Climate
Since A.D. 1500

Climate
Since A.D. 1500

Edited by

Raymond S. Bradley

University of Massachusetts, Amherst

and

Philip D. Jones

University of East Anglia, Norwich

London and New York

First published 1992
by Routledge
11 New Fetter Lane, London EC4P 4EE

Simultaneously published in the USA and Canada
by Routledge
a division of Routledge, Chapman and Hall, Inc.
29 West 35th Street, New York, NY 10001

This book was produced from authors' disks and has
been passed for press by the editors.

Typeset in 10 on 13 point Times by Phoenix Photosetting, Chatham, Kent
and printed and bound in Great Britain by Richard Clay Ltd, Bungay, Suffolk

British Library Cataloguing in Publication Data

ISBN 0–415–07593–9

A catalogue reference for this title is available
from the British Library

Library of Congress Cataloging in Publication Data
has been applied for

Contents

SECTION C: ICE CORE EVIDENCE

SECTION D: FORCING FACTORS

SECTION E: SUMMARY

Acknowledgements

As editors, we wish to thank all of our contributing authors whose names and addresses are found in the following section. We appreciate their tenacity and patience in what has turned out to be a rather long and arduous project. We hope they find the final product worthy of their efforts.

In the course of producing this book we called upon many colleagues to give us the benefit of their expertise by reviewing early drafts of each chapter. Chapters were generally reviewed by two specialists in the field, as well as by the editors. We are very grateful to all the contributing authors and to the following individuals who assisted us in this task: B. Alt, T. J. Blasing, G. Boulton, C. Caviedes, P. Damon, D. Eckstein, G. Farmer, L. Graumlich, J. Grove, R. E. Holmes, M. J. Ingram, J. Jouzel, P. M. Kelly, J. Kington, R. M. Koerner, J. Lough, N. Nicholls, S. Payette, C. Wilson, T. M. L. Wigley, M. Yoshino.

Numerous other individuals have helped with the production of the book, by proof reading, word processing, digitising and drafting figures, constructing an index and by facilitating communications around the world. In this regard we would like to thank Paul Doyle, Roy Doyon, Clem Earle, Greg Garfin, Pavel Groisman, Jean Hamilton, Christine Jeffrey, Philip Judge, Jennifer Ouimet, Marie Litterer, Alan Robock and David Taylor.

We also thank the European Geophysical Society which helped us to organise the meeting in Bologna where the seeds of this book took root. Finally, both editors wish to acknowledge the support they have received in recent years from the U.S. Department of Energy, Carbon Dioxide Research Division, for their research into climatic variability.

Note on Data sets

We had originally planned to include in this book an Appendix containing all of the data presented in the figures of climatic reconstructions. In the end, it was obvious that this would make the book too long and expensive. We did receive the data for a number of time series published in this book. These can be obtained on disk from the Paleoclimatology Program Manager, National Geophysical Data Center, 325 Broadway, Boulder, CO 80309, U.S.A. Please specify "Data Disk for Climate Since A.D. 1500" and indicate size and density of disk required.

Contributing Authors

T. F. Ball
Geography Department
University Winnipeg
515 Portage Avenue
Winnipeg, Canada R3B 2E9

V. Banzon
Instituto di Fisica dell'Atmosfera, CNR
P. Sturzo 31
00144 Roma
Italy

W. R. Baron
Center for Colorado Plateau Studies
Northern Arizona University
Flagstaff, Arizona 86011
U.S.A.

J. A. Boninsegna
Instituto Argentino de Nivologia y
    Glaciologia
Casilla de Correo 330
5500 Mendoza
Argentina

E. P. Borisenkov
Main Geophysical Observatory
Karbyshara 7
Leningrad 194018
U.S.S.R.

R. S. Bradley
Department of Geology & Geography
University of Massachusetts
Amherst, MA 01003
U.S.A.

T. F. Braziunas
Quaternary Research Center
University of Washington AK-60
Seattle, WA 98195
U.S.A.

K. R. Briffa
Climatic Research Unit
University of East Anglia
Norwich NR4 7TJ
England, U.K.

D. Camuffo
Instituto di Chimica e Tecnologia
    dei Radioelementi
Corso Stati Uniti, 4
I-35020 Padova
Italy

A. J. W. Catchpole
Department of Geography
University of Manitoba
Winnipeg, Manitoba R3T 2N2
Canada

M. K. Cleaveland
Dept. of Geography
University of Arkansas
Fayetteville, Arkansas 72701
U.S.A.

M. Colacino
Istituto di Fisica dell'Atmosfera, CNR
P. Sturzo 31
00144 Roma
Italy

E. R. Cook
Tree Ring Lab
Lamont Doherty Geological Obs.
Palisades, New York 10964
U.S.A.

R. D. D'Arrigo
Tree Ring Laboratory
Lamont-Doherty Geological Obs.
Palisades, NY 10964
U.S.A.

S. Enzi
National Research Council
ICTR
C.So Stati Uniti 4
I-35020 Padova
Italy

H. C. Fritts
Laboratory of Tree-Ring Research
University of Arizona
Tucson, Arizona 85721
U.S.A.

G. Gong
Institute of Geography
Chinese & Academy Sciences
Beijing, China

D. A. Graybill
Laboratory of Tree-Ring Research
University of Arizona
Tucson, Arizona, 85721
U.S.A.

G. P. Gregori
Istituto di Fisica dell'Atmosfera, CNR
P. Sturzo 31
00144 Roma
Italy

J. Guiot
Laboratoire de Botanique Historique et
    Palynologie
CNRS UA 1152
Faculté de St. Jerome
13397 Marseille Cedex 13
France

G. Holdsworth
Arctic Institute of North America
University of Calgary
2500 University Drive N.W.
Calgary, Alberta T2N 1N4
Canada

M. K. Hughes
Laboratory of Tree-Ring Research
University of Arizona
Tucson, Arizona 85721
U.S.A.

G. C. Jacoby, Jr.
Tree-Ring Laboratory
Lamont-Doherty Geological Obs.
Palisades, New York 10964
U.S.A.

P. D. Jones
Climatic Research Unit
University of East Anglia
Norwich NR4 7TJ
England, U.K.

T. Karl
National Climatic Data Center
Asheville, NC 28801
U.S.A.

H. R. Krouse
Dept. of Physics & Astronomy
University of Calgary
Calgary, Alberta T2N 1N4
Canada

D. M. Meko
Laboratory of Tree-Ring Research
University of Arizona
Tucson, Arizona 85721
U.S.A.

E. Mosley-Thompson
Byrd Polar Research Center
Ohio State University
125 South Oval Mall
Columbus, Ohio 43210
U.S.A.

A. Murata
Marine Department
Japan Meteorological Agency
Tokyo 100
Japan

V. T. Neal
College of Oceanography
Oregon State University
Corvallis, Oregon 97331
U.S.A.

D. A. Norton
School of Forestry
University of Canterbury
Christchurch 1
New Zealand

M. Nosal
Dept. of Mathematics & Statistics
University of Calgary
Calgary, Alberta T2N 1N4
Canada

A. E. J. Ogilvie
Climatic Research Unit
University of East Anglia
Norwich NR4 7TJ
England, U.K.

J. G. Palmer
Department of Plant Science
Lincoln College
P.O. Box 84
Lincoln, New Zealand

M. Pasqua
Instituto di Fisica dell'Atmosfera, CNR
P.Sturzo 31
00144 Roma
Italy

M. P. Pavese
Instituto di Fisica dell'Atmosfera, CNR
P. Sturzo 31
00144 Roma
Italy

D. A. Peel
British Antarctic Survey
Madingley Road
Cambridge CB3 OET
England, U.K.

C. Pfister
Historisches Institut der
Universitat Bern
Engehaldenstrasse 4
CH-3012 Bern
Switzerland

D. Portman
Atmospheric & Environmental Research Inc.
840 Memorial Drive
Cambridge, MA 02139
U.S.A.

W. H. Quinn
College of Oceanography
Oregon State University
Corvallis, Oregon 97331
U.S.A.

F. H. Schweingruber
Swiss Federal Institute of Forest, Snow
   and Landscape Research
Zürcherstrasse 11
CH-8903 Birmensdorf
Switzerland

F. Serre-Bachet
Laboratoire de Botanique Historique et
   Palynologie
CNRS UA 1152
Faculté de St. Jerome
13397 Marseille Cedex 13
France

X. M. Shao
Laboratory of Tree-Ring Research
University of Arizona
Tucson, Arizona 95721
U.S.A.

S. G. Shiyatov
Laboratory of Dendrochronology
Institue of Plant and Animal Ecology
Ural Division of the U.S.S.R. Academy
   of Sciences
Sverdlovsk
U.S.S.R.

D. W. Stahle
Dept. of Geography
University of Arkansas
Fayetteville, Arkansas 72701
U.S.A.

M. Stuiver
Quaternary Research Center
University of Washington
Seattle, Washington 98195
U.S.A.

A. Tarussov
Remote Sensing Center
Sherbrooke University
Sherbrooke, Québec J1K 2R1
Canada

L. Tessier
Laboratoire de Botanique Historique et
   Palynologie
CNRS UA 1152
Faculté de St. Jerome
13397 Marseille Cedex 13
France

L. G. Thompson
Byrd Polar Research Center
Ohio State University
125 S. Oval Mall
Columbus, Ohio 43210
U.S.A.

P. K. Wang
Dept. of Meteorology
University of Wisconsin
Madison, WI 53706
U.S.A.

W. C. Wang
Department of Atmospheric Science
State University of New York
Albany, N.Y.
U.S.A.

X. D. Wu
Institute of Geography
Academia Sinica
Beijing 100012
People's Republic of China

D. Zhang
Academy of Meteorological Science
State Meteorological Administration
Beijing, People's Republic of China

P. Zhang
Institute of Geography
Chinese Academy of Sciences
Beijing, People's Republic of China

Preface

The idea for this book arose from a two day session on *"Climatic Fluctuations in the Instrumental and Historical Period"* organised by the editors at the European Geophysical Society meeting in Bologna, Italy during March, 1988. Flushed with enthusiasm from the meeting, and a couple of bottles of Chianti, we foolishly decided that what the world really needed was an up-to-date assessment of how climate had varied over the last few hundred years. After more protracted thought, we decided that it would be best to focus on the period since about A.D. 1500. This was the interval for which the most data could be assembled, and it was also a period of particular interest, encompassing the so-called "Little Ice Age", as well as a period of unusual solar activity (the Maunder minimum) and several exceptionally large explosive volcanic eruptions. Our principal criterion for selecting types of data was to consider only those sources which had annual (or better) resolution, which largely meant a focus on historical documentary records, dendroclimatic data and ice cores. The book thus contains three main sections divided along these disciplinary lines. In addition, we have included a fourth section dealing with potentially important forcing factors (causes of climatic change) which have relevance for understanding the records described in the preceding sections.

Contributions were sought from authors throughout the world to provide as wide a geographical dimension as possible (see Chapter 1, Figure 1). However, assembling a truly comprehensive set of papers proved to be a daunting task and certain omissions will be readily apparent. No chapters, for example, deal explicitly with the important studies of ice cores from Greenland and the Canadian Arctic simply because the experts in those matters were not able to contribute due to other pressing commitments. In other cases there is simply not enough known to produce a worthwhile review. Thus, large regions of the earth, such as Africa and South Asia, remain relatively unknown for this period. In fact, tropical regions in general are poorly represented by high resolution paleoclimatic data. We hope that book will stimulate research in those areas about which so little is known. Obvious candidates for future study are historical records from South and Southeast Asia, South America and the Middle East, further dendroclimatic research in Asia, South America and Africa, and additional studies of ice cores from the high mountain ice caps of the world. We also hope that the book will lead to the application of new techniques to reconstruct paleoclimatic conditions of the past, and to the further assessment of reconstructions using many different lines of evidence.

Amherst

Section A: DOCUMENTARY EVIDENCE

1 Climate since A.D. 1500: Introduction

R. S. Bradley and P. D. Jones

1.1 The need for perspective

It is now common knowledge that today's climate is unlikely to prevail into the 21st century. An increase in "greenhouse gases", as a result of human activity, is likely to perturb the world's energy balance, leading to higher surface temperatures and a redistribution of precipitation patterns (National Research Council 1982; Bolin *et al.* 1986; I.P.C.C., 1990). The magnitude of any climatic change, and its distribution both geographically and seasonally, is estimated by the use of computer models of the general circulation. Simulations of equilibrium climatic conditions with early twentieth century CO_2 levels are generally compared with those resulting from doubling of CO_2 to obtain the differences that might be expected in the future (Schlesinger 1984). More recently, transient climate models have been developed to simulate the gradually changing climate as CO_2 and the other greenhouse gases increase from year to year. The focus of all these experiments is, of course, to isolate the impact of human activities on climate over the next century or so. However, whatever the anthropogenic climatic effects may be in the future, they will be superimposed on a climatic system which also responds to "natural" forcing factors. Unless we improve our understanding of what these factors are, and how the climate system has responded to them in the past, there is little prospect of interpreting, or anticipating, future climatic changes. Therefore, in order to understand how climate may vary in the future we must understand how and why it has varied in the past. With such knowledge we may be able to place our contemporary climate in a longer term perspective and identify any underlying trends or periodicities in climate upon which future climatic changes might be superimposed. With such knowledge we may be able to isolate the causes of past climatic fluctuations, causes which may continue to operate in the future and influence the course of forthcoming climatic events (Bradley 1990).

In this book we focus on the most recent period of climate history, the last 500 years. This is an important interval for a number of reasons. Firstly, we can construct a fairly comprehensive picture of climatic variations during this period, and we can also document variations in potentially important forcing factors. There is thus the opportunity to develop and test hypotheses about how the climate system responds to these factors. Secondly, climatic variability on a decadal to century time-scale is of most relevance to concerns about future climate and the extent to which "natural" variability will amplify or subdue anthropogenic effects. Thirdly, in the last 500 years, world population has increased by a factor of 12, at least; our society has changed from one which produced local, or perhaps regional environmental impacts, to one which now produces environmental impacts of global extent. Understanding how human activity may already have altered climate locally, regionally and perhaps globally is of great significance as we enter a new century facing even more rapid increases in world population and continuing environmental degradation.

1.2 Climates of the past

Although the focus of this book is the last 500 years, it is appropriate to consider briefly how this period fits in with the longer term changes that we know have affected the earth's climate system. Twenty thousand years ago the world was experiencing a period of major continental glaciation, the most recent of a series of such events which have occurred with some regularity over the last 2 million years (Imbrie and Imbrie 1979). These events were brought about by small changes in the position of the earth relative to the sun and the consequent redistribution of solar radiation across the earth (Berger 1980). Changes in atmospheric composition and of atmospheric aerosol loading may also have played a role in the evolution of climate from glacial to interglacial periods and back again to glacial periods (Barnola *et al.* 1987; Genthon *et al.* 1987; Chappellaz *et al.* 1990). The last continental ice sheets in the Northern Hemisphere had largely disappeared by 6000-7000 years B.P. (Before Present) and this induced dramatic changes in the distribution of plants and animals and of the world's coastlines as the sea returned to higher levels.

There is evidence that some parts of the world experienced quite warm summers around 5000 to 6000 years ago, a period sometimes referred to, rather loosely, as the mid-Holocene Optimum (cf. Webb and Wigley 1985). Whether this was a globally extensive warm period, or if it was warm in other seasons is not yet known. However, it is thought that glaciers in many parts of the world reached their post-glacial minima around this time, a condition made all the more significant by the fact that glaciers subsequently expanded over the course of the next few thousand years. Such periods of glacier expansion are known collectively as Neoglacial episodes, times of renewed glacier activity (Porter and Denton 1967). It has been argued that such periods have a certain regularity in time, suggesting some periodic forcing, but the evidence is quite weak and the causes of these glacier advances, whether regional or global in extent, remains obscure. For our purposes, the important point is that the most recent of these neoglaciations occurred during the last 500 years, and there is abundant evidence that this most recent episode was the most significant of all the periods of glacier expansion that have occurred since the last Ice Age. The period since A.D. 1500 is thus of extraordinary scientific interest.

Although there is voluminous indirect evidence that climatic conditions in the past 500 years were often quite different from our contemporary experience (e.g. von Rudloff 1967; Lamb 1982; Grove 1988) the precise nature of these differences, and what caused them, remains elusive. Certainly, there were a number of cold intervals which had dramatic environmental consequences. In almost every glacierised mountain region of the world, glaciers grew and advanced down-valley, often to positions as extensive as at any time since the last Ice Age, more than 10,000 years ago. These changes in alpine glaciers are so characteristic of the period that it is often referred to as "the Little Ice Age". However, among those who use this term there is little consensus on when it began, or when it ended.

Furthermore, the term suggests a period of uniformly cold conditions which obscures the fact that relatively warm intervals did occur (see Chapter 33). It also focuses attention on those parts of the world where snow and ice are common phenomena; how climate changed in tropical and sub-tropical regions during the last 500 years is far less well-documented. Only by careful paleoclimatic reconstruction can we hope to unravel the sequence of events which occurred during this "Little Ice Age" and to understand how extensive the changes really were geographically. With this information, it may then be possible to isolate the causes of such climatic variations.

1.3   The world in A.D. 1500

The beginning of the 16th century was somewhat of a watershed in the history of civilisation. Between 1492, when Columbus reached the Caribbean, and 1532, when Pisarro arrived in Peru, the map of the known world had irrevocably changed. Vasco de Gama rounded the southern tip of Africa in 1497 and reached as far as Calicut in India, opening up an entirely new trading route between Europe and the East. By 1519, Magellan's expedition had circumnavigated the globe. The stage was set for the emergence of colonial states, ruled from small but powerful countries, geographically isolated from their remote territories. Only Australasia and the extreme polar regions were to remain beyond the reach of explorers for a further 200 years or more.

Over the course of the next 500 years extraordinary changes in society took place. These occurred against a background of environmental changes, which may have played a critical role in some of the events which occurred. However, until we can document climatic variations of the last 500 years, the extent of such influences will remain controversial.

1.4   Sources of high resolution data for paleoclimatic reconstruction

Meteorological measurements (from which we can assess climatic conditions) are only available for relatively short periods (generally a century or less) from most parts of the world. Although observations have been maintained in a few locations for two centuries or more (see Jones and Bradley, Chapter 13) to obtain a broader picture of past climatic variations we must rely on additional non-instrumental records, from which climatic conditions can be deduced. Such records of climatically sensitive natural phenomena are surrogate or proxy measures of past climate; they contain climatic information which must be extracted and separated from the non-climatic matrix in which it is embedded. The analyst must isolate the climatic signal from the extraneous noise. As more detailed and geographically extensive records are built up, the possibility of identifying causes and mechanisms of climatic variation is increased and so the prospects of understanding future climate are enhanced.

Although there are numerous approaches to the reconstruction of past climates (Bradley 1985) only a few types of evidence have the potential of providing a record which can be resolved to the annual or seasonal level (Table 1.1). Of these, quantitative estimates of past

3

Table 1.1 Sources of high resolution paleoclimatic data.

Type of record	Main distribution	Potential Information:
Historical Documents	All continents	Almost all aspects of climate.
Tree rings	Continental areas*, excluding desert and tundra regions	Temperature precipitation pressure patterns, drought runoff.
Ice cores	Polar and high mountain regions	Temperature, precipitation atmospheric aerosols, atmospheric composition.
Varved sediments	Continents and some coastal basins	Temperature, precipitation solar radiation.
Corals	Tropical oceans	Sea surface temperatures, adjacent continental rainfall.

* Studies of tree growth in tropical areas have not yet provided useful paleoclimatic reconstructions

climate from studies of varved sediments, and corals, have not yet been widely carried out. These natural archives have great potential for paleoclimatology, but a number of problems have yet to be resolved. Two approaches to climatic reconstruction have the potential of providing wide geographic coverage: the analysis of documentary records (discussed in Section A) and of tree growth (dendroclimatic) indices (Section B). These can be supplemented in high latitude and high altitude regions by the analysis of ice cores (Section C). Figure 1.1 shows the distribution of records discussed in the various chapters of this book. By combining these different approaches, the aim is to construct a picture of past climatic variations on larger and larger spatial scales in which the whole is greater than the sum of its individual parts. At the same time, we need records of those phenomena which are thought to have played a role in causing climatic variations of the past. The most likely candidates for climate forcing on this time scale include explosive volcanic eruptions, solar activity variations and El Nino-Southern Oscillation (ENSO) events of varying magnitude. As with the paleoclimatic record itself, proxy records can be used to reconstruct the history of these climatic forcing factors. Such records are discussed in Section D.

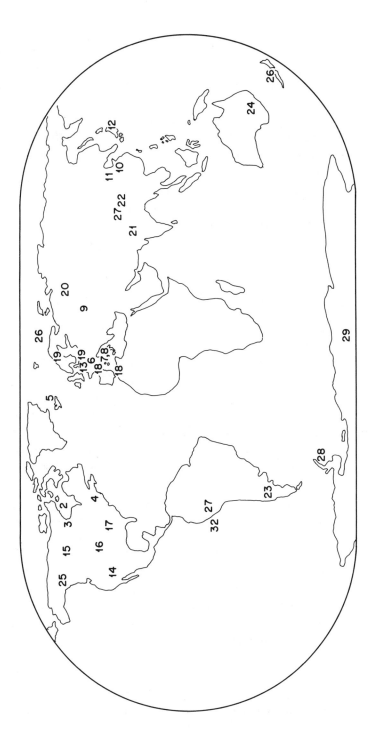

Figure 1.1 Approximate location of studies discussed in text. Numbers refer to chapters in this volume.

1.5   Methods of paleoclimatic reconstruction

1.5.1   Documentary records

Historical data can be grouped into three major categories. First, there are observations of weather phenomena *per se*, for example, the frequency and timing of frosts, or the occurrence of rainfall or snowfall recorded by early diarists. Secondly, there are records of weather-dependent natural phenomena (sometimes termed parameteorological phenomena) such as droughts, floods, lake or river freeze-up and break-up, etc. Thirdly, there are phenological records, which deal with the timing of recurrent weather-dependent biological phenomena, such as the dates of flowering of shrubs and trees, the timing of harvest (both fruit and grain) or the arrival of migrant birds in the spring. Within each of these categories there is a wide range of potential sources and an equally wide range of possible climate-related phenomena.

Potential sources of historical paleoclimatic information include: (a) ancient inscriptions, (b) annals, chronicles, etc., (c) government records, (d) private estate records, (e) maritime and commercial records, (f) personal papers, such as diaries or correspondence, (g) scientific or quasi-scientific writings, such as (non-instrumental) weather journals and (h) fragmented early instrumental records (Ingram *et al*. 1978). In all these sources, the historical climatologist is faced with the difficulty of ascertaining exactly what the qualitative description of the past is equivalent to, in terms of modern-day observations. What do the terms "drought," "frost," "frozen over," really mean? How can qualifying terms (e.g. "extreme" frost) be interpreted? Baker (1932) for example, notes that one 17th century diarist recorded three droughts of "unprecedented severity" in the space of only five years!

An approach to solving this problem has been to use content analysis (Baron 1982) to assess in quantitative terms, and as rigorously as possible, climatic information in the historical source. Historical sources are examined for the frequency with which key descriptive words were used (e.g. "snow," "frost," "blizzard," etc.) and the use which the writer may have made of modifying language (e.g. "severe frost," "devastating frost," "mild frost," etc.). In this way an assessment can be made of the order of increasing severity as perceived by the original writer. The ranked terms may then be given numerical values so that statistical analyses can be performed on the data. This may involve simple frequency counts of one variable (e.g. snow) or more complex calculations using combinations of variables. The original qualitative information may thus be transformed into more useful quantitative data on the climate of different periods in the past. Perhaps the most comprehensive work of this kind has been carried out for Switzerland by Pfister who discusses his approach in Chapter 6.

Non-climatic information is often required to interpret the climatic aspects of the source. Where did the event take place (was the event only locally important; was the diarist itinerant or sedentary?) and precisely when did it occur and for how long? This last question may involve difficulties connected with changing calendar conventions as well as trying to define what is meant by terms such as "summer" or "winter," and what time span might be represented by a phrase such as "the coldest winter in living memory." Not all of these problems may be soluble, but content analysis can help to isolate the most pertinent and unequivocal aspects of the historical source (Moody and Catchpole 1975).

Historical sources rarely give a complete picture of former climatic conditions. More commonly, they are discontinuous observations, often biased towards the recording of extreme events, and even these may pass unrecorded if they fail to impress the observer. Furthermore, long-term trends tend to go unnoticed since they are beyond the temporal perspective of one individual. In a sense, the human observer acts as a high-pass filter, recording short-term fluctuations about an ever-changing norm (Ingram *et al.* 1981).

It is worth noting that not all historical sources are equally reliable. It may be difficult in some cases to determine if the author is writing about events of which he has first-hand experience, or if events have been distorted by rumor, or the passage of time. Ideally, sources should be original documents rather than compilations; many erroneous conclusions about past climate have resulted from climatologists relying on poorly compiled secondary sources which have proved to be quite erroneous when traced back to the original data (Bell and Ogilvie 1978, Wigley 1978, Ingram *et al.* 1981). The problem of dealing with these types of fragmentary evidence is discussed for Italy by Camuffo and Enzi and Pavese *et al.* in Chapters 7 and 8, respectively, and for the Soviet Union by Borisenkov in Chapter 9.

As with all proxy data, historical observations need to be calibrated in some way, in order to make comparisons with recent data possible. This is commonly done by utilizing early instrumental data which may overlap with the proxy record, to develop an equation relating the two data sets. Thus, Bergthorsson (1969) was able to calibrate observations of sea-ice frequency on the north Icelandic coast with mean annual temperatures during the 19th century and then use this equation to assess long-term temperature fluctuations over the last 400 years from sea-ice observations. This work has been re-assessed by Ogilvie in Chapter 5 using historical information on both sea-ice extent and climatic conditions inland. Sea-ice observations (from the Hudson's Bay Company trading ships) are also used by Catchpole in Chapter 2 to compare sea-ice conditions during the late eighteenth and early nineteenth century with modern records.

Some observations may not need direct calibration if recent comparable observations are available. This applies to such things as rain/snow frequency, dates of first and last snowfall, river freeze/thaw dates etc., providing urban heat island effects, or technological changes (such as river canalization) have not resulted in a non-homogeneous record. For example, rain day counts are used to reconstruct monthly precipitation data for eastern China by Wang and Zhang in Chapter 11 and by Murata in Chapter 12 for Japan. The relationship between precipitation occurrence and temperature in Beijing during the eighteenth century is used by Wang *et al.* in Chapter 10 to estimate summer temperatures. Early weather diaries and instrumental measurements are used by Baron in Chapter 4 to estimate climate changes over New England during the seventeenth to nineteenth centuries. Observations of weather conditions by employees of the Hudson's Bay Company are discussed by Ball in Chapter 3.

1.5.2 Dendroclimatology

Variations in annual tree ring parameters from one year to the next have long been considered an important source of past climatic information. Climatic reconstructions have been made not only from measurements of a series of ring widths but more recently from measurements of maximum latewood density and from isotopic studies of the cellulose in individual tree rings (e.g. Briffa *et al.* 1988). Dendroclimatic reconstructions have been made in sub-tropical, temperate and high latitude regions of both hemispheres, using trees which can be shown to contain a primary climate-dependent signal. The technique is limited in tropical regions because most tree species there do not form distinct annual rings and growth is less susceptible to the inter-annual variability of climate.

There are three important steps in dendroclimatic reconstruction:

(1) Standardization of the tree-ring parameters to produce a site chronology.
(2) Calibration of the site chronology with instrumentally-recorded climatic data, and production of a climatic reconstruction based on the calibration equations.
(3) Verification of the reconstruction with data from an independent period not used in the initial calibration.

Here we briefly discuss these steps to provide some background for the chapters which follow in Section B.

Standardization Following the cross-dating of all series of tree-ring measurements from a site, it is generally necessary to standardize these in some way. The reason for this is that single series of tree-ring width measurements often exhibit systematic trends which are directly attributable to the aging of the tree. The best example of this is the wider widths of young rings compared to the thinner widths of older rings. In order for series from young and old trees to be amalgamated or compared it is necessary to remove the influence of this growth trend. This is generally done by fitting some form of mathematical function to the raw data and dividing or subtracting each measured value by the expected value according to the fitted curve. This procedure leads to a new series of dimensionless indices with a variance that is roughly equal through time (Fritts 1976).

A site chronology is produced by averaging the indexed series from each tree. Averaging the indices increases the expression of the common signal because variability which differs from tree to tree is lost in the averaging. Much of the common signal is climatically related. Ideally, a chronology should contain enough samples to ensure that the residual noise, left after averaging, is minimal. One way of gauging this is to calculate the Expressed Population Signal (EPS). This parameter estimates how well the site chronology approximates the population chronology. It is based on the average intersample correlation coefficient (r) (see Briffa 1984 and Wigley *et al.* 1984 for details). The lower the r value, the greater the replication required to achieve a certain level of confidence in the overall chronology.

The key point in standardization is the degree of fit between the functional form and the measured ring width series. If the fit is tight then the standardization procedure will take out more of the low-frequency variance, which may be related to climate, than desired. In some circumstances it may be necessary to trade-off the loss of low-frequency variance with the increase in statistical reliability of any reconstructions. For further information on this aspect of standardization the reader is referred to Cook (1985) and Briffa *et al.* (1987).

Dendroclimatic studies of trees from the arid southwestern United States tend to use modified exponential functions of the form $Yt = ae^{-bt} + k$, where a, b and k vary according to the tree-ring measurements (Fritts 1976). This approach is conservative in that it minimized the loss of potential climatic information on relatively long timescales. The use of such 'functions' is justified by the particular growth habitat, lack of competition and the relatively long-lived nature of semi-arid tree species.

The extension of dendroclimatology to many other parts of the world, most notably to Europe and eastern North America, involved the use of deciduous trees often growing in closed-canopy situations. Their growth curves are poorly modelled by the negative exponential function as they often contain periods of growth enhancement and suppression related to non-climatic factors such as competition, management, insect infestation etc. Several functional forms have been suggested to overcome these problems. Examples include orthogonal polynomials (Fritts 1976) Gaussian filters (Briffa 1984) and cubic splines (Cook and Peters 1981). Whichever functional form is used, it is important to decide how tight to fit the function to the ring width series. For the most complete discussion of this often neglected issue in dendroclimatology, the reader is referred to Cook and Briffa (1990).

Calibration   Once a master chronology of standardized indices of some tree growth parameter (e.g. ring width, maximum latewood density) has been produced, the next step is to relate this to variations in climatic data. Mathematical and/or statistical procedures are used to derive an equation relating the tree indices to a climate variable. The equation is developed over a period known as the calibration period. If the equation adequately describes instrumental climatic variability over the calibration period, then it can be applied to past tree growth data to reconstruct the climate variable back as far as the beginning of the chronology series. As with all mathematical-statistical relationships of this kind, it is preferable to retain some climate data outside the calibration period. These data can then be used to assess the performance of the fitted equation with independent data from a verification period. The importance of this is discussed in the next section.

Calibration methods can be classified according to a hierarchical structure of complexity (see Table 1.2). The following is a brief outline of the statistical approaches illustrated with some recent examples from the dendroclimatic literature. More complete descriptions of some of the statistical procedures are given by Fritts (1976) Hughes *et al.* (1982) and Kairiukstis and Cook (1990).

Table 1.2   Calibration methods.

Level	Number of variables of: Tree Growth	Climate	Statistical Technique
1	1	1	Simple regression analysis
2a	n	1	Multiple linear regression analysis (MLR)
2b	nP	1	Principal component analysis (PCA)
3a	nP	nP	Orthogonal spatial regression (PCA + MLR)
3b	n	n	Canonical regression analysis

nP = number of variables after discarding unwanted ones from PCA

SIMPLE REGRESSION (LEVEL I)   This is the simplest level of climate reconstruction. Variations in growth indices at a single site are related to a single climatic parameter, such as mean summer temperature or total summer precipitation. In mathematical terms the equation might be:

$$t_i = c + bw_i$$

where t_i and w_i are the mean summer temperature and tree ring widths in year i, and c and b are the coefficients estimated by simple regression. An example of this approach is the work of Jacoby and Ulan (1982) where the date of the first complete freezing of the Churchill River estuary on Hudson Bay is reconstructed from a single chronology located near Churchill, Manitoba. Graybill and Shiyatov in Chapter 20 use this approach to reconstruct summer temperatures in the northern Ural mountains.

Increasing complexity might involve using additional predictors in the regression such as the previous year's growth, or using the average of a number of separate site chronologies. For example, Duvick and Blasing (1981) reconstructed annual (August-July) precipitation totals for Iowa from a single regional chronology which was an average of three site chronologies of the same species. Similarly, in Chapter 21, Hughes averages together ring width chronologies from eight sites and maximum latewood density chronologies from seven sites to reconstruct early spring and late summer temperatures and precipitation totals in the Kashmir region of the Himalayas.

MULTIPLE LINEAR REGRESSION (LEVEL 2)   Most dendroclimatic reconstructions make use of multivariate statistics relating a single climate variable to an array of tree ring chronologies. An example is

$$t_i = c + b_1 w_{i1} + \ldots, + b_n w_{in}$$

where there are n regression coefficients b_j for each of the n chronologies w_{ij}. This approach is simply an expansion of the level 1 technique, but bringing in more predictors generally accounts for progressively more climate variance. Incorporating too many terms (coefficients) into the equation is not to be recommended, even though it would be possible to explain all the variance in the climate parameter if enough terms were used. Adding too many coefficients simply widens the confidence limit about the reconstruction estimates.

Two procedures have been proposed for deciding how many of the possible n chronology predictors should be retained in the prediction equation. The first uses stepwise multiple regression to select the number of climate variables. In this technique, the chronology which accounts for most of the climate variables is selected, followed successively by the chronology that accounts for most of the remaining variance. Predictor chronologies are continually added until the point is reached where the addition of another chronology explains an amount of variance below some pre-determined threshold (e.g. 5%). An example of this approach is the reconstruction of drought in Southern California by Meko et al. (1980). Variations of the basic approach are used by D'Arrigo and Jacoby, Wu, and Boninsegna to reconstruct paleoclimatic conditions in northern North America, western China and southern South America, in Chapters 15, 22 and 23, respectively.

A major problem with stepwise regression is that intercorrelations between the tree ring predictors can lead to instability in the predictor equation. In statistical terms this is referred to as multi-collinearity. The second procedure attempts to overcome this by expressing the variance of the tree ring data in terms of principal components (PCs) (also called empirical orthogonal functions or EOFs) and using these in the regression procedure. Principal components analysis (PCA) is a statistical transformation of the original (intercorrelated) variables to produce a set of orthogonal (uncorrelated) principal components (for a review, see Richman 1986). The first PC is the mode of variation of the data that explains the most variance. The second is that which explains most of the remainder, but which is orthogonal to the first, and so on. Although there are as many PCs as there were original variables, the transformation means that most of the variance of the original data set is explained by a few PCs. A selection criteria will then be needed to decide how many PCs to retain. Various methods have been proposed (see e.g. Preisendorfer et al 1981).

Principal components analysis, apart from reducing the number of potential predictors, also considerably simplifies the multiple regression. It is not necessary to use the stepwise procedure because the new potential predictors are all orthogonal. An example of this kind of procedure is the reconstruction of July drought (Palmer Indices; Palmer 1965) in the Hudson Valley in New York by Cook and Jacoby (1979). Meko, Serre-Bachet et al. and Norton and Palmer (in Chapters 16, 18 and 24, respectively) use modifications of the principal component regression to reconstruct the paleoclimatic history of the U.S. Plains, southwestern Europe and northwest Africa and Australia and New Zealand.

SPATIAL REGRESSION (LEVEL 3)   The procedures used in this level are the most complex of all reconstructions attempted in dendroclimatology. A spatial array of tree-growth data is used to reconstruct the spatial field of some climatic parameter. Details of this complex statistical procedure are given by Fritts (1976) and Blasing (1978). The earliest spatial dendroclimatic reconstruction used canonical regression analysis to establish the relationship between patterns of tree ring growth over western North America and mean sea-level pressure patterns over the North Pacific and adjacent continental regions of eastern Asia and North America (Fritts et al. 1971). This approach is extended in chapter 14 by Fritts and Shao by mapping temperature and precipitation patterns over the United States.

Related procedures have been used by other workers. Briffa et al. (1988) for example, use orthogonal spatial regression techniques to reconstruct April to September temperatures over Europe west of 30°E using densitometric information from conifers over Europe. In their procedure both the spatial array of temperature and the spatial array of densitometric data are first reduced to their principal components. Only significant components in each set are retained. Each retained PC of climate is then regressed in turn against the set of retained densitometric PCs. This procedure can be thought of as repeating the level 2 approach m times, where m is the number of retained climate PCs. Having found all the significant regression coefficients, the set of equations relating the climate PCs to the tree growth PCs are then transformed back to original variable space, resulting in an equation for each temperature location in terms of all the densitometric chronologies. In Chapter 19, Briffa and Schweingruber use this approach to present temperature reconstructions from northern Europe, and in Chapter 17 Cook et al. reconstruct Palmer (drought) Indices for the eastern United States using a spatial version of the level 2 principal components regression. In this technique the Indices are reconstructed point-by-point over the study region. It differs from the level 3 spatial regression approach by not reducing the drought indices to principal components.

Verification No matter how complex the calibration procedure is, it is essential that the empirically-derived equation(s) be tested with independent data from the verification period. The statistical procedures described in the previous section ensure that the maximum amount of climate variance is explained by the tree growth parameters over the calibration period. It is likely that this amount of explained variance will be reduced when the prediction estimates are tested against independent data. The performance of the equation over the verification period gives the best guide to the likely quality of the reconstruction for periods prior to instrumental measurement. All the authors in section 3 pay particular attention to this aspect of dendroclimatic reconstruction.

There is really no substitute for independent data to assess the reconstruction performance. At least one-third to one-half of climate data should be retained for verification. Many reconstructions split the climate data into two halves calibrating on one half and verifying on the other. The process may then be repeated, reversing the calibration and verification periods.

A number of statistical comparisons between reconstructed and independent climate data have been proposed (see, for example, Fritts 1976). These include comparisons of the mean, standard deviation, correlation coefficient, reduction of error and the Durbin-Watson test for series correlation on the residuals. For further discussion of verification procedures the reader is referred to Gordon (1982) where there is discussion of the many sub-sample replication techniques proposed by Mosteller and Tukey (1977).

1.5.3 *Ice core records*

The accumulation of past snowfall in the remote polar and alpine ice caps and ice sheets of the world provides an extraordinarily valuable record of paleoclimatic and paleoenvironmental conditions. These conditions are generally studied by detailed physical and chemical analyses of ice and firn (snow which has survived the summer melt season) in cores recovered from the highest elevations on the ice surface, where snow melt and sublimation are essentially zero. On ice caps where melting may accasionally occur, the re-freezing of meltwater within the underlying snowpack produces an identifiable crystal fabric which can be studied as an indicator of warm summer conditions in the past (Koerner 1977). In Chapter 26, Tarussov discusses such a record from Spitsbergen and the Soviet Arctic Islands.

Snowfall provides a unique record, not only of accumulation amounts *per se*, but also of air temperature at the time the snow formed in the atmosphere, atmospheric composition (including gaseous composition and aerosol content) and the occurrence of explosive volcanic eruptions. Although many ice cores have now been recovered, the detailed record of the last 500 years has been studied in relatively few of them. In Section 4 of this volume, details of recent research on high elevation ice cores from the Alaska-Yukon border, from Peru and from Tibet are discussed, together with records from Antarctica, the Soviet Arctic Islands and Spitsbergen.

A primary source of paleoclimatic information in ice cores is the variation of oxygen isotopes in the water molecules which made up the original snow crystals. These variations arise because the vapor pressure of $H_2^{16}O$ is 1% higher than that of $H_2^{18}O$. Evaporation from a water body thus produces a vapor which is relatively depleted in the heavier isotope, ^{18}O, than the original water. However, when condensation occurs, the lower vapor pressure of $H_2^{18}O$ results in this heavier molecule passing from the vapor to the liquid state more readily than the lighter $H_2^{16}O$ molecules. As condensation proceeds, this fractionation causes the remaining vapor to become increasingly depleted in the heavier isotope, and so precipitation from the remaining vapor contains less and less $H_2^{18}O$. The greater the fall in temperature, the more condensation will occur and the lower will be the heavy isotope concentration in the precipitation, compared to that in the original water source. Similarly, falls in temperature lead to lower proportions of deuterium (^2H) relative to hydrogen (^1H) in precipitation. Isotopic concentration in precipitation can therefore be considered as a function of the temperature at which the condensation occurs. Some of the problems of establishing the precise nature of this relationship are discussed by Peel in Chapter 28. The relative proportions of ^{16}O and ^{18}O in a sample are expressed in terms of departures ($\delta^{18}O$) from a standard, known as SMOW (Standard Mean Ocean Water); a $\delta^{18}O$ value of -10 indicates that a sample has an $^{18}O/^{16}O$ ratio 1% (10 per mil) less than SMOW (Bradley 1985).

Because of the relationship of oxygen (and hydrogen) isotopes in snowfall to temperature, there is a strong seasonal variation of $\delta^{18}O$ (and δD) which enables seasonal layers in annual accumulation to be detected in ice cores. Such variations are a primary means of identifying annual layers in the ice core record, enabling a chronology to be established in that part of the core where distinct layering can be detected (see Mosley-Thompson, Chapter 29). At greater depths, the accumulation of overlying snow and molecular diffusion between layers causes the seasonal signal to become less distinct and other means of dating have to be employed. However, in most of the sites discussed in Section 4, such problems are only encountered at significantly greater depths than those occupied by the record of the last 500 years. When confusion arises as to whether a clear sequence of seasonal layering occurs, other signals may be used to help establish the correct chronology; for example, in many ice core records, a seasonal variation in aerosols is found, commonly peaking in spring or early summer months and reaching a minimum in winter months (see, for example, records from the Quelccaya Ice Cap, discussed by Thompson in Chapter 27). Cross-checking between the isotopic variations and the aerosols generally resolves any uncertainties in the chronology (cf. Hammer *et al.* 1978 and Mosley-Thompson, Chapter 29). As an additional check, distinct horizons with elevated acidity content are often detectable by measuring the electrolytic conductivity of the ice. These layers represent times of explosive volcanic eruptions, many of which are known from historical sources, often to the day of the eruption. Holdsworth *et al.*, (Chapter 25) demonstrate the value of such chronostratigraphic markers in their study of an ice core from Mount Logan, Alaska. The overall record of explosive volcanism is discussed in more detail in Chapter 31 by Bradley and Jones.

Once a reliable stratigraphy has been established, it may be possible to determine variations in accumulation rate over time, assuming losses due to ablation and deformation are minimal. Such information provides important insights into atmospheric circulation, as illustrated by studies of the Quelccaya Ice Cap ice cores (Chapter 27) and of the Mount Logan ice core (Chapter 25).

1.6   Causes of Climatic Variations

Ice cores provide a unique opportunity to study some of those factors which may have caused (forced) climatic variations (such as explosive volcanic eruptions) in direct comparison with a proxy climatic record which may also record the *impact* of the forcing factor. This enables the rate of climate change to be established, at least locally, without any uncertainties about dating and comparisons of separate, independent records. However, not all important forcing factors are recorded in ice cores and past variations in climate may still require additional records of forcing to understand the paleoclimatic record. Section D provides details of three important factors which may be responsible for large scale climatic variations in the past – solar activity variations (Stuiver and Braziunas, Chapter 30) explosive volcanic eruptions (Bradley and Jones, Chapter 31) and El Nino-Southern Oscillation (ENSO) events (Quinn and Neale, Chapter 32).

References

Baker, J. N. L., 1932. The climate of England in the 17th century. *Quarterly J. Royal Meteorological Society*, 58, 421-436.

Barnola, J. M., Raynaud, D., Korotkevich, Y. S. and Lorius, C. 1987. Vostok ice core provides 160,000 year record of atmospheric CO_2. *Nature*, 329, 408-414.

Baron, W. R., 1982. The reconstruction of eighteenth century temperature records through the use of content analysis. *Climatic Change*, 4, 385-398.

Bell, W. T. and Ogilvie, A. E. J., 1978. Weather compilations as a source of data for the reconstruction of European climate during the Medieval Period. *Climatic Change*, 1, 331-348.

Berger, A., 1980. The Milankovitch astronomical theory of paleoclimates. A modern review. *Vistas in Astronomy*, 24, 103-122.

Bergthorsson, P., 1969. An estimate of drift ice and temperature in Iceland in 1000 years. *Jokull*, 19, 94-101.

Blasing, T. J., 1978: Time series and multivariate analysis in paleoclimatology. In: *Time Series and Ecological Processes* (H. H. Shugart, Jr. ed.). SIAM-SIMS Conference Series, No. 5, Society for Industrial and Applied Mathematics, Philadelphia, 213-228.

Bolin, B., Doos, B. R., Jaeger, J. and Warrick, R. A. (eds.) 1986. *The Greenhouse Effect, Climatic Change and Ecosystems: a Synthesis of Present Knowledge*. J. Wiley, New York.

Bradley, R. S., 1985. *Quaternary Paleoclimatology: methods of paleoclimatic reconstruction*. Allen and Unwin, Boston, 472pp.

Bradley, R. S., 1990. Pre-instrumental climate: how has climate varied over the past 500 years? In: *Greenhouse Gas-induced Climatic Change: A Critical Appraisal of Simulations and Observations*. M. E. Schlesinger (ed.) Kluwer, Dordrecht (in press).

Briffa, K. R., 1984: *Tree-climate Relationships and Dendroclimatological Reconstruction in the British Isles*. Unpublished PhD dissertation, University of East Anglia, Norwich, U.K., 285pp plus Appendices.

Briffa, K. R., Wigley, T. M. L. and Jones, P. D., 1987: Towards an objective approach to standardization. In: *Methods of Dendrochronology* (L. Kairiukstis, Z. Bednarz and E. Feliksik, Eds). IIASA/Polish Academy of Sciences, Warsaw, 69-86.

Briffa, K. R., Jones, P. D., Pilcher, J. R. and Hughes, M. K., 1988: Reconstrucing summer temperatures in northern Fennoscandinavia to 1700 A.D. using tree-ring data from Scots pines. *Arctic and Alpine Research*, 20, 385-394.

Chappellaz, J., Barnola, J. M., Raynaud, D., Korotkevich, Y. S. and Lorius, C., 1990. Ice-core record of atmospheric methane over the past 160,000 years. Nature, 345, 127-131.

Cook, E. R., 1985: *A time-series analysis approach to tree-ring standardization*. Unpublished PhD dissertation, University of Arizona, Tucson, U.S.A. 161pp.

Cook, E. R. and Jacoby, G. C., 1979: Evidence for quasi-periodic July drought in the Hudson Valley, New York. *Nature*, 282, 390-392.

Cook, E. R. and Peters, K., 1981: The smoothing spline: a new approach to standardizing forest interior tree-ring width series for dendroclimatic studies. *Tree-ring Bulletin*, 41, 45-53. Cook, E. R. and Briffa, K. R. 1990: Chapter 2 on standardization. In *Methods of Tree-Ring Analysis: Application in the Environmental Sciences*. (L. Kairiukstis and E. R. Cook, Eds) Kluwer Academic Publishers.

Duvick, D. N. and Blasing, T. J., 1981: A dendroclimatic reconstruction of annual precipitation amounts in Iowa since 1680. *Water Resources Research*, 17, 1183-1189.

Fritts, H. C., 1976: *Tree-rings and climate*. Academic Press, London, 567pp.

Fritts, H. C., Blasing, T. J., Hayden, B. P. and Kutzbach, J. E., 1971: Multivariate techniques for specifying tree-growth and climate relationships and for reconstructing anomalies in paleoclimate. *Journal of Applied Meteorology*, 10, 845-864.

Genthon, C., Barnola, J. M., Raynaud, D., Lorius, C., Jouzel, J., Barkov, N. I., Korotkevich, Y. S. and Kotlyakov, V. M., 1987. Vostok ice core: climatic response to CO_2 and orbital forcing changes over the last climatic cycle. *Nature*, 329, 414-418.

Gordon, G. A., 1982: Verification of dendroclimatic reconstructions. In: *Climate from Tree Rings* (M. K. Hughes, P. M. Kelley, J. R. Pilcher and V. C. Lamarche, Jr., eds) Cambridge University Press, 58-61.

Grove, J. M., 1988. *The Little Ice Age*. Methuen, London, 498pp.

Hughes, M. K., Kelly, P. M., Pilcher, J. R. and LaMarche, V. C., Jr., (Eds) 1982: *Climate From Tree Rings*, Cambridge University Press, 223pp.

Imbrie, J. and Imbrie, K. P., 1979. *Ice Ages: solving the mystery*. Macmillan, London.

Ingram, M. J., Underhill, D. J. and Farmer, G., 1981. The use of documentary sources for the study of past climates. In: *Climate and History*, T. M. L. Wigley, M. J. Ingram and G. Farmer (eds.) Cambridge University Press, Cambridge, 180-213.

I.P.C.C. (Inter-Governmental Panel on Climatic Change) 1990. *A Scientific Assessment of Climatic Change*. World Meteorological Organisation, Geneva (in press).

Jacoby, G. C. and Ulan, L. D., 1982: Reconstructions of past ice conditions in a Hudson Bay estuary using tree rings. *Nature*, 298, 637-639.

Kairiukstis, L. and Cook, E. R., (Eds) 1990: *Methods of Tree-Ring Analysis: Applications in the Environmental Sciences*. Kluwer, Dordrecht (in press).

Koerner, R. M., 1977. Devon Island Ice Cap: core stratigraphy and paleoclimate. *Science*, 196, 15-18.

Lamb, H. H., 1982. *Climate History and the Modern World*. Methuen, London.

Meko, D. M., Stockton, C. W. and Boggess, W. R., 1980: A tree-ring reconstruction of drought in southern California. *Water Resources Bulletin*, 16, 594-600.

Moody, D. W., and Catchpole, A. J. W., 1975. *Environmental data from historical documents by content analysis: freeze-up and break-up of estuaries on Hudson Bay, 1714-1871*. Manitoba Geographical Studies No. 5, University of Winnipeg, Winnipeg.

Mosteller, F. and Tukey, J. W., 1977: *Data analysis and regression*. Addison-Wesley, Reading, Massachusetts, 588pp.

National Research Council, 1982. *Carbon Dioxide and Climate: a second assessment*. National Academy Press, Washington D.C. 72pp.

Palmer, W. C., 1965: *Meteorological Drought*. U.S. Weather Bureau Research Paper 45, U.S. Weather Bureau, Washington D.C. 58pp.

Porter, S. C. and Denton, G. H., 1967. Chronology of neoglaciation in the North American Cordillera. *American J. Science*, 265, 177-210.

Preisendorfer, R. W., Xwiers, F. W. and Barnett, T. P., 1981: *Foundations of Principal Components Selection Rules*. SIO Reference Series 81-4, Scripps Institute of Oceanography, La Jolla, California. 192pp.

Richman, M. B., 1986: Rotation of principal components. *Journal of Climatology*, 6, 293-335.

Rudloff, H. von, 1967. *Die Schwankungen und Pendelungen des Klimas in Europa seit dem Beginn der regelmässigen Instrumenten-Beobachtungen (1670)*. Vieweg, Braunschweig, 370pp.

Schlesinger, M. E., 1984. Climate model simulations of CO_2-induced climatic change. *Advances in Geophysics*, 26, 141-235.

Webb, T. III, and Wigley, T. M. L., 1985. What past climates can indicate about a warmer world. In: *Projecting the Climatic Effects of Increasing Carbon Dioxide* (eds. M. C. MacCracken and F. M. Luther). U.S. Department of Energy, Washington D.C., 237-257.

Wigley, T. M. L., 1978. Climatic change since 1000 A.D. In: *Evolution des atmosphères planétaires et climatologie de la terre*. Centre National D'Etudes Spatiales. Toulouse, 313-324.

Wigley, T. M. L., Briffa, K. R. and Jones, P. D., 1984: On the average value of correlated time-series, with applications in dendroclimatology and hydrometeorology. *Journal of Climate and Applied Meteorology*, 23, 201-213.

2 Hudson's Bay Company ships' log-books as sources of sea ice data, 1751-1870

A. J. W. Catchpole

2.1 Introduction

There was a symbiosis between the winds and sailing ships. These vessels sailed within a framework of limitations set by the winds and, in turn, their log-books were replete with descriptions of wind direction and force. It is mainly this wind information that has been used in studies of sailing ships' log-books as sources of historical climatic data (Landsberg 1985). Two centuries ago, they revealed the global wind belts and the effects of the earth's rotation on wind velocity (Hadley 1735, Kutzbach 1987). These log-books were among the sources used by Lamb and Johnson (1966) in their reconstruction of July and January surface pressure patterns over the North Atlantic in each year since 1750. Several other studies employed this wind information to reconstruct the structure and development of historic storms (Landsberg 1954, Oliver and Kington 1970, Douglas *et al.* 1978).

Sea ice was a hazard to be avoided by sailing ships and it was described in their log-books only on the rare occasions when they were forced to venture into ice. Unlike the winds, however, ice was visible in forms that could be described in intricate detail and vivid reconstructions of the extent, fragmentation and motion of the ice can be made from the log-book descriptions. Sporadic use has, therefore, been made of sailing ship log-books as sources of historic sea ice information. These log-books, together with land-based records of the presence of off-shore ice, have provided several records of sea ice in the historical period. A detailed inventory of these records, giving their geographical location, temporal resolution, period of record and data source has been assembled by Barry (1986). The longest ice record, extending back to ~AD 865, gives the incidence of East Greenland ice off the coast of Iceland (Koch 1945). However, it should be noted that some of Koch's information, particularly prior to ~AD 1600 is not reliable (Ogilvie 1984). In her contribution to this volume, Astrid Ogilvie presents indices of sea ice off the coast of Iceland from 1601 to 1800. East Greenland ice that has rounded Cape Farewell to drift up the west coast of Greenland is termed *storis*. Speerschneider (1931) compiled a record of the annual extent of *storis* since 1821. A companion record from the western side of the Labrador Sea gives annual indices of the amount of ice drifting south in summer off the coast of Labrador in the 19th century (Newell 1983). Indirect indications of ice in Baffin Bay are given by a record of dates of penetration of whalers into the North Water, a polynia in the north of the bay (Dunbar 1972). This record extends from 1817 to 1870. Historical sea-ice records are also available for the Baltic Sea. These include records of winter maximum ice extent reconstructed annually since 1720 and of the date of opening of the port of Riga since 1530 (Lamb 1977, p.586-9).

This chapter examines a collection of log-books that has yielded indices of summer ice severity in Hudson Bay, Hudson Strait and the western margin of the Labrador Sea. These

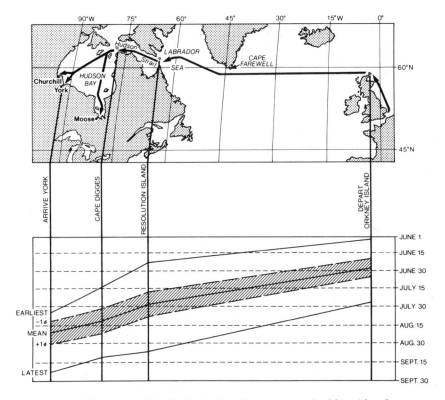

Figure 2.1 The route of the Hudson's Bay Company supply ships. Also shown are the times of departure from the Orkney Islands and the times of arrival at selected places on the outward voyages in the period 1751 to 1870. The times shown at each location include the mean date, the two standard deviations range and the extreme range.

log-books were kept on the supply ships of the Hudson's Bay Company (HBC) that sailed annually between London and the Company's trading posts on the bay. The route of the supply ships is shown in Figure 2.1. Usually they sailed from the Thames estuary to Stromness in the Orkney Islands where the Company's servants were embarked. Thence, by latitude sailing at 57° to 59°N, the ships were able to safely round Cape Farewell in Greenland. Hudson Strait was entered immediately to the south of Resolution Island. At the entrance to the bay the route diverged with one branch leading directly south to Moose Factory and the other striking across the bay to York Factory or Churchill Factory. The sailing routine compressed the outward voyage, the sojourn in the bay and the return voyage into the brief interval of summer ice break-up. This required the outward passage to be made early in the break-up period when severe ice was frequently encountered in three portions of the route (Danielson 1971, Crane 1978, Markham 1988). These portions were:

1) the western margin of the Labrador Sea where the ships traversed the Labrador current transporting pack ice and icebergs south from Baffin Bay into the Atlantic Ocean;

2) Hudson Strait in which high tidal amplitudes and strong tidal currents keep in motion pack

18

ice that drifts into the strait from Foxe Basin and Hudson Bay in the west and from the Labrador Sea in the east;
3) the southern margin of Hudson Bay between Churchill and James Bay where the prevailing northwesterly winds cause pack ice to accumulate throughout the summer (Figure 2.2).

The log-books written in these portions of the outward voyages contain detailed descriptions of the ice encountered. These descriptions have been interpreted in the reconstruction of four summer ice severity indices in the period 1751 to 1870:

1) *Hudson Strait*. The derivation of this index and its values were initially presented in Faurer (1981) and a revised method and index are now published in Catchpole and Faurer (1983). Information contained in the *Book of Ships' Movements* (HBCA, PAM, c4/1) enabled this record to be extended to 1889 (Catchpole and Hanuta 1989).
2) *Hudson Bay – approach to Moose Factory*. The method used to derive this index and its annual values are given in Catchpole and Halpin (1987).
3) *Hudson Bay – western approach*. This study applied the method developed in Catchpole and Halpin (1987) to the log-books written on ships sailing to York and Churchill. The annual indices are given in Catchpole and Hanuta (1989).

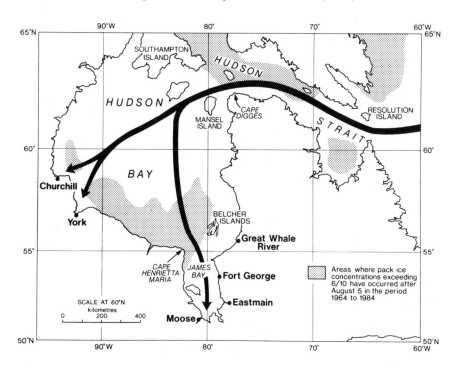

Figure 2.2 Location map of Hudson Bay and Hudson Strait showing the areas where ice concentrations exceeding 6/10 have been observed after 5 August in the period of the modern ice record. Sailing ships could not penetrate through pack ice with these concentrations. The ice data were derived from weekly ice summary maps published from 1964 to 1973 (Canada, Atmospheric Environment Service 1966 to 1981) and from unpublished manuscript maps for 1974 to 1984.

4) *Labrador Sea*. Teillet (1988) derived these indices using a modification of the method applied in the bay by Catchpole and Halpin (1987).

This research was undertaken in two phases with the Hudson Strait indices being derived before work commenced on the log-books kept in Hudson Bay and the Labrador Sea. The method applied to the Hudson Strait log-books was based on the logistics of sailing ship penetration through the strait. Its major variables were times of arrival at the entrance to the strait, durations of the westward passages through the strait and frequencies with which the ships grappled with or were fast in ice during the westward passage. At the outset of the second phase of the research, it was determined that the method developed for the derivation of ice indices in the strait is uniquely applicable to the strait and not universally applicable to all seas in which the HBC ships encountered ice. This prompted a reconsideration of the principles to be applied in the derivation of numerical ice indices from sailing ships' log-books. The outcome of this was a universally applicable method which assigns to each daily ice encounter a numerical index based on the seasonal lateness and the severity of the ice. It is this revised method that is examined in this chapter. The indices derived by the application of this method to the log-books kept in Hudson Bay are given in Tables 2.1 and 2.2.

2.2 The Hudson's Bay Company

On 2 May 1670 King Charles II of England granted a charter to a company of 18 merchants investing them with monopoly rights of trade and colonization in the lands draining into Hudson Bay. At that time this territory had not been explored and its extent was unknown, but it was eventually revealed that the HBC's chartered domain embraced 23% of the land surface of North America. Shortly after its incorporation the Company's charter was challenged by French fur traders who embarked on hostilities against its bayside posts in the late

Table 2.1 Annual indices of summer ice severity in Eastern Hudson Bay (the *Approach to Moose Factory*) 1751-1870.

Year	0	1	2	3	4	5	6	7	8	9
1750	*	-17	-2	-2	-12	-4	8	1	63	80
1760	0	4	-6	0	-5	-3	19	33	2	0
1770	*	21	*	*	74	*	*	126	-12	*
1780	-6	-20	3	0	2	124	0	17	41	*
1790	-12	-6	12	56	-10	-20	31	11	-15	-6
1800	-6	5	7	30	-5	-13	-10	4	2	-9
1810	8	0	23	180	-3	4	113	108	-8	-8
1820	-4	22	-1	2	43	1	0	-1	36	94
1830	122	-6	*	-1	-6	-2	12	16	-6	*
1840	*	*	34	288	85	150	-7	-2	9	-8
1850	0	46	0	0	-6	2	0	-1	-3	24
1860	2	-10	-4	-2	*	85	68	1	9	85
1870	0									

* Log-book not preserved in HBC archives.

Table 2.2 Annual indices of summer ice severity in Western Hudson Bay (the *Western Approach*) 1751-1869.

Year	0	1	2	3	4	5	6	7	8	9
1750	*	-4	6	0	0	0	26	0	10	88
1760	16	0	27	127	13	20	-5	8	9	3
1770	25	15	16	10	7	44	12	-1	3	174
1780	0	-4	4	0	4	40	3	-4	2	20
1790	0	0	15	2	0	64	35	61	20	0
1800	0	0	86	20	4	-9	0	0	84	0
1810	19	*	19	0	0	105	0	*	13	0
1820	38	65	2	22	5	10	0	83	15	-8
1830	38	*	-2	0	0	83	399	11	8	*
1840	*	*	-3	21	30	43	9	0	15	13
1850	5	*	12	0	5	207	0	7	5	4
1860	0	0	4	2	0	0	147	-4	3	3

* Log-book not preserved in HBC archives.

17th century. During the early decades of the 18th century this competition widened to include domestic rivals in England. This rivalry came to a head in 1749 when a Committee of the House of Commons was appointed to consider petitions to end the Company's monopoly. The petitioners were unsuccessful and the charter remained in force until 1870 when the HBC transferred its territories to the Canadian government (Williams 1970).

The climatic severity of the Company's lands was thus matched by the hostile commercial climate which prevailed, especially in its first century of business. The Company responded to this by developing a strategy of secrecy whereby its servants were forbidden to convey to outsiders information about its affairs or about the geography of the bay. While the Company thus restricted the flow of information to its enemies, it strengthened its own hand by instituting policies that ensured this flow to the Governor and Committee in London. These required its servants to keep detailed written records and to submit them regularly to the Governor. These policies, sustained over three centuries, brought into being and preserved the rich collection of written correspondence, reports, journals and account books that exists today as the HBC archives (Craig 1970). Tim Ball's contribution to this volume demonstrates the wealth of climatic information that has been extracted from these sources. This consists primarily of reconstructions of the frequencies of occurrence of weather phenomena, dates of first occurrences of events that mark the passage of the seasons and daily measurements of air temperature made over lengthy but sporadic time periods.

2.3 The log-book record

An element of the information that was to be carefully acquired and guarded was that bearing upon the navigation of the ice-congested waters of Hudson Strait and Hudson Bay. The Company's success in concealing information gathered by its mariners over a period of 147 years is revealed in a journal written by Lieutenant Edward Chappell. He was an officer on HMS *Rosamond*, a man-of-war which escorted the HBC ships *Prince of Wales* and *Eddystone* in 1814. On entering Hudson Strait, the Royal Navy officers found that 'nothing can be more incorrect than the Chart supplied by the *Admiralty* for the guidance of a man-of-war in *Hudson's Straits*: it absolutely bears no resemblance to the channel of which it is intended to be an exact delineation' (Chappell 1817, p.175-6). In commenting on this, Chappell noted that 'the officers of the *Hudson's Bay* ships have a motive in concealing from the public the knowledge which they actually possess relative to the navigation of the Northern Seas . . . this illiberal concealment has its origin in the Company themselves, who (as I am told by their own officers) have issued the strictest and most peremptory commands to the people in their employment, "that they take especial care to conceal all papers, and every other document, which may tend to throw light upon the Company's fur trade" . . .' (Chappell 1817, p.174-5).

One aspect of the practice of 'illiberal concealment' is that the Company undertook the preservation of copies of its ships' log-books and these survive today in its archives. The sailing ship log-book record extends from 1751, immediately following the parliamentary attack on the Company's charter, to 1870. The 119-year record has one interruption spanning 1839 to 1841 and the record includes a total of 485 log-books kept on 316 ships. There is a discrepancy between the number of log-books and the number of years of record because in most years the supply voyages involved two or more ships and quite often more than one log was kept on a ship. The archives also contain a large collection of log-books written on the Company's steam ships in the 20th century. These have not been used in this study because of the fundamental differences between the logistics of sailing ship and steam ship navigation.

2.4   The log-books

The general aspects of log-book keeping in the late 18th century and early 19th century are summarized in Falconer's *Marine Dictionary*. The first edition of this was published in 1769, shortly after the commencement of the HBC log-book record, and the revised *New Universal Dictionary of the Marine* appeared in 1815, roughly at the mid-point of the record. Falconer (1815, p.206) commented that 'There are various ways of keeping journals, according to the different notions of Mariners concerning the articles that are to be entered.' The 485 log-books in the HBC collection were written by 79 ships' officers over a period of 119 years. These numbers imply considerable scope for 'the different notions of mariners' to be revealed in idiosyncrasies in the format and contents of the individual log-books. However, the HBC log-books do not display such irregularities; they are highly uniform in terms of their style and their substance.

The origin of this homogeneity is not to be found in written instructions to the ships' captains from the Company. The *Sailing Orders and Instructions* issued by the Governor and Committee to the captains on the occasion of each voyage gave detailed orders on the route, the handling of the ship, the conduct of the crew and the protection of the Company's trade. However, these instructions were silent on the need for and nature of the ship's log-book. A scrutiny of the authorship of the log-books indicates that their uniformity probably arose from practices that were informally developed and shared by the officers themselves.

The HBC log-book collection is largely the work of a small number of captains and mates, each of whom was employed for a lengthy period of time. Almost half of the entire collection is credited to the ten most prolific log-keepers. Three officers authored 87 log-books, approximately one fifth of the whole. Figure 2.3 identifies the 19 officers who each kept ten or more log-books, showing the period during which the officer served and the number of log-books that he kept. It is apparent that the Company endeavoured to retain cadres of four to six experienced officers who served concurrently for periods of two or more decades. In the 1750's George Spurrell and William Coats completed their long service, and were relieved by Joseph Spurrell, William Norton and Jonathan Fowler Sr. These officers were, in turn, replaced by William Christopher, Joseph Richards and Jonathan Fowler Jr. who continued into the 1780s. This pattern of employment continued with only brief interruptions in the late 1830's and 1860's. Such a system of employment afforded excellent opportunities for the long-serving officers to work together, sharing knowledge and handing down that knowledge to their successors.

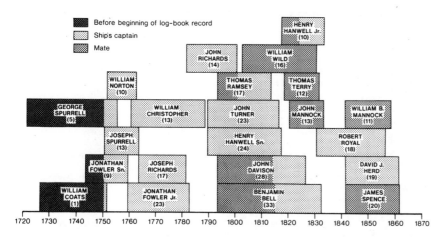

Figure 2.3 The Hudson's Bay Company ships' officers who each kept ten or more log-books. Also included are three long-serving officers who kept less than ten log-books and whose service commenced before the beginning of the log-book collection in 1751. Indicated for each officer are the years on which his service began and ended, the year of his promotion to captain and the number, in parenthesis, of his log-books preserved in the archives.

The format of the log-book page remained virtually unchanged throughout the period of the record. It accommodated routine observations made at two-hour intervals and irregular comments recorded in a remarks column on the right of the page. The routine observations were time of day, speed in knots, depth in fathoms, course direction, wind direction and weather. The seaman's day ended at noon when the errors of navigation by dead reckoning were generally corrected by solar observation. Hence midnight is in the middle of the page and noon at the bottom. Beneath the noon entry, there was a space for recording the compass variation and navigational information. The latter usually included the course direction and distance run over the past 24 hours together with the latitude and longitude based on dead reckoning and, when available, celestial observations. The information on sea ice conditions was recorded in the remarks column with additional comments occasionally appearing in the noon summary at the bottom of the page. Although sea ice descriptions were not part of the routine information to be recorded at two-hour intervals, there is an element of regularity in their incidence throughout the day. The surveillance of ice was the responsibility of the officer of the watch. During the course of the day there were six watches, each of four hours duration. Consequently, there were six occasions during the day when an officer of the watch commenced his duties and was apt to examine the sea surface and take note of any changes in ice conditions warranting description in the log-book. These times were 12.00h, 16.00h, 20.00h, 00.00h, 04.00h and 08.00h. It was at these times that the ice descriptions tended to be made. However, they were not distributed among the watches with equal frequency, but were generally more common during the daylight watches and were concentrated at noon. Figure 2.4 gives frequencies of ice descriptions made during the 775 days on which ice was encountered by ships sailing the approach to Moose Factory in Hudson Bay. Over 80% of these log-book pages contained at least one ice description made in the forenoon watch and 36% contained at least one made in the middle watch. An average of five ice descriptions per day appeared on log-book pages written when the ships encountered ice.

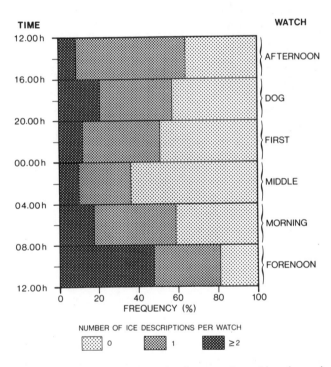

Figure 2.4 Division of the day into watches. Also shown is the frequency of occurrence of ice descriptions written during each watch on days when the ships were in the presence of ice when sailing the approach to Moose Factory in the period 1751 to 1870.

2.5 Range of visiblity from a sailing ship

The maximum range of vision from a ship's masthead to the horizon determines the field of direct observation of the sea surface under optimum light conditions. There is a substantial discrepancy between the theoretical value of this range and the values occasionally reported in the log-books. The range of vision depends on the height of the mast and estimates of the latter can be derived from information on ships' dimensions and displacements given in the *New Universal Dictionary of the Marine* (Falconer 1815). During the 18th and 19th centuries, the Company's vessels were primarily frigates, brigs and schooners of 100 to 250 tons burden. These ships carried mainmasts of 32 to 38 m in height from which the range of vision to the horizon was 23 to 25 km (Falconer 1815, p.198). The maximum ranges of vision reported in the log-books were of the order of 6 to 7 leagues (33 to 39 km). On 24 August 1824 when the *Camden* was beset in ice between Cape Henrietta Maria and the Belcher Islands, her log-book commented: 'no water to be seen from the masthead and can see 6 or 7 leagues all round' (HBCA, PAM, C1/224, 1824). On some occasions low-lying islands were reported visible at these distances. The log-book of the *Emerald* reported on September 15, 1816: 'at 6 saw the N. Belchers from the Masthead bearing 6 lgs ESE' (HBCA, PAM, C1/324, 1816). These apparently excessive ranges of vision may have been real effects of the arctic mirage. This occurs when relatively warm air is superimposed over a cold surface. The resultant temperature inversion may cause a curvature in light rays allowing direct vision of objects located below the horizon (Sawatzky and Lehn 1976). The enhancement of vision by the arctic mirage was, however, a rare phenomenon because it depended upon a specific combination of meteorological conditions.

A second, much commoner, optical effect extended the effective range of vision of ice conditions well beyond the horizon. This is the process of multiple reflection of light between the surface and the atmosphere. In arctic and mountain regions this causes a bright illumination of the sky over ice and snow that is termed *ice-blink*. The arctic explorer William Scoresby described it as 'a stratum of a lucid whiteness which appears over ice in that part of the atmosphere adjoining the horizon' (Scoresby 1820, p.299). Scoresby reported that 'when the ice-blink occurs under the most favourable circumstances, it affords to the eye a beautiful and perfect map of the ice, twenty or thirty miles [35 km to 55 km] beyond the limit of direct vision, but less distant in proportion as the atmosphere is more dense and obscure. The ice-blink not only shows the figure of the ice, but enables the experienced observer to judge whether the ice thus pictured be field or packed ice: if the latter, whether it be compact or open, bay or heavy ice . . .' (Scoresby 1820, p.299-300). There is no statement in the log-book collection referring explicitly to ice-blink but many comments imply that the state of the sea surface was being inferred from the appearance of the sky on the horizon. Thus, when the *Prince of Wales* was sailing in James Bay, on 14 September 1837 her log-book reported 'sailing through streams of heavy ice. Icy sky to No'ᵈ' (HBCA, PAM, C1/833, 1837).

Opportunities for the visual inspection of the sea surface in Hudson Bay and Hudson Strait are enhanced by the long hours of daylight and twilight in these latitudes in summer. The *Smithsonian Meteorological Tables* (List 1966) contain durations of daylight and civil twilight tabulated according to latitude and time of year. These have been used in the analysis of light conditions given in Figure 2.5. Daylight is the interval between sunrise and sunset. Civil twilight occurs before sunrise and after sunset to the time when the stars and planets of the first magnitude are just visible. Light fades during civil twilight but there is still a downward flux of diffuse solar radiation to illuminate ice on the sea surface. Consequently, the network of isopleths in Figure 2.5 gives the daily total duration of daylight and civil twilight. Superimposed on these isopleths are the dates on which the ships arrived at Resolution Island,

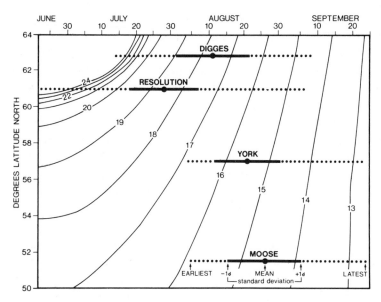

Figure 2.5 Total daily durations (hours) of daylight and civil twilight at latitudes 50°-64°N between 23 June and 25 September. Also shown are the sailing dates of the outward-bound ships at Resolution Island, Cape Digges, York Factory and Moose Factory. The mean dates, extreme ranges and standard deviations are based on the period 1751 to 1870.

Cape Digges, York Factory and Moose Factory on their outward voyages from England. The mean dates and one standard deviation ranges from the means show that at Resolution Island the sea surface was usually visible for roughly 18 to 19 hours per day. This period was 17 to 19 hours at Cape Digges, 15 to 16 hours at York Factory and 14 to 15 hours at Moose Factory.

Figure 2.5 demonstrates that, with the exception of the ships which arrived very early at Resolution Island and Cape Digges, all of the ships sailed in darkness during part of the day. The distance covered in darkness will be termed the *night traverse*. Its length was a function of latitude, time of year and speed of the ship. At each latitude and date the duration of darkness was constant and the length of the night traverse was directly dependant on the speed of the ship. The visual inspection of the sea surface was not interrupted along the entire length of the night traverse. At sunset the crew could look forward into the night traverse and at dawn they looked backward over sea surfaces which had recently been in darkness. Since the maximum range of vision from the masthead was 25 km, it is apparent that, under optimum circumstances, 50 km of the night traverse was made visible in this way. In these conditions, a portion of the night traverse would have been totally shrouded in darkness only when its length exceeded 50 km. Using the information given in Figure 2.5 it is a simple matter to calculate the sailing speeds required to produce night traverses exceeding this length at each latitude on each day. In July these speeds ranged from 7 to 15 km h^{-1}. In August the range was from 5 to 8 km h^{-1} and in September they ranged between 4 and 6 km h^{-1}. Ice was a major cause of the reduced sailing speeds that allowed direct vision into the entire night traverse.

2.6   Determination of ships' locations

The information in sailing ships' log-books can be fixed in time with much higher resolution and accuracy than it can be located in space. The log-book entries were made on twelve occasions, at two-hour intervals throughout the day. However, the determination of the ship's position was made only once in 24 hours. The spatial resolution of the log-book information is thus an order of magnitude less than its resolution in time. In addition, the accuracy of the estimates of the ships' locations was degraded by several deficiencies in navigation in the 18th and 19th centuries.

The main difficulty, which endured into the mid-19th century, was the measurement of longitude. The log-books kept during the 1750's indicate that celestial observations for latitude were made using both Hadley's and Elton's quadrants; thereafter, Hadley's quadrant alone was apparently used. Thus, at the outset of the period of study, the ships' captains observed latitudes with the instrument that continued in use on British merchant ships until the beginning of the 20th century (Hewson 1951, p.82). Harrison's invention of the chronometer had in principle solved the problem of the measurement of longitude as early as 1761. However, the great cost of this instrument restricted its early use to scientific voyages and long delayed its acquisition by captains of merchant ships (Taylor 1956, p.262). Chronometers first became standard equipment on HBC ships in 1834 and there are only two log-book entries mentioning their use before that date. Dead reckoning provided the alternative to navigation by celestial observation, but it could not rectify the errors in the estimation of longitude. Though simple in principle, dead reckoning was plagued by a multitude of errors involving the measurements of course directions and distances run.

These uncertainties prompted the development of two tests of the accuracy of the latitude and longitude coordinates given in the noon entries in the log-books. These are the *close consort test* and the *landmark sighting test*. A standing instruction in the Company's sailing orders required the captains in each voyage to keep their ships in close company until they entered the bay. The close consort test exploited the principle that when ships were sailing together their log-books should have recorded similar, though not identical, concurrent latitudes and longitudes. Discrepancies are estimates of the errors in these measurements. The test was applied by selecting 17 outward-bound voyages in which the log-books confirmed that the ships did keep close consort from the Orkney Islands to Cape Farewell. At least one such voyage was available in each decade from the 1750s through the 1840s. In each voyage the ships were paired and a total of 40 pairs obtained from the 17 voyages. For each day during the voyage to Cape Farewell the following discrepancies were then tabulated using the navigational information in the paired log-books: latitude by account, latitude by observation and longitude by account. The average differences between all 40 pairs of ships tested were: latitude by account, $00°\ 09'$, or 17 km; latitude by observation, $00°.04'$, or 7 km; longitude by account $00°.55'$, or approximately 45 km at 60°N.

The landmark sighting test (Teillet 1988) was applied when the ships made their first sighting of land, on Resolution Island after sailing across the Atlantic Ocean. This test involved a comparison of the position of the ship determined from the noon latitude and longitude recorded immediately before the sighting with the position derived from the bearing and distance of the landmark. The test was applied to landmark sightings that satisfied stringent criteria regarding the identity of the landmark sighted, the time of the sighting and knowledge of the compass variation. The average discrepancies between the positions based on landmark sightings and on the noon coordinates were: latitude, $00°.08'$, or 15 km; longitude prior to 1834, $01°.04'$, or 54 km; longitude after 1834, $00°.11'$, or 9 km. The results of the landmark sighting test are thus consistent with those of the close consort test. They indicate that, as expected, the locational data in the log-books cannot be used to determine precise daily positions of the ships, but they can provide an estimate of these positions. Accordingly, a network of marine sectors was established each comprising a rectangle with dimensions $00°.15'$ latitude and $01°.00'$ longitude. Each ship's position at noon was established within this network of marine sectors and the ice descriptions on a log-book page were treated as descriptions of the ice conditions in a particular sector on a specific day.

2.7   Derivation of the ice severity indices

A detailed explanation of the derivation of the ice severity indices is given in Catchpole and Halpin (1987), and the following is only an outline of the method. It involved four discrete stages of analysis:

1) The ice descriptions written in each sector on each day were evaluated against ordinal scales of *relative lateness*, **d** and *severity*, **s**.
2) These were combined by multiplication to yield a *daily ice index*, **d s**.
3) In each year the daily ice indices were combined by summation to give an *unadjusted annual ice index*.
4) The index obtained in Stage 3 was then converted to the *annual ice index* by applying a correction based on the modern record of ice conditions.

The relative lateness **d** of each daily ice description in each sector was evaluated by comparing the date of the ice description with the median date on which ice was last observed in this sector in the period of modern observations. These median dates are now available for the period 1964 to 1984 (Markham 1988) but when this research was undertaken the record of median dates extended from 1964 to 1973 (Sowden and Geddes 1980). The values of **d** assigned to ice descriptions ranged from one to eight. The lag between the date of the ice description and the corresponding median date was one to ten days for **d** = 1 and 71 to 80 days for **d** = 8.

An individual description provides information on the severity of the ice conditions in a particular sector on a specific day. The severity index **s** was based upon a content analysis of a randomly selected 10% sample (2460) of these descriptions. This content analysis determined the various word roots which, in different semantic forms, appeared in the descriptions and it enumerated the frequencies of occurrence of each word root. A total of 102 word roots was identified but these varied greatly in frequency and 50% of all descriptions were based on the 11 most common word roots. A classification of the word roots was undertaken to determine the types of information that they conveyed on sea ice conditions. This classification first recognized a basic dichotomy between word roots that described the appearance or behaviour of ice and those that described the handling and manoeuvering of the ship in the presence of ice. Slightly over half (51%) of the entries described ice, 33%

described the handling of the ship, and 16% were vague references to ice that could not be classified. The word roots that described ship-handling in the presence of ice consisted almost entirely of verbs referring to the navigation of the ship or the action of ice on the ship. Most word roots were common English words but some were nautical terms, interpreted with the aid of Falconer's 18th century marine dictionary (Falconer 1815). They indicated a wide range of sailing conditions imposed by the presence of ice, and content analysis subdivided these conditions into three categories: (1) course changes to avoid ice ahead, (2) penetration through ice, and (3) ship's passage blocked by ice. A fourth category was adopted to include occasions on which the ship was damaged by ice. Figure 2.6 lists the root words that were assigned into each category and ranks them in order of frequency of occurrence. The lists are complete, except for root words that occurred only once in the sample analysed.

The word roots that described the appearance and behaviour of the ice consisted mainly of adjectives, though nouns and verbs appeared also in this class. The majority described the degree of fragmentation and amount of pack ice. Smaller categories of descriptive words

ICE CATEGORIES

WORD ROOTS AND PHRASES		PACK ICE: Open (SHIP MAKING EASY PASSAGE)	PACK ICE: Closed 1 (FORCED PASSAGE)	PACK ICE: Closed 2 (PASSAGE BLOCKED)	ICEBERGS	ICE PRESENT	ICE ABSENT
ICE APPEARANCE AND BEHAVIOUR	State of Pack Ice: OPEN PACK	OPEN PIECE STRAGGLE LOOSE SHATTERED SLACK SMALL				ANY REFERENCE	
ICE APPEARANCE AND BEHAVIOUR	State of Pack Ice: CLOSE PACK		CLOSE HEAVY PACKED BODY FLOE LEDGE THICK FAST ICE FIELD NO WATER			ANY REFERENCE	
ICE APPEARANCE AND BEHAVIOUR	Icebergs				ICE ISLANDS ICEBERGS		
ICE APPEARANCE AND BEHAVIOUR	Ice Motion	STREAM SAIL DISPERSE				ANY REFERENCE	
ICE APPEARANCE AND BEHAVIOUR	Ice Free						CLEAR OPEN NO ICE
SHIP HANDLING AND MANOEUVRES	Avoiding Ice	TACK TURN WORE SHIP	SAIL ALONG ROUND HAUL UP ALTER COURSE BEAR AWAY			ANY REFERENCE	
SHIP HANDLING AND MANOEUVRES	Penetrating Ice	SAIL THROUGH RUN IN PASS THROUGH STEER THROUGH TRAVERSE	FORCE WORK AMONG ROW AND TOW WARP DRIVE THROUGH BORE THROUGH FALL THROUGH FORGE THROUGH BREAK THROUGH			ANY REFERENCE	
SHIP HANDLING AND MANOEUVRES	Passage Blocked			GRAPPLE BESET (SET) FAST IN CAST OFF BRING TO CANNOT MOVE STOPPED MOORED ANCHORED		ANY REFERENCE	
SHIP HANDLING AND MANOEUVRES	Damage by Ice			PRESS STOVE BLOWS STRIKING		ANY REFERENCE	
VAGUE REFER-ENCES						PASSED SAW APPEAR FALL IN WITH OBSERVE CLOSE TO COME UP TO MEET	

Figure 2.6 Content analysis of the word roots and the ice categories based on this analysis (Catchpole and Halpin 1987, p 240).

referred to icebergs, to the movement of the ice and to ice-free water. Within the log-books the term *ice island* was used to describe icebergs. The commonest word roots used to describe ice were 'open' and 'closed'. These word roots were used as the basis for a subdivision of the state of pack ice into three categories: Open, Closed 1 and Closed 2. Open Ice was assumed to exist when the ice was described as being highly fractured and dispersed, when it was in rapid motion, or when the ship easily sailed through it. When the ice was described as being heavily packed it was designated Closed Ice 1. This condition was also assumed to exist if the course of the ship was changed to avoid the ice or if it could make way through it only by forcing a passage. This could be done with all sails set, by towing the ship from its boats (rowing and towing) or by attaching the ship to ice using grapnels and pulling it forward (warping). The second category of closed ice was specified on occasions when the ship was unable to make headway or when the captain chose to grapple to an ice floe to secure his ship.

These definitions of open and closed ice differ from the present-day internationally accepted terminology established by the World Meteorological Organization (WMO) and applied in Canada (Canada, Hydrographic Service 1988). The WMO definitions of 'Open Water,' 'Very Open Pack,' 'Open Pack,' and 'Close Pack' are based on the extent of pack ice measured in tenths of the surface covered. However, there is no basis for assigning numerical limits to the ice concentrations through which a sailing ship could sail easily or proceed by forcing its passage. The definitions of 'Open,' 'Closed Ice 1,' and 'Closed Ice 2' will therefore remain qualitative. For the purposes of calculating daily and annual ice indices, numerical values of ice severity s were arbitrarily assigned to each ice category. These values were zero for 'Ice Absent,' one for 'Open Ice' and 'Ice Present,' two for 'Closed Ice 1,' and three for 'Closed Ice 2.'

The daily ice indices were obtained by multiplying **d** and **s**. The unadjusted annual ice indices were obtained by summing the daily ice indices. Since there was no overlap between the era of sailing ship navigation in the bay and the period of modern ice observations, there is no basis for directly calibrating these indices against the modern records. However, an adjustment was applied to the ice index, which permits comparisons between the historical ice conditions and those that existed in the most severe ice year in the period 1964 to 1984. A scrutiny of the annual *Ice Summary and Analysis* reports for 1964 to 1973 and of the subsequent weekly ice charts showed the most severe late summer ice in the period 1964 to 1984 occurred in 1969 in the approach to Moose Factory and in 1973 in the western approach. A procedure was applied to compare the severity of the 1969 ice with that encountered by the sailing ships in the approach to Moose Factory in each year between 1751 and 1870. This was repeated for the western approach using 1973 as the base for comparison. The procedure involved plotting each of the annual routes of the sailing ships onto the weekly ice charts for 1969 (or 1973). To facilitate this, the grid of marine sectors was superimposed on each of the 1969 (or 1973) weekly ice charts. This enabled the estimation of hypothetical daily values of ice severity s' and lateness d' that would have been observed if each of the ships' voyages had been made in 1969 (or 1973). These values of s' and d' were combined into hypothetical daily and annual ice indices. The procedures used here were the same as those applied previously, except for the determination of s'. In the historical period, s was based on the descriptions in the log-books and the ice categories given in Figure 2.6. In the 1969 (or 1973) study, s' was arbitrarily based on the two WMO definitions of ice concentrations: 'Open Ice' was equated with the WMO's 'Very Open Pack' (one tenth to three tenths covered by ice); 'Closed Ice 1' was equated to the WMO's 'Open Pack' (four tenths to six tenths); 'Closed Ice 2' was equated to the WMO's 'Close Pack' (seven tenths to eight tenths) and greater concentrations; 'Ice Absent' was equated to the WMO's 'Open Water' (less than one tenth). These definitions assume that a sailing ship could make a forced passage through four tenths to six tenths of ice. Dunbar (1985, p.110) designated ice covers of five tenths in the most congested areas as being the upper limit for penetration by a sailing ship while acknowledging that sailing ships could occasionally force through denser ice concentrations. In this way, series of hypothetical annual ice indices were obtained for the 1969 (or 1973) ice conditions. Each was derived by superimposing an annual sailing route on the 1969 (or 1973) ice patterns. The final correction was made by subtracting the hypothetical ice index from each corresponding unadjusted annual ice index. This yielded the annual ice indices given in Tables 2.1 and 2.2.

2.8   Nature of the ice severity indices

The adjustment applied in the last stage in the derivation of the ice indices provides a yardstick for comparing each year between 1751 and 1870 with the year which experienced the most severe summer ice in the bay during the period 1964 to 1984. However, these indices have not been calibrated against modern ice conditions. The 94 years between the end of the sailing ship log-book record and the commencement of the modern ice record remains a void in terms of knowledge of summer ice conditions in the bay. Glimpses of these conditions during this void may be revealed by other historical sources, particularly in the period following the opening of the port of Churchill in 1929. For example, records of the dates of 'opening and closing the harbour at Churchill to navigation' extend back to 1928 (Allen and Cudbird 1971). These dates are, however, primarily determined by the thawing and freezing of the freshwater discharging into the Churchill River estuary, rather than the thawing and freezing of the saltwater along the sailing route to Churchill. There has not yet been a systematic inventory of the historical evidence of sea ice conditions in the bay during the past century. Therefore, there is no basis at this time for speculating whether or not these sources may yield numerical summer ice severity indices for the period 1871 to 1963.

Although a direct approach to the calibration of the ice severity indices is impractical, there are good grounds for anticipating that this may eventually be undertaken indirectly using proxy evidence of summer ice conditions in the bay. Two features of the Hudson Bay region have endowed it as a rich source of proxy evidence of climatic change during the past three centuries. The earliest HBC posts were located at the mouths of rivers draining into the bay and these have furnished the longest and most continuous post journal records. A substantial body of historical evidence of past climates has been derived from these sources (Moodie and Catchpole 1975, Catchpole and Ball 1981, Ball and Kingsley 1984, Wilson 1988). The bay region is also emerging as an important source of dendroclimatological data. The North American forest-tundra ecotone is displaced southward in the eastern part of the continent. It intersects the bay in the vicinity of Churchill, parallels the south west coast to James Bay and then intersects the coast again adjacent to the Belcher Islands. This ecotone is a fruitful zone for dendroclimatological research where tree growth is sensitive to summer temperature and where the fragmentation of the forest cover into isolated pockets offers refuges from the ravages of forest fires and opportunities for enhanced longevity of tree growth (see, for example, D'Arrigo and Jacoby, Chapter 15, this volume). Dendroclimatological research in the bay region has been focussed in the vicinity of Churchill (Jacoby and Ulan 1982, Scott and Hansell 1987, Scott *et al.* 1988) and in the vicinity of James Bay and Great Whale River (Parker *et al.* 1981, Filion *et al.* 1986). Guiot (1987) has integrated these historical and tree-ring data in his reconstruction of annual and seasonal temperatures in central Canada in each year between 1700 and the present. These data may provide a means for the indirect calibration of the ice severity indices but this potential has not yet been investigated.

Since the ice indices are uncalibrated they function as ordinal not interval data. As such, the indices rank the years on the basis of summer ice severity, but they are not numerical measures of the quantities of ice present in each summer. Furthermore, the rank orders are only approximate since they are dependant on several arbitrary decisions made during the content analysis of the log-book descriptions. This applies, for example, to the choice of weightings applied to the various degrees of ice lateness and severity. The decisions to combine **d** and **s** first by multiplication and later by addition were also arbitrary. Another feature of the ice indices is that their frequency distributions are highly skewed. For example, over 50% of the indices in Tables 2.1 and 2.2 are below five and less than 10% exceed 100. This property implies that the indices discriminate more accurately between the ranking of the few severe ice years than between that of the large number of moderate and light ice years. Therefore, it is recommended that the ice indices should be primarily used in studies concerning the incidence of exceptionally severe late summer ice.

2.9   Occurrence of severe summer ice, 1751 to 1889

To facilitate this, Figure 2.7 identifies the years having the ten highest ice indices in each of the ice severity records derived in western Hudson Bay, eastern Hudson Bay and Hudson Strait. The method used to reconstruct the indices in the strait is different from the method applied in the bay and outlined in this chapter. For this reason the numerical values of the Hudson Strait indices are not given here. However, the severe ice years in the strait are included in Figure 2.7 to extend the regional coverage of these data and to provide information for the two decades following 1870. The severe ice record from the Labrador Sea is not included in Figure 2.7 because the sailing route across the sea is connected, at Resolution Island, with the route through Hudson Strait. Consequently, the log-book sources used by Teillet (1988) to reconstruct the ice indices in the Labrador Sea included some of the log-books used in the Hudson Strait reconstruction.

Figure 2.7 identifies only two years, 1816 and 1836, when severe ice indices were derived in two of the records and there was no year in which all three records yielded indices ranking among the highest ten values. There were several periods in which severe ice indices occurred consecutively, as in 1813 to 1817, 1835 to 1836, 1843 to 1845, 1854 to 1855, 1858 to 1859, and 1865 to 1866. Catchpole and Hanuta (1989) have identified a tendency for severe summer sea ice in Hudson Strait and Hudson Bay to follow major volcanic eruptions in the period 1751 to 1889. This condition occurred after the eruptions of Lakagigar in 1783, Tambora in 1815, Coseguina in 1835, Sheveluch in 1854, Askja in 1875, and Krakatau in 1883. The virtual lack of coincidence between the severe ice indices in the three records may cast doubt on the validity of these data but it is consistent with the nature of summer ice dispersal processes in

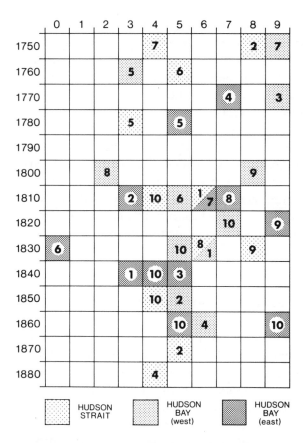

Figure 2.7 Years with the ten highest ice severity indices in each record. The numerical values give the ranks on a scale which allocates Rank 1 to the greatest severity. There are no log-books available for 1871 to 1889 but the ice severity in Hudson Strait in that period can be derived from information given in the Book of Ships' Movements (HBCA, PAM, c.4/1).

the bay and the strait (Crane 1978, Danielson 1971, Markham 1988). These are dynamic processes, responding to the seasonal regimes of radiation and temperature, but strongly regulated by winds and sea currents. It is beyond the scope of this chapter to compare the atmospheric and sea circulation patterns prevailing in the bay with those in the strait. However, a brief comparison of ice dispersal in the eastern and western parts of Hudson Bay, traversed by the HBC sailing routes, is appropriate. The first parts of the bay to clear of ice in summer are normally the waters adjacent to Southampton Island in the extreme north west. The ice is cleared from here by the prevailing northwesterly winds which drive it south to a melting zone between Churchill and James Bay (Danielson 1971). It is within this zone that the last remnants of ice are normally found in late summer (Figure 2.2). In years with a zonal atmospheric circulation this last ice is driven towards the east and accumulates in the path of the sailing route to James Bay. A meridional circulation permits the last ice to remain in the west in the vicinity of the sailing route to York and Churchill. The two ice records derived in the bay should not, therefore, be considered as separate entities indicating the incidence of severe ice in different water bodies. It is more appropriate to treat these records as indicators of severe ice at the lateral limits of a single water body. This is the portion of the bay in which the last remnants of ice are normally found. This implies that the records of severe ice from western and eastern Hudson Bay should usually be used together and not separately.

The combination in Figure 2.7 of the severe ice years in the strait with those in the bay is justified on grounds that relate to the nature of the log-book sources rather than to ice dispersal processes. Severe ice in Hudson Strait retarded the entry of the ships into the bay to such a degree that they first encountered the bay ice in September. Normally this occurred in the middle of August (Figure 2.1). In September even very late summer ice is generally cleared from the bay. Severe ice in the strait therefore limited the opportunitites for the log-books to describe any late ice that may have occurred concurrently in the bay. For this reason the records of severe ice in the strait should be consulted when decisions are made about the incidence of severe ice in the bay.

2.10   Comparison between ice indices and Wilson's thermal indices

This chapter will conclude with an application of the ice records which compares the incidence of severe summer sea ice with measures of relative summer warmth in the Hudson Bay region derived by Wilson (1985, 1988) from other historical sources. The objectives of this application are to identify the main features of summer thermal variations in this region in the 19th century and to make tentative observations on the validity of the sea ice indices. Wilson's data comprise a series of summer thermal indices based on weather observations and proxy weather data from the HBC post journals, correspondence and annual reports for Great Whale River, Fort George and Eastmain (Figure 2.2). The weather observations used in this reconstruction were daily temperature measurements that were made sporadically at the trading posts. The proxy data range from descriptions of snowfalls, rainfalls, frosts, frozen ground, snow and ice covers to descriptions of a wide array of botanical and faunal phenological events. Also used were dates of commencement of various stages in gardening activities at the posts. The methods used in this derivation are given in Wilson (1985) and the indices appear in Wilson (1988). The thermal indices were derived for four seasonal intervals: May to June, July to August, September to October, and May to October. Figure 2.8 selects for comparison with the ice indices only the May to June thermal indices. This choice was made because there is no possibility that the sea ice descriptions in the log-books of the ships sailing to Moose Factory could have contributed to the May to June thermal indices. Therefore, the two sets of data in Figure 2.8 are derived from independent sources.

The outstanding feature apparent in Figure 2.8 is the coincidence between exceptionally low thermal indices and a sequence of severe ice indices during the second decade of the 19th century. The two anomalies do not coincide exactly since the severe ice commenced in 1813, one year before the plunge in the thermal index. It is noteworthy that Wilson (1988, p.7) designated 1815 to 1821 as the only period in which the quality of the thermal indices is excellent. During this period weather diaries at several posts recorded air temperatures twice or thrice daily. During the period from 1829 to 1845 there is a fairly close concurrence between severe summer ice and low thermal indices and Wilson (1988, p.8) described the

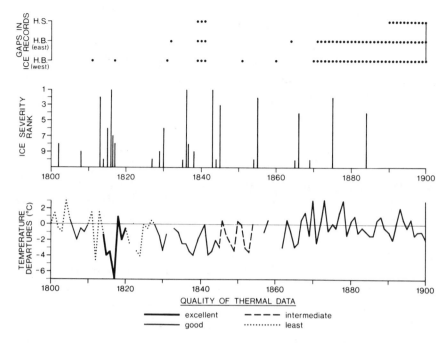

Figure 2.8 Thermal indices for the southeast coast of Hudson Bay (Wilson 1988) and years with severe summer ice in Hudson Strait and Hudson Bay during the 19th century. The thermal indices are estimates of departures from the 1941 to 1970 normals of temperatures in May and June. The quality of these indices was assessed by Wilson (1988, p. 7-8). Severe ice years are those identified in Figure 7. Gaps in the ice record are shown for Hudson Strait (H.S.) and the two parts of Hudson Bay (H.B.).

quality of the thermal indices in this period as good. After 1845 there are but few concurrences between severe ice and low thermal indices. However, in the latter half of the century opportunities for concurrences are reduced by gaps in the two records, particularly in the ice record after 1869. In conclusion, it can be stated that during the periods when the two records are available and when the quality of the thermal index is described as excellent or good, there are fairly good agreements between these records. Since cold summer weather in this region may be a cause or a result of severe summer sea ice, the agreements revealed in Figure 2.8 provide a partial confirmation of the validity of the two sets of data.

2.11 Summary

Shortly after its incorporation in 1670, the Hudson's Bay Company adopted policies which led to the keeping and preservation of detailed written records of its fur trade. These ranged in scope from letters, district reports and business accounts to daily journals of events at the trading posts and log-books kept on the Company's supply ships. A remarkably complete collection of these documents is preserved in the Company's archives in Winnipeg. The documents were written in a harsh subarctic climate and made frequent reference to weather conditions. Thus, they comprise a valuable source of written historical climatic information.

The sailing ship log-book record extended from 1751 to 1870 with only one interruption in 1839 to 1841. It comprises a total of 485 log-books. These are relatively homogeneous in terms of the time of year in which they were written, the sailing routes where they were written, their format and their substantive contents. The substantive homogeneity is largely attributable to the nature of their authorship. The Company retained the services of its ships' officers for long periods of time and the majority of the log-books were kept by a very few officers. Almost one fifth of the entire collection was kept by three people and one half is the work of the ten most prolific log-keepers.

It was necessary for the outward voyages to commence early in the summer and severe ice was often encountered in the Labrador Sea, in Hudson Strait and in Hudson Bay. This ice was frequently described in the log-books and there are grounds for maintaining that these descriptions were closely representative of the ice encountered along a broad corridor in the path of the ships. The direct range of vision from the masthead of a sailing ship was about 23 to 25 km and the phenomenon of ice-blink rendered sea ice visible as reflected light for a distance of up to 55 km beyond the horizon. Summer nights are brief in these northern waters. Indeed, it is concluded that when sailing speeds were low to moderate no part of the route was completely obscured by darkness. Two tests were devised for assessing the accuracy of the daily latitudes and longitudes recorded in the log-books. These revealed that the point-locations of the ships cannot be accurately determined but that the ships can be located within a network of rectangles with dimensions $00°.15'$ latitude by $01°.00'$ longitude.

The derivation of the ice severity indices commenced with the identification of the location at which each daily log-book page was written. When sea ice was encountered at this location numerical estimates were made of its severity and its seasonal lateness measured in relation to modern records of the summer ice dispersal process. These estimates of severity and lateness were combined to give daily and annual ice severity indices. The calculation of the latter involved an adjustment based on the ice conditions observed in the year having the most severe late summer ice in the period of the modern ice record.

It was stressed that these indices comprise ordinal, not interval data. They serve to rank the years according to summer ice severity, but do not provide numerical measures of the quantities of ice present in each summer. It was also stressed that this rank ordering of years is only an approximation and that it is more sensitive in ranking the severe ice years than in discriminating between the moderate and low ice years. The years with the ten highest ice indices in each of three ice records were identified. The distribution of these in the 19th century was compared with a concurrent record of thermal indices for the east coast of James Bay and adjacent southeast coast of Hudson Bay. It was concluded that the similarities between these two records provide a partial confirmation of the validity of the two sets of data.

References

Allen, W. T. R. and B. S. V. Cudbird 1971. *Freeze-up and break-up dates of water bodies in Canada.* CL1-1-71. Toronto: Canadian Meteorological Services.

Ball, T. F. and R. A. Kingsley 1984. Instrumental temperature records at two sites in central Canada: 1768 to 1910, *Climatic Change*, 6,39-56.

Barry, R. G. 1986. The sea ice data base. In: *The Geophysics of Sea Ice*, N. Untersteiner (ed.), 1099-1134, New York: Plenum Publishing Corporation.

Canada, Atmospheric Environment Service 1966-1981. *Ice summary and analysis 1964-1973: Hudson Bay and approaches*. Toronto. Canada, Hydrographic Service 1988. *Sailing directions Labrador and Hudson Bay*. Ottawa.

Catchpole, A. J. W. and T. F. Ball 1981. Analysis of historical evidence of climatic change in western and northern Canada. In: *Climatic change in Canada 2, Syllogeus* 33, C. R. Harington (ed.), 48-96.

Catchpole, A. J. W. and M. A. Faurer 1983. Summer sea ice severity in Hudson Strait, 1751-1870. *Climatic Change*, 5, 115-39.

Catchpole, A. J. W. and J. Halpin 1987. Measuring summer sea ice severity in eastern Hudson Bay, 1751-1870. *The Canadian Geographer*, 31, 233-44.

Catchpole, A. J. W. and I. Hanuta 1989. Severe summer ice in Hudson Strait and Hudson Bay following major volcanic eruptions, 1751 to 1889 A.D. *Climatic Change*, 14, 61-79.

Chappell, E. 1817. *Narrative of a voyage to Hudson's Bay in His Majesty's Ship Rosamond*. London: J. Mawman.

Craig, J. 1970. Three hundred years of record. *The Beaver*, 301, 65-70.

Crane, R. G. 1978. Seasonal variations of sea ice extent in the Davis Strait – Labrador Sea area and relationships with synoptic-scale atmospheric circulation. Arctic, 31, 434-47.

Danielson, E. W. 1971. Hudson Bay ice conditions. *Arctic*, 24, 90-107.

Douglas, K. S., H. H. Lamb and C. Loader 1978. *A meteorological study of July to October 1588: the Spanish Armada storms*. Publication 6, Norwich: Climatic Research Unit, University of East Anglia.

Dunbar, M. 1972. Increasing severity of ice conditions in Baffin Bay and Davis Strait and its effect on the extreme limit of ice. In: *Sea Ice*, Proceedings of an International Conference, T. Karlssen (ed.) 87-93. Reykjavik: National Research Council of Iceland.

Dunbar, M. 1985. Sea ice and climatic change in the Arctic since 1800. In: *Climatic change in Canada 5: Critical periods in the Quaternary climatic history of northern North America, Syllogeus* 55, C. R. Harington (ed.) 107-19.

Falconer, W. A. 1815. *A new universal dictionary of the marine, modernized and enlarged by William Burney*. Reprinted 1974. London: Macdonald and Jane's.

Faurer, M. A. 1981. *Evidence of sea ice conditions in Hudson Strait, 1751-1870, using ships' logs*. M.A. Thesis. Winnipeg: University of Manitoba.

Filion, L., S. Payette, L. Gauthier and Y. Boutin 1986. Light rings in subarctic conifers as a dendrochronological tool. *Quaternary Research*, 26, 272-9.

Guiot, J. 1987. Reconstruction of seasonal temperatures in central Canada since A.D. 1700 and detection of the 18.6- and 22-year signals. *Climatic Change*, 10, 249-68.

Hadley, G. 1735. Concerning the cause of the general trade winds. *Philosophical Transactions of the Royal Society of London*, 39, 58-62.

Hewson, J. B. 1951. *A history of the practice of navigation*. Glasgow: Brown, Son and Ferguson.

Jacoby, G. C. and L. D. Ulan 1982. Reconstructions of past ice conditions in a Hudson Bay estuary using tree-rings. *Nature*, 298, 637-9.

Koch, L. 1945. The east Greenland ice. *Meddelelser om Grönland* 130.

Kutzbach, G. 1987. Concepts of monsoon physics in historical perspective: the Indian monsoon (seventeenth to early twentieth century). In: *Monsoons*, J. S. Fein and P. L. Stephens (eds.) 159-209. New York: Wiley.

Lamb, H. H. 1977. *Climate: present, past and future*. Volume 2: *Climatic history and the future*. London: Methuen.

Lamb, H. H. and A. I. Johnson 1966. Secular variations of the atmospheric circulation since 1750. *Geophysical Memoirs* 110, London: Meteorological Office.

Landsberg, H. E. 1954. The storm of Balaklava and the daily weather forecast. *Scientific Monthly*, 79, 347-52.

Landsberg, H. E. 1985. Historical weather data and early meteorological observations. In: *Paleoclimate analysis and modelling*, A. D. Hecht (ed.), 27-70. New York: Wiley.

List, R. J. 1966. Smithsonian meteorological tables. *Smithsonian Miscellaneous Collections* 114. Washington: Smithsonian Institution.

Markham, W. E. 1988. *Ice atlas: Hudson Bay and approaches*. Ottawa: Canada, Atmospheric Environment Service.

Moodie, D. W. and A. J. W. Catchpole 1975. Environmental data from historical documents by content analysis: freeze-up and break-up of estuaries on Hudson Bay, 1714-1871. *Manitoba Geographical Studies* 5, Winnipeg: Department of Geography, University of Manitoba.

Newell, J. 1983. Preliminary analysis of sea-ice conditions in the Labrador Sea during the nineteenth century. In: *Climatic change in Canada 3, Syllogeus* 49, C. R. Harington (ed.) 108-29.

Ogilvie, A. E. J. 1984. The past climate and sea-ice record from Iceland, part 1: data to A.D. 1780. *Climatic Change*, 6, 131-52.

Oliver, J. and J. A. Kington 1970. The usefulness of ships' log-books in the synoptic analysis of past climates. *Weather*, 25, 520-8.

Parker, M. C., L. A. Jozsa, S. G. Johnson, and P. A. Bramhall 1981. Dendrochronological studies on the coasts of James Bay and Hudson Bay. In: *Climatic change in Canada 2, Syllogeus* 33, C. R. Harington (ed.), 129-88.

Sawatzky, H. L. and W. H. Lehn 1976. The arctic mirage and the early North Atlantic. *Science*, 192, 1300-5.

Scoresby, W. 1820. *An account of the arctic regions, with a history and description of the northern whale-fishery*. Vol. 1, Edinburgh: Archibald Constable Co.

Scott, P. A., D. C. F. Fayle, C. V. Bentley, and R. I. C. Hansell 1988. Large-scale change in atmospheric circulation interpreted from patterns of tree growth at Churchill, Manitoba, Canada. *Arctic and Alpine Research*, 20, 199-211.

Scott, P. A. and R. I. C. Hansell 1987. Establishment of white spruce populations and responses to climatic change at the treeline, Churchill, Manitoba, Canada. *Arctic and Alpine Research*, 19, 45-51.

Sowden, W. J. and F. E. Geddes 1980. *Weekly median and extreme ice edges for eastern Canadian seaboard and Hudson Bay*. Ottawa: Ice Climatology and Applications Division, Environment Canada.

Speerschneider, C. I. H. 1931. The state of the ice in Davis-Strait 1820-1930. *Meddelelser* 8. Copenhagen: Danish Meteorological Institute.

Taylor, E. G. R. 1956. *The haven-finding art: a history of navigation from Odysseus to Captain Cook*. London: Hollis and Carter.

Teillet, J. V. 1988. *A reconstruction of summer sea ice conditions in the Labrador Sea using Hudson's Bay Company ships' log-books, 1751 to 1870*. M.A. Thesis. Winnipeg: University of Manitoba.

Williams, G. 1970. Highlights of the first 200 years of the Hudson's Bay Company. *The Beaver*, 301, 4-59.

Wilson, C. V. 1985. *The summer season along the east coast of Hudson Bay during the nineteenth century: Part III summer thermal and wetness indices, A. Methodology. Report 85-3*. Downsview, Ontario: Canadian Climate Centre, Atmospheric Environment Service.

Wilson, C. V. 1988. *The summer season along the east coast of Hudson Bay during the nineteenth century: Part III summer thermal and wetness indices, B. The indices 1800 to 1900. Report 88-3.* Downsview, Ontario: Canadian Climate Centre, Atmospheric Environment Service.

Manuscript sources cited

Canada, Atmospheric Environment Service. Manuscript maps of ice conditions in Hudson Bay and its approaches, constructed at weekly intervals, 1974-1984. Ottawa: Ice Forecasting Central.

Hudson's Bay Company Archives (HBCA), Provincial Archives of Manitoba (PAM). Log-book of *Emerald*, 1816, c.1/324.
HBCA, PAM. Log-book of *Camden*, 1824, c.1/224.
HBCA, PAM. Log-book of *Prince of Wales*, 1837, c.1/833.
HBCA, PAM. Book of Ships' Movements, c.4/1.

3 Historical and instrumental evidence of climate: western Hudson Bay, Canada, 1714-1850

T. F. Ball

3.1 Introduction

The reconstruction of past climates is essential to any understanding of the scope and mechanisms of global weather. A major problem in any reconstruction is the transition from one type of information to another to produce a meaningful and continuous record. Such a record is essential if we are to produce a paradigm that will link together the multiple 'red' noises that result in the 'white' noise that is climate.

There are few places in the world where historical and early instrumental records were maintained. It is even more unusual to find such records that span long periods of time. Ironically, the remote and harsh region of central and northern Canada has such records. The diaries, journals, accounts books, district reports and instrumental weather records of the Hudson's Bay Company provide an extended and detailed picture of environmental conditions. The richness and extent of the record required techniques that would be applicable to all sites at which similar observations were made. This chapter outlines the methods used to create an objective, quantitative procedure for combining proxy, phenologic and instrumental records into a continuum. It concludes with analysis of the results.

Three sites, York Factory, Churchill Factory, and Cumberland House were chosen as representative of the records and used to establish a methodology (Figure 3.1). Two of the sites, York Factory and Churchill Factory located on the southwest shore of Hudson Bay, then became the basis for creating a long, continuous, quantified record of climate. Cumberland House is located in the heart of the boreal forest on the Saskatchewan River west of its entry into Lake Winnipeg. It was included to ensure that any terminology different from the coastal stations be incorporated into the coding system. York Factory has the longest and most complete record. Churchill Factory has a record almost comparable in length, but also has a statistically significant modern record for comparison.

The region is important to the reconstruction of global climates. The Rocky Mountains and Hudson Bay affect the Westerlies and planetary waves thus influencing the pattern of weather in the northern hemisphere. The period covered is important because it spans a major portion of the Little Ice Age and the transition to the warmer conditions of the 20th century. It allows for construction of a continuous record incorporating historical and instrumental sources. The data has been used for comparison with tree ring studies allowing reconstruction of conditions prior to the period of record.

This chapter provides details of the sites, the source and type of information available, the method of analysis, analysis of the results and an attempt to put the information into a hemispheric context.

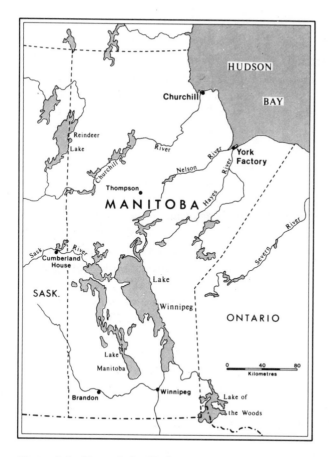

Figure 3.1 Sites of the Hudson's Bay Company trading posts.

3.2 The sites

Churchill Factory with a record which commences in 1718 and York Factory with a record beginning in 1714, are situated close enough for comparison and far enough apart to determine relative homogeneity. Churchill today continues as a seaport at the mouth of the Churchill River in the southwest corner of Hudson Bay.

York Factory, situated approximately 140 km southeast of Churchill on the isthmus between the Hayes and Nelson Rivers, was abandoned by the Hudson's Bay Company in 1952. Ironically, it was the most important of the two posts during the fur trade era, but quickly declined when it was determined that Churchill had a superior harbour for modern shipping. The most complete records were kept at this post, especially the instrumental records which began in 1770.

Both sites are in low coastal areas at the northern limit of trees. The flatness and lack of trees, other than low willow shrub, make ideal locations for climate observations.

3.3 The records

The Hudson's Bay Company was controlled by a Board of Governors sitting in London. This remoteness required that they receive as much accurate information as possible; as a result very detailed records were kept and sent to London. Three important things result. The records are long, detailed, and consistent. All are desirable for climate reconstruction, but the latter is the most important. The sheer volume of the data required very basic analysis some of which is presented here.

The Company specified what information should be recorded. They chastised employees who did not provide adequate detail or were remiss in making entries. Most officers were trained in navigation and all were very familiar with weather. Familiarity came from the almost total control the weather had upon their lives. It determined the number of animals and thereby the success of the fur trade. It dictated the ease or difficulty of transport, whether by snow, land, or water. Most importantly, it affected their food supply. Their perception of weather subtleties, is exemplified by the distinction they made between snow melting in the shade and melting in the sun.

Unfortunately there are periods of incomplete record at some sites, the most extensive at Churchill and York Factory occurring from 1805 to approximately 1820. This is unfortunate because the period from 1809 to 1820 is of unusual climatological interest. The gap is a combination of several things including harsh weather conditions. These conditions seriously aggravated a conflict between the Hudson's Bay Company and the North West Company; as a result, records were neglected or often destroyed. Fortunately, there are very few years in which records are not available at one or more sites.

Statistical tests for consistency of the records were carried out. These included tests for consistency of terminology, variation from one observer to another, variation between the records at both sites, random checks on coding and manual checking of apparently anomalous records. It is essential to ensure that variations are due to the climate and not a change of observer.

3.3.1 Instrumental records

There were many incentives for accuracy in the recording of meteorological information. The Company, desirous of reducing costs by creating self-sufficiency, encouraged the planting of European crops and the use of local foodstuffs. They wanted to assess the climate to determine the viability of European vegetables.

They also had a very close connection with the Royal Society. Samuel Wegg, who was Treasurer of the Royal Society for thirty-four years and a member of the Board (including being Governor of the Company for six years) constantly encouraged the collection of scientific information. The meteorological instruments used by the Company were provided and calibrated by the Society, including those brought out by the astronomers William Wales and Joseph Dymond in 1768. The Society, caught up in the growing intellectual curiosity, provided several instruments in 1773 and again in 1811, including thermometers and barometers.

Knowledge about the construction, quality and calibration of the instruments is important. Thomas Hutchins wrote in the preface to his journal of 1771 that:

The instruments used in taking these observations are a barometer and thermometer of Nairne's construction, and we have great reason to think them both very good as Mr. Wales the astronomer (who remarked the last transit of Venus at Churchill) was commissioned to send them. The thermometer is that termed the standard with Fahrenheit's scale; the freezing point is at thirty-second degree above the Cypher. (P.A.M., H.B.C.A. B239/a/67, p.1)

Detailed studies of the instruments maintained by the Company at posts on the east side of Hudson Bay was carried out by Cynthia Wilson (1982). At those sites the type of instruments used was not recorded creating some difficulty in analysing the data. Despite these limitations it was determined that if

The historical series were subjected to the modern AES (Atmospheric Environment Services) quality control procedures [they] passed without difficulty, in spite of what later appeared to be exaggerated mean daily maximum and mean noon values. (Wilson 1982, p. 206)

The type of instruments, their siting and aspects are known in most of the records maintained at Churchill Factory and York Factory. All analysis showed that these records are comparable in accuracy and diligence of observation to modern records maintained at the same locations.

The regular journals were maintained by the officers, but many of the instrumental journals were kept by the surgeons. All of the journals follow the format specified by Dr. James Jurin (1722). Notable are the works of Thomas Hutchins and Andrew Graham. The former carried out experiments on the freezing of mercury at Fort Albany on James Bay and published his results in the Royal Society Transactions of 1788.

Hutchins, who was instructed in the use of the instruments by William Wales, notes that the thermometer was placed on a north facing wall away from direct sunlight (PAM, HBCA, B239/a/67, f.10). The walls were made of unbarked logs which would ensure little reflected radiation. All subsequent records indicate that instruments were placed in similar appropriate locations. Dr. John Rae was especially precise: "The thermometers were suspended within a couple of inches of each other, under a tunnel-like covering of stout canvas, facing north and protected as much as possible from the sun's rays at the same time quite detached from any building. Height of thermometers from the ground, four feet six inches." (PAM, HBCA, B239/A/164, F.1).

One record has, so far, been excluded from any analysis. This was kept by W. Jefferson at Churchill in 1791 and records a temperature of 108°F. It is possible that the reading is valid because Jefferson located the instrument appropriately, however, it is far enough above the modern record maximum of 91°F to be questionable.

The instrumental records maintained by Peter Fidler are fascinating but frustrating. Fidler was a self-taught scientist who was given thermometers by the Company. His readings are important because he travelled through the interior of the prairies. Unfortunately, many of his readings were taken while travelling and we have no way of determining his precise location. Fidler's travelling record has another problem because he did not record the time of the observation. It is the only such record; all others followed the instructions of Jurin and recorded the time of the observation. Many of the records have only a single noon reading, but a majority have at least three, usually morning, noon and evening. The hours of observation vary from record to record and from month to month, probably as a result of the daily work schedule. This created difficulty in determining a daily average.

The temperature records were reduced to the best approximations of daily averages (Ball and Kingsley 1986). Weightings were calculated for each hourly reading in the modern Churchill record so that the daily average could be calculated for any combination or number of readings. The original readings were taken in degrees Fahrenheit, they were converted to Celsius for comparison with the modern records. The reader is referred to Ball and Kingsley (1986) for a detailed description of the data and analysis of the temperature record; general results are presented later.

Most of the instrumental records were sent to the Royal Society, but there is no evidence that any scientific analysis was carried out prior to the 20th century. The only records published prior to 1961 were those maintained by Wales and Dymond at Churchill in 1768/69 and printed in the Transactions of the Royal Society for 1771. A.B. Lowe (1961) was the first in the modern period to examine the climatological data. He examined the instrumental records, but focused primarily on the anecdotal weather comments.

3.3.2   Non-instrumental records

R. Minns (1970) used the records for a study of air mass frequencies. He concluded that outbreaks of arctic air were more frequent in the 19th century than at present. Mackay and Mackay (1965) published a study of the recorded date of freeze-up and break-up of rivers. Moodie and Catchpole (1975) applied rigorous testing to freeze-up and break-up data using the technique of content analysis to achieve systematic and objective results. The efficacy of this technique made it appropriate for converting the daily weather entries into statistical form. All testing procedures recommended in their work were applied to the records used in the following reconstructions.

Five major concerns shaped the methodology:

1 The volume of non-quantitative material.
2 The occurrence of instrumental records that increase in number through time.
3 The need to establish uniformity throughout the length of the record.
4 The need to establish meaningful measures of variables from which prevailing atmospheric conditions can be inferred.
5 The need to establish a technique that would apply to the full range of weather commentaries.

The preliminary research had to establish a long baseline record. The records for York Factory and Churchill Factory were ideal. Three records, York Factory, Churchill Factory and Cumberland House were examined in detail to determine the complete range of weather information. The record is remarkable for its consistency of terminology and amount of detail. The record changed in 1752 when the Gregorian Calendar reform was introduced but all adjustments were made in the subsequent computer record. A classification into the following categories was undertaken.

Instrumental Temperature   was recorded in degrees Fahrenheit. There are a total of 63,230 individual readings for York Factory and 22,601 for Churchill, not including the modern record.

Wind Direction   recorded by observing a flag set on a tall mast. Andrew Graham notes in the introduction to his meteorological journal of 1770 that the mast was 50 feet high and well clear of the influence of buildings. This exposure was enhanced by the low-lying terrain and vegetation at both sites. All the observers were skilled mariners trained by the Company which insisted on procedure and completeness of records. The terminology is consistent throughout and wind is recorded to 32 points of the compass. It is debatable that wind can be observed that accurately from a flag, but it does mean that a reduction of the data to 8 quadrants is going to yield a more accurate representation of the true wind.

Most of the record only has one entry per day, however it was noted that any significant change in weather, particularly wind direction was recorded. Several tests were run on the modern record for Churchill to determine the validity of using a single reading as representative of a 24 hour or 12 hour period. Studies were also done to determine whether the wind was backing or veering. Information on these are not presented here, but will appear in future publications. The computer was programmed to determine the number of times that the wind blew from each sector for each month and year of the record. Actual and percentage frequency counts were determined.

The general or typical pattern of upper air circulation over Hudson Bay is determined by the counterclockwise flow around a cyclone centred on northern Baffin Island. Flow is strong in the winter and weak in the summer creating a predominant north and northwest flow across the Bay. The strength of the flow limits the penetration of southerly air. A measurement of increased southerly flow and cyclonic activity in southern Hudson Bay would reflect a weakening of the persistent pressure regime over Hudson Bay.

The single season of winter, measured by the period with permanent snow cover, is six months which collectively equals the other three seasons. The latter are marked by the influx of cyclonic systems from the west and south which are frequently preceded by the advection of warm, moisture-laden air resulting in unsettled conditions despite the stabilizing effect of the cold waters of the Bay.

Wind Strength and Type   There is no record of actual measurement of wind speed. Some numerical records exist related to observable natural events. The most common is that of Jurin, but the Beaufort scale also appears. Examples of strength are, "smart, stiff, strong" and type, "calm, breeze, gale." In the original record strength and type were usually combined.

Precipitation   Three major divisions were established; rain events, snow events, and miscellaneous moisture events. There are no measurements of precipitation and only very infrequent estimates of snowfall occur.

Cloud Cover   Seven categories occur including, Clear, Cloudy, Overcast, Flying Clouds, Part Clear, Part Cloudy, Part Clear/Part Cloudy.

Thunder   It is impossible to have thunder without lightning and vice versa but they can be observed independently. This variable is very important because changes in the frequency of thunderstorms are a direct indication of changes in the stability of the atmosphere. There is a possibility of cannon fire and ice cracking being mistaken for thunder, but analysis of the data suggests this is not a problem.

Non-instrumental Temperature   Fourteen categories of subjective references to weather occur. These have only been used as secondary support for analysis and only those that indicate extremes. A system for using these data has yet to be developed. Two comments are of particular interest; "Warm for time" and "Cold for time."

General Weather   Subjective comments that were also used as secondary support. They are: Pleasant, Fine/Good, Mild, Moderate, Stormy, Close/Thick, Fair, Sultry, Variable.

Melting   This category indicates the level and sophistication of observation and understanding of weather conditions. Distinction was made between snow melting in the sun and in the shade thus distinguishing between ambient air temperature and radiation balance.

Frost   Distinction was made between the type of frost and the time of occurrence, for example, "Hoar Frost", and "Froze Hard at Night."

Drift   Care was necessary because the term 'drifting weather' was occasionally used for low scudding clouds and generally unsettled conditions. Low drift is a distinctive subarctic and arctic condition that should allow an estimate of wind speed.

Remarks   The disadvantage of most diaries is that they only comment upon extreme events. These diaries have daily weather information and extreme events combined. There are also comments on a wide variety of phenological information, all of which were noted and recorded.

3.4   Calibration methodology

The inherent consistency, due to the explicit directions of the Hudson's Bay Company (HBC) made calibration easier. The major problem was that no definitive methods existed for such historical climatological data. Catchpole and Moodie (1978) established distinct techniques for content analysis using HBC documents, but there are subtle variations and problems from one type of information to the next. The major concerns and guidelines that they delineate served as an excellent basis, but modifications are always necessary. The objective was a system and results that would allow for comparison with other records by other researchers.

All of the coding for York Factory and Churchill Factory was carried out by the author to assure consistency. However, test coding of the same selected section was carried out by two other recorders. Statistical analysis of the results confirmed the efficiency and consistency of the method.

Despite the fact that diarists were instructed in the ways of the Company there are likely to be variations from one individual to the next. Several tests were performed on various portions of the record. There were two objectives. First, to determine if a change in diarist could be detected by statistical examination of the results. Second, to determine if significant changes in the record were a result of a change in observer.

It is apparent from the general tone of the diaries and from specific comments made that additional consistency is provided because a new diarist was instructed by his predecessor. They also referred back in the record for guidance, direction, and comparison. The record kept by John Newton served as a good test of observer consistency. Newton joined the Company in 1748 after serving on merchant ships in the Mediterranean for thirty years. Obviously, his comment that it was "the coldest November ever he knew" must be taken in context. Despite his lifetime experiences which were so different from employees who had risen from apprenticeship to command within the Company, his record cannot be distinguished from the continuum.

Two tests were carried out to ensure that transcription from the record and accurate input to the computer was achieved. A statistically significant number of randomly generated samples were checked. In addition, any data that were extreme, unusual, or inconsistent were identified by the computer and checked.

The instrumental record required different techniques. Problems associated with the instruments, their calibration and location were discussed earlier. Different hours of observation were dealt with as follows. Average daily temperature (T) was calculated for the modern record at Churchill using 24 hourly readings. A regression was run on the modern record to achieve the estimate (T') for all combinations of hourly readings that occurred in the historic record. It was assumed that the correction factor calculated for the Churchill record would apply to the York Factory data.

3.5   Results

Two categories of data are defined. Primary data comprise the dates of occurrence or frequency of specific climatic events. Some of those analyzed so far include:

a) First day of snow in the fall
b) First day of rain in the spring
c) First day of frost in the fall
d) First day of melting in the spring
e) First day of thunder/lightning in the spring
f) Number of days with snowfall, October through May.
g) Number of days with rainfall, January through December.
h) Number of days with thunder/lightning each year
i) Number of days with each of the eight wind directions
j) Monthly mean temperatures

The secondary data attempt to synthesize different types of information and create synoptic analyses. These studies include, spectral analysis of precipitation data, cross-tabulation of the date of arrival of geese with the onset of spring, and synoptic analysis of annual weather at both stations.

It is impossible to present the results of all the analyses in a single chapter. Only very general results, many not available in other publications, are discussed. A year-by-year synopsis of notable conditions in each year for which complete data are available is presented for a sample decade. The reader is referred to the bibliography for published results produced from this large data base.

3.5.1   Date of first rain in summer (Figure 3.2)

Rain is recorded in every month over the whole period of record, however March 1 was taken as the cutoff date. Overall the two curves are similar indicating homogeneity of the record. From approximately 1765 to 1800 Churchill generally has a later date of first rain. The variance in date, from year to year is noticeable throughout the record, but appears more extreme at Churchill in the period 1815 to 1840.

3.5.2   Date of first snowfall in winter (Figure 3.3)

The date of first snowfall is usually related to the first southerly advance of the Arctic Front. This means that it is also a strong indicator of the onset of winter.

The first date of snowfall includes all events after August 1st. Every month has a record of snow events, but August has the least. Early winter is more harmful to wildlife and the human condition therefore an early date was chosen. The first snowfall of any type was used.

There is less variability in the date at both sites compared to rainfall. No divergence of the curves is evident. There is homogeneity between the curves, although they seem to be slightly out of phase in the period from 1720 to 1740. Five trends are noteworthy.

1 A gradual shift in the entire record from the first week in September in 1715 to the first week in October in 1820 and then back to the middle of September by 1850.
2 A shift from the first to the last week in September between 1720 and 1735.
3 In 1737 the first event occurs in the last week of August, the earliest period in the record after which it moves to a progressively later date until 1765.
4 Between 1765 and 1815 the date moves from the second week of September to the first week of October.
5 From 1815 to 1850 the date of first snowfall shifts back to the second week of September, but the range of variability increases slightly.

3.5.3   The date of first thaw in the spring (Figure 3.4)

The base date used was the 1st of March although thawing events occurred before that date in rare years. The level of sophistication of the observations is indicated by the thawing categories. A distinction was made between thawing in the lee and thawing in the sun thus distinguishing ambient air temperature from radiant energy.

The plot of dates does not distinguish between the various types of melting. A more detailed analysis will be carried out in a future study. The general points to observe about the curves for Churchill and York Factory are as follows:

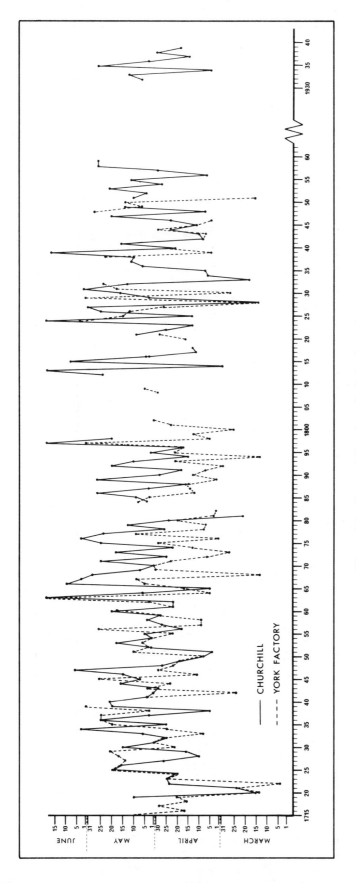

Figure 3.2 Dates of first day of rain in the spring.

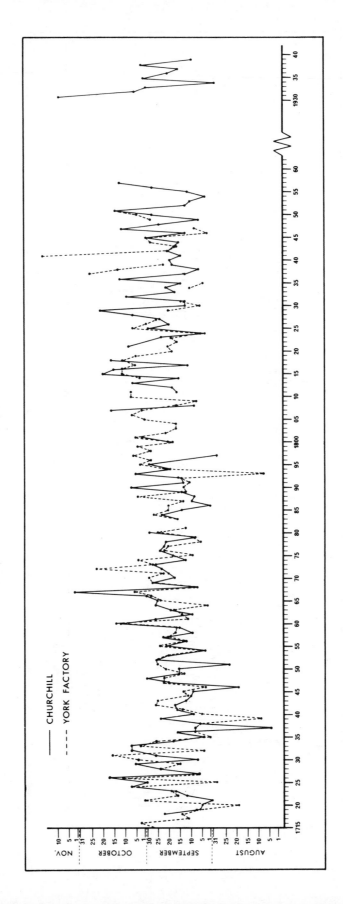

Figure 3.3 Dates of first day of snow in the fall.

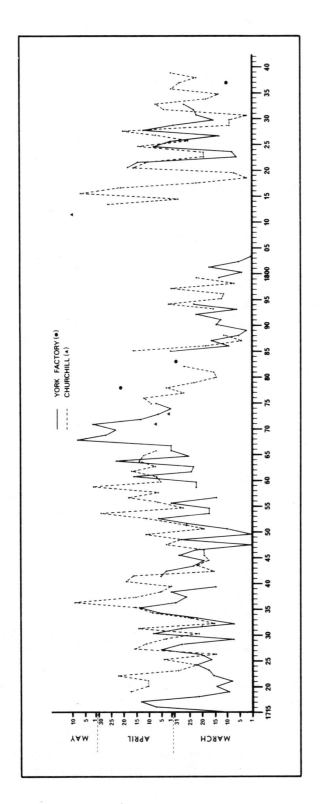

Figure 3.4   Dates of first day of melting in the spring.

1 There is an apparent homogeneity between the curves as they tend to follow each other.
2 The date of first thaw is generally slightly earlier at York Factory, the more southerly of the two sites.
3 The date of first thaw tends to fall between March 15th and April 15th from 1715 to 1750.
4 From 1750 to 1770 the date shifts to a later period falling between March 25th and April 25th.
5 From 1770 to 1800 there is a very distinct trend to a much earlier date of first frost, generally between March 1st and the 25th.
6 Between 1820 and 1840 the date has reverted to those seen in the early part of the 18th Century.

3.5.4   The date of first frost in the fall (Figure 3.5)

The computer was programmed to select the first day of recorded frost after August 1st. This was the most problematic of the parameteorological indicators, but the data are presented because they appear to show a slight trend towards earlier dates of frost. There is a wider range of dates from 1770-1790, which is consistent with other evidence of extreme variability of weather in this period. There is potential for further analysis of these data, but problems with the terminology require much additional work.

The major problem is a lack of distinction between visible frost and the term 'froze.' The former is also classified as precipitation. It is an event that occurs frequently with relatively warm arctic sea water frequently open to extremely cold air. The latter term refers to cooler thermal conditions.

3.5.5   Date of first thunder in spring (Figure 3.6)

A journal entry for the 23rd of August 1746 (New Calendar) reports hailstones 4.75 inches (12 cm) in circumference. This means a stone approximately 1.5 inches (3.8 cm) in diameter, a considerable size and indicative of strong vertical development for this latitude. The frequency of thunder and lightning is a good indicator of changes in the stability of the atmosphere. Once again York Factory's southerly location is reflected in the higher frequency of events.

March 1st was used as the base date. The following observations of the curves can be made.

1 Homogeneity in the general trend between the curves of the two sites is evident.
2 Churchill has a very wide range of dates from the 7th of March to the 1st December. York ranges from the 14th of March in 1790 to the 16th of September in 1778.
3 There is a progressive shift in the curves from a date in the latter part of May in the 1720s, to a date in the middle of July in the 1760s. From 1760 to 1790 the curve declines back to the latter part of May. The curves are incomplete from 1800 to 1850, however the date appears relatively stable around the second week of June.

3.5.6   Number of days with snow events, October through May (Figure 3.7)

The graph indicates the total number of days on which snow was recorded. There is apparent homogeneity between the two sites, but there should be less than with rainfall because of the

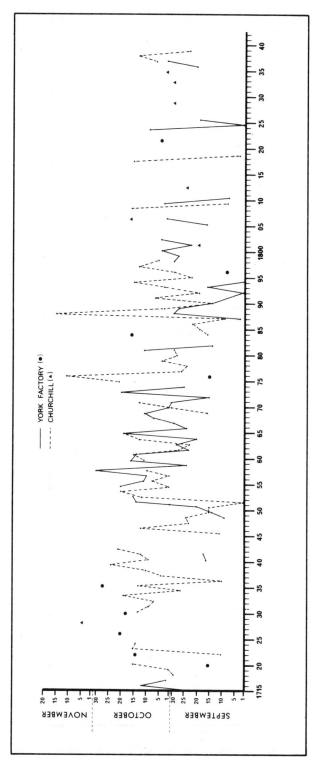

Figure 3.5 Dates of first day of frost in the fall.

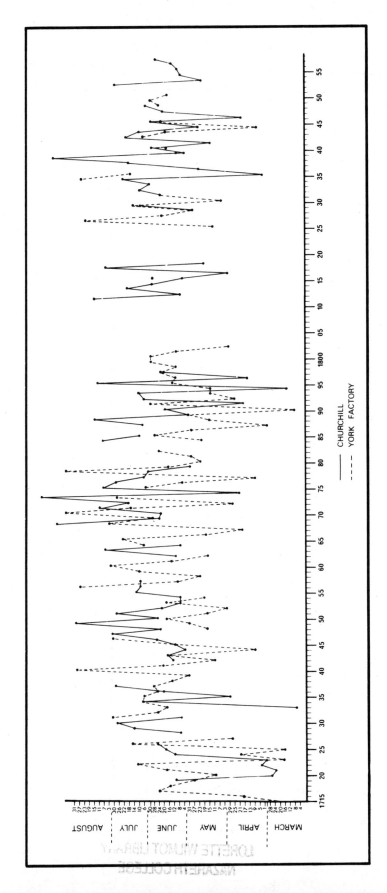

Figure 3.6 Dates of first day of thunder in the spring.

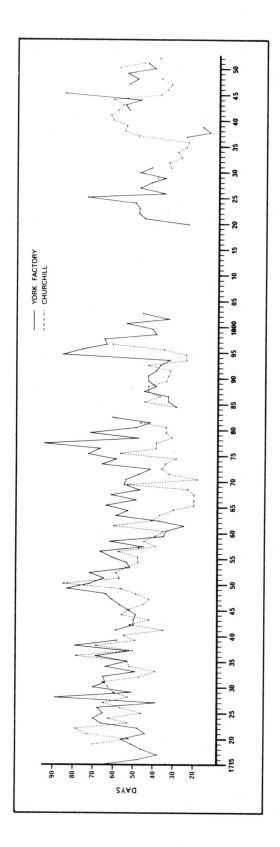

Figure 3.7 Number of days with snow events, October through May.

very localized occurrence of snow showers. This would be more noticeable at York Factory than at Churchill. In a description of the climate of Hudson Bay, H. A. Thompson writes:

Not only is there a noticeable southward increase in snowfall over Hudson Bay, but during the open-water months of autumn snowstorms are characteristically heavier and more frequent along the east coast than on the west coast. (Thompson 1970 p.269)

Other important points include:

1 There is a gradual decline in the number of days from approximately 60 per year in 1715 to 50 per year in 1850. This would be consistent with the global warming coming out of the Little Ice Age.
2 There is a much greater variation in the number of days of snowfall than for rainfall. This is consistent with the nature of snow events.
3 The general trend of the curves is synchronous up until approximately 1762. The curves are divergent until 1780 but then, unlike the rainfall curves, are synchronous again.
4 On a smaller scale there is a stable pattern from 1715 to 1740 when a brief decrease to 1745 is followed by a rapid increase to a peak in 1750. From 1750 to 1760 there is a rapid decline followed by a gradual increase to 1780. The period from 1780 to 1793 has reduced variability around a very low average of 75 events per year. 1795 to 1850 sees a return to the levels of the early part of the 16th Century.

3.5.7 Number of days of rainfall (Figure 3.8)

A synoptic study of climate in the Canadian Arctic shows that York Factory has more days of rain per year than Churchill in this century (Fletcher and Young 1976). Figure 3.8 shows that from 1718 to 1760 the two stations recorded approximately the same number of days of rainfall. Today the two sites lie on either side of the mean summer position of the Arctic front. The implication of this and other evidence is that both places were under similar climatic conditions in the first half of the 18th Century. The author has substantiated this contention elsewhere arguing that this situation is consistent with colder conditions of the Little Ice Age (Ball 1985). The issue will be discussed in more detail later in this chapter.

Other features of the curve are as follows:

1 The general synchronicity of the curves indicates relative homogeneity.
2 Between 1715 and 1765 there is a gradual decrease in the number of days of rain from 35 to approximately 20.
3 There is a rapid increase from 20 to 40 between 1765 and 1780.
4 A gap in the record from 1800 to 1820 is followed by partial curves to 1850. The average appears to be close to that of the 1740s, but the variation is much greater than any period in the 18th Century.

3.5.8 Number of days of thunder and lightning (Figure 3.9)

Modern records for the period 1941-1970 show a yearly average of 5.6 days with thunder at

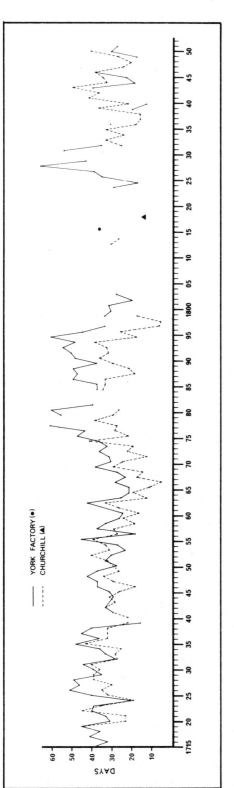

Figure 3.8 Number of days with rain events, May through October.

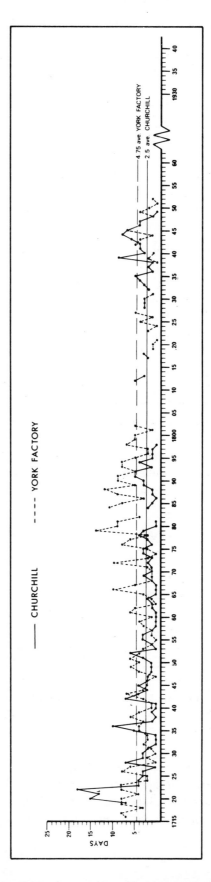

Figure 3.9 Number of days of thunder and lightning.

Table 3.1 Average number of days with thunder in the modern and historical records.

| Month | Churchill 1941–70 | Churchill 1718–98 | | York Factory 1715–1808 | |
	Average per month	Events	Average per month	Events	Average per month
April	0.0	19	0.24	14	0.15
May	0.2	13	0.16	32	0.36
June	1.2	34	0.43	70	0.79
July	2.1	64	0.81	136	1.54
August	1.7	44	0.55	118	1.34
September	0.3	11	0.13	23	0.26
October	0.1	2	0.02	7	0.07
Totals	5.6	187	2.4	400	4.5

Churchill. Table 3.1 provides a comparison of York Factory and Churchill Factory with the modern record.

Thompson (1970, p.282) notes that "The stabilizing effects of the cold water surface show up in the low frequency of thunderstorms in summer . . . Southern locations such as Churchill and Great Whale River report two to six thunderstorms per year." As far as is known there are no records of water temperatures of Hudson Bay taken by employees of the Hudson's Bay Company. It is most likely that temperatures would have been lower during the Little Ice Age. Detailed studies of ice conditions by Catchpole (see elsewhere this volume) using Company ship's logs indicate that this was the case. The lower number of days of thunder compared to the modern record is partially due to colder water temperatures and longer periods of ice cover. The second factor is a change in the frequency of influx of southerly air. Figure 3.9 shows a similar numbers of events in the record up to 1760 after which the curves diverge. From 1760 to 1800 York Factory has distinctly more days with thunder than Churchill. This supports the hypothesis that there was a shift in the mean summer position of the Arctic Front marking the transition out of the Little Ice Age.

3.5.9 Frequency of wind from 8 points of the compass

Further evidence for changes in climate should be found in changes in the direction of wind. The W.M.O. Technical Note on Climatic Change states that "Valuable early records of wind directions are available from some exposed places, including the log-books of ships in port or patrolling stretches of coast" (W.M.O. 1966, p.11). There is a general scepticism of early wind observations, which is probably justified in most cases. The observations of the Hudson's Bay Company are as detailed and as reliable as possible from historic records.

Wind direction, when combined with the few barometric pressure readings, temperature information, and general weather characteristics, offers the best opportunity for producing synoptic weather maps. An attempt is made later in this chapter to connect this information with Lamb and Johnson's (1977) work on the Eastern United States and the Western Atlantic.

North Winds: (Figure 3.10a) The curves are coincident from 1729 to 1752 after which the Churchill frequency increases while York decreases. In 1763 the curves cross so that there is a higher percentage at York than Churchill. The period from 1763 to 1776 marks the highest percentage of north winds at York Factory for the whole period of record. Between 1776 and

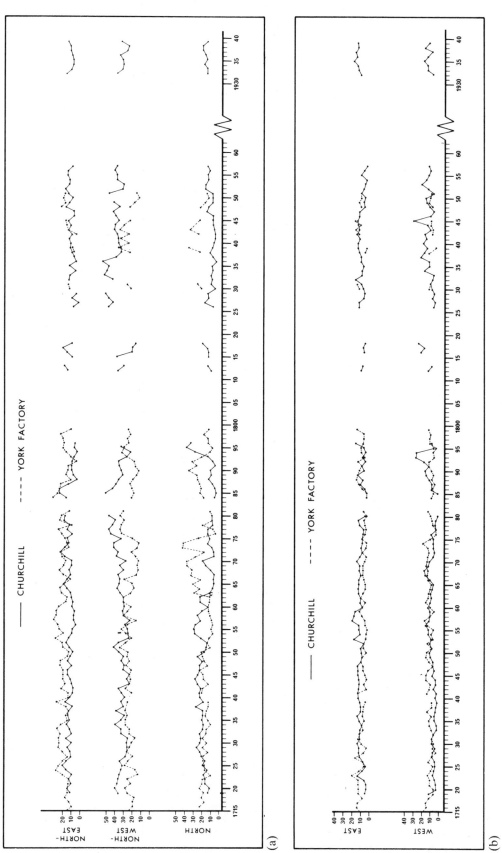

Figure 3.10 (a) Percentage frequency of winds from the North, Northwest and Northeast. (b) Percentage frequency of winds from the West and East.

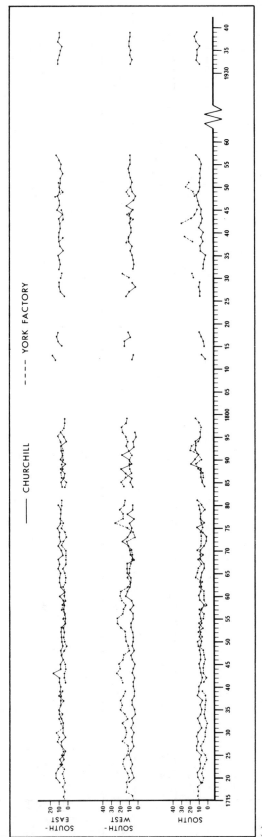

Figure 3.10 (c) Percentage frequency of winds from the South, Southwest and Southeast.

(c)

1782 the curves are coincident after which Yorks shows an increase. They cross in 1794 again with a decline at York and an increase at Churchill.

The scanty record suggests that the crossing pattern continues at least until 1850. Starting in 1750 the sequence of years between the intersection points of the curve are as follows:

$$1750\text{-}1763 = 13$$
$$1763\text{-}1776 = 13$$
$$1776\text{-}1783 = 7$$
$$1783\text{-}1794 = 11$$

Herman and Goldberg write that,

Atmospheric circulation changes result from a spatial redistribution of atmospheric pressure, and, since the latter has been observed to correlate with solar activity, especially in terms of stationary planetary wave oscillations in the surface and 500 mb pressures, one might conclude that some circulation changes be indirectly responsive to solar influence. (Herman and Goldberg 1978 p. 245).

Later in discussing the location of the main trough of the ionosphere they state,

It is noteworthy that the latitudinal position of the trough places it in the vicinity of the polar vortex that is so well known to meteorologists. A southward movement of polar vortex in association with the geomagnetic storm time behaviour of the main trough would tend to force the cold polar air equatorward . . . during solar maximum years when geomagnetic storms occur often there would be frequent southward movement of the trough, polar vortex, and the jet stream. (Herman and Goldberg 1978, p.245).

Earlier in the same book the authors refer to the Schwabe cycle which they point out varies from 8.5 to 14 years between minima and 7.3 to 17 years between maxima. With a 13 year span in the epoch 1784-1797. (Herman and Goldberg 1978, p. 13). The recent works of Labitzke and Van Loon (1988) and Bucha (1988) have demonstrated correlations between solar activity and global wind patterns.

Northwest: (Figure 3.10a) The highest percentage of winds throughout the record are from the northwest at both locations. The most significant feature of these curves appear to be as follows:

1 There is a strong degree of homogeneity between the curves.
2 The percentages at Churchill are consistently higher than for York Factory.
3 The curves are separated from 1720 to 1725, coincident from 1725 to 1763, and divergent from 1763 to 1795.
4 Longer trends appear to be present but the data are limited.

Northeast: (Figure 3.10a) York Factory has a higher percentage of northeast wind. It appears that these curves are the inverse of those for the northwest winds. From 1720 to 1725 the curves are coincident, divergent from 1725 to 1765 and coincident from 1765 to the end of the record.

East: (Figure 3.10b) The percentage of east winds is very low at both stations therefore significant or recurrent trends are not immediately apparent. From 1720 to 1725 York Factory has a higher percentage than Churchill, but from 1725 to 1780 Churchill is higher. From 1785 to 1797 York is dominant again.

West: (Figure 3.10b) There is a gradual increase in the percentage of west winds at Churchill from 1740 to 1780 and again from 1825 to 1857. Both these periods experienced a gradual and general warming and a shift to more westerly flow would have created warmer conditions in this region.

South: (Figure 3.10c) Churchill had a higher percentage of south winds from 1720 to 1749 after which the curves are coincident to 1796. There is evidence of a slight increase in south winds at both stations between 1788 and 1795. The period from 1825 to 1852 is incomplete but appears to show a much higher percentage of south winds at York Factory while Churchill percentages remain average. Thus it appears to be a period of zonal flow.

Southwest: (Figure 3.10c) York has the higher value for the entire period of record except from 1765 to 1775 when they are coincident. 1773 to 1795 is marked by greater variability in the Churchill record. The period from 1718 to 1760 at Churchill is remarkable for its uniformity.

Southeast: (Figure 10c) The overall percentage of southeast winds is low at both stations, although Churchill Factory has a slightly higher percentage than York Factory. This is probably due to a sea breeze effect at Churchill, an hypothesis that is supported by the increase of southeast winds in the summer months.

3.6   Monthly mean temperatures (Figure 3.11)

Figure 3.11 presents the plots of monthly mean temperatures for York Factory and Churchill. The former has a more extensive record than the latter, but it is convenient to show them together although each will be discussed separately. (Modern mean values for Churchill are shown in the text, below, in brackets). A major feature of both records is that summer maximum temperatures are relatively constant, whereas winter temperatures vary a great deal. There are insufficient data to determine trends, but there are some general features. The greatest range of seasonal temperatures occurs in the latter part of the 19th century. This is consistent with the official records maintained at Winnipeg which commence in 1876. Most of the record for the middle part of the 19th century indicates a narrower range, especially from 1844 to 1858.

More specific details for each record are as follows:

1) At York Factory the period from 1775 to 1779 had below average summer temperatures.

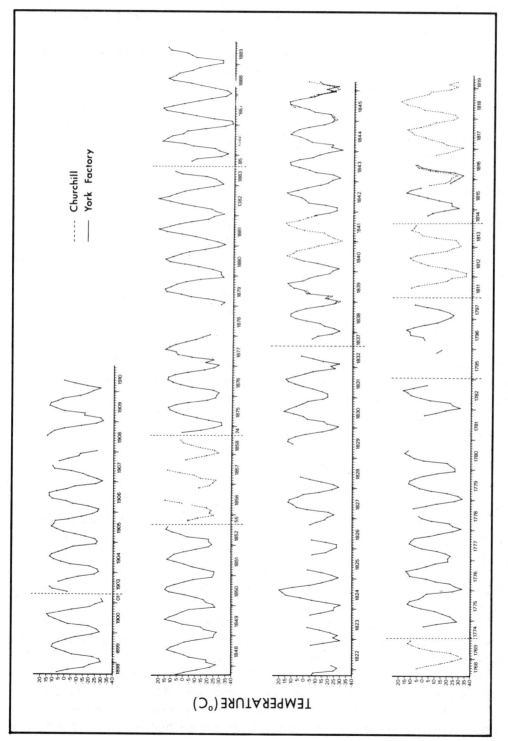

Figure 3.11 Monthly mean temperature for York Factory and Churchill.

2) The winter of 1776-1777 was very mild with virtually no snowfall at Churchill until March 1777 when very heavy snowfall occurred. York Factory mean temperature for January was -21.0°C (-27.4°C) and February -22.8°C (-27.0°C). These data are interesting when compared with the conditions recorded in the eastern United States for the same years.

3) The cold period beginning in 1809 and reaching a nadir in 1816/17 recorded in other regions, such as reported by Borisenkov in chapter 8 (this volume) is also manifest. July 1813 with a mean of 11.7°C (12.6°C) was noted as being the warmest summer for three years but is still over 1°C below the modern mean.

4) A break in the aforementioned cold occurred in 1818. The York diarist noted that "November very mild, mildest for 30 yrs or more." Churchill recorded -4.7°C (-12.4°C) for the same month.

5) February 1831 had winds from the south 30% of the month resulting in a temperature of -19.5°C (-27.0°C).

6) The spring of 1841 was the latest since 1822. Churchill recorded a monthly mean in April of -15.9°C (-11.9°C).

7) June 1842 was cold at both sites. Churchill recorded 4.2°C and York 2.6°C (5.8°C).

8) Mean temperature for November 1849 at York was -5.8°C (-12.4°C) supporting the written observation that there was a, "Very late Fall. Interior rivers and lakes unfrozen until the 10th of December."

9) Support for the position that the Little Ice Age had definitively ended circa 1850 is seen in the York summer temperatures which were relatively constant and close to the modern July average of 12.6°C.

Despite the incompleteness of the record it provides a valuable series of absolute measures for comparison with other regions. It is an acceptable record that supports and extends the evidence of climate change. It also serves as a basis for confirming other secondary or less precise measures such as tree rings.

3.7 Comparisons with european conditions

There has been considerable debate over the dates of onset and cessation of the Little Ice Age. The Hudson's Bay Company records do not cover the onset but they do provide information for comparison during that period. One of the main arguments used to define the onset of the main phase of the Little Ice Age are the different percentage frequencies of wind direction given in Tycho Brahe's diaries. Similar changes in wind directions and other climatic variables suggest that the nadir of the Little Ice Age occurred around 1769 in central Canada.

Manley has said that climatic fluctuations "…should not be expected to occur at the same time or with the same intensity in all longitudes in either the North American or the Eurasian continent, even in those critical latitudes from 45 degrees north to 60 degrees north in which such changes may have their maximum effect" (Manley 1974 p.173). However, Lamb writes, "It (a delay in onset of the Little Ice Age) is not contradicted by the observation that the cooling of climates setting in after the 1940s took effect almost a decade sooner in the Arctic – and in tongues extending from the Arctic towards the heart of the northern continents – than it did in much of Europe and the U.S.A. Its amplitude was also much greater in the Arctic and in northern parts of the continental interiors than elsewhere" (Lamb 1977, p.477) Churchill and York Factory are affected by the situations described by Manley and Lamb.

It is difficult to compare regions when there are no specific measurements, indeed many argue that there is little point in comparing such widely separated areas. The objective of the following comparisons is to define the large scale patterns of climate that probably reflect hemispheric changes. It is also important to determine whether conditions were zonal or meridional.

The early period, that is up to 1768, must rely upon subjective ratings for comparison with other regions. C. Easton (1928) rated the winters of western Europe using a subjective index value ranging from warm to very severe (Tables 3.2 and 3.3). Lamb argues that although Easton's data have been superseded the extreme years would be valid in any system of classification because they are so different. Easton's subjective classification is shown in brackets.

Ironically the year 1716 was very mild at both North American stations, while Easton lists it as a severe winter. The only year that is warm at all three locations is 1807. There are 4 warm years in the eighty years of record in the 1700s and 7 in the fifty years of the 19th century. The movement of the mean summer position of the Arctic Front after 1760 previously postulated, seems to be confirmed. Churchill has far fewer mild winters than York.

The predominance of cold winters in the 1700s is apparent at all locations. The 1740s was a decade of cold winters, especially at the North American stations. The period from 1786 to

Table 3.2 Mild winters* (identified by the year in which the January fell).

Western Europe	York factory	Churchill Factory
	1716	1716
1717(warm)	----	----
----	----	1727
1725(warm)	----	----
----	1744	1744
----	1757	1753
1764(mild)	----	----
----	1777	1777
1796(warm)	----	----
1807(warm)	1807	1807
----	1818	----
1822(warm)	----	----
----	1825	1825
----	1831	----
1834(very warm)	----	----
----	----	1835
----	1838	----
1846(warm)	----	----
----	1849	----

* Defined by a combination of comments in the journals, frequency of south wind, and actual temperatures when available. Data for western Europe from Easton (1928).

Table 3.3 Cold winters*.

Western Europe	York Factory	Churchill Factory
1716(severe)	----	----
1729(severe)	----	----
----	----	1731
1740(very severe)	----	----
----	1741	1741
1742(severe)	----	----
----	1748	1748
----	1749	1749
1757(severe)	----	----
1763(severe)	----	----
----	1786	1786
1784(very severe)	----	----
----	1791	1791
----	1792	1792
----	1797	1792
1795(very severe)	----	----
----	1796	1796
1799	1799	1799
----	1801	1801
----	----	1803
----	----	1804
----	1805	1805
----	1806	1806
----	18..	1822
1823(severe)	----	----
----	----	1826
1870(very severe)	----	----
1878(severe)	----	----
----	1879	----
1841(severe)	----	----
1845(severe)	----	----

* Defined by comments in the journals, frequency of
north winds, and actual temperatures when available.
Data for western Europe from Easton (1928).

1796 was extremely cold at York and Churchill. In western Europe it was also cold in the 1780s, however it did not extend beyond 1795.

The Company journals are replete with comments on the hardship that weather conditions created in the 1780s and 1790s. Cold and extreme snow conditions that fluctuated between no cover and extreme depths led to a lack of game for food. The natives and Company men suffered from shortages and references to starvation dominate the journals of Joseph Colen. (He was the Chief Factor at York Factory from 1788 to 1802). A detailed study of journal entries for the period 1789 to 1800 by Wilkerson (1986) shows a dramatic increase in the references to hunger, malnutrition and starvation. Scurvy was constant and great stress resulted. A comment for the 5th of February 1792, is indicative of the conditions and reads as follows; "They (Indians) also inform me that the winter set in so early upwards that many

Swans and other waterfowl were froze in the lakes and they found many of the former not fledged, they likewise say that the snow is remarkably deep" (HBCA PAM, B239/a/97).

The failure to find correlations between regions so far apart is not surprising. Although single years do not correlate there are parallels for decades and in the overall trend from colder to warmer conditions. Similarities also occur in the precipitation patterns. For example, the 1780s were a period of increased precipitation on both continents.

A classic approach to the problem of determining variability of precipitation without measured records has been the summer wetness/dryness index. In this technique months that are very dry are assigned a value of zero, normal months a value of one-half and exceptionally wet months a value of one. Lamb (1977, p.562) presents data for each decade from A.D.1100 to 1960 for England, Germany and Russia. A comparison of the decades from 1710 to 1800 is given in Table 3.4.

Table 3.4 High summer (July and August) wetness/dryness index.

Decade	England	Germany	Russia	Churchill	York
1710-19	10.0	8.5	10.5	---.--	--.--
1720-29	11.5	8.0	13.5	11.0	11.5
1730-39	10.0	10.5	13.5	11.5	10.5
1740-49	7.0	6.5	10.5	11.5	9.5
1750-59	13.5	12.0	10.0	12.0	9.0
1760-69	11.5	12.0	9.0	6.0	8.9
1770-79	11.5	11.0	6.0	11.0	13.0
1780-89	12.0	10.0	10.0	11.0	12.0
1790-99	10.5	12.5	10.0	7.5	14.5

The European stations are at 50° N latitude. Churchill and York Factory are approximately 56° and 55° N respectively. The Russian station at 35° E longitude best reflects the continentalism of the North American stations. The months of July and August were used to determine the values. The 1760s and 1790s were dry at Churchill while the 1740s, 50s and 60s were dry at York Factory. The 1760s were also dry in Russia, but beyond that there are no similarities between any of the stations. The record supports the contention that York Factory was wetter than Churchill after the 1760s. It suggests that Churchill remains in the subarctic region and had similar precipitation patterns throughout the Little Ice Age, while York Factory becomes boreal after 1760.

The lack of correlation with the European record could have been due to the fact that August and September are the wettest months at York and Churchill. The use of data for those months showed no change in the pattern. Further research is necessary to determine the mechanisms that create these patterns of wet and dry decades. Undoubtedly the patterns are a function of the mean summer location of the Arctic Front and the amplitude and positions of the Rossby waves. The 18th century was a period of climatic extremes as these waves varied greatly in amplitude. Support for this argument is provided by Borisenkov in Chapter 9 (this volume).

Barry attributes cooler summers and increased passage of cyclonic systems ". . . with a westward displacement of the mean 700mb trough over eastern North America . . ." (Barry *et al*. 1978 p.89) while Locke and Locke speculate that,

The depression of glaciation thresholds and equilibrium-line altitudes of about 400 m along the coast of Baffin Bay may indicate increased precipitation during the Little Ice Age due to an increase in the frequency of low pressure centred over Davis Strait and Baffin Bay. (Locke and Locke 1977 p.299)

Extrapolating from the maps created by Brinkmann and Barry (1972) to the various locations of the 700 mb trough on the east side of North America yields the pattern given in Table 3.5 for York Factory and Churchill.

Table 3.5 Precipitation conditions at Churchill and York Factory in relation to position of 700mb trough over eastern North America.

Location of 700mb trough	Precipitation
Winter	
1. Westward, shallower trough	Normal precipitation
2. Westward normal trough	Below normal precipitation
3. Eastward	Well above normal precipitation
Summer	
1. Deep trough over eastern Hudson Bay	Normal to low precipitation
2. Westward normal trough	Normal to low precipitation
3. Eastward deeper than normal	Normal to low precipitation

Dey studied the synoptic conditions that occur during summer dry spells in the Canadian Prairies. He concluded that,

. . . there was a general eastward migration of all pressure systems and that the North Pacific high and the North Atlantic high tended to extend north eastward from their normal positions, whereas the Hudson Bay low moved southeastward during the dry spells. (Dey 1973 p.169)

At the 700 mb level he observed an eastward displacement of the high pressure ridge that formed on the lee side of the Rocky Mountains. This would result in an eastward displacement of the 700 mb trough. Eastward movement of the North Pacific high and eastward movement of the 700 mb trough in eastern North America result in below normal precipitation in Prairie Canada and above average precipitation in the southern region of Hudson Bay. The cause of this increased precipitation would be the location of the Rossby waves in a low zonal index, meridional flow, thus increasing the frequency of cyclonic systems tracking through the western and southern portion of Hudson Bay.

3.8 Annual analysis of significant weather

This synopsis of annual conditions will be of value to other researchers attempting to compare and explain the pattern of weather and the associated winds in their region. It also validates the quality of the records because the observer comments and the quantified data correlate extremely well.

Precipitation days refers to the number of days on which rain or snow of any sort was recorded. To save space the words snow and rain refer to snowfall days and rainfall days respectively. The number of days of events per month is shown in brackets. It does not refer to the amount of precipitation, although it is reasonable to assume that a higher number of days would mean a greater volume. Comments refer to written comments recorded by the original diarist. Only the records from 1715 to 1725 are included here, as an example of the information available. Readers interested in the remainder of the record to 1852 should contact the author.

1715 York: Snowfall days (snow) above average; especially high in March and April. Rain on February 16. Rainfall days (rain) average. Comment on cold September; it was also wettest month and 65% of winds from north.
Churchill: No record.

1716 York: Comments on mild weather; ice 2 feet less than previous year. Below average snowfall and rain. Thunder and lightning in April with 69% of winds from south and southwest.
Churchill: No record.

1717 York: Well below average snow. None in February. Average rain.
Churchill: No record.

1718 York: Average snow except October with 11 days, also higher than normal NE (20%) and E (46%) winds.
Churchill: Record begins in September. 12 days snow in October with 55% E winds. Comment on deep snow in October and December.

1719 York: Above average rainfall days. Below average snow. High percentage E winds (20%) and comment on mild February.
Churchill: Above average rainfall and snow. October with 14 days of rain. High percentage of S winds (30%) and comment on mild February.

1720 York: Average year except high percentage of NE winds in May (33%).
Churchill: Average year except high percentage of NE winds in May (43%).

1721 York: Slightly below average snow.
Churchill: Above average snow, especially January and February. High percentage S winds in March (29%).

1722 York: High percentage of SW winds in November (42%). Comment on mild October (39% S winds).
Churchill: Well above average rain, especially June and August. Well above average snow, especially January and March. Winds above average; NW in January 61%, February 68% and March 45%. Comment on moderate December.

1723 York: Slightly above average snow; April 12 days. Thunder and lightning in April. Above average NE and E winds for the year.

Churchill: Slightly above average rainfall and snow. High percentage of NE and E winds in April (NE 14% E 10%) May (ENE 13% E 32%) and June (NNE 24% E 28%). Comment on mild winter.

1724 York: Well below average rain. Well above average snow, especially January, March and May. Above average NE winds for year (27%) below average NW winds (15%) high percentage of NE winds in every month except October, November and December with April highest (35%). Written comments on snow melting in April; heavy snowfall in May and December extremely cold.

Churchill: Well below average rain. Well above average snow, especially January and March. Slightly above average NE winds for the year. High percentage of E winds in April (37%) and very high percentage of NW winds in October, November and December. Comments on strong winds in May.

1725 York: Slightly above average rain for year, well above for July (10) and August (14). Slightly above snow but well above in April (14). High percentage of NE winds in April (46%). Comments on heavy snow in April, heavy rain in July and wet fall.

Churchill: Slightly above average rain for year but above average for July (10) and August (11). High percentage of NE winds in June (34%) and August (44%).

3.9   Conclusion

A great deal of unique climate data is available in the journals of the Hudson's Bay Company. Two stations were analysed to establish relative homogeneity of climate change over the period from 1714 to 1850. A coding system established a uniform method of quantifying written records so that they could be integrated with instrumental data.

The study showed that Churchill and York Factory were in the subarctic zone up to 1760. After that date the mean summer position of the Arctic Front moved north apparently marking the turning point of the Little Ice Age in this part of the world. A study by the author (Ball 1986) confirmed this northward movement and the onset of warming by identifying the northern limit of trees in 1772 and demonstrating how much they had moved in the intervening 200 years.

Other research also supports the existence of colder conditions that led to movement of the treeline. Guiot (1986) produced a temperature curve for the period 1700-1979 using a variety of sources including break-up and freeze-up of rivers, tree-rings and pollen data. A relationship between the principal components of temperature and the proxy data for the period 1925-1979 was established and the results used to extrapolate back to 1700. Further statistical refinement produced temperature and pressure reconstructions for all four seasons.

Peter Scott of the University of Toronto led a team that has produced extensive and detailed tree-ring analysis for the Churchill region using cross-section analysis of a tree at fixed intervals over its entire height (Hansell 1984; Scott *et al.* 1988). This produces what the researchers have called a 'growth layer analysis' from which 'growth trends' can be determined. These results also show 1760 as pivotal in climate trends for that region. They further show that the period from 1760 to 1820 was one of variability as the climate oscillated from temperate to arctic conditions. A more recent study by Diaz *et al.* (1989) supports the contention of cold conditions before the middle of the 18th century, however they point out that it was cold again in the first half of the 19th century.

The Churchill journals of the last 20 years of the 18th century are replete with comments on the extreme variability of conditions. York Factory experienced three consecutive decades, from 1770 to 1800, of very wet and variable conditions. A notable feature of this period was winters with either a great deal of snow or virtually none. This seriously reduced the wildlife population with terrible impact on people dependent on these resources for food and economy.

The records are not as complete for the first decade of the 19th century as the Company struggled with the changes created by the harsh conditions of the 1780s and 1790s. The cold conditions of the period from 1809 to 1820, with a nadir in 1816-1817, were recorded by instruments and observations of the journal keepers. Extensive details of this period are provided in other publications. The reader is referred to Ball (1988) Catchpole (1985) and Wilson (1982, 1985). These studies support the evidence of Diaz *et al.* (1989) that it was cooler in the first half of the 19th century, but they indicate that the fairly cold average conditions were due to the excessive cold of one decade. The difference is partly one of sensitivity of measurement rather than disagreement. It is also partly due to the regional differences between the east and west sides of Hudson Bay. Regardless, all the studies combine to provide a continuum that is essential if we are to understand the natural variations and mechanisms of climate.

The overall pattern of climate for these two sites indicates similar trends to those noted for other regions of the northern hemisphere, but individual extreme years and seasons cannot be expected to coincide directly. With the variations between zonal and meridional flow that are associated with Rossby waves, it is not surprising that these stations do not have sequences comparable to those of Europe.

Ironically the vast, remote, inhospitable region of central Canada has a rich legacy of climate data. The dependence of the fur traders on the environment for survival, sustenance and profit made them acutely aware of its condition. Their masters in London required that detailed records be kept of those conditions thus producing a unique legacy for climatic reconstruction.

References

Allsopp, T. R. 1977. *Agricultural Weather in the Red River Basin of Southern Manitoba over the Period 1800-1975*, Environment Canada Report CLI-3-77.

Alt, B. T. 1983. Synoptic Analogs: a Technique for Studying Climatic Change in Canadian High Arctic in *Syllogeus*, C. R. Harington (ed.) Volume 3, National Museum of Canada, 70-107.

Ball, T. F. 1985. A Dramatic Change in the General Circulation on the West Coast of Hudson Bay in 1760 A.D.: Synoptic Evidence Based on Historic Evidence, in *Syllogeus*, C. R. Harington (ed.) Volume 5, National Museum of Canada, 219-228.

Ball, T. F. 1986. Historical Evidence and Climatic Implications of a Shift in the Boreal Forest Tundra Transition in Central Canada. *Climatic Change*, 8, 121-132.

Ball, T. F. and Kingsley, R. G. 1984. Instrumental Temperature Records at Two Sites in Central Canada. *Climatic Change*, 6, 39-56.

Barry, R. G. 1967. Seasonal Location of the Arctic Front over North America. *Geographical Bulletin*, 9, 79-95.

Brinkman, W. A. R. and Barry, R. G. 1972. Paleoclimatological Aspects of the Synoptic Climatology of Keewatin, Northwest Territories, Canada. *Palaeogeography, Palaeoclimatology, and Palaeoecology*, 11, 77-91.

Bucha, V. 1988. Influence of Solar Activity on Atmospheric Circulation Types. *Annales Geophysicae*, 6 (5) 513-524.

Catchpole, A. J. W. 1985. Evidence from Hudson Bay Region of Severe Cold in the Summer of 1816, in *Syllogeus*, C. R. Harington (ed.) Volume 5, National Museum of Canada, 121-146.
Catchpole, A. J. W. and Moodie, D. W. 1978. Archives and the Environmental Scientist. *Archivaria*, 6, 113-137.

Dey, B. 1973. *Synoptic Climatological Aspects of Summer Dry Spells in the Canadian Prairies*, Unpublished Ph.D. University of Saskatchewan.
Diaz, H. F., Andrews, J. T. and Short, S. K. (1989). Climate Variations in Northern North America (6000 BP to present). Reconstructed from Pollen and Tree-Ring Data. *Arctic and Alpine Research*, 21 (1) 45-59.

Easton, C. 1928. *Les Hivers dans L'Europe Occidentale*, Leyden: Brill.

Fletcher, R. J. and Young, G. S. 1976. *Climate of Arctic Canada in Maps*, Boreal Institute for Northern Studies, Edmonton, Occasional Publication Number 13, 1-85.

Guiot, J. 1986. *Reconstruction of Temperature and Pressure for the Hudson Bay Region from 1700 to the Present*, Canadian Climate Centre, Report No. 86-11.

Hansell, R. (ed.). 1984. *Study on Temporal Development of Subarctic Ecosystems 1. Determination of the Relationship Between Tree-ring Increments and Climate.* Report to Department of Supply and Services OSU84-00041 for Environment Canada, Canadian Climate Centre.
Herman, J. R. and Goldberg, R. A. 1978. *Sun, Weather and Climate.* Washington D.C., NASA, Sp-426.

Jurin, J. 1722. Invitatio ad Observationes Meteorologican Communi Consilio Instituendos. *Philosophical Transactions of the Royal Society*, 32, 422-427.

Labitzke, K. and Van Loon, H. 1988. Associations between the 11 year solar cycle, the QBO and the atmosphere. Part I: the troposphere and stratosphere in the northern hemisphere in winter. *Journal of Atmospheric Terestrial Physics*, 50, 197-206.
Lamb, H. H. 1977. *Climate, Present, Past and Future: Climate History and the Future*, Volume 2, London: Methuen and Co. Ltd.
Lamb, H. H. and Johnson, A. I. 1966. Secular Variation of the Atmospheric Circulation Since 1750. *Geophysical Memoirs*, No. 110, H.M.S.O., London.
Locke, C. W. and Locke, W. W. 1977. Little Ice Age Snowcover Extent and Paleoglaciation Thresholds: North-Central Baffin Island, N.W.T., Canada. *Arctic and Alpine Research*, 9, 291-300.
Lowe, A. B. 1961. Canada's First Weatherman. *Beaver*, Summer, 4-7.

MacKay, D. K. and Mackay, J. R. 1965. Historical Records of Freeze-up and Break-up on the Churchill and Hayes Rivers. *Geographical Bulletin*, 7, 7-16.
Manley, G. 1974. Central England Temperatures: Monthly Means 1695 to 1973. *Quarterly Journal of the Royal Meteorological Society*, 100, 389-405.
Minns, R. 1970. *An Air Mass Climatology of Canada during the Early Nineteenth Century: An Analysis of the Weather Records of certain Hudson's Bay Company Forts*, Unpublished Masters Thesis, University of British Columbia.
Moodie, D. W. and Catchpole, A. J. W. 1975. *Environmental Data from Historical Documents by Content Analysis.* Manitoba Geographical Studies 5, Department of Geography, University of Manitoba.

Rannie, W. F. 1983. Break-up and Freeze-up of the Red River at Winnipeg, Manitoba, Canada, in the 19th century and some Climatic Implications. *Climatic Change*, 5, 283-296.

Scott, P. A., Fayle, D. C. F., Bentley, C. V., Hansell, R. I. C. 1988. Large Scale Changes in Atmospheric Circulation Interpreted from Patterns of Tree Growth at Churchill, Manitoba, Canada. *Arctic and Alpine Research*, 20 (2) 199-211.

Thompson, H. A. 1978. The Climate of Hudson Bay. In: *Science, History and Hudson Bay*, Volume 1, Queen's Printer, Ottawa, 263-286.

W.M.O. Climatic Change. *Technical Note* No. 79, World Meteorological Organization Geneva.

Wilson, C. V. 1982. *The Summer Season Along the East Coast of Hudson Bay During the Nineteenth Century. Part 1: General Introduction: Climatic Controls: Calibration of the Instrumental Temperature Data 1814 to 1821*. Canadian Climate Centre, AES, Report Number 82-4.

Wilson, C. F. 1985. Daily Weather Maps for Canada, Summers 1816 to 1818 1a. Pilot Study, In: *Syllogeus*, Volume 5, C. R. Harington (ed.) National Museum of Canada, 191-218.

4 Historical climate records from the northeastern United States, 1640 to 1900

W. R. Baron

4.1 Introduction

The northeastern United States is endowed with the richest cache of old climate records available for anywhere in the Western Hemisphere, apart from a few locations in Latin America. Within the region are some of the earliest points of European settlement. Some of the earliest arrivals were well educated for their day and had an interest in natural philosophy, the predecessor of today's science. They immediately began to make a record of the New World's environment so that they could compare their findings to the world they had left behind. Among their notes can be found the grist for the preparation of climate records. By the beginning of the eighteenth century a thermometer and barometer were in place at Harvard College, Cambridge, Massachusetts and a record of observations was being sent to the Royal Society in London. Much of this data gathering was in the form of qualitative entries in diaries and journals, as well as instrument observations. This resulted from a growing interest in the role that climate played in deciding what crops could be grown, how soon rivers would be open to navigation in the spring, and what if any impact settlement was having on the occurrence of cold winters, droughts and floods.

Prior to the mid-twentieth century various authors attempted to assemble these data in order to understand the climate and to assess perceived changes, usually in the form of milder winters and warmer summers (see, for example, Webster 1843 and Blodget 1857). Unfortunately, their studies were based on a collection of instrumental records taken under greatly varying conditions with a wide assortment of instrument designs and calibrations. In the United States, standardization of instruments and observation times only came in 1872 with the centralization of record keeping under the U.S. Army Signal Service. Only recently have researchers undertaken work to develop techniques designed to reconstruct long instrumental observation series from data recorded prior to the 1870s. In addition historical climatologists have worked to design procedures to quantify qualitative diary and journal records containing weather and climate information. The purpose of this chapter is to describe these techniques as they have been developed for use on northeastern U.S. materials and to present the results of some reconstructions that may help to increase our understanding of seventeenth through nineteenth century climate in New England.

4.2 Northeast instrumental data and reconstructions

4.2.1 Instrumental record development

There are no instrumental records for the seventeenth century and observations for the the eighteenth century are very rare and extremely limited in their geographic and chronological distribution. The earliest of these records is one kept by John Winthrop of Harvard College for the period 1742 through 1779 which was assembled by Landsberg *et al.* (1968). Later records of varying lengths include those for Salem, Massachusetts, 1754-1829, (Holyoke 1833); Philadelphia, 1758-1759, 1767-1777, 1798-1804, (Schott 1876); New Haven, Connecticut., 1778-1795, (Landsberg 1949); Nazareth, Pennsylvania., 1787-1792, (Schott 1876); Cambridge, Massachusetts, 1790-1812, (Farrar 1809); and Morrisville, Pennsylvania., 1790-1845, (Schott 1876). J.M. Havens (1958) has completed a very useful, annotated bibliography of all available records. Based on these data and other shorter fragments, there are long reconstructions compiled for New Brunswick, New Jersey, 1738-1976, (Reiss *et al.* 1980) and for the eastern United States adjusted to Philadelphia, 1738-1967, (Landsberg *et al.* 1968).

During the early nineteenth century, records were begun at several other northeastern locations (see Table 4.1). All of these records were kept by amateur observers who earned their livings as doctors, farmers, or teachers. Data from some of these records appear in Forry (1842) Blodget (1857) and Schott (1876, 1881).

Table 4.1 Early nineteenth century instrument records for the Northeastern United States.

Location	Starting Date	Record Source
Albany, N.Y.	1795	Schott 1876
Belleville, N.J.	1803	Rutgers 1829
Brunswick, Me.	1807	Cleaveland 1867
Williamstown, Ma.	1811	Milham 1950
New Bedford, Ma.	1813	Rodman 1905
Portland, Me.	1816	Moody 1852
Dennysville, Me.	1816	Lincoln 1861
Newport, R.I.	1817	Gould 1857
Boston, Ma.	1818	Rumford Foundation 1839
Germantown, Pa.	1820	Schott 1876
New York City, N.Y.	1822	Schott 1876
Burlington, Vt.	1828	Clayton 1927
Randolph, Vt.	1828	Nutting 1863
Providence, R.I.	1831	Caswell 1882
Hanover, N.H.	1834	Dartmouth College 1905
Gettysburg, Pa.	1839	Schott 1876

The Surgeon General's Office was the first agency of the federal government to organize a meteorological network. Record keeping began at various military posts in 1819 and continued until the 1870s (U.S. Surgeon General's Office 1855). During the nineteenth century several federal meteorological networks were organized to collect data. The most important of these early networks was that administered by the Smithsonian Institution which ran its own network from 1848 through 1872 (Langley 1894) and assisted the Patent Office (U.S. Patent Office, 1848-1858) and the Bureau of Agriculture (U.S. Bureau of Agriculture 1863-1870) in operating a special network for farmers. The Navy was another organization which supported the collection of maritime meteorological data. In the early 1870s, Congress tired of funding several agencies to run networks and ordered the consolidation of all meteorological data gathering under the U.S. Army Signal Service (U.S. Army 1872-1892). Political pressure to bring meteorological activities under civilian control in the late 1880s eventually led to the creation of the United States Weather Bureau. During the nineteenth century, at one point in time or another, over 350 different Northeastern locations were represented in one or more of these early networks.

In addition to the federal government's meteorological activities, each of the Northeast states sponsored limited record keeping. Usually these records were connected with agricultural pursuits and were kept at state farms and, later, at state experiment stations. Other meteorological observations were taken at local colleges, new land grant universities, near drinking water reservoirs and at canal lock stations.

During this century, researchers have published a number of nineteenth and twentieth century record reconstructions. The most useful set of data is the Smithsonian's *World Weather Records* (Clayton, 1927) which provides records for mostly urban locations. A number of early precipitation records appear in the New England Water Works Association's publications (Goodnough, 1915). There are also state compilations for Connecticut and Maine which list some of the longer records (Kirk 1939, Baron *et al* 1980).

Many of the nineteenth century federal network records are available on microfilm from the U.S. National Archives in Washington (Darter, 1942). Records for the last 96 years are available through a number of publications and on magnetic tape obtainable from NOAA's National Climate Center at Asheville, N.C. A large number of digitised records are available at the U.S. Department of Energy's Carbon Dioxide Information Analysis Center in Oak Ridge, Tennessee.

A final source of early American meteorological records (1600-1920) of all types is the Historical Climate Records Office, part of the Center for Colorado Plateau Studies at Northern Arizona University, Flagstaff. The Office has on file and computer disk a large number of U.S. records collected for the seventeenth through nineteenth centuries. There is a particularly strong record group for the northeastern U.S. The Office was founded by the author with the intention of making these climate materials available to other researchers. Record collection was undertaken by the members of the now disbanded Northeast Environmental Research Group centered at the University of Maine and, after 1985, by the staff of the Historical Climate Records Office.

Despite the relatively large number of northeast nineteenth century instrument records, there are many that are unusable for climatic reconstruction. There are a variety of reasons for this: many records are simply too short, or the instruments were changed or moved frequently. Instrument sites were often poorly chosen and the instruments were inadequately maintained and went out of adjustment over time. In other cases, observation times were varied, sometimes on a daily basis. While these records are not without value, their use is compromised because they do not represent mathematically distinct sets but are instead a series of separate, short heterogeneous records for a given location. Reconstruction based on a heterogeneous record can at times be worse than no record at all as it is impossible to distinguish between trends and changes caused by climate or caused by problems with instrumentation and exposure.

In order to reconstruct long "homogeneous" time series, a climatologist must locate other long reliable records from nearby locations with similar geophysical features (Baron *et al.* 1980; Baron and Gordon, 1985). In practice this is a very complex operation requiring a large number of records to complete a reconstruction for a single location. The techniques used for temperature reconstructions are different from those for precipitation. The publications of

Thom (1966) Craddock (1977) and Bradley (1976) are useful introductions to these methodologies.

4.2.2 Instrument record reconstructions

Figures 4.1 and 4.2 are plots of yearly mean temperatures for some of the the longest surviving instrumental records of the eighteenth and nineteenth centuries. The most complete record, that by Landsburg *et al* (1968) for Philadelphia, exhibits a cooling trend running from the 1740s which bottoms out sometime between 1760 and 1776. The warming trend that follows stretches into the first decade of the nineteenth century and is followed by a

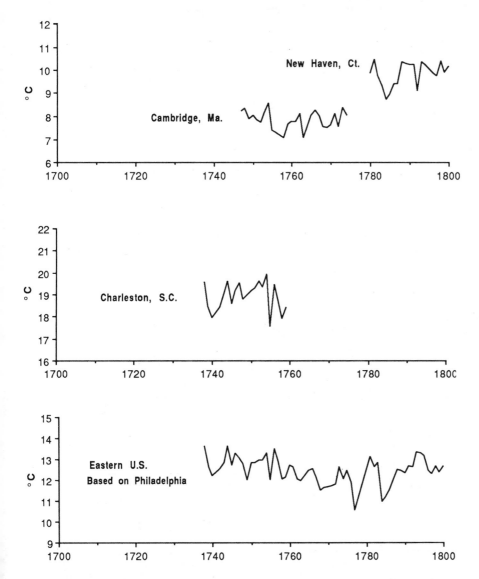

Figure 4.1 Mean annual temperature for locations in the eastern United States during the eighteenth century.

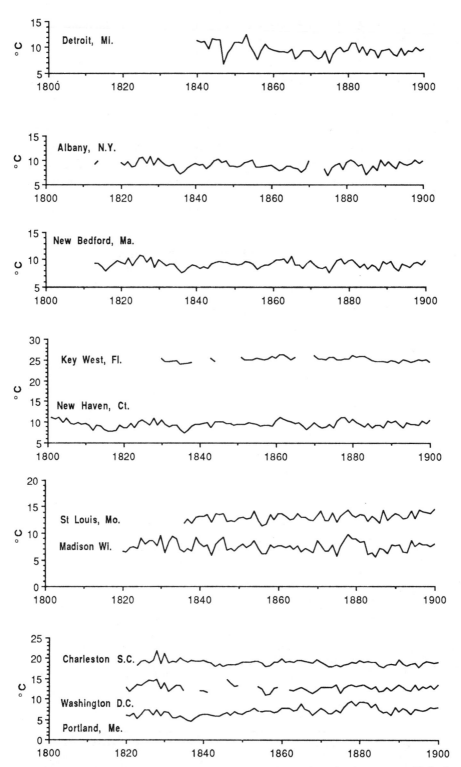

Figure 4.2 Mean annual temperature for selected locations in the eastern United States during the nineteenth century.

cool second decade. For most records the warming trend leading into the early twentieth century began sometime during the 1870s.

Precipitation records for the eighteenth century are very fragmentary (Figure 4.3) as are those for the early nineteenth century (Figure 4.4). Data for the last three quarters of the nineteenth century show considerable variability. For at least coastal locations such as New Bedford and, during the 1830s, Charleston, the 1820s and 30s were apparently somewhat wet. However, for at least one interior location, the Mohawk Valley, the same period was relatively dry. Some records exhibit a gradual increase in precipitation from 1850 through the mid-1870s. Thereafter most locations experienced a decline in precipitation.

While there are a fair number of instrumental records available from the second half of the

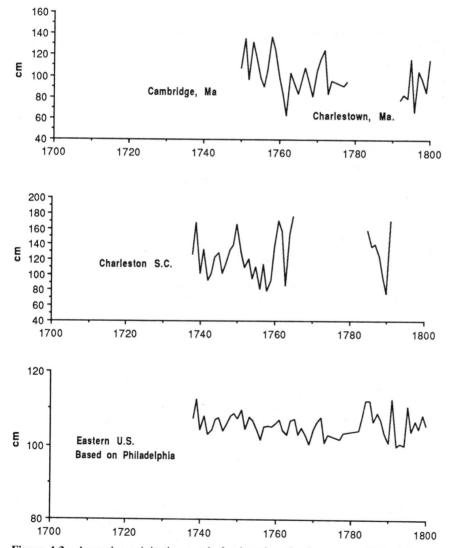

Figure 4.3 Annual precipitation totals for locations in the eastern United States during the eighteenth century.

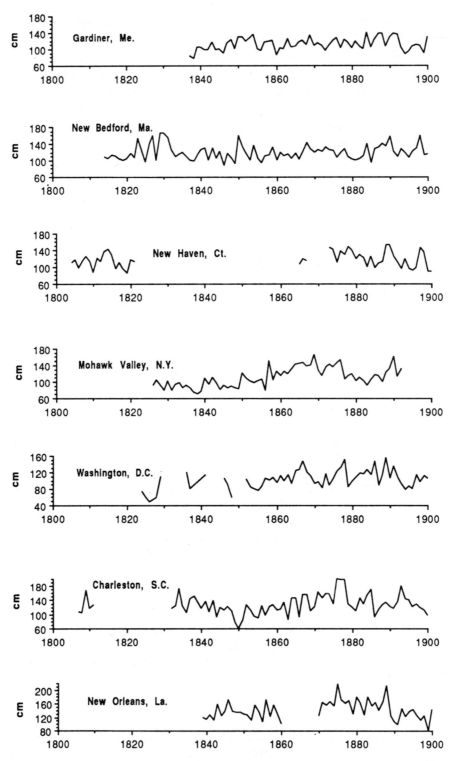

Figure 4.4 Annual precipitation totals for selected locations in the eastern United States during the nineteenth century.

nineteenth century, our understanding of the climate for earlier periods must depend on additional data derived from non-quantitative sources.

4.3 Northeast qualitative data and reconstructions

4.3.1 Sources and early reconstructions

Historical documentation such as that found in diaries, almanacs, old newspapers and periodicals is often richer in detail and has a better geographic distribution than early instrument records. The author, through a survey of over 150 archives with holdings of northeastern manuscript materials, located over 2,500 diaries found to contain weather-related information. Of this number there are at least 500 diaries that have detailed, daily descriptions of the weather and a much smaller number have multiple daily descriptions. A large number of others have monthly or seasonal weather descriptions. The chronological length of these diaries ranges from a few months to over 60 years. Most seventeenth century materials describe meteorological events only in terms of months, seasons, or years while diaries with daily entries are most frequently found for the eighteenth and nineteenth centuries.

Additional qualitative materials are found in old newspapers and periodicals. Newspapers first appeared at Boston in 1704 and from their earliest issues contained references to storms, droughts and other climatic phenomena. By the mid-nineteenth century, over 250 newspapers were published throughout the Northeast. Periodicals first appeared in the mid-eighteenth century. Periodicals concerned with agriculture, science and medicine are most useful for climatological record reconstruction. Articles on meteorology and climatology, old weather records, and, by the very early nineteenth century, instrumental records were published fairly regularly.

Attempts to reconstruct New England's climate using qualitative materials are not new. There were impressionistic overviews written in the late eighteenth and early nineteenth centuries (Williams 1790, Currie 1792, Webster 1843 and Thompson 1981). Early in the present century Brooks (1917) Weber (1930) and Smith (1934) used frequency counts and general descriptions of colonial snowfall, rainfall and winter weather in an effort to reconstruct early climatic patterns. Ludlum (1966, 1968, 1976, 1983) has produced perhaps the most complete overview of northeastern weather based on both early instrumental records and published diaries, newspapers and periodicals. Unfortunately much of his work was reported in qualitative form.

4.3.2 "Departure" methodology and reconstructions

To reconstruct quantitative records from qualitative source materials requires a sometimes complex and varied methodological approach. Advances in communications research dealing with the content analysis of historical documents provide a method that can be employed by climate researchers working on qualitative source materials.

The author modified a methodology used by Lamb (1965) on medieval materials to reconstruct a temperature and precipitation record based on seasonal and yearly weather descriptions. This procedure takes advantage of the fact that the diarists who recorded weather observations came from very similar educational and regional backgrounds. In other words, they employed the same vocabulary, nouns, verbs and descriptors in their writing. This commonality of expression contributed greatly to the compilation of a homogenous data set. Using dictionaries for the time period and region to insure an historically accurate definition of all words, the author designed rank-ordered word scales that delineated various degrees or departures from what was *perceived* by the diarists as "normal" temperatures and precipitation.

These word scales were then fitted to an arbitary numerical index. As the temperature word scale contained 29 divisions which allowed for adjectives such as "cold" or "cool" and their modifiers ("very", "little" etc.) the author designed a scale ranging from +14 ("very hottest") to -14 ("very coldest") with 0 considered as "normal." A scale for precipitation was arranged in a similar fashion (+10 to -10) to allow for the 21 divisions of the precipitation word scale. In instances, where diarists made no report conditions were assumed to be "normal."

Reconstructions using the "departure" method depend on the use of reports for a given season or year from a number of observers. By taking the consensus of a set of observations, individual idiosyncrasies due to ill health, or poor or inaccurate memory of previous weather experienced can be offset. While tests by Lamb (1965) and the author show that resulting reconstructions correlate in a "general" way to data derived from other record types, recent research has shown that personal long term impressions of past weather are sometimes skewed or misleading and tend to emphasize the catastrophic (Oliver 1975). Therefore the departure reconstruction method should be used only for time periods in which instrumental records are not readily available.

A further consideration when using the departure method is to determine the frame of reference for all observers. Observers who have recently moved into a geographic area may not be familiar with what weather is "normal" for their new locale. A historical background check on all observers is imperative. For example while New England source materials are available for the period 1620 through 1640, a reconstruction for this time is problematical as most observers evaluated the weather based on their English weather experience and on misconceived but generally accepted ideas concerning North American climate (Kupperman 1982). Fortunately by the 1640s there were a sufficient number of diarists who had spent at least ten years in New England. By the 1650s the first native-born generation started to produce observations. Furthermore, significant immigration into New England had greatly diminished as a result of the Puritan Revolution, leaving the region's society to expand mostly through natural reproduction.

Figure 4.5 is a reconstruction of temperature departures, based on yearly and seasonal weather descriptions drawn from over 300 diaries and journals representing eighty locations throughout New England from 1640 through 1820. The temperature reconstruction, representing exclusively daylight observations, shows that observers preceived that most of the seventeenth century had near "normal" temperatures with most departures from "normal" falling on the cold side. The decades of the 1650s and 60s appear to have been more variable than most. After 1730 there was a considerable time period, stretching into the mid-1780s, when the temperature was extremely variable. From 1750 to the early 1780s there were several very cold years. Several of these appear to have been more significant to observers

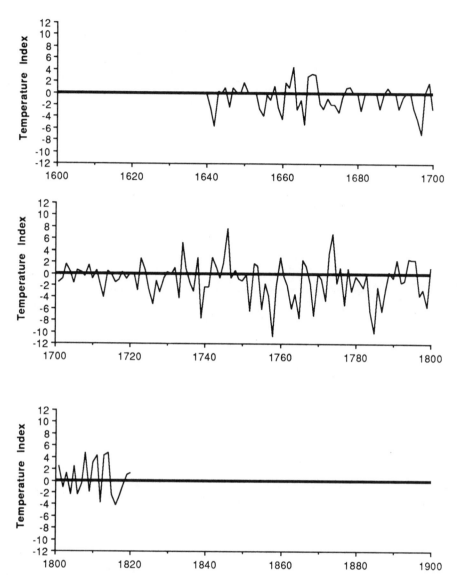

Figure 4.5 Reconstruction of New England temperatures, 1640-1820, using the departure method.

than either of the better known cold years of the early nineteenth century, 1812 or 1816. This situation is partly explained by an analysis of seasonal records which revealed considerably greater seasonal agreement for the late eighteenth century than for the period 1800 through 1820. For example in 1757 and 1784, two of the coldest years, all four seasons were reported as colder than normal where in both 1812 and 1816 falls were described as somewhat above normal in temperature.

The reconstruction of precipitation departures (Figure 4.6) is based on the same source materials as those used to produce Figure 4.5. The precipitation record for the seventeenth century shows both dry and wet periods. The 1660s were drier while the early 1670s and the

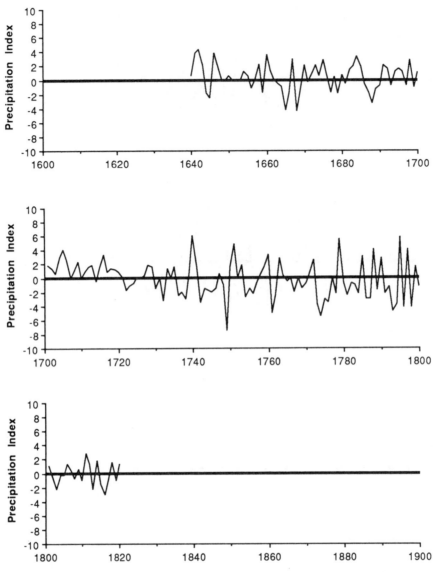

Figure 4.6 Reconstruction of New England precipitation, 1640-1820, using the departure method.

1690s were wet. The 1740s through 90s was a period of considerable variability with mostly drier conditions punctuated fairly frequently with individual years that were very wet.

4.3.3 Two-tiered methodology and reconstructions

The detailed daily diaries available for the eighteenth and nineteenth centuries were analyzed utilizing a two-tiered content analysis methodology. To use this methodology, weather entries must be sorted by type and then assigned to word scales in a manner similar to that described for the "departure" method. However, the weather descriptions employed are much more detailed than those used for departure reconstructions and information is available for each day of the year. Not only modifiers relating to quantity or degree, but also the time duration of each weather event are weighed (Baron, 1982). The arbitary numerical scale designed to fit the resulting complex qualitative scale is constructed with a range of 0 (for "no precipitation") to 30 (for "wettest day ever remembered").

This method makes use of a considerably more sophisticated differentiation between qualitative degrees of change and relies much more heavily on an individual diarist's selection of descriptors than does the "departure" analysis. As a result one can obtain a very detailed chronological resolution down to and including day to day reconstructions which are usually homogenous because they are based on data produced by the same diarist. Regression analysis in which the results of these reconstructions have been compared with independently derived instrumental records has yielded positive correlations of 0.6 or higher (Baron 1982)

It is very important that a background history of the diarist be completed to determine their period of residence, general health, occupation (whether in-door or out-of-doors work) age, and education. Fortunately most diaries and journals include much of this information. All of these factors can have a significant impact on temperature observations. For example ill health or advancing age often affects the body's ability to properly sense temperature, while unfamilarity with the local environment will affect an individual's ability to perceive "normal" conditions.

One significant disadvantage of this method is that it is very difficult to construct long master records made up of several diaries because some diarists were more demonstrative in their use of descriptors than others. Still another problem with this method is that for best results it requires relatively long diaries of from ten to forty years that have few daily entries with missing weather observations. Such diaries are indeed rare!

Figure 4.7 shows yearly precipitation reconstructions based on four daily diaries kept during the eighteenth century for locations in or near Boston. These records demonstrate considerable interannual variation. However, until more diaries are located that overlap the ones presently reconstructed it is difficult to conclude how representative these records may be.

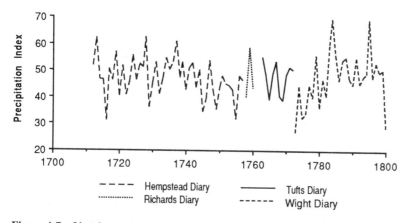

Figure 4.7 Yearly precipitation records from four daily diaries from southern New England using two-tiered content analysis.

4.3.4 Frequency count methodology and reconstructions

Records of drought frequency and growing season lengths based on qualitative materials require a different methodological approach. Interpretation and analysis of these qualitative data demand a precise set of definitions. For example what is meant by such terms as a "white frost" or a "killing frost"? The specific species of plants killed by frost help determine temperature limits because some plants are more resistant to freezing temperatures (Baron *et al.* 1984). Using these data, regional dates for the beginning and end of growing seasons can be determined. The same sort of methodology can be used to produce records of droughts with agricultural implications.

Figure 4.8, based on 143 diaries and 99 agricultural records, is a reconstruction of the frequency of droughts during growing seasons. There was an increase in droughts during the 1740s, 60s, 70s, and 80s as well as a relatively high number of droughts for the period 1850 through 1889. Records of growing season lengths such as the one for southern New Hampshire (Fig 4.9) illustrate the great variability that existed throughout most of the

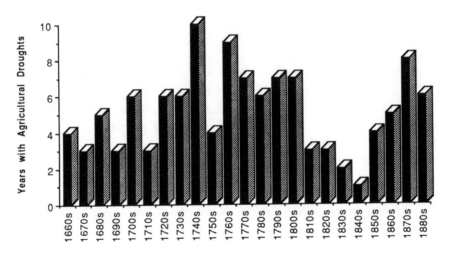

Figure 4.8 Droughts in southern New England during the growing season, by decade.

eighteenth and nineteenth centuries. Only with the onset of the warming period in the 1880s through 1890s does the degree of variability decrease. Between 1740 and 1840, it was not uncommon for the growing season lengths to vary as much as one month from one year to the next. These records, drawn from over 200 agricultural journals, also exhibit a systematic relationship between the timing of spring and fall killing frosts (Baron *et al.* 1984).

The simplest form of frequency reconstruction employs counts of various weather events such as thunder and lightning. The number of occurrences of each of these phenomena are recorded by month, season and year. They are then reported as aggregates or in index form using a specified year or decade as a base period (Baron 1982).

Figure 4.10, based on 39 diaries and 20 observation records, demonstrates the great range of frequency in the number of thunder storms from year to year. The period 1740 through 1780s was one of increased thunderstorm activity. From 1800 to 1815 there appears to have

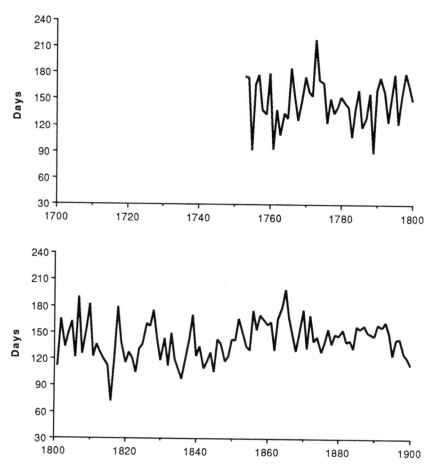

Figure 4.9 Length of the growing season in southern New Hampshire, 1752-1900.

been a decrease in activity with another increase during the period 1816 through 1828. From 1828 to 1861, there was a fairly steady decline in activity with the exception of an eleven year period from 1835 through 1845. In the 1870s and 1880s there was an increase in activity exceeded in aggregate by only the 1750s.

4.4 Conclusions

The climatological picture that emerges from these reconstructions shows a cool, somewhat wet three centuries marked by occasional decades of drought, and considerable interannual variability that does not decrease until the last two decades of the nineteenth century. These results correspond well with reconstructions for other geographic regions. Lamb (1977) observed that the period from 1200-1400 to 1850-1890 was characterized by a trend toward colder climates in the Northern Hemisphere. The circumpolar vortex expanded and the zone of heightened cyclonic activity shifted south. Yet despite the overall trend toward cooler and wetter regimes, there were intermittent episodes of warmth and drought. This climatic

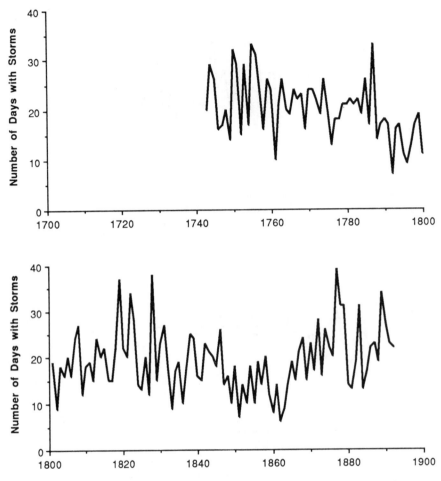

Figure 4.10 Days per year with thunderstorms in eastern Massachusetts, 1743-1892.

episode, often called the Little Ice Age, continued until the middle to latter part of the nineteenth century when a gradual warming trend began. Lamb's description is based primarily on European records but adequately describes conditions in the Northeast as well.

The methods described in this chapter can be used for any time period and geographical locale for which there is, at a minimum, good qualitative descriptions of the weather. The development of these new techniques has broken the researcher's total dependence on instrumental records as the sole source of detailed climatic records. The techniques allow for the use of detailed qualitative weather descriptions from a large number of geographical locales. The long, quantitative record reconstructions that result provide considerable additional information that will help us better understand the operation of the earth's atmospheric system as well as the ways in which that system changes over time.

References

Baron, W. R. 1982. The reconstruction of eighteenth century temperature records through the use of content analysis. *Climatic Change*, 4, 385-398.

Baron, W. R. and G. A. Gordon 1985. A reconstruction of New England climate using historical materials, 1620-1980. *Syllogeus* 55, 229-245.

Baron, W. R., G. A. Gordon, H. W. Borns, and D. C. Smith, 1984. Frost-free season record reconstruction for eastern Massachusetts, 1733-1980. *Journal of Climate and Applied Meteorology*, 23, 317-319.

Baron, W. R., D. C. Smith, H. W. Borns, J. Fastook, and A. E. Bridges, 1980. *Long Time Series Temperature and Precipitation Records for Maine, 1808-1978*. Maine Agricultural Experiment Station Bulletin 771. Orono: University of Maine.

Blodget, L. 1857. *Climatology of the United States*. Philadelphia: J. B. Lippincott.

Bradley, R. S. 1976. *Precipitation History of the Rocky Mountain States*. Boulder: Westview Press.

Brooks, C. F. 1917. New England snowfall. *Geographical Review*, 3, 222-240.

Caswell, A. 1882, Meteorological observations made at Providence, R.I. from Dec.1831 to Dec. 1876. *Smithsonian Contribution to Knowledge* 24 (443) 1-31.

Clayton, H. H. 1927. Burlington. In: *World Weather Records, Smithsonian Miscellaneous Collections* 27, 817-819.

Cleaveland, P.. 1867. Results of meteorological observations made at Brunswick, Maine, between 1807 and 1859, reduced and discussed by C. A. Schott. *Smithsonian Contributions to Knowledge* 16 (204) 2-25.

Craddock, J. M. 1977. A homogeneous record of monthly rainfall totals for Norwich for the years 1836 to 1976. *Meteorological Magazine*, 106, 267-278.

Currie, W. 1792. *An Historical Account of the Climates and Diseases of the United States of America*. Philadelphia: T. Dobson, (reprinted in 1972 edition by Arno Press and The New York Times, New York).

Darter, L. J. Jr. 1942. *List of Climatological Records in the National Archives*. Washington D.C.: U.S. Archives.

Dartmouth College, 1905. Monthly, annual and average temperature and precipitation at Hanover, N.H., 1834-1904, *Climate and Crops: New England Section*, Annual Summary 1905:11.

Farrar, J., 1809. Abstract of meterological observations made at Cambridge, New England. *Memoirs of the American Academy of Arts and Sciences*, old series 3 (1):361-412.

Forry, S. 1842. *The Climate of the United States and Its Endemic Influences*. New York: J. and H. G. Langley (reprinted in 1978 by American Meteorological Society Press, New York).

Goodnough, X. H. 1915. Rainfall in New England. *Journal of the New England Water Works Association*, 29, 237-425.

Gould, S., 1857. Ms. Meteorological Journal for Newport, R.I., 1817-1857. Newport Historical Society, Newport, R.I.

Havens, J. M. 1958. *An Annotated Bibliography of Meteorological Observations in the United States, 1715-1818*. Key to Meteorological Records Documentation No. 5.11. Washington, D.C.: U.S. Government Printing Office.

Holyoke, E. A., 1833. A meteorological journal from the year 1786 to the year 1829 inclusive. *Memoirs of the American Academy of Arts and Sciences*, new series 1:107-216.

Kirk, J. M. 1939. *The Weather and Climate of Connecticut*. State of Connecticut Geological and Natural History Survey Bulletin 61. Hartford: State Geological and Natural History Survey.

Kupperman, K. O., 1982. The puzzle of the American climate in the early colonial period, *American Historical Review* 87:1262-1289.

Lamb, H. H., 1977. *Climate History and the Future*, vol. 2 of *Climate: Present, Past and Future*. Methuen, London.

Lamb, H. H., 1965. The early medieval warm epoch and its sequel. *Palaeogeography, Palaeoclimatology, Palaeoecology*, 1:13-37.

Landsberg, H. E., 1949. Climatic trends in the series of temperature observations at New Haven, Connecticut. *Geografiska Annaler*, 1-2:125-132.

Landsberg, H. E., Yu, C. S., and Huang, L., 1968. *Preliminary Reconstruction of a Long Time Series of Climatic Data for the Eastern United States*. Technological Note B14-571, Institute for Fluid Dynamics and Applied Mathematics, University of Maryland, College Park.

Langley, S. P., 1894. The meteorological work of the Smithsonian Institution. *United States Weather Bureau Bulletin*, 11:216-220.

Lincoln, T., 1861. Ms. Record Book of Weather Reports for Dennysville, Maine, 1816-1861. at: Maine Historical Society, Portland.

Ludlum, D. M., 1983. *The New Jersey Weather Book*. Rutgers University Press, New Brunswick.

Ludlum, D. M., 1976. *The Country Journal New England Weather Book*. Houghton Mifflin, Boston.

Ludlum, D. M., 1968. *Early American Winters II, 1821-1870*. American Meteorological Society, Boston.

Ludlum, D. M., 1966. *Early American Winters, 1604-1820*, American Meteorological Society, Boston.

Milham, W., 1950. *Meteorology in Williams College*. McClelland Press, Williamstown.

Moody, L., 1852. Mss. Thermometrical Records for Portland Observatory, Maine, 1816-1852. 2 vols. at: Maine Historical Society, Portland.

Nutting, W., 1863. Mss. Meteorological Observations for Randolph, Vt., 1828-1863. at: Vermont Historical Society, Montpelier.

Oliver, J. E., 1975. Recollection of past weather by the elderly in Terre Haute, Indiana. *Weatherwise*, 28:161-171.

Reiss, N. M., Groveman, B. S., and Scott, C. M., 1980, Seasonal mean temperatures for New Brunswick, N.J., *Bulletin of the New Jersey Academy of Science*, 1980:1-10.

Rodman, T. R., 1905. Monthly, annual and average temperature and precipitation at New Bedford, Mass., 1813-1904, *Climate and Crops: New England Section, Annual Summary* 1905: 9.

Rumford Foundation, 1839. Mss. Meteorological Observations for Boston, 1818-1839 at: Harvard University Libraries, Cambridge.

Rutgers, G., 1829. Mss. Meteorological Observations, 1803-1829 at: Rutgers University Library, New Brunswick.

Schott, C. A., 1881. *Tables and Results of the Precipitation in Rain and Snow in the United States and Some Stations in Adjacent Parts of North America, and in Central and South America*. In: *Smithsonian Contributions to Knowledge*, 2d. ed., no. 353. Smithsonian Institution, Washington, D.C.

Schott, C. A., 1876, *Tables, Distributions, Variations of the Atmospheric Temperature in the United States and Some Adjacent Parts of North America*. In: *Smithsonian Contributions to Knowledge*, no. 277. Smithsonian Institution, Washington D.C.

Smith, Jr., F. H., 1934. Some old-fashioned winters in Boston. *Massachusetts Historical Society Proceedings*, 65:269-305.

Thom, H. C. S. 1966. *Some Methods of Climatological Analysis*. Technical Note 81, WMO No. 199. TP. 103. Geneva: World Meteorological Organization.

Thompson, K., 1981. The question of climatic stability in America before 1900, *Climatic Change*, 3:227-241.

United States Army, 1872-1892. *Department of War Annual Reports*, 1872-1892, U.S. Army, Washington, D.C.

United States Bureau of Agriculture, 1863-1870. *Annual Reports*, 1863-1870, Bureau of Agriculture, Washington, D.C.

United States Patent Office, 1848-1858. *Annual Reports*, 1848-1858, Patent Office, Washington, D.C.

United States Surgeon General's Office, 1855. *Meteorological Register*, U.S. Government, Washington, D.C.

Weber, J. H., 1930. The rainfall of New England: Historical statement, *Journal of the New England Water Works Association*, 44:1482-1485.

Webster, N., 1843. On the supposed change in the temperature of winter. In: *A Collection of Papers on Political, Literary and Moral Subjects*, n.p., New York (reprinted in 1968 by Burt Franklin, New York) pp. 119-162.

Williams, S., 1790. Change of Climate in North America in: Mss. Williams Papers at: University of Vermont Special Collections Library, Burlington.

5 Documentary evidence for changes in the climate of Iceland, A.D. 1500 to 1800

A. E. J. Ogilvie

5.1 Introduction

An understanding of the nature and causes of climatic change is of particular relevance today when it appears highly probable that the activities of humankind may have far-reaching consequences for global climate (Jones *et al.* 1988). In order to further our understanding of changes in climate it is necessary to look, not just at present day climate or the recent past, but to consider long timescales. As quantitative meteorological observations have only a limited span backwards in time, "proxy" data must be used to estimate changes in climate beyond this time. One type of proxy evidence that is potentially particularly useful is documentary data. Although written accounts of the weather and climate are qualitative in nature they may yield accurate and detailed information.

For much of its thousand-year-long history, Iceland is unusually rich in both the quality and quantity of its documentary information on climate. This information includes descriptions of the incidence of the Arctic drift ice off the coasts of Iceland. As a positive correlation exists between temperature and sea-ice incidence (Bergthórsson 1969, Ogilvie 1981, 1984a) the descriptions of sea ice provide an additional and useful way of gauging past temperature variations. (With regard to Icelandic spelling, the two Icelandic characters for "th", Þ, pronounced as in "throne", and ð, pronounced as in "clothe", are retained in personal and place names except where these have already been anglicised, as in "Bergthórsson".)

In this chapter, the major historical sources containing weather and sea-ice information are evaluated for the period A.D. 1500 to 1800. Thermal and sea-ice indices for the years 1601 to 1780 have already been developed (Ogilvie 1981, 1984a). These are continued here to 1800, and are also estimated for the two decades in the sixteenth century when there is sufficient information to do this, 1561 to 1580. Variations in temperature and sea ice during the period 1500 to 1800 are considered in the light of these indices. Prior to this discussion of the sources and their evidence, some of the principal features of the climate of Iceland will be noted.

5.2 Principal features of the climate of Iceland

Continuous meteorological observations were begun in the west of Iceland, at Stykkishólmur, in 1846 (Sigfúsdóttir 1969). By ~1900, some twenty observational sites had come into existence and many more soon followed. As data on the climate of the twentieth century have increased, so has knowledge of the basic climate systems affecting Iceland. Of prime importance is the location of the country close to two major air masses (cold Polar air and warmer

Atlantic air) and to two principal ocean currents (the warmer Irminger current and the colder East Iceland current).

Iceland's position within the range of the Arctic drift ice is also of major importance to the country's climate. Using data from the twentieth century, Eythórsson and Sigtryggsson (1974) define a "normal" ice year as one when the ice edge is about 90 to 150 kilometres away from Straumnes in northwest Iceland. A "mild" year is when the distance of the ice from this point is about 200 to 240 kilometres. "Severe" years occur when the ice reaches the coast of Iceland. It must be noted, however, that in the seventeenth, eighteenth and nineteenth centuries, "severe" sea-ice years occurred much more frequently than in the twentieth century.

Sea ice most commonly affects the northwestern, northern and northeastern coasts. The ice reaches the coasts of Iceland as a result of a complex amalgam of winds, weather patterns and ocean currents. Thus, the unusual event of sea ice off the south coast can only partially be associated with a very cold weather pattern. However, when the ice does occur close to the land its main effect is to lower temperatures, and, as stated above, there is a good positive correlation between land temperatures and the presence or absence of the ice.

As a consequence of Iceland's maritime location, the annual temperature range is small compared to continental locations at the same latitude. Winters are fairly mild and the summers cool. Thus, during the period 1931 to 1960, mean temperatures in Reykjavík were 0.4°C in January and 11.2° in July (Einarsson 1976). However, strong winds may make temperatures seem colder than they are.

Temperatures have varied considerably in Iceland during the twentieth century. In the early part of the century, the trends paralleled those of the Northern Hemisphere as a whole, but the recent (post-1970) warming is less evident in Iceland. Winter and summer temperature anomalies during the period 1851 to 1988 are shown in Figure 5.1. Historical data suggest considerable fluctuations in temperatures on an annual, decadal and century time scale. Regional variability is also of importance.

Precipitation in Iceland is very frequent due to the North Atlantic depressions moving across or close to the country. These can move at great speed, often at 50 to 100 kilometres per hour (Eythórsson and Sigtryggsson). However, distribution is unequal due to the varying terrain and orographic features. For example, far less precipitation falls in the north and northeast than in the south (Einarsson 1976). During the period 1931 to 1960, the place with the highest precipitation was Kvisker in the southeast, with an average of over 3300mm per year. In contrast, Reykjahlíð, near Mývatn in the north, only had an average of 394mm per year (Eythórsson and Sigtryggsson 1971, Einarsson 1976).

5.3 Documentary sources A.D. 1500 to 1800

5.3.1 Introduction

Many attempts have been made to explain the existence of the numerous and diverse historical sources written in Iceland. Lack of space makes any detailed discussion of this topic inappropriate here. However, a few comments on the origin of the Icelanders, and their way of life, may go some way towards explaining why they wrote so copiously about the weather.

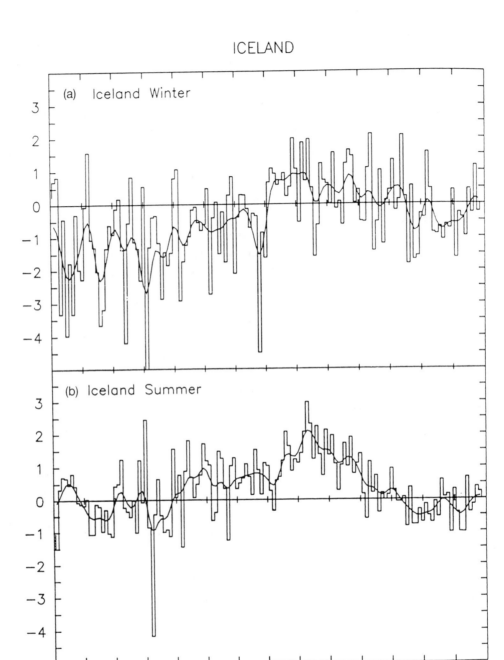

Figure 5.1 Winter and Summer Temperature Anomalies for Iceland in degrees Celsius from 1851 to 1988 shown as deviations from the 1951 to 1970 mean. Winter is defined as December, January and February, with the winter dated according to the month of January. Summer is defined as June, July and August. The smoothed curve is a Gaussian filter designed to show variations on decadal and longer timescales. These temperature data are by courtesy of Dr P. D. Jones of the Climatic Research Unit.

Iceland was first settled, principally by Norwegian Vikings, in the period ~A.D. 870 to 930. Throughout most of Iceland's history, the principal economic occupation has been animal husbandry. A farming life, shared by all members of society, fostered a keen interest in the weather. This interest was further augmented by the marginal nature of Icelandic farming, with variations in the weather playing a significant role in people's lives (Thórarinsson 1956, Friðriksson 1969, Ogilvie 1981 and 1984b, Bergthórsson et al. 1988). The pivot on which Icelandic farming hinged was the annual hay crop. This was needed to provide supplementary fodder for livestock over the winter when weather conditions became such that it was impossible for them to graze. Thus, for example, if a cold wet summer resulted in a poor hay crop, and was subsequently succeeded by a severe winter, food shortages, and even deaths by starvation, could follow for both the animal and human populations.

This interest in weather and climate, plus a high level of literacy, and a tradition of writing, has resulted in the existence of a wide variety of documents in which the nature of the weather over a given year, season, month or day is recounted in some detail. It must be noted, however, that it is only after ~A.D. 1600 that these documents are extensive. For the period from the late twelfth century to 1430, there is sufficient information on the weather in Iceland to give some suggestions as to how the climate may have varied, but that is all. For the years 1430 to ~1560, few sources are available. The major reason for this may have been the ravages of epidemics which hit Iceland in the early and late fifteenth century and seriously disrupted society. These epidemics may have been the bubonic plague experienced in other parts of Europe. For the timespan 1561 to 1600 there is some contemporary evidence. The sources and evidence for changes in climate in Iceland from ~A.D. 865 to 1598 are discussed in Ogilvie (1990).

For the purpose of climate reconstruction, the most useful kinds of documents are annals, early newspapers, journals, private and official correspondence, descriptions of the country by native Icelanders, and travel books written by foreign visitors. Some of the most important of these sources are discussed below. Prior to this discussion, a few words may be said on the topic of source analysis.

5.3.2 Source Analysis

As the need to analyse historical sources in order to ensure reliability has already been discussed in detail elsewhere (Vilmundarson 1972, Bell and Ogilvie 1978, Ingram et al. 1981) this topic will be raised only briefly here. All the sources used were carefully evaluated in order to establish their reliability. The primary purpose of this was to ascertain as clearly as possible that the information in any given source represented an accurate picture of a weather event. In particular, details of authorship and the date and place of writing were noted. This established whether or not a source was written close in time and space to the events described. All the eighteenth-century sources, and the majority of the seventeenth-century sources used here are contemporary. Some sixteenth- and early seventeenth-century sources are not. These are used, with caution, where no other sources are available, and because they do have some claim to reliability.

5.3.3 The sources

Annals One of the most useful types of historical source for the period ~1500 to 1800, and particularly for the years ~1600 to ~1730, is the annal. Some 33 annals exist, all written by different individuals. The annals, together with their authors, and the time periods they cover, are listed in Table 5.1. The majority of these annals come under the category "later" annal, to distinguish them from the medieval annals (see below). The later annals have been published, by various editors, in the recently completed six-volume series *Annálar 1400-1800 (1922-88)*. The annals differ from diaries in that they relate events of a general rather than a personal nature and, as their name suggests, their entries are on an annual rather than a daily basis. A few words on how these annalistic writings developed will make their nature clearer.

During the medieval period in Iceland a number of annals were compiled, mostly by unknown authors. These annals give brief notices of the major events of the year. This phase of annal-writing ceased in 1430 and does not concern us here, except in so far as it formed the pattern on which the seventeenth and eighteenth century annals were built. The later annals differ from the medieval annals in that the authorship and inception of the later annals is well documented. They also generally give quite detailed information. A third collection of annals, consisting just of three works, does not fit precisely into the category of either medieval or later annal and may be termed "sixteenth-century annals".

The writers of the different annals were frequently priests or men involved in the legal profession, but the structure of society was such that they were usually farmers as well, or at least had an interest in the running of a farm. This ensured their preoccupation with the weather. The annals were written in many different locations in Iceland, but mainly in the north and west where the population was more concentrated. Although the annals are, on the whole, excellent sources, they vary as to reliability and quality, and therefore must all be evaluated individually. Many of the annals are reliable only in part, with the first section deriving from earlier works, and only the latter part forming an independent and original source. (Those annals containing little or no contemporary information, such as *Setsberg-sannáll, Ketilstaðaannáll, Hrafnagilsannáll* and *Vatnsfjarðarannáll hinn yngri* have been deliberately excluded from this analysis.)

For the early part of the period under consideration, that is, from A.D. 1500 through to the early part of the seventeenth century, there is very little weather information available. Much of what exists for this time comes from the three "sixteenth-century annals" known as *Gottskálksannáll, Annalium in Islandia farrago* and *Biskupa-Annálar Jóns Egilssonar*.

Gottskálksannáll, which covers the years A.D. 1 to 1578, was compiled by Gottskálk Jónsson (1524-1590) and his son Jón Gottskálksson (died 1625). The early part is a copy of an older annalistic work. The latter part derives from the contemporary knowledge of Gottskálk and Jón. The entries in this work are brief, like those of the other medieval annals, but, particularly for the years 1561 to 1578, several interesting weather descriptions are given. This work is published in Storm (1977).

The *Biskupa-Annálar Jóns Egilssonar* or "Annals of Bishops" by Jón Egilsson (1548-1636) is primarily a history of Iceland bishops from the eleventh century to the late sixteenth century. The work also mentions notable events that occurred during the bishops' lifetimes, including severe seasons and other natural phenomena. Jón's major sources were oral accounts, particularly from his parents and grandfather. Weather descriptions from this work, together with weather extracts taken from *Gottskálksannáll*, are discussed in detail in Ogilvie (1990).

The third work in this group is the *Annalium in Islandia farrago* by Bishop Gísli Oddsson (1593-1638) or *Íslenzk annálabrot Gísla Oddssonar*. This was first published in 1917 in an

	ANNAL	AUTHOR	PLACE OF ORIGIN	TIMESPAN	CONTEMPORARY
1.	Skarðsárannáll	Björn Jónsson	Skarðsá, Skagafjord district	1400-1640	1636-1640
	Continuation	Brynjólfur Sveinsson	Skálholt, Árnes district	1641-1645	1641-1645
2.	Seiluannáll	Halldór Þorbergsson	Seila, Skagafjord district	1641-1658	1652-1658
3.	Vallholtsannáll	Gunnlaugur Þorsteinsson	Vallholt, Skagafjord district	1626-1666	1626-1666
4.	Vatnsfjarðarannáll Hinn Elzti	Jón Arason	Vatnsfjord, N. Ísafjord district	1395-1654	1645-1654
	Continuation	Sigurður Jónsson	Vatnsfjord/Ögur N. Ísafjord district	1655-1661	1655-1661
5.	Vatnsfjarðarannáll Hinn Yngri	Guðbrandur Jónsson	Vatnsfjord, N. Ísafjord district	1614-1672	1656-1672
6.	Annálsgreinar frá Holti	Sigurður Jónsson	Holt, W. Ísafjord district	1673-1705	1673-1705
7.	Ballarárannáll	Pétur Einarsson	Ballará, Dala district	1597-1665	c.1620-1665
8.	Vallannáll	Eyjólfur Jónsson	Vellir, Eyjafjord district	1659-1737	1690-1737
9.	Mælifellsannáll	Ari Guðmundsson	Mælifell, Skagafjord district	1678-1702	1678-1702
		Magnús Arason	Mælifell, Skagafjord district	1703-1738	1703-1738
10.	Annáll Páls Vídalíns	Páll Vídalín	Víðidalstunga, W. Húnavatn district	1700-1709	1700-1709
11.	Fitjaannáll	Oddur Eiríksson	Fitjar, Borgarfjord district	1400-1712	1700-1712
	Continuation	Jón Halldórsson	Hítardalur, Mýra district	1713-1719	1713-1719
12.	Kjósarannáll	Einar Einarsson	Garðar, Kjós district	1471-1687	1670-1687
13.	Hestsannáll	Benedikt Pétursson	Hestur, Borgarfjord district	1665-1718	1694-1718
14.	Hítardalsannáll	Jón Halldórsson	Hítardalur, Mýra district	1724-1734	1724-1734
	Continuation	Gísli Bjarnason?	Melar, Borgarfjord district?	1735-1740	1735-1740
15.	Hvammsannáll	Þórður Þórðarson	Hvammur, Dala district	1707-1738	1717-1738
16.	Eyrarannáll	Magnús Magnússon	Eyri, N. Ísafjord district	1551-1703	1673-1703
17.	Annálsgreinar Árna á Hóli	Árni Magnússon	Eyri, N. Ísafjord district	1632-1695	1662-1695
18.	Grímsstaðaannáll	Jón Ólafsson	Grímsstaðir, Snæfellsnes district	1402-1764	1734-1764
19.	Setbergsannáll	Gísli Þorkelsson	Setberg, Snæfellsnes district	1202-1713	-
20.	Sjávarborgarannáll	Þorlákur Magnússon	Sjávarborg, Skagafjord district	1389-1729	1727-1729
21.	Ólfusvatnsannáll	Sæmundur Gissurarson	Ölfusvatn, Árnes district	1717-1762	1740-1762
22.	Ketilsstaðaannáll	Pétur Þorsteinsson	Ketilsstaðir, S. Múla district	1724-1784	-
23.	Höskuldsstaðaannáll	Magnús Pétursson	Höskuldsstaðir, E. Húnavatn district	1730-1784	1734-1784
24.	Húnvetnskur annáll	unknown	E. Húnavatn district	1753-1776	1765(?)-1776
25.	Hrafnagilsannáll	Þorsteinn Ketilsson	Hrafnagil, Eyjafjord district	1717-1754	-
26.	Íslands Árbók	Sveinn Sölvason	Munkaþverá, Eyjafjord district	1740-1782	1740-1782
	Continuation	Jón Sveinsson	Eskifjord, S. Múla district	1782-1792	1782-1792
27.	Espihólsannáll	Jón Jakobsson	Espihóll, Eyjafjord district	1768-1799	1768-1799
28.	Þingmúlaannáll	Eiríkur Sölvason	Þingmúli, S. Múla district	1663-1729	1700-1729
29.	Desjarmýrarannáll	Halldór Gíslason	Desjarmýri, N. Múla district	1495-1766	-
30.	Vatnsfjarðarannáll Hinn Yngsti	Guðlaugur Sveinsson	Vatnsfjord, N. Ísafjord district	1751-1793	
31.	Sauðlauksdalsannáll	Björn Halldórsson	Sauðlauksdalur, Barðastrandar district	1-1778	c.1740-1778
32.	Annáll Eggerts Ólafssonar	Eggert Ólafsson	Sauðlauksdalur, Barðastrandar district	1700-1759	1752-1759
		Tómas Tómasson	Stóra-Ásgeirsá, W. Húnavatn district		
33.	Djáknaannáll	Hallgrímur Jónsson	Sveinsstaðir, E. Húnavatn district	1731-1794	1780-1794

Table 5.1 Seventeenth and eighteenth century Icelandic Annals.

edition by Halldór Hermannsson. It has recently been republished in the collection of later annals *Annálar 1400-1800*. This useful new edition has been prepared by Guðrún Ása Grímsdóttir. Gísli's annal covers the period 1106 to 1636. For our purposes here, the only section of interest is that based on Gísli's own contemporary knowledge from 1591 onwards. *Annalium farrago* is unique amongst the annals in that it places particular emphasis on natural phenomena (Grímsdóttir, in *Annálar 1400-1800 5*, p.472). Gísli wrote one other work, *De mirabilibus Islandiæ*, discussed below.

To Björn Jónsson (1574-1655) may be accredited the writing of the first of the later annals proper. Björn's annal, named *Skarðsárannáll* after Skarðsá in the north of Iceland where he lived, runs from 1400 to 1640 but, as most of his work on the annal was done during the years 1636 to 1639, its contemporary span is short. One of his known sources is *Gottskálksannáll* and, for the 1570s, Björn quotes this annal almost word for word. However, he accidentally missed the year 1570 and went straight on to 1571 with the result that, between 1570 and 1578, when *Gottskálksannáll* ends, the weather information that he had copied is one year out. Apart from this unfortunate error, and some other random errors in dating, there is evidence to suggest that Björn's work is reliable before 1636, at least from ~1580. For example, apart from the information taken from *Gottskálksannáll*, Björn includes few weather references before the early 1600s, suggesting that he only included information that he considered to be accurate. Björn's work must have been based partly upon oral tradition, but he also had many documents at his disposal since he had access to the archives at the see of Hólar, the northern bishopric, and he often worked there for long periods (Þorkelsson 1887, Benediktsson 1948). Thus, although *Skarðsárannáll* does not become contemporary until 1636, some of its information from 1600 onwards has been included in the climate analysis presented here and earlier in Ogilvie (1984a).

Three other particularly useful seventeenth-century annals are *Vallholtsannáll*, *Mælifell-sannáll* and *Ballarárannáll*. These are all good sources, and provide long, continuous records. *Mælifellsannáll* was written by a father and son, and spans sixty years, from 1678 to 1738. *Vallholtsannáll*, which covers the years 1626 to 1666, is one of the most important and reliable annals for the seventeenth century (Þorsteinsson *Annálar 1400-1800 1*, p.320). *Ballarárannáll*, as well as being one of the most original and entertaining of the annals, gives fuller details regarding the weather than many of the others. *Ballarárannáll* covers the period 1597 to 1665, and although it does not become contemporary until ~1620, some weather references before this date are included here. This seems justified as they invariably refer to the author's personal circumstances and seem unlikely to have been falsified. Furthermore, there is no reason to suspect any dating errors.

Usually, annals which were not written contemporaneously are of limited value. However, as with *Skarðsárannáll* and *Ballarárannáll*, there may be exceptions to this general rule. For example, some annals incorporated earlier contemporary writings into their own works. Thus, Þorlákur Markússon, the author of *Sjávarborgarannáll* (which spans the period 1389 to 1729) began to write his annal in 1728 and therefore, except for the last year, it is clearly not written contemporaneously. However, Þorlákur seems to have used several reliable sources for the earlier years. For example, between 1609 and 1627 he gives extracts from a lost annal which appears to have been written contemporaneously in the south of Iceland (Sjávarborg is in the north). Also, for 1645 to 1650 and 1668 to 1671 the author takes extracts from another annal, *Gufudalsannáll*, written by Halldór Teitsson (died ~1685) from Gufudal in the northwest (Jóhannesson, *Annálar 1400-1800 4*, p. 228-230). The original version of this annal was burned in the great fire of Copenhagen in 1728 which destroyed many manuscripts.

Vallaannáll (1659-1737) does not become contemporary until 1701, but there is a case for it being reliable from 1690 as, from that time, it seems likely that the annal was based on a diary, written by the author or someone else, which described the weather on particular days, such as Christmas Day and New Year's Day (Þorsteinsson, *Annálar 1400-1800 1*, p.380).

In the eighteenth century the annals become more numerous. *Hítardalsannáll*, which covers the period 1724 to 1734, was written by Jón Halldórsson, one of the best historians of his time, and is an important and reliable work (Þorsteinsson, *Annálar 1400-1800 2*, p.591). *Hvammsannáll* was written by Þórður Þórðarson who was a pupil of Jón Halldórsson. This annal is not regarded as being particularly important, but some things in it are not found elsewhere, including some very good weather descriptions, such as an account of the severe weather in 1737. *Grímsstaðaannáll*, whose author was Jón Ólafsson of Grímsstaðir in Breiða- vík, was little known by other annalists and contains much original material. *Grímsstaðaann- náll* is useful in that it is more "localized" than many of the other annals, referring almost entirely to matters of relevance for the Snæfellsnes district. Its contemporary span is ~1734 to 1764. Two other annals which provide long and useful weather records are *Höskuldsstaðaann- náll* and *Íslands Árbók*, both written in the north of Iceland.

Three other eighteenth-century annals, previously only available in manuscript form, *Djáknaannáll*, *Sauðlauksdalsannáll* and *Annáll Eggerts Ólafssonar* have recently been published in the collection *Annálar 1400-1800 6* (edited by Guðrún Ása Grímsdóttir) and their weather information has been included in this analysis.

Private and official correspondence Many Icelanders corresponded with friends and collea- gues at home and abroad. In their letters they frequently commented on the weather. The large collection of letters to the Danish naturalist Ole Worm (1588-1654) for example, contains many descriptions of the weather. These letters cover the period from 1626 to 1649 and were written by such learned Icelanders as Arngrímur Jónsson (1568-1648) and Gísli Magnússon (1621-1696). Arngrímur is mentioned below with regard to the geographical treatises he wrote on Iceland. Of particular interest is a seventeenth-century account of Iceland, written by Sir Thomas Browne (1605-1682) of the Royal Society, based on corre- spondence sent to him by an Icelandic priest, Þórður Jónsson (~1609-1670) in 1651, 1656 and 1664. Other notable Icelandic letter-writers were Árni Magnússon (1663-1730), Páll Vídalín (1667-1727) and Eggert Ólafsson (1726-1768).

Such private correspondence, while important, forms a minor source in comparison with the official correspondence available from the early eighteenth century onwards. The origin of this type of correspondence may be explained as follows. From 1380 up to the time of the Second World War, Iceland was a colony of the Danish Crown and ruled over by the central authority in Copenhagen. In the 1680s a "Governor" or "Prefect" (*Stiftamtmand*) was appointed over Iceland. Up to 1770, the Governor was resident in Copenhagen, thereafter in Iceland. The Governors were all Danish until 1752, when it became more common for Icelanders to be appointed. The Governor was required to give an annual report on the situation in Iceland to the Danish government, and lesser officials in Iceland, the sheriffs (*Sýslumaðr*) in each district (*Sýsla*), were similarly required to send annual reports on their district to the Governor. The sheriffs were usually Icelanders and there were generally about

Districts (Sýsla)

1 Gullbringusýsla
2 Kjósarsýsla
3 Borgarfjarðarsýsla
4 Mýrasýsla
5 Snæfellsness – og Hnappadalssýsla
6 Dalasýsla
7 Austur – Barðastrandarsýsla
8 Vestur – Barðastrandarsýsla
9 Vestur – Ísafjarðarsýsla
10 Norður – Ísafjarðarsýsla
11 Strandasýsla
12 Vestur – Húnavatnssýsla
13 Austur – Húnavatnssýsla
14 Skagafjarðarsýsla
15 Eyjafjarðarsýsla
16 Suður – Þingeyjarsýsla
17 Norður – Þingeyjarsýsla
18 Norður – Múlasýsla
19 Suður – Múlasýsla
20 Austur – Skaftafellssýsla
21 Vestur – Skaftafellssýsla
22 Rangárvallasýsla
23 Árnessýsla

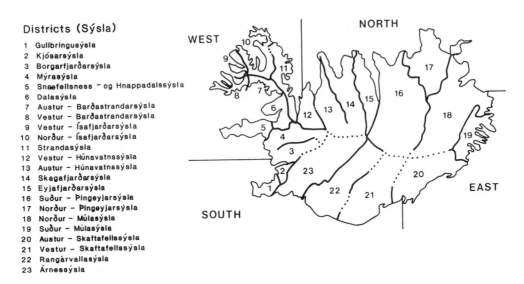

Figure 5.2 Administrative Divisions into Districts, (Sýsla, plural Sýsslur) and Geographical Regions.

22 in office at any one time during the 1700s. The location of all the diferent districts may be seen in Figure 5.2.

The sheriffs' reports were written in the form of a letter to the Governor. They contain information on legal, economic and social matters, and make specific references to problems with regard to trade, the hay-harvest and fishing. Also included are descriptions of weather, climate and sea ice. The attention given to weather and climate in the letters varies; some are brief, others give considerable detail, some even include daily data or early temperature observations. These letters are excellent contemporary records and they form the major eighteenth-century source of climatic data for the analyses presented here.

The time periods covered by the letters naturally vary in accordance with the length of office of a particular sheriff. Once appointed, a person generally held this position for life. Occasionally, a gap of one or more years occurs between the retirement of one sheriff and the appointment of another. There are very few letters from before 1720, and up to 1729 they are sporadic. After that they are plentiful. The geographical coverage provided by the sheriffs' letters is excellent, although confined to the mainly coastal centres of habitation. The places at which many of the letters were written may be seen in Figure 5.3. The letters were written in Danish, in Gothic script. They are unpublished and are held in the national archives (Þjóðskjalasafn), Reykjavík. The sheriffs' letters may be found in the references in the section on unpublished sources in the series *Bréf til Stiftamtmanns* and *Islands Journal*.

Descriptions of Iceland by Icelanders   Towards the end of the sixteenth and beginning of the seventeenth century it is possible to detect a new interest among Icelanders in natural history and the physical conditions of the country. This had its origins in the new ideas heralded by the Reformation and the Renaissance. One more direct cause may have been the influence of two individuals on the Icelanders. One of these was Tycho Brahe, the brilliant Danish

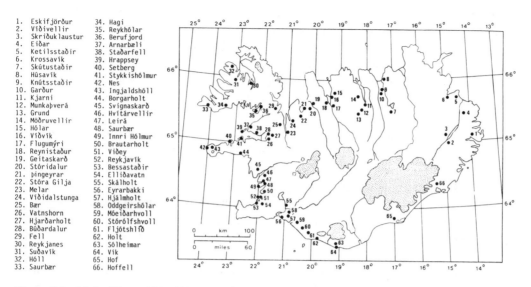

1. Eskifjörður
2. Víðivellir
3. Skriðuklaustur
4. Eiðar
5. Ketilsstaðir
6. Krossavík
7. Skútustaðir
8. Húsavík
9. Knútsstaðir
10. Garður
11. Kjarni
12. Munkaþverá
13. Grund
14. Möðruvellir
15. Hólar
16. Víðvík
17. Flugumýri
18. Reynistaður
19. Geitaskarð
20. Stóridalur
21. Þingeyrar
22. Stóra Gilja
23. Melar
24. Víðidalstunga
25. Bær
26. Vatnshorn
27. Hjarðarholt
28. Búðardalur
29. Fell
30. Reykjanes
31. Suðavík
32. Hóll
33. Saurbær
34. Hagi
35. Reykhólar
36. Berufjord
37. Arnarbæli
38. Staðarfell
39. Hrappsey
40. Setberg
41. Stykkishólmur
42. Nes
43. Ingjaldshóll
44. Borgarholt
45. Svignaskarð
46. Hvítárvellir
47. Leirá
48. Saurbær
49. Innri Hólmur
50. Brautarholt
51. Viðey
52. Reykjavík
53. Bessastaðir
54. Elliðavatn
55. Skálholt
56. Eyrarbakki
57. Hjálmholt
58. Oddgeirshólar
59. Móeiðarhvoll
60. Stórólfshvoll
61. Fljótshlíð
62. Holt
63. Sólheimar
64. Vík
65. Hof
66. Hoffell

Figure 5.3 Major Sites of Sheriffs' Letters.

astronomer, and the other was Ole Worm, the Danish naturalist, with whom many learned Icelanders corresponded (see above).

In the latter part of the sixteenth century two particularly interesting works were written. These were the *Qualiscunque descriptio Islandiæ* or *Íslandslýsing* by Oddur Einarsson (1559-1630) and the *Brevis Commentarius de Islandia* by Arngrímur Jónsson (1568-1648). These early works of geographical description give particularly interesting and accurate accounts of such natural phenomena as sea ice and the Icelandic glaciers. Both works were written primarily in order to refute a number of inaccurate and scurrilous foreign accounts of Iceland.

Arngrímur, known as "the learned", wrote his work in 1592. It was then published in Copenhagen in 1593. An English translation was included in Hakluyt's *The Principal Navigations* of 1598. Although it is likely that Oddur wrote the greater part of his work earlier, probably during the winter of 1588-89 (Benediktsson 1956) it remained unpublished. Perhaps he felt that it had become superfluous upon the publication of Arngrímur's work. In later years, all manuscripts of *Íslandslýsing* were believed lost. It was not until the early part of this century that a manuscript was fortuitously found in a library in Hamburg. It was published there in 1928 (by Burg) and subsequently in Iceland in 1971. Although Arngrímur's work is still of interest, Oddur's *Íslandslýsing* is now regarded as the more important of the two.

Oddur Einarsson was Bishop of Skálholt, Iceland's southern bishopric, from 1589 to 1630. In 1631 this position was filled by his son, Gísli Oddsson. Gísli, author of *Annalium farrago*, discussed above, also compiled the first geographical treatise on Iceland to be written in the seventeenth century. In compiling this treatise, known as *De mirabilibus Islandiæ*, Gísli used

101

his father's work, *Íslandslýsing*, as a source. This latter work is superior, but the *De mirabilibus* also has much to commend it. It contains forty brief chapters on different topics relating to natural history. One of these is on sea ice. The work was written in 1638, over a period of two weeks, shortly before Gísli's death. However, as is evident from a letter he sent to Ole Worm, he had been planning such a work for a long time (Hermannsson 1917 p. viii). Like *Íslandslýsing*, Gísli's *De mirabilibus Islandiæ* was not published until the present century (Hermannsson 1917, Oddsson 1942).

An early work on Iceland's natural history, *Um Islands Aðskiljanlegar Náttúrur* ("Natural History of Iceland") was written by Jón Guðmundsson the Learned (1574-1658). It was first published in this century and was written sometime between 1640 and 1644 (Hermansson 1924 p.xiii). Unlike most other seventeenth- and eighteenth-century Icelandic authors he was uneducated in the sense that he did not follow the usual course of formal education at home and in Copenhagen. However, he must have been a gifted and clever man. Jón's work is a mixture of superstition and pertinent observation and is the first natural history of Iceland written in the vernacular.

The two other most important seventeenth-century descriptions of Iceland were written by Bishop Þorlákur Skúlason (1597-1655) and Bishop Brynjólfur Sveinsson (1605-75). These were not published until the twentieth century (Benediktsson 1943). The treatises, both written in 1647, resulted when the Secretary to the Danish Chancellery, Otte Krag, sent the Bishops some specimens of foreign writing on Iceland and asked their opinions on them. As these writings were mostly grossly inaccurate, the treatises were produced in order to present a more correct view of the country. Regrettably, Arngrímur Jónsson's *Brevis Commentarius*, written with this same aim, had not prevented the writing of inaccurate accounts. The treatises are both quite short, but contain some useful details.

An important eighteenth-century description of Iceland is by Magnús Stephensen (1762-1833) who, in addition to being Chief Justice of Iceland, was active in politics, especially with regard to improvements in the Icelandic economy, and a prolific writer on many subjects. However, his major work was *Eftirmæli átjánda aldar* ("Iceland in the Eighteenth Century") first published in 1806. In this work Stephenson concentrates on social and economic matters with sections on, for example, farming, fishing, trade, population, and government officials. There is also a section on natural history. This includes a discussion of "good and bad years" and their causes (Stepensen believed the major cause of "bad" years to be severe winters and sea ice). Stephensen concluded that out of the hundred years of the eighteenth century, forty-three of these had been dearth-years with severe seasons and that, in fourteen of these, there was considerable loss of human life, as well as of domestic animals. Stephensen's book is one of the most important sources on eighteenth-century Iceland.

Another work which may be metioned here is the *Ferðabók* (Travel Book) by Sveinn Pálsson, (1762-1840) a doctor, and one of Iceland's most brilliant naturalists. Pálsson travelled around Iceland during the years 1791 to 1795 on behalf of the Danish Natural History Society and subsequently submitted an account of his observations. These laid particular stress on the geology of Iceland. Pálsson was one of the first to understand many of the phenomena associated with glaciers. His account was published in Icelandic in 1882/3. Sveinn Pálsson is also well-known for his careful meteorological observations, made in the south of Iceland, from 1798 to 1840.

The Jón Jónssons' weather diary from Eyjafjord During the nineteenth century many individuals kept weather diaries but this was not common in earlier times. However, one such early diary of excellent quality is the weather diary of Jón Jónsson senior (1719-1795) and his son, Jón Jónsson junior (1759-1846), who were both priests. This diary, which has not been published, was kept from 1747 to 1846 and thus spans virtually a whole century. The observations were all made in one locality, in Eyjafjord, and although they were made at various sites (depending on where the Jón Jónssons lived) they were all fairly close together.

The observations are not instrumental but they are detailed, careful and very full. They include daily reports on the wind, weather and state of sky, and also weekly weather summaries. There is also an annual summary of the weather during 1748 to 1768. This includes details on fishing and the harvest. Other matters are described when the Jónssons considered them relevant. For example, the summary for 1766 includes a detailed account of the sandfall in the Eyjafjord region from the Hekla eruption of that year, and there is also an extensive description of the effects of the 1783 Skaftáreldar eruption on Eyjafjord. Details of the presence of sea ice, especially the times of its arrival and departure, are also included. The Jón Jónssons' weather diary is thus an extremely valuable document, both because of the quality and extent of the observations, and also because of the length of time that it covers. For further discussion of these diaries see Bergþórsson (1957) and Kington and Kristjáns-dóttir (1978).

5.4 Methods of data analysis

The information on weather and ice contained in the above sources is, by its nature, qualitative. Because of this, its use immediately raises the problem of subjectivity, both on the part of the writer, whose personal perception may have influenced his comments on the weather, and on the part of the reader of these comments. While it is not possible to completely overcome these subjective elements, it is possible to reduce them. Firstly, with regard to the subjectivity of the writer, this may, for example, be overcome in part by comparing several different accounts of a particular weather event. For the greater part of the time period considered here there are generally several sources available for each year. With regard to the subjectivity of the researcher, the data may, as far as is possible, be quantified in order to produce a clearer pattern of weather events.

With this latter aim in mind, information on weather and sea-ice events in the sources discussed above were categorized according to season and type. For the purposes of this analysis, the seasons were classified as follows: winter, mid-October to mid-April; spring, mid-April to mid-June; summer, mid-June to the end of August; autumn, September to mid-October. These seasonal divisions correspond closely to those understood by the writers of the sources used here. Iceland was also divided into four geographical regions: north, south, east and west. These areas may be seen in Figure 5.2. The weather described during the seasons was then noted for all regions. Initially, the full description was extracted. Following this, the seasons were classified according to their most salient features. Thus, a winter might be described as severe, mild, average or variable etc. In order to give a figurative representation of possible temperature variations, a decadal winter/spring thermal index was devised. This covered the period mid-October of one year to mid-June of the next, eight months in all. For this index, the seasonal characteristics of the winter and spring seasons were reduced to their simplest level. Where these were described as very cold they were given a value of minus one. Mild seasons were accorded plus one, and average or variable seasons, a value of zero. Because Iceland was divided into four regions, a total of four "points" was possible for each season, winter or spring. This could result in a total of minus four, if all four regions reported very cold weather, plus four points if the weather were mild everywhere, or no points at all if all regions described average or variable weather. Other numbers would obviously result from different combinations. If no information was available from a particular region for a particular year then that was given a value of zero. The number of mild and cold seasons in each decade were then counted, and the cold seasons subtracted from the mild seasons. The resulting number was taken to represent that particular decade. Thus, during the decade 1781 to 1790, which appears to have been particularly cold, forty-two cold seasons occurred, as opposed to twenty mild seasons. This decade thus has a final value of -22. Using this method, the value for each decade during the period 1601 to 1800 is shown in Figure 5.4a. The greater part of this index, from 1601 to 1780, has already been published (Ogilvie 1984a). Here it is extended forwards and backwards in time to include data not given in the earlier analysis. Unfortunately, with the exception of the 1560s and 70s, sixteenth-century data are not sufficient to be represented by such an index (Ogilvie 1990).

A similar quantification of sea ice data is shown in Figure 5.4b. Here, one "sea-ice point" is awarded for each region and each season which reported ice in any given year. Thus, in 1782, a very heavy ice year, twelve sea-ice points are accumulated. These may be explained thus: the east reported ice during all four seasons of the year; the south during two seasons; the north and northwest during three seasons each. Further details on the methods of data analysis used here may be found in Ogilvie (1981).

In the section below, possible variations in temperature and sea ice during the period ~1500 to 1800 are considered in the light of the indices discussed above. Some consideration is given here to the spatial variability of climate in Iceland. However, major emphasis is placed on the temporal variability of the climate of Iceland as a whole. For a discussion of regional variations in Iceland during the period 1601 to 1780, see Ogilvie (1984a).

With regard to these discussions, it is important to note that the data used here present certain spatial limitations. The greater part of the interior of Iceland consists of mountains and glaciers and is, consequently, unsuitable for human habitation. The means that the variations in climate discussed below are based primarily on data from the coastal, lowland regions. There is, furthermore, a bias towards the more populous areas of the north, west and southwest. Large parts of the east and southeast are uninhabitable. The map shown in Figure 5.3, although only illustrating data points from one type of source, gives a good indication of the spatial coverage of the data in general.

5.5   Temperature and sea-ice variations in Iceland ~A.D. 1500 to 1800

5.5.1   The sixteenth century

As outlined above, there are only a handful of sources available for the sixteenth century. The most important of these are: *Gottskálksannáll*, relating mainly to the north; *Biskupa-*

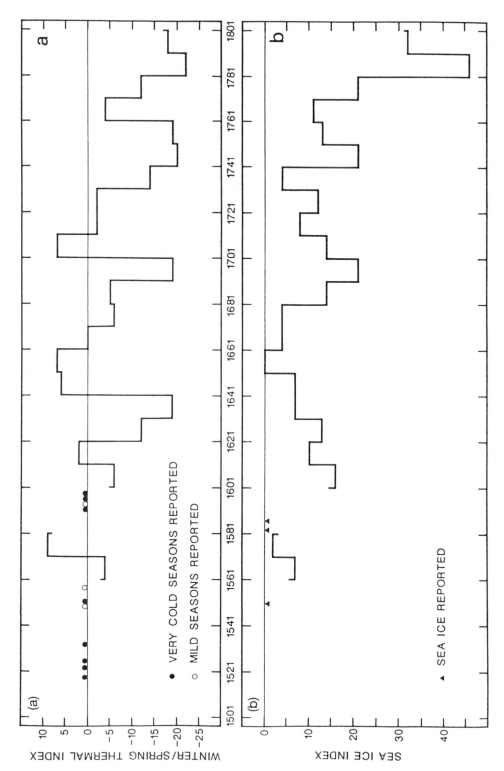

Figure 5.4 (upper panel) Winter/Spring Thermal Index for Iceland 1561 to 1580 and 1601 to 1800. Isolated reports of very cold and mild seasons also shown. (lower panel) Sea Ice Index for Iceland 1561 to 1580 and 1601 to 1800. Isolated reports of sea ice also shown.

Annálar Jóns Egilssonar, written in the south; and the descriptions of Iceland, *Qualiscunque descriptio Islandiæ* or *Íslandslýsing*; and *Brevis Commentarius de Islandia*, both written by northerners. Sixteenth-century sources and their climate evidence are discussed in detail in Ogilvie (1990).

Only a few isolated weather descriptions exist for the period ~1500 to 1561. The first of these is for the winter of 1518-19. This is noted as severe (Egilsson 1856 p.62). Another severe winter may have occured in 1520-21 as *Gottskálksannáll* reports heavy losses of livestock (Storm 1977 p.374). This account seems to be confirmed by a contemporary letter written at Gunnarsholt in Hlíðarendi in the south on 29 June 1521 (*Diplomatarium Islandicum 8* 1906-1913 pp. 793-794). Other severe winters are recorded in 1525, 1532 and 1552 (Egilsson 1856 pp. 64, 117).

The description for 1525 is particularly interesting as this severe winter, lasting from 11 January to ~23 April, appears to have had considerable social effects. (All dates given have been converted to the Gregorian or "New Style" calendar.) The source of information is said to be Jón Egilsson's grandfather, Einar, and the account refers to Grímsnes in Árnes district in the south. People there were prosperous until this disaster occurred. The great "livestock-death winter" as Jón Egilsson calls it, is said to be the most notable event during the early years of Bishop Ögmundur Pálsson, Bishop of Skálholt, the southern diocese, from 1519 to 1540.

Livestock deaths occurred again during Bishop Ögmundur's rule, in 1532 (Egilsson 1856 p.117). These were caused by a great snow during the Icelandic month *einmánuður* or "ein" month (22 March to 20 April). The same happened in 1552 with alternating thaws and severe snowfalls. This "hard winter" (Egilsson 1856 pp.101-102) lasted from 23 December 1551 to Easter (27 April 1552). Prior to the severe winter of 1552, there occurred a very good year for grass in 1551, and then another in 1557. Jón states that his source for this information is Jón Bárðarson who, in the autumn of 1575, told Jón that the three best grass years he had experienced on his farm at Sólheimar, in the south were during 1551, 1557 and 1575 (Egilsson 1856 p.117).

As might be expected from someone living in the south, where sea ice was a rare phenomenon, Jón Egilsson does not mention the ice himself at all. However, a note has been inserted in his work describing sea ice off the south coast, a relatively unusual event. This is said to have occurred early in the days of Bishop Marteinn Einarsson who was Bishop from 1548 to 1556. The note reads:

> . . . it stretched for more than a sea-mile to the shore and far beyond Þorlákshöfn (in Árnes district) and there was much seal hunting on it; it came before the end of the fishing season (~24 April) at the beginning of summer . . . (Translated from Egilsson 1856 p.100).

In a footnote to this text, the editor of Jón Egilsson's work, Jón Sigurðsson, suggests that the description of ice is likely to have been added by Bishop Oddur Einarsson (see the discussion of his work, *Íslandslýsing*). Oddur was a reliable writer but, as he was not born until 1559, this is not his eye-witness account.

For the 1560s and 70s, annual data, mainly from *Gottskálksannáll*, make it possible to give a figurative representation for these years (Figs 5.4a and b). From this it is clear that, on the whole, the 1560s were mainly cold, but the 1570s, by contrast, were mild. Sea ice was described several times in the 1560s. In 1564, there was said to be "much sea ice and countless seals were caught in all of the northern quarter". In 1566 sea ice appeared after Easter (24 April) and lay off the north coast until Midsummer (24 July). Sea ice was also present off the northern coasts during 1567 and 1568. During the mild decade of the 1570s, sea ice was only referred to in one year, 1572, when the ice seems to have been present off the north coast from around 20 April to 2 June.

The first winter of the next decade, 1580-81, was described as good up to the last day of *góa* (23 March). Then great quantities of snow fell, and this lay past the beginning of summer (23 April). Apart from this, there are no specific weather descriptions from the 1580s at all. However, some indication of the climate of the years ~1580 to 1592 may be gleaned from the excellent treatises on Iceland by Arngrímur Jónsson and Oddur Einarsson; the *Brevis Commentarius de Islandia*, and the *Qualiscunque descriptio Islandiæ* or *Íslandslýsing*. As these were written in 1592 and 1589 respectively, it seems reasonable to assume that they may have most relevance for the decade of the 1580s.

Arngrímur comments on the variability of the sea ice, and states that the ice is not present *every* year. In the year he was writing, for example, 1592, there was no sea ice off the coasts of Iceland. The fact that he mentions the lack of ice in this particular year implies that this was somewhat unusual. Certainly, the overriding impression gained from Arngrímur's work is of a time of relatively severe climate with much sea ice. He is familiar with the movements and duration of the sea ice, and describes them thus: "for the most part the ice melts in April or May and is driven towards the west. It does not then return before January or February and very often even later" (Jónsson 1957 p.16). He also describes the common practice of catching seals on the ice.

Oddur's detailed and knowledgeable description of sea ice may be taken as a clear indication that there was a high incidence of sea ice around the time that he was writing (1589). However, even more than Arngrímur, he emphasises the variability of the ice. It is interesting that his remarks reflect what we already know about sea-ice incidence in the latter half of the sixteenth century, namely that there was much sea ice during the 1560s, but little during the 1570s.

The Icelanders who have settled on the northern coasts are never safe from this most terrible visitor. The ice is always to be found between Iceland and Greenland although sometimes it is absent from the shores of Iceland for many years at a time . . . (Translated from Einarsson 1971 pp.34-35).

There are only a few weather descriptions available for the 1590s. As noted above, Arngrímur Jónsson stated that there was no sea ice in 1592 (Jónsson 1957 p.16). According to *Skarðsárannáll*, a severe winter occurred everywhere in the north in 1594. The winter of 1595 was mild, but in 1596 a great snow winter occurred. The summer of 1596 was good for grass in the south (Egilssson 1856 p.117). In 1598 there was a hard winter with losses of livestock everywhere except in the north (*Ballarárannáll*).

5.5.2 *The seventeenth century*

The major sources for the seventeenth century are the annals, in particular *Skarðsárannáll*, *Ballarárannáll*, *Sjávarborgarannáll*, *Kjósarannáll*, *Eyrarannáll*, *Annálsgreinar Árna á Hóli* and *Mælifellsannáll*. Also of importance are the letters written to Ole Worm, and the various treatises on Iceland. In order to save space, and as entries in the annals are easily found according to the year, full bibliographical details for references from the annals are generally not given in the text. Complete details for each year may be found in Ogilvie (1981).

On the winter/spring thermal index (Figure 5.4a) the first decade of the seventeenth century does not appear to have been extremely cold; it was on a par with the 1670s and 80s and the 1770s. However, the early years of the century included a time of great dearth, similar to that which occurred in the 1690s, 1750s and 1780s (see below). The severity began in 1602 when all seasons were harsh throughout Iceland. The dearth and severe weather continued through 1603 and 1604 and caused the deaths through hunger of around nine thousand people (*Skarðsarannánnáll, Ballarárannáll, Annalium farrago*). There were several severe ice years during this decade. Sea ice is recorded in the north in 1602 "far into summer"; in 1604 when "ice came"; in 1605 when "much ice came" and in 1608 when it lay from mid-April to mid-June (*Skarðsárannáll*). In 1602, *Ballarárannáll* refers to ice in Hvammsfjord being crossed on horseback in early June. However, this is likely to refer, not to drifting sea ice, but to the sea itself being frozen *in situ*. In 1605 *Skarðsárannáll* states that sea ice also occurred off the eastern coast, and was present off Grindavík in the south at the end of the spring fishing season (June). Neither *Skarðsárannáll* nor *Ballarárannáll* mention sea ice in 1610, yet *Sjávarborgarannáll* states that it came to the south coast. If it affected the south this year it would have occurred in the north and east as well, and drifted around the coast to the south. Between the years 1601 and 1690, this decade experienced more sea ice than any other (Figure 5.4b).

In Figure 5.4a, the 1610s appear as mild on the whole, and the 1620s as rather cold. These decadal divisions mask the fact that, at least in the north, mild winters occurred one after the other from 1618 to 1624 (*Skarðsárannáll*). It should be noted, however, that these were offset by several cold springs. Years of particularly cold weather during this period were 1615 and 1625. Björn Jónsson, author of *Skarðsárannáll*, wrote a poem about the latter winter entitled *Svellavetur* or "ice-winter". There were three severe sea-ice years this decade. In 1612 "ice came" to the north. In 1615 it came to the north during *Þorri*, the Icelandic winter month (23 January to 22 February) and stayed until June. In 1618 the ice visited the north "during the spring" (*Skarðsárannáll*). In 1615, the ice also reached the eastern and southern coasts as far as Reykjanesröst, the most southwesterly point of Iceland (*Skarðsárannáll, Annalium farrago*).

During the 1620s, six severe sea-ice years were recorded in the north: in 1622 "ice came", in 1624 it was present "in the spring" and in 1625 when the ice came during the month of *góa* (23 February to 25 March) and remained until ~2 July. Sea ice off the northwestern coast is said to have hindered rowing until early June (*Sjávarborgarannáll*). In 1626, "seals were caught at Skaga on the ice"; in 1628 it came again during *góa*, and stayed until 12 August; in 1629 sea ice came to the north but did not remain (*Skarðsáránnáll*).

The 1630s appear to have been especially cold, similar to the 1690s, 1740s, 1750s and 1780s. Particularly severe years all over Iceland were 1633 and 1639. The author of *Ballarárannáll* compared 1633 with the very cold year of 1602. Sea ice occurred off the northern coasts from *Þorri* (21 January to 20 February) to late June (*Skarðsárannall*) or late July (*Valholtsannáll*). The ice was also present in 1636, when, according to *Skarðsárannáll*, six people were drowned in Skagafjord. They were hunting seals on the ice when their over-loaded boat capsized. Sea ice drifted in to the northern coasts around early November 1638, and remained throughout the winter of 1639. It also reached the eastern and southern coasts as far as Suðurnes (the southern part of the Reykjanes peninsula). According to *Sjávarborgarannáll* the sea ice came to the east three weeks after Easter, and then lay fast to the land all along the south coast to past Reykjanes. It continued to drift around until late June.

A distinct mild period occurred from ~1641 to ~1670 (Figure 5.4a). It is interesting to note from Figure 5.4b that these decades coincided with the period of least ice in the seventeenth century (from ~1631 to ~1681). There is no comparable period in the eighteenth century. The only equally ice-free decade then was during the 1730s. Sea ice was recorded twice during the years 1641 to 1650, in 1648 and 1650. For the latter year, *Vallaannáll*, a northern source, simply records "there was sea ice". This is likely to have occurred sometime during February and/or March. The former year, 1648, was a very severe sea-ice year, and a very cold year on land; an exception to the mild years of the period. The sea ice lay off the northern and northwestern coasts during the winter and right through to the summer.

It is noteworthy that, during these decades of fairly mild climate, an enlightened individual, Gísli Magnússon (1621-1696) managed to grow barley and a number of other plants on his farm at Fljótshlíð in the south (Ólason 1942). Barley had been grown during early settlement times but this practice subsequently ceased, possibly as a result of a colder climate (Thórarinsson 1956).

The 1670s and 1680s saw a return to a colder régime although Iceland's shores remained relatively ice free. During the 1670s, sea ice was only recorded during one year, in 1672. *Vatnsfjarðarannáll Yngri* states that "a lot of ice" entered all the fjords in the northwestern district of Ísafjarðarsýsla, and that there was much ice in the area around Rauðisandur, in Vestur-Barðastrandarsýsla, also in the northwest, and to the north "all around Skjálfandi" off Suður-Þingeyjarsýsla. The author does not say exactly when the ice came, but it was still present during the summer when a ship was wrecked in the ice off Eyjafjord. The sources recount four occurrences of sea ice during the 1680s: in 1683, 1684, 1685 and 1688. The ice came to the northern and nothwestern coasts during all these years. The worst ice years seem to have been 1684 and 1685. In the former year the ice was present from January or February to Midsummer (~24 July) and in the latter year from March to September.

Severe winters during the 1670s were: 1671, which was "harsh" in the south, 1673, which was severe in the north and east but good in the south, and 1674 and 1680 which were severe everywhere. Cold springs were recorded in 1672, 1674, 1675 and 1678. Except for 1688 and 1690, the winters during the 1680s were not particularly severe. However, several cold springs were recorded: in 1682, 1685, 1687 and 1689 (*Vatnsfjarðarannáll Yngri, Kjósarannáll, Eyrarannáll, Annálsgreinar Árna á Hóli, Mælifellsannáll*).

The last decade of the seventeenth century was extremely cold. It was also the decade with the most sea ice during the years 1601 to 1700 (Figs 5.4a and b). Severe winters occurred in 1691, 1692, 1694, 1695, 1696, 1697, 1699 and 1700. Only two winters were recorded as "good": 1693 and 1698. In 1692 sea ice affected the north during winter and spring. In 1694 the sources record a heavy sea-ice year with the ice present off the northern, northwestern and eastern coasts during spring and summer. The ice is said to have penetrated as far as the Westman Islands and Eyrarbakki in the south. In 1695 a sea-ice year of even greater severity occurred. The sources for the sea ice this year, together with other heavy sea-ice years in the seventeenth century are discussed in detail by Vilmundarson (1972). During this year of 1695, sea ice was present off the north and northwest coasts from winter, probably around

Christmas, to late June. The ice drifted around the eastern and southern coasts as far as Faxa Bay (i.e. past Reykjavík) on 24 April (*Vallaannáll*). It is very rare indeed for ice to penetrate this far west. Another source commented on the ice in Borgarfjord and wrote that people estimated that it would be possible to walk on the sea ice from Akranes to Reykjavík (*Hestsannáll*). Sea ice was also present in 1697, in winter, off the northern coast, and in 1699 when it affected the north and northwest during the springs. The major sources for the 1690s are: *Mælifellsannáll, Vallaannáll, Eyrarannáll, Fitjaannáll, Hestsannáll, Annálsgreinar frá Hólti* and *Annálsgreinar Árna á Hóli*.

5.5.3   The eighteenth century

Temperature and sea-ice variations during the decades 1701 to 1780 have been discussed in some detail in Ogilvie (1986). These will therefore be dealt with relatively briefly here. More attention will be paid to the decades of the 1780s and 90s which have not been considered previously.

Major eighteenth-century sources are the annals, travel descriptions and the official correspondence written by Icelandic sheriffs. As with seventeenth-century sources, full bibliographic details for each year up to 1780 may be found in Ogilvie (1981). The major sources for the 1780s and 90s are the sheriffs' letters, as listed in the *Islands Journal* series given in the references.

The first three decades, 1701 to 1730, were the mildest of the century (see Figure 5.4a). The decade 1701 to 1710 was particularly mild, and similar to the years ~1640 to 1660. Sea ice did occur on several occasions during this decade, however. In 1701 it affected the north in the summer and autumn. In 1703 it lay off the eastern and northwestern coasts in spring (*Vallaannáll* states specifically that there was no sea ice in the north). 1703 was a heavy ice year with northern and eastern sources reporting ice in spring and summer. Northwestern sources state that it was present in the spring. Around 7 May this year it reached Eyjafjöll in the south, the area of land which faces the Westman Islands to the southwest. In 1706 ice came to the north and east in spring, and in 1708 to the north in winter through to the summer.

There were only three recorded instances of sea ice in the decade 1711 to 1720. These were in 1714, 1717 and 1718. In 1714 the ice lay off the northern and northwestern coasts in spring and summer. The eastern coast was affected during the winter. In 1717 the ice "came late" to the north, presumably in late summer (*Viðauki Fitjaannálls*). In 1718 sea ice came to the north during the spring.

There were four severe sea-ice years in the 1720s. In 1726 the ice came to the north and west in spring. In 1727 the north was affected during the spring. More prolonged ice occurred in 1728, lying off the northern coast during winter, spring and summer, and off the northwestern coast during spring and summer. In 1729 the sea ice lay off the western coast during spring and summer. Major sources for the years 1701 to 1730 are: *Mælifelsannáll, Eyrarannáll, Vallaannáll, Fitjaannáll, Hestsannáll, Annáll Páls Vídalíns, Þingmúlaannáll, Hvammsannáll* and, from 1720 onwards, the sheriff's letters.

The 1730s were cold on the whole, although there was considerable year-to-year variability. The most severe years of the decade were 1737 and 1738, especially the former. One source, *Ölfusvatnsannáll*, written in the south, suggests that the winter of 1737 was the worst since the winter of 1633. It is interesting to recall that 1633 was compared by a contemporary source, *Ballarárannáll*, to the severe winter of 1602 (see above). No sea ice was recorded during 1737 and 1738. During the 1730s sea ice is only mentioned in the sources for 1732 and 1733. In the former year the ice came to the north during the spring and lay until the end of August. In 1733 a northern source commented "cold near the sea because of drift ice which lay off the north coast all summer" (sheriff's letter from Skagafjarðarsýsla in the north).

The 1740s and 1750s were very cold, similar to the 1630s and 1690s. In Figure 5.4a, the 1740s appear as slightly colder than the 1750s. It was during the 1750s, however, that human beings suffered most as a result of the harsh weather. The reason for this may have been a higher proportion of unfavourable summers during the 1750s. To give an idea of the difficulties endured by people at this time, we may quote Jón Jónsson senior, the weather diarist, who, during the extremely severe year of 1756, commented: "There was terrible hardship amongst people . . . Poor people left their homes in many places . . . and there were people travelling around but few people had food to give them. As a consequence people died of hunger". A discussion of climate impact in Iceland during the seventeenth and eighteenth centuries may be found in Ogilvie (1981 and 1984b).

As may be seen in Figure 5.4b, the 1740s were years of considerable sea ice, similar to the 1690s, with sea ice recorded for seven years out of ten. In 1741, sea ice was present off the northern coast during the spring, and the eastern coast during the summer. The north was affected in 1742 from winter through to summer, and in 1743 during spring and summer. The year 1745 was a particularly heavy ice year. The ice came early to the north and east and stayed to around June. It eventually affected the whole country, with the exception of the Faxaflói and Breiðafjord area. This year was also severe with regard to the weather. The main feature of the winter seems to have been the intensity of the frost and a lack of snow. Sea ice affected the north in the spring of 1748 and the winter of 1749. In 1750 ice lay off the northern coasts from winter through to summer, and off the northwestern coasts from winter to the following autumn.

In some previous works on sea ice off the coast of Iceland, such as that by Koch (1945) the 1750s are represented as years of higher ice incidence than the 1740s. However, this is due to the inclusion of spurious information (Vilmundarson 1972, Ogilvie 1984a). According to the reliable sources consulted here, sea ice was present off the Icelandic coasts during five years in the 1750s. In 1751, the ice lay off the northern coast during winter and spring. In 1756, it affected the north in spring and the northeast in summer. One southern source records that the ice came as far as Reykjanes on 24 June (Ölfusvatnsannáll). The Sheriff of Árnessýla in the south wrote: "Drift ice came to the south on 24 June and began to break up on 6 July". In 1757 the ice lay off the northern coast in spring and summer, and in 1759 a similar distribution pattern to that of 1756 occurred: "The sea ice surrounded all the eastern and southern parts of the country up to Reykjanes for two to three weeks during June and July" (sheriff's letter from Rangárvallasýsla). Sea ice also came to the north in the spring of 1760.

The 1760s saw a return to a milder climatic régime. However, sea ice distribution in the 1760s was similar to the 1750s, with ice reported during five years. In none of these years did the ice reach the south coast, however. In 1764 the ice came only briefly to the north, probably during June. In 1766 it was more prolonged, remaining off the northern coast from winter through to summer. The ice also affected the northwest during spring. Sea ice came to the north in the spring of 1767, and in the winter of 1769. In 1770 the northern and eastern coasts were visited by ice in the spring and summer.

The decade of the 1770s appears to have been somewhat colder than the previous decade, and sea ice was reported for every year except 1779 and 1780. In only one of these years was the ice extensive (1772), and it did not occur in southern waters during this decade. Nevertheless, the incidence of sea ice in the 1770s is as great overall as during the 1690s and 1740s. In 1771 the ice was present off the northern coast during the spring only. In 1772 it remained off the north, northwest and east from winter through to summer. In 1773 it returned to the north during the spring. The next year, 1774, saw the presence of sea ice off the northern coast in winter, and the northwestern coast in spring. In 1775 the north and northwest were both affected by summer ice. The ice came to the northern and northwestern coasts in the spring of 1776. In 1777 and 1778 the ice came to the north in the spring.

Other events in Iceland during the 1780s are overshadowed by the great Skaftáreldar (Laki) eruption of 1783. The volcanic ash which fell over virtually the entire country poisoned vegetation and, through this, the livestock. In the famine that followed approximately 9,000 people died (out of a total population of around 50,000). The (mainly cold) weather during 1783 is discussed in detail in Ogilvie (1986). See also Bjarnar (1965) and Gunnlaugsson et al. (1984). The difficulties caused for people by the Skaftáreldar eruption were clearly compounded by the harsh weather during this decade and also the extensive sea ice which hindered fishing and prevented merchant ships from reaching their harbours.

As may be seen from Figure 5.4a, the 1780s were the coldest decade in the entire series 1601 to 1800, possibly even from 1501. It is also the decade with the greatest extent and duration of sea ice, with the ice present every year (Figure 5.4b). The years 1781, 1786, 1787, 1788 and 1789 were what we have come to regard as "average" years of sea ice. The ice was present in the north and northwest over one or two seasons. In 1790, there was only a little ice in the north. It is the years 1782, 1783, 1784 and 1785 which make this decade so spectacular as regards sea-ice incidence. In all of these years the ice was extensive off the northern, northwestern and eastern coasts, and present from the winter through to the summer. In 1782, the worst sea-ice year of the decade, the ice also reached the south coast, and was present from May onwards. In some places in the south it remained to August. The Sheriff of Gullbringusýsla, who travelled from the south to Copenhagen on 27 August, wrote that the ice then lay so densely off the Westman Islands that the ship was forced to change its course from southeast to southwest for a period of eight hours in order to get past it.

Mainly severe weather continued into the 1790s. However, this decade does not appear to have been as cold as the 1780s or, indeed, the 1630s, 1690s, 1740s or 1750s (see Figure 5.4a). Several winter and spring seasons were characterised as mild. In most regions, for example, both the winter and spring of 1794 and 1797 were mild, the winter of 1793 was mild in the northwest, east, west and the north, and the winter of 1795 was mild in the southwest, the southeast and the west. The spring of 1799 and the winter of 1800 were favourable in most areas.

From Figure 5.4b, it may be seen that sea-ice incidence over the decade of the 1790s was extensive. Over the period 1601 to 1780, this decade comes second only to the 1780s for quantity of ice. Sea ice was present off the Icelandic coasts every year except for 1795, 1797 and in 1800. Years of comparatively little ice were 1793, 1794 and 1799. The very heavy ice years during this decade occurred in 1791, 1792, 1796 and 1798. Of these, the worst was 1791, with ice off the northern, northwestern and eastern coasts from winter through to summer. In June there was said to be much ice off Eyrarbakki in the south. In 1792 the ice was present off the northwest, north and east from December. Accounts from different regions reflect the retreat of the ice in spring and summer from the east around the northern coasts. In the eastern part, in southern Múlasýsla, the ice was present until the end of April. In northern Múlasýsla, the ice is said to have remained far into May. The Sheriff of Þingeyjarsýsla states that the ice remained until the end of June. The ice was still densely packed off Eyjafjarðarsýsla on 8 July and the Sheriff of Húnavatnssýsla wrote that the ice first left on around 10 August.

5.6 Summary

It is a commonly-held belief that when historical sources such as annals do comment on climate, they tend to dwell on cold, severe and generally extreme weather, and ignore average or mild seasons. That this is a misconception, at least with regard to the sources used here, is attested to by the significant proportion of mild seasons described. Particularly striking are the mild decades of the 1570s and the 1640s, 1650s, 1660s and the early 1700s (see Figure 5.4a). It is interesting that, out of all these years, the 1570s appear to have been the mildest. However, it should be remembered that there is only one major source for this decade, *Gottskálksannáll*, written in the north. The decade of the 1570s (and the 1560s) cannot therefore be directly compared with later decades when data are available from all regions of Iceland.

The charge of subjectivity, often levelled at historical records of climate, may be partially answered by the use of statistical methods. Although only a rudimentary form of statistical analysis is used here, it is interesting to note that when such an exercise is performed, the final climate picture that appears may be quite different from that which first presents itself. An example of this is the decade 1701 to 1710. A cursory reading of the accounts of the weather written during this decade suggests that it was very cold. There were numerous complaints about the difficulties that people were suffering and these were frequently ascribed to poor weather. Yet when the actual weather descriptions of cold and mild seasons are analysed it appears that the decade 1701 to 1710 was the mildest of the entire eighteenth century.

It is interesting to speculate upon when the phenomenon that has come to be known as the "Little Ice Age" occurred in Iceland. This has frequently been ascribed a general date of around A.D. 1550 to 1850. Lamb suggests 1550 to 1700 as a reasonable time for the main phase "in most parts of the world" (Lamb 1977 p. 463). As research in different countries has continued, however, it has become apparent that the timing and duration of cold periods within and around this general timeframe has varied considerably between different regions (see, for example, Grove 1988). It has also become clear that there was not necessarily one distinct period of cooling. This has been observed, for example, in Switzerland (Pfister 1980 and Chapter 6, this volume) and it is certainly true for Iceland. Indeed, what is most noticeable from the study of the climate of Iceland from 1500 to 1800 is its profound variability on both an annual and a decadal timescale.

The Icelandic sources used here do suggest a cooling trend around the end of the sixteenth century and the beginning of the seventeenth, but this is interrupted by the mild decade of the 1610s. The cooling of the 1620s and 1630s was brought to an abrupt halt by the mild period ~1641 to 1670. The 1670s and 1680s remained relatively mild. The 1690s were undoubtedly very cold, in Iceland and elsewhere (Lamb 1977 pp. 471-72, Grove 1988 pp. 417-18). However, the first decade of the eighteenth century showed a return to a milder régime and this continued until ~1730. Three decades of very cold climate, 1731 to 1760, were followed by the milder 1760s. The 1770s and 1790s do not appear to have been as cold as the 1730s, 1740s or 1750s, but the 1780s stand out as the coldest decade in the entire series. It was also during the 1780s that the greatest sea-ice incidence occurred.

The mainly positive correlation between sea-ice incidence and temperature variations may be seen in Figure 5.4a and 5.4b. The years 1561 to 1580 agree well, as do 1601 to 1630, and 1641 to 1680. The 1690s are in marked agreement. The 1630s are anomalous, with cold weather and little sea-ice, and so are the 1680s, this time with a reversed pattern of pronounced sea-ice but only "averagely cold" climate. The 1710s show a marked decrease in sea ice compared with previous decades, but the climate is colder after the milder years 1701 to 1710. The 1730s are also anomalous. The decades from 1741 to the end of the eighteenth century show agreement.

A discussion of glacier oscillations has been outside the scope of this paper. However, while considering the "Little Ice Age" it is appropriate to mention them. A summary of the evidence from Icelandic glaciers suggests that by around the mid-eighteenth century most of them were larger than at any time since the settlement of Iceland. Although minor fluctuations occurred, they remained generally enlarged throughout the next 150 years (Grove 1988, p.27-54). This picture accords fairly well with the evidence from the climate data discussed here, and with unpublished research by Ogilvie on the climate of the nineteenth century. It is thus perhaps primarily during the period ~1750 to ~1900 that the "Little Ice Age" must be sought in Iceland. Further research on the climate of the nineteenth century will clarify the longer term picture. There can be no doubt, however, that the climate of the twentieth century has been far more clement than that of the nineteenth, eighteenth and much of the seventeenth century.

Acknowledgements

This chapter was re-written while the author was visiting the *Stofnun Árnamagnússonar* (Manuscript Institute) in Reykjavík during September 1989. I am grateful to the Director, Professor Jónas Kristjánsson, and his staff, for their hospitality. I should particularly like to thank Guðrún Ása Grímsdóttir for her valuable comments on the manuscript. The existence of the sheriff's letters was first brought to my attention by Professor Þórhallur Vilmundarson of the *Örnefnastofnun* (Place-Name Institute) Reykjavík. Part of the research for this paper was supported by grant GR3/7013 from the Natural Environment Research Council (U.K.)

References

Manuscript Sources

Þjóðskjalasafn (National Archives), Reykjavík

Bréf úr Árnessýslu til Stiftamtmanns 1711-1750, 1751-1785
Bréf úr Austur-Skaftafellssýslu til Stiftamtmanns 1709-1786
Bréf úr Barðastrandarsýslu til Stiftamtmanns 1709-1785

Bréf úr Dalasýslu til Stiftamtmanns 1704-1785
Bréf úr Húnavatnssýslu til Stiftamtmanns 1711-1735, 1736-1792
Bréf úr Ísafjarðarsýslu til Stiftamtmanns 1707-1803
Bréf úr Múlaþingi til Stiftamtmanns 1708-1768
Bréf úr Rangárvallasýslu til Stiftamtmanns 1711-1750,1751-1785
Bréf úr Skagafjarðarsýslu til Stiftamtmanns 1708-1740,1741-1790
Bréf úr Snæfellsnessýslu til Stiftamtmanns 1711-1803
Bréf úr Þingeyjarsýslu til Stiftamtmanns 1707-1751,1752-1803

Islands Journal 5 1780-1782, fylgiskjöl nr. 375: Rtk. 40.5; fylgiskjöl nr 618: Rtk. 40.90
Islands Journal 6 1782-1786, fylgiskjöl nr. 218: Rtk. 41.4; fylgiskjöl nr. 420; Rtk. 41.6; fylgiskjöl nr. 566: Rtk. 41.8; fylgiskjöl nr. 1093: Rtk. 41.13; fylgiskjöl nr. 1408: Rtk. 41.17
Islands Journal 7 1786-1788, fylgiskjöl nr. 490: Rtk. 42.6; fylgiskjöl nr. 552: Rtk. 42.7; fylgiskjöl nr 1402: Rtk. 42.16; fylgiskjöl nr. 1522, 1528, 1529: Rtk. 42.18
Islands Journal 8 1788-1792, fylgiskjöl nr. 536: Rtk. 43.7; fylgiskjöl nr. 609: Rtk. 43.8; fylgiskjöl nr. 1009: Rtk. 43.2; fylgiskjöl nr. 1246: Rtk. 43.6; fylgiskjöl nr. 1952: Rtk. 43.24 fylgiskjöl nr. 2151: Rtk. 43.27; fylgiskjöl nr. 2410: Rtk. 43.29
Islands Journal 9 1792-1796, fylgiskjöl nr. 332: Rtk. 44.6; fylgiskjöl nr. 567: Rtk. 44.9; fylgiskjöl nr. 617: Rtk 44.10; fylgiskjöl nr. 1065: Rtk. 44.15; fylgiskjöl nr. 1475, 1476: Rtk. 44.20 fylgiskjöl nr. 1530, 1540: Rtk. 44.21; fylgiskjöl nr. 1669, 1670, 1672-4, 1685, 1687: Rtk.44.23
Isands Journal 10 1796-1804, fylgiskjöl nr. 387: Rtk. 45.5; fylgiskjöl nr.398: Rtk. 45.6; fylgiskjöl nr. 638: Rtk. 45.9; fylgiskjöl nr. 806, 809-813, 815-16, 818-19: Rtk. 45.11; fylgiskjöl nr. 1115, 1120: Rtk. 45.13

Landsbókasafn (National Library), Reykjavík

The Jón Jónssons Weather Diary, Lbs 332 8vo, IBR 82 8vo

Published Works

Annálar 1400-1800 *(Annales Islandici posteriorum sæculorum) 1-6.* 1922-88. Reykjavík: Hið Íslenzka Bókmenntafélag

Bell, W. T. and A. E. J. Ogilvie 1978. Weather compilations as a source of data for the reconstruction of European climate during the medieval period. *Climatic Change 1*, 331-48

Benediktsson, J. (ed.) 1943. Two treatises on Iceland from the seventeenth century. Þorlákur Skúlason: Responsio subitanea. Brynjólfur Sveinsson: Historica de rebus Islandicis relatio. In *Bibliotheca Arnamagnæana* Vol.3, J. Helgason (ed.). Copenhagen: Ejnar Munksgaard

Benediktsson, J. (ed.) 1948. Ole Worm's correspondence with Icelanders. In *Bibliotheca Arnamagnæana* Vol.7, J. Helgason (ed.). Copenhagen: Ejnar Munksgaard

Benediktsson, J. 1956. Hver samdi Qualiscunque descriptio Islandiæ? *Nordæla Afmæliskveðja til Sigurðar Nordals*, 97-109. Reykjavík: Helgafell.

Bergþórsson, P. 1957. Veðurathuganir í Eyjafirði 1747-1846. *Veðrið 2*, 25-27

Bergthórsson, P. 1969. An estimate of drift ice and temperature in 1000 years. *Jökull 19*, 94-101

Bergthórsson, P., H. Björnsson, Ó. R. Dýrmundsson, B. Guðmundsson, A. Helgadóttir and J.V. Jónmundsson 1988. The effects of climatic variations on agriculture in Iceland. In *The impact of climatic variations on agriculture, Vol. 1*, M. L. Parry, T. R. Carter and N. T. Konijn (eds), 383-509. Dordrecht, Boston, London: Kluwer Academic Publishers

Bjarnar, V. 1965. The Laki eruption and the famine of the mist. In *Scandinavian studies*, C. F. Bayerschmidt and E. J. Friis (eds), 410-21. Washington: American-Scandinavian Foundation, University of Washington Press

Browne, Sir T. 1835. *Works Vol 2* S. Wilkins (ed.) London: Bohn's Antiquarian Library

Burg, F. (ed.) 1928. *Qualiscunque descriptio Islandiæ*. Hamburg: Veröffentlichungen aus der Hamburger Staats und Universitäts Bibliothek, Band I.

Diplomatarium Islandicum. *Íslenzkt Fornbréfasafn 8. 1906-1913*. Copenhagen and Reykjavík: Hinu Íslenzka Bókmentafélagi

Egilsson, J. 1856. Biskupa-annálar Jóns Egilssonar með formála, athugagreinum og fylgiskjölum eptir Jón Sigurðsson. In *Safn til Sögu Íslands og Íslenzkra bókmenta að fornu og nýju* Vol 1. 15-117. Copenhagen: Hinu Íslenzka Bókmentafélagi

Einarsson, M. A. 1976. *Veðurfar á Íslandi*. Reykjavík: Iðunn

Einarsson, O. 1971. *Íslandslýsing. Qualiscunque descriptio Islandiæ*. Introduction by J. Benediktsson. Reykjavík: Bókaútgáfa Menningarsjóðs

Eythórsson, J. and H. Sigtryggsson 1971. The climate and weather of Iceland. In *The Zoology of Iceland* Vol. 1, pt. 3, S.L. Tuxen (managing ed.). Copenhagen and Reykjavík: Ejnar Munksgaard

Friðriksson, S. 1969. The effects of sea ice on flora, fauna and agriculture. *Jökull 19*, 146-57

Grove, J. M. 1988 *The Little Ice Age*. London and New York: Methuen

Gunnlaugsson, G. A., G. M. Guðbergsson, S. Þórarinsson, S. Rafnsson and Einarsson 1984. *Skaftáreldar 1783-1784. Ritgerðir og heimildir*. Reykjavík: Mál og Menning

Hermannsson, H. (ed.) 1917. Annalium in Islandia farrago and De mirabilibus Islandiæ by Gísli Oddsson. *Islandica 10*. Ithaca, N.Y: Cornell University Library

Hermannsson, H. (ed.) 1924. Jón Guðmundsson and his natural history of Iceland. *Islandica 15*. Ithaca, N.Y: Cornell University Library

Ingram, M. J., D. J. Underhill and G. Farmer 1981. The use of documentary sources for the study of past climates. In *Climate and history. Studies in past climates and their impact on Man*, T. M. L. Wigley, M. J. Ingram and G. Farmer (eds), 180-213. Cambridge: Cambridge University Press

Jones, P. D., T. M. L. Wigley, C. K. Folland, D. E. Parker, J. K. Angell, S. Lebedeff, J. E. Hansen 1988. Evidence for global warming in the past decade. *Nature 332*, 790

Jónsson, A. 1928. Brevis commentarius de Islandia. In *Hakluyt's Voyages*, Vol. 1 of the Foreign Voyages and Vol. 9 of the Principal Navigations. Originally published 1598. London: J. M. Dent and Sons Ltd., New York: E. P. Dutton and Co

Jónsson, A. 1957. Arngrimi Jonæ Opera Latine Conscripta Vol. 4, J. Benediktsson (ed.) In *Bibliotheca Arnamagnæana*, Vol. 12, J. Helgason (ed.). Ejnar Munksgaard, Hafniæ

Jónsson, B. 1913. Svellavetur. *Andvari. Tímarit hins Íslenzka Þjóðvinafélags 38*, 104-110

Kington, J. A. and S. Kristjánsdóttir 1978. Veðurathuganir Jóns Jónssonar eldra og yngra og gildi þeirra við daglega veðurkortagerð eftir sögulegum gögnum. *Veðrið 21*, 42-51

Koch, L. 1945. The East Greenland ice. *Meddelelser om Grönland 130 (3)*. Copenhagen: Udgivne af kommissionen for videnskabelige undersögelser i Grönland

Lamb, H. H. 1977. *Climate present past and future. Volume 2. Climatic history and the future*. London: Methuen. New York: Barnes and Noble Books

Oddsson, G. 1942. *Íslenzk annálabrot (Annalium in Islandia farrago) og Undur Islands (De mirabilibus Islandiæ)*. J. Rafnar (trans.). Akureyri: Þorsteinn M. Jónsson

Ogilvie, A. E. J. 1981. *Climate and society in Iceland from the medieval period to the late eighteenth century*. Unpublished Ph.D. dissertation, University of East Anglia, Norwich, U.K.

Ogilvie, A. E. J. 1984a. The past climate and sea-ice record from Iceland, part 1: data to A.D. 1780. *Climatic Change 6*, 131-52

Ogilvie, A. E. J. 1984b. The impact of climate on grass growth and hay yeild in Iceland. In *Climatic changes on a yearly to millennial basis*, N.-A. Mörner and W. Karlén (eds), 343-52. Dordrecht: D. Reidel

Ogilvie, A. E. J. 1986. The climate of Iceland, 1701-1784. *Jökull 36*, 57-73

Ogilvie, A. E. J. 1990. Climatic changes in Iceland c.A.D. 865 to 1598. *Acta Archaeologica* (in press)

Ólason, P. E. 1942. *Saga Íslendinga: Seytjánda öld höfuþðættir 5*. Reykjavík: Menntamálaráð og Þjóðvinafélag

Pálsson, S. 1945. *Ferðabók, dagbækur og ritgerðir 1791-1797*. Reykjavík: Snælands Utgáfan

Pfister, C. 1980. The climate of Switzerland in the last 450 years. In Geography in Switzerland. *Geographica Helvetica* 35, no.5 (special issue), F. Müller, L. Bridel and L. Schwabe (eds), 15-20

Schepelern, H. D. (trans.) 1965-68. *Breve fra og til Ole Worm 1-3*. Copenhagen: Udgivet af Det Danske Sprog og Litteraturselskab, Munksgaard

Sigfúsdóttir, A.B. 1969. Temperature in Stykkishólmur 1846-1968. *Jökull 19*, 7-10.

Stephensen, M. 1806. *Eftirmæli átjándu aldar eptir Krists hingaðburð, frá ey-konunni Íslandi*. Leirárgörðum: Íslands opinbera vísinda-stiptun

Storm, G. 1977. Islandske annaler indtil 1578. Reprinted Oslo: Norsk-Historisk Kjeldeskrift-Institutt. Originally published 1888 Christiania: det norske historiske Kildeskriftfond

Thórarinsson, S. 1956. *The thousand years struggle against ice and fire*. Reykjavík: Museum of Natural History, Dept. of Geology and Geography, Misc. Papers no. 14

Vilmundarson, 1972. Evaluation of historical sources on sea ice near Iceland. In *Sea Ice*, Proceedings of an international conference, M. Á. Einarsson (ed.) 159-69. Reykjavík: National Research Council

Þórkelsson, J. 1887. Þáttur af Birni Jónssyni á Skarðsá. *Tímarit hins Íslenzka Bókmenntafjelags 8*, 34-96.

6 Monthly temperature and precipitation in central Europe 1525-1979: quantifying documentary evidence on weather and its effects

C. Pfister

6.1 From weather accounts to climate history

The growing consciousness that the world is undergoing a period of significant warming has triggered a wave of research activity into past climates. Within this context observations in historical sources are an essential ingredient of any study dealing with the reconstruction of climate in the period between the Middle Ages and the beginning of instrumental network observations in the 19th century.

Proxy data from natural archives cover periods up to hundreds of thousands of years. However, their time resolution is restricted to individual years or to periods of several months in the best case. Moreover they respond to a variety of meteorological variables making it difficult to disentangle the climatic information. Documentary material from anthropogenic archives, that is highly specific with regard to time, place and meteorological parameter, covers a period of roughly seven hundred years. This provides a safer basis for assessing the natural variability of climate than the shorter instrumental period. In particular this holds for the interpretation of exceptional events, such as the very mild winter 1989-90 in Central Europe (Pfister 1990).

Seasonal and monthly estimates of temperature and precipitation for the pre-instrumental period are obtained from the interpretation of documentary material. They provide a broad basis for calibrating natural proxy information and a new testing ground for climatic models, that need still to be much improved regarding regional and seasonal scenarios.

Assessing and modelling impacts of past climates upon past societies requires a very detailed record of past changes which is in fact a history of weather. Among agrarian scientists and demographic historians there is agreement on the point that monthly temperature and precipitation data are needed in order to quantify the influence of meteorological factors upon yield formation and upon diseases in a meaningful way. (Hanus and Aimiller 1978; Georgelin 1979). This holds also for the climatic interpretation of demographic patterns (Flinn 1981:100).

The types of data used for the reconstruction of past climates have been described in great detail by Lamb (1981). Within the same volume Ingram, Underhill and Farmer (1981) have provided a survey of documentary sources. The classification scheme shown in Figure 6.1 groups the evidence firstly into natural and anthropogenic data according to their origin. The data fall into descriptive and proxy data with respect to the kind of information which they contain. The term 'proxy' is used to denote any material, that provides an indirect measure of

ORIGIN / INFORMATION	NATURAL			MAN-MADE		
			Documentary Sources		Descriptive Reports	Instrumental Observations
Direct: weather patterns and meteorological parameters					– extreme events – rough sequence of weather situations – daily weather	– barometric pressure – temperature – precipitation – water-gauge
Indirect (proxy data): phenomena governed or affected through meteorological parameters	Geophysical – isotopes – sediments – moraines etc	Biological – marine plankton – pollen – tree rings etc			Geophysical / para-meteorological – water levels – snow falls – duration of snowcover – freezing-over of water bodies	Biological – time of blossoming and ripening of plants – yield and sugar content of vine – time of grain harvest and vintage
				Material Sources	paintings, prints and photographs; maps and charts buildings, settlements, roads, waterways abandoned farms and fields archeological remains	

Figure 6.1 A survey of evidence for reconstructing past weather and climate.

climate. It comprises both natural and man-made evidence. One kind of proxy data is related to geophysical and para-meteorological phenomena, mostly snow-cover and the freezing of water bodies, the second one refers to phenophases or other signs of biological activity. Anthropogenic data may also be grouped into documentary sources and material sources according to their form and to the place where they are found. Written sources, manuscript or printed, are preserved in libraries and archives or owned by private individuals. Moreover inscriptions referring to climatic anomalies are sometimes painted on the front of houses. In some locations the height of floods is marked on buildings, or the level of low water at certain times is engraved on rocks. Objective data are found in museums or in the field, pictorial data are stored in libraries, archives and displayed in museums or on private property.

The present approach attempts to demonstrate, how to bridge the gap between climatic history and weather history by cross-dating different kinds of documentary proxy data with descriptive evidence. The aim is to produce a combined record which provides both the quantitative estimates of temperature and precipitation needed by the scientist, the economist and the policy maker and the detailed weather account which the historian requires for reconstructing the past. The procedure starts by collecting the smallest bits of evidence available, weather observations, early instrumental data and proxy information. The resulting 'weather history' is coded, homogenized and calibrated according to the type of data and stored in a data-bank (Figure 6.2). All the evidence is then boiled down to a numerical wetness and temperature index for each month. In the last step this data is converted into a 'climate history' by computing transfer functions for estimating decennial temperature and precipitation. The results of the reconstruction of climate and the bearing that climatic fluctuations have upon the economy and demography of Switzerland are discussed in considerable detail in the 'Klimageschichte der Schweiz' (Pfister 1984) from which this article is mainly drawn.

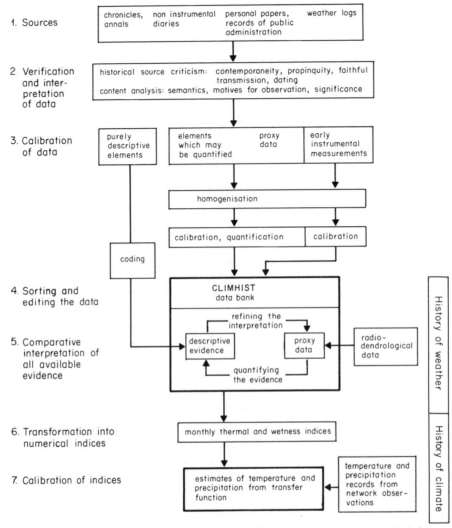

Figure 6.2 Flow diagram illustrating the procedures used in obtaining climatic information from historical documents.

6.2 Survey of sources and data

The human evidence which documents the study has originated principally from chronicles, from personal papers (diaries, calendars) containing intermittent meteorological entries, from non-instrumental weather diaries to weather logs, that contain both instrumental measurements and visual observations. More than 118 manuscripts and over 150 printed sources with weather evidence were found in the Swiss libraries and archives.

Some of the main observers and sources are briefly described in the following: Wolfgang Haller (1525-1601), Arch-deacon of the Cathedral of Zürich, kept a non instrumental weather diary for thirty years (1545-1576) in which he describes the daily weather with one or two words such as 'rather warm', 'rain', 'snow', 'dull' etc. Flohn (1949) has shown the decline of winter temperatures after 1560 from this source. Renward Cysat (1545-1613), Chancellor in Lucerne, has quite regularly observed a broad variety of meteorological and environmental data, mentioning snowfalls in summer on the summits surrounding the city. Joseph Dietrich (1645-1704) and Sebastian Reding (1667-1724), who were monks in the monastery at Einsiedeln, have included an incredible wealth of detailed lengthy weather reports in their diaries an the end of the seventeenth century. They already distinguished the basic form of clouds. Heinrich Fries (1639-1718), who was a professor at Zürich, kept a non-instrumental diary in which he noted down quite regularly the formation and the melting of snow-cover. The physician Johann Jakob Scheuchzer (1672-1733), the pioneer of natural history in Switzerland, was the first in Central Europe to measure precipitation in his garden in Zürich (from 1708). Hans Rudolf Rieter (1665-1748), a baker, has left an extremely detailed account on the weather from 1721 to 1738 which even includes many observations for the night hours. Most remarkably, he has regularly noted down the time of blossoming and ripening of eight cultivated plants and trees (beech tree, sweet cherry, pear, rye, barley, strawberry, spelt [a variety of wheat] and grapes). His record is somewhat older than that of the Marsham family in Norfolk, England (Kington 1974). Johann Jakob Sprüngli (1717-1803), a parson, made some 4000 phenological observations between 1760 and 1802. He also collected precise data on the formation and melting of snow cover in his surroundings and on the Alps.

The data-bank includes more than 33,000 records. The time resolution of a record may vary from five or ten day periods to a month or a season. The northern and eastern regions in the Swiss lowlands are better represented than those in the French-speaking part and in the Alps. About 80,000 daily entries in weather logs were integrated to form monthly sums of days with precipitation, snow, thunderstorms, fog or clear sky. The data-bank includes some 3000 phenological (i.e. plant) observations and about the same number of observations concerning snow-cover at various altitudes.

The evidence increases in volume, density and diversity over time. For the period 1525-1549 the entries originate mainly from chronicles and annals. Accordingly, weather sequences are mainly described at a seasonal level; information is missing for 43% of the months and the emphasis is on *anomalous* rather than on *ordinary weather*. In the second period 1550-1658 monthly data from weather diaries and personal papers are abundant. The few missing monthly data (7%) are concentrated in the months from October to December. From 1659 monthly meteorological data are continuous. From 1684 monthly precipitation is almost thoroughly quantified either by counting entries in weather diaries or from measurements (after 1708) and the thermal character of each month is derived from a body of reliable para-meteorological and phenological proxy data.

The thermometric evidence dates from 1755 and is based on the Basel series. In order to document conditions in mountain environments, the thermometric series from the Great St. Bernard pass (2460 m above sea level) which originates in 1817, was included. Scattered series of rainfall measurements (e.g. Zürich, Bern) are available from 1708. The Geneva precipitation series (from 1778 to the present) is the longest continuous record of this kind in Switzerland. The creation of the national weather service in Switzerland in 1864 was chosen as a dividing line between the periods of the 'historical' and the 'modern' data.

6.3   Data verification

Only a decade ago it was discovered, that documentary sources of information about past climates are not equally reliable. Compilations often contain a mishmash of valuable and worthless data. Their main flaws are inaccurate or uncertain dating of particular events, acceptance of accounts which are distortions or amplifications of original observations and spurious multiplications of events through mis-dating (Alexandre 1987; Bell and Ogilvie 1978; Ingram *et al*. 1981, Pfister 1984). An example of mis-dating comes from Central Europe. The well-known chronicle of Thann (Alsace) which Klemm (1974, p.409) takes to be "extremely valuable" has been used again and again as a source of information for reconstructing past climate. In vain, the editor himself, Abbé Merklen, had cautioned, that the chronicle contains a lot of contradictory evidence and incorrect dating.

Historians have developed a standardized methodology for evaluating sources and rejecting unreliable information. The most important critical tests are those based upon the principles of contemporaneity, propinquity and faithful transmission. Recorded statements cannot be regarded as reliable and valuable, unless it can be shown either, that the writer lived close in time and space to the events he purports to describe, and recorded his observations immediately or within a short space of time after these events had taken place, or that he had access to first hand oral or written reports and can be presumed to have accurately transmitted the information derived from them. (Ingram *et al*. 1981). In the context of the *Klimageschichte der Schweiz* (Pfister 1984, 1985a) non-contemporary material was not completely rejected. Where a reliable picture of weather patterns had already been obtained from contemporary data, second hand reports which contributed to the understanding of weather situations were included. In order to clearly mark their lower quality, the name of the author was omitted in the printout of the data-bank (see Section 6.6).

Accurate dating was one of the thorniest problems. Up to the end of the eighteenth century the Gregorian and Julian calendars were simultaneously in use, often within the same village or district. In the data-bank all the dates have been converted to Gregorian style and every dating correction is made explicit. Apparent contradictions arise, when weather reports from catholic and protestant cantons refer to the same month. In those cases a time resolution of the individual records from ten days to ten days, such as it is provided in the CLIMHIST data-bank, becomes indispensable for an appropriate interpretation. Moreover the reference to ten day-periods is very convenient for handling the dating related to Saint's days, which do not fit into the scheme of calendar months.

For events in 'winter', dating becomes a problem particularly when the source gives just one year for identification. Then it must be derived from the context or from other sources, whether the 'old' or the 'new' year is meant in order to prevent spurious multiplication of events. The term of 'Winter' itself was related to the duration of snow-cover rather than to specific months. Likewise 'Herbst' (autumn) described the time of the vintage.

6.4   Calibration of proxy data

The following types of data are included in the data-bank:

1 Miscellaneous observations from descriptive sources

    1.1 Purely descriptive data
    1.2 Non-serial proxy data

        1.2.1 Freezing of water bodies
        1.2.2 Snowfall and snow-cover
        1.2.3 Phenological data
        1.2.4 Floods and low water levels

2 Daily non-instrumental observations
3 Monthly frequencies of rainy, sunny, foggy, snowy days etc. obtained from weather logs
4 Instrumental records for precipitation
5 Instrumental records for temperature
6 Serial proxy-data:

    6.1 Dates of auctions of tithes paid in grain
    6.2 Grapevine harvest dates
    6.3 Grapevine harvest yields
    6.4 Phenological series
    6.5 Dendroclimatic data

The descriptive evidence (without the 'serial' data) was standardized by the use of a numerical code (Pfister 1981a). Calibration has to be done separately for each type of data.

6.4.1 *Freezing of water bodies*

From the cases documented with thermometric measurement it has been established, that lakes in the alpine borderland freeze in a specific rank order, according to their surface, depth and individual characteristics. The freezing is primarily a function of the sum of below-freezing-point daily mean temperatures, plus such other factors as wind-speed. Frequently it is specified in the sources, whether the ice was thick enough to carry men and cargoes and how long the ice-cover remained. This provides an additional clue. For the Lake of Zürich, which has the highest number of entries in the CLIMHIST data-bank, it is known that a sum of at least 350°(C) freezing degree days is needed to form an ice cover which is thick enough for a safe walk.

    The freezing of rivers cannot be properly calibrated, because in the past these events are not adequately documented with thermometric measurements. Moreover river-beds have been changed because of river regulation schemes in the nineteenth and twentieth centuries. However, it may be hypothesized from the scanty thermometric evidence, that such events did occur, when temperatures were between -25° and -30°C. Temperatures in this range were also associated to the formation of frost cracks in trees. This produces repeated claps similar to gunfire which are described in some sources as symptoms of a bitter cold.

6.4.2 *Snowfall and snow-cover*

The ratio of the number of snowy days to the number of rainy days may be computed from weather-logs and used as a proxy for temperatures in winter (Flohn 1949). Moreover frequent snowfalls reported for October, November, March and April point to below average temperatures in those months.

Snow cover is an eye-catching meteorological element used in chronicles to describe extremely long winters. From the end of the seventeenth century the formation and melting of snow-cover has been quite regularly described in weather logs. For the Mittelland (250 m to 600 m a.s.l.) there is at least one observation for 271 out of 306 winters from 1684 to the present; from the mid-eighteenth century the date of thaw is known for higher elevations, too (Pfister 1985b, 1990).

The persistence of snow-cover is related to the sequence and duration of weather situations favouring accumulation or ablation. In most cases this may be drawn from weather reports. A very small number of days with snow cover is mostly related to warm winters, occasionally also to dry and cold anticyclonic situations. For the pre-instrumental period the hypothesis of warmth needs, therefore, to be supported with observed signs of vegetation activity. Altitude and orographic factors (such as windward or leeward slope) of the locations of observation were considered in interpreting the data (Schüepp 1980; Witmer 1986).

For calibrating the series from the Mittelland a separate model has been developed for every winter month, which yielded fair estimates for temperatures (Pfister 1977). The date of alpine thaw, which depends primarily on temperature and radiation patterns, was compared among stations in the Swiss alps located at similar altitudes. It turned out to be highly correlated and closely related to phenological data, such as flowering and ripening of the vine, (Pfister 1985b).

6.4.3 Phenological data

In order to describe temperature patterns during the vegetative period observers in the pre-instrumental period frequently referred to stages in the growth and maturity of cultivated plants, which were known to be more accurate yardsticks of warmth and coldness than impressions of individuals.

The evidence enables the development of long phenological series, that cover almost the entire vegetative period: blossoming of sweet cherries (from 1721), the bloom of grapevines (from 1702), the start of rye harvest (from 1557), the ripening of early varieties ("Aeugstler") and ordinary varieties of the red burgundy grape (from 1721), the beginning of the wine harvest (from 1370) (Pfister 1988).

Two conditions must be met for interpreting and calibrating pre-instrumental observations.

1 Corresponding phenophases should have been regularly observed for at least ten years close to a meteorological station, where temperature, precipitation and duration of sunshine were measured.
2 Phenophases for the pre-instrumental and for the instrumental period should be compared in order to determine, whether significant shifts did occur in the averages. Moreover the knowledge of averages within the pre-instrumental period allows for the interpretation of isolated observations for anomalous years.

Calibration of the historical data was based upon series of corresponding phenophases carried out at three meteorological stations in the Canton of Schaffhausen from 1875 to 1950.

It turned out that most phenophases are significantly correlated with temperature patterns: the beginning of cherry blossom depends on conditions in March and April, the beginning of vine flower is primarily an indicator of temperatures in May, the beginning of the rye harvest is tied to temperatures in May and June, the ripening of both varieties of the red burgundy grape is a function of temperatures in June and July.

In the central frame of Figure 6.3 the average dates of ten phenophases observed in the pre-instrumental period are given below the line. The corresponding averages from the Schaffhausen series are represented above the line. The days of the year are marked on the x-axis. There is a good agreement between the two sets of observations, when differences in altitude are considered. The advanced blooming of vine in the eighteenth century is bound to the specific climatic conditions of the period.

Above the frame, the earliest phenophases known from 1525 are provided. For those observations which fall into the instrumental period the deviation of temperatures from the 1901-60 mean in the months preceding the phenophases is marked by means of horizontal bars. The bars indicating positive deviations point to the left. The merger of evidence from the two periods allows us to compare the earliest known springs and summers within the pre-instrumental period with those which are documented with thermometrical series.

Below the frame, the latest phenophases and thus the chilliest springs and summers known from 1525 are shown in the same way. The horizontal bars pointing to the right indicate negative deviations from the 1901-60 mean in the months preceding a given event.

In the following those extremes are discussed.

Cherry flowering: advances by two weeks or more are mostly connected to summerlike temperatures in February or March (1822, 1897, 1794). In 1990 sweet cherries began flowering around March 20 which is equivalent to the earliest springs which are known (Pfister 1990). A delay of flowering by three weeks or more suggests that the March-April period may have been more than 5°C below the 1901-60 average. March 1785 was 8°C below this average!

Vine flowering: we should consider the observations which were carried out in an open vineyard only. Plants which are sheltered by the wall of a house will flower considerably earlier. Differences in varieties, however, can be neglected prior to the late nineteenth century. The first flower is advanced or delayed mainly according to temperatures in May, an early flower may also follow a very warm April (e.g. 1811, 1893) and, according to descriptive evidence, in 1723. The times of the full bloom and the last flower vary with temperatures in both May and June. Extreme delays (1542, 1627, 1628, 1642, 1675, 1740) were much more frequent and much more pronounced than extreme advances.

Start of the rye harvest: While the three-field system was in use (i.e. until the early nineteenth century) agreement was reached jointly by the farmers of a village to begin the harvest. The maturity of rye is controlled by temperatures in June and, to a lesser extent, by those in May.

Wine harvests: In all those years for which an early flowering of vine or an early cereal harvest was reported (e.g. 1540, 1616, 1636-38, 1660, 1718 1719, 1811, 1822) wine harvests throughout Central and Western Europe were also very advanced. On the other hand the latest wine

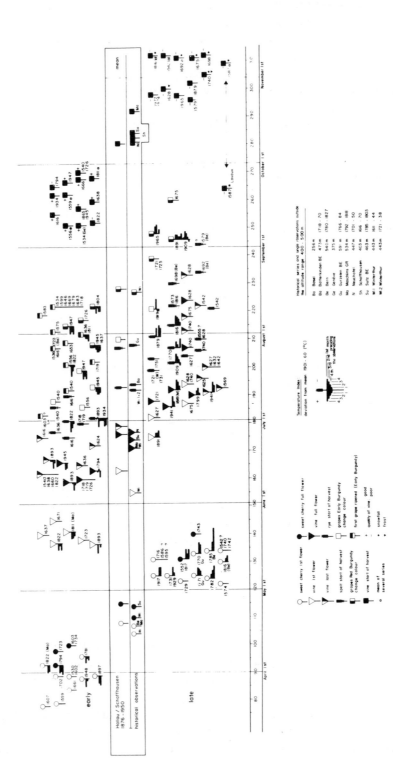

Figure 6.3 Dates of phenological events in Switzerland. Central panel shows the average dates of ten phenophases observed in the historical period (below center line) compared with the corresponding averages in the period 1876–1950 (above center line) based on observations in the Canton of Schaffhausen. Symbols above central panel indicate the earliest phenological events since 1525; those below the line indicate the latest events. For further explanation of symbols, see text.

harvests go along with a delayed bloom of the vine or a delayed cereal harvest (1542, 1555, 1573, 1627, 1628, 1675, 1698, 1740, 1816, 1879, 1980). This corroborates the results of statistical analyses that show the time of wine harvest is controlled by temperatures in May and June to a large degree (Pfister 1984; Flohn 1985). The timing of the latest wine harvests shows almost no variation, because at the end of October or at the beginning of November grapes had to be picked before the onset of winter, even if they were not mature yet.

In addition to focusing upon the analysis of specific phenophases, the whole pattern of phenophases from spring to autumn may be examined. A large shortening and lengthening of the period between phenophases, as compared to ordinary years, may be interpreted in terms of temperature anomalies. To take an example, a drastic shortening of the interval between the first vine flower and the start of the rye harvest (about two weeks compared with thirty-three days on average) occurred in 1616. The rye reached maturity even six days earlier than in 1822, which was the hottest June since 1755. This early date suggests that the heat-wave in June 1616, which is impressively described in the sources, was the most severe since 1525 at least.

On the other hand major delays between phenophases point to temperatures far below average. In 1628, to take an extreme case, the duration of vine flowering was some thirty-five days instead of nineteen, on average. This suggests an extreme cold spell in midsummer which is consistent with the high number of snowfalls reported from the alpine pastures.

6.4.4   Serial proxy data

Flohn (1985) has carefully investigated and cross-correlated long proxy data series from Switzerland, southern Germany and France. He advises that a careful examination of historical series would require the following checks:

1 the homogeneity of all records by comparison with adjacent records
2 the spatial representativity of the data. Area averaged series are preferable if their coherency can be warranted.
3 the time and the type of response to climate parameters, which they may represent.

In the following, the three types of proxy data – tithe auction dates, grapevine harvest yields and tree-ring densities are discussed. For wine harvest dates the reader is referred to Legrand (1979) Le Roy Ladurie and Baulant (1980) and Flohn (1985).

Tithe auction dates   The date on which the tithes paid in grain were sold by auction is regularly listed in some records from the mid-sixteenth century. It has been shown, that the date of the auction was closely related to the time of maturation of rye which was the earliest winter grain. In most cases the rye harvest immediately followed the auction of the standing crops. 42 local series, mainly from central and eastern Switzerland, were combined into an area averaged series from 1611. An analysis of both modern phenological observations on the beginning of the rye harvest (the Schaffhausen data) and the historical series of tithe auction dates has revealed that both primarily reflect the temperatures in May and June. The correlation of the tithe auction dates with the Basel temperature series over the period 1755-1825 is $r=-0.8$ (significance <0.001). A comparison with the Central England temperature series over the period 1659-1758 has still yielded an astonishing $r=-0.68$ (significance <0.001) despite the large distance between the two countries. It follows from the model that an advance or a delay of the mean tithe auction dates of seven days corresponds roughly to a positive or a negative deviation of temperatures of 1°C in May and June taken together.

The area averaged series of Swiss tithe auction dates is highly correlated (r=>0.70) with the area averaged series of wine harvest dates (Le Roy Ladurie and Baulant 1980) which shows again that the latter are primarily tied to temperature patterns in May and June (Flohn 1985).

Grapevine harvest yields   Until very recently (Pfister 1981b; Lauer and Frankenberg 1986) very little research has been devoted to fluctuations in grapevine harvest yields in historical times. This may be due to the fact, that modern investigations focus upon conditions in single vineyards, where local weather conditions, in particular late frosts, prevail over the large scale effects of temperature and rainfall. Moreover, reliable data from the pre-instrumental period are difficult to find and their interpretation is controversial. For the analysis long area-averaged series had to be established as a first step. Most of the secular series found in Swiss archives refer to vineyards owned by public institutions and authorities, which were cultivated by tenants according to share-cropping agreements. The production was divided between the tenant and the landlord, the latter's part being listed in the document. Some series give also the acreage of the vineyard. The earliest long record begins in the early sixteenth century; a sufficient number of series are available for the period after 1550.

Four regional series for different parts of the Swiss plateau, compiled from 17 local series, for which the acreage is not known show highly significant correlations (from r=0.56 to r=0.76) between each other. A Swiss area-averaged series was compiled from those four regional series. Likewise an area-averaged series was established from 20 local series, for which the acreage is known. In turned out that the two main series were highly correlated (r=0.85) (Pfister 1981b). Thus we may conclude that the former series, which goes farther back in time and which represents a larger volume of wine harvested, also reflects yields per acre for the most part. A stepwise regression analysis of this multisecular area-averaged series with the Basel temperature record yielded that July had the greatest weight, with $R^2=0.29$, June had $R^2=0.16$ whereas July of the previous year had $R^2=0.13$. The coefficients for August are not significant.

According to present day knowledge yearly fluctuations in wine yields are chiefly related to weather patterns in summer. High yields can be expected if temperatures in midsummer (June-July) are high, unless the water supply is deficient. On the other hand, cool and wet weather during and after flowering will hamper fertilization and the flowers may drop in the following. It must be stressed, however, that the effects of a widespread late frost, if it occurs repeatedly can almost annihilate the harvest, even when temperatures are favorable for growth and maturity in the following period. The statistical models should therefore always control for years with severe late frosts which are well known from the descriptive record. Flohn (1985) who from his models advised to use wine yields only cautiously as climatic indicators, did not exclude the years with heavy late frost. This was the reason while his results were not that convincing.

To conclude, the grapevine has three major advantages as a proxy:

1   The plant remains the same for twenty to fifty years. No annual planting is required.
2   The entire length of the growing season from March-April to September-October is needed to bring the grapes to maturity.

3   Harvest date, yield per acre and sugar content can be used as climatic proxy evidence for three different periods of the growing season: late spring-early summer, midsummer and late summer-early autumn (Pfister 1981b).

Dendroclimatic data   Representative results can be expected from trees at the alpine timberline, where the temperature of the short vegetative period controls the growth rate. Significant progress has been made through the Roentgen density measurements of wood, which allow the evaluation of a quantitative parameter such as the density of late wood which is produced mainly during late summer. This data is a good proxy for temperatures in July, August and September (Schweingruber 1983; Flohn 1985). The well known series from Lauenen (Bernese Oberland) is included in the CLIMHIST data-bank (Schweingruber *et al.* 1978). A cross correlation with the Basel temperature series has revealed a weakness of these data. Whereas low densities are very reliable indicators for cool summers, some of the hottest summers in the last 450 years (1616, 1719, 1947) do not stand out in the record (Pfister 1985c). This suggests that cross-checking with anthropogenic evidence would certainly be helpful in the interpretation of tree-ring density data in this region.

6.4.5   *Survey of proxy-data used as substitutes for thermometric measurement*

Table 6.1 gives a survey of the proxy-information which has been used as substitutes for thermometric measurements. The data type taken as the best indicator for the temperature of an individual month is underlined. The winter months are primarily documented on the basis of snow and ice features, while plant indicators are the best evidence for the vegetative period. Most of the biological proxy-data used reflect the temperature pattern over a period of several months. The longer the period involved, the higher the precision of the estimate. No equivalent proxy exists for October and November. Moreover even descriptive observations are sometimes missing for those months before 1658.

6.5   Proxy substitutes for measured precipitation

6.5.1   *Weather logs*

Weather diaries can easily be quantified by counting the frequencies of events such as rain, snow, thunderstorms etc. Whether it pays to attempt the cumbersome operation of counting depends on the quality of the observation. Comparing frequencies based on qualitative data with frequencies based on measurements provides a useful check of reliability. We may rely on the assumption that the meteorological framework does not change dramatically over time. Yearly averages of days with precipitation obtained from the weather-logs of careful observers differ only insignificantly from those which are based upon measured daily precipitation of >0.3 mm. The quantified historical data may therefore be compared to 1901-60 statistics from the same or from a neighboring station. However, considering the diary of Wolfgang Haller (1545-1576) it turned out that the average was even somewhat below the days with measured precipitation of >1mm. In order to get at least a source-specific rough criterion for assessing the wetness of the individual months from this unique evidence the

Table 6.1 Survey of indicators used for the determination of the thermal charater of individual months.

Month	Cold	Warm
Dec., Jan.	Uninterrupted snow cover freezing of lakes	Scarcity of snow cover signs of vegetation
March	Long snow cover, high snow frequency	Sweet cherry first flower ($\pm 1.3°$)
April	Snow cover and snow frequency, beech tree leaf emergence, sweet	beech tree leaf emergence (tithe auction date)
May	Tithe auction dates ($\pm 0.6°$) Vine first flower ($\pm 1.2°$) Barley harvest beginning	
June	Tithe auction dates ($\pm 0.6°$) Vine full flower ($\pm 1.2°$) Vine last flower Coloration of first grapes	
July	Vine yields ($\pm 0.6°$) Coloration/maturity of first grapes	
April-July	Wine harvest dates ($\pm 0.6°$)	
August	Wine yields ($\pm 0.6°$) Tree ring density ($\pm 0.8°$)	
September	Vine quality Tree-ring density ($\pm 0.8°$)	
October	Snow cover, snow frequency	Reappearance of spring vegetation (cherry flowering etc.)
November	Long snow cover, high snow frequency freezing of lakes	No snowfall, cattle in pastures

The figures give the standard error of estimate in °C.

duodecile distribution (cf. Table 6.4) of the days with observed precipitation in the diary was computed for every calendar month.

6.5.2 Floods and low water levels

Evidence that bears on floods and low water marks is found in written sources and also in form of marks on buildings. From the eighteenth century water-gauges have been installed on the major rivers and lakes. In order to exclude local events the analysis has focussed upon the large rivers (Rhine, Rhone, Aare) and on the major lakes in the Swiss lowlands.

A quantitative estimate on floods may be obtained by cross checking marks of historical floods on buildings with measured discharge or precipitation. In Basel, where the Rhine drains almost two thirds of the surface of Switzerland, monthly discharge of the river has been measured since 1808. Some hundred meters upstream from the water-gauge the level of some of the major floods within the last few centuries is marked on the building located at Oberer

Rheinweg 93. A comparison of this evidence allows 'calibrating' the descriptive data with the measured discharge record. Moreover, the old observers often compared the observed floods to previous ones, sometimes in quantitative terms or with reference to marks on buildings which have since been pulled down. Combining all the existing evidence has allowed the rank order of the most severe floods at Basel since the mid sixteenth century to be established.

In interpreting flood records it is essential to consider the time of the year and the nature of the drainage basin. Heavy floods of the major Swiss rivers in early summer are sometimes the result of the melting of unusually large amounts of snow in the mountains (e.g. 1817). Therefore we must also consider the snow record in order to assess the relative proportion of snow-melt and heavy rainfall for the event. The interpretation must also allow for corrections of the river and for changes in the level of buildings which might have occurred over time.

Low water levels are quite reliable indicators of precipitation patterns. They occur mostly in winter and spring during long spells of anticyclonic weather. Because they had almost no economic impact they are less frequently described in the sources than floods. Sources specify for some cases, how far one could walk in a river-bed which had partially dried up. Occasionally, extremely low water levels were marked on rocks which emerged, when the water was very low. This evidence is again 'calibrated' by comparing the older marks to more recent ones and to that of water-gauges.

6.6   Coding and editing the evidence: the CLIMHIST data-bank

The collected material was analyzed by a set of computer programs which had been specifically developed for this purpose. The descriptive evidence was converted into numerical form by means of a comprehensive code. Those observations which couldn't be expressed by the code, were literally written on a specific file to be included into the final version of the output in form of footnotes. Moreover, each type of serial proxy-data had to be homogenized and calibrated by means of standard software, before it was included into the data pool. Finally the entire evidence was pre-processed and sorted according to time, type of data and region. Subsequently the resulting data-bank was reconverted into a weather chronology called CLIMHIST-CH. It lists each of the 33,000 records including all the attributes, that are necessary for the interpretation, such as place and altitude of observation, name of the observer (if the data is contemporary) the style of the original observation etc. (Pfister and Schwarz-Zanetti 1986). Weather patterns and their impacts upon the hydrosphere, the biosphere and society (agricultural prices, diseases etc.) are given in a temporal frame of ten day periods, months or seasons. Serial information such as the number of days with precipitation or early instrumental measurements is compared to that of the 1901-60 reference period. The date of phenological observations is converted to days of the year. Serial proxy-data are given as deviations from their average or in terms of standard deviations.

The program sequence conveniently allows for error-correction or updating when new evidence is discovered, even including new footnotes (Pfister 1985a). At present four routines are availble that reconvert the numerical code into English, French, Italian or German. Routines for any additional language might be set up from corresponding translations of the code-book. Originally the program was tailored to Swiss data. In the meantime improved versions have been devised that handle data from all over Europe.

Table 6.2 provides the information for August 1723 as an example. Each record is explained by a number of attributes: **R** indicates to the meteorological region in which the place of observation is located. If the name of the observer is listed, this means that the data is contemporary. **S** is related to the source number in the bibliography. **F** followed by a number would refer to a footnote in the appendix.

Table 6.2 Example of an entry in the CLIMHIST data bank (Pfister 1985a) (see discussion of coding in text).

1723 August

1st ten day period
Warm/Hot. Preponderantly sunny (shorter periods of rain). **R**:2 Winterthur: 442m (Rieter, **S** 90).
Continuing rain (no statement). **R**:10 Stans:452m (Buenti, **S** 146)

2nd ten day period
Warm. Preponderantly sunny (isolated thunderstaorms). **R**:2 Winterthur: 442m (Rieter, **S** 90)
Preponderantly sunny (no statement). **R**:10 Stans: 452m (Buenti, **S** 146).
First red berries 8 11th (223) R:2 Winterthur:442m (Rieter, **S** 90)

3rd ten day period
Hot. Preponderantly sunny (isolated rain). **R**:2 Winterthur:442m (Rieter, **S** 90).
Preponderantly sunny (no statement). **R**:10: 452m (Buenti, **S** 146)

Entire month
Hot. Continuing rain. **R**:8 St. Blaise NE:433m (Peter, **S** 220)
Warm. Preponderantly sunny. **R**:5 Baetterkinden, Be: 473m (Wieniger, **S** 268).
Dry 87mm (mean:132mm). **R**:6 Zuerich:408m (Schuechzer, **S** 237)
Dry 10 days of PR (M:14 days) Warm **R**:2 Winterthur:442m (Rieter, **S** 90)

Temperature Index: 2, Warm. **Precipitation Index**: -2, Dry.

The first three paragraphs in Table 6.2 give information which is available according to ten day-periods (five day-periods are separated with a slash). The last paragraph is related to the data that refer to the entire month. In the example, rainfall at Zürich measured by Scheuchzer, and the number of days with precipitation computed from the Rieter diary, are compared to the duodecile statistic for the period 1901-60 at the same place. The thermal and wetness indices listed on the last line are estimates for temperature and precipitation derived from all the evidence by the interpreter (see the following section).

6.7 Deriving monthly indices

Estimates for temperatures and precipitation are estimated from regression models, that include series of proxy data and instrumental measurements. Significant results are obtained for periods of two or three months only. The models cannot account for shorter periods, because the variety of the underlying pattern of cold and heat spells is too large. However, this missing information is obtained from descriptive sources. To take an example, the opening of the first vine flower in a vineyard is bound to temperatures in April and May. An early flowering at the end of May may occur after an exceptional heat-wave in April followed by an average temperature in May, or after a long warm spell in May preceded by average weather in April. Based upon detailed descriptive evidence we might be able to decide which of the two patterns is more likely to have occurred. This then allows a monthly temperature profile, i.e. the 'excess of warmth' to be estimated for the entire period from the regression attributed to an individual month in an interpretative way. This takes into account all the quantitative and qualitative information available for the period. From this example it has been demonstrated, how the two kinds of evidence, proxy data and descriptive sources, control and complement each other. We may conclude that they form a coherent body of information, from which estimates of monthly temperature and precipitation patterns are obtained, based upon a synthesis of statistical reasoning and contextual interpretation.

Two types of indices, a weighted and an unweighted index, have been derived. For the weighted index, the frequency distribution of monthly means for the period from 1901-1960 was adopted as the standard of comparison (Table 6.3).

Table 6.3 The definition of the weighted temperature and precipitation index values over the period 1901-1960.

	lowest <------					------> highest	
	8.3%	16.6%	16.6%	16.6%	16.6%	16.6%	8.3%
Duodecile	1	3	5	7	9	11	
Index	-3	-2	-1	0	+1	+2	+3

The value of 0 was used for 'normal' weather conditions and for all months for which the evidence is missing. The values of +3 and -3 were applied to those cases which can unmistakably be considered as "extreme" by twentieth century standards. The values of +2 to -2 were adopted for the less marked gradations. The values of +1 to -1 were applied to all months for which only descriptive evidence is available as well as to those months which according to proxy information fall in the corresponding range of temperature and precipitation (cf. Table 6.3).

For precipitation two separate indices have been computed. The first, a frequency index, draws from the number of rainy days in Basel and Geneva, while the second, a rainfall index, is based upon measured precipitation at Geneva, Zürich and Bern. Subsequently, the two indices have been merged (cf. Pfister 1984, Table 1.26). Within the instrumental period the indices were computed from measured temperature and precipitation according to the duodecile distribution of the values (cf. Table 6.4). The unweighted index downgrades all the positive or negative weights to three gradations: +1, 0 and -1. Accordingly this index is more homogeneous than the weighted one, but is does not fully exploit the informative potential of the data. Which of the two indices is more 'realistic' depends on the quality of the evidence.

Table 6.4 Standard errors in estimating temperature (°C) and precipitation (%) from indices for the period 1864-1979.

Month/ Season	Temperature Index		Precipitation Index	
	unweighted	*weighted*	*unweighted*	*weighted*
January	0.4°	0.26°	12.8%	6.5
February	0.34°	0.26°	12.6%	11.4%
March	0.28°	0.18°	11.6%	9.2%
April	0.21°	0.11°	9.8%	9.6%
May	0.29°	0.16°	8.2%	7.1%
June	0.22°	0.16°	7.5%	6.7%
July	0.21°	0.11°	8.0%	6.6%
August	0.22°	0.14°	3.8%	5.0%
September	0.25°	0.14°	7.0%	11.7%
October	0.27°	0.13°	18.9%	18.1%
November	0.34°	0.19°	18.8%	13.7%
December	0.55°	0.35°	13.8%	12.6%
Spring	0.15°	0.1°	7.7%	7.2%
Summer	0.15°	0.1°	4.2%	2.9%
Autumn	0.15°	0.13°	9.4%	7.5%
Winter	0.25°	0.2°	9.5%	7.5%
Year	0.12°	0.08°	6.4%	5.2%

An earlier version of the indices has been published in Pfister (1981a). However, the indices have been considerably improved since, as new evidence has been found. The reader is therefore referred to the values published in Pfister (1984) and to the CLIMHIST data-bank in which all the basic evidence can be scrutinized (Pfister 1985a).

6.8 Estimates of temperature and precipitation from transfer functions

In order to bridge the gap between the history of weather and the history of climate, monthly temperature and precipitation patterns for the pre-instrumental period have been estimated from the indices. For every calendar month and for every season a model has been set up which, for the period of network observation from 1864, compares the weighted and unweighted indices (computed from the record) with the record itself. This has yielded a set of transfer functions for estimating temperature and precipitation patterns on the basis of the indices.

For temperatures, the deviations of the Basel series (1755-1979) from the 1901-60 average were included in the regression model . Precipitation series covering the whole country for the entire period, were not available in sufficient number. Instead the analysis was done by means of six series from the Swiss 'Mittelland' – Geneva, Bern, Zürich, Basel, Einsiedeln and St. Gallen – where precipitation has been measured from 1864 to 1979. It turned out that the models yielded excessively large standard errors for individual months. Thus estimates were attempted for decennial averages only. In interpreting the standard errors for the decennial averages (Table 6.4) we should bear in mind, that they are obtained from the transformation of measured data. For the pre-instrumental period, the 'true' standard deviations are certainly larger for the weighted indices, because the nature of the evidence and the process of interpretation involves additional biases. On the other hand, they are probably smaller than those obtained for the unweighted indices (Table 6.4), because the evidence allows for more than just simply distinguishing between 'warm', 'cold' and 'average' months (provided that we ignore the period prior to 1550 and the data for autumn prior to the late seventeenth century).

6.9 Patterns of temperature and precipitation since the time of the Reformation

The results of the reconstruction will only briefly be discussed. Figure 6.4 shows deviations from the means of the 1901-1960 reference period in form of 11 year moving averages. Positive temperature deviations are shown above the line, negative deviations below. Precipitation is shaded and shown in the same way as temperature.

Considering the long time-scale temperature and precipitation patterns, no clear pattern emerges for the four seasons and the year. A 'Little Ice Age', that might be associated with the known fluctuations of glaciers, is hard to detect. In contrast to the twentieth century three features are common to the preceding centuries :

1   Winter and spring months tended to be colder and drier. This holds especially for March.
2   The climate was more variable – in particular around 1600 – and in many cases the extremes were more marked than those registered in the instrumental period.
3   Fluctuations of the same type occurred repeatedly and often simultaneously in winter, spring and summer (e.g. 1569-1574, 1586-1589, 1688-1694, 1769-71).

However, summer periods were not significantly colder in the 'Little Ice Age' than in the present century, although it is well-known that glacier fluctuations are generally related to temperature in summer. A tendency towards increased precipitation is apparent, in particular for the late 16th century and for the 18th century. Quite often a warm and dry period in August followed after a rainy and cool spell in June and July.

Temperature patterns in the autumns are rather balanced up to 1670, and this season was rather dry over this period. Up to the early twentieth century autumns were colder than since and they included both wet and dry phases.

If changes over time are considered, a marked contrast stands out between the second third and the last third of the 16th century. Apart from the winters, the weather conditions that predominated from 1530 to 1560 proved nearly as favorable as those that have prevailed during the climatic optimum of the twentieth century. Over the following decades, then, annual mean temperatures declined by more than 0.6°C. Summers became about 0.8°C colder and more than 20% wetter. A marked increase in summer wetness was also observed in northern Germany (Lenke 1968). By the 1580s the broad Denmark Strait between Iceland and Greenland was often found entirely blocked by pack-ice during the summer (Lamb 1982). In Switzerland the frequency of severe floods in the last third of the 16th century

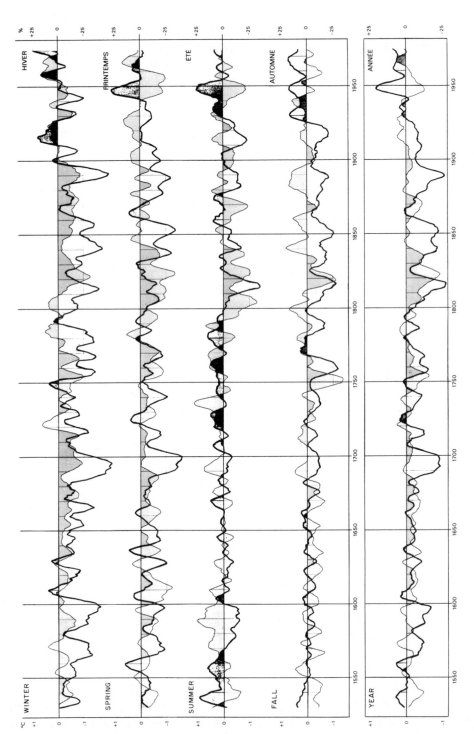

Figure 6.4 Temperature and precipitation estimates for Switzerland (11 year moving averages) expressed as departures from the mean for 1901-60.

increased fourfold compared with the second third of the century. Some alpine glaciers advanced several hundred meters within the span of a few years. Also, the variability of the climate became more marked during the last third of the 16th century.

Drought was the most noticeable characteristic of the 17th century. Wet summers still occurred, however, particularly during 1600-1630, the 1670s and 1685-1699. Summers were moderate from 1645 to 1684, and mean temperatures were comparable to those witnessed in the present century. If winters are disregarded, the amplitude of the temperature anomalies between 1630 and 1688 was rather small. On the whole, the seventeenth century was not as cold as Le Roy Ladurie (1971) suggests from wine harvest dates and evidence of glacier fluctuations.

After the year 1685 temperatures dropped sharply in all seasons. From 1690 to 1699 mean annual temperatures were almost 1.0°C lower than in the 1901-1960 reference period, in Central Europe. In Central England they were 0.8°C lower during this period (Manley 1974). Based upon reports of sea ice and the cod fishery, Lamb (1982) concludes that the ocean surface temperature between Iceland and the Faroe Islands was probably 5°C lower than it is today (see also Chapter 5, this volume). This suggests that Arctic cold water had spread far south to the Norwegian Sea. In the Swiss mountains the vegetative period was noticeably shortened over this period.

During the first two decades of the 18th century, mean seasonal temperatures rose, except for winters. For the decade of the 1720s mean annual temperatures were at the level of the twentieth century. Springs and autumns became cooler in the 1730s. In summer, warmth persisted and precipitation increased between 1760 and 1790.

The 19th century proved to be the coldest hundred-year period since 1500. This holds true for all seasons, in particular for autumn and it is also revealed from the known extreme cold fluctuation from 1812 to 1817. In the second half of the century warm and dry summers prevailed from 1855 to 1875. Spring temperatures rose above the 20th century means in the 1860s. In winter, temperature and precipitation trends became positive from the end of the 19th century. In the last few decades this season has changed dramatically. Compared to the long term average of the past half millennium this season has been 1.3°C warmer and 25% wetter since 1965. The recent succession of three warm winters (1987-88 to 1989-90) with almost no snow-cover in the lowlands is unique in the last seven hundred years (Pfister 1990).

6.10   Comparing temperature trends in Central England and in Switzerland since 1659

Lauterburg (1990) has examined the spatial variability of climatic change in Europe over the period 1780-1960. Using cluster analysis (Ward's method) he defined areas of similar year-to-year fluctuations of climatic parameters for the three sub-periods 1781-1840, 1841-1900 and 1901-1960. The regions that emerge from the procedure turn out to be relatively stable over time from one sub-period to another for spring and summer, whereas they change somewhat for autumn and winter.

Based upon the Swiss temperature indices and the Central England temperature series (Manley 1974) this comparison was extended back to 1659. From this year the Manley series begins, and monthly weather in Switzerland is described continuously, i.e. the series of temperature indices doesn't include any missing months. The entire period of three hundred years was split into three sub-periods of approximately equal length for the analysis. In the first one (1659-1754) the Swiss indices are based upon descriptive and proxy data. The second period (1755-1863) is covered by instrumental measurements. The third period (1864-1960) is that of the detailed observational network.

The Central England series was transformed into indices to adjust it to the level of measurement of the Swiss series (cf. Table 6.5). The Gamma coefficient (which is a symmetrical measure for association of two ordinal variables) was used for the correlation, because of the Pearson correlation coefficient is not appropriate for ordinal data. Gamma can achieve limiting values of -1.0 or + 1.0 regardless of the number of ties (Loether and McTavish 1974).

All coefficients between the two series are highly significant, even those for the first period, which is not covered by instrumental measurement (Table 6.5). This confirms the reliability of the Swiss data. However, the seasonal and yearly correlations (autumn excepted) are somewhat lower for the first sub-period 1659-1754. This is almost certainly due to the semi-quantitative evidence on which the Swiss indices are based. On the other hand it might also be due to changes in the frequency of weather patterns to some extent. Considering correlations of monthly variables, the difference between the first sub-period and the two others is particularly large for July and August. This might be related to the fact that phenological indicators for summer temperatures are more related to June than to July and August (tree-ring densities excepted).

Table 6.5 Gamma correlation coefficients between the Central England temperature series and the Swiss temperature series 1659-1960.

Month	1659-1980	1755-1754	1864-1863	1959-1960
January	0.669	0.594	0.734	0.668
February	0.645	0.578	0.648	0.677
March	0.612	0.544	0.667	0.653
April	0.667	0.619	0.678	0.678
May	0.609	0.501	0.682	0.635
June	0.471	0.452	0.490	0.540
July	0.658	0.467	0.664	0.786
August	0.540	0.396	0.560	0.619
September	0.571	0.502	0.650	0.600
October	0.498	0.483	0.478	0.566
November	0.461	0.467	0.391	0.510
December	0.548	0.588	0.564	0.589
Winter(12-2)	0.552	0.509	0.582	0.564
Spring	0.544	0.496	0.598	0.588
Summer	0.461	0.374	0.487	0.530
Autumn	0.438	0.437	0.436	0.455
Year	0.484	0.431	0.554	0.499

Significance: < 0.0001 for all pairs of variables

Surprisingly, correlations for October, November and December are somewhat higher in the first sub-period, when the Swiss series is based on semi-quantitative data, compared to the second one, when it is based on measured temperature. This might be related to prevailing weather patterns that differ somewhat from those prevailing in the preceding and the following period.

6.11 Beyond time series analysis: establishing historical weather maps

Most reconstructions of climate from natural or man-made archives are presented in the form of time series. The climatic variations shown refer to areas, that have the dimension of a German Bundesland, an Italian Province or a small nation such as Switzerland. The focus is primarily on improving the time resolution of the findings down to seasons and month and on interpreting the results by the means of statistical techniques that become more and more sophisticated.

Compared to time-series analysis the investigation of the spatial dimension of climatic change has been neglected so far. It has not been sufficiently realised that regional time series are not isolated pieces of evidence, which may be interpreted for their own sake, but that they should be related to a larger entity which is the global climate system. A comparison cannot be done just by comparing fluctuations in time series as it has been done so far. In order to provide a coherent picture of climatic change in space and over time we need to focus upon

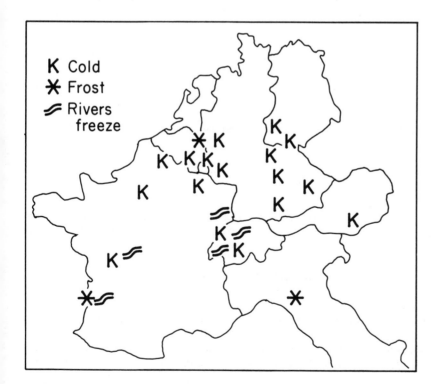

Figure 6.5   Weather map of the cold winter 1407-08 in Europe. This winter was similar to that of 1962-63. (Schwarz-Zanetti and Pfister, in preparation).

the analysis of this system. All reconstructions of climatic change that have already been obtained at a regional scale need to be compared on a common basis. This requires a common standard of representation for scholars from all over Europe. On this continental level, analysis of climatic change should be represented in terms of historical weather maps, that show the spatial dimension of outstanding anomalies (Figure 6.5) and trends related to coherent regions (Lauterburg 1990).

In order to initiate international cooperation directed towards this goal, a workshop was held recently by the European Science Foundation (E.S.F.) at the Academy in Mainz. Scholars from 14 European countries and from Japan discussed the standardization of historical climatology and the creation of an international data-bank for the history of climate which might become the basis for a new spatial image of climatic change (Frenzel and Pfister, in preparation).

References

Alexandre, P., 1987: *Le Climat en Europe au Moyen-Age. Contribution à l'histoire des variations climatiques de 1000 1425, d'après les sources narratives de l'Europe occidentale.* Paris Ed. de l'Ecole des Hautes Etudes en Sciences Sociales.

Bell, W. and Ogilvie, A. E., 1978: Weather Compilations as a Source of Data for the Reconstruction of European Climate during the Medieval Period. *Climatic Change*, 1 (4) 331-348.

Flinn, M., 1981: *The European Demographic System 1500-1820.* Baltimore Md.

Flohn, H., 1949: Klima und Witterungsablauf in Zürich im 16. Jahrhundert. In: *Vierteljahrsschrift der Naturforschenden Gesellschaft.* Zürich, 94 (1) 28-41.

Flohn, H. 1985: A critical assessment of proxy data for climatic reconstruction. In: *The Climatic Scene*, ed. by M. J. Tooley & G. M. Sheail. Allen & Unwin, London, p.93-103.

Georgelin, J., 1979: L'écologie du froment en Europe occidentale. *Cahiers des Etudes rurales* IV. S. 569-582.

Hanus, H. and Aimiller, O., 1978: Ertragsvorhersage aus Witterungsdaten. *Beiheft z. Zeitschrift für Acker- und Pflanzenbau* Jg. 5, Berlin.

Ingram, M., Underhill, D., Farmer, G., 1981: The Use of Documentary Sources for the Study of Past Climate. In: *Climate and History*, ed. T. M. L. Wigley *et al.*, Cambridge University Press, Cambridge, p. 180-213.

Kington, J. A., 1974: An application of phenological data to historical climatology. *Weather*, 29 (9), 320-328.

Klemm, F., 1974: Die Entwicklung der meteorologischen Beobachtungen in der Schweiz bis zum Jahre 1700. In: *Vierteljahrsschrift der Naturforschenden Gesellschaft* Zürich, 119 (4), 405-454.

Lamb, H. H., 1981: An approach to the study of the development of climate and its impact in human affairs. In: *Climate and History*, ed. T. M. L. Wigley *et al.*, Cambridge University Press, Cambridge, p. 291-310.

Lamb, H. H., 1982: *Climate, history and the modern world.* Methuen, London.

Lauer, W. and Frankenberg, P., 1986: *Zur Rekonstruktion des Klimas im Bereich der Rheinpfalz seit*

Mitte des 16. Jahrhunderts mit Hilfe von Zeitreihen der Weinquantität und Weinqualität. Stuttgart (Fischer).

Lauterburg, A., 1990: Klimaschwankungen in Europa. Raum-zeitliche Untersuchungen in der Periode 1841-1960. *Geographica Bernensia* G 35. Bern Geographisches Institut der Universität, Bern.

Legrand, J. P., 1979: L'expression de la vigne au travers du climat depuis le Moyen-Age. In: *Revue Française d'Oenologie*, 75, 23-50.

Lenke, W., 1968: Das Klima Ende des 16. und Anfang des 17. Jahrhunderts nach Beobachtungen von Tycho de Brahe auf Hven im Sund DK, Leopold III. Treuttwein in Fürstenfeld Oberbayern und David Fabricius in Ostfriesland. *Berichte des dt. Wetterdienstes*, 15. Offenbach.

Le Roy Ladurie, E. 1971: Times of Feast, Times of Famine. A history of climate since the year 1000. London (Allen & Unwin). (Translation of the 1967 French edition: L'histoire du climat depuis l'an mil. Paris).

Le Roy Ladurie, E. and Baulant, M. 1980: Grape Harvests from the fifteenth through the nineteenth Centuries. In: *J. of Interdisciplinary History* 10 (4) 839-849.

Loether, H. J. and McTavish D. G., 1974: *Descriptive Statistics for Sociologists. An introduction.* Allyn and Bacon, Boston.

Manley, G., 1974: Central England temperatures: monthly means 1659 to 1973. *Quarterly J. Royal Meteorological Society* 100, 389-405.

Pfister, C., 1977: Zum Klima des Raumes Zürich im späten 17. und frühen 18. Jahrhundert. In: *Vierteljahrsschrift der Naturforschenden Gesellschaft Zürich*, 122 (4) 447-471.

Pfister, C., 1979: Getreide-Erntebeginn und Frühsommertemperaturen im schweizerischen Mittelland seit dem frühen 17. Jahrhundert. *Geographica Helvetica*, 34, 23-35.

Pfister, C., 1980: The Little Ice Age: Thermal and Wetness Indices for Central Europe. *Journal of Interdisciplinary History*, 10 (4) 665-696.

Pfister, C., 1981a: An analysis of the Little Ice Age climate in Switzerland and its consequences for agricultural production. In: *Climate and History*, ed. T. M. L. Wigley *et al.*, Cambridge University Press, Cambridge, p.214-247.

Pfister, C., 1981b: Die Fluktuationen der Weinmosterträge im Schweizerischen Weinland vom 16. bis ins frühe 19. Jahrhundert. Klimatische Ursachen und sozio-ökonomische Bedeutung. *Schweiz Zeitschrift für Geschichte*, 31 (4) 445-491.

Pfister, C., 1984: *Klimageschichte der Schweiz 1525-1860. Das Klima der Schweiz von 1525-1860 und seine Bedeutung in der Geschichte von Bevölkerung und Landwirtschaft.* 2 volp. Bern. 3rd. ed. 1988.

Pfister, C. 1985a: *CLIMHIST – a weather data bank for Central Europe 1525 to 1863.* May be ordered from METEOTEST, Fabrikstr. 29a, CH 3012 Berne.

Pfister, C. 1985b: Snow Cover, snow lines and glaciers in Central Europe since the 16th century. In: *The Climatic Scene*, ed. by M. J. Tooley and G. M. Sheail. Allen and Unwin, London, 154-174.

Pfister, C., 1985c: Veränderungen der Sommerwitterng im südlichen Mitteleuropa von 1270-1400 als Auftakt zum Gletscherhochstand der Neuzeit. *Geographica Helvetica* 4, 186-194.

Pfister, C. 1988: Variations in the Spring-Summer Climate of Central Europe from the High Middle Ages to 1850. In: Wanner, H. and Siegenthaler, U. (eds.) *Long and Short-term variability of Climate*, Springer, Berlin p. 57-82.

Pfister, C. 1990: Einmalig in 700 Jahren: Drei warme Winter hintereinander. Unser Wetter im Lichte der historischen Klimaforschung. *Neue Zürcher Zeitung* 82, 7(8). April, p. 23-4. Pfister, C., Schwarz-Zanetti W. und G., 1986: Ein Programm- und Methodenpaket zur Rekonstruktion von Klimaverhältnissen seit dem Hochmittelalter. In: *Datenbanken und Datenverwaltungssysteme als Werkzeuge historischer Forschung*, Historisch-sozialwissenschaftliche Forschungen, ed. Bd. 20, p. 75-92.

Schüepp, M., 1980: Methoden und Probleme der Bearbeitung langjähriger meteorologischer Beobachtungsreihen. In: *Das Klima, Analysen und Modelle, Geschichte und Zukunft*, ed. H. Oeschger, B. Messerli, M. Svilar. Berlin, p. 191-206.

Schweingruber, F., *et al.*, 1978: The X-Ray Technique as Applied to Dendroclimatology. *Tree-Ring Bulletin*, 38, 61-91.

Schweingruber, F., 1983: *Der Jahrring. Standort, Methodik, Zeit und Klima in der Dendrochronologie.* Bern (Haupt).

Witmer, U., 1986: Erfassung, Bearbeitung und Kartierung von Schneedaten in der Schweiz. *Geographica Bernensia* G 25. Bern.

7 Reconstructing the climate of northern Italy from archive sources

D. Camuffo and S. Enzi

7.1 Introduction

Starting from the second half of the 1800s, much information on historic changes in Italian and European climate has been published. Some publications, however, are not wholly reliable and inexact dating has resulted in much confusion. For example, Toaldo (1781) through a reading error changed an X into II, transforming the year MDXI into MDIII, said that the artillery of Pope Giulio II crossed the frozen River Po (instead of the protected moat of the castle). The fact was then propagated by Easton (1928) as an independent event, in addition to the correctly reported event of 1511. The event thus created seemed to have all the characteristics of a 'Great Winter' (much worse than it actually was) because the frozen river supported the crossing troops.

The Italian archives supply a large quantity of information for the last millennium (Roccatagliata 1976; Veggiani 1986; Pavese and Gregori 1985; Camuffo 1987, 1990) and a smaller amount of data for almost another millennium and a half. At present, research is underway (some 200 chronicles have been analyzed) in a joint programme between the Commission of the European Communities and the National Research Council, in order to reconstruct the climate of the last millennium. Information about both 'average' climate and 'extreme' meteorological situations is being collected, validated and analyzed. The type of information depends on the sources, for example, diaries may furnish daily data whereas letters or reports may give desultory information. Chronicles and annals give a good overall view of climatological features and hazards. Extreme events are described by several sources, while normal weather conditions are mainly found in diaries.

The aim of this chapter is to discuss the period beginning with the onset of the Little Ice Age.

7.2 Comments on sources

After the Medieval period and the age of Municipalities, the Renaissance Humanism in the 15th and 16th centuries rediscovered and revalued the classic and local antiquities. Essays appeared that 'reconstructed' local history, even going as far back as 'the origins of the world' or to that of Rome or the City under consideration by the chronicler. Existing chronicles were enhanced and brought up to date using original manuscripts, many of which have been subsequently lost. Such works were of great importance in the transmission of ancient data. The erudite spirit of the time caused the authors to pay particular attention to the historical reconstruction, in view of the extremely complex dating systems then in existence. However, many errors remained or were introduced due to the change in the type of calendar used, the latter often associated with the name of the Consul, the Podestà or the local governor.

As well as the scholarly essays of the Renaissance type, the 17th century saw the beginnings of work of a scientific and experimental nature arising from direct investigation, of which Galilei was one of the first exponents. This new spirit was then spread by the establishment of the Academies, the first of which was the Lincei Academy in Rome (1603-1630) of which Galilei was a member. The Cimento Academy was next, founded by Leopoldo de' Medici. In collaboration with this Academy, the Grand Duke of Tuscany, Ferdinando II de' Medici established, between the end of 1654 and the beginning of 1655, the first network of meteorological observations in Italy and at other sites in Europe i.e. Florence, Vallombrosa, Cutigliano, Bologna, Parma, Milan, Innsbruck, Osnabruck, Paris and Warsaw. Only part of these data, which were sent back regularly to Florence, has survived the vissicitudes of time (Cantù 1985). Some of the original instruments can still be seen and have recently been calibrated against modern standards (Vittori and Mestitz 1981). The seventeenth century, therefore, was the beginning of many of the instrumental series taken in Italy (Cantù and Narducci 1967). Some are of extreme interest in that they document the Little Ice Age, even though some of them only lasted a few years.

By the 18th century the scientific spirit was well established and many instrumental time series were initiated. Details of the instrumentation, the calibration and measurement techniques have been documented. The Academy of Science was founded at Bologna and in 1716 began taking a series of instrumental measurements in the city (Comani 1987). At Padua, towards the end of 1724, Poleni initiated the series that was then continued by other scholars at the University (Camuffo 1984). These were then followed by series of measurements taken in Milan (1763) and Parma (1778) (Menella 1956; Santomauro 1957). In Rome, the Collegio Romano began a new series of meteorological observations in 1782, one year after the foundation of the Societas Meteorologica Palatina in Mannheim, Germany (Trevisan 1980; Colacino et al. 1983; 1986). Many academies, especially in cities with universities, studied meteorological conditions, because it was believed that these were responsible for many illnesses and epidemics. Many of the observations were made by doctors, for example, those recorded by Morgagni at Padua.

It was also believed that meteorological phenomena were induced by astronomical factors. The most important exponent of this opinion was Toaldo, and his *Giornale Astrometeorologico* was widely read throughout Europe. This theory originated from the Caldei through the writings of Aristotle (*Meteorologica*) and Ptolemy (*Tetrabiblos*). This confusion was, in part, caused both by the study of Aristotle's treatise at university courses on astronomy and meteorology and by the diatribe on the origins of comets and the frequent polar aurorae which had a great effect on public opinion.

A negative aspect, introduced on the recommendation of the Royal Society, London (Jurin 1723) was the practice of taking temperature readings inside a north-facing room, preferably in which a fire was never lit. This measuring technique, in which the thermic diffusion of the walls behaved like a low-pass filter, caused the loss or the attenuation of the extreme temperatures and the data were affected by internal heat sources, wind cooling etc. Fortunately, Poleni sometimes disobeyed this rule (Poleni 1731) and took measurements outdoors. All the places at which the measurements were taken at his house (which is still standing) were noted.

In 1836 the first routine meteorological network composed of many stations was established. The network increased further after the unification of Italy (1861) and the data were sent to a Forecast Office which was, at first, in Florence and later (1876) in Rome. The wealth of instrumental data meant that notes by chroniclers and diarists came to be of only passing interest, except when special phenomena occurred.

Historiography was, by now, a mature science; critical revisions and reprinting of many codes and other medieval sources were brought together into large collections, noted and commented upon: *Rerum Italicarum Scriptores*; *Fonti per la Storia d'Italia*; *Regia Deputazione Storia Patria*; *Monumenta Historiae Patriae*; *Archivio Storico Italiano*; *Monumenta Germaniae Historica*. However, only a part of the enormous archival heritage has been reprinted.

7.3   Validation of data

Each period should be considered individually, as the historical and scientific interests and aims changed with time. The number of Italian chronicles is impressive; in order to give some idea of the quantity for the Middle Ages, 35% of all documents that Alexandre (1987) found in Europe were taken from northern and central Italy.

Dating styles are not homogeneous and often very complex, but usually there are no problems in recognizing the correct date. Sometimes local chronicles may leave doubts, especially when dealing with winters when it is often difficult to decide in which of two consecutive years the event occurred. Making an exact critical evaluation of the reliability of the data requires careful analysis (Alexandre 1987; Bell and Ogilvie 1978; Ingram *et al.* 1978, 1981a,b; Pfister 1981). The large quantity of manuscripts generally makes cross comparison of the sources possible, thus verifying the accuracy of the data and the reliability of each author. The validation of data is very important and is based both on a historical-philological and a physical point of view. The physical problem of how representative a report may be, arises in that different weight is given to descriptions which may have been conditioned by non-climatic factors.

A critical examination of the chronicles, comparing one with another, helps in the recognition of any eventual exaggeration, especially when individual events may appear somewhat doubtful. Sometimes, the sequence of daily information which in itself may appear insignificant, such as 'rain today' or 'drizzle today' can, if taken together, supply a wider general picture, so that a comparison can be made between the monthly precipitation of that time and the present day. In this sense, daily news such as that given by Sanudo (16th century) or Gennari (18th century) have a rather different meaning, in that they give an exhaustive picture of a certain period.

The importance of non-climatic factors is often difficult to discern; factors such as the abandonment of land in time of war, enemy raids, systematic felling of trees during a siege or for the construction of fleets, reduction in manual labour during or after an epidemic, indiscriminate pasturing or the invasion of locusts may produce effects similar to those which may result from climatic changes, and are not easily checked. All of these events may, however, contribute to the desertification of an area when its hydrological balance is delicate, or when the type of cultivation was unsuitable for the ecosystem.

An example of meteorological phenomena which may depend on the state of the land are tornadoes which occur in the Venice area in late summer. In this season the plain in the lee of the Alps remains warmer than the surrounding regions, being shielded from the cold air masses coming from the north. When the Azores anticyclone becomes weaker, there are invasions of cold air through the alpine passes, resulting in marked instability with the formation of tornadoes which travel along the frontal line. The distribution of vegetation and marshes may locally affect the frontal contrast and hence the severity and path of these tornadoes.

In other cases technology was important. Storms may have been recorded objectively in terms of their occurrence, but reports on the resulting damage were less reliable due to the state of the building techniques of the time. A similar problem arises with roofs collapsing due to heavy snow. In the case of sea storms the effects were conditioned by the state of vessels and harbours. Famine was highly affected by storage ability and transport. Drought was compounded by the limited irrigation technology, and fires, by the use of inflammable materials.

Any event must also be verified objectively by taking into account psychological conditioning due to the culture and challenges of the time. For example, comets were regarded as being unlucky and their passage, of necessity, was followed by negative events so meteorological hazards may have been exaggerated. That climate greatly affected man is witnessed by the copious documentation regarding events which were not severe but, all the same, difficult to endure. In the case of events not influenced by anthropogenic factors, it is also possible to evaluate the intensity of the forcing climatic factors on the basis of their effects and these may be classified on a scale of severity.

Certainly, each epoch was characterized by different climatic vicissitudes but their impact on the population depended on various factors. This would have led to different concepts of hardships due to the weather, which were cited with greater accuracy and prominence in particular periods, so that it is not always easy to compare the severity of the climatic conditions, unless they were associated with specific effects.

Moreover, even the effects may depend on factors which vary in time. Floods for example, depend on: the quantity of rain and the rate at which it falls; the extent of forests, undergrowth and crops which can attenuate the runoff; the state of the river embankments (often rudimentary, sometimes consisting only of sacks filled with earth) or other interventions by man; the depth of the river-bed with respect to the altitude of the countryside that may range from a pluvial to a dry regime (Veggiani 1986). The floods of the past must be evaluated differently from the present day ones and it is very difficult to compare accurately past and modern events. The problem also occurs when landslides are considered.

Besides events which have been influenced landscape changes over time, there are others which have been less affected. Among these, those less susceptible to outside influences should be looked for, and used as reliable indicators. Hot summers, great winters, freezing of rivers, severe sea floods in Venice, and to a lesser extent the frequency of rainfall, can all be considered objective data of undoubted interest. In the case of the freezing of rivers, it is possible to establish a scale in terms of severity from the weight which could be supported, such as the passage of small animals, people, men on horseback, carriages drawn by oxen, troops with cannons, etc.

146

7.4   Some climatic features of the Little Ice Age

7.4.1   Locusts

Invasions of locusts (mainly *Locusta migratoria*) in northern Italy may be related to some climatic factors although their appearance was also determined by the ecosystem and the means of destroying them (Camuffo and Enzi 1990). From the archive sources it is evident that invasions occurred in the warm season (maximum frequency in August) when the wind blows from the northeast. It has been possible to ascertain that locusts were periodically transported by the sirocco from the Near-East along the Danube to the Pannonian Plain (Hungary). Once this region was infested the bora wind then transported swarms towards Italy. These invasions virtually ended when Hungary was more intensively cultivated, making this region unsuitable for locust reproduction. Invasions were particularly frequent in 14th century and then in the 16th and 17th, in the heart of the Little Ice Age. This shows that, for locust invasions, wind circulation was more important than temperature. In addition, it can be deduced that the northeast wind (which now presents its minimum frequency between late spring and early summer) was fresh and lasted for some days, thus allowing for the transport of swarms which may be airborne only a few hours a day. In fact, they come to land either during the heat of the day or at night, when air temperature is too high or too low.

7.4.2   Freezing of the Venetian lagoon

The analysis of the freezing of the Venetian lagoon (Camuffo 1987) raises problems in terms of data representativity; the enormous hydraulic works carried out from the 14th century onwards, progressively reduced the amount of water entering from the open sea. The critical conditions for the freezing-over of the lagoon changed accordingly, making it difficult to compare the severity of the events.

7.4.3   Sea flooding at Venice

Sea floods due to exceptional tides at Venice (the well known 'high water') are better climatic indicators. Flood tides depend on the combined action of various factors. First, when the sirocco blows along the Adriatic sea from southeast, the sea level rises in the northern part of the basin, due to the mass transport and at the same time, for dynamic reasons (Camuffo 1981) a 'dark bora' wind is generated from northeast and this causes further transport of sea water towards Venice. Many chronicles report that the high water followed a fresh sirocco and heavy rainfall, which is associated with the 'dark bora'. Second, the Mediterranean sea is substantially closed, and a non-homogeneous distribution of atmospheric pressure over the sea surface causes variations in sea level which are dependent on the atmospheric forcing. A variation of 1 hPa equals nearly 1 cm change in water level. Third, rapid changes in the atmospheric pressure pattern cause free oscillations in the Adriatic sea (*sessae*) which have a period of about 24h. This is the same order of the tide, and lasts for several days. Fourth, when moon, earth and sun lie along the same line (syzygy) the tide is maximum. Especially during winter, the presence of depressions over the Mediterranean, or the formation of low pressure in the lee of the Alps enhanced by air-sea interactions (cyclogenesis) are generally

147

associated with the sirocco and formation of *sessae*. Although high water in Venice results from several factors, it can be considered as an index of the persistence of depressions generating sirocco. Flood tides were documented in several archive sources with instrumental records beginning in 1867. Floods typically occur between October and December with a marked maximum in November, when the Mediterranean is warmer than the air masses and the release of heat and moisture from the sea deepens the depressions. They are quite rare in spring and summer.

The occurrence of extreme events may be represented with marks on the time axis, but this monodimensional representation is not very convenient to show trends. A two-dimensional graph may be obtained plotting along the ordinate the number of events which occurred over a certain time interval. Histograms are subject to a principle of indetermination: in order to contain on average more than one event, the time interval should cover several years, and the wider the interval, the lower the resolution. Another possibility is to plot the cumulative frequency against time; the slope of the graph shows the actual distribution of frequency, slope constancy means persistence of a given frequency and periods with particular trends are easily identified. The distribution of the frequency of occurrence of abnormally high tides in Venice and their trend in the past are clearly shown in Figure 7.1. The frequency was almost constant except for a few periods: in the first half of the 16th century (0.44 per year); between 1740 and 1800 (0.35 per year) and after 1920. The excavation in 1920 of a new deep channel for tankers changed the hydraulic equilibrium due to the greater exchange of sea water with the lagoon. For this reason we terminated the series with the famous flood in 1966.

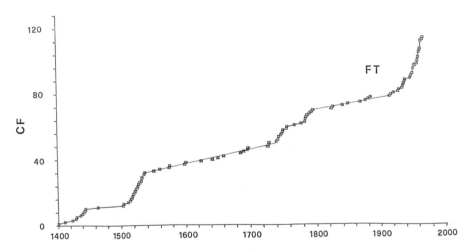

Figure 7.1 Cumulative frequency (CF) of flood tides (FT) at Venice.

7.4.4 Thunderstorms and hailstorms

The effect of thunderstorms and hailstorms depends not only on meteorological factors, but also on the state of the territory. Small differences in interannual frequency or weak trends are in no way significant. However, there is no doubt that these events occurred very frequently in the 15-16th centuries and especially in the 18th. Thunderstorms were associated with synoptic instability which was, among other things, responsible for the low temperatures typical of the period. Some of Gennari's notes are particularly interesting. He was an accurate observer of weather, and he reported in his diary that he does not remember any rainfall without thunder and that such storms were frequent also in winter. He also wrote that at the beginning of the 18th century, when he was a boy, after any hailstorms people used to go to offer their condolences to the owners of crops. In the second half of the century, hail was so frequent that this habit was discontinued. This is a very good indication of the instability which characterized both the beginning and the end of the Little Ice Age. The same can be demonstrated from the variability in teleconnections of rainfall in northern Italy in the last two centuries, given by long instrumental series (Camuffo 1984). This also appears from the series of droughts in archival sources; several droughts are well documented in certain periods by reliable authors, but their extension was very limited because they are not mentioned by chroniclers living in other cities.

7.4.5 Winters

Winters were particularly important in the Little Ice Age and were more objective indicators. They were classified on the basis of their effects. A 'great winter' was defined when the cold was particularly severe and lasted for a relatively long time over a large area, i.e. when reliable and widespread descriptions of large water bodies (e.g. great rivers, Venetian lagoon) froze over and ice supporting people were found. Often, this information was also associated with the mention of wine frozen in butts and other effects. A 'severe winter' was defined when the cold was severe but lasted only a few days, causing the death of animals or plants but not the complete freezing of large water bodies. Winters claimed by witnesses to be 'very cold', 'severe', or 'very severe' without citing particular effects, were classified as 'cold', by taking into account the possible subjective impression, especially in the case of cooling due to wind. Mention of normal, mild or warm winters can also be found.

The cumulative frequency of great (GW) and severe winters (SW) and GW alone are reported in Figure 7.2. The total number of very cold winters in the 15th century was very high, 8 GW and 9 SW. This was an impressive climatic change in comparison with the preceding century, when the frequency of cold winters was very low, 1 GW and 2 SW. The occurrence of GW and SW had a marked increase of frequency during two periods, 1450-1514 and 1570-1614. A milder interval occurred in the first half of the 17th century, as already noted (Camuffo 1987).

This series ends in 1800, which is the upper limit of our research into archive sources except for some phenomena, such as the freezing of large water bodies, which define the GW. It was considered useless to extend the analysis of manuscripts to a period covered by instrumental measurements. A comparison of SW deduced from effects on the ecosystem and SW deduced from temperature alone was not very satisfactory, as these two series are not homogeneous. For ecosystem effects other factors, such as the wind speed and the number of consecutive days below zero are important. Further work is necessary to compare the two series obtained with different criteria.

Great winters have had an almost regular frequency distribution except for the end of the 15th century. In Italy the cold of the Little Ice Age began in the mid-15th century, with the second half of that century even colder. Lamb (1969, p.185) suggested an earlier date (1430)

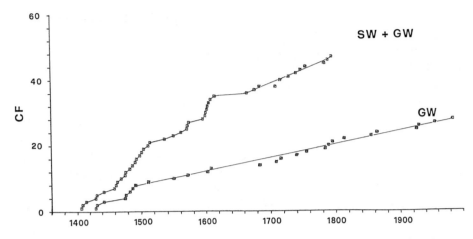

Figure 7.2   Cumulative frequency (CF) of severe and great winters (SW+GW) and great winters alone (GW) in northern Italy.

for the onset of the cold in western Europe. Surprisingly, after the onset of the Little Ice Age, the frequency of the GW was practically unchanged.

Table 7.1 shows that there is only partial agreement between the results of this work and the Lamb (1977) and Buisman (1984) classifications. Only 7 winters (1571, 1595, 1608, 1709, 1740, 1789, 1795) were universally classified as 'great' or 'very severe', and 5 others (1432, 1443, 1511, 1684, 1784) 'very severe'. Lake Bodensee was frozen over only 4 times in such winters. Three of our severe winters were classified 'mild' (1406, 1430, 1605) by Buisman

Table 7.1   Great and severe winters in northern Italy, compared with other works. Winters dated by January.

GW	Great Winter
SW	Severe Winter
WBF	Water bodies frozen
PL	Freezing of plants
O	Others
NS	Number of different sources
L	Lamb's (1977) classes for western Europe
B	Buisman's (1984) classes for western Europe
LB	Bodensee frozen over from Lamb (1977)
P	Pfister's (1985) temperature index for Switzerland

Other codes: see Table footnotes.

YEAR	GW	SW	WBF	PL	O	NS	L	B	LB	P
1406		+				1		m		
1408		+	+			1		vs		
1414		+				1				
1430	+		+			1		m		
1432	+		+			3	s	vs		
1443	+		+	+		6	s	vs		
1459		+			a	2				
1462		+		+	c	2		s		

Year										
1464		+		+		1				
1470		+	+			2		s		
1476	+		+			1		s		
1477	+		+			3		vs		
1483	+		+			1		c		
1487	+		+	+		3		c		
1491	+		+	+		7	s	s		
1493		+	+			2				
1498		+	+			1		c		
1501		+	+		e	2		c		
1504		+			b	2		c		
1511	+		+			2	s	vs		
1514(?)		+	+			3				
1536		+			e	1		s		-1
1549	+		+	+		4		s		0*
1561		+	+			2		s		-3
1570		+	+			2		s		-1
1571	+		+			1	vs	s	+	-3
1573		+	+	+	b	2	s	s	+	-3
1595		+	+	+		2	vs	vs		-3
1599		+		+	ab	3		c		-1
1600		+				2		s		-3
1602		+				1		c		-1
1603	+		+	+		3		c	+	-3
1605		+	+	+	d	2		m	+	0
1608	+		+		c	4	g	vs	+	-3
1614		+		+		1				-2
1665		+	+			1	s	s		-2
1677		+				1	s	s		-3
1684	+		+	+	ab	2	s	vs	+	-3
1709	+		+	+		7	g	vs		-3
1716	+		+			2	s	s		-3
1729		+				1	s	s		-2
1740	+		+			3	vs	vs	+	-3
1747		+	+			1				-1
1755	+		+			4		c		-3
1784	+		+			2	s	vs		-3
1789	+		+			2	vs	vs		-3
1795	+					2	vs	vs		-3
1814	+		+			s		c		-3
1855	+		+			c	vs	vs		-3
1864	+		+			f				
1926	+		+			f		c		
1929	+		+			f	c	s		
1956	+		+			f	c	s		
1985	+		+			f		s		

m = mild; c = cold; s = severe; vs = very severe (in Buisman 1984 this is the maximum degree of severity); g = great winter (Lamb only); w = warm.

The Pfister classes are: -3 (very cold); -2 (cold); -1 (subnormal); 0 (average); positive values are for mild months.

a = death of people; b = death of animals; c = mills locked with ice; d = wine frozen in butts; e = glaze; f = the years after 1800 refer to freezings of the Venice Lagoon, after Camuffo (1987).
* Pfister (1985 p. 59) notes for this year: 'Signs of strong coldness: larger lakes partly frozen'.

(1984) while others were not classified at all. Further studies are necessary in order to determine whether these discrepancies are due to the pattern of the dominant synoptic situation of those years which makes the teleconnections between distant regions weaker, or to incorrect classification based on unsound data.

A comparison between our classification and the Pfister (1985) temperature classes in Switzerland for the period 1525-1863, however, was most successful. As our winter severity is based on the effects of the cold for a period of the order of a month, we used the most severe index for the three winter months reported by Pfister (1985). The 14 winters classified 'great' in northern Italy were all classified 'very cold' in Switzerland except for 1549. At Venice some chronicles state that in this year the lagoon froze over, supporting people, and one at Verona reports that several trees and birds died because of the very intense cold. All the Italian chronicles note that the cold began on January 20 and lasted for a short time. Pfister (1985) defined this month as 'normal' although he noted for the last ten days of December: 'signs of strong coldness: larger lakes partly frozen'. The difference in classification arises from the fact that our scale of severity was defined on the basis of the effects, whereas Pfister defined this period as having an 'average' temperature index due to the predominant temperature. The 14 winters classified as 'severe' in northern Italy were, in Switzerland, respectively: 5 very cold, 3 cold, 5 subnormal, 1 normal. The winter of 1604-5 was very cold both at Venice and in Pedemont with the lagoon and rivers frozen over; Pfister defined the temperature index of all the winter months as 'average', without further details.

It appears that the teleconnection between northern Italy and Switzerland was very good for great winters, and still good, but with a greater variance for severe winters. The variance in the latter case may be due either to real climatic differences between the two regions, or a less homogeneous definition of severity for the intermediate cases. Severity is evaluated on the basis of the effects, and only in some cases can the alpine region and the Po Valley be compared. In particular, the Alps may experience warm winds, such as the föhn or the sirocco, when the Po Valley is immersed in fog or is subjected to cold bora inflows from the eastern side.

The onset of the Little Ice Age was signaled by changeable conditions which did not appear in phase. From several factors it appears that the 18th century had frequent climatic anomalies of different types which concluded the Little Ice Age. This seems to indicate that the weather types which characterized this period of extreme events differed in frequency and predominance.

7.5 Summary

A review has been made of the existing archival sources in northern Italy, showing the different cultural approach which characterized each period: the new Renaissance spirit; the beginnings of work of an experimental nature and the foundation of Academies which led to the first European meteorological network; the progress of historiography and the critical editing of old manuscripts. A semi-quantitative analysis of the data is possible after the factors which influence climate have been evaluated on the basis of their effects. It is thus necessary to distinguish between events which are rather objective indicators and those which are more influenced by the territory or the technology of the time.

The onset and the end of the Little Ice Age were characterized by phenomena of instability which caused very frequent thunderstorms and hail, even during the cold season. Increased spatial variations of both rainfall and drought in northern Italy showed the dominance of local effects. The flood tide at Venice is an index of the persistence of the sirocco when a depression lies over the western or central Mediterranean. The analysis showed that exceptionally high tides were particularly frequent in autumn, during the first half of the 16th century and in the period between 1750 and 1800.

In Italy the Little Ice Age began in the middle of the 15th century with a series of great and severe winters. Later, their occurrence was more regular, especially for the great winters. Surprisingly, after the end of the Little Ice Age, their frequency was practically unchanged. The coldest winters in Italy were compared with those in western Europe (Lamb 1977; Buisman 1984; Pfister 1985) agreeing well with the Swiss data.

Acknowledgements

This work was carried out under the Climatology and Natural Hazards Programme of the Commission of the European Communities DGXII (contract EV4C-0082-I-A) and with the contribution of the National Research Council (CNR). The authors are grateful to Drs L. Gentile, L. Megna, L. Mettifogo, M. Sghedoni and A. Ongaro for their important contributions in analyzing the manuscripts and computer programming.

References

Alexandre, P., 1987. *Le Climat en Europe au Moyen-Age.* Contribution à l'histoire des variations climatiques de 1000 à 1425, d'àpres les sources narratives de l'Europe occidentale. Ed. de l'Ecole des Hautes Etudes en Sciences Sociales, Paris.

Bell, W. and Ogilvie, A. E. 1978. Weather compilations as a source of data for the reconstruction of European climate during the medieval period. *Climatic Change*, 1, 331-348.
Buisman, J., 1984. *Bar en Boos, Zeven Eeuwen Winterweer in de Lage Landen, Bosch and Keuning.* Baarn, Netherlands.

Camuffo, D., 1981: Fluctuations in Wind Direction at Venice, Related to the Origin of the Air Masses. *Atmospheric Environment, 15*, 1543-1551.
Camuffo, D., 1984. Analysis of the Series of Precipitation at Padova, Italy. *Climatic Change*, 6, 57-77.
Camuffo, D., 1987. Freezing of the Venetian Lagoon Since the 9th Century AD, in Comparison to the Climate of Western Europe and England. *Climatic Change, 10*, 43-66.
Camuffo, D., 1990. *Clima e Uomo.* Garzanti, Milan, in press.
Camuffo, D. and Enzi, S., 1990. Invasioni di cavallette e fattori climatici dal Medioevo al 1800. *Bollettino Geofisico*, in press.
Cantù, V., 1985. Alla ricerca di documenti sul clima passato. *Accademie e Biblioteche d'Italia*, 53, 103-110.
Cantù, V. and Narducci P., 1967. Lunghe serie di osservazioni meteorologiche. *Rivista Meteorologia Aeronautica*, 27, 71-79.
Colacino, M. and Purini, R., 1986. A Study on Precipitation in Rome from 1782 to 1978. *Theoretical and Applied Climatology*, 37, 90-96.

Colacino, M. and Rovelli, A., 1983. The Yearly Averaged Air Temperature in Rome from 1782 to 1975. *Tellus*, 35 A, 389-397.

Comani, S., 1987. The historical temperature series of Bologna (Italy): 1716-1774. *Climatic Change*, 11, 375-390.

Easton, C., 1928. *Les hivers dans l'Europe Occidentale*. Brill, Leyde.

Ingram, M. J., Underhill D. J. and Wigley T. M. L., 1978. Historical Climatology. *Nature*, 276, 329-334.

Ingram, M. J., Farmer, G. and Wigley, T. M. L., 1981a. Past climates and their impact on Man: a review. In: (T. M. L. Wigley, M. J. Ingram and G. Farmer, Eds.) *Climate and History*, Cambridge University Press, Cambridge, 3-50.

Ingram, M. J., Underhill, D. and Farmer, G., 1981b. The use of documentary sources for the study of past climates. In: (T. M. L. Wigley, M. J. Ingram and G. Farmer (Eds.) *Climate and History*, Cambridge University Press, Cambridge, 180-213.

Jurin, J., 1723. Invitatio ad observationes meteorologicas communi consilio instituendas a Jacobo Jurin M.D. Soc. Reg. Secr. et Colleg. Med. Lond. Socio. *Philosophical Transactions 379*, 422-427.

Lamb, H. H., 1969: Climatic Fluctuations. In: H. Flohn (Ed.) *General Climatology*, 2, Elsevier, Amsterdam, 173-249.

Lamb, H. H., 1977. *Climate Present, Past and Future, Vol.2*. Methuen, London.

Menella, C., 1956. *L'andamento annuo della pioggia in Italia nelle osservazioni ultrasecolari*. Mareggiani, Bologna.

Pavese, M. P. and Gregori, G. P., 1985. An analysis of six centuries (XII through XVII century A.D.) of climatic records from the Upper Po Valley. In W.Schröder (Ed.) *Historical events and people in geosciences*, Peter Lang, Frankfurt, 185-220.

Pfister, Ch., 1981. An analysis of the Little Ice Age climate in Switzerland and its consequences for agricultural production. In: T. M. L. Wigley, M. J. Ingram and G. Farmer (Eds) *Climate and History*, Cambridge University Press, Cambridge, 214-248.

Pfister, Ch., 1985. *CLIMHIST: Climate-History Data Bank*. Meteotest, Bern.

Poleni, G. M., 1731. Viri celeberrimi Johannis Marchionis Poleni, R.S.S. ad virum doctissimum Jacobum Jurinum, M.D.R.S.S. epistola, quacontinetur summarium observationum meteorologicarum per sexennium Patavij habitarum. *Philosophical Transactions, 421*, 201-216.

Roccatagliata, A., 1976. Variazioni climatiche, pestilenze e vita sociale nel territorio alessandrino nei secoli XIII-XVII. In *Rivista di Storia, Arte e Archeologia per le province di Alessandria e Asti, 85*, 185-216.

Santomauro, L., 1957. Lineamenti climatici di Milano. *Quaderni della città di Milano*, Milano.

Toaldo, G., 1781. *Della vera influenza degli astri sulle stagioni e mutazioni del tempo. Saggio meteorologico*. 1st and 2nd edition, Stamperia del Seminario, Padova.

Trevisan, V., 1980. *Meteorologia Romana – La serie storica delle osservazioni al Collegio Romano (1782-1978)*. CNR AQ/5/28, Rome.

Veggiani, A., 1986. Clima, uomo e ambiente in Romagna nel corso dei tempi storici. In *Romagna vicende e protagonisti*, Vol.I, Edison, Bologna, 3-19.

Vittori, O. and Mestitz, A., 1981. Calibration of the 'Florentine little thermometer'. *Endeavour*, 5, 113-118.

8 Three historical data series on floods and anomalous climatic events in Italy

M. P. Pavese, V. Banzon, M. Colacino, G. P. Gregori and M. Pasqua

8.1 Introduction

Historical-climatological data are measurements irregularly recorded in time and space of an experiment that cannot be repeated. Therefore the analysis of historical data series can give unique and valuable information, provided that it is done carefully and by suitable specific algorithms. It is well known, in fact, that several contradictory conclusions and lines of evidence can be found in an extensive literature dealing with historical data series analysis (e.g. Herman and Goldberg 1978).

Any kind of quantitative historical investigation is basically and critically influenced by two factors (and by their combined effect): (i) the reliability of the data base, and (ii) the method of statistical analysis.

A historical climatological data base relies on several heterogeneous sources, including human accounts in the documentation, and natural records of environmental parameters ("proxy" data). Errors, however, can be large and often difficult to quantify. In the case of written documents, exaggerations of chronicles, uncritical reference to previous sources, misprints, etc., play a relevant role. Literature of the professional-historian is normally not concerned with either the reporting or the problems of robustness in the analysis of environmental data records. Geophysicists, on the other hand, when attempting to extract by themselves historical environmental information, often lack the expertise and time-tested methodology of the professional historians. Hence, some historical environmental data bases may contain a large percentage of erroneous data (Pfister 1988).

In this respect, there appears to be some controversy among different authors, about the degree of rigour to be exercised when deciding what information to reject for a certain data series. This is clearly subjective, albeit mathematical, as it depends on the analysis to which the data will be used. A non-robust method can suffer dramatically from the inclusion of a few spurious events, while a robust method can safely tolerate a small percentage of misdated or erroneous events. Therefore, in general, any data set should strictly contain only reliable information. Inclusion of some less reliable data may be unavoidable but, in this case, their limited reliability should be clearly stated, emphasized, and also, if possible, quantified. Data series are therefore incomplete and their error is not homogeneous, in the sense that reports from different sources or epochs have different errors. This must be taken into account when performing data analysis. Since there is apparently no clearly assessed and generally agreed-upon methodology for such kind of data, authors generally rely on some intuitive or original procedures. Otherwise, they use standard statistical tools that can be found in the literature, but these generally require, in order to obtain meaningful results, a complete data series without gaps, or a uniform error bar all along the data series, etc. On the other hand, intuitive methods are generally not robust enough with respect to data gaps, or uneven error bars, or the eventual insertion within the data series of some percentage of false information, etc.

Seldom have such relevant aspects been recognized in the literature (Pfister 1988; see also Chapter 1). As soon as a greater awareness of the importance of the reliability of the data base is reached, specific papers dealing only with the data collection and evaluation *per se* should be found in the literature, independent of those dealing only with specific data handling or analysis. At present, the literature mostly emphasizes the climatological and geophysical results, overlooking the importance of the data base. However, the actual geophysical implications should only be of concern later, when several accurate data sets become available, and when suitable multivariate statistical analyses are developed so as to allow for significant statistical inferences to be made.

This paper contains a description of a few data bases collected at the Instituta di Fisica dell'Atmosfera in Rome. The reliability of every series is discussed in detail, and data logs are included. The analysis of such data sets was also carried out by means of procedures reviewed in Gregori (1989 and 1990) and have appeared elsewhere (Pavese and Gregori 1985; Gregori *et al.* 1988; Banzon *et al.* 1989; Banzon *et al.* 1990).

8.2   The data

Three historical data sets are presented (see Figure 8.1). They are of different types: one deals with the floods of the River Tiber, a second with the anomalous climatic events in north-western Italy, and a third with spates and floods of a minor river in the same region.

The Tiber flood data series (that has been listed and discussed in detail by Gregori *et al.* 1988) is listed in Table 8.1. It is based on information extracted from a wide range of literature. Unfortunately, a systematic reference procedure to obtain the very first historical record of every event would require a much greater time-consuming effort. The present data

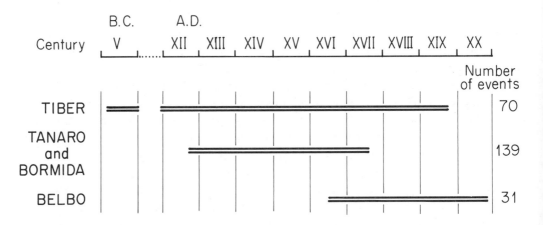

Figure 8.1   Span of the data series presented in this work.

base is the result of a cross reference and comparison of several different sources that have been available; the work, however, could be completed by scanning a few additional sources that are not, as yet, available to the authors. Several events have been rejected, when it was possible to recognize a specific cause of error (e.g., a printing error in one source, which was not recognized by a later author, produced a duplication of the event in a later listing, etc.). All floods were recorded in Rome. The first known flood dates back to 414 B.C., and the last one to A.D. 1870 (new embankments were built beginning in 1876 in Rome along the Tiber).

In general, notwithstanding the accuracy of the scanning, the Tiber flood data series has a non-uniform reliability all along its time span, depending on the different kinds of sources

Table 8.1   Dates of Tiber floods, A.D. 1230-1870 (from Gregori *et al.*, 1988).

Year	Date
1230	February 1st
1277	November 7th
1310?	Autumn-winter
1379	November 8th or 9th.
1415?	November 26th
1422	November 30th
1438	November
1467	September 29th
1475	November
1476	January 8th
1480?	
1493?	Between September 20th and October 21st
1495	December 5th
1498?	October
1514	November 13th
1530	October 8th
1557	September 14-15th
1572	December 31st
1589	November 4th and 10-11th
1598	December 24th
1606	January 23rd
1628*	(presumably wrong datum)
1637	February 22nd
1647	December 24th (less probably November 24th, or December 7th)
1660	November 5th
1686	November 6th
1695	
1702	December 22nd
1742	
1750	December 6th
1772	
1780	
1805	From January 31st through February 2nd
1809	December 21st
1843	First days of February
1846	December 10th
1855	February 17-19th
1858	December 3rd
1870	December 28th

* Not used in the authors' analyses

Note: In 1876 new embankments were built along the river. A flood occurred in 1900 when they were almost completed and the damage was limited. However, during the recovery a minor part of the walls was ruined (from the right branch of the Isola Tiberina to the Lungo-tevere dell'Anguillara). The best test of the embankments occurred in 1937 when an exceptional spate was reported (a picture of the time shows the arches of the Castel Sant'Angelo bridge to be completely submerged). All the region downstream to Rome had a large flood, but Rome suffered no damage.

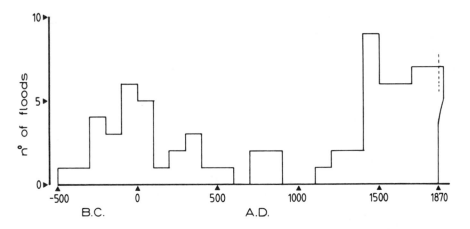

Figure 8.2 Number of reported Tiber floods per century in Rome. The largest fraction of the events dates back either to the classical republican and imperial times, or from Renaissance time onward. (From Gregori *et al.* 1988).

(classic authors, Medieval chroniclers, compilation sources from Renaissance onward, official documents).

The flood frequency vs. century is shown in Figure 8.2. Notice the wealth of information clustered within the classical republican and imperial times, plus from Renaissance onward.

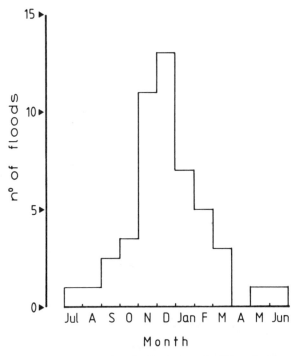

Figure 8.3 Seasonal cycle diagram of the Tiber floods. The winter maximum is typical of the Mediterranean climate of Rome. (From Gregori *et al.* 1988).

A seasonal cycle diagram (Figure 8.3) clearly shows the maximum in winter, a typical feature of the Mediterranean climate (compare this with the cycle diagrams of the Po valley series here below).

The Tanaro/Bormida data set is concerned with anomalous climatic events of several different kinds almost always recorded in the city of Alessandria, in north-western Italy, southern Piedmont, located at the confluence of the Bormida river with the Tanaro river, a main right affluent of the upper Po valley (Figure 8.4). A detailed listing, formerly given by Gregori *et al.* (1988) is shown in Table 8.2. Analysis of the data is discussed in the same paper, as well as in Pavese and Gregori (1985). A substantial fraction of this data series was based on

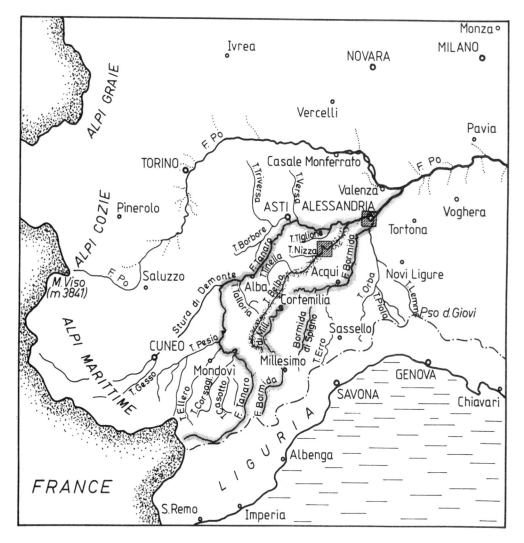

Figure 8.4 Catchment basin of the Tanaro/Bormida river system (darker shade) and of the Belbo torrent (lighter shade). The historical data series deal with observations collected at the two sites, each indicated by a square: upper square is Alessandria and lower is Incisa Scapaccino (located on the horseshoe-shaped meander). **F** is for "fiume", the Italian word for river, and **T** is for torrent. (Adapted and redrawn from Cati 1981, and Gregori *et al.* 1988).

159

Table 8.2 Anomalous climatic events in the Tanaro valley (A.D. 1500-1659) (after Gregori *et al.*, 1988).

Key to coding

Class	Specifications
a	Floods of Tanaro and Bormida rivers
b	Floods not specifically referred to either Tanaro or Bormida, but reported by the sources
c	Time intervals with heavy rain, not followed by reported floods
d	Severe winter
e	Typical winter phenomena occurred out of season
f	Hail
g	Drought and/or heat wave
h	Time intervals with mild winter and/or without snowfall

Symbols

B	Bormida flood	I	Ice (rivers or ponds)	S	Snowfall
C	Cold	K	Hoarfrost	T	Tanaro flood
D	Drought	M	Mild winter	V	Very cold
F	Flood	N	No snowfall	W	Heat wave
H	Hail	R	Rain		

Year A.D.	a	b	c	d	e	f	g	h	Comments
1500							W		Spring-summer
				V					Autumn-winter
1501							D		Spring-summer
1508			R						Summer
1510				VSI					last months?
						H			?
1511				S					earliest months
	TB								March 25th
			R			H			May
1514						H			August 5th
	TB								Half of November
1515				SC					May 20th
		F							Over the year
1516				KV					February 2nd
1517							DW		Summer?
								NM	End of the year
1518				VS					End of the year?
1519	T								April 2nd
							W		Spring-summer?
1520	TB								July 28-29th
1521	TB								May 29-30th
								NM	End of the year
1523				VI					November 11th and subsequent days (continuing also in 1524)
1524			R			H			Over the year
1540							W		Summer
1541	TB								September
1545	T								May 23rd (22nd according to Ghilini)
1550							D	NM	Earliest months
			R						Spring

Year A.D.	a	b	c	d	e	f	g	h	Comments
1552			R						Autumn?
				C					Last months
1555			R				W		August?
1556			R						Over the year (several storms)
1557			R	S					Earliest months
1562							D		From February to October
1564				S					December 23-24th
1565				SI					January-March
1567	TB								October 30th
				S					December 25th
1568				C					Earliest months
			R						Spring
1569				VS					Earliest months
		F							Over the year
1570				VSI					Last months
1571				VSI					Earliest months?
			R						Spring ("*maxima intemperies*" according to Schiavina, with snow lasting on ground until end of May)
1573				C					Earliest months
					S				April 17th
1578				C					April 10th and subsequent days
						H			May 1st
							D		Spring-summer?
1579				KC					April 26th
	TB								October 10-11th
1580			R						Spring
1584		F							July 6th (Cevetta and Bormida rivers near Cortemilia)
1590			R						April-May
							D		Summer
1593						H			May 31st (30th according to Ghilini)
								M	Last months (Ghilini reports flowering of meadows at beginning of February; Schiavina reports the event at the beginning of January)
1594				C					July
							W		August (heat wave with cold nights)
1595				VI			D		Earliest months
					S				April 22nd (Schiavina); April 3rd (Ghilini)
				C					April-May (much snow visible on the Alps, lasting for a long time)
	TB								October 27-28
1596								NM	Earliest months
	T								Half of April
				C					From June-July 11th (a snowfall on the Cozie Alps [between the Maritime and Graian Alps] occurred in July, according to Schiavina, June 9th, according to Ghilini)
			R						Summer (after cold)
								WD	Summer (after the rainy period)

(continued)

Year A.D.				Class					Comments
	a	b	c	d	e	f	g	h	
	T								December 1st
1597						DW			From February-October
1598					H				July 25th (24th according to Ghilini)
			R						Before end of June
						W			Around end of June
			R						From mid-September to end of year
1599				CI					March 28th (snowfall on Alps)
					H				June 24th
				C	H				June 29th (hail on Apennines and Alps at border between Piedmont and Liguria)
					H				July 10th
1600				VS					Earliest months
					C				Snow lasting on ground until May
	T								Between May and June
				S					December 1st
1601			R	VSK	H				Earliest months
			R	CSK	H				Until mid-June (until mid-July according to Ghilini who states that the sky was almost always overcast throukghout 1601 [possibly related to Kamchatkan erution; see Lamb 1970; Hammer et al., 1980; LaMarche and Hirschboeck 1984]).
				C					End of the year
1602								M	End of the year
1603						W			June, July, August
					H				August 6th
	T								October 3rd
1604			R						March-April
						W			August-September
		TB							September 29th
				VKI		D	N		From October 15th (but Schiavina, contrary to Ghilini, says that winter was without ice)
1605							N		Until February 6th
						D			Throughout first half of May
					CIK				Half of April
		TB							October-November
1606				VSI					January
			R						February 27th (winter storm)
		TB							May 14-15
1607							N		Earliest months?
					CK				End of March
								M	Earliest or last months
1608				VI					Earliest months
1609							N	M	Earliest months?
					H				July 16th
1610				V			N		Until February 22nd
				VS					From February 22nd
1611						D			Mid-April to end of June
							N	M	Last months?
1612				V					January

Year A.D.	a	b	c	d	e	f	g	h	Comments
						H	D		Before end of June
	TB								October-November
1613						H			July 19th
					C				July 19th-21 (hail stayed on ground for two days)
				VSI					End of December
1614				VS					Earliest months, snow lasting on ground until mid-March; between October 1613 and mid-May 1614 the sky was almost always perturbed with frequent rain or snowfall
					C				April-May
					S				May 10th
	TB								Around November 25th
1615				S					January
			R						Snow lasting on the ground until mid-March
						W			July-August
						D		N	Last months
1616						D		N	Until January 18th
				S					January 19th
	T								June 17th (18th according to Ghilini)
1620	TB								October 31st-November 2nd
1622				S					January 2nd
1626	T								April 27th
1627	T								October 16-17th
1635					CI				April 27th
						H			July 22nd
1638					S				April 12th
1639							D		From June through August 11th
1644						H			May 20th
					CSI				October 18-19th
					S				October 25th
1646	B								Around August 20th (caused by tributary rivers Orba and Lemme)
	TB								November 19th
								M	December (according to Ghilini, the year was 1647)
1647								M	Earliest months
	TB								April 9th
						H			June 28th
			R						September-October (rain caused a flood of Bormida in earliest days of October)
		B							November 8th
1649			R						From February through June
1650					CK				May 11th
					C				October 21st (snowfall on hils around Tortona, as on May 11th also)
1653			R						May-June
1654	T								Around May 20th
		B							October 19-20th

(continued)

Table 8.2 (*continued*)

Year				Class					Comments
A.D.	a	b	c	d	e	f	g	h	
1655				CIK					April 18th and subsequent days
		T							April 26th
1656				CIK					May 24th
1657	T								November 27th

Addendum: As a matter of curiosity (in view of the major eruption of Tambora in April 1815) on June 8th, 1816 the hills around Voghera (~35km from Alessandria, see Figure 4) "*were white like snow but they said it was hail*". This information was found by one of the authors (M.P.P.) on a hand-written notebook kept in a parish church . Further study of the manuscript is in progress.

the direct testimony of coeval chroniclers. The data base was obtained by a direct scanning of the original first printing of each chronicle, as explained in detail in Pavese and Gregori (1985). The time span is between A.D. 1174 (the historical flood that compelled Frederick Barbarossa to raise the siege) until A.D. 1659, that is, the last year of information given by G. Ghilini, the last of the chroniclers. The series contains some gaps, mostly in earlier times, and is continuous and reasonably homogeneous only within the life-spans of the four chroniclers concerned.

Figure 8.5 shows a composite plot of the different kinds of information provided by this series, compared with the life-span of the chroniclers. It helps for a quick-look evaluation of the continuity of the information. Figure 8.6 shows the number density of all-kind events per decade. Figure 8.7 shows the seasonal cycle diagrams for different phenomena. Notice the maxima of floods and heavy rain at equinoxes, a well known fact from the modern climatology of this region (continental-type climate of the Po valley). Notice however, the higher relative maximum for floods in autumn, as compared with the higher relative maximum for heavy rain in spring time. Compare this with the analogous results for the Tiber shown above.

The historical sources are listed in an appendix to the reference list. A few biographical sketches of the chroniclers can help in evaluating the data reliability. Giovanni Antonio Claro was a notary who lived in Alessandria between the second half of the 15th century through the first years of the subsequent century (A.D. 1516?). He wrote a *Chronica Alexandriae* in Latin which covers the time interval between A.D. 1154 through A.D. 1499. He briefly reports on the history of Alessandria, and also on a few events which preceded its foundation. Owing to his business, as a notary, he was presumably acquainted with handling old documents and with critically considering their actual reliability.

Raffaele Lumelli was a jurisconsult who lived in Alessandria in the second half of the 16th century. He wrote *Chronologica descriptio de origine civitatis Alexandriae, ab anno suae fondationis MCLXVIII (et successive usque ad anno 1586)* in Latin. He relied also on Claro's Chronica.

Guglielmo Schiavina spent all of his life in Alessandria, where he was born in A.D. 1542. He was a canon in the cathedral and died in A.D. 1616. He spent a long time to write his *Annales Alexandrini* in Latin. He reports on events which occurred from the foundation of the town (A.D. 1168) until A.D. 1616. He did a remarkable job using literary and docu-

FLOODS OF THE RIVER TANARO AND REMARKABLE CLIMATIC EVENTS AT ALESSANDRIA (PIEDMONT, ITALY) AND ITS COUNTRYSIDE FROM XII CENTURY THROUGH XVII CENTURY

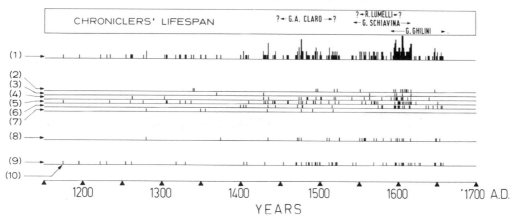

Figure 8.5 Chronological display of the available information.
1: Remarkable unusual climatic events (sum of all kinds of events);
2: Mild winter or lack of snow;
3: Drought (but, not during summer time);
4: Heat wave or summer drought;
5: Severe winter;
6: Cold out of season;
7: Hoar-frost;
8: Heavy rain without reported floods;
9: Floods;
10: This is the historical flood which occurred in A.D. 1174 during the siege of Alessandria by Frederick Barbarossa, and compelled him to raise the siege.

Notice that the data do not cluster into warm or cold epochs, or into dry and wet epoches, etc. On the contrary, one can recognize epochs of "perturbed" climate, when either drought or flood, severe winter or a heat wave etc. can equally occur, and epochs of "quiet" climate. The data sequence is obviously more reliable during the life-spans of the chroniclers than during the years for which they had to rely on previous information. A vague indication of a periodicity of ~40 years can possibly be envisaged. (From Pavese and Gregori 1985).

mentary sources. He was particularly careful and clever in reporting climatic phenomena. He relied also on the two chroniclers mentioned above.

Girolamo Ghilini was born in 1589 in Monza (an old town close to Milan and shown in Figure 8.4) but his father was born in Alessandria. He chose the ecclesiastical state after the death of his wife. Later he became titular of St. James Abbey in the village of Cantalupo in Molise (in southern Italy) but it seems he never went there. In subsequent years, he was Protonotary Apostolic and canon of the collegiate basilica of St. Ambrose in Milan. He died in Alessandria in A.D. 1668. He wrote several literary works, the most important being the *Annali di Alessandria* in Italian. He reports on events from the foundation of the town until A.D. 1659. He based his research mostly on Schiavina's *Annales*. However, he also included several additional data he got from other sources, as well as information subsequent to A.D. 1616 which was the upper time boundary of Schiavina's report.

165

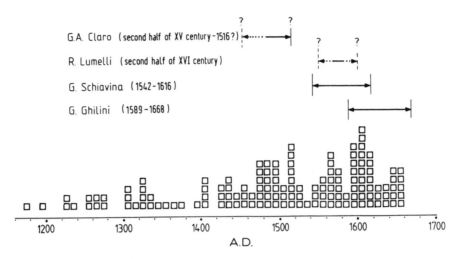

Figure 8.6 Total number of anomalous climatic events (all kinds) per decade, reported from the Tanaro valley. The chroniclers' life-spans are shown on top of the Figure. Note the apparent ~40 year modulation occurring after the middle of the 16th century (when the series is presumably continuous). (From Gregori *et al.* 1988).

The data series of the torrent Belbo (see Figure 8.4) is concerned with spates and floods. The list is given in Table 8.3. The time span covers the period A.D. 1551 until present. The information was recorded at the settlement of Incisa Scapaccino, located on a horseshoe-shaped meander of the river (see Figure 8.4). All data were directly read from official documents, like the minutes of municipal meetings which are still stored either in the archives of the local town hall or in the archives of the "Region" in Turin. The record of more recent events was based on reports contained in some major newspapers. Therefore, all these data are highly reliable, and the series is likely to have no gaps, except, maybe, in the interval from 1554 through 1645 (only four events reported). The record of spates is affected by some subjective judgements: in particular, spates, unlike floods, are not usually reported by a newspaper. For this reason, the data series contains less spates in more recent than in older times. The seasonal cycle diagram is shown in Figure 8.8.

8.3 Concluding remarks

Historical proxy data series cannot be expected to provide, in general, any final and very clear geophysical conclusion. Albeit, they can give some important indications that are of a unique value in the fact that they refer to a unique experiment, the evolution of the environment, an experiment that cannot be repeated, and that was monitored only very irregularly both in space and time. Moreover, proxy data series are strictly complementary to instrumental data series, which cover only too short a time span to allow for the investigation of some relevant long range trends or periodicities. Errors in proxy data series can play a paramount role, depending on the robustness of the actual mathematical analysis, but the general methodology has not as yet been assessed (Gregori 1989 and 1990). It is important to avoid as much as

166

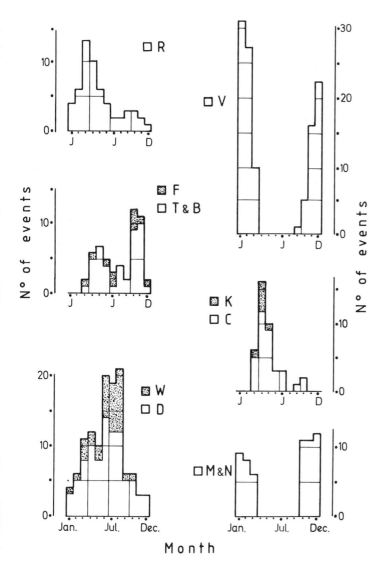

Figure 8.7 Seasonal cycle diagrams for the Tanaro/Bormida series, with events selected according to the classes defined in Table 8.2 or in Figure 8.5. (From Gregori *et al.* 1988).

possible including unreliable data in the data series. In addition, one should clearly emphasize what data could not be fully correct, in order to make the effort of the mathematician more effective.

Such relevant items are likely to be solved only when: (i) a large multivariate set of correct historical data series is available; (ii) some mathematical methods of analysis are selected as being specifically suited for them; and (iii) some good physical modelling is achieved. Such progress is likely to occur only by means of a joint step by step interdisciplinary effort, beginning with a careful and critical collection of several historical data sets.

Table 8.3 Historical data series of the spates and floods of the torrent Belbo recorded at Incisa Scapaccino (province of Asti). The * following a year means spate occurrence without any mention, by the source, of subsequent flood. Table based on direct scanning of original sources and on the manuscript of one of the authors: Michele Pasqua, *Incisa e il Belbo – Cronaca delle alluvioni del Belbo ad Incisa con cenni storici e geografici della sua valle* (Incisa and Belbo – A chronicle of the floods of Belbo at Incisa with historical and geographical notes on its valley) stored in the archive of the parish church of Saints Vittore and Corona at Incisa Scapaccino (province of Asti).

Year	Date
1511	May
1553*	December 25-26
1646*	February
1648*	August and December
1649*	April
1651*	January
1671*	May
1680*	November 15-16
1684*	February
1698*	
1709*	Autumn
1742*	September
1744*	October 3-4
1776*	September
1792*	Autumn
1801	Autumn
1803	March
1839	Autumn
1840	May 30-31
1857	October 21
1859	
1879	May 26-28
1914	October 30-31
1926	May 15
1948	September 4 and 12
1951	November 10
1957	April 10
1958	April 14
1959*	December 2
1960*	December 19
1968	November 23

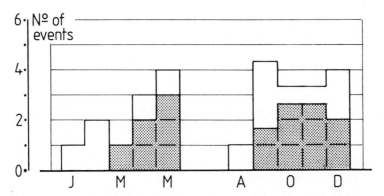

Figure 8.8 Seasonal cycle diagram of the Belbo floods (stippled) and spates (white). The trend is similar to the Tanaro and Bormida data base, and different from the Tiber series.

Acknowledgements

This investigation has been carried out in partial fulfilment of a CEE sponsored programme on climatic research and of a programme supported by the CNR Finalized Project "Climate, environment and territory in southern Italy". We thank M. R. Valensise for kindly providing us with the paper by Chodzko (1831). We also thank E. Lo Cascio for the careful preparation of the Figures.

References

Banzon, V. F., M. Colacino, G. de Franceschi, L. Diodato, G. P. Gregori, M. P. Pavese, and R. Santoleri 1989: Tiber floods, anomalous climatic events in the upper Po valley, and volcanic activity, (in print) in Schröder (1989) *op. cit.*

Banzon, V. F., M. Colacino, G. de Franceschi, L. Diodato, G. P. Gregori, M. P. Pavese, and R. Santoleri 1990: A study on the Tiber floods, on the anomalous climatic events of the Tanaro valley (upper Po valley) and on their relation with volcanic activity, (in preparation) to be submitted to *Nuovo Cimento C* 1990.

Cati, L., Idrografia e idrologia del Po, *Pubblicazione n. 19 dell'Ufficio Idrografico del Po*, 310 pp., Istituto Poligrafico e Zecca dello Stato, Roma 1981.

Chodzko, L. 1831: *Relazione storica, politica, geografica, legislativa, scientifica, letteraria, etc. della Polonia antica e moderna.* Prima traduzione italiana, Tomo I, (Ch. 3) Livorno dalla tipografia di G. P. Pozzolini.

Gregori, G. P., R. Santoleri, M. P. Pavese, M. Colacino, E. Fiorentino, and G. de Franceschi 1988: The analysis of point-like historical data series. In: Schröder (1988) *op. cit.*, pp. 146-211.

Gregori, G. P. 1989: A few mathematical procedures for the analysis of incomplete historical data series, (in print) in Schröder (1989) *op. cit.*

Gregori, G. P. 1990: Multidimensional statistical analysis of incomplete point-like historical data series, (in print) in Martellucci and Gregori (1990) *op. cit.*

Hammer, C. U., H. B. Clausen, and W. Dansgaard 1980: Greenland ice sheet evidence of post-glacial vulcanism and its climatic impact, *Nature*, 288, 230-235.

Herman, J. R., and R. A. Goldberg 1978: *Sun, Weather and Climate.* National Aeronautics and Space Administration, Washington, D.C., 360 pp.

LaMarche Jr., V. C., and K. K. Hirschboeck 1984: Frost rings in trees as records of major volcanic eruptions, *Nature*, 307, 121-126.

Lamb, H. H. 1970: Volcanic dust in the atmosphere; with a chronology and assessment of its meteorological significance, *Phil. Trans. R. Soc. London*, 266A, 425-533.

Martellucci, S., and G. P. Gregori (eds) 1990: Interpretation methods and their test by available data sets, *Proceedings of the First HEFEST workshop* held at Erice (Sicily) February 27 – March 1 1989, (in print). World Laboratory, Project HEFEST, Lausanne.

Pavese, M. P., and G. P. Gregori 1985: An analysis of six centuries (XII through XVII century A.D.) of the climatic records from the upper Po valley. In: Schröder (1985) *op. cit.* pp. 185-220.

Pfister, C. 1988: Variations in the spring-summer climate of central Europe from the high middle ages to 1850. In: Wanner and Siegenthaler (1988) *op. cit.*, pp. 57-82.

Schröder, W., (ed.) 1985: *Historical Events and People in Geosciences.* Peter Lang, Frankfurt am Main etc., 220 pp.

Schröder, W., (ed.) 1988: *Past, Present, and Future Trends in Geophysical Research.* Interdivisional commission on the history of IAGA, Bremen-Roennebeck, 342 pp.

Schröder, W., (ed.) 1990: *Advances in geosciences*, Proceedings of two symposia held by the Interdivisional Commission on History of IAGA, Exeter, UK, IAGA General Assembly, August 1989, in print, Interdivisional Commission on History of IAGA.

Wanner, H., and U. Siegenthaler, (eds) 1988: *Long and Short Term Variability of Climate.* Springer-Verlag, Berlin etc. 175 pp.*

Historical sources for the Tanaro/Bormida data series

Joannis Antonii Clari Chronica Alexandrina (in Latin) in Memorie politiche, civili e militari della città di Alessandria di G. O. Bissati e vecchi cronisti alessandrini, ed. by L. Màdaro 167-186, Casale Monferrato 1926; based on a previous edition ed. by G. B. Moriondo, Monumenta Aquensia, II, columns 725-738, Taurini (i.e. Turin) 1790.

Raphaelis Lumelli de origine civitatis Alexandriae ab anno suae fundationis MCLXVIII chronologica descriptio, (in Latin) in *Memorie politiche, civili e militari della città di Alessandria di G. O. Bissati e vecchi cronisti alessandrini*, ed. by L. Màdaro, 204-322, Casale Monferrato 1926; based on a previous edition ed. by G. B. Moriondo, Monumenta Aquensia, I, columns 545-626, Taurini (i.e. Turin) 1789.

Guillelmini Schiavinae *Annales Alexandrini*, (in Latin) ed. by Ferrerus Ponzilionius, Augustae Taurinorum (i.e. Turin) 1857; reprinted in *Historiae patriae monumenta*, XI, scriptores IV, columns 1-688, Augustae Taurinorum (i.e. Turin) 1863.

Girolamo Ghilini, *Annali di Alessandria, annotati e documentati da A. Bossola fino all'epoca quarta (i.e. up to A.D. 1559) e da G. Jachino dall'epoca quarta in avanti* (in Italian) vol. I 1903, vol. II 1906, vol. III 1908, and vol. IV 1915. First edition: Girolamo Ghilini, *Annali di Alessandria*, ovvero le cose accadute in essa città, nel suo e circonvicino territorio dall'anno dell'origine sua sino al MDCLIX. I fatti memorabili de' suoi cittadini. Alcuni avvenimenti notabili altrove occorsi nell'istesso tempo. Et un breve trattato delle terre che concorsero alla fabbrica di essa città, (in Italian) Milano, Gioseffo Marelli, al segno della Fortuna 1666.

English translations of the titles of the historical sources

J. A. Claro, Alexandrinian chronicles, in *Political, civil and military memoirs of the town of Alexandria by G. O. Bissati and old Alexandrinian chroniclers . . .*

R. Lumelli, Chronological description on the origin of the town of Alexandria from the year of its foundation A.D. 1168 and later up to the year A.D. 1636, in *Political, civil and military memoirs of the town of Alexandria by G. O. Bissati and old Alexandrinian chroniclers . . .*

G. Schiavina, *Alexandrinian Annals, . . .*

G. Ghilini, *Annals of Alexandria*, with notes and documents (collected) by A. Bossola up to the fourth epoch (i.e up to 1559) and by G. Jachino from the fourth epoch onward. . . . First edition: G. G., Annals of Alexandria, or what happened in that town, and in its own and in the nearby territory from the year of its origin until 1659. The memorable facts of its citizens. Some remarkable events occurred elsewhere in the same time interval. And a brief treatise on the lands which concurred to the assemblage of the same town (in Italian) Milan, Joseph Marelli (this is the name of the printer) "at the sign of Fortune" (this is the address) 1666.

9 Documentary evidence from the U.S.S.R.

Ye. P. Borisenkov

9.1 Introduction

The territory of the Soviet Union spans many climatic zones and experiences large temporal and spatial variations in climate. Studying the climate of these territories is not only of scientific but also of great practical importance. Suffice it to say that during the last millenium of Russian history there were more than 350 "hungry" years, most of them precipitated by unfavorable climatic conditions (Borisenkov and Pasetsky 1983).

The first climatic descriptions of Russia may be found in works by Anuchin (1983) Vild (1883) Shchepkin (1886) Veselovsky (1887) Voeikov (1902) Bogolepov (1907) Buchinskii (1957) and others. According to historical chronicles, extreme climatic anomalies always strongly affected crop yields in Russia, and were often accompanied by famine and social unrest (Karamzin 1816 to 1829; Lezhkov 1854; Romanovich-Slovatinsky 1892; Gadzatsky 1907; Stanislavsky, 1926; Berg 1947; Sreznevsky 1957; Pashuto 1964; Shvets 1971; Kes *et al.*, 1980). An analysis of unusual natural phenomena during the 15th to 19th centuries was undertaken for the first time in a study by Dugov (1883) who related such adverse phenomena in the Russian state to certain social upheavals.

Notwithstanding the seemingly greater independence of modern society from adverse climatic conditions, droughts, recurring cold spells during the season of crop ripening, severe winters and exclusively humid summers, still heavily strain the agricultural services, power production facilities (especially hydroelectric complexes) and other branches of national economies. The severe droughts of 1972 and 1975 in the U.S.S.R. (which were particularly harmful) serve to illustrate such effects; in 1975 the overall countrywide loss in grain yield amounted to approximately 40% (Borisenkov 1982).

Using historical chronicles, Borisenkov and Pasevsky (1983) analyzed climatic extrema and related them to social events in Russian history throughout the 11th to 17th centuries. Another monograph by the same authors (1988) treats the same events in more detail for the millenium spanning the 9th to 20th centuries. It includes a chronological, century-by-century compendium of such extreme natural phenomena.

9.2 Principal climate reconstruction techniques (16th to 19th centuries)

Routine magnetic and meteorological observations started in Russia in 1832, following the establishment of the Normal Observatory in St. Petersburg (now Leningrad) under the Department of Mining [locations of places mentioned in the text are shown in Figure 9.1]. In 1849 the Main Physical Observatory was founded (now known as The Voeikov Main

Geophysical Observatory) evolving from the Normal Observatory. The history of organization of the regular instrumental meteorological observations in Russia may be found in a monograph by Pasetsky (1978).

Since regular meteorological observations cover only a limited time span, various sources of proxy data have to be used when reconstructing the climate from the 16th century on. Of primary significance amongst these are the historical chronicles. Some of the earliest records are of unusual natural phenomena in the ancient Kiev Rus, mentioned in chronicles dating from the rule of Prince Yaroslav the Wise. The most ancient Kiev historical chroncle was compiled in 1037-1038 (Likhachev 1947; Rybakov 1963). Longhand chronicling started in Velikii Novgorod in the 11th century and went on until the beginning of the 18th century. These chronicles contain not only the most ancient but also the most detailed data on extreme natural phenomena, from the Baltics to the Northern Urals and from the Black Sea to the shores of the Arctic Ocean.

By the mid-12th century, chronicles were kept in Rostov Velikii and during the first half of the 13th century in Galich, Pskov and Tver (now Kalinin). This proliferation of Russian chronicles was somewhat hampered by the Tartar invasion. However, chronicling did not stop. It became particularly active in Novgorod, Pskov, Rostov Velikii, Galich, Tver and Smolensk. The best traditions of Russian chronicling developed further during the 14th century in the Moscow, Lvov and Vologda-Perm chronicles.

During the 17th century, Russian chronicling suffered a certain crisis, though it still went on in Novgorod, Pskov, Moscow and in some northern and Siberian cities and cities of South Western Russia. During the second half of the 16th century and the first half of the 17th

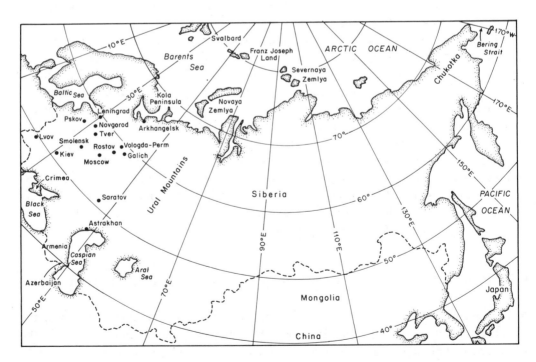

Figure 9.1 Locations of places mentioned in the text.

century, the data on weather became more scarce, and the chronicles themselves gradually lost their former importance. By that time, however, the Russian lands were already joined into a centralized state, so that data on adverse weather phenomena since then may be found in governmental documents and Grade books (from 1575-1605).

Unique notes on Moscow weather from 1657-1675 may be found in a study by Belokurov (1909). From 1695-1724 Peter the Great took weather notes in his "Hike Journals". Starting in 1720, following his orders, the Russian Navy was required to keep "watch journals" with observations taken six times a day. State archives hold meteorological notes from the late 17th century to the first half of the 19th century.

To reconstruct the history of climate, 37 volumes of the complete set of Russian chronicles (mainly originating from churches and monasteries) were used, together with longhand chronicle compendiums not included in the published set. Studies by historians and natural science researchers were also involved. More than 2,000 weather notes taken in the Moscow Kremlin during the 17th century, as well as meteorological notes by Peter the Great and his courtiers, were studied. The overall result is a compilation in the form of a 1,000 year summary of extreme weather phenomena and an analysis of climate dynamics during that period. Of particular note, are the highly objective descriptions of natural phenomena recorded in the chronicles. Historical chronicles stand out for their clear, simple expressions and unpretensiousness of entries.

9.3 General description of the climate of the U.S.S.R. for the pre-instrumental period (starting in the 16th century)

The analysis of historical chronicles and a comparison with other climatic data reveals a certain scheme of climate dynamics in the Russian territories. It is apparent that a distinct transition from a minor climatic optimum to the "Little Ice Age" had started here by the late 12th to early 13th centuries.

The Little Ice Age featured highly variable climate and frequent climatic extremes. Starting from the 13th century, seasonal contrasts became stronger, with dry summers followed by rainy autumns, etc. Such an increase in climate instability led to heavy crop failures and famines. For example, in the beginning of the 13th century famines occurred for 17 years in a row, thus significantly reducing the population of Russia. Similar instability was typical of the 14th century. All in all, more than a hundred climatic extremes were recorded in the 14th century chronicles. These extremes resulted in 30 "hungry years" in Russia; it is notable that four of these events were pan-European and did not only affect Russia.

The 15th century saw more than 150 climatic extremes. Heavy rains and droughts, severe winters and summer cold spells brought about more than 40 hungry years, of which 15 were very severe.

The 16th century was noted for similar climatic contrasts: severe winters became much more frequent, particularly by the end of the century, and summer cold spells recurred more often. Although in general, the 16th century (especially the first half) featured higher temperatures, the recurrence of climatic extrema during that period was quite high.

Unusually dry, hot weather was noted in 1512, 1525, 1533, 1541, 1542, 1561, 1571 and 1585. In contrast, during that century, extremely heavy rains were listed 17 times in summer, and six times in autumn. The years 1518, 1524, 1529 and 1557 were especially rainy. Climatically, the first two decades of the 16th century were favorable. However, starting around 1524 the incidence of climate extremes sharply increased. This period of climate instability lasted until 1570. The 1573-1582 decade may be characterized as the most favorable. From 1583 on, the picture once again changed sharply and unusually cold, mild, extremely snowy and snowless winters alternated.

During the first half of the 17th century there were fewer climatic extremes, except for the very beginning of the century. Unfavorable climate at that time precipitated the "Great Famine" of 1601-1603, and riots by angry people followed (the well known "Godunov Hungry Riot"); according to witnesses, 120,000 people perished from hunger in Moscow alone. Contemporaries of this great famine believed that "one third of the Moscow Tzardom" died out during these years. Favorable climate in the Northern Rus domain during the first half of the 17th century is testified to by data on the unprecedented navigation in the Arctic seas. In the second quarter of that century, Russian vessels reached Chukotka and passed through the Bering Straits, thus opening the Northeastern Route from the shores of the Kola Peninsula to the Pacific.

Russian chronicles from that time contain few notes on climatic extremes. It is hard to say whether this fact reflects their scarcity. Testimonies from certain years (1604-1608, 1619, 1623 and 1625) do describe some adverse phenomena. From the mid-17th century on, universal climate cooling set in for almost 200 years, until the middle of the 19th century, resulting in heavier ice cover in the Arctic seas.

The second half of the 17th century was also characterized by drier conditions. Droughts were particularly common from 1640-1659 (11 dry years) and 1680-1699 (9 dry years). Rainy summers, with recurring early season cold spells and late season frosts, became common. Severe winters alternated with extremely mild ones. During one such winter (1696-7) 35 ships froze into the ice near Arkhangelsk. All in all, 33 years out of the second half of the seventeenth century were considered dry by the chronicles.

The 18th century was characterized by climate contrasts and instability, accompanied by more frequent climatic extremes. In that century, 39 years were dry and 19 were rainy; there were 40 severe winters and 22 mild ones, 33 unprecedented spring floods and 22 great storms. Cold springs, summer cold spells and late summer frosts became more frequent than during the preceding centuries. Frosts were particularly severe during the winters of 1708-9, 1709-10 and 1739-40. According to the chronicles, extremely severe frosts were also noted in Western Europe from December 19, 1739 to February 15, 1740.

At the start of the third decade of the 18th century, Russia suffered several severe famines due to droughts and recurring summer cold spells. The noted Russian historian, V.O. Klyuchevsky, stated: "In 1733, peasant mobs flooded cities, begging alms." Many villages were deserted.

The second half of the century started with a short respite. The fifties and sixties of the 18th century were considered climatically favorable. However, later the climate instability increased again (1765-1775) embracing not only Russia, but Western Europe as well. Many Russian provinces suffered from droughts in 1772, 1773 and 1776. Epidemics and epizootics started and bread prices soared. In Moscow alone 56,672 people perished from hunger and epidemics during the period 1770-1773. Severe droughts happened in 1791, 1793 and 1794. All in all, during the 18th century Russia had suffered 68 extremely hungry years, and most of them were due to unfavorable climate.

The first half of the 19th century had a high proportion of severe winters: 37 out of 50. Particularly cold were the winters of 1802-1803 and 1808-1812, when records indicate that mercury froze in several Russian cities. Although a few mild winters did occur during this cold spell (for example, in 1814-1815, 1816-1817 and 1821-1822) the first half of the century stands out for its severe frosts. Such conditions were also typical for Western Europe.

Several examples may be mentioned. During the severe winter of 1819-20, 33,426 trees, 62,275 horses, 3,654 camels and 28,367 heads of cattle died due to temperatures below freezing in the Astrakhan province alone. In 1827-28 more than one million sheep, 280,500 horses, 78,450 head of cattle and 10,500 camels froze to death in the Saratov province. The winter of 1844-45 was exceptionally severe, when approximately 100,000 sheep perished in the Crimea, and 22,960 head of cattle froze in Armenia and Azerbaijan.

During the period of severe winters, summers alternated between extreme droughts and cold, rainy conditions. Historical sources of the first half of the century listed 35 droughts and 25 rainy summers. 23 recurrent cold spells were noted in late spring to early summer and 21 frosts occurred not only in late summer, but mid-summer as well. 44 years out of 50 were considered "hungry".

The second half of the 19th century was warmer, in concert with the general global warming. The frequency of climatic extremes puts this 50 year span among the highly unstable climatic periods. 27 cold winters and 31 years of cool, rainy summers were registered. Simultaneously, 44 drought years were listed in various provinces of the country. This is why 36 years out of those 50 were noted for poor crops and famines. However, extremely cold winters became rarer compared with those of the second half of the 18th and the beginning of the 19th centuries. Notwithstanding some severe winters, recurring spring frosts (1891-1892) and droughts (1891) the century definitely ended in a warmer climate, which continued into the 20th century, peaking during the 1920s-1940s.

On a larger scale, dividing the period from 1201-1980 into 30 year intervals, Lyakhov (1988) identified seven episodes of lowest air temperatures for Mid-Russia: 1411-1440, 1441-1470, 1471-1500, 1621-1650, 1651-1680, 1861-1890 and 1891-1920. Using historical chronicles he reconstructed seasonal temperature variations, calibrating them against the instrumental data period (Figure 9.2). A period of comparatively warm conditions, in all seasons, during the first half of the 16th century catches one's attention. According to Lamb (1982) the situation was approximately the same in Western Europe. It should be noted that climatic descriptions of the first half of the 16th century are in many respects similar to those during the first half of the 20th century.

The precipitation regime in European Russia over the last millennium is outlined in Table 9.1. For each 30 year period it gives the number of those years which were either wet or dry (judging by the amount of precipitation during the respective spring/summer seasons). One may see that dry spring/summer seasons were common in the first 30 years of the 13th century, and again, variously from 1351-1380, 1411-1440, 1801-1830, 1831-1860, 1891-1920 and 1951-1980.

9.4   Some features of the climate of the U.S.S.R during the instrumental observational period

Instrumental climate observations are fairly complete for the last 100-150 years. Basically, this was a period of world-wide climate warming, starting from the end of the last century. Such a

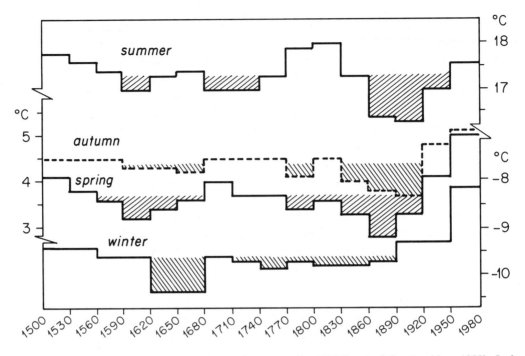

Figure 9.2 Thirty year seasonal mean temperatures for Mid-Russia (after Lyakhov 1988). Left scale for spring and fall; upper right scale for summer; lower right scale for winter. Long-term averages for each season designated by dotted horizontal lines.

warming was quite apparent in the Soviet Union. According to the discussion above, the most noticeable consequences of climatic changes in the Soviet Union were not always so much those resulting from climatic warming or cooling but from a higher or lower frequency of extreme climatic phenomena (droughts, severe winters, moist cool summer seasons, etc.)

Table 9.1 Tri-decadal spring/summer precipitation extremes in European Russia, 1201-1980 (after M. E. Lyakhov, 1988).

Tri-decade	Season Dry	Moist	Tri-decade	Season Dry	Moist
1201-1230	8	4	1591-1620	3	4
1231-1260	1	1	1621-1650	3	0
1261-1290	3	3	1651-1680	5	1
1291-1320	4	3	1681-1710	2	4
1321-1350	4	1	1711-1740	2	1
1351-1380	9	1	1741-1770	2	1
1381-1410	3	3	1771-1800	1	2
1411-1440	7	3	1801-1830	7	*
1441-1470	4	3	1831-1860	9	*
1471-1500	4	3	1861-1890	5	2
1501-1530	3	4	1891-1920	9	6
1531-1560	3	3	1921-1950	5	5
1561-1590	3	4	1951-1980	7	6

* extremely moist periods not identified

176

and from the periods of moist or dry climate. Average temperatures, which only reflect climatic trends, are hardly indicative in that respect.

Table 9.2 (after Mesherskaya and Blazhevich 1978) lists generalized data on the extremely dry or moist, warm or cold seasons during the instrumental observational period. Periods when the average precipitation for a given area exceeded 110% of the norm (for 1891-1975) were classified as moist. If the average amount of precipitation was below 90% of the norm, the period was classified as extremely dry.

Table 9.2 Extremely dry or moist, cold and warm seasons in the U.S.S.R. territory.

REGION	MOIST PERIODS Seasons		DRY PERIODS Seasons	
	Cold	Warm	Cold	Warm
European USSR:	1913-1917 1966-1970	1925-1933 1929-1930	1917-1936	No
Asiatic USSR:	1899-1919	1945-1947 1958-1960	1931-1935	No

One can see that in the warming climate not only was the warm season precipitation trend positive in the European territory of the country, but the deviations from this trend were also positive, i.e. large positive anomalies in precipitation prevailed. Cold seasons of the warmer climate spell (1927-1936) mainly deviated negatively from the precipitation trend, i.e. dry anomalies prevailed. Such a feature was typical also for the pre-instrumental period. Apparently, higher summer evaporation and lack of proper winter precipitation may explain the 20th century retreat of the Caspian Sea.

To assess the degree of climate extremes during the instrumental period, Sazonov and Kanayeva (1985) compiled a catalogue for the years 1891-1985. According to that study the complete range of moisture extremes (E) is scaled from +10 to -10 with the positive values corresponding to higher dryness, and vice versa. The same scaling was done for winter severity (± 10). Cold winters were assigned negative values and warm summers positive.

The data from 500 stations of the Northern hemisphere having long observational series were catalogued. Sazonov and Kanayeva considered that all the years for which the moisture index, E, lies in the range 6-10 should be classified as dry, and the years with E= 8-10 as extremely dry. The years with E= -5 to +5 are close to normal. Similarly the years with E= -6 to -10 are considered excessively moist, and those having E= -8 to -10 as extremely moist. In exactly the same way the temperature extremes may range from +6 to +10 (warm years) +8 to +10 (extremely warm years) -6 to -10 (cold) and -8 to -10 (extremely cold).

Such a classification permits certain conclusions to be reached on the dynamics and extremes of climate during the current century. Noticeably, during the warmer period of the 1930s and 1940s the number of "normal" winter decreased and the winters tended to be both warm and cold extremes. The percentage of warm or cold winters (i.e. E $\geq$ +6 or $\leq$ -6) climbed from 48-50% in the beginning of the century, to 62-65% in these two decades. Later such extremes dropped to a minimum, coinciding with the generally colder 1960s; however, in the 1980s they sharply increased once again.

Such a significant rise in the frequency of both wintertime extremes during the warming spell of the 1930s and 1940s indicates that such a spell cannot be envisaged as simply a uniform period of gradual temperature build-up, to be followed by a similar decline. Actually, it occured against the background of the occurrence of major climatic anomalies of both signs. We once again draw attention to similarites between this period and the warming of the first half of the 16th century.

It should be noted that hemispherically the frequency of normal conditions had, on the average, decreased during the 1930s and 1940s, while the extremes correspondingly increased. More than half of all the extremes in various regions of the hemisphere (90 out of 170) were strong or severe droughts, while extremely wet conditions were rare.

During the warming of the 1930s and 1940s the number of major droughts was almost twice the norm for practically all regions. These events were particularly frequent in the U.S.A. Certain ideas on climate variability in various regions may be formed from examining Table 9.3 compiled by the author of the present study.

Table 9.3 (and other data which are not cited for lack of space) vividly illustrate high temporal and spatial inhomogeneity in climatic extremes. The higher frequency of extremes in both the U.S.S.R. and Europe as compared to other regions is quite obvious. A similar pattern is found in the temperature regime. It is also apparent that climatic extremes are regional in nature.

One of the important characteristics of climate in the territory of this country is winter severity. The index usually used to characterize it is the sum of negative temperatures. A winter is considered cold if its sum of the negative temperatures exceeds 500°C, in absolute terms. Recent research identifies four periods of differing winter-time conditions in Central Europe (Borisenkov 1988): 1) from the start of observations in the mid-18th century until 1828-29; 2) 1829-30 to 1894-95; 3) 1895-96 to 1927-28; 4) 1928-29 to 1970s. Of these the first and the last were known for their mild winters. Extremely cold winters mainly occurred during the second period. The analysis by Yefanova (1976) demonstrated that 90-95% of cases with stable, cold winter spells (December excluded) in the European U.S.S.R. and Western Siberia occur during circulation regimes classified by Vangengeim (1952) as either meridional or eastern atmospheric circulations.

In Eurasia, one of the severest winters between 1891 and 1971 was the winter of 1928-29. Several more extremely severe winters took place in 1890-91, 1939-40, 1941-42, 1946-47, 1953-54, 1962-63 and 1978-79, though cold conditions extended over comparatively small areas. Winters of 1892-93, 1899-1900, 1916-17, 1918-19, 1928-29, 1929-30, 1931-32, 1932-33, 1949-50, and 1968-69 may be classified as cold in large parts of Eurasia.

Clearly, most of these 18 cold and extremely cold winters fall in the period of maximum warming in the 1920s-1940s. However, winter severity is not a sufficient characteristic in itself. For example, it is impossible to interpret the character of icing in the Baltic Sea if one disregards the phase of the cooling period. Indeed, the sums of the negative temperatures for the winters of 1921/22, and 1928/29 reached 711 and 747°C, respectively, differing by only 4%, yet in the first of these winters the frozen area of the Baltic was 260,000 km^2, compared to 390,000 km^2 during the second, that is, they differed by a factor of 1.5.

Sazonov and Kanaeva (1985) compared the number of extremely cold winters from two

Table 9.3 Chronology of extremely dry or humid summer seasons in various regions of the U.S.S.R. and the Northern Hemisphere, 1891-1985.

Region	Period	Extremely humid years (-10 ≤ E ≤ -8)	Total humid extremes
European U.S.S.R.	Apr-May	1896, 1900, 1904, 1908, 1913, 1923, 1941, 1942, 1945, 1964	10
Altai mountains	May-Jul	1894, 1897, 1903, 1912, 1919, 1946, 1947, 1954, 1958, 1959, 1960, 1972	12
Northern Europe	Jun-Aug	1891, 1903, 1910, 1912, 1922, 1924, 1927, 1931, 1948, 1956, 1958, 1960, 1965	13
Southern Europe	Jun-Aug	1900, 1901, 1906, 1914, 1915, 1940, 1959, 1975, 1979	9
U.S.A.	Apr-Jun	1892, 1908, 1944, 1945, 1947, 1983	6
U.S.A.	Jun-Aug	1891, 1908, 1915, 1943, 1945, 1947, 1951	7
U.S.A. and Canada	Apr-Jul	1892, 1902, 1909, 1927, 1929, 1943, 1945, 1947, 1961, 1974, 1980, 1983	12
India	Jun-Aug	1909, 1956, 1967, 1978	4
Mexico	Jul-Sep	1906, 1916, 1919, 1932, 1936, 1974, 1975, 1978	8
		Extremely dry years (+10 ≤ E ≤ +6)	Total dry extremes
European U.S.S.R.	Apr-May	1901, 1906, 1920, 1921, 1924, 1948, 1951, 1957, 1967, 1975, 1984	11
Altai Mountains	May-Jun	1893, 1900, 1901, 1904, 1923, 1929, 1931, 1935, 1940, 1945, 1951, 1955, 1962, 1963, 1965, 1974, 1982	17
Northern Europe	Jun-Aug	1893, 1899, 1901, 1904, 1911, 1921, 1934, 1937, 1947, 1949, 1952, 1959, 1973, 1975, 1976, 1983, 1984	17
Southern Europe	Jun-Aug	1894, 1904, 1922, 1927, 1928, 1945, 1946, 1952, 1958, 1962, 1965, 1977	12
U.S.A.	Apr-Jun	1910, 1931, 1934, 1936, 1937	5

(continued)

Table 9.3 (*continued*)

Region	Period	Extremely dry years (+10 ≤ E ≤ +6)	Total dry extremes
U.S.A.	Jun-Aug	1931, 1933, 1934, 1935, 1936, 1937, 1970	7
U.S.A. and Canada	Apr-Jul	1895, 1898, 1911, 1913, 1941, 1955, 1964, 1965, 1971	9
India	Jun-Aug	1905, 1928, 1941	3
Mexico	Jul-Sep	1891, 1892, 1930, 1940, 1956, 1962, 1964, 1977	8

45-year intervals (1881-1925 and 1926-1970) demonstrating that during the second of them, the number of cold months in Europe grew by a factor of 5-10; in western Siberia, however, this factor only reached 2. The situation was similar in other regions of the globe. It follows from the above that the dynamics of cold extremes (in this case, of severe winters) differ considerably from that of the average background indices of hemispheric climate. It appears too risky to try to characterize the climate in general, and the climate of the instrumental period in particular, using only the averaged trends. Such a conclusion is of practical importance; though temperature, on the average, increased in the 1980s, this increase was accompanied by more frequent climatic extremes. A more detailed catalogue of cold winters for the continent of Eurasia, based on the data of A. V. Yefanova is presented in Borisenkov (1988).

9.5   The formation of climate anomalies: Circulation patterns

Climate extremes of both signs always formed under certain circulation patterns and were definitely regional in character. Lamb (1982) demonstrated that the mildest European winters correspond to periods of prevailing westerly and south-westerly air flow (e.g., the beginning of the 16th century and 1920-1929) while the warm seasons and droughts fall during anticyclonic circulation periods (e.g., 1940-1949, 1976, etc.). As a rule, decades of cold European winters are associated with either cold Arctic intrusions or with the development of winter anticyclones over the continent. Anticyclonic situations in the Arctic entail easterly and north-easterly winds, leading to severe winter conditions in Europe. Such easterly winds prevailed during the winters of 1560-1569, 1699, 1820-29, and 1890-1899. The above holds for the current climate too. Several Soviet studies (Borisenkov 1988) have shown that cyclonic activity in the Atlantic and in the Western and Central Arctic plays a key role in the formation of extended Eurasian temperature anomalies. Climate oscillations in this region, lasting from years to several decades correspond to variations in the general atmospheric circulation.

During the last hundred years there have been three circulation patterns, each one differing in character. The first (1899-1915) was primarily cold, with winters dryer than summers. The extreme decade of that period was predominantly dry. According to B. L.

Dzerdzeevsky's classification (Dzerdzeevsky 1975) this peak was dominated by the northern meridional circulation.

The second period (1916-1956) was of predominantly zonal circulation, though both of the northern and southern meridional flows were frequent too. In most regions of the country and over the Northern hemisphere as a whole, this period was warm and humid. However droughts exceeded their normal frequency in the principal grain-growing areas of the country.

The third (current) period is meridional in nature, with a minimal zonal component. In contrast to the first period, this time the southern meridional circulation dominates. As of 1989, this period appears to be the one most disturbed; it started in 1956 and is not over yet. Frequent changes in the type of circulation processes, especially during the latter decade, makes this an epoch of meteorological extremes.

Table 9.4 lists the data on drought frequency in various regions of the country (after the All Union Scientific Research Institute of Hydrometeorology, World Data Center B (WDC-B) Catalogue, and another catalogue by Yu. L. Rauner [see Borisenkov 1988]). One may see that in the principal grain areas of the country the drought frequency was higher during the second circulation epoch. Since climate perturbations result from given circulation patterns, studying the major climatic anomalies in terms of circulation processes, is a vital task of climatology, particularly in view of the current period of increasingly unstable climatic conditions.

Table 9.4 Principal agricultural regions of the USSR: drought frequency vs. circulation epochs.

Region	Circulation epoch					
	1 1899–1915 17 years		2 1916–1955 40 years		3 1956–1972 17 years	
	A	B	A	B	A	B
After the VNIIGMI-MCD (WDC) catalogue						
Kazakhstan	5	29	27	68	9	53
Western Siberia	6	35	18	45	3	18
Ukraine	5	29	15	38	5	29
European Russia	7	41	17	42	6	35
After Yu. G. Rauner's catalogue						
Ukraine	6	35	18	45	8	34
Volga basin	6	35	19	48	9	36
Central chernozem provinces	2	12	8	20	7	28
Northern Caucasus	3	18	8	20	7	28
Altai province	6	35	22	55	10	40
North and Central Kazakhstan	5	29	19	48	10	40

A = number of droughts; B = ratio of dry years to epoch duration, per cent.

9.6   Conclusions

Studies in reconstructing the Russian climate of the last millenium from historical chronicles, and instrumental data have made it possible to reproduce its basic features and effects upon the economic and social life of the society.

An overview of Russian climate from the 10th century onwards, demonstrates that the country shifted from a minor climatic optimum to the Little Ice Age, approximately in the 13th century, much earlier than in Western Europe. In the climatic sense the Little Ice Age was highly variable both spatially and temporally. The main feature of the last millenium was the frequent recurrence of climatic extremes, during which Russia suffered 350 "hungry years" as a result of unfavorable climatic conditions. Of these 225 fell during the 16th-19th centuries, with extremes and famine frequent during the 17th, 18th and, particularly, the 19th centuries.

An examination of cooling or warming climatic trends is not sufficient to characterize the principal features of climates at different times. Such characteristics as the frequency of climatic anomalies (droughts, rainy and cool summers, severe winters, etc.) are more informative in this respect. Periods of climate warming (in particular, the first half of the 16th and 20th centuries) and the cold periods as well, were characterized by climatic extremes of both signs.

The spatial variability of climate was also quite high, for example, whole settlements perished in the Caspian basin and in Middle Asia due to drought, which started there in the 15th century. Now these areas are desertified.

Climate was highly variable during the period of instrumental observations. For example, cold, wet seasons in the European territory of the U.S.S.R. were registered in 1913-1917 and 1966-1970, and warm, wet seasons in 1925-1933 (excluding 1929/30). The Asiatic territories experienced such humid periods during the winter seasons of 1899-1919, and the summer seasons of 1945-1947 and 1958-1960. Dry, winter seasons in the European territory of the U.S.S.R. were observed in 1917-1936 and in the Asiatic territories, from 1931 to 1951.

Formation of climatic anomalies decisively depends on the pattern of the general atmospheric circulation. In view of the increasing sensitivity of national economies and the activities of society to climatic extremes, and the instability of our current climate, understanding the causes of climatic changes (with a view to eventually forecasting such anomalies) should be given top priority.

References

Anuchin, D. N. 1883. On the question of climatic changes. *Russkiye Vedomosti* 70.

Berg, L. S. 1947. *Climate and Life*. Moscow: OGIZ.

Belokurov, S. A. 1909. *Daily notes of the Prikaz of Secret Deeds*. Moscow, p. 368.

Bogolepov, M. P. 1907. On the climate oscillations in the Eurasian Russia in the historical epoch. Moscow: *Zemledeliye*, Vol.3, p.1-188, Vol. 4, p. 242-264.

Borisenkov, Ye. P. 1982. *Climate and Man's Activities*. Moscow: Nauks.

Borisenkov, Ye. P. (ed.) 1988. *Climate Oscillations of the Last Millennium*. Leningrad, Gidrometeoizdat.

Borisenkov, Ye. P. and V. M. Pasetsky, 1983. *Extreme Natural Phenomena in Russian Chronicles of the 11th to the 17th Centuries*. Leningrad, Gidrometeoizdat.

Borisenkov, Ye. P. and V. M. Pasetsky, 1988. *The Millenium of Unusual Natural Phenomena: A Chronicle*. Moscow: Mysl.

Buchinskii, I. Ye. 1957. *On the Past Climates of the Russian Plains*. Leningrad, Gidrometeoizdat.

Dugov, A. V. 1883. *Geographical Environment and the History of Russia. The End of the 15th to the Mid-19th Centuries*. Novosibirsk: Nauka.

Dzerdzeevsky, B. L. 1975. General Atmospheric Circulation. *Selected Studies*. Moscow: Nauka, p. 285.

Gadzatsky, G. 1907. *Struggling with Famine in the 11th to the 13th Centuries*. Saint Petersburg, p. 241.

Karamzin, N. M. 1816-1829. *History of the Russian State*. Saint Petersburg, Vol.1-14.

Kes, A. S., V. P. Kostyuchenko and G. N. Lisitsina, 1980. *History of Population and Ancient Irrigation in the South-West Turkmenia*. Moscow: Nauka.

Klyuchevsky, V. O. 1956-1980. *Collected Studies*. Vol. 1-8. Moscow.

Lezhkov, V. N. 1854. *On the People's Provision in Ancient Russia*.

Lamb, H. H. 1982. *Climate: Present, Past and Future*. Vols. 1 and 2, Methuen, London.

Likhachev, D. S. 1947. *Russian Chronicles*. Moscow-Leningrad.

Lyakhov, M. Ye. 1988. In: Borisenkov, Ye.P. (Ed.) *Climate Oscillations of the Last Millenium*. Leningrad: Gidrometeoizdat, p. 408.

Mesherskaya, A. V., and V. G. Blazhevich. 1978. Interannual variability of general humidity in principal agricultural regions of the European Soviet Union, Northern Caucasus, and Western Siberia. *Trudy GGO* (MGO Proceedings). issue 400, p. 87-98.

Pasetsky, V. M. 1978. *Meteorological Center of Russia: History of Foundation and Establishment*. Leningrad: Gidrometeoizdat. Pashuto, V. G. 1964. Hungry years in ancient Russia. *Agricultural Yearly of the Eastern Europe*. Minsk: Ac.Sci. Byelorussian SSR.

Romanovich-Slovatinsky, L. V., 1892. *Famine in Russia*. Kiev.

Rybakov, B. S. 1963. *Ancient Russia: Legends, Sagas, Chronicles*. Moscow: Ac.Sci. U.S.S.R.

Sazonov, B. I. and A. D. Kanayeva, 1985. On the increasing repetition frequency of cold months in Eurasia. *Trudy GGO* (MGO Proceedings). p. 486, Leningrad: Gidrometeoizdat.

Shvets, G. I. 1971 *Outstanding Hydrological Phenomena in the South-West Russia*. Leningrad: Gidrometeoizdat.

Shchepkin, V. N. 1886. Weather in Russia. *Istoricheskii Vestnik*, Vol. 24, p. 39-51.

Sreznevsky, B. I. 1902. *Climate Effects on Man*. Yuriev (Tartu).

Stanislavsky, L. I. 1926. Chronology of hungry years in 14 centuries: meteorological aspect. *Problems of Crop Yield*. p. 309-337, Moscow.

Vangengeim, G. Ya. 1952. Foundations of macrocirculation methods for the Arctic long range weather forecasts. *Trudy AANII* (Proceedings Arct. Antarct. Res. Inst.).

Veselovsky, K. S. 1887. *On the Climate of Russia*. Saint Petersburg.

Vild, G. I. 1883. *On Air Temperature in the Russian Empire*. Saint Petersburg.

Voeikov, A. I. 1902. On the problem of climate oscillations. *Meteorologicheskii Vestnik*. Saint Petersburg.

Yefanova, A. V. 1976. *Cold Winters in the Continents of the Northern Hemisphere*. Leningrad: Gidrometeoizdat, p. 114.

10 Reconstruction of 18th century summer precipitation of Nanjing, Suzhou, and Hangzhou, China, based on the Clear and Rain Records

P. K. Wang and D. Zhang

10.1 Introduction

Of all the nations in the world, China is one of the few that is blessed with a long and continuous history. In addition, thanks to the relatively early development of a writing system and numerous antiquary scholars in the past, historical studies have played a central role in classical Chinese literature. One of the more extreme of these scholars even stated that, "Although a country can be eliminated, the history should not be erased". While such a philosophy did not necessarily help the survival of the political systems at the time, it did help to create a lot of documents related to history.

Many of these historical documents contain information about weather and climate. For example, as a tradition, private diaries of scholars usually contain daily weather information. Poets wrote about the scene of different seasons in their romantic observations and unwittingly left behind information about the climate of their time. Military reports, grain price records, agricultural notes, and numerous other documents were related to weather and climate in some way (see Wang 1979, 1980; Wang and Chu 1982, for some examples).

The government, of course, kept records about the weather records of the country also. The weather records served many purposes. Should a city be waived of certain taxes because of a recent flood? Should the grains of South China be shipped to the North because of a widespread and prolonged drought that caused famine? What had the Emperor done wrong such that an unusually cold winter had fallen upon his country? All these were serious matters to the government and the imperial astronomers, and sometimes other officials who were ordered to make careful observations and report to the throne. More often than not, these reports contain weather information that surpasses other sources in terms of accuracy, continuity, and completeness.

In this study, we utilized a set of weather records preserved in the First National Historical Archives in Beijing, China to reconstruct the precipitation of the lower Yangtze region in the 18th century. This set of records is called the *Qing Yu Lu* which we translate as *Clear and Rain Records* (CRR). The First National Historical Archives is part of the former Imperial Palace and is now used for storing various documents. The Clear and Rain Records are the focus of study in this and the following chapter by W. C. Wang *et al*.

10.2 The Clear and Rain Records

The Clear and Rain Records have been recently discussed by Wang and Zhang (1988). They are basically daily weather records of the Qing Dynasty (1636-1910) which were compiled by local officials in various places and submitted to the emperor monthly or occasionally bi-monthly. The exact purpose of the order to submit these reports is not yet clear but it is believed to be related to taxation, for weather is closely related to grain production and grain was the basis of taxation in China for a long time.

The earliest known order for the submission of Clear and Rain Records was dated in the 10th month (of the Chinese lunar calendar) of the 24th year of the Kangxi period of the Qing Dynasty (i.e. A.D. 1685). Each county and prefecture was asked to submit such reports monthly in the form of a "yellow book". However, this practice changed over time and therefore the earlier records were probably incomplete. In addition, as time went on, some records were destroyed or simply lost. Consequently, only a few sets of these records have survived until today. Among the existing sets, the Beijing, Nanjing, Suzhou, and Hangzhou Records are the longest and most continuous. Beijing's Clear and Rain Records have already been studied (Central Meteorological Bureau 1976). The other three sets have not been

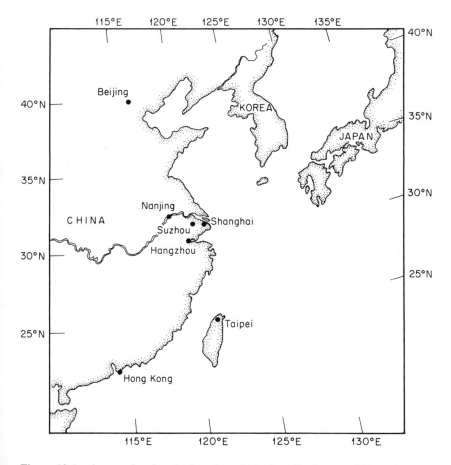

Figure 10.1 A map showing the location of Nanjing, Suzhou, and Hangzhou.

subjected to analysis. Our goal in this study is to use these 3 sets of Clear and Rain Records to reconstruct the summer (May-August) monthly precipitation of Nanjing, Suzhou, and Hangzhou in the 18th century. These three cities are located in the lower Yangtze River region as shown in Figure 10.1.

The contents of the Clear and Rain Records consist of the year, month, day, and the weather conditions. The weather conditions usually include the sky conditions (clear, overcast, fog, or rain, etc.) wind directions (N, S, E, W, NE, SE, NW, SW) precipitation type and intensity (rain, snow, light rain, heavy rain, thunderstorm, etc.) and duration (the beginning and ending hours). The duration was reported using the traditional time system in China: a day is divided into 12 intervals, called 12 *shichens* and one *shichen* equals two hours. By carefully reading the entries, it is possible to obtain a time resolution of one hour. An example of the Clear and Rain Records is shown in Figure 10.2. For some reasons not yet clear, the Records of Beijing did not include the wind information, but those of Nanjing, Suzhou, and Hangzhou did include all the weather information described above.

The lengths of these four sets of records are different. Beijing's records are the longest, beginning in 1724 and ending in 1903. The lengths of the other three are: Nanjing 1722-1798; Suzhou 1723-1821; and Hangzhou 1723-1773. As indicated above, Beijing's records have been analyzed and used for reconstructing the precipitation series of Beijing in this period. The reconstructed precipitation series have been used for studying the characteristics of wet and dry periods (Feng 1982) and in the reconstruction of summer temperature series (Zhang and Liu 1986; W. C. Wang *et al.*, this volume). On the other hand, the records of the other three cities have not been studied. Beijing's records were studied first, not because Beijing is the national capital but because they can be quantified with relative ease. This is because during the period from 1840 to 1903, both the traditional Clear and Rain Records and

蘇州晴雨錄

六月

初一日晴東南風戌時微雨即止夜陰
初二日晴東北風夜晴
初三日晴東北風夜晴
初四日晴東南風夜晴
初五日陰東北風未時細雨起亥時未止
初六日雨東北風午時雨止夜陰
初七日晴東南風巳時細雨即止夜晴
初八日晴東南風夜晴
初九日子時細雨起東北風酉時雨止夜陰
初十日晴東南風夜晴
十一日晴東南風夜晴

Figure 10.2 A page of the Clear and Rain Records of Suzhou, June 1742.

instrumental observations were maintained. One can simply use the instrumental data to calibrate the descriptions in the Clear and Rain Records, and then use the calibration to convert the descriptive information in the records into quantitative precipitation series. This method is unfortunately not applicable to the other three cities, for there were no such convenient overlapping periods of instrumental records. In order to utilize the precipitation information in the Clear and Rain Records of these three cities to reconstruct the 18th precipitation series for this region, some other reliable conversion techniques have to be found. This is addressed in the following sections.

10.3   Digitization of the Clear and Rain Records

The first step in the reconstruction work is to check whether there were errors in these records. Here we are speaking of errors such as mistakes in dates, lost paragraphs, wrongly written or missing words, and other written errors. As far as the original records are concerned, these errors did not seem to happen. This is probably due to the fact that these reports were to be submitted to the emperors. Any obvious error would result in stiff punishment of the reporter. In addition, no contradictory entries such as two mutually exclusive reports (e.g. clear and rain occurring in the same hour) were found. The only error that needed correction was the disorder of the pages. This probably occurred when the records were moved from one storage place to the other and sheets of records were piled together without due regard for the correct order. Since the year and the month were only written at the beginning and the end of the document, those pages in the middle showed only the day. Thus if some pages in the middle were misplaced, no year and month information can be checked. Fortunately, by carefully matching the style of the caligraphy, it was possible to return these pages back to their proper order. More difficult to deal with are the missing records. They may have been destroyed during the many wars in the 19th and early 20th century, or displaced and stored in other places, or simply lost. Fortunately, the amount of missing records in these three sets is relatively small and therefore does not pose a serious problem.

In order to facilitate computer processing and analysis, we designed a code system to transform the verbal information in these records into digital series. The dates in the original records were given in the traditional Chinese lunar calendar system. In the process of digitization, they were converted to Gregorian dates.

10.4   Rainy day information in the Clear and Rain Records and modern precipitation data

Since there were no instrumental data to calibrate the Clear and Rain Records of Nanjing, Suzhou, and Hangzhou, there is no way to be absolutely sure about the reliability and data quality. But at least we can make some tests to see if the statistical properties of a certain series derived from these records show reasonable characteristics. One way to do this is to compare the statistics of these series with those from modern observational records. The simplest series for comparison purposes is the dry/rainy day series (a 0 1 series).

From the three sets of digitized Clear and Rain Records we extracted information on rainy days and non-rainy days for the 30-year period of 1741-1770 and called the resulting series the CRR series. Similar information was extracted from modern data from the same three cities and formed the rainy/non-rainy day series for the 30-year period of 1951-1980. The t-test was then performed for these two groups of series. First, the mean (x_1) and the standard deviation (s_1) of the CRR series and the corresponding x_2 and s_2 of the modern series were calculated. The t-value was then calculated. The limiting t value, t_c is 3.46 at 0.1% significance limit. The calculated t-values for the frequency of rainy days for each month and each location are tabulated in Table 10.1. The fact that all t values are smaller than t_c indicates that the CRR and the modern precipitation series can be considered as independent samples of the same normally-distributed population. Thus data derived from the Clear and Rain Records can be regarded as reliable and comparable with modern data.

Table 10.1 Frequency of rainy days in the 18th century from Clear and Rain Records (CRR) and in modern times from instrumental observations.

	Nanjing		Suzhou		Hangzhou	
	CRR	Modern	CRR	Modern	CRR	Modern
May	8.0	10.5	11.4	13.0	13.6	16.9
Jun	9.7	11.4	13.5	12.5	15.0	14.4
Jul	9.5	12.3	11.3	12.0	10.9	11.7
Aug	9.5	11.3	10.9	10.7	11.3	12.5
Jun-Aug	36.7	45.5	47.2	48.2	50.8	55.5

Table 10.1 further reveals that the geographical variations of the rainy day statistics are consistent in both the CRR and modern precipitation series. In both series, Hangzhou has the highest rainy day frequency, followed by Suzhou and Nanjing. The difference in rainy day frequency certainly represents the difference in local climates at a fixed time. On the other hand, at each location there is also evidence of climatic change. For example, the mean rainy day frequency of May was smaller than that of June in all three locations in the 18th century. The opposite is true in the modern data. This reflects a shift in the rainy season in this region. More detailed analyses of the climatic change will be discussed in another paper.

In studying Table 10.1, we discovered that while the monthly rainy day frequencies of Hangzhou and Suzhou show consistent temporal behavior, that of Nanjing shows systematic bias. For instance, in the CRR series the mean rainy day frequency was more in May and less in June and July, than the corresponding modern values in the case of Suzhou and Hangzhou. The frequencies are about the same in August. On the other hand, the Nanjing case, shows that the mean monthly rainy day frequency in the 18th century was always less than the modern value in all three months. Since Nanjing is geographically adjacent to Hangzhou and Suzhou, one would expect them to have similar, if not identical, climatic trends. It is therefore felt that the CRR data of Nanjing contains a systematic bias.

One possible cause of this bias is the omission of trace rain in Nanjing's Clear and Rain Records. Perhaps officials in Nanjing at the time did not feel that it was necessary to bother with such small amounts of rainfall. Although this omission would not result in significant error in the total precipitation, it would result in substantial error in monthly rainy day counts. In order to check whether such omissions occurred or not, we made a simple comparison between the statistics of precipitation of Suzhou and Nanjing as revealed in the Clear and Rain Records. Combining 3 intensity categories (light rain, rain, heavy rain) and 6 duration categories (1 hour, 1-4 hours, 5-8 hours, 9-12 hours, 13-16 hours, 17-24 hours) we formed 18 combinations (1: light rain [1 hour]; 2: light rain [1-4 hours]; . . ., 18: heavy rain [17-24 hours]). We then compared the monthly frequency of these 18 combinations for Nanjing and Suzhou. The results are shown in Figure 10.3. The premise of such a comparison is that since Nanjing and Suzhou are close to each other (within a distance of 200 km) the patterns of the distributions of these frequencies should be at least qualitatively similar. From Figure 10.3 we see that while the ratios of the frequencies of most combinations are similar in both cities, the ratios of the combinations of light rain with short durations show a marked difference. Suzhou had a high frequency of these combinations but Nanjing had very low frequency. We also made such a comparison for the modern precipitation data of Suzhou and Nanjing, and found that there weren't any substantial differences. The only reasonable explanation of this is that Nanjing reporters had consistently omitted many short duration, light rains.

Since the monthly rainy day frequency is one of the predictors for our regression relations (to be derived below) it is necessary to take care of this bias so that its effect can be minimized. The treatment of this problem will be addressed later.

10.5 Reconstruction of the 18th century precipitation series

We are now ready to describe the method used in reconstruction of the monthly precipitation series in the 18th century based on the Clear and Rain records. Although descriptions of the precipitation intensity were given in these records, they are merely qualitative statements and cannot be directly equated to definite precipitation rates. It is clearly useless to simply 'guess' the actual intensity based on these descriptions. We need to look for an objective method so that these qualitative statements can be converted to quantitative numbers and the errors, if any, can be assessed objectively.

The method we adopted for this study is decribed as follows. First, we use modern precipitation data to search for some predictors that are well-correlated to precipitation. We then study the Clear and Rain Records to see if the time series of these predictors can be constructed. If so, then we can establish a set of regression relations between the preci-pitation and the predictors using modern data. Assuming that the modern relations are applicable to the 18th century case, we can retrieve the 18th century precipitation by using the historical predictor series in the regression relations. Needless to say, the results obtained should be carefully tested.

The first predictor we tested was the total monthly frequency of rainy days (f_m) because it is already known that the monthly precipitation (R_m) is well-correlated with f_m in the lower Yangtze region where the three cities are located. From modern precipitation data, we established the regression relations of R_m and f_m for these three cities. Next the 18th century f_m series were constructed from the Clear and Rain Records. Using the ancient f_m series with the regression equations, the 18th century monthly precipitation series were obtained.

The results from this first attempt were not very satisfactory. The correlation coefficients between the modern observed precipitation series and the regression-derived series range

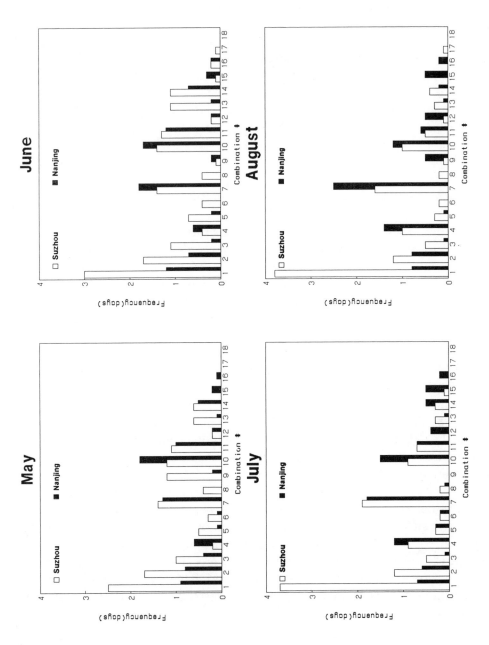

Figure 10.3 Comparison of the mean monthly frequencies of 18 rain class/duration combinations of Nanjing and Suzhou in the Clear and Rain Records. The combinations are (1) light rain (< 1 hour) (2) light rain (1-4 hours) . . . (7) rain (< 1 hour) . . . (18) heavy rain (17-24 hours).

from 0.4 for the Suzhou June series to 0.8 for the Hangzhou May series, with 0.6 to 0.7 the most common situation. While these correlations are not necessarily bad, they are far from satisfactory. The situation can become worse when applied to the 18th century case, considering other uncertainties involved in the data.

The main difficulty in the above univariate approach is in the fact that each single rainy day was assigned an equal weight in the regression. In reality, the precipitation in a rainy day can vary from 0.1 mm to more than 100 mm. It is more reasonable to give days with greater precipitation more weight. To achieve this goal, we have to apply a multiple regression method to our problem. The predictors will now consist of several rainy day frequencies of different intensity categories, instead of a single total monthly frequency.

The first thing to decide is how many intensity categories should be considered in the regression. Too few categories would result in poor performance similar to the univariate approach. Too many categories, on the other hand, makes the treatment of data very cumbersome. Besides, it is impossible to form that many categories from the Clear and Rain Records anyway. We tried from 2 to 8 categories and finally decided that a 4 category scheme is the optimum choice. These four rain intensity categories are:

Category 1: $0.1 \leq R < 2.0$ mm/day
Category 2: $2.0 \leq R < 10.0$ mm/day
Category 3: $10.0 \leq R < 25.0$ mm/day
Category 4: 25.0 mm/day $\leq R$

where R is the daily rainfall rate. These categories are essentially the same as the descriptive rainfall intensities from daily weather reports currently used by the Central Meteorological Station of China: light rain (0.1-10 mm/day) medium rain (10-25 mm/day) and heavy rain (>25 mm/day) except we further divided the light rain category into two. The monthly frequencies of these 4 categories plus the total monthly frequency form 5 predictors x_1, x_2, . . ., x_5, respectively. Let y_i be the precipitation of month i (e.g., y means the precipitation of May). From the modern data, we determined a set of relations between y_i and x_i's using a stepwise regression technique. The lengths of data used for the regression are: Nanjing 1905-1980; Suzhou 1930-1980; Hangzhou 1951-1970. These were the data available to us at the time of the analysis.

The regression equations determined in this way are listed in Table 10.2, where y_{678} is the total summer precipitation from June to August. The correlation coefficients are very high in all cases, indicating the regression equations simulate the data very well. Note that not all regression coefficients are present in these equations. In general, the weights of x_1 and x_2 are smaller, sometimes vanishing, compared to those of of x_3 to x_5 although there are some exceptions.

In order to apply these regression relations to retrieve the 18th century precipitation, we need to determine the 5 predictor series x_1 to x_5 from the Clear and Rain Records. The determination of the x_5 series presented no difficulty at all since the total monthly frequency of rainy days can be counted easily. The other four series, on the other hand, were not so straightforward. The main difficulty was: what rain intensity do statements such as "light rain" or "heavy rain" in the Clear and Rain Records stand for? Without such a knowledge, it was impossible to determine the intensity categories and hence the x_1 to x_4 series.

Table 10.2 Regression equations for precipitation derived from modern precipitation data of Nanjing, Suzhou, and Hangzhou. The definitions of x_1 to x_5 are given in the text.

(1) Nanjing

$y_5 = -2.228 + 1.348x_1 + 5.316x_2 + 14.797x_3 + 43.385x_4$ $r = 0.97$
$y_6 = -24.673 + 7.782x_3 + 48.037x_4 + 5.935x_5$ $r = 0.83$
$y_7 = -29.263 + 81.872x_2 + 17.080x_3 + 65.991x_4$ $r = 0.92$
$y_8 = -14.298 + 7.770x_2 + 14.971x_3 + 52.280x_4$ $r = 0.91$
$y_{678} = -20.153 + 56.070x_4 + 4.846x_5$ $r = 0.88$

(2) Suzhou

$y_5 = 7.753 + 4.953x_2 + 16.213x_3 + 32.094x_4$ $r = 0.96$
$y_6 = 32.103 + 16.504x_3 + 39.362x_4$ $r = 0.89$
$y_7 = -14.822 + 18.666x_2 + 16.361x_3 + 53.248x_4 - 4.010x_5$ $r = 0.96$
$y_8 = 3.053 + 7.732x_3 + 41.567x_4 + 4.072x_5$ $r = 0.88$
$y_{678} = -84.407 - 11.703x_1 + 31.442x_4 + 13.560x_5$ $r = 0.91$

(3) Hangzhou

$y_5 = 104.458 - 5.968x_1 + 9.829x_3 + 3.870x_4 + 11.645x_5$ $r = 0.98$
$y_6 = 26.367 + 20.600x_3 + 40.582x_4$ $r = 0.90$
$y_7 = 10.282 + 23.084x_3 + 37.919x_4$ $r = 0.98$
$y_8 = -0.809 + 51.412x_4 + 5.374x_5$ $r = 0.91$
$y_{678} = -56.089 + 21.364x_3 + 31.906x_4$ $r = 0.94$

In searching for a reasonable way to determine the equivalent rain intensity for these qualitative statements, we noticed that there exists a special characteristic of the rain intensity frequency distribution in modern precipitation. The distribution curve of monthly frequencies of rain intensity at a location seems to possess a relatively constant shape. Figure 10.4 illustrates the example of Nanjing for the months of May to August. Modern daily precipitation data for Nanjing from 1905 to 1980 are used. What are shown here are the mean monthly frequencies (in days) of rainy days with intensity greater than the values indicated in the abscissa. The values in the abscissa are selected to be 1.0, 5.0, 10.0, 15.0, . . ., 100 mm/day. It is obvious that the frequency of light rain days is higher than that of heavy rain days. The frequency of rainy days with an intensity of 5 mm/day represents 51% to 58% of the total. This feature, and the general slopes of the curves, are very similar for the same month for three different periods: 1905-1930, 1935-1980, and 1905-1980. On the other hand, there are differences between months. For example, the frequency of days with a rain intensity of >100 mm/day is high in June and July, less in August, and nearly zero in May. We also found that the frequency of days with 1.1 to 5 mm/day intensity is consistently higher than that with 0.1-1.0 mm/day intensity in June and July (diagrams not shown here). The frequency distributions of Suzhou and Hangzhou show essentially the same features.

If we plot the frequency of the 4 predictor categories intead of using a 5mm/day interval, we obtain the results as shown in Figure 10.5. The simplification to 4 categories is desirable since we would like to apply the conclusions to the 18th century case where it is impossible to have detailed categories as in Figure 10.4. Figure 10.5 shows that the general characteristics described in the preceding paragraph remain valid, although there are some discrepancies between the frequencies of Category 1 and 2 in June and July in two different periods. The mean frequency of Category 1 in June in the period 1905-1930 was somewhat less than that of

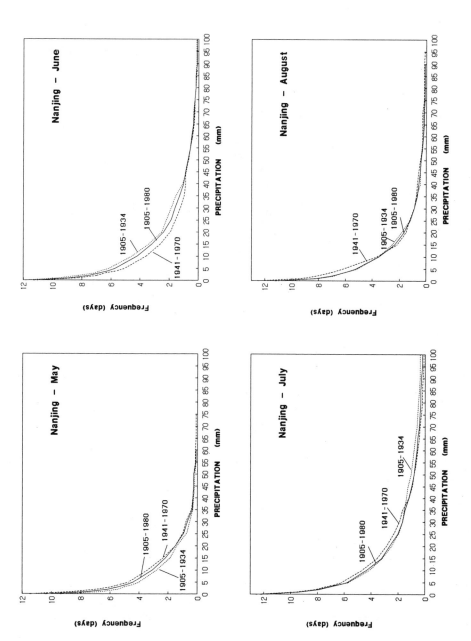

Figure 10.4 Mean monthly frequency of precipitation intensity of Nanjing larger than a certain precipitation rate for May–August. For example, the point corresponding to 15 mm means the number of days in a month with a precipitation rate > 15 mm/day.

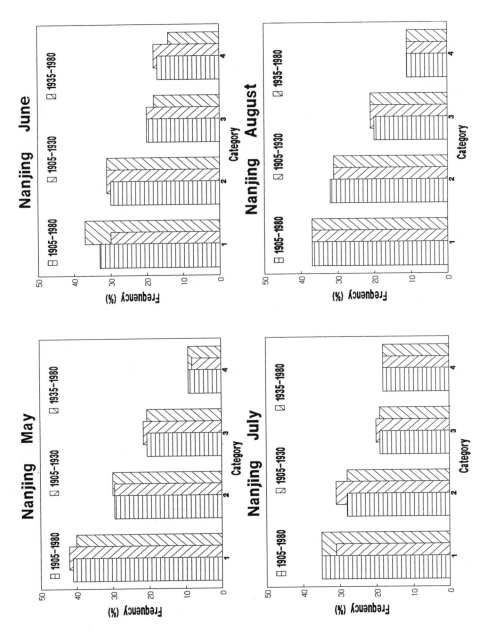

Figure 10.5 Mean monthly percentage frequency of rainfall grade for Nanjing (modern data).

Category 2. The distribution in the period 1935-1980 was the opposite. A similar situation exists in July. However, this would not cause any severe problem in retrieving the monthly precipitation since the weight of Category 1 is zero in the June and July regression equations of Nanjing. Indeed, as can be seen from the previous regression equation sets, Category 1 frequency is absent in all equations except for Nanjing and Hangzhou in May.

The above observation suggests that the frequency distribution of rainfall rates may be a fairly constant characteristic of local climate in the lower Yangtze region. It therefore seems to be reasonable to consider that the 18th century summer precipitation of Nanjing, Suzhou, and Hangzhou had the same characteristic frequency distributions. Specifically, we can think of the distributions shown in Figure 10.5 as *the* distributions of 18th century precipitation and require that the data compiled from the Clear and Rain Records should also show similar, if not exactly the same, distributions.

The last statement enables us to design a scheme to convert the qualitative intensity descriptions in the Clear and Rain Records to more quantitative rainfall rate categories. The steps to be taken are as follows:

(1)　Form various rain intensity/duration combinations from the Clear and Rain Records.
(2)　Determine in which one of the 4 rainfall rate category a particular combination belongs.
(3)　Count the monthly frequency of the 4 categories formed in (2).
(4)　If the distributions of frequencies obtained from (3) are significantly different from that of Figure 10.5, go back to step (2) to revise the category assignment.

The above procedure is repeated until satisfactory results are obtained. Obviously, the core of the above procedure is step (2). In the following, we use the case of Suzhou as an example to illustrate how it is done.

We first form 24 rain intensity/duration combinations from the Clear and Rain Records of Suzhou. Each combination is given a number as shown in Table 10.3. These 24 combinations

Table 10.3　Designation of combination categories from Clear and Rain Records at Suzhou. The rainfall class descriptions are taken from the language of the original records.

Class	Duration in hours							
	≤ 1	1-2	3-4	5-6	7-8	9-12	13-17	17-24
light rain	1	2	3	4	5	6	7	8
rain	9	10	11	12	13	14	15	16
heavy rain	17	18	19	20	21	22	23	24

are then divided into 4 categories. Thus, for example, combinations 1, 2, and 3 in May are considered as equivalent to rainfall rate category 1, combinations 4-11 are equivalent to rainfall category 2, etc. Each month the conversion scheme is somewhat different from other months. The designation of a combination to a certain category was not made arbitrarily but had to satisfy the following rules:

(1)　Each category must occupy a contiguous area in a chart like Table 10.4.

Table 10.4 The scheme used to convert the intensity/duration combinations into 4 precipitation grades for Suzhou.

Rainfall Class	Duration (hours)							
	< 1	1-2	3-4	5-6	7-8	9-12	13-16	17-24
May								
light rain	1	1	1	2	2	2	2	2
rain	2	2	2	3	3	3	4	4
heavy rain	2	3	3	3	4	4	4	4
June								
light rain	1	1	1	2	2	2	2	3
rain	2	2	2	3	3	3	4	4
heavy rain	2	3	3	3	4	4	4	4
July								
light rain	1	1	2	2	2	2	3	3
rain	2	2	3	3	3	4	4	4
heavy rain	2	3	3	3	4	4	4	4
August								
light rain	1	2	2	2	2	3	3	3
rain	2	2	3	3	3	4	4	4
heavy rain	2	3	3	4	4	4	4	4

(2) In a specific row, categories of higher rainfall rate must be located to the right of categories of lower rainfall rate.

(3) In a specific column, categories of high rainfall rate must be located lower than categories with low rainfall rates.

The ultimate choice of a conversion scheme depends on how much the resulting frequency ratios between these categories resemble the modern ratios. Exact similarity was not expected but an attempted was made to make these ratios as close as possible. Extensive trial-and-error routines had to be gone through before a satisfactory scheme was finally attained. Table 10.4 was obtained this way and represents the satisfactory scheme for Suzhou. The comparison of the ratios between the 18th century and the modern cases is shown in Figure 10.6. Although the ratios are not exactly the same in the two cases, they are fairly close. The comparison for the cases of Hangzhou and Nanjing are shown in Figures 10.7 and 10.8, respectively.

The treatment of Nanjing records was somewhat different because of the previously-mentioned bias in Category 1 frequency. It was decided to maintain good relative ratios for Categories 2, 3, and 4, but not for Category 1. It is seen in Figure 10.8 that the frequencies of

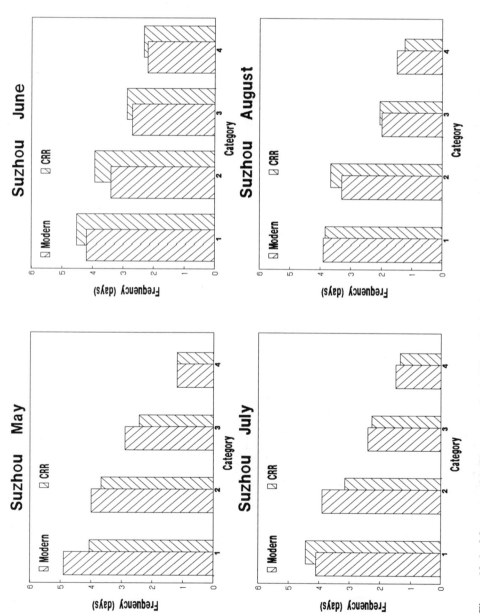

Figure 10.6 Mean monthly frequency of rainfall grades for Suzhou compiled from CRR and modern data using the conversion scheme in Table 10.4.

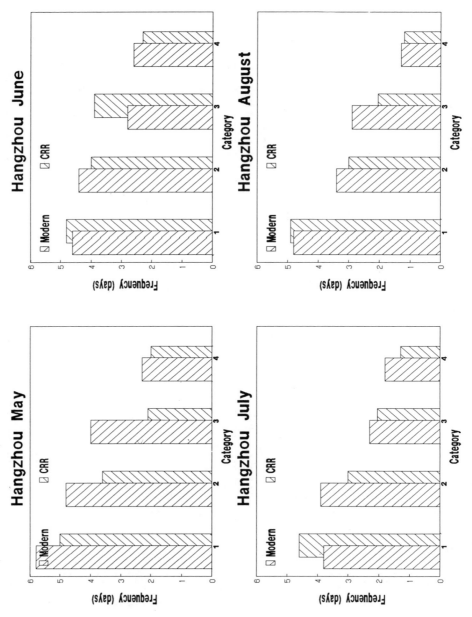

Figure 10.7 Mean monthly frequency of rainfall grades for Hangzhou compiled from CRR and modern data using the conversion scheme in Table 10.4.

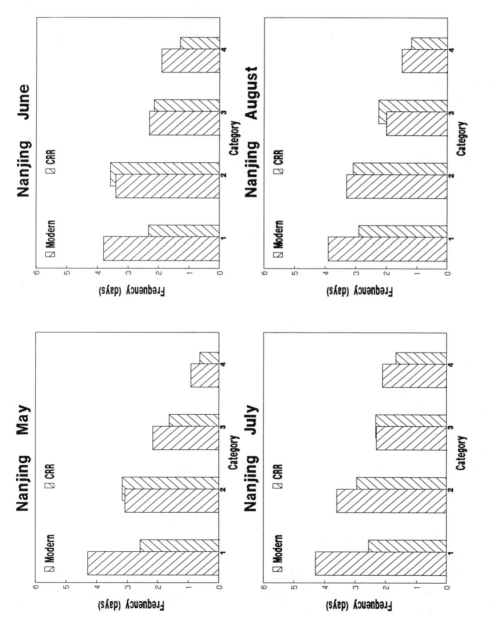

Figure 10.8 Mean monthly frequency of rainfall grades for Nanjing compiled from CRR and modern data using the conversion scheme in Table 10.4.

Category 1 in the 18th century are low in all months when compared with the modern frequencies, but the comparisons of the other 3 categories are quite satisfactory. Except in the May case, the influence of the lower Category 1 frequency is completely negligible. In May, due to the lower Category 1 frequency and the non-zero regression coefficient, the rainfall rate retrieved from the regression equation might be slightly over-estimated *on the average*. It is important to keep in mind that the comparisons of the frequency ratios and the determinations of the regression equations are all based on the *mean* frequency over a certain period. The actual effect of the above bias on a particular month in a particular year might be somewhat different from the others.

After such conversions were done, the x_1 to x_5 series of the three cities were established. Then, by means of the regression equations, monthly precipitation in the 18th century was calculated. Figure 10.9 shows a flow chart that explains the methodology more concisely.

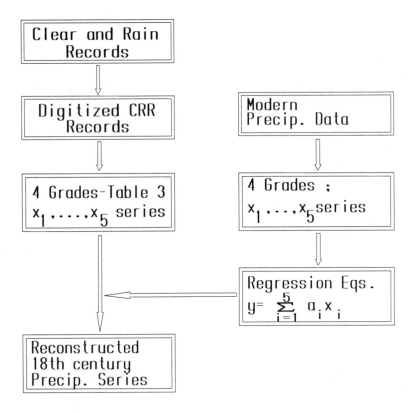

Figure 10.9   A flow chart summarizing the reconstruction methodology.

10.6   Results and discussions

Figures 10.10 to 10.12 show the reconstructed monthly precipitation of May-August in Nanjing, Suzhou, and Hangzhou in the 18th century based on the Clear and Rain Records as described in the previous section. Since these reconstructions were based on the regression

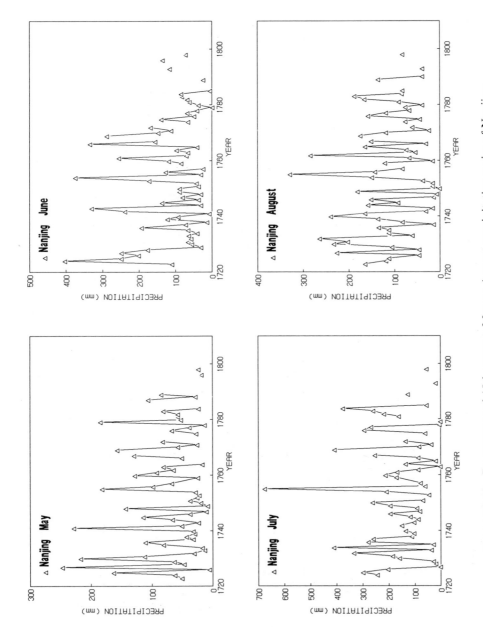

Figure 10.10 Reconstructed 18th century May to August precipitation series of Nanjing.

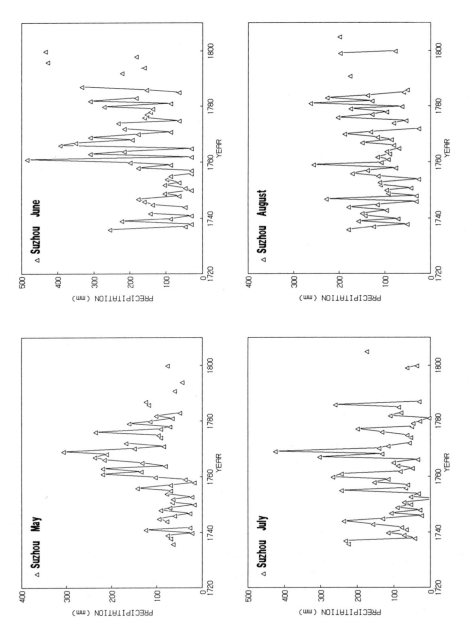

Figure 10.11 Reconstructed 18th century May to August precipitation series of Suzhou.

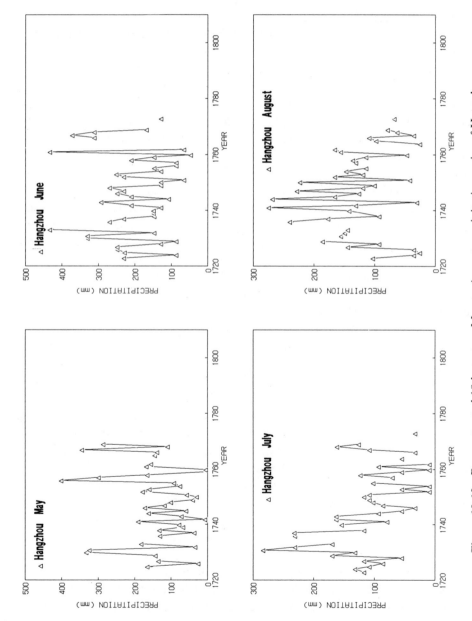

Figure 10.12 Reconstructed 18th century May to August precipitation series of Hangzhou.

relations derived from modern data, it is necessary to examine the influence of modern data in more detail before one accepts these results. We first note that the lengths of modern data used to derive the regression equations are different for these three places. This is entirely due to the limitation of the available data at hand. Potentially, the regression equations based on a longer data series and those based on a shorter series may be different. In addition, two data series of the same length but of different periods may also result in different regression relations. The question here is: would the data length and the selection of data affect the reconstructed results? In the following, we use the case of Nanjing to obtain an answer.

The regression equations for Nanjing were derived from the precipitation data of 1905-1980. There were 5 years of missing data so the actual data length is 71 years. We selected two subsections of this series: (A) 1905-1934 and (B) 1951-1980, for our checking purpose. It was thought that 30 years of data is sufficiently long for the regression purpose. Using these two distinct data series and going through the same procedure as described in the previous section, we derived two sets of regression equations, A and B. Applying these two sets of equations to the x_1 to x_5 series of Nanjing for the period of 1905-1980, we get two fitted precipitation series for the same period. It was found that the two fitted series based on two different data subsets are very similar to each other, the cross-correlation coefficient being easily significant at the 0.1% significance level (>0.93). This test indicated that using data of different periods to derive regression equations has little effect on the reconstructed results.

To test the effect of data length on the reconstruction, we compared the two fitted series, A and B, with the original fitted series (called series C here) of Nanjing (using the 1905-1980 data as done in last section). We also compared the new reconstructed 18th century series Q_A and Q_B with the original reconstructed series Q_C. The correlation coefficients of these series are greater than 0.97 (again easily significant at the 0.1% significance level). This test indicates that the length of data has little effect on the reconstructed results.

The above two tests revealed that neither the length of data nor the selection of a particular sub-set has any appreciable effect on the reconstructions, as long as the sub-set has reasonable length and quality. Thus, even though we only used 20 years of data to derive the regression equations for Hangzhou, the reconstructed results are probably acceptable. Suzhou's regression equations were derived based on 50 years of data and should be quite adequate.

The above tests do not guarantee the reliability of the reconstructed results, however. The only absolute proof of the reliability of these results is to show that they agree with the contemporary instrumental data which are unfortunately non-existent. The next best proof then is to see if there was other evidence, preferably derived from completely different data sources, that show similar characteristics as the present results. For this purpose, we can compare the present results with the 500 year flood and drought (F-D) series of China which includes the 3 cities considered in this study (Central Meteorological Bureau 1982). The F-D grade series was established from climate information contained in local records, which are completely different from the Clear and Rain Records. The techniques of constructing the F-D series are completely different from the present technique. The two data sets are therefore independent series.

The F-D series are represented by moisture grade indices. From historical flood and drought records, a moisture grade index was determined for a station in a certain year. There are 5 grade indices: grade 1 represents the wettest condition, grade 5 the driest, and grade 3 the normal condition. The moisture index is defined as: [6-(F-D grade)]. Thus the higher the index, the more humid the year was. The F-D series reflect not the summer precipitation but rather the *annual* moisture conditions. They are therefore not really equivalent to our reconstructed summer precipitation series and exact correspondence of the two sets of series is not expected. Nevertheless, a large percentage of the annual precipitation in this region should be accounted for by summer precipitation.

Figures 10.13a to c show a year-to-year comparison of the F-D and the reconstructed 18th century summer precipitation series for Nanjing, Suzhou, and Hangzhou. While we are not expecting exact correspondence of the two, we see that there is a general consistency of the two series in all three places. The features of Figures 10.13a and 10.13b for Nanjing and Suzhou are basically the same as described in the previous paragraphs. Figure 10.13c shows the comparison for Hangzhou. For those years where the comparisons can be made, the two series are mostly in phase.

There are 6 points in Figures 10.13a to c where the two series show opposite trends that can be explained. These are: (1) Nanjing 1733 (2) Nanjing 1742 (3) Suzhou 1751 (4) Suzhou 1762 (5) Hangzhou 1724 and (6) Hangzhou 1762. After examining the original records, we found that these inconsistencies can be explained as follows. (1) and (2) are caused by errors in the F-D series. Both flood and drought occurred in the same year in these two years. In both years, drought occurred first, then flood. The original records were not sufficiently detailed to allow correct judgement of the relative importance of flood and drought. (3) is caused by the missing Suzhou local records in that year. The F-D grade index was obtained by interpolating indices from neighboring areas and the interpolation might be incorrect. (4) and (6) are caused by over-estimates of precipitation from a severe storm that caused a flood in this region. (5) is caused by the fact that abnormally high rainfall occurred in the spring of 1724 which was, of course, not reflected in the summer rainfall series.

Figures 10.14a and b show a comparison of the moving averages of the two sets of series. Figure 10.14a shows that the two 5-year moving average series of Suzhou have quite similar characteristics. The dominant peaks and troughs in the middle sections of the two series are in phase, except the 1771 peak of the moisture index which remains to be studied. This suggests that: (1) summer rainfall was the dominant component of annual precipitation in Suzhou as it is now (as can be attested by modern data) and (2) the data derived from the Clear and Rain Records of Suzhou agree fairly well with that from local records. In fact, it is more appropriate to state the point (2) conversely, namely, local records conform with the Clear and Rain Records since the latter contain more detailed and accurate descriptions of the weather. Figure 10.14b shows the comparison for Nanjing. Unfortunately, the F-D series has missing data for a period after 1748, so only the section from 1723 to 1748 can be compared. It is seen that the general trends of decreasing precipitation in the two series are similar. Minima in the F-D series are also consistent with that in the precipitation series. It thus appears that the two conclusions for Suzhou are also applicable to Nanjing. These two figures demonstrate that the reconstructed series are at least consistent with series obtained from completely different data sources.

The comparison for Hangzhou is not made because both F-D and precipitation series contain missing data which occurred in different years. The moving averages cannot be formed. However, the following comparison should make it clear that in Hangzhou, too, the data series derived from the Clear and Rain Records and local records are consistent. Many

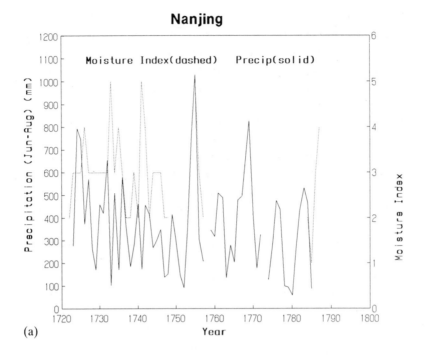

(a)

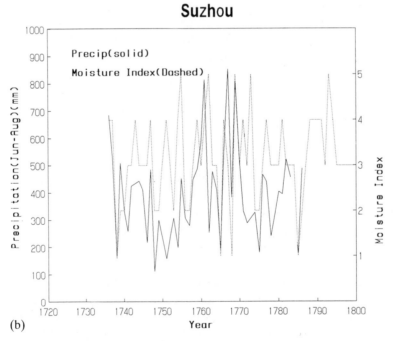

(b)

of the descriptions in local records can be verified directly by the entries in the Clear and Rain Records. Statements like "heavy rains for several days", "No rain in summer. Severe drought" and "x Month, x Day: severe storm caused a flood" can be cross-checked by statements in the Clear and Rain Records. Only the Clear and Rain Records gave more detailed, semi-quantitative descriptions than the local records.

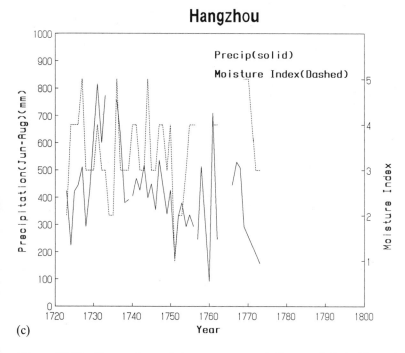

(c)

Figure 10.13 Comparison of the reconstructed Clear and Rain Record precipitation series with the moisture index series derived from the flood-drought records. (a) Nanjing (b) Suzhou and (c) Hangzhou.

10.7 Conclusions

In the above sections, the Clear and Rain Records of Nanjing, Suzhou, and Hangzhou were described and analyzed. Reconstructions of the 18th century monthly precipitation series from May to August were made based on these records. The results of the reconstruction are generally consistent with the flood and drought grade index series derived from local records. On the one hand, this says that these two historical series are both reliable sources of past climate. On the other hand, we have to realize that the data quality of the Clear and Rain Records is higher than that of the local records. The time resolution of these records is one hour whereas the best resolution one can get from the local records is essentially one season. The Clear and Rain Records can be used as a more precise calibration for the local record series.

In this study, we have only discussed the Clear and Rain Records themselves and the reconstructed precipitation series derived from them. We have not discussed the detailed climatic features of these series, which will be examined in another article. In addition, there is other useful information in the Clear and Rain Records that are currently under study such as the wind and sky conditions. The wind series, especially, will be of great value in determining the relations between summer precipitation in this region (the *meiyu*) and the Northwest Pacific high pressure circulation regime.

207

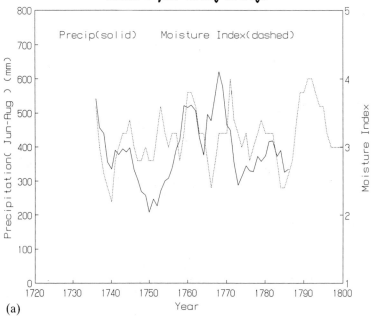

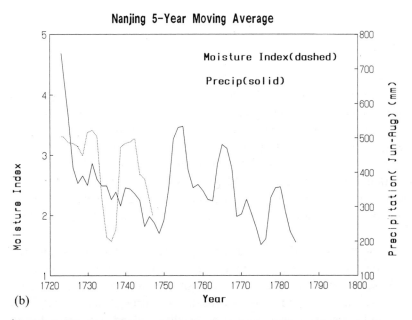

Figure 10.14 Comparison of the 5-year moving averages of the reconstructed Clear and Rain Record precipitation series with the moisture index series. (a) Suzhou. (b) Nanjing.

Acknowledgements

This work was partially supported by U.S. National Science Foundation Grant ATM-8511905 (Climate Dynamics Program, Division of Atmspheric Science)

References

Central Meteorological Bureau 1976: *250 Years of Beijing's Precipitation*. Central Meteorological Bureau, Beijing, China. (in Chinese)

Central Meteorological Bureau 1982: *Atlas of Flood and Drought in China in the Last 500 Years*. Map Press, Beijing, China. (in Chinese)

Feng, Liwen 1982: The drought characteristics of growing seasons in Beijing in 1724-1979 and their interannual variations. *Acta Geographica*, 37 194-205. (in Chinese with English abstract)

Wang, P. K., and De'er Zhang 1988: An introduction to some historical governmental weather records of China. *Bulletin American Meteorological Society, 69*, 753-758.

Wang, P. K. 1979: Meteorological records from ancient chronicles of China. *Bulletin American Meteorological Society, 60*, 313-317.

Wang, P. K. 1980: On the possible relationship between winter thunder and climatic changes in China over the past 2200 years. *Climatic Change*, 3, 37-46.

Wang, P. K., and J. H. Chu 1982: Some unusual lightning events reported in ancient Chinese literature. *Weatherwise, 35* 119-122.

Zhang, De'er, and Chuangzhi Liu 1986: Reconstruction of summer monthly temperature series of Beijing in 1724-1903. *Kexue Tongbao, 31*, 597-599. (in Chinese)

11 Beijing summer temperatures since 1724

W. C. Wang, D. Portman, G. Gong, P. Zhang and T. Karl

11.1 Introduction

Quantitative temperature measurements with records longer than one hundred years exist in a few places in China such as Beijing and Shanghai, but for most stations in China the temperature records start during the middle of this century (cf. Domros and Peng, 1988). Consequently, to study temperature fluctuations for the last few hundred years we will have to rely on proxy data from sources such as historical documents (Zhu 1973, Zhang 1988).

One of the most valuable historical documents is the *Qing Yu Lu* ("Clear and Rain Records") kept in the National First Historical Archive at the Palace Museum in the Forbidden City in Beijing. These documents (also discussed in Chapter 10) contain the *daily* weather conditions, somewhat like today's weather reports except that no instruments were used. For Beijing, the *Qing Yu Lu* were reported with little interruption from the second year of Emperor Yongzheng (1724) to the 30th year of Emperor Guangxu (1904). In China, this weather information has been used to estimate historical precipitation amounts (State Meteorological Administration 1981, Feng 1980) and to study patterns of droughts and floods (Wang and Zhao 1981, Ronberg and Wang 1987).

For Beijing, the surface temperature during summer is greatly affected by clouds and rainfall, because direct solar radiation is reduced on overcast days and continuous rainfall is often associated with a cold air mass. Consequently, it is expected that there exists a negative correlation between surface temperature and rainfall. Such a negative correlation also exists over much of North America (Karl *et al.* 1990) and England and Wales (Tout 1987).

Recently, Zhang and Liu (1986) used the Beijing precipitation information reported in the *Qing Yu Lu* to reconstruct the Beijing June and July monthly mean and maximum temperatures for the period 1724-1903. They first used the 1951-1982 temperature and rain-day data to establish the statistical correlation between the two variables, and then applied the empirical relationship to the rain-day data in the *Qing Yu Lu* to reconstruct the temperature. The temperature reconstruction was found to be reasonably good, as demonstrated in the comparison with instrumental temperature for the period 1897-1903.

Here, we study further the use of precipitation in Beijing for summer temperature reconstruction. This study differs from Zhang and Liu's (1986) work in two aspects. First, instead of using instrumental temperature and precipitation data of recent decades to establish the empirical relationship, we employed some of the 19th century instrumental temperature records which overlap with the *Qing Yu Lu*. Second, the effect of the timing of precipitation (the starting and ending hours of the rainfall) has also been considered and included in our correlation and regression analysis.

Below, we briefly describe data sources for the Beijing summer temperature reconstruction used in the present study. Next, the method for the temperature reconstruction is discussed and the complete time series are presented. Finally, major characteristics of the reconstructed temperature record are discussed.

11.2 Data sources

11.2.1 Qing Yu Lu

The *Qing Yu Lu* are kept in the National First Historical Archive in Beijing; the earliest record is for Beijing and dates back to 1672 (Zhang 1982). These records describe whether the sky was sunny or rainy or snowy. When it first started, this system of reporting weather conditions was limited to Beijing. The system was implemented countrywide in November 1685, as reported in a March 1686 memo to the Kangxi Emperor by *Hsieh Zu-Du* [the *Shunfu* (administrator) of the Anhui province at that time]. However, for unknown reasons, starting March 1686, the level of importance of the weather report was downgraded. Instead of a stand-alone, special report, it was reduced to an appendix in the normal memos to the Emperor (Gong *et al.* 1983). This change of reporting status made the weather report less important and resulted in losses of many earlier weather records.

At present, there are only four cities with long and relatively continuous *Qing Yu Lu*. Beijing (39.9°N, 116.4°E) has the longest, covering the period 1724-1904 while records at Nanjing (32.0°N, 118.7°E) Suzhou (31.3°N, 120.6°E) and Hangzhou (30.2°N, 120.1°E) are shorter, covering 1722-1798, 1723-1810 and 1723-1773 respectively. There is evidence that many other cities such as Yangzhou (29.1°N, 115.3°E) maintained *Qing Yu Lu*, but these records have not been found.

The *Qing Yu Lu* in Beijing were observed at the site located outside of the Jianguo Gate. The observation was the duty of *Xin-Tian-Jian* (title of the official) who reported to the Emperors about the weather conditions. In addition, the National First Historical Archive also has a few reports (covering the period from 24 September 1730 to 12 June 1742) which were written by personnel living inside the Forbidden City.

Two kinds of observations were taken: (1) irregular observations of special weather phenomena and (2) regular daily observations similar to the current daily weather report. The *Qing Yu Lu*, which belong to the second category, contain reports of sky conditions (overcast, clear, fog, rain) wind direction (N, S, E, W, NE, SE, NW, SW) and precipitation (light rain or snow, heavy rain or snow). The date, time, and duration of each weather event (i.e. starting and ending hours) were also recorded. The date, which is according to the Chinese lunar calendar, can be converted to the Gregorian calendar date. The time, which is recorded in terms of the traditional Chinese *Shi Chen*, can also be converted to western hours.

For example, one description for precipitation translates as follows:

On July 16, the third year of Emperor Yongzheng, it drizzled from *Xu Shi* through *Zi Shi*. On July 17, it drizzled from *Zi Shi* to *Yin Shi*, rained from *Mao Shi* to *Si Shi* and drizzled from *Wei Shi* through *Zi Shi*.

In other words, there were four hours of drizzle on 23 August 1725, followed by fifteen hours of drizzle and six hours of light rain on 24 August. Therefore, for each month the total

211

number of rain-hours and the number of rain-days (i.e. days with rain) can be easily computed. A more detailed description of the *Qing Yu Lu* and some examples were given by Wang and Zhang (1988).

The *Qing Yu Lu* in Beijing contain 15,951 daily precipitation records for the period 1724-1903. Some records are missing for 1810, 1831, 1844, 1851, 1893 and 1900. As many as 6 occurrences of precipitation are reported on a single day. In general, the number of rain-days in July is about 2-3 times higher than in May. Some years are of special interest. For example, in 1801, there were 27 rain-days in July (out of 31 possible) but less than 10 in May and June and in 1877, only 8 rain-days were observed for July. The rain-hour data, in general, have similar characteristics as rain-day data except for a few occasions. The most striking one was in July 1890 when the total number of rain-hours was 133 during daytime hours and 104 during the nighttime. Note that relatively more rain-hours also occurred in July 1801 when rain-days had their maximum value. On the other hand, minimum values of daytime and nighttime rain-hours in July occurred in 1755. We have also plotted these data (not shown here); it appears that there is no visible trend in the number of rain-days.

11.2.2   Beijing instrumental data

A French priest, Pater Gaubil, started keeping climate records in Beijing during the eighth year of the reign of Emperor Qianlong (1743). Unfortunately, his records were lost except for a few statistics. From early 1757 to 1763, Jesuit Father Amiot recorded temperature, atmospheric pressure, cloud amount, and wind directions in Beijing and the data were published in *Memoirs de Mathematique et de Physique*, published by Messier of Paris. However, it is difficult to assess the data quality because of the lack of information. Amiot's record of Beijing monthly mean January to December temperatures averaged over the seven-year period are shown in Figure 11.1. The modern temperatures averaged over 1951-1980 are also shown for comparison. It is interesting to note that although the annual mean temperatures for the two periods are nearly identical, the recent period has a colder winter and warmer summer and thus a larger seasonal cycle. No other records are known to exist until 1830, when G. von Fuss (a Russian) resumed the record-keeping in Beijing for just half a year. The Russian Church started systematic observations around 1840, but no records were kept for 1856-1858 and 1862-1867. In 1867, the Russian Academy of Sciences in St. Petersburg sent H. Fritsche to Beijing to take meteorological observations. Since then, instrumental temperatures have been measured continuously up to the present, although no records have been found for 1885-1888, 1890, 1900-1904, 1909, 1912-1913, 1926-1929 and 1937-1939.

11.3   1724-1903 Summer temperature reconstruction

Zhang and Liu (1986) used the rain-day data from the *Qing Yu Lu* to reconstruct the June and July monthly mean and maximum temperature. Here, we re-examine the statistical correlations between temperature and precipitation to study how we can further improve the correlations. We have used instrumental temperature data recorded during the nineteenth century and have included the timing of the precipitation, which Zhang and Liu did not consider.

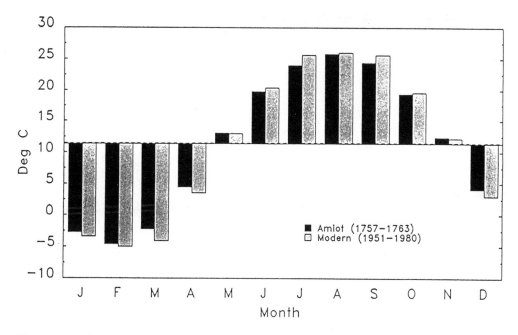

Figure 11.1 Comparisons of Beijing monthly mean temperatures (°C) between Father Amiot's data for 1757-1763 (annual mean 11.4 °C) and modern values for 1951-1980 (annual mean 11.5 °C) from Domros and Peng (1988).

Correlation coefficients were calculated between the monthly mean temperature and monthly total number of rain-days based on (1) the 1951-1985 instrumental temperatures and rain-days and (2) the 1841-1897 instrumental temperatures and the (*Qing Yu Lu*) rain-days. In the 1841-1897 period, we chose 35 years which had 12 months of complete data; the May, June and July monthly mean temperatures for these years are shown in Figure 11.2. The correlation coefficients, summarized in Table 11.1, suggest that using the temperature data in 1841-1897 can provide a much better correlation for months March through August. The main reason for the better correlation is that for 1841-1897 the number of (*Qing Yu Lu*) rain-days includes the days with rain amounts of less than 0.1mm while for 1951-1985 the number of (instrumental) rain-days only considers the days with rain amounts larger than 0.1mm. Differences between these two records are illustrated in Table 11.2, which summarizes in each record the mean number of rain-days and its standard deviation for months April through July. Two other periods were also used for comparison. In general, the number of rain-days is larger in the *Qing Yu Lu* than in the instrumental data while the standard deviations for both are about the same. For example, for July the total number of rain-days is about 16 in the *Qing Yu Lu* but only 14 in the instrumental data. This aspect was considered by Zhang and Liu (1986) who adjust their correlation by considering the variance of the rain-days with rainfall less than 0.1mm.

The duration of the precipitation will affect the correlation. For example, the effect on temperature will be different between 2 July, 1762 with one hour of rainfall and 15 July, 1762 with 21 hours of rainfall, although they are both considered to be one rain-day. In addition, the timing of the rainfall will also affect the temperature. In general, rainfall at night will tend

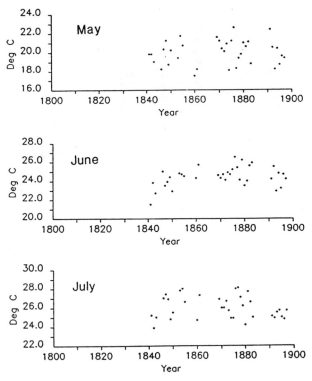

TEMPERATURE MEASUREMENTS

Figure 11.2 Beijing May, June and July mean instrumental temperature for 1841-1900.

Table 11.1 Correlation coefficients between the monthly mean temperature and monthly total number of rain-days calculated for two periods 1951-1985 and 1841-1897.

| | Correlation Coefficient | |
Month	1951-1985	1841-1897
1	-0.270	-0.416
2	-0.420	-0.393
3	-0.440*	-0.521*
4	-0.401	-0.599*
5	-0.245	-0.673*
6	-0.537*	-0.577*
7	-0.613*	-0.636*
8	-0.249	-0.435*
9	-0.160	-0.346
10	-0.230	0.116
11	-0.080	-0.249
12	0.100	-0.021

*Statistical significance $\geq$ 99%; N = 35 years.

Table 11.2 Mean (R) and standard deviation (σ) of the number of rain-days during different periods for April, May, June and July.

Month		Qing Yu Lu		Instrumental Data	
		1724-1897	1841-1897	1920-1985	1951-1985
April	R	6.4	6.5	4.2	4.7
	σ	2.86	2.90	2.92	3.17
May	R	9.1	9.9	6.0	6.0
	σ	6	2.37	2.56	3.04
June	R	12.6	13.6	9.2	9.7
	σ	3.68	3.38	3.28	3.10
July	R	16.7	16.3	13.6	13.9
	σ	3.46	3.57	3.95	4.01

to be associated with more cloudiness and thus prevent night-time radiation loss. Therefore, rainfall during the daytime should have a better negative relationship with temperature than at night. To address these issues, we categorize the correlation between temperature and rain-hours using rainfall at different times of the day. These results are shown in Table 11.3. Note first that the correlation between the monthly mean temperature and rain-hours is generally highest for May, June and July. Second, the results suggest that, for these three months, the correlations between monthly mean temperature and the number of daytime rain-hours are even higher than those between mean temperature and the total number of rain-days shown in the last column of Table 11.3.

Given these considerations, we have developed a new empirical equation relating the monthly mean temperature to precipitation information. The functional form is:

$$T = C_1 + C_2 R_1 + C_3 R,$$

where T is the monthly mean temperature, R_1 is the monthly total number of rain-hours during daytime, and R is the monthly total number of rain-days. The empirical coefficients C_1, C_2, and C_3 obtained for May, June, and July are summarized in Table 11.4. For comparison we also show the empirical coefficients using R alone (two-parameter model). The results indicate that using the three parameters substantially improves the correlation coefficients. Three points are noted here. First, in the temperature reconstruction, we explain at a maximum only two-thirds of the interannual variance. This implies that there can be large trends in the temperature records unrelated to precipitation, which we would never detect. Second, the correlation between the maximum temperature and precipitation is expected to be better than the mean temperature, as found in Zhang and Liu (1986). Third, it would also be valuable to examine directly the instrumental precipitation data to study its interannual variations.

Table 11.3 Correlation coefficients for 1841-1897. Monthly mean temperatures are correlated with (1) the monthly total number of rain-hours at different times of the day and (2) the monthly total number of rain-days.

Month	Number of Rain-Hours						Night & Day	Number of Rain-Days
	0-6 hrs	6-12	12-18	18-24	Night	Day		
1	-0.378	-0.536*	-0.436*	-0.217	-0.334	-0.509*	-0.450*	-0.521*
2	-0.253	-0.368	-0.387	-0.305	-0.302	-0.392	-0.372	-0.398
3	-0.524*	-0.583*	-0.499*	-0.511*	-0.557*	-0.559*	-0.575*	-0.517*
4	-0.564*	-0.260	-0.405	-0.404	-0.564*	-0.369	-0.493*	-0.600*
5	-0.543*	-0.704*	-0.684*	-0.478*	-0.618*	-0.781*	-0.782*	-0.676*
6	-0.659*	-0.685*	-0.639*	-0.345	-0.635*	-0.770*	-0.774*	-0.600*
7	-0.519*	-0.574*	-0.619*	-0.559*	-0.583*	-0.683*	-0.683*	-0.627*
8	-0.177	-0.314	-0.426	-0.208	-0.209	-0.409	-0.346	-0.188
9	-0.078	-0.097	-0.353	-0.271	-0.156	-0.242	-0.213	-0.328
10	0.061	0.237	-0.106	-0.005	0.042	0.088	0.075	0.122
11	-0.269	-0.369	0.069	-0.212	-0.284	-0.184	-0.248	-0.249
12	-0.161	-0.083	-0.056	-0.094	-0.153	-0.104	-0.136	-0.056

*Statistical significance $\geq$ 99%; N = 35 years.

Table 11.4 Regression coefficients for the functional form $(T = C_1 + C_2R_1 + C_3R)$ relating monthly mean temperature (T) to the monthly total number of daytime rain-hours (R_1) and rain-days (R).

Month	C_1	C_2	C_3	Correlation* coefficient
A.	Three-parameter model			
May	22.74	-0.067	-0.138	0.80
June	26.20	-0.052	0.000	0.77
July	29.11	-0.034	-0.091	0.71
B.	Two-parameter model			
May	23.56	--.---	-0.363	0.68
June	26.94	--.---	-0.184	0.60
July	29.36	--.---	-0.205	0.63

* Values are between proxy temperature and instrumental temperature for 1841-18

11.4 Beijing summer temperature characteristics

Using the empirical coefficients given in Table 11.5 together with the *Qing Yu Lu* we can reconstruct the May, June and July monthly mean temperatures. These temperatures together with the 3-month mean temperature are shown in Figure 11.3. Note that the missing temperatures during the years mentioned at the end of section 2 have been filled in with data from the neighboring station of Tienjing (39.1°N; 117.2°E) by the Beijing Weather Bureau (1982). There are several interesting features in these temperature time series. First, for all three months the peak temperatures during the 1930s were comparable in magnitude to those during the first part of the eighteenth century. Second, the temperatures during the decade of the 1970s were the coldest on record, especially for May and June. Third, the fluctuations exhibit regularity.

Table 11.5 Average (T) standard deviation (s) and range (r) of Beijing summer temperatures.

Month	1724-1854	1855-1986	1724-1986
May			
T	20.27	20.11	20.19
s	1.05	1.19	1.12
r	5.0	6.3	6.6
June			
T	24.79	24.62	24.71
s	0.89	1.15	1.03
r	5.0	6.0	6.9
July			
T	26.16	26.06	26.11
s	0.91	1.02	0.97
r	5.1	5.6	5.6

We also examine the statistics of the Beijing summer temperature shown in Figure 11.3. In addition to the statistics for the whole period 1724-1986, we have compared the statistics between 1724-1854 and 1855-1986, the individual periods with over 90% proxy and instrumental data respectively. Table 11.5 summarizes the statistics of mean, standard deviation, and range (difference between the maximum and minimum) of the temperatures. The results indicate that, for the earlier period, the temperatures for all the three months were about 0.1-0.2°C warmer and had smaller values of standard deviation and range. These features were particularly evident in June.

Power spectrum analysis was performed to study the temperature fluctuations. The 1724-1986 spectra for the three months are shown in Figure 11.4. The most distinct features in the power spectra are the peaks at the periods 18.9-21.3 years, all significant at 95% confidence level. This periodicity has also been identified in the analysis of Beijing precipitation data (Hameed *et al.* 1983, Clegg and Wigley 1984). Other notable common peaks are the 7.4, 3.2 and 2.5 years.

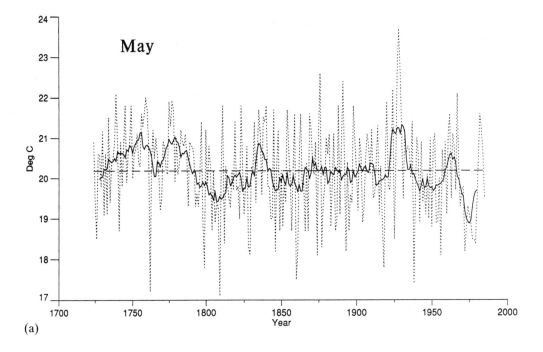

(a)

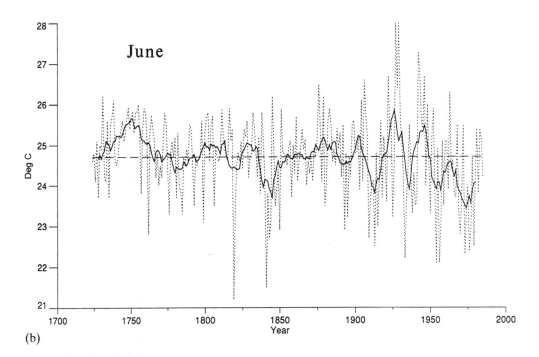

(b)

Figure 11.3 Beijing May, June, July and three-month mean temperatures for the period 1724-1986. These temperature were derived from a combination of proxy and instrument sources. Both yearly values (dotted line) and 10-year running mean (solid line) are shown.

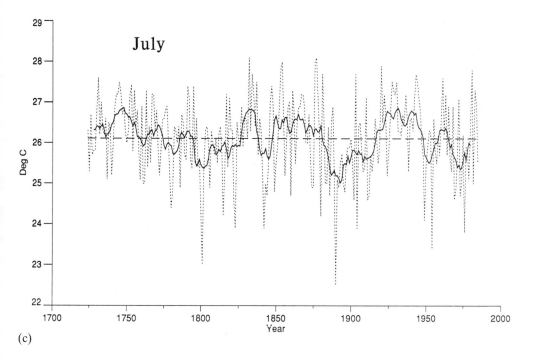

(c)

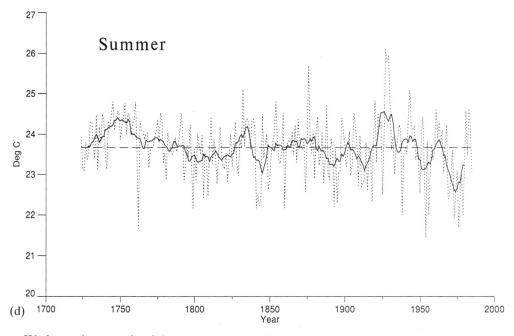

(d)

We have also examined the power spectra of the two individual periods mentioned above. In general the characteristics between the two periods are different. For example, the June spectra, depicted in Figure 11.5, indicate that the inflated variances at 21.5 years appeared in the latter instrumental period while the 7.2 year periodicity occurred in the earlier proxy

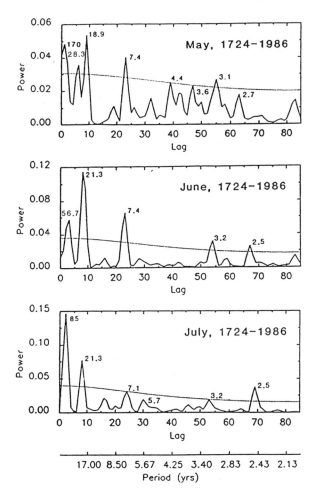

Figure 11.4 Power spectrum of 1724-1986 Beijing temperature during May, June and July. Dashed line indicates 95% level of statistical significance.

period. On the other hand, the July spectra (Figure 11.6) show significantly high variance at 21.5 years in the proxy data. The 7.2 and 2.5 year periodicities are also found only in the earlier period.

The present study utilizes the precipitation information contained in the Beijing *Qing Yu Lu* to reconstruct the summer temperature. As the *Qing Yu Lu* also includes other climate observations such as cloudiness, it would then be worthwhile to use the cloudiness information to study the consistency between precipitation and cloudiness as an indirect means for validation of the temperature reconstruction. In addition, comparative study of the temperatures in Beijing versus other regions such as in Europe is also warranted so that a global temperature pattern for the last two to three hundred years can be provided. Studying temperature variability in the last few centuries will help define the natural variability of

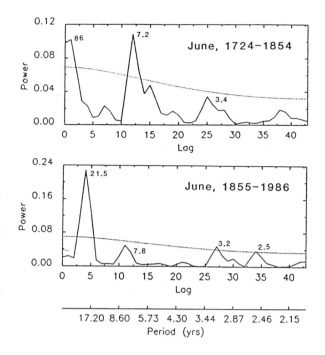

Figure 11.5 Power spectrum of Beijing June temperature for 1724-1854 and 1855-1986. Dashed line indicates 95% level of statistical significance.

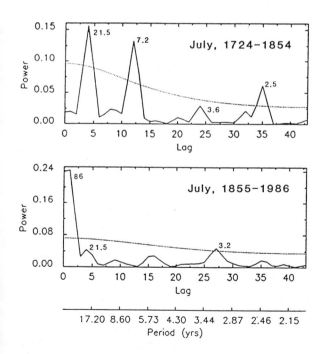

Figure 11.6 Power spectrum of Beijing July temperature for 1724-1854 and 1855-1986. Dashed line indicates 95% level of statistical significance.

climate, a necessary pre-requisite for identifying the anticipated climate warming signal caused by increases of atmospheric greenhouse gases.

Acknowledgement

We thank M. Riches of the Atmospheric and Climate Research Division, Office of Health and Environmental Research, Department of Energy for his encouragement and comments. This research was supported by the United States' Department of Energy and the People's Republic of China's Chinese Academy of Sciences Joint Research on the Greenhouse Effect.

References

Beijing Weather Bureau 1982. *Beijing meteorological records*. Beijing Weather Bureau.

Clegg, S. L. and T. M. L. Wigley 1984. Periodicities in precipitation in northeast China, 1470-1979. *Geophysical Research Letters*, 11, 1219-1227.

Domros, M. and G. Peng 1988. *The Climate of China*. Berlin. Springer.

Feng, L. W. 1980. The rainy season and its variation in Beijing during the last 255 years. *Acta Meteorologica*, 38, 342-350. (in Chinese)

Gong, G. F., P. Y. Zhang, C. D. Wu and G. R. Zhang 1983. *The methodology to study climate change during historical times*. Beijing: Science Publ.

Hameed, S., M. Li, W. Yeh, R. Cess, and W.-C. Wang 1983. An analysis of periodicities in the 1470 to 1974 Beijing precipitation record. *Geophysical Research Letters*, 10, 436-439.

Karl, T., W.-C. Wang, M. E. Schlesinger, R. W. Knight and D. Portman 1990. A method of relating general circulation model simulated local climate to the observed local climate. Part I. Central tendency and disposition. Submitted to *Journal of Climate*.

Ronberg, B. and W.-C. Wang 1987. Climate patterns derived from Chinese proxy precipitation records: an evaluation of the station networks and statistical techniques. *Journal of Climatology*, 7, 391-416.

State Meteorological Administration 1981. *Annals of 510 years precipitation Record in China*. Beijing: Cartographic Publ. (in Chinese, English summary).

Tout, D. G. 1987. Precipitation-temperature relationships in England and Wales summers. *Journal of Climatology*, 7, 118-184.

Wang, P. K. and D. Zhang 1988. An introduction to some historical governmental weather records of China. *Bulletin of American Meteorological Society*, 69, 753-758.

Wang, S. and Z. Zhao 1981. Droughts and floods in China. 1470-1979, in *Climate and History*. Wigley, T. M. L., Ingram, M. J. and Farmer, G. (eds.) p. 271-288. Cambridge: Cambridge University Press.

Zhang, G. R. 1982. *The weather information in historical records during Ching Dynasty*. Report on Historical Records, National First Historical Archive, Beijing.

Zhang, J. C. (ed.) 1988. *The reconstruction of climate for historical times*. Beijing: Science Press

Zhang, D. and C. Liu 1986. *Reconstruction of summer temperature series (1724-1903) in Beijing*. Annual Report, Academy of Meteorological Science, State Meteorological Administration, Chinese Meteorological Press.

Zhu, K. C. 1973. A preliminary study on the climate fluctuations during the last 5000 years in China. *Scientia Sinica*, 16, 226-256.

12 Reconstruction of rainfall variation of the *Baiu* in historical times

A. Murata

12.1 Introduction

Baiu is a distinct rainy season in early summer in Japan. The rainfall amount during this season influences strongly water resources in western Japan, where the *Baiu* rainfall constitutes 30-40 per cent of the annual total. It is, therefore, very important to understand the nature of rainfall variation in the *Baiu* season. Since instrumental observations started in Japan around the end of the nineteenth century, the data set available for the study of long-term change is limited. Therefore, reconstruction of rainfall variation in the historical time can provide useful information in this regard. In addition, the reconstruction would be useful for the understanding of atmospheric circulation over East Asia in historical time because the *Baiu* appears as a seasonal change of the general circulation of the atmosphere.

The purposes of the present study are:

(1)   Reconstruction of monthly rainfall variation in the *Baiu* season, and
(2)   Investigation of the conditions of the *Baiu* since 1700.

In the following sections, we first describe the reconstruction procedure in detail. Second, reconstructed rainfall variations are discussed in comparison with other studies indicating the conditions of the *Baiu* during this time.

This study revises previous studies (Yoshino and Murata 1988, Murata and Yoshino 1988) in order to obtain more reliable reconstructed data.

12.2 Data sources

With a view to reconstructing monthly rainfall variations in the historical time, weather descriptions written in diaries, documents, archives, etc. are utilized as proxy data. In Japan, paleoclimatic reconstruction based on data of this kind can be found in many studies (see for example Yamamoto 1971a,b). However, the results obtained in these studies were not good enough for comparison with climatic conditions in the instrumental period, because the data were both spatially and temporally fragmentary. Nowadays, the development of a "Historical Weather Data Base" (Yoshimura and Yoshino 1988) enables us to resolve many of these problems. This data base is now being produced with the aim of collecting daily weather information for most of Japan during historical time. In the near future, each researcher will be able to access easily the data base through a computer network system. In the present study, five documentary sources were taken from the data base. The details of the documentary sources used in the present study are shown in Table 12.1 and Figure 12.1.

Table 12.1 List of documentary sources used in the present study. HWD in the table means "Historical Weather Data Base" and "Original data" means a documentary source originally collected by the author. Numbers in parentheses in the last column show the last three digits of index numbers given in WMO No. 9, Volume A-Stations.

Symbol	Name of documentary source	Data collection source	Observation station used for estimation equation
A	Hachinohe-han-nikki	HWD	Hachinohe (581)
B	Hirosaki-hancho-nikki	Maejima et al. (1983)	Aomori (575)
C	Nanbuhan-karoseki-nisshi	HWD	Morioka (584)
D	Hirosaki-hancho-edo-nikki	Original data	Tokyo (662)
E	Sekiguchi-nikki	Original data	Yokohama (670)
F	Okyo-zakki	HWD	Kanazawa (605)
G	Kakuson-nikki	HWD	Kanazawa (605)
H	Shiryokan-nikki	HWD	Tsu (651)
I	Ikota-jinjya-nikki	Original data	Osaka (772)
J	Inatsukake-nikki	Original data	Osaka (772)
K	Usukihan-kaisyo-nikki	HWD	Ooita (815)
L	Isahayahan-nikki	HWD	Nagasaki (817)

Documentary source D: Original data collected in the City Library of Hirosaki were used.
Documentary source E: A publication Sekiguchi-nikki Vol. 1-20 by the Education Committee of Yokohama City was used.
Documentary source I: A publication Ikeda-shishi(shiryo-hen) (The History of Ikeda City (historical materials)) was used.
Documentary source J: Same as Documentary source I.

Figure 12.2 shows an example of weather descriptions written in a diary. The data recorded in a diary used the old calendar. This was converted to the Gregorian calendar used today. Days with the description such as "rainfall" were counted for June and July, respectively. The total number of such days was taken as the number of days with precipitation for each month. Months with more than 6 days of missing weather descriptions were excluded. The same procedure was performed for other documentary sources.

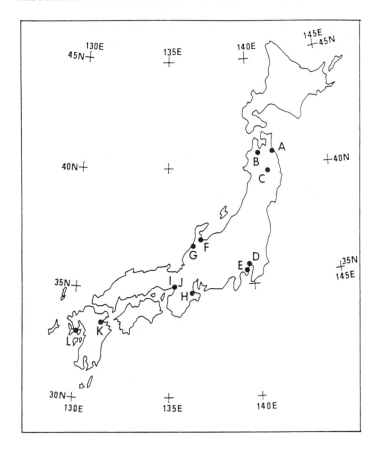

Figure 12.1 Location of documentary sources. Letters in the figure correspond to those listed in Table 12.1.

12.3 Reconstruction procedure

Figure 12.3 shows a flow chart of the reconstruction procedure used in the present study. First, regional division is carried out based on monthly rainfall amounts for June, July and June plus July in order to arrange stations with similar rainfall variations into a region. Second, the Rainfall Variation Index (RVI) which represents monthly rainfall variations in each region, is defined. Third, estimation equations for the reconstruction of the RVI in the historical time are established using the method of least squares. With the estimation equations, RVIs in the historical time are calculated and graphically displayed. In a region, at least two documentary sources are necessary to have a valid reconstructed RVI. If two curves of the RVI in the same region, which are depicted as a function of time, give a similar picture for long-term change, it means that the reconstructed RVI is valid and can be used for further studies. Otherwise, the region is not used in this study. The methods used in the reconstruction are stated in the following subsections.

```
 1  曇   巳ノ剋比方雨未ノ剋前止夕方方雨　　夜中大雨
 2  陰晴　申ノ剋前方雨
 3  晴
 4  晴   未ノ剋過方曇夕方東風
 5  雨   辰ノ剋前方止
 6  晴   未ノ剋比方陰晴
 7  陰晴
 8  雨   未ノ剋比方止
 9  晴
10  陰晴　午ノ剋過方晴
11  雨   折々少雨
12  曇□方折々雨
13  雨   入梅　巳ノ剋前方止
14  曇   巳ノ剋前方雨　　午ノ剋前止
15  曇
16  曇   午ノ剋方陰晴　　夕方戌ノ下剋過方雨□雷
17  快晴
18  晴
19  雨
20  曇
21  曇   午ノ剋比方陰晴南風夕方東風
22  曇   午ノ中剋比方少雨夜中大雨
23  雨   北風未ノ剋過方雨止晴
24  晴
25  曇   巳ノ剋前方曇　未ノ剋過方少雨夜中北風烈
26  大雨　巳ノ剋過止午ノ剋前方晴申ノ剋前方曇同中剋比雨夕方止夜入晴
27  快晴
28  晴   午ノ剋比方曇　夕方方雨夜中降
29  雨   終日　夜中
30  雨   南風　夜中雨降
```

Figure 12.2   Daily weather descriptions in June, 1836 written in documentary
source D. Solid circles indicate a description such as "rainfall".

12.3.1   *Regional division*

The results obtained in this subsection have already been stated in the studies of Murata and
Yoshino (1988) and Yoshino and Murata (1988) but the outline is given here because the
regional division technique constitutes the principal feature of the present study.

Regional division was carried out based on June, July and June plus July rainfall data of an
84-year period from 1901 to 1984. The rainfall data were taken from 50 Japanese stations,
whose locations are given by dots in Figure 12.5. The varimax rotated Principal Component
Analysis (PCA) was applied to the rainfall data in order to obtain groupings into a small
number of regions. Actual calculation was performed using the correlation matrix after the

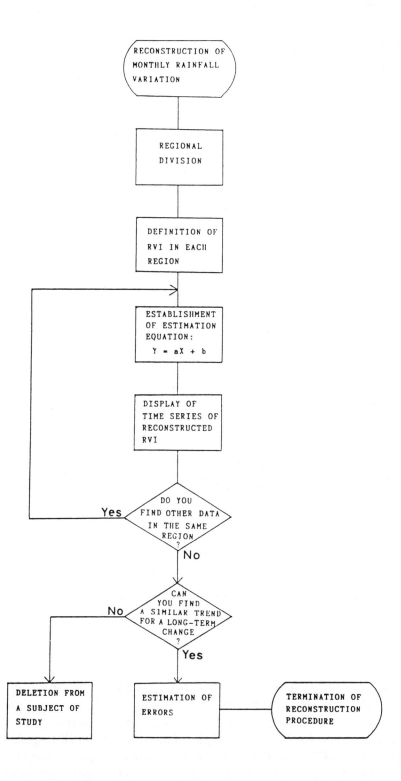

Figure 12.3 The flow chart of the reconstruction procedure for monthly rainfall variation.

data was normalized using the natural logarithm transformation. The number of components rotated was eight, nine and nine for June, July and June plus July, respectively because these components had an eigenvalue of at least one in the initial solution before the rotation.

Figure 12.4 shows the spatial patterns of component/factor loading in the higher two components for each month. In the case of June, the maximum value is seen in the northern Kyushu District for the first component and in the Kinki-Chubu Districts for the second component. For both July and June plus July, the maximum values are seen in the Kinki-Chubu Districts for the first component and in almost the whole of Kyushu for the second component. Here, for each component pattern, the larger the value of the component/factor loading, the more dominant the rainfall variation related to the component is at each station. For other PCs, maximum values were also observed for each component. The positions of these maximum values do not overlap owing to the inherent nature of the varimax rotated PCA. Regional division was made using the component/factor loading.

Figure 12.5 depicts the regional division by the above method. The capital F used to represent regional names represents the initial letter of 'Factor'. (Correctly speaking, 'P' should be used instead of 'F' because PCA is applied in this study. Here, 'F' is used just as a symbol). The smaller and the capital numbers represent the month and the order of components, respectively. For example, 'F61' represents the region which is determined in accordance with the spatial pattern of the first component for June rainfall through the Varimax rotated PCA. Each region consists of stations with component/factor loading of more than 0.6 for each component. This value was chosen in order to obtain regions as large as possible but without any overlap.

The validity of the regional division proposed here has already been discussed in comparison with other studies (Murata 1990). Also Murata and Yoshino (1988) and Murata (1990) give some climatic characteristics inherent in each region.

12.3.2 Calibration

The first component score was defined here as the Rainfall Variation Index (RVI). This variable is sufficiently representative of rainfall variations in each region since the first component comprises most of the total variance (Table 12.2). The next step is to reconstruct the RVIs in historical time using the monthly number of days with precipitation inferred from documentary sources. The RVIs in historical time were estimated by the following procedure.

First, simple linear regression equations were derived by using the RVI and the number of days with precipitation as the dependent and independent variables, respectively. The period 1943-84 (42 cases) was used for the calculation. The number of days with precipitation was taken from the nearest station to the location of the documentary sources (Table 12.1). Here, the category of daily rainfall amount must be decided before calculation of the regression coefficients. It is necessary to decide which daily rainfall category (e.g. 1mm, 5mm or 10mm) the count of precipitation days obtained from documentary sources is related to. These categories were adopted here due to data availability. In the present study, the category was decided using the Student's t-statistic, which judges significance of the difference of averages between two groups, on the assumption that the average of the number of days with

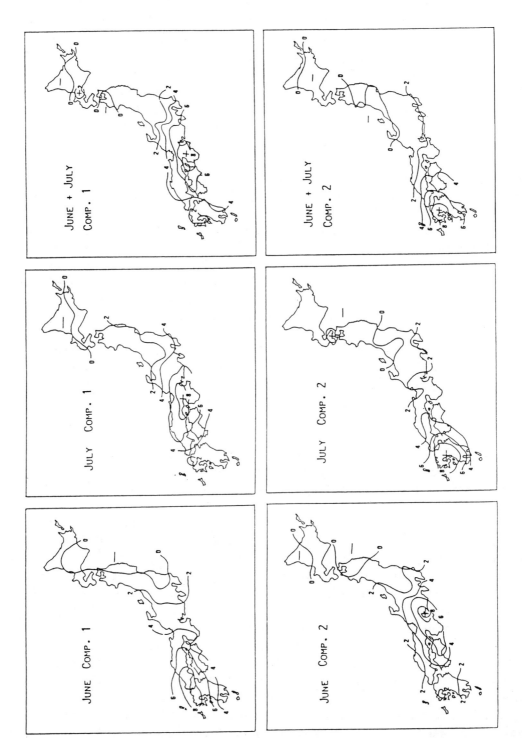

Figure 12.4 The geographical patterns of the varimax rotated component/factor loading. The values are multiplied by ten.

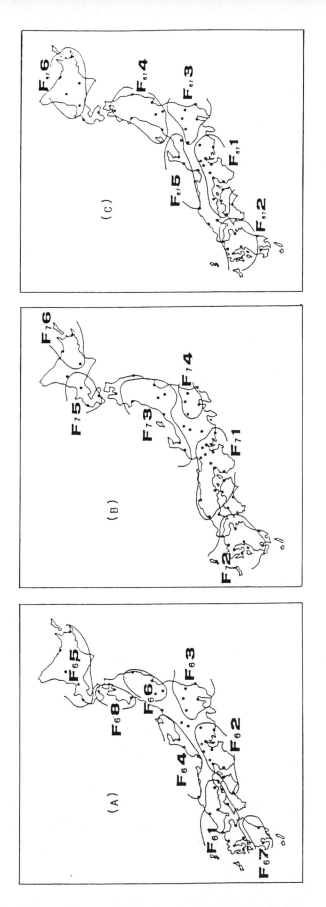

Figure 12.5 Regional divisions by the varimax rotated component/factor loading. (A) June (B) July (C) June plus July.

Table 12.2 Percent variance of the first (P1) and second (P2) components in each region.

Region	Percent variance explained	
	P1	P2
F_61	74.4	9.5
F_62	72.5	8.9
F_63	71.7	13.4
F_64	82.8	8.8
F_65	61.6	14.5
F_66	73.2	13.0
F_67	89.2	10.8
F_68	76.7	14.2
F_71	68.7	8.4
F_72	74.6	9.7
F_73	72.0	10.6
F_74	68.4	15.2
F_75	70.5	17.1
F_76	69.6	17.8
F_671	71.6	8.5
F_672	73.5	11.0
F_673	68.5	11.4
F_674	68.5	11.4
F_675	76.4	11.4
F_676	63.3	15.1

(a)

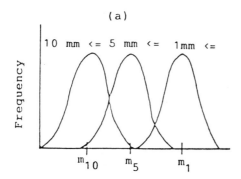

(b)

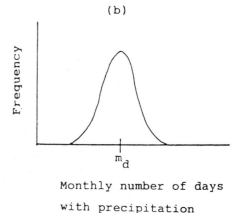

Monthly number of days with precipitation

Figure 12.6 Schematic figure of the way to decide the number of days with precipitation obtained from documentary sources. Figure (a) indicates the frequency of monthly days with precipitation for three categories, more than 1mm, more than 5mm and more than 10mm, which are calculated based on instrumentally observed data. Figure (b) indicates the same as figure (a) but based on a documentary source. The category of the number of days with precipitation obtained from a documentary source is decided with the help of the t-statistics by judging which is closest to the documentary mean. The "m" indicates mean for the cases of more than 1mm (subscript 1) more than 5mm (5) more than 10mm (10) and a documentary source (d) respectively.

precipitation obtained from a documentary source should coincide with the average of that of one of the three categories mentioned above. Figure 12.6 illustrates this idea schematically.

Table 12.3 shows the results of the analysis. It can be judged that the number of days with precipitation obtained from documentary source A belongs to the category of "daily amount

Table 12.3 The results of Student's t-tests. Letters in the table correspond to those listed in Table 12.1. Level of significance (1) shows the reults of F-test for homogeneity of variance. Level of significance (2) shows the results of T-test for the case of equal variance (left) and that of unequal variance (right).

Documentary source	Category of daily rainfall	Level of significance (1) %	Level of significance (2) %	
A	1mm <=	34.55	0.01	0.01
	5mm <=	32.80	41.10	41.85
	10mm <=	0.05	0.01	0.01
B	1mm <=	0.83	0.01	0.01
	5mm <=	0.01	0.01	0.01
	10mm <=	0.01	0.01	0.01
C	1mm <=	55.83	0.01	0.01
	5mm <=	12.64	14.22	17.59
	10mm <=	0.01	0.01	0.01
D	1mm <=	3.82	0.11	0.26
	5mm <=	0.01	0.01	0.01
	10mm <=	0.01	0.01	0.01
E	1mm <=	13.10	0.01	0.01
	5mm <=	38.66	15.99	17.16
	10mm <=	0.07	0.01	0.01
F	1mm <=	31.73	1.72	1.96
	5mm <=	87.95	0.01	0.01
	10mm <=	4.85	0.01	0.01
G	1mm <=	19.10	79.80	80.89
	5mm <=	78.00	0.01	0.01
	10mm <=	16.33	0.01	0.01
H	1mm <=	3.97	3.18	5.07
	5mm <=	0.01	0.01	0.01
	10mm <=	0.01	0.01	0.01
I	1mm <=	34.53	0.82	1.01
	5mm <=	0.63	0.01	0.01
	10mm <=	0.01	0.01	0.01
J	1mm <=	33.06	29.58	31.43
	5mm <=	0.50	0.01	0.01
	10mm <=	0.01	0.01	0.01
K	1mm <=	84.09	0.25	0.24
	5mm <=	3.18	0.01	0.01
	10mm <=	0.01	0.01	0.01
L	1mm <=	98.29	0.01	0.01
	5mm <=	39.14	71.72	72.41
	10mm <=	0.81	0.01	0.01

more than 5mm" because the difference of the averages is rejected at the 41.10 or 41.85% level. In a case when the category of daily rainfall cannot be assigned (documentary source B) the category of daily amount more than 1mm was adopted. By this procedure, we can decide the category of daily rainfall amount for each documentary source and minimize the bias originating from the use of the number of days with precipitation based on documentary sources.

Estimation equations used for reconstruction are listed in Table 12.4 with correlation coefficients and standard errors. Correlation coefficients in Table 12.4, which are all significant at the 99% level, indicate that the estimation equations provide a reasonably good fit to

Table 12.4 Estimation equations ($Y = aX + b$) used for reconstruction. Correlation coefficients (r) and standard errors are also given. Alphabets in the column of DS correspond to those listed in Table 12.1.

Region	DS	a	b	r	standard error
F_61	K	0.213	-2.809	0.742	0.698
	L	0.253	-2.394	0.851	0.548
F_62	I,J	0.183	-2.120	0.624	0.679
	H	0.207	-2.598	0.647	0.765
F_63	D	0.217	-2.475	0.671	0.729
	E	0.224	-1.785	0.632	0.801
F_71	I,J	0.190	-1.700	0.752	0.643
	H	0.176	-1.767	0.724	0.620
F_72	K	0.171	-1.875	0.751	0.685
	L	0.210	-1.625	0.807	0.612
F_73	G,F	0.187	-2.199	0.745	0.754
	C	0.224	-1.674	0.769	0.722
F_74	D	0.162	-1.669	0.674	0.624
	E	0.186	-1.110	0.484	0.740
F_671	I,J	0.138	-2.875	0.744	0.640
	H	0.144	-3.300	0.735	0.653
F_672	K	0.142	-3.396	0.700	0.769
	L	0.149	-2.537	0.751	0.711
F_673	D	0.121	-2.593	0.690	0.670
	E	0.156	-2.144	0.645	0.697
F_674	A	0.194	-2.146	0.685	0.769
	B	0.186	-3.200	0.713	0.740
	C	0.176	-2.424	0.777	0.665

the observations and may be used to make reconstructions. Additionally, the residuals were examined using the Durbin-Watson D-statistic. The results show that the residuals were normal and independent, which gives us more confidence in the results.

In order to confirm the reliability of the estimation equations, the following diagnostic procedure was used. First, we examined whether or not the estimation equations can predict the RVIs for 1901-42. Second the RVI and the number of days with precipitation for 1901-84, were divided into two groups by the generation of random numbers. It was then examined whether or not the estimation equations derived from one group could predict the RVI of the other group. This diagnostic procedure was repeated ten times for each estimation equation. The results of this diagnostic procedure for regions F_32, F_71 and $F_{67}1$ are listed in Table 12.5. Both correlation coefficients and standard errors in Table 12.5 do not show remarkable differences to those in Table 12.4. Therefore it can be judged that positive correlations observed between the RVI and the number of days with precipitation are statistically stable and the estimation equations established here are useful for reconstruction.

Table 12.5 Correlation coefficients (r) and standard errors (se) for other combinations of data. The bottom row shows the standard error based on the date of 1901-42.

		F_62			F_71			$F_{67}1$	
	N		se	N	r	se	N	r	se
1	35	0.742	0.736	39	0.796	0.655	34	0.797	0.655
2	43	0.667	0.688	37	0.814	0.612	40	0.756	0.643
3	46	0.707	0.758	43	0.766	0.684	39	0.767	0.572
4	44	0.554	0.577	37	0.817	0.752	39	0.719	0.662
5	45	0.714	0.757	35	0.713	0.705	39	0.752	0.716
6	44	0.742	0.749	41	0.788	0.482	40	0.715	0.611
7	42	0.591	0.719	40	0.793	0.625	34	0.825	0.707
8	44	0.692	0.693	38	0.738	0.673	39	0.775	0.593
9	40	0.706	0.682	39	0.823	0.538	46	0.788	0.638
10	30	0.596	0.712	42	0.740	0.570	40	0.730	0.616
			0.679			0.643			0.640

Although the correlation coefficients in Table 12.4 indicate that the estimations equations can reconstruct the RVI in historical time, the predictability of the equations is not satisfactory enough because the standard errors are rather large. However, this shortcoming will be lessened if we deal with long-term changes. Figure 12.7 shows the squared coherence between the RVI and the number of days with precipitation for regions F_62, F_71 and $F_{67}1$. The highest coherence is observed at low frequency (below 0.1 cycles per year) or at periods equal to 10 years or more for all the regions. This implies that the reconstructed RVIs from the number of days with precipitation can represent rainfall variations over more than 10 years. This fact could be also confirmed for other regions.

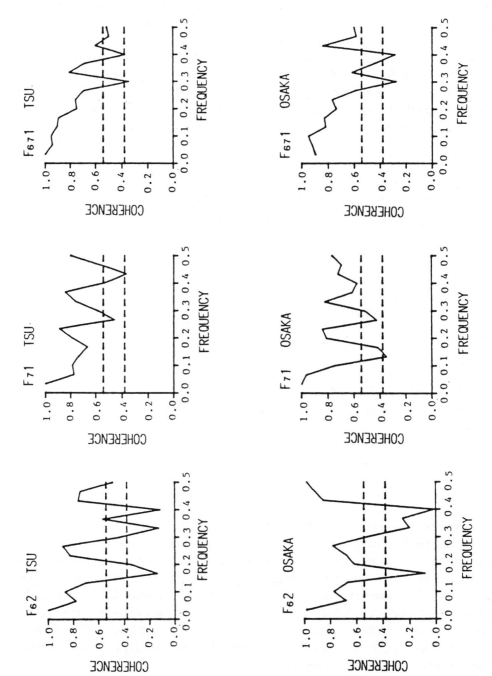

Figure 12.7 Squared coherence between the RVI and the number of days with precipitation. Dashed lines show the 95 per cent (lower) and the 99 per cent (upper) confidence levels.

12.3.3 Estimation of errors

Errors due to both the statistical method and the proxy data are inevitably contained in the reconstructed RVIs for historical time. With a statistically determined error ϵ, the regression model can be rewritten as follows:

$$Y_i = aX_i + b + \epsilon_i \tag{1}$$

where X_i is the number of days with precipitation in the ith year, Y_i, the "true" RVI and both a and b are the parameters decided by the method of least squares. Equation (1) means that the estimated RVI for a given point in time has a difference of ϵ_i to the true RVI. The variance of RVI, Var(Y) is calculated as follows:

$$\begin{aligned} Var(Y) &= Var(aX + b + \epsilon) \\ &= a^2 Var(X) + Var(\epsilon) \end{aligned} \tag{2}$$

X in Equation (2) is thought to be composed of X_t and X_e where X_t is the true number of days with precipitation and X_e is the error due to documentary sources. Consequently, Equation (2) can be rewritten as follows:

$$Var(Y) = a^2 Var(X_t) + a^2 Var(X_e) + Var(\epsilon) \tag{3}$$

The last two terms in the right side of Equation (3) represent the variance due to the errors. The value of Var (ϵ) can be easily obtained because it accords with the residuals of a regression model. On the other hand, the value of $Var(X_e)$ can be estimated using documentary sources from the instrumental period. That is, the error X_e can be defined as the difference between the monthly number of days with precipitation based on instrumental observation and that based on documentary sources kept in the instrumental period. Figure 12.8 shows the relative frequency of the difference. The distribution is characteristic of the

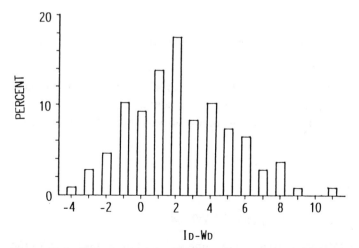

Figure 12.8 Frequency of the difference between the number of days with precipitation based on instrumental observation (I_D) and that on documentary sources (W_D).

normal distribution, which has a mean of 2.3 ($=E[X_e]$ expressed in days) and a variance of 8.6 ($=Var[X_e]$). Thus the variance due to the error, which seems to be random, can be calculated. In the case of June plus July, a mean of 4.6 and a variance of 19.3 was substituted for $E(X_e)$ and $Var(X_e)$ respectively. These values were used for documentary sources whose categories of daily rainfall amount were identified as being more than 1mm in the previous sub-section. For documentary sources classified into the category of daily amount more than 5mm, the error X_e was calculated as follows:

First, the days with precipitation more than 5mm were found for June and July. Second, the days with a description such as "rainfall" were counted for each month among the selected days. Thus we had data for the number of days with precipitation more than 5mm based on instrumental observations and that based on documentary sources. The difference between the number was defined here as the error, X_e. Consequently, a mean of 1.7 and variance of 2.2 were obtained for the separate cases of June and July. A mean of 3.4 and a variance of 4.9 were used for the case of June plus July.

The means and variances calculated above were commonly used for corresponding documentary sources.

When taking the errors into consideration, a mean of the RVI ($E(Y)$) is transformed as follows:

$$
\begin{aligned}
E(Y) &= E(aX + b + \epsilon) \\
&= aE(X) + b + E(\epsilon) \\
&= aE(X_t) + aE(X_e) + b
\end{aligned}
\tag{4}
$$

Equation (4) means that the reconstructed RVIs have a small estimated error. In addition, as a filter is used to delete short-term variations, the variance of the errors is transformed by the weight of the filter. The low-pass filter used in this study was constructed so that we can observe long-term changes above 10 years. The weights (0.004, 0.013, 0.034, 0.071, 0.121, 0.166, 0.182, 0.166, 0.121, 0.071, 0.034, 0.013, 0.004) were adopted here for the purpose.

12.3.4 Assumptions

In the present study, two assumptions pertaining to the reconstruction procedure are as follows:

(1) The relationship between rainfall variation and the number of days with precipitation in historical time is the same as that observed in the instrumental period.
(2) Regional divisions presented in the instrumental time are also as observed in the historical time.

12.4 Results and discussions

12.4.1 The conditions of the Baiu in the period 1700-1900

Although reconstruction of the RVIs was carried out for June, July and June plus July, only the result of June plus July is shown in this section because the rainfall variation of these months are representative of the *Baiu* season (Murata 1990).

Figure 12.9 shows the reconstructed RVIs in region $F_{67}1$. Since three documentary sources were available in this region, three time series of the RVI are shown. As these three curves resemble each other with regard to long-term changes, it is considered that the reconstructed RVIs represent actual rainfall variations. In this figure, the centres of the 95% confidence interval are not on the reconstructed RVI, but shifted upward due to $aE(X_c)$ which was calculated in the previous section. These curves should be drawn so that the centre of the intervals are on the curves. However, that is not performed here because the aim was to clarify the difference between the reconstruction, taking the error into consideration.

From Figure 12.9, it is observed that increases in RVI occurred during the period 1730-50 and again between 1830-50. In contrast to this, relative decreases occurred during the period 1760-1820. The difference between the driest and wettest periods are significant because the 95% confidence intervals for the two periods do not overlap. According to Yamamoto (1971b) Japan experienced the peak of the Little Ice Age during the period 1750-1850. In addition, it is said that it was wetter than normal during this period in the summer season.

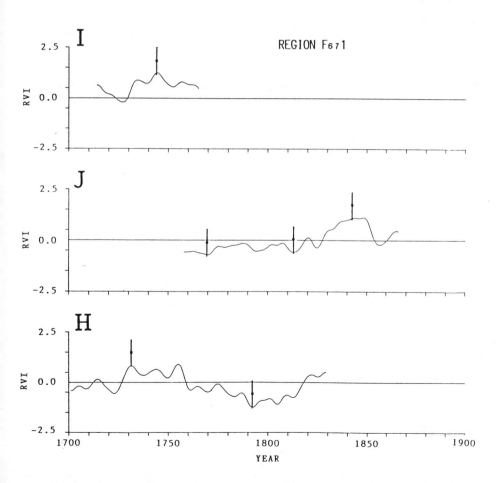

Figure 12.9 The reconstructed RVIs in region $F_6 1$. The vertical lines on each curve indicate the confidence interval of 95 per cent. The centres of the intervals are shifted upward by $aE(X_c)$. Alphabets in the figure correspond to those listed in Table 12.1.

Wetness cannot be identified in the present study. Instead, dryness or little rainfall is identified during this period. This tendency is also confirmed by Mizukoshi (1986) who reconstructed *Baiu* precipitation (rainfall amount between the beginning and the end of the *Baiu* season). Furthermore, the reconstructed rainfall anomaly in Beijing as shown in Figure 12.3 of Wang *et al.* (1981) supports the result because a weak negative correlation (-0.36 for N = 30) is observed between the RVI in region $F_{67}1$ and the rainfall anomaly in Beijing. According to Murata and Yoshino (1988) rainfall variation of this region is related to variation of the North Pacific High. In fact, from the reconstructed variations of the Subtropical High by Huang and Wang (1985) it is recognized that rainfall variation in region $F_{67}1$ is related to longitudinal changes of the Subtropical High (compare Figure 12.9 of the present study with Figure 12.7 of their study).

Figure 12.10 shows the reconstructed RVIs in region $F_{67}2$. Since the two series do not agree over the long-term, it is not possible to know whether rainfall has varied over this region. The dissimilarity is probably attributable to the inaccuracy of weather descriptions in the documentary sources used for reconstruction. More documentary sources are necessary for reconstruction in this region.

Figure 12.11 shows the reconstructed RVIs in region $F_{67}3$. From these two curves, wetter than normal conditions are observed during the entire period. In addition, relative maxima are seen in the periods 1730-50, 1780-90, 1820-40 and 1865-75. Among these four periods, the years 1780-90, 1820-40 and 1865-70 have been identified as famine periods due to cool summers in northern Japan. They are referred to as *Tenmei no Daikikin*, *Tempo no Daikikin*

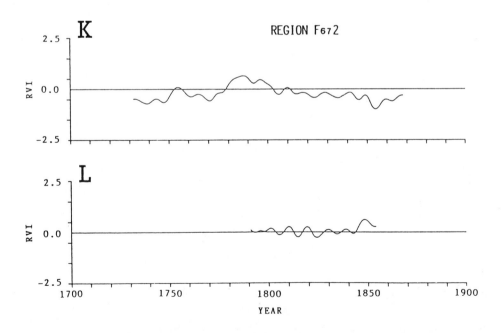

Figure 12.10 The reconstructed RVIs in region $F_{67}2$. The vertical lines on each curve indicate the confidence interval of 95 per cent. The centres of the intervals are shifted upward by $aE(X_e)$. Alphabets in the figure correspond to those listed in Table 12.1.

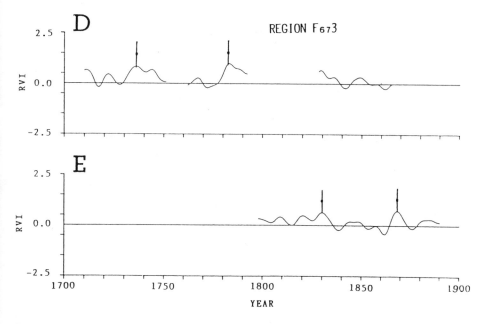

Figure 12.11 The reconstructed RVIs in region $F_{67}3$. The vertical lines on each curve indicate the confidence interval of 95 per cent. The centres of the intervals are shifted upward by $aE(X_c)$. Alphabets in the figure correspond to those listed in Table 12.1.

and *Keio-Meiji no Kyosaku*, respectively. According to Murata and Yoshino (1988) rainfall variation in region $F_{67}3$ is in good agreement with the development of the Okhotsk High, which frequently stagnates over the Okhotsk Sea during the *Baiu* season and brings a cool northeasterly wind called *Yamase* over Japan. Cool summers in northern Japan are often attributable to abnormal stagnation of this high which coincides with the occurrence of a blocking high in the upper circulation. This relationship explains the relative maxima observed in rainfall variation in region $F_{67}3$.

Figure 12.12 shows the reconstructed RVIs in region $F_{67}4$. For this region, three documentary sources were available for reconstruction, however, the reconstructed RVI for documentary source B does not display a similar long-term change compared with the other sources. Therefore the characteristics of this region are inferred from the other two reconstructed RVIs. Relative maxima are seen in the periods 1740-50 and 1780-90, which accords with the case of region $F_{67}3$. In addition, a relative minimum is observed in the period 1800-20. In contrast to the case of region $F_{67}3$, a distinct maximum cannot be seen in the years 1820-30, but rather, a relative decrease is observed. The rainfall variation in this region is mainly ascribed to the stagnation of the *Baiu* front over northern Japan. Relative maxima and minima of the RVI observed in the present study perhaps reflect the activities of the *Baiu* front over northern Japan. Furthermore, since the rainfall variation in this region is parallel with that in the Peninsula of Korea (Murata and Yoshino 1988) the results of the present study can be verified with Korean proxy data.

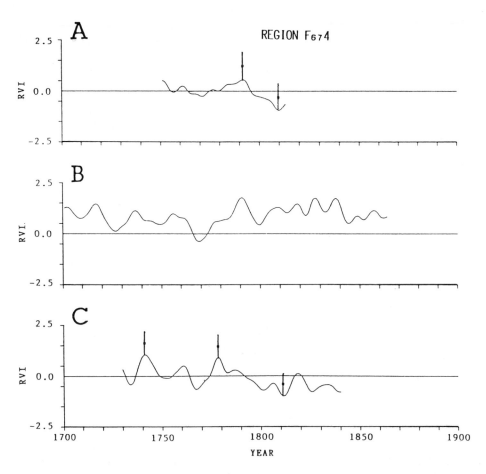

Figure 12.12 The reconstructed RVIs in region $F_{67}4$. The vertical lines on each curve indicate the confidence interval of 95 per cent. The centres of the intervals are shifted upward by $aE(X_e)$. Alphabets in the figure correspond to those listed in Table 12.1.

12.4.2 *Uncertainties for reconstruction procedure*

In the present study, two assumptions were made for reconstruction. The first of these is that the relationship between rainfall variation and the number of days with precipitation in the historical time is the same as that in the instrumental period. In general, we observe a significant positive correlation between rainfall and the number of days with precipitation. As shown earlier, monthly rainfall variation was reconstructed on the basis of the correlation over the recent period. However, we should pay attention to uncertainty pertaining to this relationship. It is possible that the amount of precipitation per day may have changed. For example, we observed a relative decrease in region $F_{67}1$ in the period 1760-1820. However, this change will be lessened if precipitation amount per day during the period was large. In fact, Mizukoshi (1988) revealed that heavy rainfall occurred frequently in the years 1770-80 and after 1820. More detailed treatment of weather descriptions in documentary sources are necessary to help eliminate this uncertainty.

The other assumption is that the regional division presented in the instrumental period is also observed in the historical time. It is extremely difficult to confirm whether this assumption holds or not. Uncertainty must remain concerning this assumption. In the reconstruction procedure, it was determined that the reconstructed RVI should be ignored, for further studies, if the long-term change is not similar to that of other RVIs. According to this criterion, region $F_{67}2$ failed to reconstruct rainfall variation and the RVI based on documentary source B in region $F_{67}4$ was judged to be noisy. The inability to produce consistent reconstructions possibly means that the regionality of rainfall variation has changed, especially for region $F_{67}2$. This might be possible for this region if air flow systems have changed. This is because southwestern Japan including this region often experiences local heavy rain which is closely related to southwest air flow and is topographically controlled (Mizukoshi 1962). One way to overcome this problem might be to divide $F_{67}2$ into two smaller regions such as the eastern part and the western part of the territory. Abundant proxy data would be necessary to test this.

In the case of region $F_{67}4$, the accuracy of descriptions perhaps relates to the dissimilarity of the RVI based on documentary source B. Documentary source B could not be classified into any category based on the number of days with precipitation (Table 12.3). This is likely because the writers of documentary source B recorded daily weather in great detail and as a result, they recorded as "rainfall" days with precipitation less than 1mm which do not contribute much to the monthly total. In fact, the upward shift of the RVI, compared to other RVIs, demonstrates that it is incorrectly evaluated. It can be concluded that the dissimilarity observed in region $F_{67}4$ has nothing to do with the regional division proposed in the present study unlike the case in region $F_{67}2$.

In this study, we devised a way to reconstruct the monthly rainfall variations using "Historical Weather Data Base", which contains several kinds of written source data such as diaries, formal records of old federal clans, etc. Since they are very diverse in their data length and contents, previous arrangement is necessary to utilize these data for reconstruction, especially when using them at the same time. For this purpose, we decided the category of daily rainfall amount related to the number of days with precipitation obtained from documentary sources. Furthermore, it is desirable that the reconstructed results can be validated by comparing them with instrumentally-observed data. Unfortunately, we do not yet have written source data covering the historical and the instrumental period, except for a few examples.

A specific documentary source can sometimes provide very fruitful information about the climate in historical time if it contains weather descriptions in detail and also as a continuous record. In fact, Wang and Zhang (this volume) successfully reconstructed the eighteenth century summer monthly precipitation in three Chinese cities with this type of written source data. Even in Japan, a documentary source with detailed weather descriptions has been found: documentary sources B and D in this study. For these examples, however, the contents of weather descriptions change from month to month, year to year or decade to decade, which is probably due to differences between various writers. Thus attention should also be paid to this point.

In this study, we did not deal with individual documentary sources in detail, rather using them simply, but with statistical inference and two clear assumptions. The confidence interval in consideration of the error due to documentary sources will be useful in improving the reliability of reconstructed data. Also, regional division will be helpful to arrange diverse data in advance when using multiple documentary sources at the same time. Nevertheless, we had to give up the reconstruction of year to year variations by this method because the correlation coefficients are not so large. This is the shortcoming of the method.

By these procedures, we could obtain some reliable reconstructed data for the *Baiu* in the period 1700-1900. Long-term changes from the historical to the instrumental period will be discussed elsewhere.

12.5   Summary

In the present study, monthly rainfall variation in the *Baiu* season has been reconstructed using documentary sources as proxy data. In order to produce reliable reconstructed results, a systematic approach has been devised. The analysis methodology and findings are as follows:

(1)   After dividing Japan into some regions on the basis of rainfall variation in the instrumental period, reconstruction was made for each region.
(2)   By using a number of documentary sources, the validity of the results could be verified.
(3)   The error due to documentary sources was quantitatively estimated using documentary sources during the instrumental period.

Using this approach, reliable rainfall variations could be successfully reconstructed. As a result, distinct regional differences were found during the eighteenth and nineteenth centuries. The results indicate that Japan does not appear to have been wetter than normal throughout this period. Finally, uncertainties pertaining to reconstruction procedures were also discussed.

Acknowledgement

The author is grateful to Prof. M. Yoshino of the Institute of Geoscience, the University of Tsukuba, for his suggestions and encouragement throughout the study. Thanks are also due to Prof. M. Yoshimura of Yamanashi University, for permission to use documentary sources collected in the "Historical Weather Data Base".

This study was done as a part of the World Climate Research Program in Japan.

References

Huang, J.Y. and S.Y. Wang 1985. Investigations on variations of the subtropical high in the western pacific during historical times. *Climatic Change* 7, 427-440.

Maejima, I., M. Nogami, S. Oka and Y. Tagami 1983. Historical weather records at Hirosaki, northern Japan, from 1661 to 1868. *Geographical Reports of Tokyo Metropolitan University* 18, 113-152.
Mizukoshi, M. 1962. Distribution of rainfall related to the activities of *Baiu* fronts. *Geographical Review of Japan*, 35 (1) 35-44 (in Japanese with English abstract).

Mizukoshi, M. 1986. Kinki-Tokai chiho ni okeru *Baiu* tokusei no choki hendo keiko (Long-term change of *Baiu* in Kinki-Tokai district, central Japan). *Human Science Report of Mie University* 5, 95-102 (in Japanese).

Mizukoshi, M. 1988. Kinsei igo ni okeru Kinki chiho cyubu no *Baiu* no choki keiko (Long-term change of the *Baiu* in central Kinki district, central Japan after Kinsei, 1700). *Mizu shigen kenkyu senta kenkyu hokoku (Reports of water resource center)* 65-78 (in Japanese).

Murata, A. and M. Yoshino 1988. Reconstruction of rainfall variations in the *Baiu*, a rainy season in Japan. *Geographical Review of Japan* 61(Ser. A) 8, 643-656 (in Japanese with English abstract).

Murata, A. 1990. Regionality and periodicity observed in rainfall variations of the *Baiu* season over Japan. *International Journal of Climatology* (in press).

Wang, S. H., Z. C. Zhao and S. H. Chen 1981. Reconstruction of the summer rainfall regime for the last 500 years in China. *Geojournal*, 5(2) 117-122.

Yamamoto, T. 1971a. On the climate change in XV and XVI centuries in Japan. *Geophysical Magazine*, 35 (2) 187-206.

Yamamoto, T. 1971b. On the nature of the climatic change in Japan since the "Little Ice Age" around 1800 A.D. *Journal of Meteorological Society of Japan* 49 (special issue) 798-812.

Yoshimura, M. and M. M. Yoshino 1988. Outline of a historical weather data base for Japan. In *Recent Climatic change*, London and New York, S. Gregory (ed.) 267-271. London and New York: Belhaven press.

Yoshino, M. and A. Murata 1988. Reconstruction of rainfall variation in the *Baiu* rainy season in Japan. In *Recent Climatic Change*, S. Gregory (ed.) 272-284. London and New York: Belhaven press.

13 Climatic variations in the longest instrumental records

P. D. Jones and R. S. Bradley

13.1 Introduction

Routine observations of surface air temperature, precipitation and surface pressure began in western Europe during the late seventeenth and early eighteenth centuries, and gradually spread to most of the rest of the world by the twentieth century. The speed with which instrumental recording began in other parts of the world was not gradual but tended to occur around certain key dates. For example many countries set up meteorological agencies after the Vienna Meteorological Congress of 1873. Even now, coverage is sparse in both polar regions and recording did not start in parts of northern Canada and the Antarctic until the 1940s and late 1950s respectively.

One of the problems encountered in taking European-made thermometers elsewhere, particularly to the Soviet Union and Canada was that mercury froze at about -38°C (see, for example, Ball and Kingsley 1984). Alcohol thermometers were developed to overcome this problem. Other problems with early instruments and scales are summarised by Middleton (1966, 1969) and Lamb and Johnson (1966). Most of the early observers up to the middle of the last century were professional people such as doctors, physicists and astronomers. In some cases, the observations were published, but in many cases the original measurements have been lost. Climatology is fortunate that more data might have been lost if it were not for the efforts of the German meteorologist Heinrich Wilhelm Dove who collected and published, under the auspices of the Prussian Academy of Sciences, as much monthly mean air temperature data as he could obtain (Dove 1838 and later). During the 1860s he had built up a network of nearly 2000 stations, but his coverage was particularly sparse over the interior parts of Africa, Asia, South America and Australia. This data set allowed the Austrian meteorologist Wilhelm Köppen (1873) to attempt to assess whether mean global temperature had changed since 1750. Although Dove's data compilations are important, the lack of spatial coverage meant that the early hemispheric analyses are not of great importance.

The most comprehensive compilation of long-term instrumental climate (temperature and precipitation) data currently available is that of Bradley *et al.* (1985). This compilation built and improved on earlier data sets such as *World Weather Records and Reseau Mondial* by searching meteorological and other archives for published and manuscript sources of early data. The most important aspect of the compilation is that it contains details of the sources of all the station datasets and, where possible, details of the long term homogeneity of the climate time series (see technical reports, Bradley *et al.* 1985; Jones *et al.* 1985, 1986a). Clearly, if one is to study climatic change, it is vital to ensure homogeneity of time series data.

Station time series are said to be homogeneous if the variations are caused only by variations of the weather and climate (Conrad and Pollak 1962). Factors such as changes in instrumentation, exposure, location, methodological practices and environmental changes around the station can all cause inhomogeneities in station time series (Bradley and Jones 1985). All can seriously affect the station record, although, with the exception of environmental changes, the effects do not produce any consistent bias because they can act either to increase or decrease the measured temperature. Growth in towns and cities around stations would generally be expected to increase temperatures through the development of urban heat islands.

The homogeneity of many of the long European air temperature records has been assessed by the climatologists who have developed the series (e.g. Manley 1974 for Central England, Schaake 1982 for Berlin). Homogeneity of the other temperature data in the Bradley *et al.* (1985) data set was assessed by comparing neighbouring stations records. Often, each record was compared with a number of other records from sites from a few tens to a few hundred kilometers distant. Jumps and trends in the comparisons can generally be related to one or more of the above causes. Table 13.1 lists some of the longest air temperature records by continent (see Jones *et al.* 1985 1986a for details of their sources and homogeneity).

13.2   Climate from 1700 to 1850

13.2.1   Temperature

From Table 13.1, twelve of the longest and most spatially extensive surface air temperature records (indicated by asterisks in Table 13.1) were selected. All the time series come from the Northern Hemisphere. The only station with temperature data available before 1850 in the Southern Hemisphere is Rio de Janeiro. The results from this study therefore relate exclusively to the Northern Hemisphere. Furthermore, the long time series are all located between 40° and 64°N.

Seasonal and annual time series are shown for the 12 sites in Figures 13.1a-e. It is impossible to show here all the variations on the year-to-year timescale. Instead low-frequency variations are shown by season. The 10-year Gaussian filter used illustrates variations on decadal and longer timescales. For each site the seasonal and annual time series are plotted as departures from the 1901-50 reference period average. All but one of the time series was considered homogeneous over its entire station length by Jones *et al.* (1985). Corrections were applied to three series Geneva, Leningrad and New Haven. The only series which was not considered homogeneous was Toronto. Urban warming is evident in this record since the 1880s when compared with other nearby Canadian stations. Recently, however, an additional 70 years of data, extending the data back to 1770, has been found for this site (Crowe and Masterton 1990). It is included here because the record now represents the earliest instrumental time series from interior North America.

The four seasonal and one annual figure show a number of common features. Variability tends to be greatest in winter followed by spring, autumn and summer. Most of the longer time-scale variations evident in the annual data comes from the winter, spring and autumn series. In general, all annual series (Figure 13.1e) show some indication of warming from the beginning to the end of the record, although this is clearest in the case of the records which start during the nineteenth century. The very long European records from Central England,

Table 13.1 Early instrumental temperature records.

	Site	Country	Latitude	Longitude	First Year
Europe	Trondheim	Norway	63.4°N	10.5°E	1761[*]
	Stockholm	Sweden	59.4°N	18.1°E	1756[*]
	Central England	UK	~52.5°N	~2.0°W	1659[*]
	Stykkisholmur	Iceland	65.0°N	22.8°W	1846
	Godthaab	Greenland	64.2°N	51.7°W	1866
	Copenhagen	Denmark	55.7°N	12.6°E	1768
	De Bilt	Netherlands	52.1°N	5.2°E	1706
	Geneva	Switzerland	46.2°N	6.2°E	1753[*]
	Paris	France	48.8°N	2.5°E	1757
	Berlin	West Germany	52.5°N	13.4°E	1701[*]
	Munich	West Germany	48.1°N	11.7°E	1781
	Vienna	Austria	48.2°N	16.4°E	1775
	Prague	Czechoslovakia	50.1°N	14.3°E	1771
	Warsaw	Poland	52.2°N	21.0°E	1779
	Budapest	Hungary	47.5°N	19.0°E	1780
	Milan	Italy	45.5°N	9.2°E	1763
	Arkhangel	USSR	64.6°N	40.6°E	1813
	Vilnjus	USSR	54.6°N	25.3°E	1777
	Leningrad	USSR	60.0°N	30.3°E	1743[*]
	Kiev	USSR	50.5°N	30.5°E	1812
	Astrahan	USSR	46.4°N	48.0°E	1837
	Orenburg	USSR	51.8°N	55.1°E	1832
	Tbilisi	USSR	41.7°N	44.8°E	1844
	Fort Sevcenko	USSR	44.6°N	50.3°E	1848
Asia	Tobolsk	USSR	58.2°N	68.2°E	1832
	Irkutsk	USSR	52.3°N	104.3°E	1820
	Sverdlovsk	USSR	56.8°N	60.6°E	1831[*]
	Barnaul	USSR	53.3°N	83.8°E	1838[*]
Africa	Cape Town	South Africa	33.9°S	18.5°E	1857
North America	Sitka	USA	57.1°N	135.3°W	1832[*]
	Toronto	Canada	43.7°N	79.4°W	1770[*]
	Charleston	USA	32.8°N	79.9°W	1823
	New York	USA	40.7°N	74.0°W	1822
	New Haven	USA	41.3°N	72.9°W	1781[*]
	Boston	USA	42.4°N	71.0°W	1743
	Blue Hill	USA	42.1°N	71.2°W	1811
	Minnesota	USA	~45.0°N	~93.0°W	1820[*]
	Albany	USA	42.7°N	73.8°W	1820
	Rochester	USA	43.1°N	77.7°W	1830
South America	Rio de Janeiro	Brazil	22.9°S	43.2°W	1832
	Santiago	Chile	33.5°S	70.8°W	1861
	Buenos Aires	Argentina	34.6°S	58.4°W	1856
	Bahia Blanca	Argentina	38.7°S	62.3°W	1860
Australasia	Auckland	New Zealand	36.8°S	174.8°E	1853
	Wellington	New Zealand	39.5°S	174.8°E	1862
	Dunedin	New Zealand	45.9°S	170.5°E	1853
	Adelaide	Australia	35.0°S	138.5°E	1857
	Sydney	Australia	34.0°S	151.1°E	1859
	Djakarta	Indonesia	6.2°S	106.8°E	1866

[*]Time series included in Figure 13.1.

Figure 13.1 Low frequency variations of temperature at 12 selected sites with the longest records. The time series shown result from the application of a 10 year Gaussian filter. (a) Winter [DJF] (b) Spring [MAM] (c) Summer [JJA] (d) Autumn [SON] (e) Annual. Each tick mark on the vertical scale represents 1°C.

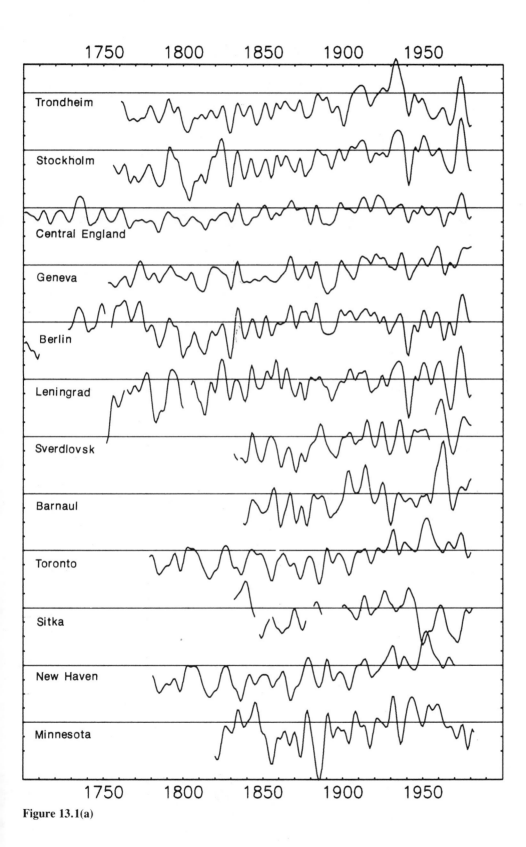

Figure 13.1(a)

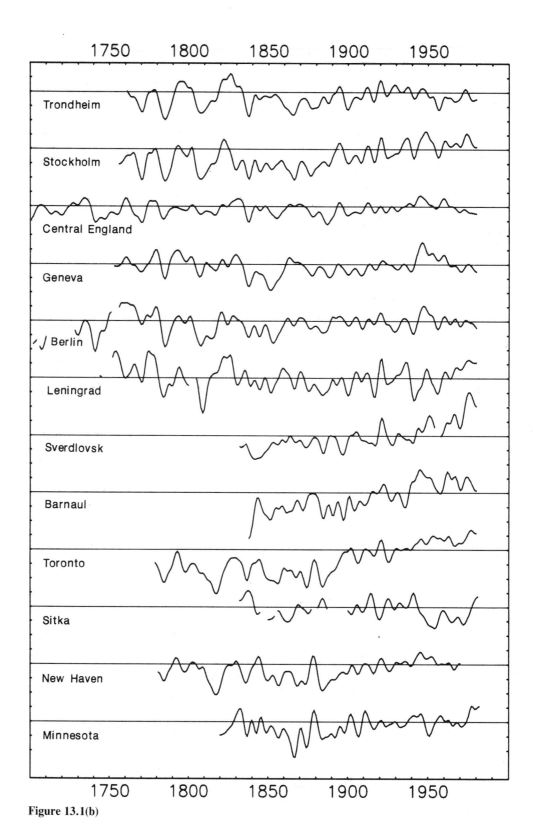

Figure 13.1(b)

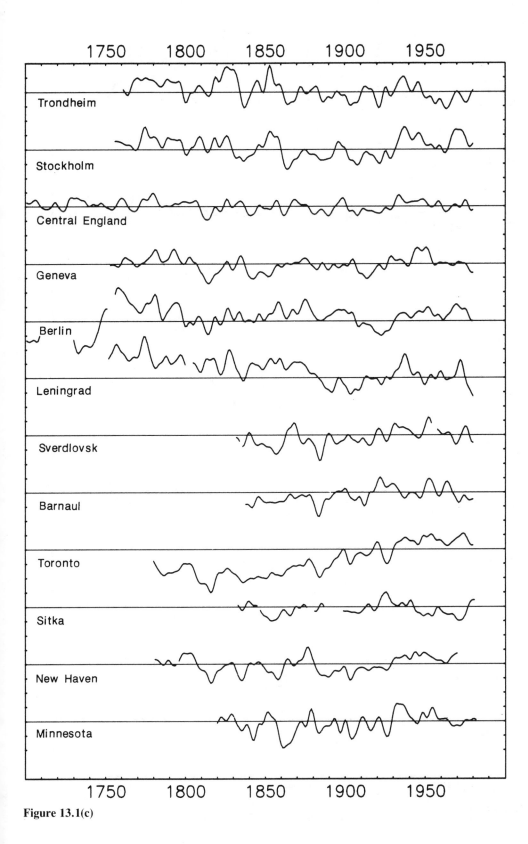

Figure 13.1(c)

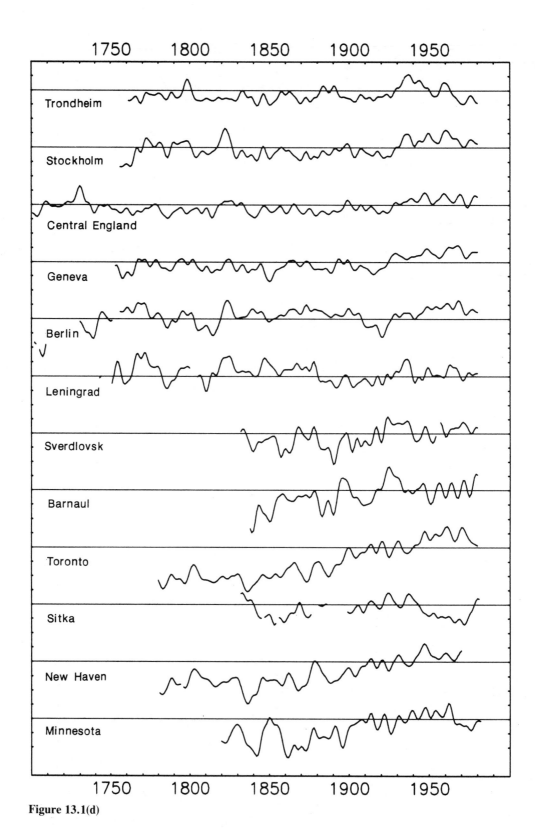

Figure 13.1(d)

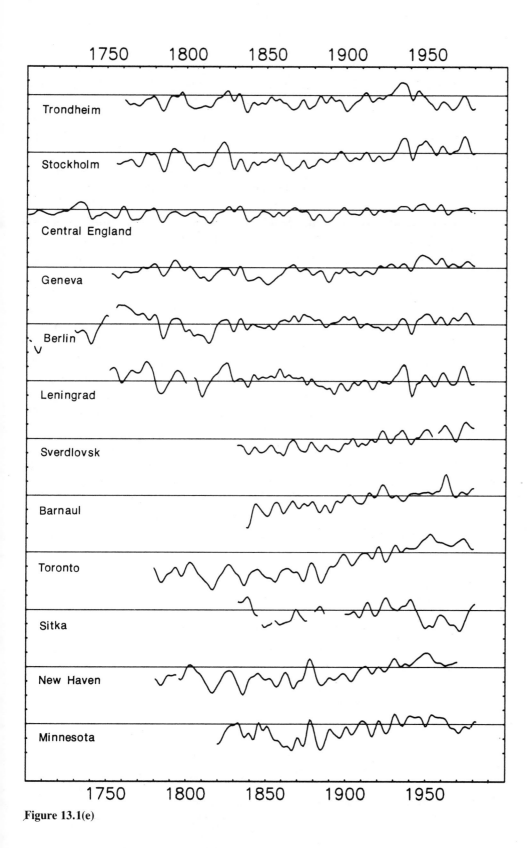

Figure 13.1(e)

Berlin and Leningrad hardly show any warming between the early 1700s and 1980. Long term warming is least evident during the summer season. Indeed both Berlin and Leningrad show a tendency towards cooler summers now than during the eighteenth century. The amount of annual temperature change explained by a linear trend is listed in Table 13.2 for various starting dates (e.g. 1700 1750 etc.). All sites show warming over the period 1851 to 1980, however a slight cooling is indicated in the longer central European records which begin in the early eighteenth century.

Comparison of the long European records shows a number of consistently cool and warm decades since 1700. Cool decades are evident during the 1740s, 1780s, 1800s and 1810s and the late 1830s and 1840s. Warm decades are limited to the 1820s only. The two Asian records show cool decades during the 1830s and 1840s as in Europe. The four North American records exhibit cold periods during the 1780s, 1810s and 1830s, again showing some similarities to the European record, though warmth is indicated during the 1800s, 1820s and to some extent during the 1840s.

Although there appears some agreement between warm and cool decades over the major northern land masses, more comprehensive studies of twentieth century temperature variations over the Northern Hemisphere have shown that no one region can be said to be representative of hemispheric-wide conditions (Jones and Kelly 1983). These and other workers have shown that the only way of producing a truly representative series for the Northern Hemisphere is to include station data from as many regions as possible. There is no short-cut method using only a few stations, such as has been tried by Groveman and Landsberg (1979). Clear examples of the differing trends are apparent here. North American stations indicate warmth during the 1800s and 1840s while European stations indicate cool conditions. Similar differences between decadal temperature trends in Europe, North America and Asia are also apparent since 1850. The most noted example concerns the early 1940s which were warm over North America and Asia but exceptionally cool over most of Europe (Jones et al. 1986a). This is particularly evident during winter (see Figure 13.1a).

Despite the problem of inferring hemispheric trends from a small number of stations it would appear that the Northern Hemisphere was cool during the 1780s, 1810s and the late

Table 13.2 Annual temperature change accounted for by the linear trend for various starting dates.

	1701–1980	1751–1980	1801–1980	1851–1980
Trondheim	-	0.23	0.31	0.15
Stockholm	-	1.02	1.27	1.40
Central England	-0.19	0.51	0.49	0.49
Geneva	-	0.66	1.05	1.07
Berlin	0.01	-0.04	0.71	0.10
Leningrad	-	-0.40	-0.09	0.08
Sverdlovsk	-	-	-	1.60
Barnaul	-	-	-	1.62
Sitka	-	-	-	0.00
Toronto[1]	-	-	3.01	3.00
New Haven	-	-	-	1.75
Minnesota	-	-	-	1.53

[1]The record for Toronto is clearly affected by an urbanization influence after about 1880.

1830s, with warmth dominating the 1820s. The cold-warm-cold oscillatory pattern between the 1810s and 1830s is particularly striking in the New Haven and Toronto records in North America and in the Stockholm and Leningrad records in Europe.

13.2.2 Precipitation

Most of the sites with long time series of temperature measurements also recorded precipitation amounts. Assessing the homogeneity of long time series is, however, generally more difficult for precipitation than for temperature. Precipitation is much more spatially variable than temperature and a denser network of stations is required to assess the homogeneity of long records. Precipitation records are particularly prone to growth of vegetation and/or building development around the site. Any interference to the wind eddies near the gauge can seriously distort the amount of rain caught (Sevruk 1986). In most cases the catch of rain is reduced from what should have been caught. The homogeneity of long time series is also affected by the methods used to assess the amount of snowfall. The ability of precipitation gauges to catch snowfall has been improved by the introduction of gauge shielding. With the shields, gauges catch more snowfall; thus precipitation totals for the winter part of the year during earlier periods must be adjusted. All gauge records in the Soviet Union must be modified for changes to gauge design during the 1960s (see Schver 1975).

Some of the longest and most continuous precipitation time series are listed in Table 13.3. Apart from Leningrad and the England and Wales, record all the European records were developed by Tabony (1980, 1981) from published sources and manuscript data held by national meteorological agencies. The England and Wales series (developed by Wigley *et al.* 1984 and updated by Wigley and Jones 1987) is a regional average based on between 5 and 35 site records spread over the region. All 35 gauge records were used after the 1820s. Averaging site precipitation data into regional series enables the underlying trends in precipitation to be more clearly seen. Local effects, specific to an individual site will tend to be smoothed out.

Table 13.3 Early instrumental precipitation records.

	Site	Country	Latitude	Longitude	First Year
Europe	Uppsala	Sweden	59.9°N	17.6°E	1774[*]
	Lund	Sweden	55.7°N	13.2°E	1748[*]
	England and Wales	UK	~52.5°N	~2.0°W	1766[*]
	Hoofdoorp	Netherlands	52.3°N	4.7°E	1735[*]
	Paris	France	48.8°N	2.5°E	1770[*]
	Marseilles	France	43.3°N	5.4°E	1748[*]
	Karlsruhe	West Germany	49.0°N	8.4°E	1779[*]
	Milan	Italy	45.5°N	9.2°E	1764[*]
	Padua	Italy	45.4°N	12.0°E	1725[*]
	Leningrad	USSR	60.0°N	30.3°E	1740[*]
Asia	Seoul	South Korea	37.3°N	127.0°E	1770
North America	Charleston	USA	32.8°N	79.9°W	1738[*]
	New Haven	USA	41.3°N	72.9°W	1804
	Philadelphia	USA	40.0°N	75.2°W	1830
	Boston	USA	42.4°N	71.0°W	1828
	Albany	USA	42.7°N	73.8°W	1826[*]
	Toronto	Canada	43.7°N	79.4°W	1841

[*]Time series included in Figure 13.2.

Figures 13.2a-e show seasonal and annual time series for 12 of the sites listed in Table 13.3 that are considered to be homogeneous. As with temperature it is not possible to show here the year-to-year variations. Low-frequency variations are shown instead. In order to compare time series which have marked differences in mean precipitation and in variability, the time series show standardized departures $(x-\bar{x})/\sigma$. The mean and standard deviation of the seasonal and annual precipitation time series were calculated over the 1901-50 period.

The longest precipitation records originate from Europe. For the period prior to the 1820s there is only an adequate time series to consider western European variations. Here decades from the 1780s to the 1810s were generally dry with the exception of the 1800s at Lund and Padua. Prior to this time, the few data sets that are available suggest that the 1760s and 1770s were wet with the 1730s 1740s and 1750s being dry. After the 1820s it is possible to consider precipitation variations over eastern North America. The available data suggest wetter conditions during the 1830s and from the 1850s to the 1880s at the two longest and most continuous records of Charleston and Albany. Over Europe the 1860s were generally dry with the 1850s also being dry over Northern Europe. Mediterranean stations were wet during the 1840s.

13.3   Changes between 1850 and the present

The last 140 years is the period for which we know most about global climate. The major expansion of instrumental recording took place with the establishment of national meteorological agencies between 1860 and 1900. Before 1850 station coverage was limited to Europe, parts of Asia and North America and some coastal areas of Africa, South America and Australasia. By 1900 there were only a few areas, in the interior of Africa and South America and the whole of Antarctica, without meteorological instrumentation.

The more complete hemispheric and global coverage allows the development of continental and hemispheric averages of temperature. Using the compilations of homogeneous stations records mentioned earlier, Jones et al. (1986a,b) have produced a grid point (5° latitude by 10° longitude) data set of surface air temperature anomalies for each month from January 1851. Interpolation of the basic station data is necessary to overcome the uneven spatial density of stations. Temperature data are expressed as anomalies from a common reference period to enable interpolation to be accomplished. Neighbouring stations, at different elevations and using different techniques to calculate monthly mean temperatures, will have different mean temperatures. Averaging the station data in absolute degrees Celsius will be affected by varying station numbers. The use of anomaly values overcomes these problems.

Time series of mean hemispheric temperature anomalies (annual and seasonal) are shown in Figures 13.3 and 13.4. The features exhibited by the two curves have been discussed before

Figure 13.2 Low frequency variations of precipitation at 12 selected sites with longest records. The time series shown result from the application of a 10 year Gaussian filter. The seasonal and annual precipitation series have been transformed to standardized anomalies by subtracting the mean and dividing by the standard deviation. The mean and standard deviation have been calculated over the 1901-1950 period. (a) Winter [DJF] (b) Spring [MAM] (c) Summer [JJA] (d) Autumn [SON] (e) Annual. Each tick mark on the vertical scale represents one standardized anomaly.

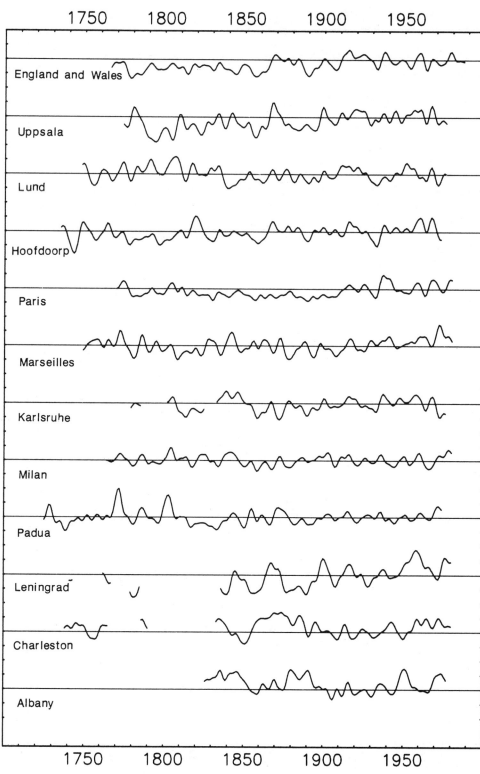

Figure 13.2(a)

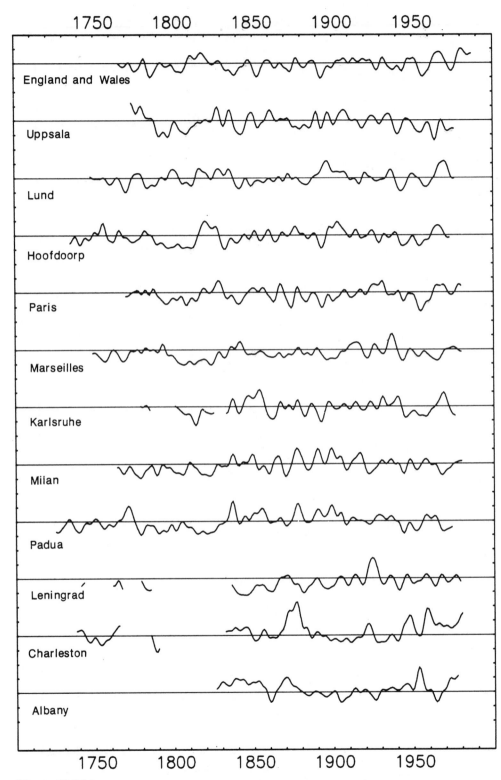

Figure 13.2(b)

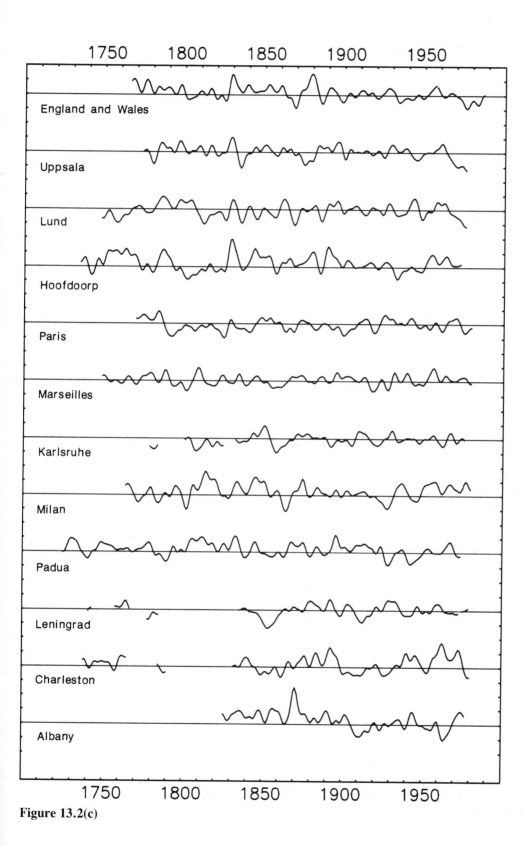

Figure 13.2(c)

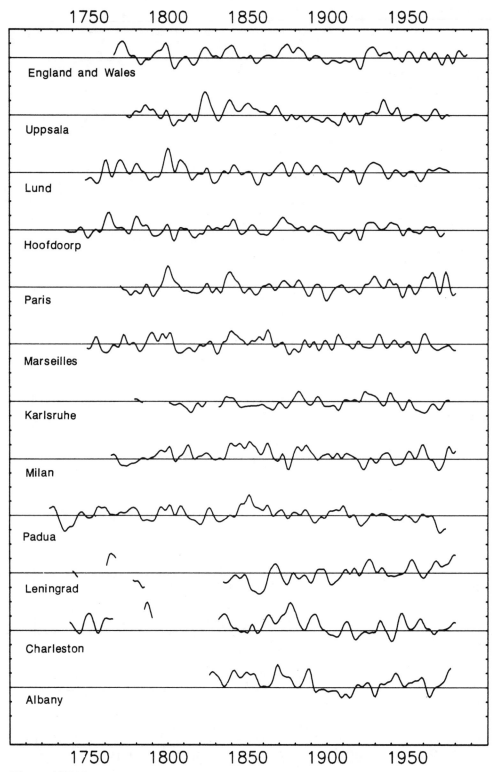

Figure 13.2(d)

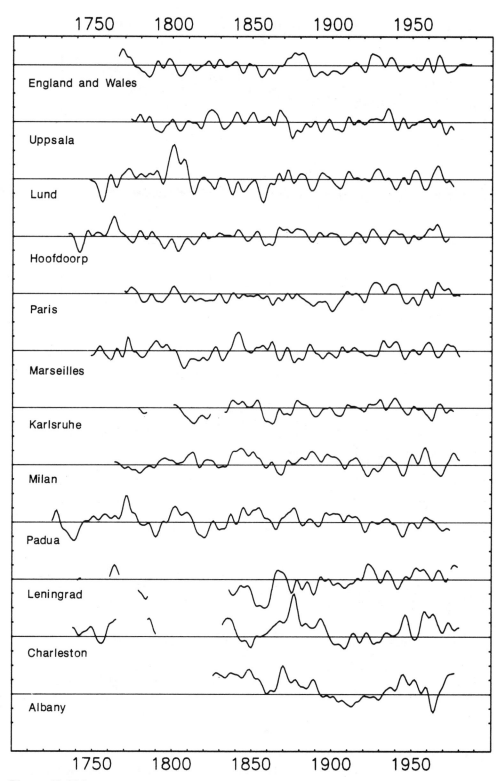

Figure 13.2(e)

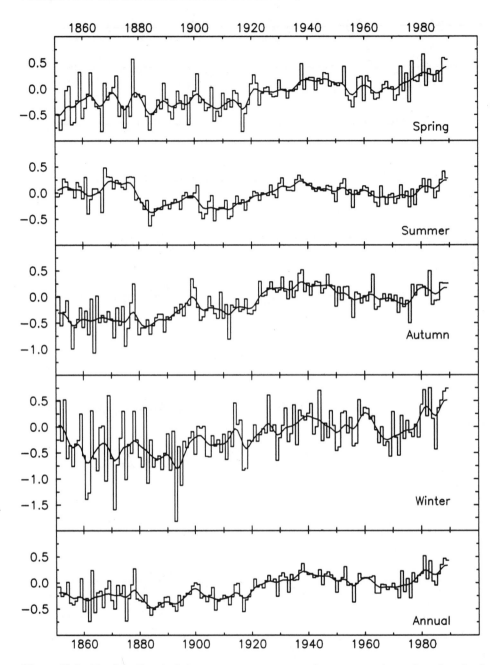

Figure 13.3 Northern hemisphere average temperatures by season and year based on land data. The smooth curve is a 10 year Gaussian filter which highlights variations on decadal and longer timescales. The temperatures are expressed as anomalies from the 1951–70 period.

(see, for example, Wigley *et al.* 1985, 1986). Both hemispheres show a warming of the order of 0.5°C since the mid-nineteenth century. The warming is more monotonic in the Southern Hemisphere with the greatest warming in the Northern Hemisphere occurring between 1920

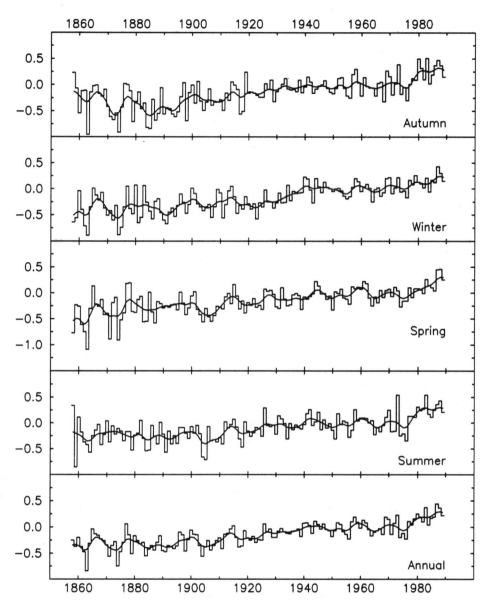

Figure 13.4 Southern hemisphere (0-60°S) average temperatures by season and year based on land data. The smooth curve is a 10 year Gaussian filter which highlights variations on decadal and longer timescales. The temperatures are expressed as anomalies from the 1951–70 period.

and 1940. Warming is evident in all seasons except summer where little if any warming has occurred since the 1850s. The 1980s are clearly the warmest decade in the Southern Hemisphere while in the Northern Hemisphere recent warming means that the 1980s are only slightly warmer than the 1940s. Comparison of the Northern Hemisphere time series with the 12 individual series in Figure 13.1a-e illustrates clearly that no region is representative of hemispheric conditions all of the time. Table 13.4 shows annual correlations between the 12

263

Table 13.4 Correlations between the 12 selected station annual temperature time series and the Northern Hemisphere annual time series. The correlations are given for three periods 1851-1915, 1916-1980 and 1851-1980. For the longer period the correlations are also calculated between low-frequency variations (resulting from a 10-year Gaussian filtered time series).

Stations	1851-1915	1916-1980	1851-1980	1851-1980
	Raw	Raw	Raw	Low
Trondheim	0.17	0.20	0.24	0.29
Stockholm	0.48	0.27	0.51	0.64
Central England	0.34	0.31	0.41	0.69
Geneva	0.12	0.32	0.43	0.71
Berlin	0.49	0.24	0.28	0.02
Leningrad	0.61	0.23	0.39	0.28
Sverdlovsk	0.59	0.24	0.57	0.73
Barnaul	0.31	0.25	0.46	0.69
Sitka	0.10	0.41	0.18	0.09
Toronto	0.38	0.35	0.62	0.76
New Haven	0.31	0.36	0.56	0.76
Minnesota	0.44	0.52	0.57	0.73

sites and the Northern Hemisphere average for the periods 1851-1915, 1916-1980 and 1851-1980. Correlations are given for both the raw series and the 10-year Gaussian-filtered time series. The correlations tend to be higher for the overall period and are markedly higher for lower frequencies. Stations with the highest correlations are generally in the continental interiors of Eurasia and North America.

The land areas represent only 29% of the Earth's surface. Until recently it was assumed that changes in air temperature over the ocean were similar to those over the land. Recent compilations of marine data taken by 'ships of opportunity' now enable this assumption to be tested. Since the 1850s ships have been obliged to take weather observations and measure the temperature of the sea surface every 6 hours. The force behind this collection was an American naval captain, Matthew Fontaine Maury, who persuaded the other major maritime nations to instruct their military and merchant navies to take measurements and record these in log books. Through his pioneering efforts in the 1830s and 1840s he helped to standardize the practice by which ships at sea collected meteorological observations, including measurements of sea surface temperature (SST) which were rare until that time. An international agreement to take, collect and exchange marine meteorological observations was signed in Brussels in 1853.

In the last twenty years, major international efforts have been made to put all of this log-book information into computer data banks. One such compilation is the Comprehensive Ocean-Atmosphere Data Set (COADS) produced by NOAA workers at Boulder, USA (Woodruff *et al.* 1987). COADS contains all the log book material that has been found so far, about 80 million non-duplicated sea surface temperature measurements between 1854 and 1988.

The marine data are, unfortunately, affected by inhomogeneities just like the land data. Sea surface temperature data were generally taken before World War II by collecting some sea water in an uninsulated canvas bucket and measuring the temperature. There was generally a few minutes delay between sampling and reading. During this time the water in the bucket generally cooled. Since World War II most readings are made in the intake pipes which take sea water on to ships to cool the engines. This change in measurement technique was quite abrupt, although there are still significant numbers of bucket measurements made today (buckets now made of plastic and thus better insulated) and some intake measurements were made prior to World War II.

Studies of the differences between the two methods indicate bucket temperatures are cooler than intake ones (James and Fox 1972). Correcting the SST data for this measurement change may, at first, seem like an intractable problem. Folland and Parker (1990) of the U.K. Meteorological Office, however, have developed a method of correcting the canvas bucket measurements based on physical principals related to the causes of the cooling. The cooling depends on the prevailing meteorological conditions, and so depends on the time of year and on location. Although the cooling is therefore a day-to-day phenomenon, the various influences are basically linear, so cooling amounts can be calculated on a monthly basis. The main free parameter is the time between reading and sampling. This is generally unknown and must be estimated from the data. The primary assumption in this estimation is that there have been no major changes in the seasonal cycle of SSTs over the period of record. Since the amount of evaporative cooling has a strong seasonal cycle in many parts of the world, an optimum exposure time can be chosen; namely that which minimises the residual seasonal cycle in the corrected data. As a check on the validity of the method, the implied optimum exposure time turns out to be quite consistent spatially.

The major problem with the technique is that it is not known with any certainty what type of buckets were used to take measurements during the nineteenth century. Assuming measurements were taken using canvas buckets rather than wooden buckets (which are

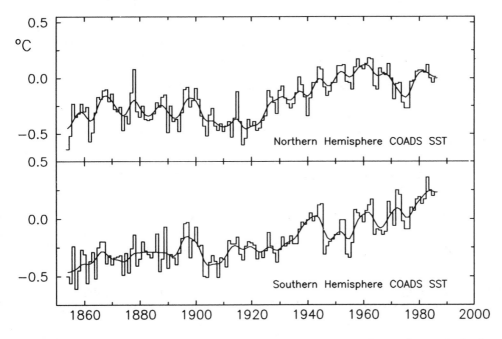

Figure 13.5 Hemispheric time series of COADS SST data corrected according to the bucket model described in the text. The temperatures are expressed as anomalies from the 1950–79 period.

better insulated) leads to corrections which result in SSTs warmer than land temperatures by about 0.2°C. The discrepancy almost disappears if wooden buckets are assumed. Although there is documentary evidence to support wooden bucket use during the mid-nineteenth century, considerable doubt remains about the transition from wooden to canvas buckets. The seasonal-cycle elimination method is not precise enough to choose between the two possibilities.

Time series of corrected COADS SST data for the two hemispheres are shown in Figure 13.5. These corrections have been derived using the wooden bucket assumption in the nineteenth century (see Jones *et al.* 1990, for details). Overall the various hemispheric land and marine time series show many consistencies. Even on the inter-annual time scale, hemispheric averages over land and ocean are strongly correlated, and virtually all longer time scale fluctuations in the marine data are reflected in the land data. Even today, however, there are still gaps in data coverage, especially over the southern oceans. Satellite data may help to resolve this problem, but only from the late 1970s.

Acknowledgements

This work was supported by the United States Department of Energy, Carbon Dioxide Research Division under Grant No. DE-FG02-89ER69017.

References

Ball, T. F. and Kingsley, R. G. 1984: (Instrumental temperature records at two sites in central Canada. *Climatic Change*, 6, 39-56.

Bradley, R. S. and Jones, P. D. 1985: Data bases for detecting CO$_2$-induced climatic change. In: M. C. MacCracken and F. M. Luther (Eds.) *Detecting the Climatic Effects of Increasing Carbon Dioxide*, U.S. Dept. of Energy, Carbon Dioxide Research Division, Washington D.C. 29-53.

Bradley, R. S., Kelly, P. M., Jones, P. D., Diaz, H. F. and Goodess, C. 1985: A climatic data bank for the Northern Hemisphere land areas 1851-1980. *Technical Report No. TR017*, U.S. Dept. of Energy, Carbon Dioxide Research Division, Washington D.C., 335pp.

Conrad, V. and Pollak, L. D. 1962: *Methods in Climatology*, Harvard University Press, Cambridge, Massachusetts, 459pp.

Crowe, R. B. and Masterton, J. M. 1990: Extension of Toronto temperature time-series from 1840 back to 1778 using United States and other data. In: C. R. Harrington (Ed.) *1816, the year without a summer*, National Museums of Canada (in press).

Dove , H. W. 1838+: Uber die geographische Verbreitung gleichartiger Witterungserscheinungen (Uber die nichtperiodischen Anderungen der Temperaturverteilung auf der Oberflache der Erde): *Abh Akad. Wiss. Berlin* I Teil (1838) 37, 285; 38, 286; 39, 345. II Tiel (1839) 40, 306-307. III Tiel (1844) 41 127. V Tiel (1852) 42, 3-4.

Folland, C. K. and Parker, D.E. 1990: Observed variations of sea surface temeprature. In: M. E. Schlesinger (Ed.) *Climate-Ocean Interaction*. NATO Workshop, Kluwer Academic Publishers (in press).

Groveman B. and Landsberg, H.E. 1979: Simulated Northern Hemisphere temperature departures 1579-1880. *Geophysical Research Letters*, 6, 767-9.

James, R. W. and Fox, P. T. 1972: Comparative sea-surface temperature measurements. Report No. 5. *Marine Science Affairs No. 366.* World Meteorological Organization, Geneva, Switzerland.

Jones, P. D. and Kelly, P. M. 1983: The spatial and temporal characteristics of Northern Hemisphere surface air temperature variations. *Journal of Climatology*, 3, 243-252.

Jones, P. D., Raper, S. C. B., Santer, B. D., Cherry, B. S. G., Goodess C., Bradley, R. S., Diaz, H. F., Kelly, P. M. and Wigley, T. M. L. 1985: A grid point surface air temperature data set for the Northern Hemisphere 1851-1984. *Technical Report TR022.* U.S. Dept. of Energy, Carbon Dioxide Research Division, Washington D.C., 251pp.

Jones, P. D., Raper, S. C. B., Bradley, R. S., Diaz, H. F., Kelly, P. M. and Wigley, T. M. L. 1986a: Northern Hemisphere surface air temperature variations: 1851-1984. *J. Climate and Applied Meteorology*, 25 161-179.

Jones, P. D., Raper, S. C. B. and Wigley, T. M. L. 1986b: Southern Hemisphere surface air temperature variations: 1851-1984. *J. Climate and Applied Meteorology*, 25 1213-1230.

Jones, P. D., Wigley, T. M. L. and Farmer, G. 1990: Marine and land temperature data sets: A comparison and a look at recent trends. In: M. E. Schlesinger (Ed.) *Greenhouse-gas-induced climatic change*, Kluwer Academic Publishers (in press).

Köppen, W. 1873: Uber Mehrjahriger Perioden der Witterung, Insbesondere Uber die 11 Jahrige Periode der Temperatur. *Zeitschrift der Osterreichischen Gesellschaft für Meteorologie*, 8, 241-248, 257-267.

Lamb, H. H. and Johnson, A. I. 1966: Secular variations of the atmospheric circulation since 1750. *Geophysical Memoirs* 110, H.M.S.O., London.

Manley, G. 1974: Central England temperatures: Monthly means 1659-1972. *Quarterly J. Royal Meteorological Society* 100, 389-405.

Middleton, W. E. K. 1966: *Invention of Meteorological Instruments.* Johns Hopkins Press, Baltimore, Maryland.

Middleton, W. E. K. 1969: *A history of the thermometer and its use in meteorology.* Johns Hopkins Press, Baltimore, Maryland.

Schaake, P. 1982: Ein Beitrag zurn Preubenjahr 1981 en Berlin: *Beilage zur Berliner Wetterkarte des Instituts für Meteorologie der Frein Universitat Berlin*, 32/82, S04/82.

Schver, T. A. 1975: Annual amount of precipitation data in the USSR, according to corrected data. *Soviet Hydrology, Selected Papers*, 3 155-157.

Sevruk, B. (editor) 1986: Correction of precipitation measurements. *Proceedings of international workshop on the correction of preciptation measurements.* ETH/WMO/IAMS Zurich 289pp.

Tabony, R. C. 1980: A set of homogeneous European rainfall series. U.K. Meteorological Office 13 Branch Memorandum 104, Bracknell, U.K.

Tabony, R. C. 1981: A principal component and spectral analysis of European rainfall. *Journal of Climatology 1*, 283-291.

Wigley, T. M. L., Lough, J. M. and Jones, P. D. 1984: Spatial patterns of precipitation in England and Wales and a revised, homogeneous England and Wales precipitation series. *Journal of Climatology*, 4 1-25.

Wigley, T. M. L., Angell, J. K. and Jones, P. D. 1985: Analysis of the temperature record. In: M. C. MacCracken and F. M. Luther (Eds.) *Detecting the Climatic Effects of Increasing Carbon Dioxide.* U.S. Dept. of Energy, Carbon Dioxide Research Division, Washington D.C., 55-90.

Wigley, T. M. L., Jones, P. D. and Kelly, P. M. 1986: Empirical climate studies: warm world scenarios and the detection of climatic changes induced by radiatively active gases. In: B. Bolin, B. R. Doös, J. Jäger and R.A. Warrick (Eds.) *The Greenhouse Effect, Climatic Change and Ecosystems.* Wiley, 271-322.

Wigley, T. M. L. and Jones, P. D. 1987: England and Wales precipitation: a discussion of recent changes in variability and an update to 1985. *Journal of Climatology*, 7, 231-246.

Woodruff, S. D., Slutz, R. J., Jenne, R. J. and Steurer, P. M. 1987: A comprehensive ocean-atmosphere data set. *Bulletin American Meteorological Society*, 68 1239-1250.

Section B: DENDROCLIMATIC EVIDENCE

14 Mapping climate using tree-rings from western North America

H. C. Fritts and X. M. Shao

14.1 Introduction

Standardized ring-width measurements from trees on spatial arrays of sites can provide continuous yearly paleoclimatic records, and this information can be transferred to estimates of climatic variables and mapped to reveal past spatial variations in climate (Fritts 1976, Fritts in press, Stockton *et al.* 1985). If the trees have the exact same response to climate, the ring-width indices theoretically need only to be averaged and plotted to reconstruct relative variations in climate (Schulman 1956; Fritts 1976; Schweingruber 1988). However, a variety of environmental conditions can limit growth, and the importance of limiting conditions can vary markedly from one season to the next depending upon the stage of growth, the conditions of the trees, microclimates and other factors of the site (Kramer and Kozlowski 1979; Fritts 1974, 1976; Fritts *et al.* manuscript). In addition, small differences in site factors can have large effects on limiting conditions particularly for species growing near their ecological limits (Fritts *et al.* 1965a, 1965b; Fritts 1969a). LaMarche (1982) suggests that differences in response to climate can be controlled by sampling trees from a particular habitat type such as arid or high-altitude sites. We propose that large differences in response still remain in selected chronology sequences from arid sites as evidenced by the large numbers of principal components (PCs) needed to reduce the chronology variance (Fritts 1976).

There are at least two strategies to deal with these differences between chronology responses from different sites. One strategy assumes that the largest and most important response is the only one of interest, and the chronology values are averaged or the first and most important PC is extracted and used to reconstruct climate. The residuals from this main pattern (lower order PCs) are little more than unwanted noise or error in the analysis (cf. Bradley and Jones, Chapter 1, Meko, Chapter 16 and Cook *et al.* Chapter 17, this volume). The other strategy extracts a relatively large number of the PCs of the growth indices (to preserve the important differences along with the main pattern in response); the most important PCs of the chronologies are calibrated with the most important PCs of climate, and the significant collective set of relationships are used to reconstruct variations in climate.

This chapter is an overview of a large project conducted for over 15 years using the second strategy. The project was a diagnostic analysis aimed at developing procedures and techniques that could capture and transfer diversity in a chronology set to estimates of different variables of climate. The details of this work are described in Fritts (in press). This chapter also utilizes three floppy disks with data and computer programs that can be used to map the tree-ring width indices and reconstructions of seasonal and annual temperature, precipitation and sea-level pressure developed in this work. The reconstructions are mapped, compared to other reconstructions of climate from the United States (see Chapters 16 and 17) and the climatic features deduced from the results.

14.2   Data

A total of 102 standardized ring-width chronologies from 1601 to present are included in the floppy disk. The first 65 chronologies were the highest quality chronologies available from arid-site ring-width chronologies when the reconstruction effort began (Fritts and Shatz 1975) and were the only tree-ring data used in that work (Fritts in press). Since ring-width growth can be statistically correlated with the growth in prior years (autocorrelation) the first order autocorrelation was removed from the original 65 chronology data set; the PCs were extracted for 1602-1963 from both the modified and unmodified chronology sets, and both were used as statistical predictors of the PCs of climate[1].

Temperature and precipitation data from the United States and southwestern Canada for 1901-1963 were examined for homogeneity and 77 temperature stations and 96 precipitation stations were selected for calibration that began in 1901 or before. The longest, most homogeneous and complete records were chosen and missing data for 1901-1970 were estimated (Fritts in press).

14.3   Calibration

Canonical regression (Glahn 1968; Blasing 1978) was used to calibrate the chronology PCs with the climate PCs for the 1901-1961 period, the chronology PCs were applied to the calibration equation to reconstruct climate for the length of the tree-ring record from 1602-1961 and all available data from the selected climatic stations that had 6 or more years of observation before 1901 were used for verification. The PCs of sea-level pressure for 96 grid points over North America and the North Pacific were also calibrated with the same tree-ring data, and sea-level pressure was reconstructed from the tree-ring data, but subsample replication techniques were used for verification because no independent pressure data were available before 1900 (Fritts in press).

The term *model* refers to one particular canonical regression equation applied to tree-ring data to make a climatic reconstruction (Fritts *et al.* 1971; Webb and Clarke 1977). The models had to be simplified somewhat by calibrating the PCs of only one climatic variable for one season at a time to allow for the differences in seasonal responses. In addition, ring-width

---

[1] At the present time, the series would be fully ARMA modeled (Jenkens and Watts 1968) and perhaps more advanced time-series techniques could be used to deal with the multivariate relationships (Benett 1979). Except for work of Meko (1981) and Cook *et al.* (1988) these possibilities have not yet been evaluated.

growth may be related to ring width in prior years or to the prior year's climate. Models used chronology PCs with and without autocorrelation and calibrated them with the chronology PCs at different lags and leads to relate growth to the preceding year's climate and to assess autocorrelated relationships. However, there were too few degrees of freedom for more than two sets of PCs to be considered in any one model; so a variety of models representing different combinations of lags and leads were calibrated.

14.4 The overall strategy

The analysis was divided into a variety of segments or different steps representing different problems and tests that were considered necessary to perform the diagnosis. The optimum and most practical procedures were selected for each segment and used throughout the remaining analysis. For lack of a well-developed theory at that time, the approach was largely empirical and used objectively derived statistics to evaluate which models and methodologies produced the most reliable reconstructions of climate.

Time-series modeling had just been introduced (Jenkins and Watts 1968) in the early 1970s when this project began, but had not been applied to multivariate analysis until 1979 (Bennett 1979). Thus stepwise canonical regression (Glahn 1968; Blasing 1978) appeared to be an optimal method to sort through, select and scale the important relationships between the modes of variance in tree-ring chronologies and the most meaningful modes of climatic variance (Fritts 1976, in press). Because a variety of responses were preserved by the PCs of the tree growth variance, individual temperature, precipitation and sea-level pressure variations were calibrated with the same tree-ring variance.

The same analyses were made for each variable, each season and each size grid. It soon became clear that a long drawn out process of experimentation, analysis and elimination would be required to complete a definitive analysis. We reduced the scope of the investigation as much as possible, but often it was a temptation to reopen and reevaluate earlier questions as new data became available or new developments and procedures were reported. However, it would have been impossible to complete the study if questions that had already been answered were reopened every time new approaches were suggested. We agreed that once a particular question had been answered and a procedure adopted, not to reconsider such questions again until all of the variables and grids had been calibrated, verified and annual values calculated.

Likewise, it was decided that the same tree-ring and climatic data sets should be used throughout the analysis and the same statistics used to evaluate technique improvements. This strategy did not allow deletion of data points once the grids were established. In general, we do not recommend that reconstructions should be retained if they cannot be validated, but the diagnostic nature of this study required that the numbers of climatic stations should be held constant when evaluating other features of the analysis. As a consequence, stations with poor verification could not be deleted; so the average statistics from the grid did not yield the highest average statistics that were possible if the poor stations had been deleted to maximize the reliability of the analysis. The average statistics that were reported should be regarded as very conservative estimates that should improve as future reconstruction efforts take advantage of new approaches, including those suggested in this work. The poorly fit data points can be regarded as the best inference possible from the available data and techniques used in this diagnostic analysis, but they should not to be considered reliable climatic reconstructions unless reliability can be confirmed by independent evidence. One type of important confirmation can come from comparison with other types of proxy data. Both the reliable and unreliable data points are preserved in the maps to allow for this type of comparison.

Three chronology grids of different density (Fritts and Shatz 1975) were initially calibrated with seasonal data of the three climatic factors and the intermediate density 65 chronology data set, which consistently produced the best reconstructions, was selected for the remaining analysis (Fritts in press).

14.5   Other questions that were addressed

1) What climatic data should be calibrated and what seasons should be used in the analysis?   Temperature, precipitation and sea-level pressure were the only data sets that were relatively continuous over space and long enough for calibration with the 20th century record beginning in 1901 and ending in 1961; so these three variables were selected as all three could be readily interpreted in the context of well understood climatic factors. Palmer drought severity indices (PDSI) were considered, but at that time the PDSI records began in 1930, which would not have provided enough degrees of freedom (greater than 30) for calibration of the spatial patterns.

The seasons were analyzed by Blasing (1975) using the variance of monthly sea-level pressure and the following seasons were chosen: winter (December-February) spring (March-June) summer (July-August) and autumn (September-November). The annual reconstructions were simply the combinations of seasons beginning with winter and ending with autumn.

In addition, the maximum distance over which significant growth-climate relationships might be present had not been determined. Two grid sizes for temperature and precipitation were used throughout the analysis to help evaluate distance relationships. The smaller grids included climatic stations in western North America covering approximately the same area as the tree-ring grid. The larger grids included eastern United States stations as well as the stations from the West. These grids were subdivided into climatic regions and the differences diagnosed to evaluate the important relationships involving space. Since the large grids appeared to have some useful information outside the area of the tree-ring grid and the estimates within the area of the tree-ring grid were almost as reliable as the estimates from the smaller grids, the estimates from all data points in the large grids were mapped for the diagnosis and comparisons with other data sets.

2) What is the maximum number of PCs of growth and climate to be modeled?   Two basic strategies were used to examine this question. The first was to determine, in some quantitative way, which eigenvectors were below the noise level and could be eliminated without affecting the reconstruction. The second was to construct and calibrate models with different numbers of PCs and use the resulting calibration and verification statistics to select the optimum combinations. This second strategy was needed because the numbers of climate PCs included as candidate predictand variables often influenced the numbers of significant canonical variates entered in regression, so that it was impossible to determine ahead of time the exact number of climate PCs needed to maximize the reconstructions of climate.

In our first strategy, four methods were used to identify the noise level PCs with very large differences in the results. The most sophisticated methods were the "scree" line approach of Tatsuoka (1974) and the simulation approach of Preisendorfer and Barnett (1977). The extreme variability in the numbers of significant PCs, both within and between methods, confirmed that there was no simple answer as to how many eigenvectors were important. It seemed best to start with a fairly high number of eigenvectors based upon the least conservative technique and to rely on objective calibration and verification tests to eliminate PCs that did not contribute to significant calibration variance. Thus we began by using 15 PCs of growth to reconstruct 15 PCs of temperature and sea-level pressure and 20 PCs of precipitation. Later the number of climate PCs was varied systematically, calibrated and verified, and the models with the highest ranking statistics were selected for reconstructing climate.

All canonical variates were entered into regression, but a systematic selection procedure using an F statistic eliminated all variates contributing insignificant variance in the regression analysis. The degrees of freedom were adjusted for autocorrelation of the residuals to avoid bias in the F statistic. This eliminated large numbers of canonical variates, which in turn eliminated the corresponding PCs that had insignificant correlations with climate.

3) How could the best models be selected? Seven calibration and 17 verification statistics, which are described by Fritts (in press) were used to test whether model estimates were better than those expected from chance relationships ($p \geq 0.95$) and to rank the significant results to determine which provided the "best" reconstructions of actual independent climatic measurements. More verification than calibration statistics were used to stress the greater importance of the independent climatic estimates in determining the reliability of reconstructions (Fritts 1976).

4) What kind of lags and leads should be modeled? Initially 13 model structures were considered that calibrated 15 PCs of climatic variables with 15 or 30 PCs of tree-ring data. These models included matrix: a) **I** corresponding to chronology PCs for the (I) immediate or current year of climate, b) *M* corresponding to the PCs for the same chronologies but (M) modeled to remove the first order autocorrelation effects, c) **B** and **MB** corresponding to the unmodeled and modeled chronology PCs (B) before the year of climate, d) **F** and **MF** corresponding to the unmodeled and modeled chronology PCs in the year (F) following climate and e) **FF** corresponding to the unmodeled chronology PCs lagging two years behind climate. The numbers of PCs that were calibrated are indicated and "T" is the climatic variable to be reconstructed (in this case winter temperature for the large grid). The ranks obtained for winter temperature are shown on the right.

Model	Rank
15T = 15B 15I	2
15T = 15MB 15I	5
15T = 15B 15M	5
15T = 15MB 15M	12
15T = 15I	4
15T = 15I 15M	1
15T = 15I 15F	9
15T = 15M 15F	10
15T = 15I 15MF	3

Model	Rank
15T = 15M 15MF	8
15T = 15F	13
15T = 15F 15MF	7
15T = 15F 15FF	11

The four highest ranking models included **I** and **M** components along with **B** and **MF** components indicating the dominance of current climate and secondary effects of prior growth and one-year lags in estimating winter temperature for the large grid. Estimates of summer climate frequently involved **F** and **FF** components suggesting that lags of one to two years were needed to estimate summer climate.

The most promising model structures were selected for each variable, season and grid, and new calibrations were made by varying the numbers of climate PCs from 2 to 20 depending upon the number of non-noise level eigenvectors in each climatic PC set. This involved more than 300 separate calibrations and verifications (see Fritts in press, for details on the various models that were considered).

5) Could reconstructions from different models be combined to reduce errors and to take advantage of differences in model structure? Bates and Granger (1969) show that the reliability of a statistical forecast may be improved by averaging statistical estimates from two or more calibrations using different combinations of predictor variables. This possibility was investigated by selecting the highest ranking models for a given variable, season and grid, averaging the reconstructions for 2 or 3 model combinations, recalculating the statistics and examining these data for significant improvements. Improvements were more marked and significant for certain combinations than for others, and the combinations with the highest ranking statistics were used for the seasonal estimates. These seasonal estimates were averaged for temperature and sea-level pressure and totaled for precipitation to obtain the annual estimates, and the statistics were recalculated for the annual data to evaluate improvements.

Since the tree-ring data are a response to several climatic factors operating over all four seasons, the calibration with climatic data for any one season was perturbed by large anomalies in climate during other seasons that had also contributed to ring-width variance. These perturbations resulted in errors in the seasonal estimates. However, these errors were not independent, as overestimates in one season were usually associated with compensating underestimates in other seasons. When the seasonal estimates were combined into annual values these errors were reduced more than would be expected if the errors were due to only random variance (Fritts in press). The grid-point estimates were combined into regional averages, or the data were filtered to emphasize the low-frequency variance. These data were compared to actual climatic data, and smaller but apparent improvements were noted (see Fritts in press).

14.6   Results and discussion

14.6.1   Reliability of the reconstructions

Merging of two or three high-ranking models appeared to improve the grid-point climatic estimates for the seasons, and these results were well above expectations due to chance variations (Livezey and Chen 1983). Annual combinations were usually more reliable than the seasonal estimates, and regional averages were more reliable than the grid-point estimates, but calibration of individual grid points for each season appeared to be necessary in order to utilize the grid diversity in response. The reliability at low frequencies varied among variables and grids.

All three climatic variables were successfully reconstructed from the same tree-ring grid. The temperature and sea-level pressure reconstructions were the most reliable. Reliability was greatest near and within the area of the tree-ring grid, but reliability was acceptable at considerable distances from the tree-ring grid for some temperature and sea-level pressure reconstructions. Except for a few grid points in the Central Plains, the verification statistics indicated that there was limited skill in reconstructing precipitation beyond the tree-ring grid. No skill was found for precipitation reconstructions east of the Mississippi. Some reconstructions of temperature appeared to be reliable in the Plains States, western Great Lakes and upper Mississippi Valley, but little or no skill for temperature was found in eastern United States and along the Gulf Coast. Sea-level pressure reconstructions were often reliable over the North Pacific, Canada and the North American Arctic.

The average calibrated variance (adjusted for degrees of freedom loss) for all individual (seasonal) model estimates at all grid points was only 12.7%. Combining the seasonal reconstructions from two or three of the best models for each variable and season boosted the average calibrated variance to 20.3%, and when these seasonal reconstructions were combined into annual values the adjusted calibrated variance increased to 30.9%. When these annual reconstructions were averaged for regions to reflect large-scale patterns and compared to regional climatic data, the average (adjusted) calibrated variance rose to 42.9%. When the reconstructions were averaged by regions and were filtered to evaluate the decade patterns, the average agreement rose to 58.5%.

Marked differences were observed between the average statistics for the three climatic variables. The adjusted calibrated variance for annual values for sea-level pressure, temperature and precipitation were 40.2, 40.7 and 16.5% respectively. When these data were averaged for regions and filtered, the corresponding values increased to 70.6, 62.1 and 48.8%. The calibration statistics were higher for some regions than for others, ranging up to 80% for filtered temperature estimates. Temperature and sea-level pressure estimates were clearly superior to precipitation estimates, and the low frequency estimates of sea-level pressure appeared superior to low frequency temperature estimates.

These are remarkable calibration percentages considering that the canonical method is conservative (Cook *et al.* 1988 and personal communications) and that the poor reconstructions of all variables in the eastern United States well outside the area of verifiable reconstruction were included in the percentages. In addition, the percentages are probably conservative estimates of what can be expected from combining and averaging individual models reconstructing seasonal temperature, precipitation and sea-level pressure from arid-site tree-ring chronologies in Western North America. The annual estimates near the tree-ring grid, averaged over climatic regions and over several years, provide the most reliable paleoclimatic estimates. The reliability diminishes with increasing distance from the tree-ring grid, but there appear to be some areas outside the tree-ring grid where some reliability exists. However, this finding does not imply that the reliability would not be improved substantially if more paleoclimatic indicators could have been included in the analysis especially for areas beyond boundaries of the tree-ring gird. In fact, the canonical analysis was used specifically because it could utilize diverse kinds of paleoclimatic information from large geographic grids. The same kind of analysis can be expanded to include more and more disparate proxy data sets from North America and other continents.

The percentage of lags and leads for the PCs of the chosen models that are plotted for the different variables and seasons (Figure 14.1) show interesting and reasonable relationships that were captured using these objective selection procedures. For the winter calibrations, the PCs for the current year (t) were dominant and those for the prior year (t-1) were second in importance. This suggests that autocorrelation is important for the winter conditions perhaps because the prior growth is an index to prior leaf set and photosynthetic potential in the response to the winter environments. Lags of t+1 and t+2 were not included in any winter relationship. For spring the leads t-1 became less important and lags of t+1 began to be important. However, by summer, lags t+1 dominate and no leads of t-1 were included in the relationship. In fact, for precipitation only lags of t+1 and t+2 were accepted. This suggests that after the growing season commences, the importance of the response in the current season diminishes while the response in the following season increases. Autumn, however, includes a mixture of lags and leads with t and t+1 being equally dominant. This may reflect the influence of autumn conditions on fruit set, bud initiation including flower production for the following season that can compete and influence cambial activity in the following summer. Temperature and precipitation models emphasize year t relationships followed by

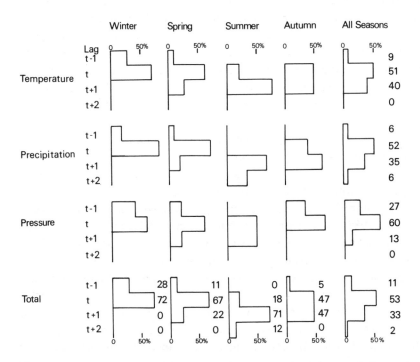

Figure 14.1 The percentage of transfer function coefficients in the selected final models associated with PCs of growth for years t-1, t, t+1 and t+2 representing different leads and lags in the calibrated relationships.

year t+1; while sea-level pressure models emphasize year t relationships with year t-1 second in importance.

While ring-width variations from trees on drought-subjected sites can be regarded as a response to local conditions of precipitation and drought, this study indicated that relationships to factors other than precipitation can and should be used to reconstruct large-scale variations in climate. Sea-level pressure can be reconstructed not because pressure influences tree-growth directly, but because spatial patterns of sea-level pressure reflect the movement of storms that in turn affect the moisture, temperature, wind, humidity and sunshine that limit ring-width growth.

The temperature estimates from trees on arid sites probably reflect the influence of temperature on water-loss, direct effects of temperature on growth and the greater collinearity of temperature over space. Because of this collinearity, the tree-ring chronologies were more highly correlated with a greater number of temperature than precipitation records, and this led to more reliable reconstructions for spatial relationships of temperature. These temperature estimates must be carefully compared to other temperature estimates using growth relationships derived from high-altitude and subarctic tree-line sites to assess the strengths and weaknesses of these two types of dendroclimatological temperature estimates.

The best reconstructions were those for the spring season, and those for winter climate were almost as good, reflecting the importance of winter and spring conditions on arid-site tree growth. Low-frequency variations in summer were fairly well reconstructed, but the high frequency variance, particularly for precipitation, was not well reconstructed. The reconstructions for Autumn were the least reliable seasonal estimates.

It was clear that a variety of information on different climatic variables could be extracted and reconstructed from a tree-ring grid on arid sites in Western North America but there were apparent limits over space. We will illustrate how maps of the different climatic estimates can be used to evaluate some of the limits, to test paleoclimatic inferences from other data sources and to deduce synoptic climatic patterns that may have caused the variations in past climate and growth.

14.6.2   Examples of the reconstructions

Figure 14.2 shows the reconstructed 17th-19th century anomalies in sea-level pressure, temperature and precipitation expressed as departures from the 1901-1970 mean values. These maps might be regarded as an estimate of the departures of the Little Ice Age climate beginning in 1602 and ending in 1900 from the 1901-1970 mean climate. The reconstructed annual temperature variations in the western United States do not show a Little Ice Age cooling but lower temperatures are clearly evident east of the Rocky Mountains. Higher precipitation was reconstructed in the Northwest and in eastern United States with dryer conditions in the Southwest. Higher sea-level pressures were reconstructed for North America and large areas of the North Pacific. This suggests that Arctic and Subtropical Highs were prominent with a possible ridge over the western United States and enhanced flow of polar air into eastern North America causing cooler temperatures there. The enhanced precipitation in the Northwest suggests that storms moved through that area more frequently than during 1901-1970.

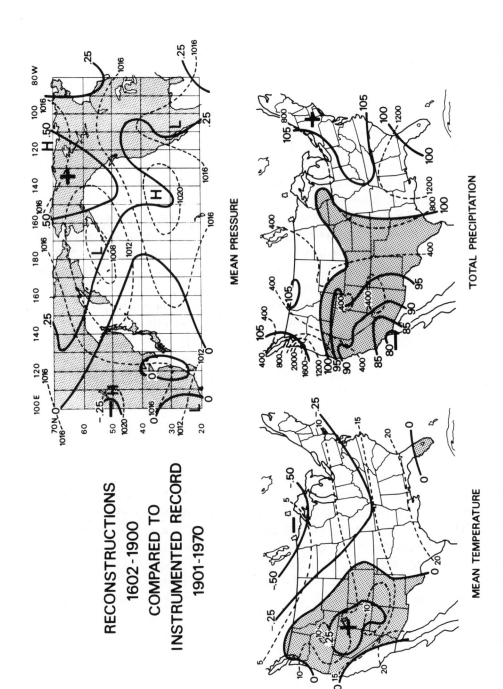

Figure 14.2 The reconstructed mean sea-level pressure, mean temperature, and total precipitation anomalies for the 1602–1900 period expressed as departures from the 1901–1970 instrumental average. The dashed lines show the mean values of the instrumental data for 1901–1970. Warm and dry anomalies are shaded.

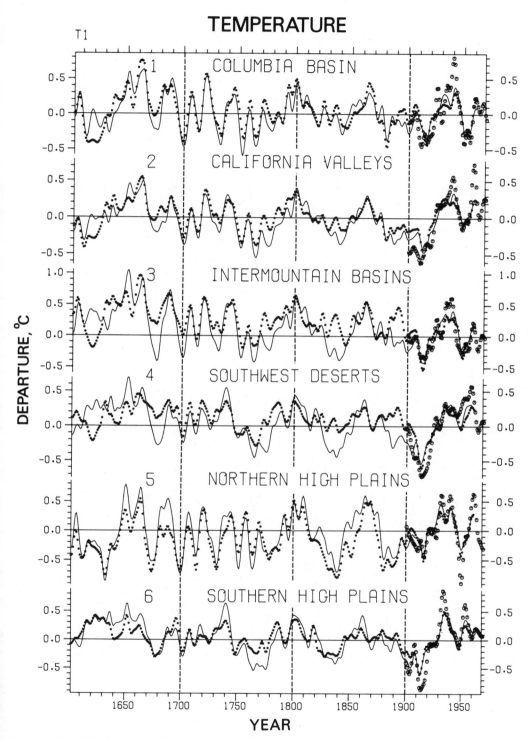

Figure 14.3 The filtered annual temperature reconstructions, averaged over all grid points in each of the 6 western regions, and expressed as anomalies from the 20th-century values. The line indicates reconstructions from the small grid, small dots indicate reconstructions from the large grid and larger dots the 20th century filtered instrumental data.

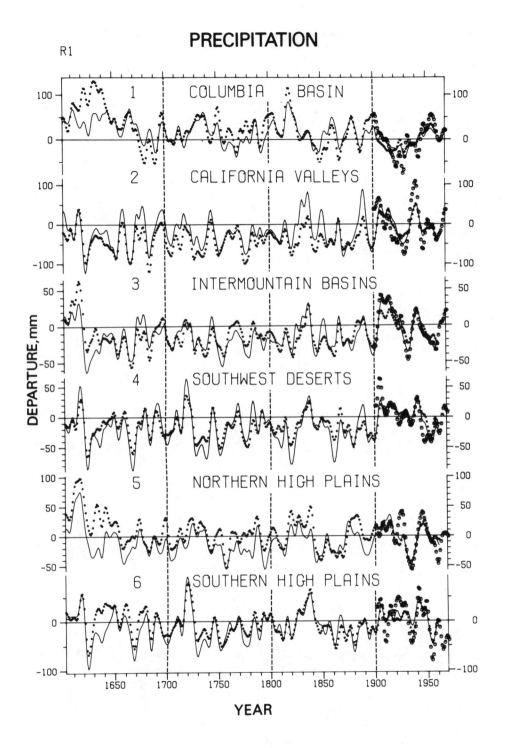

Figure 14.4 The filtered annual precipitation reconstructions, averaged over all grid points in each of the 6 western regions, and expressed as anomalies from the 20th-century values. The line indicates reconstructions from the small grid, small dots indicate reconstructions from the large grid and larger dots the 20th century filtered instrumental data.

Figures 14.3 and 14.4 are plots of the annual temperature and precipitation reconstructions averaged over six western regions and treated with a 13 weight digital filter (Fritts 1976) to emphasize the variations at time scales of seven years or longer. Figures 14.5 and 14.6 show the most important changes in spatial reconstructions of climate associated with 30-year mean temperature changes. The times of greatest change were determined by calculating 30-year running means of the grid-wide average temperature, testing the differences in non-overlapping means and selecting the years that exhibited the most significant changes. The average pattern for each period was subtracted from the average pattern of the period that succeeded it to reveal the synoptic-scale variations associated with these temperature changes.

The mean patterns of sea-level pressure, temperature and precipitation reconstructed for 1602-1636 are shown in the first three maps in row one of Figure 14.5. The second row includes the mean maps for 1637-1666. The remaining rows of maps in Figures 14.5 and 14.6 are the differences between the means for each period. For example, the 1602-1636 period was characterized by relatively low temperatures and high precipitation in the northern parts of the grids with warmer and dryer conditions to the south (note the temporal anomalies in Figures 14.3 and 14.4). High anomalies in reconstructed pressures over the U.S. and eastern North Pacific suggest storms moved more frequently through the northern than the southern part of the grid than in the 20th century. Mt. Vesuvius was the only notable volcanic eruption during this period and probably had little effect on climate in this region as it was not at a sufficient low latitude (Lough and Fritts 1987).

Around 1637 sea-level pressure was reconstructed to have declined over large areas of the eastern North Pacific but increased somewhat over the western North American continent. This appeared to have been associated with northward flow of air into the American West and warming in that area (see Figure 14.5, Row 3). A decrease in precipitation was reconstructed, which may be associated with reduced storm activity in the California Valleys, Intermountain Basins and the northern High Plains in a southwest-northeast trending area we shall call the California-Dakota storm track. Precipitation increased at more southerly localities.

Four notable volcanic eruptions that had associated dust veils were listed by Lamb (1970) in this period. Awu was the largest in 1641 and the only one at low latitudes (10°S-20°N). This was the only latitude band that Lough and Fritts (1987) found with a statistically significant relationship between great volcanic eruptions, dust veils and temperature changes over their reconstructed grids. Significant cooling was observed in the central and eastern U.S., and warming was observed in the West for 0-2 years after the eruptions. This pattern was evident in the 1637 mean changes.

Around 1667 (Figure 14.5, row 4) reconstructed sea-level pressures were higher over large areas of the North Pacific and North America, but pressures declined somewhat over the U.S. Temperature and precipitation declined more or less uniformly over the grid except for the California-Dakota storm track where precipitation was reconstructed as having increased. The West continued to be reconstructed as relatively warmer than the East (although the change in temperature was negative throughout most of the area). The general decline in temperature was consistent with greater dust-veil effects following four eruptions that occurred, including Gamkonora and Serua in 1673 and 1693, which were at low latitudes. This interval falls within the Muander Minimum (Eddy 1976) and while some cooling and drying was apparent (especially in the Central Plains and the East) the pattern of cooling did

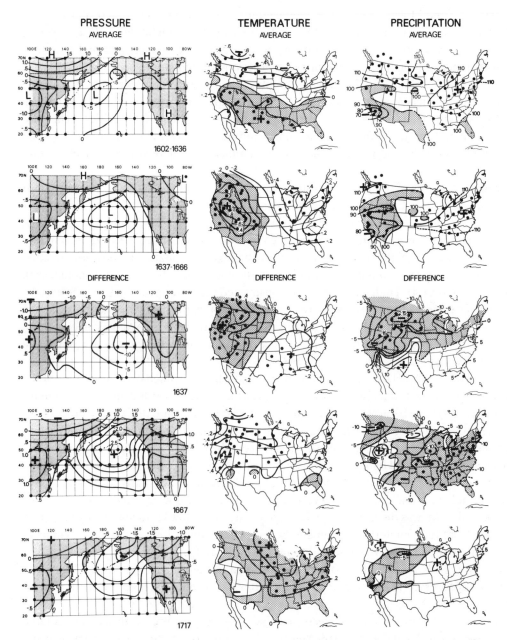

Figure 14.5 Reconstructed climatic patterns associated with the most marked and significant changes in the 30-year mean temperature. Averages in sea-level pressure, temperature, and precipitation for 1602-1636 and 1637-1666 are shown as departures from the 1901-1970 mean values in the first two rows of maps, and the differences between 1602-1636 and 1637-1666 are shown in the third row (1637). The remaining maps include successive difference maps. Dots are data points with significant anomalies ($p \geq 0.95$). Warm and dry anomalies are shaded.

not appear greatly different from patterns of cooling reconstructed for other time periods when sunspot numbers were near their maximum.

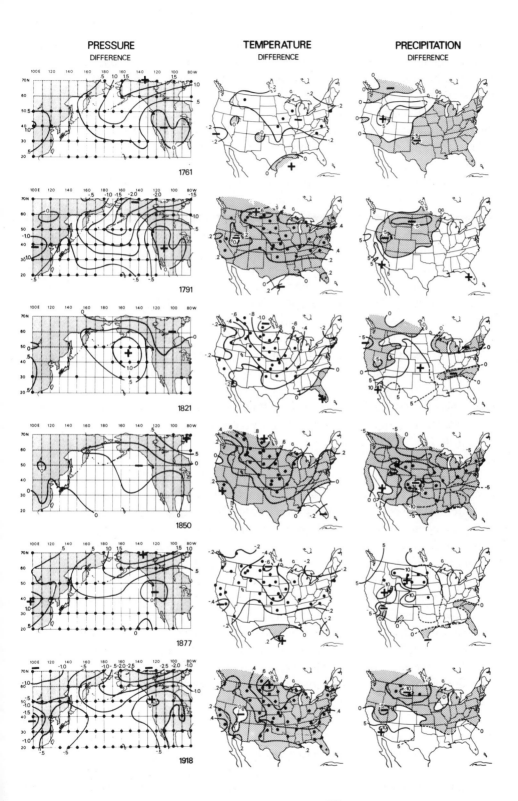

Figure 14.6 Same as Figure 14.5 except only for successive difference maps.

Around 1717 sea-level pressures declined over the North Pacific and northern regions of North America but rose slightly over western U.S. Temperatures were reconstructed to have increased, but precipitation changed very little becoming slightly lower in the area of the California-Dakota storm track and higher elsewhere on the map. Three notable volcanic eruptions occurred during this period but none were at low latitudes, which would be consistent with the reconstructed warming rather than cooling in the eastern U.S. with moderate temperatures in the West. However, considerable variability can be noted in the reconstructed temperature and precipitation during this period (Figures 14.3 and 14.4).

The change noted around 1761 (Figure 14.6) was similar to but not as marked as the change at 1667 (Figure 14.5) in that reconstructed sea-level pressures became higher, temperatures declined somewhat and precipitation was slightly lower on the average over large areas of the U.S. except near the California-Dakota storm track. There were three years of notable volcanic activity during this period, but Cotopaxi in 1768 was the only low-latitude eruption. Variable and generally rising temperatures were reconstructed in the West with variable precipitation amounts throughout this period.

The changes around 1791 mimicked the changes of 1717 (Figures 14.5 and 14.6) with declining sea-level pressures over the North Pacific and the North American North, rising temperatures over the U.S. and only slight changes in precipitation. Temperatures rose to high levels in the West at the turn of the century and gradually declined (Figure 14.3) and, except for the Columbia Basin, precipitation was reconstructed to have been low during this period but not as low as in the 1717-1760 period (Figure 14.4). The only major volcanic eruption in this interval was Tambora in 1815. It was a low latitude eruption and cooling was reconstructed, but it is superposed on top of a very warm period; so the low temperature anomalies from the 20th century means (Figure 14.3) do not appear to be particularly low after the Tambora eruption.

Around 1821 sea-level pressures were reconstructed to increase in the eastern North Pacific while they declined slightly over the North American Arctic with a greater tendency for troughs to develop over the Northern Rocky Mountains. Decade average maps (Fritts in press) show an anomalous Aleutian Low extended into the Western North Pacific in 1821-1830; this pattern was replaced by higher pressures over Alaska in 1831-1840, and this pattern was in turn replaced by low pressure anomalies in the North American Arctic in 1841-1850. The 1841-1850 pattern also showed a high pressure block in the eastern North Pacific, and this block along with low pressures in the Canadian Arctic are evident in Figure 14.6. Temperatures were reconstructed to have been low for a wide area and precipitation increased markedly especially during the first two decades of this period. The 1831-1840 decade was reconstructed to have been wetter than any other decade during the 1600-1960 interval for the area east of the Rocky Mountains. This increase in precipitation appeared to be associated with a tendency for more troughs to develop in western and central Canada with the advection of more cold air into large areas of the mid-continent.

Three major volcanic eruptions occurred in 1822, 1835 and 1845 along with Tambora that preceded them, and two were low-latitude eruptions. It is tempting to attribute the extreme conditions in these reconstructions to these eruptions, but the importance of dust veils to long-term climatic change is a controversial question and significant linkage has not been proven (Lamb 1970; Budyko 1969, 1974; Self *et al.* 1981; Kelly and Sear 1984; Bradley 1988; Mass and Portman 1989; Lough and Fritts 1987).

Around 1850 the sea-level pressure changes were not marked, but warmer and dryer conditions are reconstructed particularly in the plains states. Two notable volcanic eruptions were noted in this period but neither were at low-latitudes.

Around 1877 sea-level pressure was reconstructed to rise over the North American Arctic and the western North Pacific and to decline somewhat along the west coast of the U.S. Temperatures declined and precipitation increased especially along the California-Dakota track. Perhaps more storms entered the U.S. along the California coast and traveled in a northeasterly direction bringing more moisture and cooler weather to the West. There were two notable volcanic eruptions during this period and one, Krakatau in 1883, was from a low-latitude site. Short-term cooling was dramatic but the cause of the long-term cooling beginning in the 1870s has not been demonstrated. Cooling is reconstructed in the West early in the 20th century and cooling is seen clearly in the climatic records from the western U.S. (Figure 14.3) although the "end effect" of the smoothing used in the figure would place the first few years of the smoothed instrumental data near zero, the mean of the 1901-1970 period. The cooling was accompanied by increasing precipitation and runoff in California, the Intermountain Basins and the southwestern Deserts (Stockton *et al.*, 1985, Earle and Fritts 1986). This cooling in the western U.S. is distinctly different from the cooling in the central U.S. associated with volcanic dust veils, which was accompanied by warming in the far west.

The 1918-1960 climate reconstructions reflect the climatic data used in the calibration, but they help to place the modern record into a longer-term perspective (Figure 14.2 and 14.6). Warming was evident everywhere except in the Great Basin, sea-level pressures declined over the entire North Pacific and especially over the North American Arctic, and precipitation increased in the Southwest but diminished in more northern parts of the map. This implies that the 20th century warming was associated with an enhancement of storm activity (low pressure) in the Arctic with possible increased activity of the subtropical jet over the southern U.S. This may have been associated with less storm activity and lower moisture along the Pacific Northwest with fewer outbreaks of cold polar air inferred from lower pressures reconstructed in the North.

14.6.3 Comparisons with other dendroclimatic reconstructions

Dendroclimatic reconstructions for a particular locality, season or year can have large errors of individual estimate. While this error may be reduced by averaging the data over a number of years, filtering, or forming regional averages, it is important to compare the reconstructions to other similar data sets. In the following paragraphs we retrieve particular reconstructions from our computer disks and construct maps that can be compared to dendroclimatic reconstructions from the central and eastern U.S.

Meko (see Chapter 16) reconstructed precipitation for August-July in 10 different regions of the U.S. Plains using the first and most important PC from nearby tree-ring chronologies and explained as much as 17 to 66% of the variance. Seven years (1751, 1808, 1855, 1861, 1863, 1864 and 1893) were identified in which droughts were reconstructed for at least six out of 10 regions. Meko's study area was adjacent to our 65 tree-ring chronologies, and it coincided with our area of valid reconstructions for both temperature and sea-level pressure. In addition, some of our precipitation reconstructions appeared to be valid within Meko's

285

grid. This coincidence provided an excellent opportunity to look for similarities and to examine both data sets for evidence of larger-scale variations than could be found only looking at the individual grids.

There were 42 years between 1750 and 1900 in which Meko reported no drought. We averaged our reconstruction for these 42 years to obtain a base map of non-drought years, subtracted this average pattern from the maps of Meko's severe drought years and plotted the difference. Such difference maps removed the bias in the reconstructions associated with the changes related to the 20th century warming. Calculations were made for winter, spring and summer precipitation and spring and summer temperature. The difference maps for precipitation in winter and spring, seasons that were relatively well reconstructed, consistently show reduced amounts of precipitation for all seven years. Spring and summer temperatures were reconstructed to have been consistently warm for large areas in the West and in the Central Plains States although the specific patterns were variable.

The mean patterns for all seven years for spring precipitation and spring and summer temperature are shown in Figures 14.7, 14.8 and 14.9. On the average, spring precipitation (Figure 14.7) was reconstructed from the tree-ring data from the western U.S. as 60-80% of the 20th century values throughout the Plains States and less than 60% of the 20th century values for the extreme Southwest. Temperatures in spring (Figure 14.8) were reconstructed to have been higher than the 20th century values by as much as 1.6°C for large areas of the West centered over the Northern Rocky Mountains of the U.S. Summer temperatures (Figure 14.9) were also reconstructed to have been high, especially over the Missouri Basin during these seven years of drought. These results not only indicate excellent agreement between these two studies for years of extreme and pervasive droughts within the Plains States, but they suggest that conditions of low precipitation in the Plains were not only

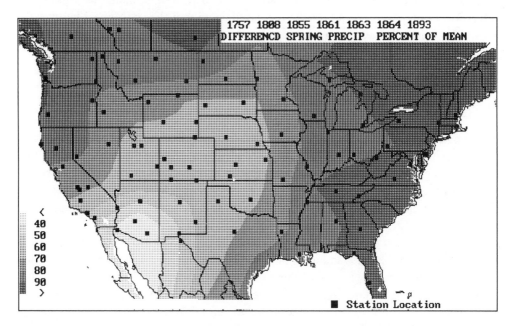

Figure 14.7 Average spring precipitation reconstructed for seven drought years expressed as departures from reconstructions for 42 years with no drought.

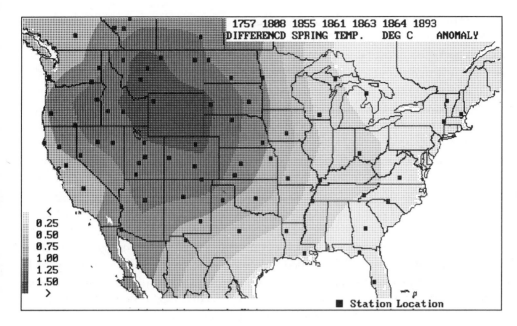

Figure 14.8 Average spring temperature reconstructed for seven drought years expressed as departures from reconstructions for 42 years with no drought.

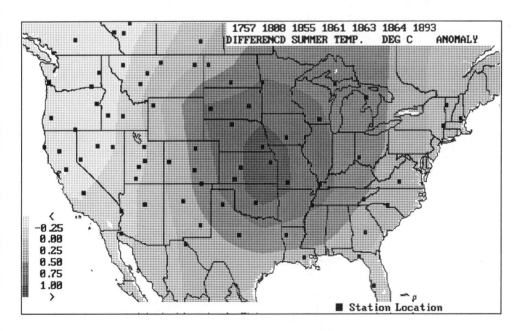

Figure 14.9 Average summer temperature reconstructed for seven drought years expressed as departures from reconstructions for 42 years with no drought.

associated with warm summers but that the summers were preceded by springs with unusual warmth in the mountains and plateaus to the west.

However, the difference maps for summer (July-August) precipitation were often inconsistent. Some summers were reconstructed to have had above average rather than below average precipitation. Possible explanations for the lack of agreement of summer precipitation is that 1) summer precipitation estimates were less reliable than those for winter and spring, 2) there is too little precipitation in summer and too much variability over space to make comparisons of this type, and 3) differences in the calibration approach. In our analysis July-August was too late in the season for precipitation to have a significant influence on the current years ring width (Figure 14.1) so our reconstructions rely primarily on the growth in the following year to reconstruct summer precipitation. Meko had averaged the annual precipitation, and his model is probably insensitive to this type of lag in growth behind summer climate.

Figure 14.10 is a difference map of spring sea-level pressure obtained by subtracting the mean of the non-drought years from the mean of the seven years of drought. A high pressure block over the U.S. is reconstructed with lower pressures in the Arctic suggesting that the Aleutian Low appeared to be well developed and storms were diverted along northerly tracks with little moisture reaching the central U.S. Figure 14.11 is a map of the mean ring-width indices from the 102 sites associated with these seven years of drought. Low growth throughout the far West was clearly associated with the seven extreme years of drought in the Central Plains of the U.S.

Cook *et al.* (Chapter 16) reconstructed PDSI for the Eastern United States from tree-ring chronologies in the East. They generated filtered drought factor scores and identified two notable dry periods and one wet period between 1700 and 1900. A second wet period was noted in 1901-1910, but we had used that period for calibration and could not regard our data for 1901-1910 as usable for an independent tests. While we recognized that we might not have any agreement because different climatic variables were reconstructed, and our reconstruc-

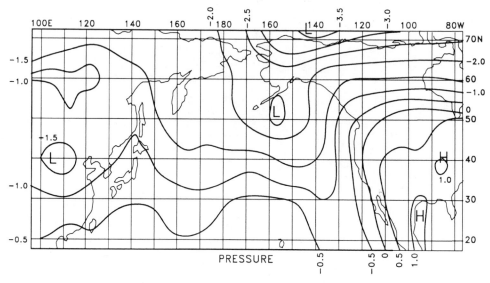

Figure 14.10 Average spring sea-level pressure reconstructed for seven drought years expressed as departures from reconstructions for 42 years with no drought.

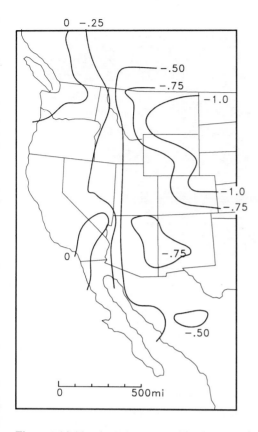

Figure 14.11 Average normalized ring-width index for the seven drought years for 102 chronologies from the western U.S.

tions appeared to be least reliable in the area of overlap with the Cook *et al.* reconstructions, we calculated the maps for the above time periods anyway and examined them for possible similarities indicating that Cook *et al.* reconstructions involved larger-scale climatic features to the West.

Cook *et al.* reported that the nine-year period spanning 1814-1822 was reconstructed to have been exceptionally dry, especially on the western margin of the Cook *et al.* grid, including areas in the Great Lakes and Ozarks. Figures 14.2 and 14.3 show high temperatures and low precipitation were reconstructed during that time period for a number of regions in the West. The average precipitation in winter, spring and summer (Figures 14.12 and 14.13) was reconstructed to have been 85 to 95% of 20th century values for substantial areas in the Southwest and U.S. Plains. The more reliable reconstructions of spring and summer temperature (Figures 14.14 and 14.15) indicate anomalous warmth throughout the West was associated with this drought in the East, and the warm anomaly extended further east than it did for Meko's data, which was confined to precipitation in the Plains states. This warmth in the West that was associated with low PDSI in the East also was associated with a high-pressure block similar to that shown in Figure 14.10, and many chronologies in the West exhibited below average growth as in Figure 14.11.

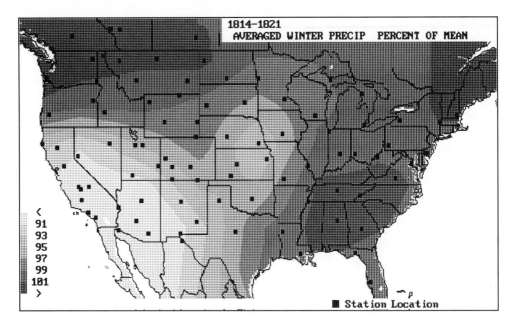

Figure 14.12   Average winter precipitation reconstructed for 1814-1821 and expressed as departures from 1901-1970 average values.

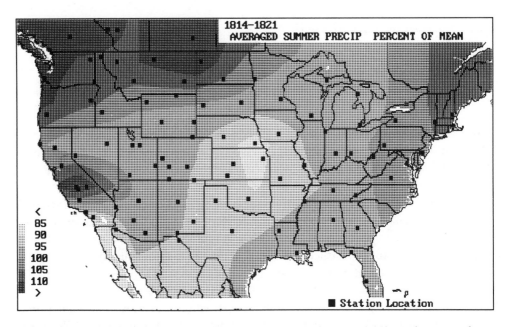

Figure 14.13   Average summer precipitation reconstructed for 1814-1821 and expressed as a percentage of 1901-1970 average values.

The interval 1827-1837 was reported by Cook *et al.* to have been exceptionally moist in the eastern U.S. Fritts (in press) had reported 1831-1840 to be the wettest decade over 370 years

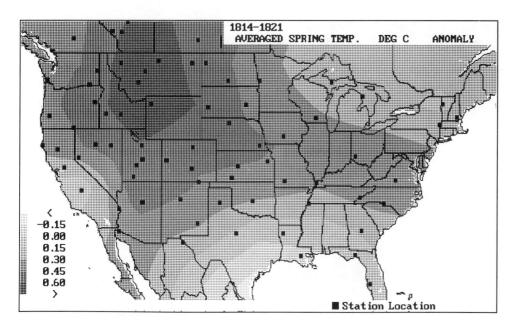

Figure 14.14 Average spring temperature reconstructed for 1814-1821 and expressed as departures from 1901-1970 average values.

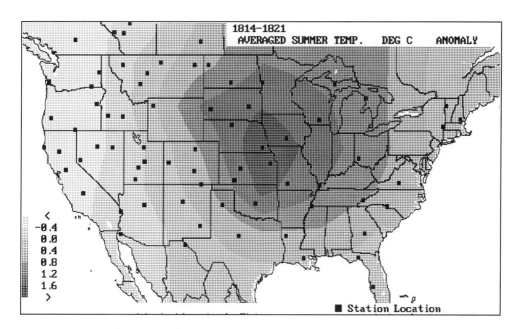

Figure 14.15 Average summer temperature reconstructed for 1814-1821 and expressed as departures from 1901-1970 average values.

and the annual precipitation map for 1827-1837 (Figure 14.16) shows high precipitation over all regions except in the far West. The spring and summer reconstructions show unusually high precipitation and cool conditions throughout the mid-continent.

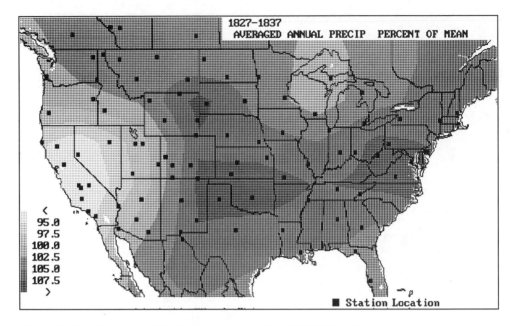

Figure 14.16 Annual precipitation reconstructed for 1827-1837 and expressed as a percentage of the 1901-1970 values.

The interval for 1769-1776 was also reconstructed by Cook *et al.* to have been dry but the most marked anomalies were in the eastern part of their grid. We reconstructed below average annual precipitation, with some low precipitation anomalies in winter, spring and summer in the central part of the grid. Temperature departures were in the right direction, but the anomalies were neither large nor significant. It appeared that, in this case, the anomalies in climate were well outside the area of reliable reconstruction from the western trees and that the anomaly could not be identified with any particular pattern of climate reconstructed in the West.

14.7 Conclusions

The agreement between the three studies during certain time periods suggests that continental-wide climatic fluctuations can be identified by comparing different paleoclimatic data sets. As more and more types of information are brought together and compared to examine such questions, we might be surprised how much information our proxy data sets contain on past climatic variations and changes. However, we must remain true to objective analysis and be willing to consider all legitimate evidence. We must not forget that all reconstructions, including our own, have error, and when two reconstructions are compared, their joint error terms can be larger, making it difficult to evaluate the true significance of the relationship.

At this time of limited funding and emphasis on single authored papers, cooperation frequently yields to competition, it is easy to loose our perspective and we may forget the importance of objective examination of diverse kinds of paleoclimatic evidence. To be truly effective, such comparisons should include all investigators working on the subject and involve open discussion of the strengths and weaknesses of the available data sets in a spirit of mutual collaboration and trust.

Paleoclimatology and more specifically dendroclimatology are not simple subjects with simple solutions to environmental questions (Fritts 1976). We have witnessed many advances in the fields in recent years, but the disciplines are still relatively young. We are becoming aware in dendroclimatology that many fascinating complex tree-growth and environmental relationships exist, but we lack the detailed knowledge of the biological growth processes, the biophysical linkages to climatic variations and the statistical procedures required to express the multiple relationships that may exist. While we can learn much from advances in other disciplines, we have accumulated a sizable intra-disciplinary expertise that we can be proud of.

Climatological and statistical models can oversimplify or misrepresent the complex responses of many paleoclimatic data sets. Paleoclimatologists need not wait for climatologists, biologists or statisticians to do more work. In such cases, objective empirical analyses provide legitimate and viable alternatives. We can utilize the best procedures that are available, analyze one question at a time using rigorous statistics, make best inferences, deduce the most likely and reasonable relationships, draw conclusions and continue developing our paleoclimatic data base and our understanding of past climate.

References

Bates, J. M. and C. W. J. Granger. 1969. The combination of forecasts. *Operational Research Quarterly* 20 (4): 451-68.

Bennet, R. J. 1979. *Spatial Time Series: Analysis-Forecasting-Control*. Pion Limited, London.

Blasing, T. J. 1975. *Methods for analyzing climatic variations in the North Pacific sector and western North America for the last few centuries*. Ph.D. Dissertation, Univ. of Wisconsin, Madison.

Blasing, T. J. 1978. Time series and multivariate analysis in paleoclimatology. In: *Time Series and Ecological Processes*, H. H. Shugart, Jr., ed., pp. 212-26. *SIAM-SIMS Conference Series*, No. 5. Society for Industrial and Applied Mathematics, Philadelphia.

Bradley, R. S. 1988 The explosive volcanic eruption signal in Northern Hemisphere continental temperature records *Climatic Change* 12:221-243.

Budyko, M. J. 1969. The effect of solar radiation variations on the climate. *Tellus* 21:611-661.

Budyko, M. J. 1974. *Climate and Life*. Academic Press, London.

Cook, E. R., K. R. Briffa and P. D. Jones. 1988. *Spatial regression methods for dendroclimatology: a review and comparison of two techniques. Part 1: the theory*. Research Report submitted to The Scientific Affairs Division, North Atlantic Treaty Organization, Brussels, Belgium.

Cook, E. R,, K. R. Briffa and P. D. Jones. personal communications.

Earle, C. J. and H. C. Fritts. 1986. *Reconstructing river flow in the Sacramento Basin since 1560*. Report to: California Department of Water Resources, Agreement No. DWR B-55395.

Eddy, J. A. 1976. The Maunder minimum. *Science* 193: 1189-1202.

Fritts, H. C. and D. J. Shatz. 1975. Selecting and characterizing tree-ring chronologies for dendroclimatic analysis. *Tree-Ring Bulletin* 35: 31-40.

Fritts, H. C., D. G. Smith, J. W. Cardis and C. A. Budelsky. 1965a. Tree-ring characteristics along a vegetation gradient in northern Arizona. *Ecology* 46(4):393-401.

Fritts, H. C., D. G. Smith and M. A. Stokes. 1965b. The biological model for paleoclimatic interpretation of Mesa Verde tree-ring series. *American Antiquity* 31(2, part 2):101-21.

Fritts, H. C., T. J. Blasing. B. P. Hayden and J. E. Kutzbach. 1971. Multivariate techniques for specifying tree-growth and climate relationships and for reconstructing anomalies in paleoclimate. *Journal of Applied Meteorology* 10 (5): 845-64.

Fritts, H. C., J. Guiot, G. A. Gordon and F. Schweingruber. 1990. Methods for calibration, verification and reconstruction. In: *Methods of Tree-Ring Analysis: Applications in the Environmental Sciences*, L. Kairiukstis and E. Cook, eds. Reidel Press, Dordrecht. p. 163-218.

Fritts, H. C., E. A. Vaganov, I. V. Sviderskaya and A. V. Shaskin. manuscript. Climatic variation and tree-ring structure in conifers: a statistical-simulative model of tree-ring widths, number of cells, cell size, cell-wall thickness and wood density.

Fritts, H. C. 1969. Bristlecone pine in the White Mountains of California: growth and ring-width characteristics. *Papers of the Laboratory of Tree-Ring Research* 4. University of Arizona Press, Tucson.

Fritts, H. C. 1974. Relationships of ring widths in arid-site conifers to variations in monthly temperature and precipitation. *Ecological Monograph* 44(4):411-40.

Fritts, H. C. 1976. *Tree Rings and Climate*. Academic Press, London, 567 pp. Reprinted in 1987 in: *Methods of Dendrochronology* Vols. II and III, L. Kairiukstis, Z. Bednarz and E. Felikstik, eds. International Institute for Applied Systems Analysis and the Polish Academy of Sciences, Warsaw.

Fritts, H. C. in press. *Reconstructing Large-scale Climatic Patterns from Tree-ring Data: A Diagnostic Analysis*. Arizona Press, Tucson.

Fritts, H. C., E. A. Vaganov, I. V. Sviderskaya and A. V. Shashkin. 1990. Climatic variation and tree-ring structure in conifers: a statistical-simulative model of tree-ring widths, number of cells, cell size, cell-wall thickness and wood density. manuscript.

Glahn, H. R. 1968. Canonical correlation and its relationship to discriminant analysis and multiple regression. *Journal of Atmospheric Science* 25(1):23-31.

Jenkins, G. M. and D. G. Watts. 1968. *Spectral Analysis and Its Applications*. Holden-Day, San Francisco.

Kelly, P. M. and C. B. Sear. 1984. Climatic impact of explosive volcanic eruptions. *Nature* 311:740-43.

Kramer, P. J. and T. T. Kozlowski. 1979. *Physiology of Woody Plants*. Academic Press, New York.

LaMarche, V. C., Jr., 1982. Sampling strategies. In: *Climate from Tree Rings*, M. K. Hughes, P. M. Kelly, J. R. Pilcher and V. C. LaMarche, Jr., (eds.) p.2-6. Cambridge University Press.

Lamb, H. H. 1970. Volcanic dust in the atmosphere; with a chronology and assessment of its meteorological significance. *Philosophical Transactions of the Royal Society of London*. 266 (1178): 425-533.

Livezey, R. E. and W. Y. Chen. 1983. Statistical field significance and its determination by Monte Carlo techniques. *Monthly Weather Review* 111:46-59.

Lough, J. M. and H. C. Fritts. 1987. An assessment of the possible effects of volcanic eruptions on North American climate using tree-ring data, 1602 to 1900 A.D. *Climatic Change* 10:219-39.

Mass, C. F. and D. A. Portman. 1989. Major volcanic eruptions and climate: A critical evaluation. *J. Climatology*. 2:566.

Meko, D. M. 1981. *Applications of Box-Jenkins methods of time series analysis to the reconstruction of drought from tree rings*. Doctoral dissertation, University of Arizona.

Preisendorfer, R. W. and T. P. Barnett. 1977. Significance tests for empirical orthogonal functions. In: *Preprint Volume Fifth Conference on Probability and Statistics*, 15-18, November 1977. Am. Meteorological Sco., Boston.

Schulman, E. 1956. *Dendroclimatic Changes in Semiarid America*. University of Arizona Press, Tucson.

Schweingruber, F. H. 1988. *Tree Rings, Basics and Applications of Dendrochronology*. D. Reidel Publishing Co., Dordrecht. 292 pp.

Self, S., M. R. Rampino and J. J. Barbera. 1981. The possible effects of large 19th and 20th century volcanic eruptions on zonal and hemispheric surface temperatures. *Journal of Volcanology and Geothermal Research* 11:41-60.

Stockton, C. W., W. R. Boggess and D. M. Meko. 1985. Climate and tree rings. In: *Paleoclimate Analysis and Modeling*, A. D. Hecht, ed., pp. 71-161. John Wiley and Sons, New York.

Tatsuoka, M. 1974. *Multivariate Analysis: Techniques for Educational and Psychological Research*. John Wiley and Sons, New York.

Webb, T., III and D. R. Clarke. 1977. Calibrating micropaleontological data in climatic terms: a critical review. *Annals of the New York Academy of Sciences* 288:93-118.

15 Dendroclimatic evidence from northern North America

R. D. D'Arrigo and G. C. Jacoby, Jr.

15.1 Introduction

Annual temperature departures for the period from 1601-1974 are reconstructed for northern North America for the general region of subarctic and arctic Canada and Alaska based on tree-ring data. Tree-ring information from the North American boreal forest zone can extend the temperature record for this region back to about A.D. 1600 with reliability. Prior to this time the data set is not adequate for large-scale coverage. There are well over 100 chronologies from sites in these northern regions from various sources (Cropper and Fritts 1981; Jacoby and Ulan 1981; Jacoby 1982). However, most of them are either short in length, end in the 1960s or earlier, or are poorly replicated. Some chronologies lack a low-frequency climatic signal due to complex forest-stand dynamics or site ecology, and some have had this signal removed by standardization methods. The reconstruction herein is therefore based on a limited number of recent ring-width chronologies that provide good spatial distribution and low and high-frequency climatic information. Long-term trends in this reconstruction are compared to other proxy and historical climatic data for the region.

The tree-ring series used here have a common period of nearly 400 years of record with which to explore theories regarding the patterns and causes of climatic change. In this paper we employ variance spectral analysis as one exploratory means of examining these long time series for periodicities which might be related to solar, lunar-nodal or other cycles.

15.2 Data

Seven tree-ring width chronologies from across boreal North America were used in this study (Figure 15.1, Table 15.1). All were standardized using negative-exponential or straight line curve fits, and other standard dendroclimatic techniques (Fritts 1976; Hughes *et al.* 1982; Cook 1985). They are part of a set of eleven such series used previously to reconstruct Northern Hemisphere and Arctic annual temperature departures (Jacoby and D'Arrigo 1989a). The seven represent more limited geographical coverage but have a longer common period than the larger data set. The chronologies are all white spruce (*Picea glauca [Moench] Voss*) with the exception of one northern white cedar (*Thuja occidentalis L.*) chronology from Gaspé, Quebec.

Temperature data for this study were derived from the averaged values of four of the 80 equal-area temperature boxes (boxes 1, 2, 6 and 7) compiled for the globe by Hansen and Lebedeff (1987) (Figure 15.2). These four boxes encompass much of the land area of arctic and subarctic North America. The temperatures compiled for each box are obtained by first

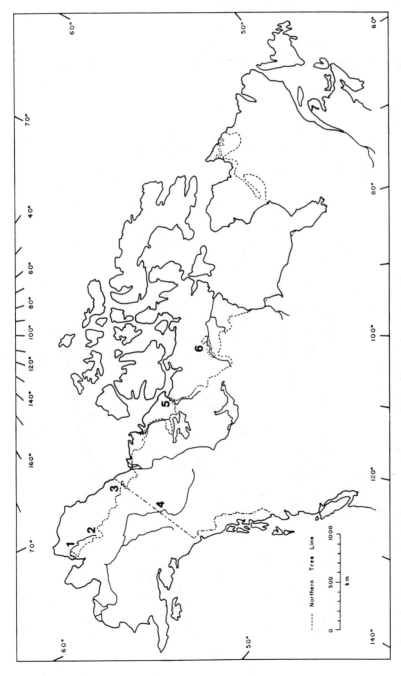

Figure 15.1 Map of Canada and Alaska showing sites of the seven tree-ring chronologies used in this study. Sites are as in Table 15.1.

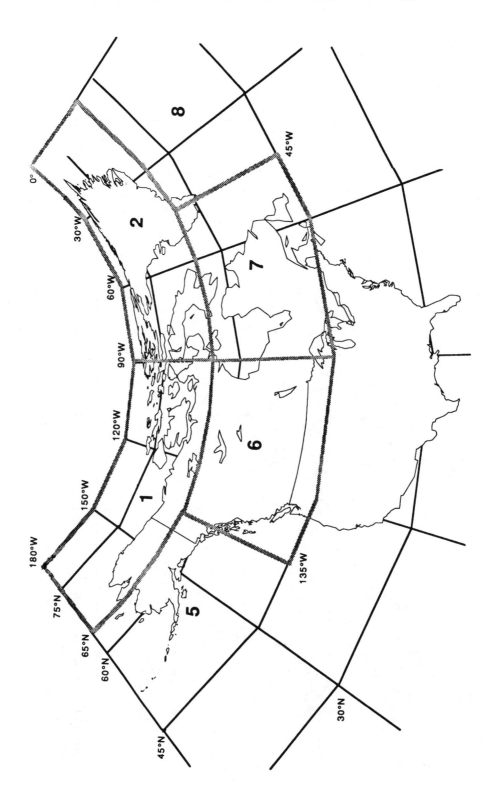

Figure 15.2 Portion of the grid of 80 equal-area temperature boxes for the globe from the Hansen and Lebedeff (1987) data set showing boxes 1, 2, 6 and 7 which have been averaged to represent the area of northern North America.

Table 15.1 Tree-ring chronology data for sites shown in Figure 15.1. This subset was selected from the 11 chronologies used in Jacoby and D'Arrigo (1989a) because of their greater length.

Site	Lat.	Long.	Yrs.	Elev. (m)	Species
1. 412, Alaska	67° 56'N	162° 18'W	1515-1977	126	*Picea glauca*
2. Arrigetch, Alaska	67° 27'N	154° 03'W	1586-1975	716	*Picea glauca*
3. Sheenjek, Alaska	68° 38'N	143° 43'W	1580-1979	808	*Picea glauca*
4. TTHH, Yukon Terr.	65° 00'N	138° 20'W	1459-1975	915	*Picea glauca*
5. Coppermine, NWT.	67° 14'N	115° 55'W	1428-1977	213	*Picea glauca*
6. Hornby Cabin, NWT.	64° 02'N	103° 52'W	1491-1983	160	*Picea glauca*
7. Gaspé, Quebec	48° 35'N	65° 55'W	1404-1982	305	*Thuja occidentalis*

averaging sub-boxes for which all stations within a 1200 km radius are included (Hansen and Lebedeff 1987). As a result there is some repetition of stations between adjacent boxes; see Hansen and Lebedeff (1987) for detailed information on the derivation of their temperature data set. The spatial coverage of these four boxes comprises a more extensive area than the boreal zone which includes the tree-ring sites (Figures 15.1 and 15.2). In particular, Box 2 is largely representative of Greenland, where climatic conditions may differ considerably from mainland Canada. However tree-ring data from the Canadian region have been shown to reflect large-scale temperature trends (e.g. Scott *et al.* 1988; Jacoby and D'Arrigo 1989a), and for this reason are believed to give an adequate estimation of temperatures for the area represented by the instrumental data.

15.3 Methodology

To reduce the number of predictors and quantify the common variance among the chrono-logies, principal components analysis [PCA] (Richman 1986; Bradley and Jones, this volume, Chapter 1) was performed on the correlation matrix of the seven time series for the period from 1600-1975. Two eigenvectors, explaining 59.1% of the variance in the data set, had eigenvalues >1 and were thus retained. The first eigenvector, explaining 42.8% of the variance, showed positive loadings for all seven chronologies, reflecting the common (presumably climatic) signal in the data. Loadings ranged from 0.26 to 0.49 except for Hornby Cabin, a site in the Northwest Territories (0.21). For the second eigenvector (accounting for 16.3% of the variance), the largest positive loading is for Hornby Cabin (Figure 15.1). The Hornby Cabin site is an outlier of trees from the northern treeline as it is usually mapped, situated north of the mean position of the polar front in summer (see Bryson 1966).

Climate-tree growth relationships often show lag effects which can influence dendroclimatic reconstructions (Fritts 1976). The first two sets of eigenvector scores and their backward and forward lags (t-1, t and t+1 for a total of six variables) were screened as candidate predictors of the averaged annual box temperatures (the single predictand or dependent variable) for subsequent multiple linear regression analysis of the 1880-1974 common interval. We thus employ levels 2A (multiple linear regression) and 2B (PCA) of the calibration methods discussed by Bradley and Jones in Chapter 1.

The chronologies were not pre-whitened prior to the PCA nor were the time series of eigenvector scores or the temperature data, although it is common practice to do so prior to regression analysis. Pre-whitening can improve regression analysis in dendroclimatic studies of semi-arid precipitation and in high-frequency studies. However when examining lower-frequency variations (3 to 10 years or more) it is inappropriate even though it makes the data satisfy certain regression analysis assumptions (see Section 15.4). Since we were primarily interested in modeling and reconstructing the low-frequency variations in the temperature record, pre-whitening of the tree-ring and temperature data was not done on this first part of the analysis (but see below).

Variance spectral analysis (Priestley 1981) was performed on the seven chronologies as an exploratory means of examining these data for periodicities which might be related to solar, lunar-nodal, Quasi-Biennial Oscillation, or other cyclical/quasi-cyclic phenomena. Standard chronologies as well as those pre-whitened (AR order 3 or 4) for the 1600-1975 interval (Box and Jenkins 1976) were each analyzed using 100 lags and a band width of 0.0126. In this case pre-whitening was performed in order to enhance the higher-frequency variations in the time series. We emphasize spectra of the individual series rather than the reconstruction because a number of studies suggest that there are regional differences in the response of the climate to atmospheric forcing factors (e.g. Herman and Goldberg 1978).

15.4   Results

The reconstruction of annual northern North American temperature departures shown in Figure 15.3 is based on the regression equation for 1880-1974 (Table 15.2). A subset of four of the original six candidate predictor variables exceeded the 0.10 probability level using a one-tailed t-test. The two non-significant predictors were then deleted and the final regression derived using the four remaining predictors: chronology values for the years t-1, t and t+1 for the first eigenvector, and chronology values for the year t+1 for the second eigenvector. Actual and estimated temperatures based on the full 1880-1974 calibration period are shown in Figure 15.4. The variance explained (adj.r^2, adjusted for degrees of freedom) is 52% for the 1880-1974 calibration model.

The 1880-1974 interval was then divided into two separate periods, 1880-1939 and 1940-1973, for calibration and verification (Fritts 1976, Hughes *et al.* 1982) using the same four predictors (Table 15.2). For the earlier calibration model the adj.r^2 is 0.45. Over the corresponding 1940-1973 verification period the RE (reduction of error statistic) is strongly positive (0.65). Any positive RE value shows that there is considerable skill in the verification period estimates as compared to the calibration period mean (Gordon and LeDuc 1981). The strength of the RE in this case, however, may partly reflect a difference in the means of the calibration and verification periods (Gordon and LeDuc 1981). The coefficient of efficiency

ACTUAL AND ESTIMATED ANNUAL TEMPERATURES FOR NORTHERN NORTH AMERICA

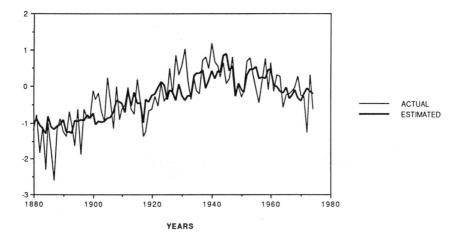

Figure 15.3 Reconstruction of annual northern North American temperature departures based on tree-ring data for the period 1601 to 1974. (Calibration on the average of four temperature boxes in Figure 15.2). Five-year smoothed values are also shown.

Table 15.2 Calibration-verification statistics for temperature reconstruction models based on temperature data from Hansen and Lebedeff (1987).

Calibration Model	1880-1974	1880-1939	1940-1973
r	.73	.69	.57
r^2	.54	.48	.32
adj.r^2	.52	.45	.23
S	--	.53***	.67***
RE	--	.65	.54
CE	--	-.003	.19
Sign Test	--	ns	35 correct* 24 incorrect
Product Means Test	--	26 correct** 9 incorrect	45 correct** 15 incorrect

r = multiple correlation coefficient

r^2 = variance explained

adj.r^2 = r^2 adjusted for loss of degrees of freedom

S = Spearman rank correlation coefficient

RE = reduction of error statistic

CE = coefficient of efficiency.

The 1880-1939 calibration model is verified over the 1940-1973 interval and the 1940-1973 calibration model verified over the 1880-1939 interval.

* Significant at <0.10 level,

** significant at <.01 level,

*** significant at <.001 level.

RECONSTRUCTED ANNUAL TEMPERATURES FOR NORTHERN NORTH AMERICA

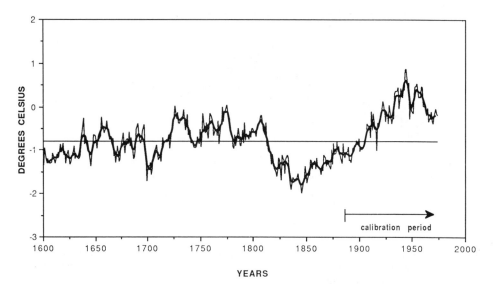

Figure 15.4 Actual (solid line) and estimated (dashed line) annual temperature departures for the 1880-1974 regression model.

(CE) statistic differs from the RE in that it compares the estimated data for the verification period to the mean for that same period (Nash and Sutcliffe 1971). The CE for the early calibration model is just slightly negative (-0.003). As additional tests of verification, the Spearman rank correlation coefficient (Steel and Torrie 1960) cross-products mean and sign test statistics (Fritts 1976) were also computed. For the earlier calibration model the Spearman correlation coefficient is 0.53 (significant at the 0.001 level). The cross-product means test measures the level of agreement of the actual and estimated values, taking into account both the sign and magnitude of the departures from the calibration average (Fritts 1976). For the earlier model the cross-product means test shows 26 correct and 9 incorrect values, which is significant at the 0.01 level. The sign test measures the first-difference direction of change, not taking into account the magnitude of the departures (Fritts 1976). The sign test results were not significant, reflecting the tendency for the model to be weaker at higher frequencies.

When the calibration-verification periods were reversed (Table 15.2), the adj.r^2 is 0.23, the RE is 0.54 the CE is 0.19 and the Spearman rank correlation coefficient is 0.67 (0.001 level). The cross-product means test showed 45 correct and 15 incorrect, significant at the .0001 level. The sign test result is 35 correct and 24 incorrect, which is significant at the 0.10 level.

Table 15.3 shows the algebraic forms of the regression equations for the three calibration periods. Although the coefficients are all positive in each case, the weights for variables 1 and 4 are somewhat unstable for the 1940-1974 calibration model. However the model passes the tests for verification (Table 15.2). The residual estimates for the 1880-1974 regression equation do not show any trends. The residuals were tested for serial autocorrelation using the Durbin-Watson and run tests (Draper and Smith 1981). The Durbin-Watson statistic for the 1880-1974 regression model is 1.583. Based on four predictors and 95 observations, any

Table 15.3 Standardized multiple regression coefficients for three calibration models.

| | Predictor variables | | | |
| | 1st Eigenvector | | | 2nd Eigenvector |
Model	t-1	t	t+1	t+1
1880-1974	.2270	.2909	.2647	.2213
1880-1939	.2651	.2678	.1715	.1781
1940-1973	.0416	.1772	.2262	.3485

value below 1.58 indicates significant (0.05 level) autocorrelation in the residual series. Thus the value obtained here shows that the residuals can be considered to be independent, albeit by a slight margin. The run test is a measure of the number of times that the residuals change sign (Draper and Smith 1981). The result was not significantly different from random, indicating no unusual series of positive or negative values in the residuals.

To summarize, a number of tests were performed in order to evaluate the regression models used in this study. PCA and preliminary screening of candidate predictors were used to minimize the number of predictor variables in the regression. Statistical tests were performed on the calibration-verification models. Finally, the residual estimates were tested for autocorrelation.

Some caveats must still be considered in examining this reconstruction. The reconstructed temperatures are most reliable in the low-frequency range, as is apparent from the sign test results which were weakly significant or not significant. The climate tree-growth model is not completely linear, as tree growth variations at temperature-limited sites tend to be more sensitive to cold than warm conditions. The mean instrumental temperature departure for the earlier calibration period is -0.56°C; it is +0.11°C for the later period. Such non-linearities have been detected in other tree-ring studies (e.g. Fritts 1976; Cook and Jacoby 1977). The cooling episode in the early to middle 1800s may not have a modern-day analog. However this cooling is corroborated by other paleoclimatic data from northern North America (e.g. Catchpole and Faurer 1983; Guiot 1987).

15.5 Discussion

15.5.1 *Northern North American annual temperature reconstruction*

Not surprisingly, the long-term trends shown are in agreement with those found previously in a reconstruction of Northern Hemisphere temperatures based partly on these same chronologies (Jacoby and D'Arrigo 1989a). The boreal reconstruction presented here provides an additional 70 years of record. The main features are below average temperatures in the early 1600s, a cooling in the early 1700s, a relative warming in the later 1700s, an abrupt transition to cold temperatures in the early to mid-1800s part of the Little Ice Age, and the general warming trend over the past century (Figure 15.3). Superimposed on this recent warming trend is a cooling episode in the 1960s to 1970s.

Although the reconstruction only extends to 1601, some of the chronologies extend back to around 1500 (Table 15.1). From these series we can infer that the early 1500s were close to the long-term mean followed by a cooling until the middle 1500s. According to the chronologies, temperatures recovered in the second half of the 1500s, punctuated by an abrupt drop in temperature in the middle 1590s.

Periods of agreement or disagreement between our reconstruction and two other large-scale temperature series (those of Groveman and Landsberg (1979) for the Northern Hemisphere and Fritts and Lough (1985) for the U.S. and southwestern Canada) are generally similar to those found previously by Jacoby and D'Arrigo (1989a). Cooler temperatures in the early 1600s in our reconstruction are in agreement with those of Groveman and Landsberg (1979) which is only based on three data sources for this period.

In addition to continental or hemispheric-scale records, there are regional-scale proxy and historical time series from northern North America available for comparison with the 400-year record of our reconstruction. One such time series is the percentage of ice-melt per annual layer in an ice core from the Devon Island ice cap in the Northwest Territories (Koerner 1977; Paterson et al. 1977) believed to be representative of summer temperatures. This record shows some similarities with our reconstructed temperatures, including a colder interval from the late 1600s to early 1700s. There is also a cooling in the early to mid-1800s in both series. $\delta^{18}O$ data developed from an ice core in Mt. Logan in the Yukon Territory is believed to be sensitive to temperature variations as well as precipitation (Holdsworth et al., this volume). This record, which comes from the vicinity of one of our tree-ring sites (TTHH, Figure 15.1, Table 15.1) also shows periods of lower $\delta^{18}O$ in the early 1700s (1715-1725) and mid-1800s (around 1850). In Quebec, a tree-ring record developed from living and dead black spruce trees for the past 600 years (Payette et al. 1985) shows periods of narrow ring width indices for three time periods within the Little Ice Age: the early 1600s, the late 1600s and the early to mid-1800s, generally corresponding to our reconstruction.

Tree-ring chronologies available from the vicinity of Hudson's Bay in central Canada were not of sufficient length to be included in our reconstruction for northern North America, with the earliest beginning in the middle to late 1600s or early 1700s (Jacoby and Ulan 1982; Scott et al. 1988). Other primary sources of paleoclimatic information in central Canada are the historical records of the Hudson's Bay Company (Ball, Chapter 3, this volume). Guiot (1987) compiled tree-ring and historical records from this area to reconstruct seasonal temperatures and sea level pressure back to 1700. He found that the 18th and 19th (including 1815-1825) centuries were colder than the 20th in this area on both an annual and seasonal basis.

15.5.2 Periodicities

The seven boreal chronologies provide long time series which extend backwards in time the limited climatic information available from instrumental records in the North American arctic and subarctic. Such records can be used to explore hypotheses regarding causes of climate variability, including solar and lunar-tidal variations, volcanic activity and the effects of CO_2 and other trace gases, as well as internal variations that might drive the climate system.

The long-term trends reconstructed for northern North America show possible relationships with hypothesized atmospheric forcings. For example, the cooling episode in the late 1600s to early 1700s coincides with the timing of the Maunder sunspot minimum (Eddy 1976). The rapid cooling of the early to middle 1800s occurs during an interval of low sunspot numbers (Eddy 1976) and several major volcanic eruptions believed to have had a cooling effect on climate (Lamb 1970). In a related study, negative tree growth departures from eastern Canada (Jacoby and D'Arrigo 1989b) coincide with elevated ice severity indices for Hudson's Bay in 1816 and 1817 (Catchpole and Faurer 1983; Catchpole and Hanuta 1989, Catchpole, this volume). These anomalies are associated with the presumed effects of the 1815 Tambora eruption (Stommel and Stommel 1983; Guiot 1987). The warming trend of the past century, also evident in our reconstruction, coincides with increasing atmospheric levels of greenhouse gases (Ellsaesser *et al.* 1986; Hansen *et al.* 1988).

Significant periodicities are identified in all seven chronologies using variance spectral analysis, although they vary across sites. One striking feature is that data from six of the sites (with the exception of Hornby Cabin in the Northwest Territories) show significant peaks at periods of about 5.5 years (Figure 15.5 a-g). As mentioned previously, the Hornby Cabin chronology shows differences from the other series in the principal components analysis. Two of the sites, TTHH from the Yukon Territory (Figure 15.5d) and Gaspé from Quebec (Figure 15.5g) show significant 11-year periods corresponding to the known sunspot cycle. Possibly the 5.5-year period represents a harmonic of this cycle. Holdsworth *et al.* (Chapter 25, this volume) note possible evidence of an 11-year cycle in ice core data from nearby Mt. Logan in the Yukon Territory. Nearly all of the sites show concentrations of variance at around 22 and 11-year (solar) and 18.6-year (lunar-tidal) cycles, although usually not significant at the 95% level. Pre-whitening of the data and the extended length of these chronologies may have enhanced detection of the signals at these sites. Most of these chronologies also show significant or nearly significant periods at about 2-3 years, possibly related to the Quasi-Biennial Oscillation (QBO) phenomenon which has been detected in numerous climate parameters (e.g. Thompson *et al.* 1986).

Spectral analyses of the non-pre-whitened tree-ring chronologies are dominated by low-frequency variance. All of the sites again showed significant periods at 2-3 years, and a few had significant spectra at around 5.5 years in the non-pre-whitened data. The results are indicated in Figures 15.6a and 15.6b for two of the chronologies (TTHH and Gaspé) which had previously shown significant 11-year periods in the pre-whitened series (Figures 15.5d and 15.5g). For the non-pre-whitened data at these two sites, the 11-year cycle is not significant.

Although spectra of the individual tree-ring series may be the most meaningful due to regional variability in the response of tree growth and/or climate to solar and lunar forcing, for completeness we present spectra of the reconstructed temperatures here. For the non-pre-whitened reconstruction, significant (95% level) peaks were found only at 2-3 years (100 lags, bandwidth = 0.0126). Following pre-whitening, (adj.r^2; Box and Jenkins 1976) the significant periods were 2.94-2.99, 4.76, 5.26-5.88 and 6.45-6.67 years.

For the period of overlap of the actual and estimated temperatures (1880-1974), both series show significant spectral estimates at periods of about 2.5 years (20 lags, bandwidth = .0629). When the series are pre-whitened, both show significant periodicities at 8.0 years, with the reconstruction also showing a peak at 2.5 years. However this time interval of less than 100 years may be too short to attach much significance to these latter results.

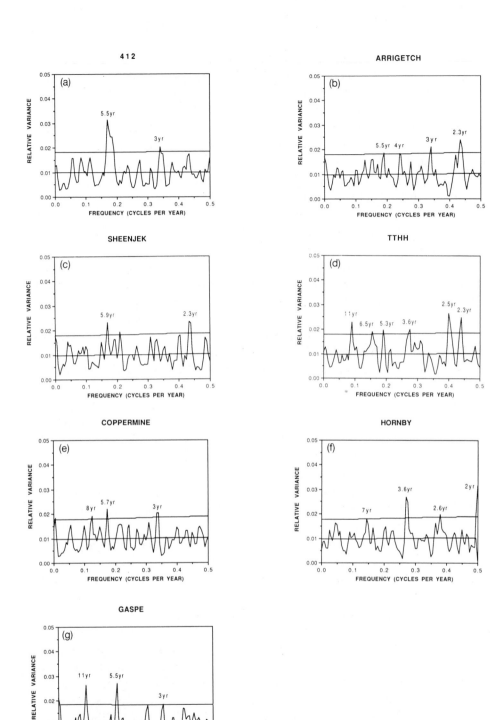

Figure 15.5(a)-(g) Variance spectra for seven pre-whitened chronologies for 1610-1975 based on 100 lags and a band width of 0.0126. The null continuum and null 95% confidence limits are also shown.

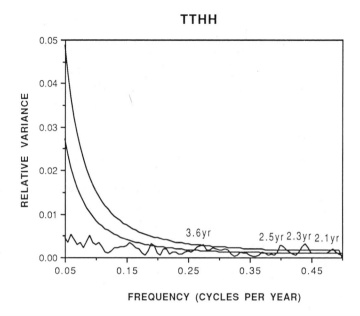

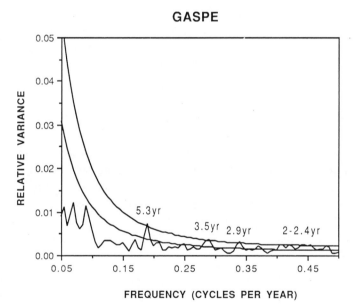

Figure 15.6(a)-(b) Variance spectra for two chronologies (TTHH and Gaspé) based on 100 lags and a band width of 0.0126, without pre-whitening of the data. The null continuum and null 95% confidence limits are also shown.

High-resolution spectral techniques such as the maximum entropy method (Priestly 1981) are needed to more adequately resolve such closely spaced periodicities as the 18.6 and 22-year cycles. Evolutionary spectral analysis is another useful technique for determining changes in spectral influences over time (Priestly 1981). These methods will be employed in later studies to more precisely assess some of the factors influencing tree growth (and hence climate) variations over the past 400 years in northern North America.

Numerous researchers have examined connections between instrumental and proxy climate data and solar or lunar-tidal cycles with conflicting results; significant relationships have been found in some analyses and not in others. Such inconsistencies have resulted in part from differences in methodology, as well as in the regions and time periods analyzed.

Early studies of possible relationships between solar variability and tree-rings date back to those of Douglass (1919, 1928, 1936). More recently, Mitchell *et al.* (1979) found a strong connection between the Hale 22-year solar cycle and reconstructed drought indices based on central and western U.S. tree-ring data back to 1700. Subsequent studies also suggest the influence of the 18.6 year cycle in tree-ring (Stockton *et al.* 1983) and instrumental data (Currie 1981) from this area (see also Meko, Chapter 16, this volume). Guiot (1987) found some evidence for 22 and 18.6 year cycles (using digital filters and a beat wave) in reconstructed temperatures based on historical and tree ring data in the Hudson Bay region, with the results varying from season to season.

Significant relationships have been found between solar and/or lunar variations and instrumental climatic data, including midwestern and northeastern U.S. precipitation and eastern U.S. temperatures (Currie 1974 and 1979; Hancock and Yarger 1979; Currie 1981; Currie and O'Brien 1989) storm tracks (Brown and John 1977; Tinsley 1988) and sea level pressure variations in the North Atlantic (Kelly 1979). However, the nature of such relationships can vary with time and location (Herman and Goldberg 1978). Competing influences related to other forcing functions (e.g. volcanism; Kelly 1979) can also obscure solar-climate relationships and vice-versa. Another possible explanation is that regional differences in the agreement between solar variations and climate reflect the long wave pattern of the atmosphere (King *et al.* 1977; Herman and Goldberg 1978). Recently, apparent linkages have been detected between solar fluctuations, climate parameters and the phases of the QBO (Labitzke 1987; Labitzke and Van Loon 1988). James and James (1989) show, however, that decadal-scale climate variations might be generated by non-linear dynamics without solar forcing. A physical explanation for these linkages might be found in the modulation of atmospheric wave patterns through feedbacks resulting from the influence of solar radiative variations on the climate system (Labitzke 1987). These connections may help explain some of the regional variations in periodicities which have been detected in instrumental and proxy climate records.

15.5.3 Other data sets

Within the last few years there have been new collections in the boreal forests of Canada and Alaska, including a recent major collection effort in the summer of 1989. These data represent a valuable new resource for climatic reconstructions. The data will be produced and/or analyzed by B. Luckman of the University of Western Ontario (presently with the Geological Survey of Canada) F. Schweingruber of the Swiss Federal Institute of Forestry Research, and G. Jacoby of Lamont-Doherty Geological Observatory in New York. New data produced by the last two workers will include density measurements. It has been shown that tree-ring density data can be a better estimator of temperature than ring-width data alone in northern regions (e.g. Hughes *et al.* 1984; Jacoby *et al.* 1988 and Briffa and

Schweingruber, Chapter 19, this volume). Analyses of reconstructions from the new data will enhance previous modeling efforts and provide improved information about climate variations over the past few hundred years.

15.6   Summary

A tree-ring based reconstruction of northern North American annual temperature departures for 1601-1974 has been presented which is primarily indicative of climatic conditions in subarctic and arctic North America. The reconstruction shows low-frequency variations including much of the Little Ice Age and a recent warming trend. The results are in general agreement with other regional and larger-scale temperature series for or within the Northern Hemisphere. There are, however, relatively few records from northern North America available for comparison with the nearly 400 years of record shown here. The data provide climatic information which can be applied to studies examining the patterns of variation and influence of forcing functions on the climate system. The spectral analysis presented here is a preliminary effort to evaluate some of the linkages between solar and other cycles on climate as reflected in tree growth variations in the North American boreal region.

Acknowledgements

This research was supported by Grants ATM87-16630, ATM85-15290 and ATM83-13789 from the Climate Dynamics Program of the National Science Foundation. We thank J. Hansen and S. Lebedeff for use of their temperature data. We also thank the Government of Canada for logistical assistance. This paper is Lamont-Doherty Geological Observatory Contribution No. 4624.

References

Box, G. E. P. and G. H. Jenkins. 1976. *Time series analysis, forecasting and control*. San Francisco: Holden-Day.
Brown, G. M. and J. I. John 1977. Solar cycle influences in tropospheric circulation. *EOS* 58, 695.
Bryson, R. A. 1966. Air masses, streamlines and the boreal forest. *Geographical Bulletin* 8, 228-269.

Catchpole, A. J. W. and M. A. Faurer 1983. Summer sea ice severity in Hudson Strait, 1751-1870. *Climatic Change* 5, 115-139.
Catchpole, A. J. W. and I. Hanuta 1989. Severe summer ice in Hudson Strait and Hudson Bay following major volcanic eruptions, 1751-1889 A.D.. *Climatic Change* 14, 61-79.
Cook, E. R. 1985. *A time series analysis approach to tree-ring standardization*. Ph.D. thesis, University of Arizona, Tucson.
Cook, E. R. and G. C. Jacoby Jr. 1977. Tree-ring drought relationships in the Hudson Valley, New York. *Science* 198, 399-401.
Cropper, J. P. and H. C. Fritts 1981. Tree-ring width chronologies from the North American Arctic. *Arctic and Alpine Research* 13, 245-260.
Currie, R. G. 1974. Solar cycle signal in surface air temperature. *Journal of Geophysical Research* 79, 5657.

Currie, R. G. 1979. Distribution of solar cycle signal in surface air temperature over North America. *Journal of Geophysical Research* 84, 753-761.

Currie, R. G. 1981. Evidence for a 18.6 year M_n signal in temperature and drought conditions in North America since A.D. 1800. *Journal of Geophysical Research* 86-C11, 11055-11064.

Currie, R. G. and D. P. O'Brien 1989. Morphology of bistable 180° phase switches in 18.6 year induced rainfall over the north-eastern United States of North America. *International Journal of Climatology* 9, 501-525.

Douglass, A. E. 1919, 1928, 1936. *Climatic cycles of tree growth*. Carnegie Institute of Washington, Publication no. 289, Parts I, II, III.

Draper, N. and H. Smith 1981. *Applied regression analysis*, 2nd edn. New York: J. Wiley and Sons.

Elsaesser, H. W., M. C. MacCracken, J. J. Walton and S. L. Grotch 1986. Global climatic trends as revealed by the recorded data. *Reviews of Geophysics* 24, 745-792.

Eddy, J. A. 1976. The Maunder minimum. *Science* 192, 1189-1202.

Fritts, H. C. 1976. *Tree rings and climate*. New York: Academic Press.

Fritts, H. C. and J. M. Lough 1985. An estimate of average annual temperature variations for North America, 1602-1961. *Climatic Change* 7, 203-224.

Gordon, G. A. and S. K. Leduc 1981. Verification statistics for regression models. In *Proceedings of a conference on probability and statistics in atmospheric sciences*, 129-133. Monterey, California, November, 1981.

Groveman, B. S. and H. E. Landsberg 1979. Simulated Northern Hemisphere temperature departures 1579-1880. *Geophysical Research Letters* 6, 767-769.

Guiot, J. 1987. Reconstruction of seasonal temperatures in central Canada since A.D. 1700 and detection of the 18.6 and 22 year signals. *Climatic Change* 10, 249-268.

Hancock, D. J. and D. N. Yarger 1979. Cross-spectral analysis of sunspots and monthly mean temperature and precipitation for the contiguous United States. *Journal of Atmospheric Sciences* 36, 746-753.

Hansen, J. and S. Lebedeff 1987. Global trends of measured surface air temperature. *Journal of Geophysical Research* 92, 13345-13372.

Hansen, J., I. Fung, A. Lacis, S. Lebedeff, D. Rind, R. Ruedy and G. Russell 1988. Global climate changes as forecast by the GISS 3-D model. *Journal of Geophysical Research* 93, 9341.

Herman, J. R. and R. A. Goldberg 1978. *Sun, weather and climate*. Washington D.C. NASA.

Hughes, M. K., P. M. Kelly, J. R. Pilcher and V. C. LaMarche, Jr.(eds) 1982. *Climate from tree-rings*. Cambridge: Cambridge University Press.

Hughes, M. K., F. H. Schweingruber, D. Cartwright and P. M. Kelly 1984. July-August temperature at Edinburgh between 1721 and 1975 from tree-ring density and width data. *Nature* 308, 341-344.

Jacoby, G. C. Jr. 1982. The Arctic. In: Hughes, M. K., P. M. Kelly, J. R. Pilcher and V. C. LaMarche, Jr.(eds) 1982. *Climate from tree-rings*. Cambridge. Cambridge University Press.

Jacoby, G. C. Jr. and R. D'Arrigo 1989a. Reconstructed Northern Hemisphere annual temperatures since 1671 based on high latitude tree-ring data from North America. *Climatic Change* 14, 39-59.

Jacoby, G. C. Jr. and R. D'Arrigo 1989b. Spatial patterns of tree-growth anomalies from the North American boreal forest treeline in the early 1800s, including the year 1816. In: Harington, C. R. (ed.). 1816: *The year without a summer?*. National Museum of Natural Sciences, Ottawa, Canada (in press).

Jacoby, G. C. Jr. and L. D. Ulan 1981. Review of dendroclimatology in the forest-tundra ecotone of Alaska and Canada, In: Harington, C. R. (ed.) *Climatic Change in Canada* 2, Syllogeus No. 33. Ottawa: National Museums of Canada, 97-128.

Jacoby, G. C. Jr. and L. D. Ulan 1982. Reconstruction of past ice conditions in a Hudson Bay estuary using tree-rings. *Nature* 298, 637-639.

Jacoby, G. C. Jr., I. S. Ivanciu and L. D. Ulan 1988. A 263-year record of summer temperature for northern Quebec reconstructed from tree-ring data and evidence for a major climatic shift in the early 1800s. *Paleogeography, Paleoclimatology, and Paleoecology* 64, 69-78.

James, I. N. and P. M. James 1989. Ultra-low-frequnecy variability in a simple atmospheric circulation model. *Nature* 342, 53-55.

Kelly, P. M. 1979. Solar influence on North Atlantic mean sea level pressure. In: *Solar-terrestrial influences on weather and climate*, B. M. McCormac and T. A. Seliga (eds), 297-298. Dordrecht: Reidel.

King, J. W., A. J. Slater, A. D. Stevens, P. A. Smith and D. M. Willis 1977. Large-amplitude stationary planetary waves induced in the troposphere by the sun. *Journal of Atmospheric and Terrestrial Physics* 39, 1357.

Koerner, R. M. 1977. Devon Island icecap: core stratigraphy and paleoclimate. *Science* 196, 15-18.

Labitzke, K. 1987. Sunspots, the QBO and the stratospheric temperature in the north polar region. *Geophysical Research* Letters 14, 535-537.

Labitzke, K. and H. Van Loon 1988. Associations between the 11-year solar cycle, the QBO and the atmosphere. Part I: the troposphere and stratosphere in the northern hemisphere in winter. *Journal of Atmospheric and Terrestrial Physics* 50, 197-206.

Lamb, H. H. 1970. Volcanic dust in the atmosphere; with a chronology and assessment of its meteorological significance. *Philosophical Transactions of the Royal Society of London Series A*. 266, 425-533.

Mitchell, J. M., Jr., C. W. Stockton and D. M. Meko 1979. Evidence of a 22-year rhythm of drought in the western United States related to the Hale solar cycle since the 17th century. In: *Solar-terrestrial influences on weather and climate*, B. M. McCormac and T. A. Seliga (eds), 125-144. Dordrecht: Reidel.

Nash, J. E. and J. V. Sutcliffe 1971. Riverflow forecasting through conceptual models. 1.A. Discussion of principles. *Journal of Hydrology* 10, 282-290.

Paterson, W. S. B., R. M. Koerner, D. Fisher, S. J. Johnsen, H. B. Clausen, W. Dansgaard, P. Bucher and H. Oeschger 1977. An oxygen-isotope climatic record from the Devon Island ice cap, arctic Canada. *Nature* 266, 508-511.

Payette, S., L. Filion, L. Gauthier and Y. Boutin 1985. Secular climate change in old-growth treeline vegetation of Northern Quebec. *Nature* 315, 135-138.

Priestley, M. B. 1981. *Spectral analysis and time series*. Academic Press, New York.

Richman, M. B. 1986. Rotation of principal components. *Journal of Climatology* 6, 293-335.

Scott, P. A., D. C. F. Fayle, C. V. Bentley and R. I. C. Hansell 1988. Large-scale change in atmospheric circulation interpreted from patterns of tree growth at Churchill, Manitoba, Canada. *Arctic and Alpine Research* 20, 199-211.

Steel, R. G. D. and J. H. Torrie 1960. *Principles and Procedures of statistics*. New York. McGraw-Hill.

Stockton, C. D., J. M. Mitchell Jr. and D. M. Meko. 1983. A reappraisal of the 22-year drought cycle. In: B. M. McCormac (ed). *Weather and climate response to solar variations*. Dordrecht: Reidel.

Stommel, H. and E. Stommel. 1983. *Volcano weather: the story of 1816: the year without a summer*. Newport, Rhode Island: Seven Seas.

Thompson, M. L., I. G. Enting, G. I. Pearman and P. Hyson 1986. Interannual variation of atmospheric CO_2 concentration. *Journal of Atmospheric Chemistry* 4, 125-155.

Tinsley, B. A. 1988. The solar cycle and QBO influences on the latitude of storm tracks in the North Atlantic. *Geophysical Research Letters* 15, 409-412.

16 Dendroclimatic evidence from the Great Plains of the United States

D. M. Meko

16.1 Introduction

Meteorological drought and its impact on agriculture in the grain belt of the interior U.S. has spurred much research into the possibility of a regular or predictable pattern in drought recurrence there. The earliest recorded tree-ring study in the Plains, by Jacob Keuchler in central Texas in 1859 (Campbell 1949) pre-dated the development of dendrochronology as a quantitative science. Tree-ring data has since played a major role in providing statistical evidence for a relationship between solar variations and drought. The first widespread tree-ring collections over western North America analyzed for cycles were those of Schulman (1956). Using an optical method of cycle analysis developed by Douglass (1936) Schulman found cycles near the 20-year wavelength in regionally averaged tree-ring series from the Banff area, British Columbia, to the Rio Grande basin. The cycles represented *average* wavelengths, however, with wide variations in time from peak to peak. Moreover, changes in amplitude, waveform, and phase were common. Schulman concluded that "in no extensive data were cyclic tendencies found of such regularity and strength as to suggest physical reality". LaMarche and Fritts (1972) found a periodicity near the 22-year double sunspot period in spectral analysis of the scores of the first eigenvector of tree-growth over the western U.S., but found no phase linkage with the sunspot series itself. A statistically significant phase linkage was later reported, however, between the double sunspot series and a 300-year tree-ring reconstruction of an index of drought area in the western U.S. (Mitchell *et al.* 1979). When the reconstructed drought-area index (DAI) and sunspot series were analyzed with a harmonic dial, a method of examining phase relationships developed by Brier (1961) drought-area was found to peak about two years after alternate minima in the sunspot series. Currie (1981) and Bell (1981) have independently examined the same DAI series and found rhythmic features near the 20-year wavelength, but argue for importance of a component near the 18.6-year lunar nodal cycle.

In the Plains, the climatic record itself has fueled speculation about cycles in drought. The rhythmic return of drought to the Plains at about 20-year intervals is well documented (Thomas 1962; Borchert 1971; Perry 1980). Whether precipitation records contain a statistically significant periodicity, however, is still questionable. Oladipo (1987) concluded from spectral analysis of growing-season precipitation at 407 stations in the interior Plains that there is no evidence for periodicity. Currie (1989) on the other hand, presents evidence for a 19-yr and 10-to-11 yr periodicity from Arkansas, Illinois, and Iowa, and attributes Oladipo's (1987) failure to find such terms to inappropriate methodology.

Tree-ring data from in and around the Plains have consistently failed to indicate a regular rhythm to drought. Inconclusive or negative evidence for cycles was provided by early tree-ring studies in Oklahoma, Nebraska, North Dakota, and the Mississippi drainage (Hawley 1941; Weakly 1943; Will 1946; Harper 1961). Much of the early tree-ring work in the Plains was reviewed by Lawson (1974) in an effort to piece together a picture of the climate of the first half of the 19th century as encountered by the Fremont, Long, and Pike expeditions. Lawson's primary conclusion regarding periodicity was that "the temporal distribution of prolonged wet or dry periods on an inter-regional basis is complex enough to resist meaningful generalizations."

One limitation to using early tree-ring studies to reconstruct the frequency of widespread drought in the Plains is that different methodologies were used by different researchers. Another is that only a small part of the Plains was represented by tree-ring coverage. Much progress has been made in overcoming both limitations in the past two decades. A standard methodology now exists for the collection and reduction of tree-ring data (Fritts 1976). Recent years have also seen an ambitious field collection effort resulting in numerous chronologies in what were formerly data voids (e.g. maps in Stockton *et al.* 1985). Annual precipitation reconstructions for various regions in and near the Plains have been published as the tree-ring database has expanded (Meko 1982; Stockton and Meko 1983; Blasing and Duvick 1984; Meko *et al.* 1985; Blasing *et al.* 1988).

Like the earlier tree-ring studies, these more recent efforts have failed to yield convincing evidence of drought periodicity in the Plains. The results appear to contradict the earlier findings of Mitchell *et al.* (1979) but differences in the spatial scales and variables used preclude a strict comparison. The DAI study essentially pointed to a near-20-year periodicity in the total area of drought west of the Mississippi, and further suggested an imperfect phase locking between the DAI and the double-sunspot cycle. These findings do not imply periodicity at specific locations or in sub-regions of the western U.S. In fact, the DAI concept was adopted largely because of the observation that the major droughts of the 1930s and 1950s were centered in different geographical locations (Stockton and Meko 1975).

The possibility remains, however, that the sparsity of tree-ring coverage in the Plains has kept previous regional tree-ring studies from discerning a regular rhythm in drought area there. This chapter presents a new regional drought history of the Plains, A.D. 1750-1964, based on tree-ring coverage considerably expanded beyond that of previous studies. The timing of major droughts relative to key dates in the double sunspot and lunar nodal cycles are examined, and inferences on drought history before A.D. 1750 are drawn from previous studies.

16.2   Tree-ring and climate data

The study area comprises the U.S. Great Plains and flanking areas to the west and east, from the Rocky Mountains to the Mississippi River. A regional approach was taken, with ten regions defined primarily by the distribution of available tree-ring chronologies. Five are along the eastern flank of the Plains, and five along the western flank (Figure 16.1).

The tree-ring data are indices of annual ring width from 58 sites collected by researchers from the University of Arizona, University of Arkansas, Oak Ridge Laboratory, and Lamont Doherty Geological Observatory between the years 1964 and 1985. Several different species are represented: *Quercus stellata* in regions 4, 6, and 8; *Quercus alba* in region 2; *Pinus*

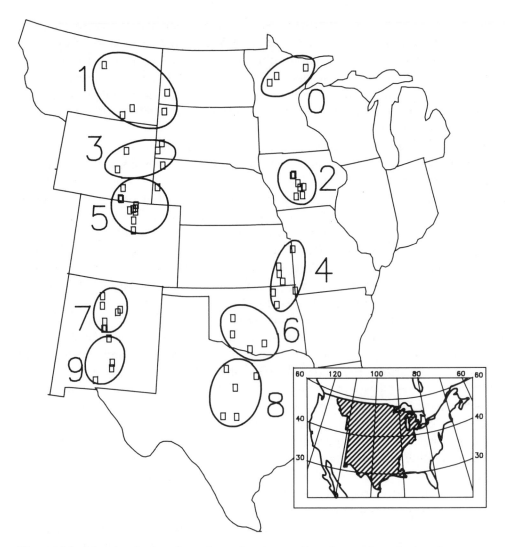

Figure 16.1 Map showing locations of tree-ring sites, and locations and numbering of ten groups defined as tree-ring regions.

resinosa and *Quercus macrocarpa* in region 0; *Pseudotsuga menziesii* and *Pinus ponderosa* in region 1; *Pinus ponderosa* in region 3; *Pinus ponderosa*, *Pseudotsuga menziesii* and *Pinus edulis* in regions 5 and 7 and *Pseudotsuga menziesii* and *Pinus edulis* in region 9. The common period covered by all chronologies limits the drought chronology to A.D. 1750-1964, although many individual chronologies date back to the mid-1600s. As shown in Figure 16.1, the number of sites per region varies from 3 in region 0 (N. Minnesota) to 11 in region 5 (N. Central Colorado).

Precipitation was used as the climatic variable for comparison with tree-ring data. Previous studies have indicated that precipitation is more strongly related than temperature to growth variations in tree-ring chronologies from both the western and eastern flanks of the Plains, and that a stronger climatic signal exists for precipitation summed over several seasons than for individual monthly or seasonal totals (Meko 1982; Stockton and Meko 1983; Blasing *et al.* 1988). These studies also suggest that negative growth departures may be accentuated when growing-season temperatures are higher than normal, presumably through the intercorrelation of temperature with other climatic variables governing potential evapotranspiration. Monthly station precipitation data covering the years 1904-1984 were taken from the Historical Climate Network of Quinlan *et al.* (1987) for all stations falling approximately within the regions defined by the ellipses around the tree-ring sites in Figure 16.1. Regions 7 and 9 have only 4 and 3 precipitation stations, respectively. Coverage is fairly uniform elsewhere, varying from 7 to 10 stations per region. Time series of annual regional precipitation were computed by averaging station monthly data over stations and summing the resulting series over the months August-July. The August-July period appears to be a reasonable compromise for grouping monthly precipitation for maximum correlation with tree-ring indices in and near the Plains (Meko 1982; Stockton and Meko 1983; Blasing and Duvick 1984; Blasing *et al.* 1988). The August-July regionally averaged precipitation is referred to as "precipitation" from here on.

16.3   Methods

16.3.1   *Principal components weighting of site chronologies*

Principal components analysis (PCA) was applied to the data at two stages in this study. The first application was to weight site chronologies into regional tree-ring series. The assumption underlying the choice of this weighting scheme is that the common ring-width variance in a region is governed by regional-scale climate fluctuations. By de-emphasizing chronologies that co-vary least with nearby chronologies, PCA weighting would therefore be expected to enhance the climatic signal in the weighted series.

Let T be an $n \times m$ matrix of tree-ring indices at m sites covering n years. Here n varies from 265 years to 281 years, depending on the common period covered by all tree-ring series in the region, while m varies according to the number of sites in a region (Figure 16.1). Uncorrelated variables C are defined that are linear combinations of the original tree-ring data at the m sites:

$$C = E'T, \tag{1}$$

where C is an $n \times m$ matrix of time series of the new uncorrelated, variables, or the eigenvector scores, and E' is transpose of the $m \times m$ eigenvector matrix. The mathematics of the transformation as applied to climatic and tree-ring data can be found elsewhere (e.g. Kutzbach 1967; Sellers 1968; Fritts 1976). The first eigenvector is the linear combination accounting for the greatest possible fraction of the variance of the data. The second eigenvector is the linear combination accounting for the greatest fraction of the remaining variance, with the constraint that it is orthogonal to the first eigenvector, and so on.

As a tool for weighting sites, only the first eigenvector was used. The analysis was run on the covariance rather than correlation matrix to retain the importance of site-to-site differences in chronology variance in the weighted series. In conversion from ring-widths to indices, tree-ring data have already been scaled into values proportional to per cent of "normal" growth for a given point on the age curve. A relatively low variance in an index series implies relatively little sensitivity to environmental factors – including climate – that vary from year to year. The time series of first-eigenvector scores for each region is an unevenly weighted average of annual growth deviations over sites in the region.

Two further transformations were made on the eigenvector-score series to facilitate inter-regional comparisons. First, if the weights of eigenvector 1 for a particular region were negative, the sign of the corresponding score series was reversed, such that a negative score in a given year implies below-normal growth for the region. This operation is merely a matter of convenience for comparison of plots. Second, the time series of scores for each region was standardized by subtracting the 1750-1964 mean and dividing by the standard deviation. The resulting ten time series, each with zero mean and unit variance, are referred to from here on as "regional tree-ring series".

16.3.2   Principal components analysis of spatial patterns

The second application of PCA was to delineate spatial modes of variation in regional tree-growth and precipitation. As discussed in Chapter 1, this type of analysis has frequently been applied in climatology and dendroclimatology. Separate runs were made on the regional tree-ring series 1750-1964, regional tree-ring series, 1904-1964, and regional precipitation, 1904-1964. The PCA for precipitation was based on the correlation matrix to avoid undue influence from regions with relatively high precipitation. The PCA of the regional tree-ring series were also based on the correlation matrix. In this case the covariance matrix could alternatively have been used with no difference in results, since the input regional tree-ring series had already been scaled to unit variance. The transformations again can be represented by an equation of the form (1). For the longer analysis period, $\mathbf{T}$ is a 215 by 10 array holding the regional tree ring series for A.D. 1750-1964, $\mathbf{C}$ is the 215 by 10 array of eigenvector scores, and $\mathbf{E'}$ is the 10 by 10 transposed eigenvector matrix. For the shorter analysis period, $\mathbf{T}$ is a 61 by 10 array of regional tree-ring or precipitation series for 1904-1964, $\mathbf{C}$ is a 61 by 10 array of scores of the new, orthogonal, variables, and $\mathbf{E'}$ is the 10 by 10 transposed eigenvector matrix.

16.3.3   Definition of regional drought

Dendrochronological drought in a particular region was defined as the occurrence of a regional tree-ring value (eigenvector score) below that level exceeded 80% of the time from 1750-1964. In other words, the lowest quintile of tree-ring values from 1750-1964 was identified as the set of drought years for a region. A "widespread" drought was arbitrarily defined as occurring when six or more of the ten regions were in regional drought in any given year.

Drought years were similarly designated for the 1904-1964 regional (August-July) precipitation series, which have been defined previously. The drought threshold here, however, is defined as the level of precipitation yielding the same number of drought years in the 1904-1964 precipitation series as in the corresponding regional tree-ring series.

16.4 Results and discussion

16.4.1 Region-by-region PCA on site chronologies

The cross-covariance matrix, 1750-1964, between all possible pairs of chronologies in each region was first examined in a preliminary step to eliminate from further analysis any chronology varying in the opposite sense from neighboring chronologies. Only in region 0 was a negative cross-covariance value found. One chronology – the only *Quercus macrocarpa* chronology in the entire study – was correlated negatively with the three *Pinus resinosa* chronologies in that region. That outlier chronology was deleted from the data set, and individual PCA were then run on chronologies from each of the ten regions, as explained in the methods section. The weights and the per cent of variance explained by the first eigenvector in each regional PCA are listed in Table 16.1.

Table 16.1 Variance explained and weights of first eigenvector of tree growth in each region.

Reg[1]	Pct[2]	Weights[3]										
0	65.0	.23	.27	.93								
2	79.0	.44	.41	.37	.39	.38	.32	.32				
4	58.2	.48	.27	.36	.39	.39	.39	.34				
6	70.0	.32	.61	.44	.58							
8	68.3	.42	.63	.25	.45	.40						
1	57.4	.68	.24	.41	.43	.35						
3	58.8	.44	.53	.40	.50	.34						
5	46.5	.47	.23	.27	.22	.39	.23	.26	.31	.22	.34	.26
7	65.5	.35	.40	.49	.31	.36	.33	.37				
9	69.8	.42	.48	.46	.62							

[1] Region number
[2] Per cent variance of tree-ring indices in region explained by first eigenvector
[3] Weights of first eigenvector, each value a weight on an individual tree-ring chronology

For some regions (e.g. region 2) weights on individual chronologies are so evenly distributed that virtually the same regional drought history would be inferred from straight arithmetic averaging of the chronologies. In other regions the regional series is very strongly dependent on one or a few of the tree-ring chronologies. The extreme latter example is region 0, Minnesota, where the highest weight is more than triple that on the other two chronologies. The variance explained by the first eigenvector ranges from 46.5% for region 5 to 79% for region 2. These percentages measure the sufficiency of a single weighted average series to represent the tree-ring variance within the region. Although the percentages for the various regions are not strictly comparable because of the differing numbers of sites per region, the weighted series appears to be most representative in region 2. Region 0 could clearly benefit most from additional sampling, as its regional tree-growth variable is highly dependent on one chronology. The per cent of variance explained for region 0 is in the medium range for regional analyses despite there being only three sites in the region.

16.4.2 Precipitation-tree ring correlations

The product-moment correlation coefficient provides a relatively simple measure of the strength of the linear relationship between tree-ring and precipitation data. The correlation coefficients, 1904-1964, between regional precipitation (August-July total) and tree-ring series derived from PCA on the period 1904-64, are listed below:

Region	r	Region	r
0	0.41	1	0.59
2	0.72	3	0.62
4	0.71	5	0.71
6	0.76	7	0.72
8	0.81	9	0.75

The left column contains correlations for regions in the western part of the Plains, the right column from the eastern part. Correlation coefficients increase from north to south along both flanks of the Plains, suggesting a gradient in strength of the precipitation signal obtainable from tree rings. The correlations indicate that if the tree-ring data were scaled into reconstructions of regional annual precipitation by simple linear regression, the explained variance would vary from perhaps 55%-65% in the south to less than 20% in Minnesota. The higher values are consistent with variance-explained estimates obtained in published regional reconstructions using relatively conservative regression models (Meko 1982, Stockton and Meko 1983, Blasing and Duvick 1984, Blasing *et al.* 1988). No previously published reconstructions are available for comparison with region 0. The value r=0.59 (implied reconstruction variance of 36%) for region 1 is considerably lower than the annual precipitation variance of 52% reconstructed for a region with nearly the same boundaries by Stockton and Meko (1983). The discrepancy (15%) represents additional explained variance gained by the design of Stockton and Meko's (1983) reconstruction model – particularly filtering out some of the noise due to non-climatic persistence in the tree-ring series by use of lagged predictors. It is reasonable to expect that lagged regression models could improve calibration statistics above those implied by the listed correlation coefficients in other regions as well, especially where tree-ring series are highly autocorrelated.

Whether the observed north-south gradient in precipitation signal is characteristic of Plains tree-ring data, or merely a function of deficiencies in the current tree-ring coverage in the north probably cannot be determined until the coverage is expanded in the band from Minnesota to eastern Montana. Possibly the northern regions are more frequently influenced by precipitation regimes that leave a less distinct signature on tree growth than further south; for example, the effect of negative precipitation anomalies may at times be mitigated by the occurrence of anomalously cool conditions (cool, dry droughts).

16.4.3 Spatial patterns

Because the regional precipitation and tree-ring variables are highly correlated, it is reasonable to expect that fields of the two types of variables should exhibit similar recurrent spatial patterns of variation. The PCAs run on the regional tree-ring and precipitation series bear out this expectation. The first three eigenvectors of annual regional tree-ring variables are very similar to those of annual precipitation (Figure 16.2). The first eigenvector represents a same-sign anomaly covering all regions, the second eigenvector a north-south contrast, and the third eigenvector an east-west contrast. Eigenvalues are greater than 1.0 only for the first three eigenvectors. The per cent of variance explained by each eigenvector is listed below:

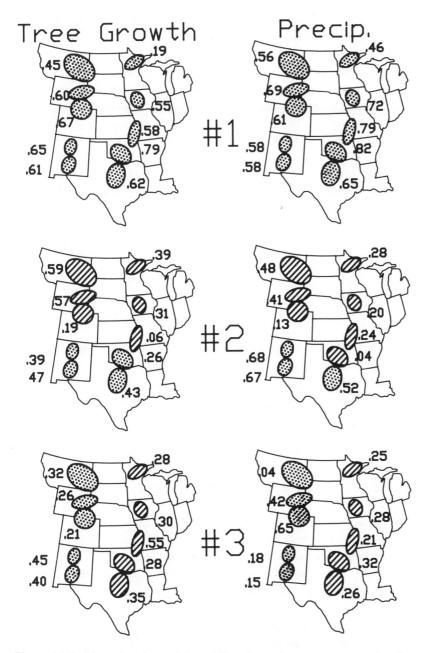

Figure 16.2 Maps showing weights of first three eigenvectors on regional tree-ring series, 1750-1964 (left) and August-July regionally averaged precipitation, 1904-1984 (right). Stippling indicates a positive sign on the corresponding eigenvector weight, hatching a negative weight.

Eigenvector	Tree-growth	Precipitation
1	35	42
2	16	18
3	12	15

That the first eigenvector weights all ten regions with the same sign indicates that the entire study area is in some sense climatologically homogeneous with respect to the direction of annual departures in both tree-growth and precipitation. This result is consistent with Borchert's (1950) observation that annual precipitation departures averaged over major drought years are negative over a large contiguous area of the Plains. Previous eigenvector studies indeed have indicated spatial coherence in spring and summer precipitation anomalies and Palmer Drought Severity Indices in the U.S. over a much greater area than the Plains (Diaz 1981; Karl and Koscielny 1981). Case studies of major droughts of the 1930s, 1950s, and 1980s (Namias 1960, 1982) also emphasize the large spatial scale of negative seasonal anomalies in precipitation over the mid-continent. A summer feature contributing to the common variance in precipitation over regions is the anomalous development of the upper-level anticyclone over the Plains (Namias 1955, 1960). The north-south and east-west contrasts of eigenvectors 2 and 3 may reflect differences in positioning of moist and dry tongues wrapping around the continental anticyclone (Namias 1955) or shifts of the storm track with the main belt of the westerlies (Diaz 1981).

16.4.4 Time variations

The change in relative importance of each pattern mapped in Figure 16.2 over time is given by the eigenvector scores of the tree-ring data (Figure 16.3). Series smoothed by an 11-weight bell-shaped filter (emphasizing low-frequency variations) are also shown. Negative departures correspond to the following inferred spatial patterns in moisture anomaly:

Eigenvector	Interpretation of Negative Scores
1	dry overall
2	relatively dry to south, wet to north
3	relatively dry to west, wet to east

Key differences between the major dust-bowl droughts of the 1930s and 1950s are evident in Figure 16.3. Eigenvector 1 suggests that the 1950s drought was more severe than the 1930s drought for the Plains as a whole. Eigenvectors 2 and 3 point to 1950s drought as more severe to the south and east, the 1930s drought more severe to the north and west. A similar contrast in the spatial locations of these two major droughts was reported previously in a rotated PCA of Palmer Drought Severity Index over the U.S. for 1895-1981 (Karl and Koscielny 1981).

Some notable shifts in the relative importance of individual eigenvectors, possibly associated with decadal-scale regimes of climate variation, can be seen in each century. From about 1910 to 1960, eigenvector 3 shows a step-like trend with the west becoming drier relative to the east. For eigenvector 2, the period from the 1770s to 1810 was characterized by a shift to drier conditions in the north relative to the south. The largest shift in overall tree-growth in

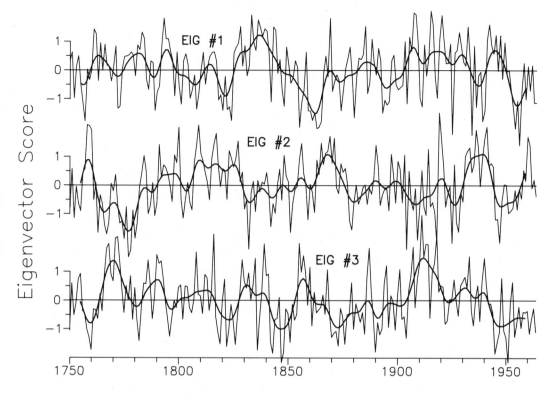

Figure 16.3 Time series plots of scores of first three eigenvectors from PCA on regional tree-ring series, 1750-1964. Also shown are series smoothed by an 11-weight raised cosine filter with a 50% frequency response at about 12 years.

the Plains as given by eigenvector 1 occurred from a high-growth period in the 1830s to a low-growth period in the 1860s.

Considering the importance of the phenomenon of drought to the Plains, it is useful to summarize the regional tree ring series in such a way that emphasizes recurrence and clustering of years of low-growth over the area as a whole. The eigenvector plots may not suffice for this purpose because a given drought may be best represented by a composite of several eigenvectors, possibly including some beyond the third. Caution must also be exercised in attributing physical meaning to patterns from PCA, as the orthogonality constraint can lead to fictitious patterns (Richman 1981).

An alternative representation of drought not prone to the pitfalls of eigenvector analysis is the simple quintile tabulation described in the methods section. A summary of this tabulation is given by the coded bar chart in Figure 16.4, which is a time series showing how many, and which, of the ten regions are in drought in each year. A "widespread" drought is defined as 6 or more of the 10 regions in drought in a given year. The coding of regions in Figure 16.4 allows a ready interpretation of the geographical location of a drought: low numbers indicate a northern location, high numbers a southern, even numbers an eastern, and odd numbers a western location. A similar tabulation based on the regional August-July total precipitation series, 1904-1984, is shown at the bottom for comparison with the tree-ring record. Arcing

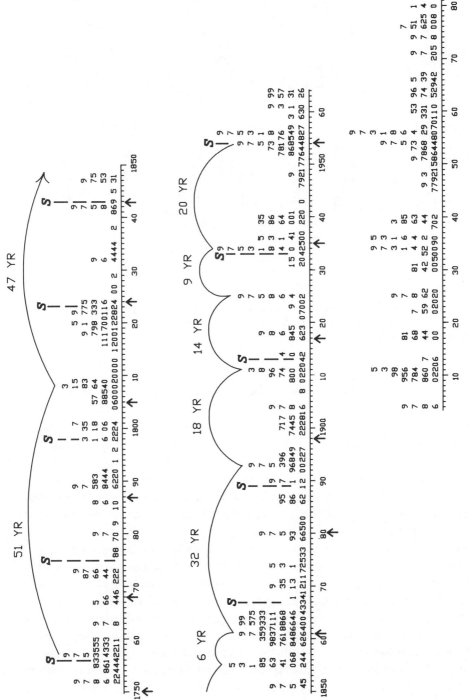

Figure 16.4 Number-coded bar plot showing how many regions, and which, are experiencing regional drought in each year, 1750-1964. Regions are numbered as in Figure 16.1. Arcing lines above each year. Arcing lines above plot join years with 6 or more regions in drought separated by at least one drought-free year. Letter "S" above plot marks years of alternate minima in double sunspot series as listed by Mitchell *et al.* (1979). Arrows below abscissa mark maxima in lunar nodal cycle, from Currie (1984). Lower panel shows a similar tabulation based on the regional August-July total precipitation series (1904-1984).

lines have been drawn above the tree-ring plot to draw attention to the intervals between years of widespread drought. Below the abscissa, years of alternate minima in the 22-year double sunspot cycle and maxima in the 18.6-year lunar nodal cycle are marked.

Several large clusters of numbers, indicating widespread drought lasting or recurring over several years, are evident in Figure 16.4. The most notable of these clusters were centered at about 1756, 1820, and 1862. By far the most severe was the 1860s drought, with 5 or more regions in drought for 6 consecutive years. Blasing *et al.* (1988) noted this period as the most severe prolonged drought from A.D. 1750-1980 in the south-central states. The most regions in drought in any single year was 9 – in years 1934 and 1956. It appears that extremes in single-year drought coverage have been more frequent in the current century than earlier, while prolonged droughts lasting several years have been less frequent.

Comparison of the two plots in Figure 16.4 for the period 1904-1964 shows that the tree-ring drought chronology tracks the actual precipitation record quite well. Droughts centered around 1910, 1934, and 1954-56 are the dominant features of both plots. The arcing lines on Figure 16.4 identify widespread droughts in the early 1890s, early 1910s, 1925, 1930s, and 1950s. All except 1925 (an isolated single-year drought) match the major Great Plains droughts identified from precipitation records by Thomas (1962). The shift from general absence of drought in the decade of the 1940s to widespread drought in the 1950s corresponds to a significant decadal-scale fluctuation in summertime average temperature and precipitation (Karl and Riebsame 1984). Their Figure 16.11 shows decreases in precipitation between the intervals 1942-51 to 1952-61 over all of the Plains except Minnesota, which is consistent with the tree-ring depiction of drought.

It may be argued that since some annual tree-ring indices are highly autocorrelated, the lengths of drought clusters shown in Figure 16.4 may be artifacts of the tree-ring data themselves. Neither the original tree-ring indices nor the eigenvector scores used in this study have been high-pass filtered to remove autocorrelation. A bias in autocorrelation is evident from a comparison of the estimated first-order autocorrelation coefficients, r_1, of regional tree-ring and precipitation series, 1904-64 (Table 16.2). Values of r_1 exceed zero by more than two standard errors in six of the 10 regions for tree-rings, but only in region 0 (Minnesota) for precipitation. The bias appears to be largest for the northwestern regions, with an extreme difference of 0.64 versus 0.18 for region 1. The focus on low-growth anomalies rather than the entire range of growth variation may act to minimize the effects of such bias in the drought chronology of Figure 16.4. The standard error in tree-ring reconstructions of annual precipitation in the Plains has been found to be generally higher in wet years than dry years (Stockton and Meko 1983). In addition, biological factors inducing autocorrelation are more likely to be active in wide rather than narrow rings (Fritts 1976). Moreover, comparison of the drought features in the tree-ring and the actual precipitation records in Figure 16.4 suggests that the distorting effects of autocorrelation on the drought chronology are small. Assuming the same biological factors introducing distortion in the post-1904 period were active in the earlier centuries, it appears that the 1850s-1860s drought was the most severe prolonged drought in the Plains since A.D. 1750. There is of course a risk in assuming that changes in the autocorrelation structure of tree ring data reflect similar changes in climate variables. Changes in the persistence structure of tree rings could presumably arise from many non-climatic factors, such as fires, changes in stand density, and insect infestations.

Table 16.2 Estimated[1] first-order autocorrelation coefficients of regional tree-growth and precipitation variables[2].

Region	Tree Rings	Precip.
1	.64	.18
3	.45	-.02
5	.45	.07
7	.33	.14
9	.12	.17
0	.53	.29
2	.05	-.13
4	.14	.15
6	.23	.04
8	.26	.17

[1] The sample period is 1904-64. The standard error of estimates is 0.13.

[2] Eigenvector scores were derived from principal components analysis on tree-ring and annual precipitation data for the period 1904-64 (see text).

16.4.5 Timing of drought relative to solar and lunar cycles

The root causes of drought in the continental U.S. are as yet unknown, although much is known about the associated atmospheric circulation features. Anomalous positioning or strength of the upper-level anticyclone over the central U.S. has been linked to major spring and summer droughts (Reed 1933; Namias 1955,1982). Low antecedent moisture conditions are thought to favor buildup of this anticyclone, as are anomalously cool sea surface temperatures in the central North Pacific (Namias 1982). It has long been speculated from statistical evidence that solar variations may also play a role in the development of drought, although a convincing theory for a physical mechanism is lacking (Siscoe 1978; Pittock 1983). Previous tree-ring evidence for drought periodicity was discussed in the introduction. In short, one study (Mitchell *et al.* 1979) reported a rhythm in total drought area west of the Mississippi near a period of 22 years, with a phase relationship to the double sunspot series, while numerous studies have failed to find drought periodicity on the smaller regional scale in various parts of the Plains (e.g., Lawson, 1974; Stockton and Meko, 1983; Meko *et al.*, 1985). The data set represented by the coded bar-chart in Figure 16.4 is intermediate in scale between the drought-area index of Mitchell *et al.* (1979) and the various regional recon-structions. Although no extensive statistical analysis is attempted here to test the relationship of this series to the solar or lunar cycles, some observations are made on the timing of drought relative to key dates in the Hale double sunspot and lunar nodal cycles. Marked on the plot are two types of key dates. First are alternate minima in the double sunspot cycle, following Mitchell *et al.*'s (1979) observation that western U.S. drought tends to peak about two years after alternate minima in the Hale series. Second are maxima in the lunar nodal series, following Currie's (1984) theory of phase locking of that series with western U.S. drought.

When the overall record is considered, no regular rhythm is evident in the drought recurrence intervals shown on Figure 16.4. The strongest hints of any drought recurrence near the double sunspot cycle are seen in the most recent 80-year record; if the isolated drought-year 1925 is ignored, the series contains a modern sequence of intervals between widespread drought of 18, 23 and 20 years. The actual precipitation record shown at the bottom of Figure 16.4 also supports such a rhythm in the 20th century, and precipitation fluctuations have been cited as possible indicators of a solar-cycle link to Plains climate (Thomas 1962; Borchert 1971; Perry 1980). However, evidence for any such rhythm appears to vanish in the years before the 1890s, when longer-wavelength variations dominate (Figure 16.4).

Figure 16.4 illustrates both a remarkably good agreement between widespread drought and sunspot minima in the past 100 years, and an equally poor agreement in previous centuries. Drought peaks in the 1950s, 1930s and near 1911 all fall within ±2 years of sunspot minima, while the earlier drought of the 1890s reached peak intensity only 4 years after the sunspot minima. The major extended drought of the late 1850s and early 1860s, however, was initiated and ran its course before occurrence of the sunspot minimum in 1867. Before 1850, major differences in the timing of sunspot minima and drought maxima are shown, most notably the absence of drought within ±10 years of the solar minima at 1775. Kelly (1979) in discussing spectra of 101 years of North Atlantic winter-mean sea-level-pressure data, noted that evidence for a solar-climate link was stronger after 1920 than before. Two possible explanations advanced by Kelly (1979) were that (1) effects of volcanic dust-loading of the stratosphere may have obscured any solar influence before 1920, and (2) the amplitude of the sunspot cycle was relatively strong after 1920. The lack of a strong phase-locking of drought with sunspots in Figure 16.4 at an earlier time of high-amplitude sunspot activity in the late 1700s (Eddy 1977) would appear to argue against the latter explanation.

The 18.6-year lunar nodal cycle also does not bear any consistent phase relationship with widespread drought as tabulated in Figure 16.4. A link might certainly be suspected from the 20th century record alone; droughts in the 1930s and 1950s both fall near lunar nodal maxima. But in earlier years several lunar maxima occur without widespread drought, and several major droughts occur without lunar maxima. The major prolonged drought of the 1860s is centered on a lunar maximum, but the drought of the 1750s is not. Some examples of lunar maxima without widespread Plains drought are 1768, 1787 and 1843.

It is likely that even if there were a solar or lunar cyclic influence on drought near 20 years, the manifestation in tree-ring or climate records would be muddled by other influences, such as the state of hemispheric sea-surface temperatures or ice and snow cover. The possibility of a weak solar/lunar related rhythm cannot therefore be rejected solely on evidence of visual examination of a time series plot such as Figure 16.4. The possibility of more complex lunar/solar influences not involving a regular phase-locking of drought with extra-terrestrial series can also not be ruled out. For example, Currie (1984) hypothesizes a "bistable flip-flop" in the phase relationship of drought to the lunar nodal series, and Bell (1981) envisions a beat phenomenon between the double-sunspot and the 18.6 year lunar nodal series to explain some features of Mitchell *et al.*'s (1979) drought-area index.

16.4.6   *Extension in time beyond A.D. 1750*

Although the availability of suitable tree-ring sites in and around the Plains restricts the period for studying widespread drought to the period beginning about the middle of the 18th century, suitable tree-ring data for specific regions are available for perhaps another 100 years. Precipitation reconstructions have been published to A.D. 1680 for Iowa (Blasing and Duvick 1984) and to the 1640s for the northwestern Plains (Meko 1982). Stockton and Meko (1983) describe regional precipitation reconstructions for four regions in or near the Plains back to A.D. 1700. The time series plots of those reconstructions are shown in Figure 16.5. These plots differ from the original versions in Stockton and Meko (1983) in showing the full lengths of the series, rather than just the post-1700 data, and in showing low-pass filtered as well as unfiltered series.

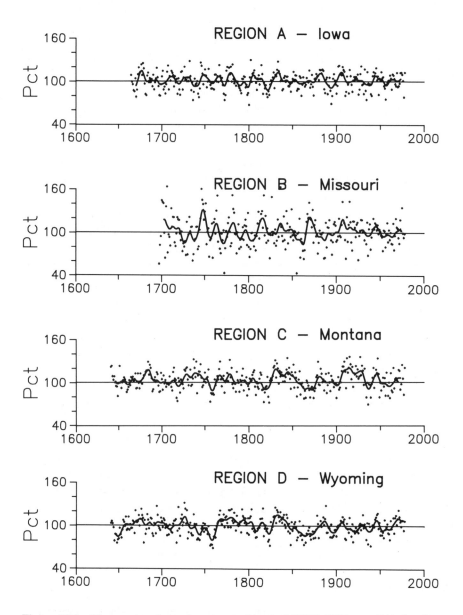

Figure 16.5 Time series plots of per cent of normal (1933-1977) precipitation for four Plains regions as reconstructed by Stockton and Meko (1983). Series smoothed by an 11-weight raised-cosine filter are also plotted.

326

A comparison of the plots in Figure 16.5 with the coded-letter plot of Figure 16.4 shows general agreement in delineating periods of extended widespread drought in the post-1750 period. This should be so, since the studies are based partly on information from the same tree-ring sites. The smoothed plots for series C and D in Figure 16.5 suggest an absence of recurrent or long-lasting widespread drought in the years 1641-1750, although a localized drought apparently affected region D around 1650. Stockton and Meko (1983) noted that longer-wavelength features (55-70 years) were more characteristic of the western regions (C and D) than the eastern (A and B). This difference shows clearly in a visual comparison of the smoothed plots in Figure 16.5. Troughs near 1750, 1820, 1870, and 1930 anchor the long-wave variations most clearly in the Montana region (region C). The smoothed plots for the two eastern regions show much greater relative variance at wavelengths shorter than about 30 years. More research is needed before the observed spatial differences in variance spectra can be interpreted as a features of climate. The operation of removing the age trend from ring-widths is itself a form of high-pass filtering, and this operation has not been carried out by uniform methods in the development of chronologies in the various regions. Lower frequencies in the chronologies in region 2, for example, very likely have been damped by detrending with relatively flexible spline functions (Meko *et al.*, 1985) a procedure not used on most of the western chronologies.

16.5 Summary

The history of widespread drought in the U.S. Plains, A.D. 1750-1964, as inferred from this study is characterized by decadal-scale and longer fluctuations not represented in the modern period covered by instrumental weather data. The historic "dust bowl" droughts of the 1930s and 1950s, while unique in the long-term record for the extent of spatial coverage of drought in single years, were less persistent or recurrent over several years than some earlier droughts. No sophisticated statistical search for periodicity has been attempted here, but visual examination of time series plots clearly indicates a lack of any strong, regular rhythm in areal drought coverage in the Plains to A.D. 1750. No evidence was found for phase-locking of Plains drought with either the 22-year double-sunspot or the 18.6-year lunar-nodal cycles.

Comparison of drought series derived by similar procedures for both tree-ring and precipitation data for 1904-1964 verifies that the tree-ring picture of drought is reasonably accurate in the current century. The long-term inferred drought history could possibly be distorted by trends in time-series properties of tree-ring data unrelated to climate. Blasing *et al.* (1988) for example, have indicated that non-climatic trends in variance may still be a problem in tree-ring data even after conventional steps have been taken to remove the growth-curve.

A gradient in strength of the signal for regional annual precipitation over the Plains was found, with the signal becoming weaker toward the far north. Additional field collections are needed to determine whether this gradient is characteristic of Plains tree-ring data, or is merely due to the paucity of modern collections in the north.

References

Bell, E. P. 1981. The combined solar and tidal influence on climate. In: *Variations of the Solar Constant* S. Sofia (ed.) 241-264. Washington, D.C.: NASA.

Blasing, T. J. and D. N. Duvick 1984. Reconstruction of precipitation history in North American corn belt using tree rings. *Nature* 307, 143-145.

Blasing, T. J., D. W. Stahle and D. N. Duvick 1988. Tree-ring based reconstruction of annual precipitation in the south-central United States from 1750 to 1980. *Water Resources Research* 24 (1) 163-171.

Borchert, J. R. 1950. The climate of the central North American grassland. *Annals of the Association of American Geographers* XL (1) 1-38.

Borchert, J. R. 1971. The dust bowl in the 1970s. *Annals of the Association of American Geographers* 61 (1) 1-22.

Brier, G. W. 1961. Some statistical aspects of long-term fluctuations in solar and atmospheric phenomena. In *Solar variations, climatic change, and related geophysical problems*, F. N. Furness (ed.) 173-187. New York Academy of Sciences.

Campbell, T. N. 1949. The pioneer tree-ring work of Jacob Keuchler. *Tree-ring Bulletin* 15, 16-20.

Currie, R. G. 1981. Evidence for 18.6-year Mn signal in temperature and drought conditions in North America since A.D. 1800. *Journal of Geophysical Research* 86 (C11) 11055-11064.

Currie, R. G. 1984. Periodic (18.6-year) and cyclic (11-year) induced drought and flood in western North America. *Journal of Geophysical Research* 89 (D5) 7215-7230.

Currie, R. G. 1989. Comments on 'Power spectra and coherence of drought in the interior plains by E.O. Oladipo' *Journal of Climatology* 9, 91-100.

Diaz, H. F. 1981. Eigenvector analysis of seasonal temperature, precipitation and synoptic-scale system frequency over the contiguous United States. Part II: spring, summer, fall, and annual. *Monthly Weather Review* 109, 1285-1304.

Douglass, A. E. 1936. *Climatic cycles and tree growth*. Carnegie Institute of Washington Publication No. 289, Vol III.

Eddy, J. A. 1977. Historical evidence for the existence of the solar cycle. In: *The solar output and its variation*, O. R. White (ed.) 51-71. Boulder, Colorado: Colorado Associated University Press.

Fritts, H. C. 1976. *Tree rings and climate*. London: Academic Press.

Harper, H. 1961. Drought in central Oklahoma from 1710 to 1959 calculated from annual rings of post oak trees. *Proceedings of the Oklahoma Academy of Science* 41, 23-29.

Hawley, F. 1941. Tree-ring analysis and dating in the Mississippi drainage. *University of Chicago Publications in Anthropology Occasional Papers* no. 2.

Karl, T. R. and A. I. Koscielny 1981. Drought in the United States: 1895-1981. *Journal of Climatology* 1, 1-16.

Karl, T. R. and W. E. Riebsame 1984. The identification of 10- to 20-year temperature and precipitation fluctuations in the contiguous United States. *Journal of Climate and Applied Meteorology* 23, 950-966.

Kelly, P. M. 1979. Solar influence on North Atlantic mean sea level pressure. In: *Solar-terrestrial influences on weather and climate*, B. M. McCormac and T. A. Seliga (eds.) 297-298. Dordrecht, Holland: D. Reidel Publishing Company.

Kutzbach, J. E. 1967. Empirical eigenvectors of sea-level pressure, surface temperature and precipitation complexes over North America. *Journal of Applied Meteorology* 6, 791-802.

LaMarche, V. C., Jr. and H. C. Fritts 1972. Tree-rings and sunspot numbers. *Tree-ring bulletin* 32, 19-33.

Lawson, M. P. 1974. The climate of the Great American Desert: reconstruction of the climate of western interior United States, 1800-1850. *University of Nebraska studies*: new series no. 46. Lincoln: University of Nebraska.

Meko, D. M. 1982. Drought history of the western Great Plains from tree rings. *Proceedings of the International Symposium on Hydrometeorology*, A. I. Johnson and R. A. Clark (eds.) 321-326.

Meko, D. M., C. W. Stockton and T. J. Blasing 1985. Periodicity in tree rings from the corn belt. *Science* 229, 381-384.

Mitchell, J. M., Jr., C. W. Stockton and D. M. Meko 1979. Evidence of a 22-year rhythm of drought in the western United States related to the Hale solar cycle since the 17th century. In: *Solar-terrestrial influences on weather and climate*, B. M. McCormac and T. A. Seliga (eds.) 125-143. Dordrecht, Holland: D. Reidel Publishing Company.

Namias, J. 1955. Some meteorological aspects of drought, with special reference to the summers of 1952-54 over the United States. *Monthly Weather Review* 83, 199-205.

Namias, J. 1960. Factors in the initiation, perpetuation and termination of drought. In *Publication no. 51 of the I.A.S.H. Commission of Surface Waters*, 81-94. (Also can be found in *Short period climatic variations*, collected works of J. Namias, volume I, 410-423, published in 1975 by the University of California, San Diego).

Namias, J. 1982. Anatomy of Great Plains protracted heat waves (especially the 1980 U.S. summer drought). *Monthly Weather Review* 110, 824-838.

Oladipo, E. O. 1987. Power spectra and coherence of drought in the interior plains. *Journal of Climatology*, 7, 477-491.

Perry, C. A. 1980. Preliminary analysis of regional-precipitation periodicity. *U.S. Geological Survey Water-Resources Investigations* 80-74.

Pittock, A. B. 1983. Solar variability, weather, and climate: an update. *Quarterly Journal of the Royal Meteorological Society* 109, 23-55.

Quinlan, F. T., T. R. Karl and C. N. Williams, Jr. 1987. *United States historical climatology network (HCN) serial temperature and precipitation data*. CDIAC Numeric Data Collection, NDP-019. Oak Ridge, Tennessee: Oak Ridge National Laboratory.

Reed, T. R. 1933. The North American high-level anticyclone. *Monthly Weather Review* 61 (11) 321-325.

Richman, M. B. 1981. Obliquely rotated principal components: an improved meteorological map typing technique? *Journal of Applied Meteorology* 20, 1145-1159.

Schulman, E. 1956. *Dendroclimatic changes in semi-arid America*. Tucson: University of Arizona Press.

Sellers, W. D. 1968. Climatology of monthly precipitation patterns in the western United States, 1931-1966. *Monthly Weather Review* 96 (9) 585-595.

Siscoe, G. L. 1978. Solar-terrestrial influences on weather and climate. *Nature* 276, 348-352.

Stockton, C. W. and D. M. Meko 1975. A long-term history of drought occurrence in western United States as inferred from tree rings. *Weatherwise* 28 (6) 245-249.

Stockton, C. W. and D. M. Meko 1983. Drought Recurrence in the Great Plains as reconstructed from long-term tree-ring records. *Journal of Climate and Applied Meteorology* 22 (1) 17-29.

Stockton, C. W., W. R. Boggess, and D. M. Meko 1985. Climate and tree rings. In *Paleoclimate analysis and modeling*, Alan D. Hecht (ed.) 71-162. New York: John Wiley and Sons.

Thomas, H. E. 1962. The meteorologic phenomenon of drought in the Southwest. *U.S. Geological Survey Professional Paper 342A*. Washington.

Weakly, H. E. 1943. A tree-ring record of precipitation in western Nebraska. *Journal of Forestry* 41, 816-819.

Will, G. 1946. Tree-ring studies in North Dakota. *North Dakota Agricultural Experiment Station Bulletin* no. 338. Fargo, North Dakota.

17 Dendroclimatic evidence from eastern North America

E. R. Cook, D. W. Stahle and M. K. Cleaveland

17.1 Introduction

Recent severe droughts in the southeastern and central United States (Bergman *et al.* 1986; Trenberth *et al.* 1988) have emphasized, once again, that a tenuous reliance of the United States on precipitation exists even for non-arid regions, because sufficient water is always required to support the needs of crops, forests, industries, and municipalities. Although technology and conservation can ameliorate some of the impact of drought on society, the need still exists to better understand the climate dynamics that determine the frequency, intensity, and duration of drought in a given region. This need is being driven, in part, by the desire for better long-range climate predictions that would be useful for planning purposes. In addition, there is the urgent need to develop a better theoretical understanding of how the climate system operates on time scales of years to centuries because of the anticipated impact of "greenhouse" warming on global and regional climates (NOAA 1989).

Karl and Koscielny (1983) studied the temporal and spatial patterns of drought in the U.S. for the period 1895-1981. Using principal components analysis (PCA) with analytical rotation, they identified nine spatial patterns of drought occurrence in a 60-point grid of monthly Palmer Drought Severity Indices (PDSI) (Palmer 1965). Each pattern reflected a different monthly precipitation climatology. In addition, Karl and Koscielny (1983) showed that the average duration of droughts was longer in the interior of the U.S. than for areas closer to the east and west coasts. However, they were unable to identify any statistically significant drought cycles or quasi-periodic rhythms in the rotated factor scores, even for regions where such behavior had been found in previous studies (Cook and Jacoby 1979; Mitchell *et al.* 1979; Currie 1981a,b). This result, and more recent Monte Carlo studies by Karl (1988) suggests that multi-year fluctuations of climate in the U.S. are primarily driven by internal stochastic processes of the atmosphere-ocean-cryosphere system that do not necessarily require any external mechanisms, such as solar (Mitchell *et al.* 1979; Currie 1981a; Labitzke 1987) or lunar tidal forcing (Currie 1981b).

A major difficulty in using the available meteorological records to model multi-year fluctuations of drought and wetness is the limited time span covered by such records. For example, Rodriguez-Iturbe (1969) showed that for a 40-year annual streamflow record the estimate of lag-1 autocorrelation could be in error by as much as 200%. Error this high is probably due to the inadequacy of short records in estimating the low-frequency persistence of climate (Stockton and Boggess 1980) and possibly short-period changes in persistence.

Insufficient time-series length affects statistical model development in other ways. If the characteristic duration of climatic fluctuations is on the order of 10-20 years (e.g. Karl and Riebsame 1984) then the number of "events" that are available for modeling persistence in the climate records will only be a small fraction of the total number of observations. This condition severely limits the statistical power of hypothesis tests used to determine the cause(s) of the observed patterns of climate.

Finally, there is often an implicit assumption that the climate records being analyzed are unbiased realizations of some underlying homogeneous stochastic process. This assumption is not easily testable given the shortness of these records. It is possible that meteorological records in the 20th century have been biased by natural and anthropogenic effects that differed in level and/or intensity in the past. Explosive volcanic eruptions (Lamb 1970; Bradley 1988) and greenhouse gases (Neftel *et al.* 1985) are two examples of possible influences on climate that are known to have varied through time. Thus, with only relatively short meteorological records, there is some danger that an incomplete or incorrect model will be developed for explaining the causes of drought in the United States.

One way to alleviate the statistical modeling limitations associated with short meteorological records is to augment these records with longer, high-quality proxy records of comparable resolution. Tree rings are ideal for this purpose. Well-dated, climatically sensitive tree-ring chronologies from old-age trees have precise annual resolution and time spans covering the past several hundred to a thousand or more years. In addition, such tree-ring chronologies are widely available throughout the temperate and subpolar latitudes of the world, making possible the comparative study of regional climatic variability. See Fritts (1976) Hughes *et al.* (1982) Cook and Kairiukstis (1990) and several companion chapters in this book for comprehensive reviews.

In this paper, we show some results of reconstructing past drought from tree rings in the eastern United States. These reconstructions approximate the coverage and spatial variance of meteorological records used by Karl and Koscielny (1983) in that region, but extend back annually to A.D. 1700 and significantly augment the sampling of processes that drive climate variability in the eastern United States.

17.2   The tree-ring data network

The tree-ring chronology network used in this study is comprised of over 150 chronologies that cover most of the eastern United States and part of southeastern Canada (Figure 17.1). This network was developed principally by dendrochronologists at Lamont-Doherty Geological Observatory, the University of Arkansas, and Oak Ridge National Laboratory for the purpose of climatic reconstruction. A brief description of methods of tree-ring chronology development is given in Chapter 1 of this book, with more complete coverage of this topic found in Fritts (1976) Cook (1987) and Cook and Briffa (1990).

Twelve tree species from the following genera comprise the network: *Tsuga* (hemlock) *Quercus* (oak) *Taxodium* (cypress) *Pinus* (pine) *Picea* (spruce) and *Juniperus* (juniper). The distributions of genera and species in the network is uneven due to the natural limits of tree species in eastern North America (Braun 1950). Consequently, hemlock predominates in the northern and eastern zones, while oak and cypress are more common in the southern and western zones. The majority of tree-ring sites are upland and well-drained, with the special exception of the lowland, riverine bald cypress sites. Most of the tree-ring chronologies showed some sensitivity to growing season drought in preliminary dendroclimatic modeling

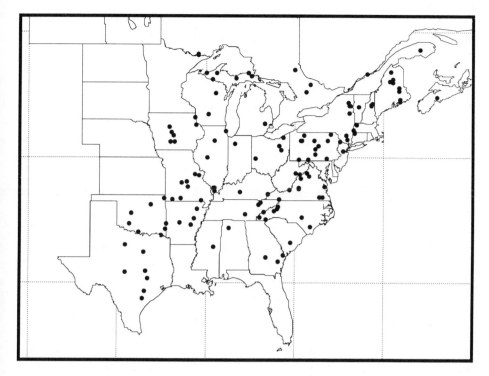

Figure 17.1 The network of tree-ring sites in eastern North America with chronologies that extend back to 1700 and earlier. The drought reconstructions reported on here were developed from a large subset of these sites.

and were subsequently used in the PDSI reconstructions reported on here. However, the *Picea* chronologies were universally rejected for failing to correlate with any measure of growing season moisture supply.

17.3 The drought index data

Past experience in reconstructing climate from tree rings in eastern North America suggests that some measure of soil moisture availability is likely to be well correlated with the tree rings (e.g. Cook and Jacoby 1977 1983; Blasing and Duvick 1984; Blasing *et al.* 1988; Stahle and Cleaveland 1988; Stahle *et al.* 1985 1988). For this reason, the Palmer Drought Severity Index (PDSI; Palmer 1965) was chosen as the climatic parameter to reconstruct. The successful reconstruction of PDSI from tree rings (Stockton and Meko 1975; Cook and Jacoby 1977; Cook *et al.* 1988; Stahle *et al.* 1985 1988) indicates that the PDSI is a realistic integrator of climate as it affects soil moisture availability and ultimately tree growth. In addition, the PDSI is a regionally independent measure of drought and wetness based on departures from regional means (Palmer 1965) which allows its use in regional studies crossing different climatic regimes.

There is no question that other climatic variables unrelated to growing season soil moisture

333

supply also influence tree growth. For example, Cook and Cole (1989; in press) described the strong positive influence of warm March temperatures on the subsequent growth of eastern hemlock throughout most of its range. Also, Cook *et al.* (1987) showed that red spruce growth is positively correlated with December and January temperatures. These dendroclimatic signals would be antagonistic to the successful reconstruction of past growing season drought if left unattenuated. This problem can be ameliorated, however, by carefully screening the tree-ring chronologies for the desired climatic signal (PDSI in this case) and by using multiple species with somewhat different "climatic windows" (Fritts 1976) in the reconstruction of past drought. Both methods of drought signal enhancement were used in this study.

The monthly PDSI data used in this study were computed from state average temperature and precipitation records spanning 1895-1983. These data were obtained from the National Climatic Data Center in Asheville, North Carolina and were used in gridded form in the Karl and Koscielny (1983) drought study. In addition, monthly PDSI's for southern Ontario, Canada were computed from regionally averaged meteorological records obtained from the Northern Hemisphere land areas climatic data bank (Bradley *et al.* 1985).

Due to large variations in state areas, areally-weighted averages of PDSI were computed for some smaller adjoining states to provide a more spatially homogeneous climatic network. Those states included in separate regional averages were Vermont-New Hampshire, Massachusetts-Connecticut-Rhode Island, and New Jersey-Maryland-Delaware. In addition, Michigan was split up into Upper and Lower Peninsulas and Texas into North Central and South Central regions (Stahle and Cleaveland 1988). As a result, drought reconstructions were made for 24 states or regions in the eastern United States and Ontario. These regions are specified in Table 17.1. However, one of these regions (Maine) was not used in our summaries because of poor reconstruction quality.

The regional nature of drought over the United States has been classified into 9 geographic factor patterns by Karl and Koscielny (1983). Based on those results, the 24 spatial units used in this study are largely covered by 5 of those factor patterns: the Central, Northeast, Southeast, East North Central, and South regions. Therefore, our drought reconstructions cover more than half of the 9 major drought regions in the United States.

Experience indicates that regional-average climatic series should be easier to reconstruct and verify than single-station data (Blasing *et al.* 1981). Presumably, this is because single-station records, with their unique microclimates and recording histories, are frequently noisier measures of climate than an average of several individual records from a given climatic division, state, or region. Perhaps as importantly, a regional climatic series will always have less variance than that in the individual records due to the averaging process. This means that there is simply less total variance to estimate and, therefore, less potential error variance in the regression model.

17.4   The method of reconstruction

PDSI was reconstructed from tree rings for each state or region through a series of 24 separate multiple linear regression analyses. In many cases, adjacent states used some (but never all) of the same tree-ring chronologies to reconstruct drought. This was necessitated by the

Table 17.1 Calibration and verification results of the drought reconstruction models. The fractional explained variance (R^2) and R^2 adjusted for lost degrees-of-freedom (R^2_a) are given for each calibration equation. The verification tests are the product-moment correlation coefficient (r) and the reduction of error (RE) statistic. All models pass the verification tests ($p < 0.05$ and $RE > 0$).

Calibration	Verification			
State or Region	R^2	R^2_a	r	RE
Arkansas	.469	.457	.658	.413
Georgia	.667	.659	.603	.300
Illinois	.539	.527	.773	.540
Indiana	.687	.671	.763	.514
Kentucky	.521	.496	.738	.541
Maine*	.066	.042	.059	-.028
Massachusetts-Connecticut-Rhode Island	.431	.401	.443	.170
Michigan (Lower Peninsula)	.220	.200	.698	.480
Michigan (Upper Peninsula)	.202	.181	.361	.046
Missouri	.604	.596	.624	.380
New Hampshire-Vermont	.361	.309	.359	.119
New Jersey-Maryland-Delaware	.493	.466	.662	.419
New York	.527	.502	.683	.390
North Carolina	.428	.398	.666	.414
Ohio	.450	.421	.674	.452
Southern Ontario	.356	.323	.517	.218
Pennsylvania	.454	.426	.682	.431
South Carolina	.452	.440	.601	.333
Tennessee	.612	.591	.654	.414
Texas (North)	.597	.589	.735	.538
Texas (South)	.613	.597	.656	.423
Virginia	.335	.318	.591	.345
West Virginia	.455	.426	.604	.334
Wisconsin	.467	.439	.630	.344
Mean	**.475**	**.454**	**.625**	**.372**
Standard Deviation	**.127**	**.132**	**.112**	**.132**

*Maine has not been used in computing the average statistics

uneven distribution of tree-ring sites over the network. Eighteen of the 24 states or regions were reconstructed by E. R. Cook. The remaining 6 (South Carolina, Georgia, Missouri, Arkansas, and North and South Texas) were independently reconstructed by D. W. Stahle and M. K. Cleaveland. The Cook reconstructions (excluding North Carolina) used principle components (PC) regression analysis (Draper and Smith 1981). This was done by applying principal components analysis (Jolliffe 1986) to the correlation matrix of the intercorrelated tree-ring predictor variables. The principal component scores for those PCs with eigenvalues greater then 1.0 were than used as the new set of orthogonal predictors of PDSI. See Fritts (1976) for an in-depth description of this procedure as applied to dendroclimatology. In contrast, the North Carolina and 6 Stahle-Cleaveland reconstructions used the untransformed tree-ring chronologies as predictors in stepwise regression analysis.

These two groups of PDSI reconstructions do differ in one notable way. The Cook reconstructions were for summer (June-July-August) average PDSI while the Stahle-Cleaveland reconstructions were for June PDSI only. In an unpublished study, Cook (n.d.) found that the large north-south range covered by his chronologies (i.e. Maine to North Carolina) resulted in a clear geographic gradient in the timing of the maximum response of those chronologies to drought. For example, tree-ring chronologies from sites south of Pennsylvania typically correlated most strongly with June PDSI while those north of Virginia correlated better with July or, less frequently, August PDSI. For this reason, Cook (n.d.) chose to reconstruct June-July-August average PDSI because it provided the best average results over his 18 states or regions. In contrast, the Stahle-Cleaveland June PDSI reconstructions were for states located roughly within the same latitude band where June PDSI always out-performed July and August PDSI as a correlate with tree rings. Phenological differences associated with the northward march of the growing season appear to be responsible for these differences in the monthly climate response of tree-ring chronologies located from southern Texas to New England. It is also possible that shorter summers (and lower heat sums) at higher latitudes delay the drying of the soil and, hence, the time when soil moisture would be limiting to growth. Although this difference in the two groups of reconstructions is not ideal, the high lag 1 autocorrelation in monthly PDSIs (typically between 0.8 and 0.9) makes them quite comparable. As will be seen in the factor analyses of these reconstructions, it is unlikely that this difference has introduced any significant bias into our analyses.

Each of the reconstructions was developed using the standard procedures of calibration and verification (Fritts 1976; Fritts and Guiot 1990). The calibration period used to develop the multiple linear regression equations relating tree rings to PDSI was based on the most recent 41 years of tree-ring and PDSI data in the Cook reconstructions and the most recent 50 years in the Stahle-Cleaveland reconstructions. However, the exact time periods used for calibration varied by a few years because of differences in when the tree-ring chronologies were developed. The remaining PDSI data not used for calibration purposes were used to verify the accuracy of the tree-ring estimates.

Considerable care was taken to avoid spurious regression relationships between tree rings and PDSI due to autocorrelation in the time series (Mitchell *et al*. 1966; Fritts 1976). This was accomplished by modeling and prewhitening the tree-ring and climate series as autoregressive processes (Box and Jenkins 1970; Meko 1981) prior to developing the calibration equations for reconstructing PDSI. In addition, care was taken to properly model any lead-lag relationships between tree rings and climate (Fritts 1976; Meko 1981) in developing the statistical transfer functions for reconstructing past drought. Then, following the method of Meko (1981) any autoregressive persistence found in the original PDSI data was added back to the prewhitened tree-ring reconstruction to reproduce the "red noise" (*sensu* Gilman *et al*. 1963) seen in the climate series.

Two verification statistics are presented here that were common to all of the reconstructions: the product-moment correlation coefficient and the reduction of error statistic. Each statistic is commonly used in dendroclimatic reconstructions. The product-moment correlation coefficient (r) is a parametric measure of association between two samples. Its use in testing for hypothesized relationships between variables is described in virtually all basic statistics texts and in Fritts (1976). The reduction of error (RE) statistic is less well known. It was developed in meteorology by Lorenz (1956) for the purpose of assessing the predictive skill of meteorological forecasts. The RE has no formal significance test, but an RE>0 is an indicator of forecast skill that exceeds that of climatology (i.e. extrapolating the the climatic mean as the forecast or prediction). See Fritts (1976) Gordon and LeDuc (1981) and Fritts and Guiot (1990) for full descriptions of this statistic, its small sample properties, and other verification tests as well.

Table 17.1 lists the calibration and verification results for 24 separate drought reconstruction models. Only the Maine reconstruction failed to verify and consequently will be eliminated from all subsequent analyses. The average (minus Maine) fractional PDSI variance (R^2) explained by tree rings and the R^2 adjusted for lost degrees-of-freedom (R^2_a) are 0.475 and 0.454, respectively. Thus, on average almost half of the PDSI variance in the calibration period was explained by the tree rings. More importantly, all but Maine passed the verification tests (p<0.05 for r and RE>0) with a mean r of 0.625 and a mean RE of 0.372 (again minus Maine). Insofar as the RE is analogous to an R^2 statistic and also measures the predictive capability of the regression equations when applied to new data, the modest difference between average R^2_a and average RE indicates that the reconstructions are of good quality on average. However, note that the verifications tend to decline in quality from south to north, especially for the upper peninsula of Michigan, southern Ontario, and the New England states. At these more northern locations, the shorter growing season coupled with generally cooler temperatures and less evapotranspiration demand apparently makes soil moisture availability less limiting to growth, thus reducing the quality of these drought reconstructions. Given this caveat, we will nonetheless proceed on the premise that the 23 verified drought reconstructions are acceptably accurate estimates of the timing, intensity, and duration of drought in the eastern United States back to 1700.

17.5   Some spatial properties of the drought reconstructions

The 23 verified PDSI reconstructions were collectively analyzed using PCA with analytical rotation for the period 1700-1972 (Richman 1986). This was done to distill the drought reconstructions into a smaller set of homogeneous spatial drought patterns that would reflect different climatic regions in the eastern U.S. (see Karl and Koscielny 1983). Although PCA alone has been used in the past to characterize the principal modes of behavior in the climate system (e.g. Kutzbach 1967; Sellers 1968; Briffa 1984) and in tree-ring networks (e.g. LaMarche and Fritts 1971; Fritts *et al.* 1971; Briffa *et al.* 1983; Briffa 1984) Richman (1986) has shown that the mathematical constraints imposed on the unrotated eigenvectors can severely distort the spatial patterns extracted from the data. Consequently, there may be little physical reality in unrotated climatic eigenvectors. Oblique rotation can effectively negate some of these constraints (especially orthogonality) so that rotated eigenvectors often express more realistic patterns.

For determining how many eigenvectors to rotate, we chose the eigenvalue-1 cutoff criterion (Kaiser 1960; Jolliffe 1986) whereby only eigenvectors having eigenvalues greater then 1.0 are retained for further analysis. Use of this criterion resulted in retaining the first six eigenvectors, which accounted for 78% of the total variance in the original data. The six retained eigenvectors were obliquely rotated using the Harris-Kaiser method (Harris and Kaiser 1964) to remove the orthogonality constraints on the original eigenvectors and make them more physically interpretable (Richman 1986). In addition, Cook (n.d.) found that oblique rotation produced reconstructed drought factors that were more congruent with similarly rotated actual drought factors, as opposed to factors derived from PCA or PCA with Varimax rotation that retains orthogonality.

Figure 17.2 shows the six reconstructed drought factor patterns with factor loadings ≥0.70. Given that the drought factors were derived from the correlation matrix, the factor loadings can be thought of as correlations between each area's PDSI record and the six drought factors. Therefore, areas are shaded where ≥49% of the variance is explained by each drought factor. We note that there is considerable agreement between the geographic placements of our drought factors and five of those identified by Karl and Koscielny (1983). For example, factors 2 to 5 (Figure 17.2) correspond well with the Northeast, Southeast, South, and East North Central factors of Karl and Koscielny (1983, Figure 17.5). Given that we did not have data for some of the states used by Karl and Koscielny (1983) to develop their factor patterns, this level of agreement is excellent and probably reflects the sub-domain stability of analytically rotated eigenvectors (Richman 1986). In contrast, factor 1 and factor 6 appear to have split Karl and Koscielny's Central drought factor. It is difficult to determine if this apparent splitting is real or an artifact of rotating too many PCs. As a test, we performed PCA on our PDSI reconstructions for the 1895-1972 period (comparable to Karl and Koscielny's 1895-1981 time period). Only five PCs had eigenvalues exceeded 1.0. After rotating those PCs, factors 1 and 6 coalesced into a factor pattern extremely similar to Karl and Koscielny's Central drought factor, while the other factors remained almost the same as seen in Figure 17.2. This suggests that only five PCs should be rotated. However, when we performed PCA on the 1700-1894 time period, six factors were even more strongly indicated by the eigenvalue trace and the resultant rotated factor patterns looked virtually identical to those shown in Figure 17.2. Therefore, it appears that the suggested need to rotate six PCs based on the eigenvalue-1 criterion is largely due to patterns in the data prior to 1900. We do not know if this represents a real difference in the drought climatology of the eastern United States prior to 1900 or is a statistical artifact in the reconstructions. Therefore, we will assume that the six drought factors shown in Figure 17.2 are "true", with the proviso that factors 1 and 6 may be manifestations of the same underlying factor.

Table 17.2 shows the correlation matrix of the six factor patterns. As expected, the obliquely rotated factors are intercorrelated, with the highest correlation of 0.53 between factors 1 and 2. These correlations are, in general, somewhat higher then those presented by Karl and Koscielny (1983, Table 1) for their obliquely rotated factors. This is probably due to the fact that we did not always use unique sets of tree-ring chronologies in each PDSI reconstruction and the regression models did not explain all of the original PDSI variance. Regardless, there is little indication for significant predictability of drought in one area over the long-term, given drought occurrence elsewhere in the network. Given the similarity of these results with those from actual climatic data (Karl and Koscielny 1983) this is probably a true feature of the climate system in this part of eastern North America. However, it is also possible that by starting with a variance-maximizing sub-set of orthogonal factors (i.e. the

Figure 17.2 The six obliquely rotated drought factors. Those areas with factor loadings greater ≥0.70 have been shaded.

eigenvectors) prior to rotation, there is a bias toward mutual independence even after oblique rotation. Investigating this possibility is beyond the scope of this paper.

The drought factor scores spanning 1700-1972 are shown in Figure 17.3. Each time series has superimposed upon it a low-pass filtered curve that highlights 10-year or longer fluctuations of drought and wetness in each region. It is apparent that fluctuations of wetness and

Table 17.2 The correlation matrix of the obliquely rotated factors (see Figure 17.2 and 17.3 for the spatial patterns of the factors and their respective scores).

	Factor 1	**Factor 2**	**Factor 3**	**Factor 4**	**Factor 5**	**Factor 6**
Factor 1	1.00	.53	.25	.00	.40	.42
Factor 2		1.00	.07	.00	.37	.18
Factor 3			1.00	.14	.00	.12
Factor 4				1.00	.06	.07
Factor 5					1.00	.41
Factor 6						1.00

dryness lasting five or more years are common within each area, but rarely occur simultaneously across many of the six regions. To illustrate this point better, the five driest and five wettest non-overlapping pentads within each of the factor scores were ranked. These ranked pentads are listed in Tables 17.3 and 17.4. The nine-year period spanning 1814-1822 is

Table 17.3 The 5 driest non-overlapping 5-year time periods in the drought factor scores. For each factor, the years are given, with the mean scores and their standard errors provided below. The mean scores are in standard normal deviates.

Factor	Rank 1	Rank 2	Rank 3	Rank 4	Rank 5
1	1835-1839	1963-1967	1803-1807	1746-1750	1770-1774
Mean	-1.38	-1.08	-1.02	-0.86	-0.86
S.E.	0.37	0.20	0.37	0.45	0.40
2	1962-1966	1769-1773	1818-1822	1910-1914	1795-1799
Mean	-1.71	-1.15	-1.13	-1.11	-1.11
S.E.	0.26	0.06	0.31	0.32	0.18
3	1816-1820	1746-1750	1877-1881	1951-1955	1894-1898
Mean	-1.05	-1.03	-0.95	-0.81	-0.80
S.E.	0.35	0.19	0.14	0.22	0.41
4	1952-1956	1859-1863	1727-1731	1772-1776	1786-1790
Mean	-1.27	-1.14	-1.06	-0.91	-0.90
S.E.	0.24	0.28	0.26	0.21	0.54
5	1818-1822	1932-1936	1870-1874	1808-1812	1770-1774
Mean	-1.60	-1.39	-1.27	-0.98	-0.97
S.E.	0.32	0.47	0.35	0.16	0.17
6	1734-1738	1799-1803	1814-1818	1870-1874	1930-1934
Mean	-1.09	-1.00	-0.98	-0.96	-0.91
S.E.	0.44	0.20	0.35	0.28	0.34

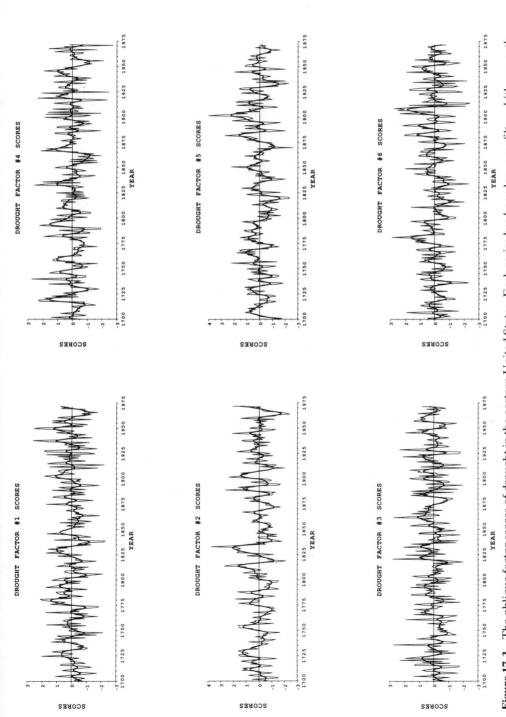

Figure 17.3 The oblique factor scores of drought in the eastern United States. Each series has been low-pass filtered (the smooth curve) to accentuate multi-year fluctuations of wetness and dryness.

noteworthy for having four of the first fifteen dry-ranked pentads (Table 17.3). Interestingly, this period of unusual drought is around the same time as the Tambora volcanic eruption in 1815 (Lamb 1970; Stothers 1984) which is believed to have affected climate over large areas of the North Atlantic sector. It also occurs at about the same time as an unusually strong El Nino event in 1815-16 reconstructed with tree rings from western North America (Lough and Fritts 1985). Historical data also indicates that a strong El Nino event began in 1814 (Quinn *et al.* 1978; Quinn and Neale, this volume). Another common period of drought is 1769-1776, which also has 4 of the driest pentads. For the wettest ranked pentads (Table 17.4) two time periods stand out as being commonly wetter-than-average: 1827-1837 with 4 of the wettest ranked pentads and 1901-1910 also with 4 ranked pentads. Beyond these particular wet and dry periods, it is difficult to see much commonality in the occurrence of multi-year periods of drought and wetness in different regions of eastern North America.

There is some limited evidence for an out-of-phase relation between low-frequency moisture anomalies reconstructed for the northern and southern sections of eastern North America. The 20th century drought episodes of the 1930s and 1950s were most intense over the northern and southern Great Plains, respectively (Warrick 1980) and the tree-ring

Table 17.4 The 5 wettest non-overlapping 5-year time periods in the drought factor scores. For each factor, the years are given, with the mean scores and their standard errors provided below. The mean scores are in standard normal deviates.

Factor	Rank 1	Rank 2	Rank 3	Rank 4	Rank 5
1	1777-1781	1827-1831	1947-1951	1901-1905	1792-1796
Mean	1.29	1.14	1.09	1.04	0.86
S.E.	0.33	0.23	0.56	0.35	0.30
2	1830-1834	1902-1906	1812-1816	1889-1893	1726-1730
Mean	2.05	1.48	1.46	1.31	1.27
S.E.	0.43	0.25	0.42	0.39	0.18
3	1821-1825	1859-1863	1832-1836	1957-1961	1721-1725
Mean	1.27	1.11	1.03	0.98	0.93
S.E.	0.50	0.47	0.34	0.17	0.60
4	1792-1796	1718-1722	1867-1871	1758-1762	1833-1837
Mean	1.54	1.48	1.14	1.12	1.12
S.E.	0.18	0.50	0.38	0.08	0.62
5	1901-1905	1881-1885	1912-1916	1765-1769	1779-1783
Mean	2.12	1.59	1.16	1.08	0.88
S.E.	0.62	0.34	0.40	0.32	0.26
6	1906-1910	1779-1783	1889-1893	1881-1885	1947-1951
Mean	1.80	1.58	1.18	1.09	0.97
S.E.	0.41	0.39	0.32	0.47	0.22

reconstructions for the Great Lakes and Texas indicate several similar latitudinal shifts in the regional focus of low-frequency moisture anomalies since 1700 (notably 1700-1740, 1855-1885 1920-1940, and after 1950; see Figure 17.3). However, this out-of-phase relation is not time stable, and the available tree-ring network may not adequately sample the geographic modes of this apparent regional variation in drought area.

17.6 Spectral properties of the drought reconstructions

Power spectra of the factor scores are described next in an attempt to characterize the frequency domain properties of drought within each area. Each spectrum was estimated from 54 lags of the autocovariance function and smoothed with the Hamming window (Jenkins and Watts 1968). The resultant spectral estimates have 12 degrees-of-freedom. In addition, a first-order Markov or "red noise" null continuum was estimated for each spectrum for the purpose of generating confidence limits for testing the statistical significance of spectral peaks (Mitchell *et al.* 1966).

Each power spectrum normalized to unit variance is shown in Figure 17.4, along with its null continuum (solid curve) and 95% confidence limit (dashed curve). For expository purposes, several of the more prominent peaks in each spectrum are identified even though some of them do not exceed the 95% confidence limits.

As has been found in other dendroclimatological studies (e.g. LaMarche and Fritts 1972) all of the spectra in Figure 17.4 exhibit varying degrees of Markovian "red noise" (Gilman *et al.* 1963) as evidenced by the rising null spectra in the lower frequencies. In addition, each spectrum has from 1 to 3 peaks that exceed the 95% confidence limit, a result that could be expected by chance alone in the absence of *a priori* hypotheses.

Drought factors 1, 4, 5, and 6, which are nearest to the Great Plains (see Figure 17.2) show some evidence of common spectral power in the 18 to 20-year bandwidth. This is consistent with results of Currie (1981b) who found evidence for the influence of the 18.6-year lunar nodal cycle on drought formation in the central United States. Although we do not see any clear evidence for the 22-year Hale sunspot cycle in our reconstructions, as found previously by Mitchell *et al.* (1979) we note that our region of study is largely on the eastern side of the Great Plains (where the 22-year cycle was identified). Therefore, our results do not neces-sarily refute the existence of the 22-year drought rhythm (cf. Meko, Chapter 16,this volume). Rather, it may be that solar forcing is not influential in our regions of study. It should also be noted that the 22-year drought rhythm identified by Mitchell *et al.* (1979) appears to operate as a spatially non-stationary and fragmented process, which makes its identification in any given sub-region of the Great Plains more difficult (e.g. Stockton and Meko 1983; Meko *et al.* 1985; Meko, this volume).

There is also some evidence for spectral power coincident with the 3 to 7-year bandwidth of the El Nino-Southern Oscillation (ENSO) (Quinn *et al.* 1978; Quinn and Neale, Chapter 32, this volume) and the 2 to 3-year bandwidth associated with the quasi-biennial oscillation (QBO). Both of these phenomena have been previously identified in meteorological data (e.g. Holton and Tan 1980; Ropelewski and Halpert 1986). However, it is not possible to say at this time if any of these spectral peaks in our drought reconstructions are related to ENSO or QBO forcing.

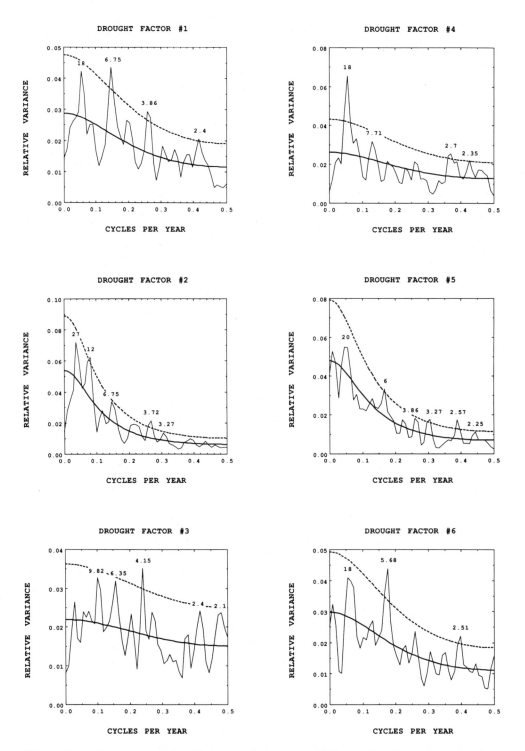

Figure 17.4 The power spectra of the factor scores shown in Figure 17.3. Each spectral estimate has approximately 12 degrees of freedom. The first-order Markov null spectrum (solid curve) and its upper 95% confidence limit (dashed curve) are superimposed on each spectrum for assessing the statistical signficance of the spectral peaks.

17.7   Conclusions

Long-term records of the Palmer Drought Severity Index for 23 states or regions in the eastern United States and southeastern Canada have been reconstructed from tree rings and verified against independent PDSI data. The temporal patterns of these 23 PDSI reconstructions can be reduced to five or six underlying drought factors that are qualitatively similar to PDSI patterns previously identified by Karl and Koscielny (1983) from actual data. This indicates that our reconstructions can be used to examine both temporal and spatial patterns of drought in eastern North America.

The reconstructions, which extend back to 1700, indicate that multi-year fluctuations of wetness and dryness are common in all of the regions of study. However, these fluctuations are rarely synchronous across all regions, and are at times strongly out-of-phase between northern and southern regions. The closest example of synchronous multi-year drought is in the 1814-1822 time period, which contains four of the fifteen driest pentads in the drought factor scores. This period is coincidental with the eruption of Tambora in 1815 (Lamb 1970; Stothers 1984) and strong ENSO events (Lough and Fritts 1985; Quinn *et al.* 1978; Quinn, this volume).

Spectral analysis of the drought factor scores indicates that there is weak quasi-periodic behavior in some of the regions that may be related to the 18.6-year lunar nodal cycle. This result is consistent with analyses of meteorological records by Currie (1981b). There is little evidence for the 22-year Hale sunspot cycle in our reconstructions. However, our region of study is mostly on the eastern side of the Great Plains region identified by Mitchell *et al.* (1979) as being influenced by the Hale cycle. Thus, we do not feel that this lack of confirmation refutes their findings. There is the suggestion of power in the bandwidths common to both ENSO events and the QBO. We have not directly tested for these associations, but believe that the well established existence of these atmospheric forcings in meteorological data is grounds for the further testing of our reconstructions.

Acknowledgements

This research was supported by the National Science Foundation, Division of Climate Dynamics, through grants ATM 87-16630 to Lamont-Doherty Geological Observatory, ATM 88-14675 to the University of Arizona, and ATM 86-12343 and ATM 89-14561 to the University of Arkansas. We thank Gordon Jacoby and Jon Overpeck for internal Lamont reviews and two anonymous external reviewers for critical reviews that improved the quality of this paper. Lamont-Doherty Geological Observatory Contribution No. 4621.

References

Bergman, K. H., C. F. Ropelewski and M. S. Halpert. 1986. The record southeast drought of 1986. *Weatherwise* 39:262-266.
Blasing, T. J., D. N. Duvick, and D. C. West. 1981. Dendroclimatic calibration and verification using regionally averaged and single station precipitation data. *Tree-Ring Bulletin* 41:37-44.

Blasing, T. J. and D. N. Duvick. 1984. Reconstruction of precipitation history in North American corn belt using tree rings. *Nature* 307:143-145.

Blasing, T. J., D. W. Stahle, and D. N. Duvick. 1988. Tree-ring based reconstruction of annual precipitation in the south-central United States from 1750 to 1980. *Water Resources Research* 24(1):163-171.

Box, G. E. P. and G. M. Jenkins. 1970. *Time Series Analysis: Forecasting and Control*. Holden-Day, San Francisco.

Bradley, R. S. 1988. The explosive volcanic eruption signal in Northern Hemisphere continental temperature records. *Climatic Change* 12:221-243.

Bradley, R. S., P. M. Kelly, P. D. Jones, H. F. Diaz and C. M. Goodess. 1985. *A Climatic Data Bank for Northern Hemisphere Land Areas, 1851-1980*. U.S. Dept. of Energy, Office of Energy Research, Washington, D.C. 335 pp.

Braun, E. L. 1950. *Deciduous Forests of Eastern North America*. Hafner Press, New York.

Briffa, K. R. 1984. *Tree-climate Relationships and Dendroclimatological Reconstruction in the British Isles*. Unpublished PhD dissertation, University of East Anglia, Norwich, U.K. 285 pp plus Appendices.

Briffa, K. R., P. D. Jones, T. M. L. Wigley, J. R. Pilcher, and M. G. L. Baillie. 1983. Climate reconstruction from tree rings: Part 1, Basic methodology and preliminary results for England. *Journal of Climatology* 3:233-242.

Cook, E. R. 1987. On the decomposition of tree-ring series for environmental studies. *Tree-Ring Bulletin* 47:37-59.

Cook, E. R. n.d. *Past drought from tree rings in the eastern United States: statistical methods and reconstruction fidelity*. Unpublished Manuscript. Tree-Ring Lab, Lamont-Doherty Geological Observatory. 27 pp.

Cook, E. R. and K. R. Briffa. 1990. Data analysis. In: Cook, E. R. and L. Kairiukstis, eds., *Methods of Tree-Ring Analysis: Applications in the Environmental Sciences*. Kluwer Academic Publishers Group, Dordrecht. pp. 97-162,

Cook, E. R. and J. Cole. 1989. Spatial patterns of climatic response for eastern hemlock and the potential impact of future climatic change. In: Noble, R. D., J. L. Martin, and K. F. Jensen (eds.) *Air Pollution Effects on Vegetation Including Forest Ecosystems*. Proceedings of the Second US-USSR Symposium, September 13-25 1988. USDA Forest Service, Northeastern Forest Experiment Station, Broomall, Pennsylvania. pp. 27-36.

Cook, E. R. and J. Cole. in press. On predicting the response of forests in eastern North America to future climatic change. *Climatic Change*.

Cook, E. R. and G. C. Jacoby, Jr. 1977. Tree-ring-drought relationships in the Hudson Valley, New York. *Science* 198:399-401.

Cook, E. R. and G. C. Jacoby. 1979. Evidence for quasi-periodic July drought in the Hudson Valley, New York. *Nature* 282:390-392.

Cook, E. R. and G. C. Jacoby. 1983. Potamac river streamflow since 1730 as reconstructed from tree rings. *Journal of Climate and Applied Meteorology* 22:1659-1672.

Cook, E. R. and L. Kairiukstis, eds. 1990. *Methods of Tree-Ring Analysis: Applications in the Environmental Sciences*. Kluwer Academic Publishers Group, Dordrecht.

Cook, E. R., Johnson, A. J. and Blasing, T. J. 1987. Forest decline: modeling the effect of climate in tree rings. *Tree Physiology* 3:27-40.

Cook, E. R., Kablack, M. and Jacoby, G. J. 1988. The 1986 drought in the southeastern United States: how rare an event was it? *Journal of Geophysical Research* (*Atmospheres*) 93(D11):14257-14260.

Currie, R. G. 1981a. Solar cycle signal in air temperature in North America: amplitude, gradient, phase and distribution. *Journal of the Atmospheric Sciences* 38:808-818.

Currie, R. G. 1981b. vidence for 18.6 M_n signal in temperature and drought conditions in North America since A.D. 1800. *Journal of Geophysical Research* 86:11055-11064.

Draper, N. R. and H. Smith. 1981. *pplied Regression Analysis, 2nd ed*. John Wiley and Sons, New York.

Fritts, H. C. 1976. *ree Rings and Climate*. Academic Press, London.

Fritts, H. C. and J. Guiot. 1990. ethods of calibration, verification, and reconstruction. In: Cook, E. R. and L. Kairiukstis, eds., *Methods of Tree-Ring Analysis: Applications in the Environmental Sciences*. Kluwer Academic Publishers Group, Dordrecht. pp. 163-218.

Fritts, H. C., T. J. Blasing, B. P. Hayden, and J. E. Kutzbach. 1971. ultivariate techniques for specifying tree-growth and climate relationships and for reconstructing anomalies in paleoclimate. *Journal of Applied Meteorology* 10:845-864.

Gilman, D. L., F. J. Fuglister, J. M. Mitchell, Jr. 1963. On the power spectrum of "red noise". *Journal of the Atmospheric Sciences* 20:182-184.

Gordon, G. A. and S. K. LeDuc, 1981. Verification statistics for regression models. In: *Proceedings of the Seventh Conference on Probability and Statistics in the Atmospheric Sciences*, Monterey, California.

Harris, C. W. and H. F. Kaiser. 1964. Oblique factor analytic solutions by orthogonal transformations. *Psychometrika* 29:347-362.

Holton, J. R. and H.-C. Tan. 1980. The influence of the equatorial quasi-biennial oscillation on the global circulation at 50 mb. *Journal of the Atmospheric Sciences* 37:2200-2208.

Hughes, M. K., P. M. Kelly, J. R. Pilcher, and V. C. LaMarche, Jr. 1982. *Climate from Tree Rings*. Cambridge University Press, Cambridge.

Jenkins, G. M. and D. G. Watts. 1968. *Spectral Analysis and Its Applications*. Holden-Day, San Francisco.

Jolliffe, J. T. 1986. *Principal Components Analysis*. Springer-Verlag, New York.

Kaiser, H. F. 1960. The application of electronic computers to factor analysis. *Educational and Psychological Measurement* 20:141-151.

Karl, T. R. 1988. Multi-year fluctuations of temperature and precipitation: the gray area of climate change. *Climatic Change* 12:179-197.

Karl, T. R. and A. J. Koscielny. 1983. Drought in the United States. *Journal of Climatology* 2:313-329.

Karl, T. R. and W. E. Riebsame. 1984. The identification of 10 to 20-year temperature and precipitation fluctuations in the contiguous United States. *Journal of Climate and Applied Meteorology* 23:950-966.

Kutzbach, J. E. 1967. Empirical eigenvectors of sea-level pressure, surface temperature and precipitation complexes over North America. *Journal of Applied Meteorology* 6(5):791-802.

Labitzke, K. 1987. Sunspots, the QBO, and the stratospheric temperature in the north polar region. *Geophysical Research Letters* 14(5):535-537.

LaMarche, V.C., Jr. and H.C. Fritts. 1971. Anomaly patterns of climate over the Western United States, 1700-1930, derived from principal components analysis of tree-ring data. *Monthly Weather Review* 99:138-142.

LaMarche, V. C., Jr. and H. C. Fritts. 1972. Tree rings and sunspot numbers. *Tree-Ring Bulletin* 32:19-32.

Lamb, H. H. 1970. Volcanic dust in the atmosphere, with a chronology and assessment of its meteorological significance. *Philosophical Transactions of the Royal Society, London* A226:425-533.

Lorenz, E. N. 1956. Empirical orthogonal functions and statistical weather prediction. *M.I.T. Statistical Forecasting Project No. 1*, Contract AF 19 (604)-1566, Massachusetts Institute of Technology.

Lough, J. M. and H. C. Fritts. 1985. The southern oscillation and tree rings: 1600-1961. *Journal of Climate and Applied Meteorology* 24(9):952-966.

Meko, D. M. 1981. *Applications of Box-Jenkins Methods of Time Series Analysis to the Reconstruction of Drought from Tree Rings*. Unpublished PhD Dissertation, University of Arizona, Tucson, Arizona.

Meko, D. M., C. W. Stockton, and T. J. Blasing. 1985. Periodicity in tree rings from the corn belt. *Science* 229:381-384.

Mitchell Jr., J. M., B. Dzerdzeevskii, B. Flohn, W. L. Hofmyer, H. H. Lamb, K. N. Kao and C. C. Wallen. 1966. *Technical Note 79, Climatic Change*. World Meteorological Organization, WMO-No. 195, T.P. 100, Geneva, Switzerland, 79 pp.

Mitchell, Jr., J. M., C. W. Stockton, and D. M. Meko. 1979. Evidence of a 22-year rhythm of drought in the western United States related to the Hale solar cycle since the 17th century. In: McCormac, B. M. and T. A. Seliga, eds., *Solar-Terrestrial Influences on Weather and Climate*. D. Reidel. pp. 125-143.

Neftel, A., E. Moor, H. Oeschger and B. Stauffer. 1985. Evidence from polar ice cores for the increase in atmospheric CO in the past two centuries. *Nature* 315:45-47.

NOAA. 1989. *Climate and Global Changes: An Integrated NOAA Program in Earth System Science*. National Oceanic and Atmospheric Administration, 24 pp.

Palmer, W. C. 1965. *Meteorological Drought*. Weather Bureau Research Paper No. 45. U.S. Dept. of Commerce, Washington, D.C. 58 pp.

Quinn, W. H., D. O. Zopf, K. S. Short, and R. T. W. Kuo Yang. 1978. Historical trends and statistics of the Southern Oscillation, El Nino, and Indonesian droughts. *Fishery Bulletin* 76:663-678.

Richman, M. B. 1986. Rotation of principal components. *Journal of Climatology* 6:293-335.

Rodriguez-Iturbe, I. 1969. Estimation of statistical parameters for annual river flows. *Water Resources Research* 5(6):1418-1426.

Ropelewski, C. F. and M. S. Halpert. 1986. North American precipitation and temperature patterns associated with the El Nino/Southern Oscillation (ENSO). *Monthly Weather Review* 114:2352-2362.

Sellers, W. D. 1968. Climatology of monthly precipitation patterns in the western United States 1931-1966. *Monthly Weather Review* 96(9):585-595.

Stahle, D. W., M. K. Cleaveland and J. G. Hehr. 1985. A 450-year drought reconstruction for Arkansas, United States. *Nature* 316:530-532.

Stahle, D. W. and M. K. Cleaveland. 1988. Texas drought history reconstructed and analyized from 1698-1980. *Journal of Climate* 1:59-74.

Stahle, D. W., M. K. Cleaveland and J. G. Hehr. 1988. North Carolinaclimate changes reconstructed from tree rings: A.D. 372-1985. *Science* 240:1517-1519.

Stockton, C. W. and D. M. Meko. 1975. A long-term history of drought occurrence in western United States as inferred from tree rings. *Weatherwise* 28:244-249.

Stockton, C. W. and D. M. Meko. 1983. Drought recurrence in the Great Plains as reconstructed from long-term tree-ring records. *Journal of Climate and Applied Meteorology* 22:17-29.

Stockton, C. W. and W. R. Boggess. 1980. Augmentation of hydrologic records using tree rings. In: *Improved Hydrologic Forecasting: How and Why*, pp. 239-265. American Society of Civil Engineers, New York.

Stothers, R. B. 1984. The great Tambora eruption in 1815 and its aftermath. *Science* 224: 1191-1198.

Trenberth, K. E., G. W. Branstator, and P. A. Arkin. 1988. Origins of the 1988 North American drought. *Science* 242:1640-1645.

Warrick, R. A. 1980. Drought in the Great Plains: a case study of research on climate and society in the United States. In: Ausubel, J. and A. Biswas, eds., *USA Climate Constraints on Human Activities*. Pergamon, pp. 93-124.

18 Dendroclimatic evidence from southwestern Europe and northwestern Africa

F. Serre-Bachet, J. Guiot and L. Tessier

18.1 Introduction

Dendroclimatic reconstructions from ring-widths available in South West Europe and North Africa mainly concern temperature (mostly summer temperature) and precipitation (over different periods of the year) (Serre-Bachet 1988). Here we present reconstructions that have been prepared over several years at the "Laboratoire de Botanique historique et Palyno-logie" (Marseille) as well as reconstructions referred to in the literature (Table 18.1). These are analysed in order to test their mutual reliability and thus make clear the main outlines of climate evolution since A.D. 1500 in the western Mediterranean regions of Europe and Africa for which hardly any climatic study has yet been made.

18.2 The reconstructions and their characteristics

18.2.1 Temperature reconstructions

Mean temperatures only were reconstructed. These reconstructions, particularly reconstructions 1 to 4 in Table 18.2, are among the longest that have ever been published (Serre-Bachet 1988); they involve very long ring-width series from living trees. The first reconstructions for the Grand St-Bernard station (Guiot 1984) (Table 18.2) are based on five sets of tree-rings from five sites, three in the Grisons, Switzerland (Munaut unpublished, Guiot *et al.* 1982) one in Savoie (Tessier 1981 and 1986b) and one in the Alpes Maritimes, France (Serre 1978). The cores were taken from larches (*Larix decidua* Mill.) except in one of the sites in the Grisons where pines (*Pinus cembra* L.) were sampled. The five master chronologies span at least 400 years. The Grisons chronologies reach back to A.D. 1600, the Savoie chronology reaches back to A.D. 1353 and the Alpes Maritimes chronology back to A.D. 933.

In the temperature reconstructions for Marseille, Rome and Grand St-Bernard (Serre-Bachet and Guiot 1987) (Table 18.1) the Savoie and Alpes Maritimes tree-ring series are combined with two additional chronologies, one for fir (*Abies alba* Mill.) from Mont Ventoux, France (Serre-Bachet 1985) the other for pine (*Pinus leucodermis* Ant.) from Calabria, Italy (Serre-Bachet 1986). These two new chronologies reach back to A.D. 1660 and A.D. 1150 respectively. The temperature reconstruction of Marseille (Guiot 1985) (Table 18.2) is based on a set of 20 cores chosen among those sampled in the Larch population of the Alpes Maritimes.

Table 18.1 Analysed reconstructions. For each one: full length of the period reconstructed; annual period reconstructed (Jan: January, Mr: March, Ap: April, Ma: May, Jn: June, Jl: July, Au: August, Sp: September, Oc: October, Nv: November, Dec: December); tree species used as predictors with meteorological station or area concerned with the reconstruction (P.: *Pinus*; P. leuc.: *Pinus leucodermis*; pub: *pubescens*).

	Time span	Annual period reconst	Predictor	Meteo.Station or area concerned
RECONSTRUCTIONS OF TEMPERATURE				
1	1590–1960	Au. Dec. Jn–Au Sp–Nv	Larix P.cembra	Grand St Bernard
2	1150–1970	Jn–Sp	Larix P.leuc. Abies alba	Marseille, Rome & Grand St Bernard
3	1200–1974	Jn–Jl	Larix	Marseille
4	1100–1970 1068–1979	Jan–Dec	Several tree species + other proxies	35°N–10°W 40°N– 0°W 40°N–10°E 50°N–20°E
5	1770–1984 1690–1984	Nv–Mr	P.nigra	NE Spain
RECONSTRUCTIONS OF PRECIPITATION				
6	1784–1980	Oc–Mr Oc–Ma Oc–Nv Ap– Ap–Sp Jn–Sp Jn–Jl	P.halepensis P.nigra P.silvestris Abies alba Quercus pub.	Marseille
7	1770–1984 1690–1984	Ma–Au	P.nigra	NE Spain
8	1100–1979	Oc–Sp	Cedrus atlantica	Morocco : Rif & Atlas

The reconstructions for five points (Table 18.2; Figure 18.1) of the temperature grid established by Jones *et al.* (1985) for the Northern Hemisphere (Guiot *et al.* 1988; Guiot 1989) are based on 23 different predictors. Eleven are tree-ring series (some of them just cited) 3 correspond to the three first principal components of the 17 longest ring-width series of cedars (*Cedrus atlantica* Endl. Carrière) from Morocco (Till 1985) and the others are historical (documentary) proxy series (such as temperature estimates and indices, grape-harvest dates, frequency of surface winds) or isotopic series.

Lastly, two reconstructions have been made by Richter (1988) for northeastern (NE) Spain

Table 18.2 Characteristics of the 27 analysed reconstructions with special focus on the statistical results given by the authors. Each reconstruction is described by four characters which refer to the annual period reconstructed and/or to the meteorological station or area. The reconstructions shown in Figures 18.3 to 18.7 are underlined. See also Figures 18.1 and 18.2.

TEMPERATURE

1- GUIOT 1984: Mean temperature reconstruction of 4 monthly periods at GRAND-St-BERNARD by (a) spectral canonical regression, (b) MARMA analysis. (Summer: June+July+August; Autumn: September+October+November).

Predictand	Predictors	Calib. period	Verif. period	Statistical Results			
				August	December	Summer	Autumn
(a) Temperature low freq. (>10 years)	20 tree ring LF ser. (indx.ser.+ 2 raw data ser. at time t=-1,0,1,2)	1874-1960 / 1818-1960	1825-1873	R2: 0.70** RE: -0.22 R2: 0.23 **AuBe**	0.81*** -0.14 0.46 **DeBe**	0.68* 0.16 0.48 **SuB2**	0.73* 0.18 0.55 **FaBe**
(b) Temperature whole spect.	12 residual series (6 res.ser. at time t=0,-2) cf.55 cores 2 sites,MARMA/indx.	1840-1949	1818-1939			R2: 0.42*** R2: 0.14 RE: 0.13 **SuB3**	

2- SERRE-BACHET & GUIOT 1987: Reconstruction of summer (june to september) mean temperature at MARSEILLE, ROME and GRAND-St-BERNARD by orthogonal multiple regression.

Predictand	Predictors	Calib. period	Verif. period	Statistical Results		
				(1)	(2)	(3)
Temperature at MARSEILLE **SuMa**	1) Rec. to 1660 AD: 12 TR ser. (4 indx. ser. at t=-1,0,1)	1851-1930 1931-1972 1851-1972		R2: 0.55*** R2: 0.49*** R2: 0.48***	0.36 -0.08 0.24***	0.28 -0.35 0.09
Temperature at ROME **SuRo**	2) Rec. to 1353 AD 9 TR ser. (3 indx. ser. at t=-1,0,1) 3) Rec. to 1150 AD 6 TR ser. (2 indx. ser at t=-1,0,1)	1851-1930 1811-1850 1951-1970 1811-1930 1951-1970		R2: 0.66*** R2: 0.29** R2: 0.50***	0.49*** -0.04 0.24***	0.37* -0.06 0.12*
Temperature at GRAND-SAINT-BERNARD **SuB1**		1851-1930 1818-1850 1931-1960 1818-1960		R2: 0.56*** R2: 0.35 R2: 0.42***	0.39 0.19* 0.27***	0.28 -0.15 0.08

3- GUIOT 1985: Mean summer (June + July) temperature reconstruction at MARSEILLE by spectral canonical regression.

Predictand (T)	Predictors (C)	Calib. period	Verif. period	Statistical Results
Temperature LF + MF	20 tree ring LF ser. cf. 20 cores, 1 site	1851-1974 (Mlle)	1818-1850 (G.St.B.)	R T/C: 0.52 R T/C: 0.31* **JJMa**

4- GUIOT 1988: Annual (January to December) mean temperature reconstruction at 5 points of JONES et al. (1985) grid by bootstrapped orthogonal multiple regression.

Predictand	Predictors	Calib. and Verif. period	Statistical Results	
LF + HF temp. at 10°W-35°N 0°W-50°N 0°W-40°N 10°E-40°N 20°E-50°N	23 with 11 TR indx. series + 3 first PC of 17 RW series from Morocco.	1950-1970 (50 simulations /1 grid point) (e/o:estimated/ observed data)	Re/o: 0.64 Re/o: 0.70 Re/o: 0.53 Re/o: 0.60 Re/o: 0.54	1W35 **0W50** **0W40** **1E40** **2E50**

Table 18.2 – *continued*

TEMPERATURE

5- RICHTER 1988: Reconstruction of winter (November to March) mean temperature for Nord-East Spain (Cuenca, Teruel, Cazorla triangle) by orthogonal multiple regression.					
Predictand	Predictors	Calib. period	Verif. period	Statistical Results (1)	(2)
Temperature from 10 meteo. stations	1) 4 residual series cf. 4 TR series from 4 sites	1902-1942	1943-1983	R : 0.40* RE: -0.18	0.49* -0.01
	2) 9 residual series cf.9 TR series from 9 sites	1943-1983	1902-1942	R : 0.44* RE: -0.10	0.59* 0.20*
		1902-1983	1902-1983	R : 0.44* RE: -0.11 <u>WiC1</u>	0.53* 0.07* <u>WiC2</u>

PRECIPITATION

6- SERRE-BACHET et al. 1988: Precipitation reconstruction of 7 monthly periods at MARSEILLE by bootstrapped orthogonal multiple regression.									
Predictand	Predictors	Calib. and Verif. years (Cy, Vy)	Statistical Results O-Mr	O-My	O-Nv	Ap.	A-Sp	Jn-S	Jn-Jl
Monthly precipitation at MARSEILLE	15 indexed TR series from 5 tree species	1861-1980 Cy (45 simulations /each monthly period) Vy	Re/o: 0.65 *** Re/o: 0.44 *** <u>OcMr</u>	0.70 *** 0.53 *** <u>OcMa</u>	0.53 *** 0.23 ** <u>OcNv</u>	0.50 *** 0.19 ** Apr	0.47 *** 0.13 <u>ApSp</u>	0.40 *** 0.00 JnSp	0.39 *** 0.07 JnJl

7- RICHTER 1988: Reconstruction of summer (May to August) precipitation for Nord-East Spain (Cuenca Teruel, Cazorla triangle) by orthogonal multiple regression.					
Predictand	Predictors	Calib. period	Verif. period	Statistical Results (1)	(2)
Precipitation from 10 meteo. stations	1) 4 residual series cf. 4 TR series from 4 sites	1902-1942	1943-1983	R : 0.50* RE: 0.01	0.52* 0.05
	2) 9 residual series cf.9 TR series from 9 sites	1943-1983	1902-1942	R : 0.48* RE: -0.03	0.49* -0.01
		1943-1983	1902-1942	R : 0.49* RE: -0.01*	0.48* -0.03
		1902-1993	1902-1983	R : 0.48* RE: -0.02 <u>MaC1</u>	0.55* 0.11* <u>MaC2</u>

8- TILL & GUIOT 1988: Annual (October to September) reconstruction of precipitation at 4 vegetational stages of MOROCCO (Rif and Moyen Atlas) by bootstrapped orthogonal multiple regression.					
Predictand	Predictors	Calib. and Verif. years (Cy; Vy)	Statistical Results Humid	Sub- Humid	Semi- Arid
Precipitation defined on 41 meteo.stations	46 indexed TR series	1924-1977 Cy (50 simulations /each veg. stage) Vy	Re/o: 0.88 *** Re/o: 0.59 *** <u>MHu</u>	0.88 *** 0.69 *** MSHu	0.85 *** 0.62 *** MSAr

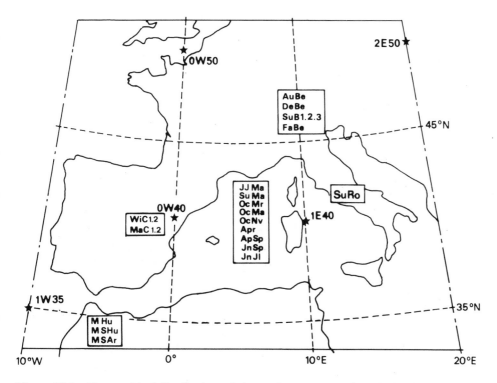

Figure 18.1 Geographical distribution of the stations or areas for which climatic reconstructions are analysed (Stars = points of Jones *et al.* temperature grid). See also Table 18.1 and 18.2.

(Table 18.2). These are based on two supra-regional chronologies, one established from nine different chronologies starting in A.D. 1770, the other from four chronologies starting in A.D. 1690. In all the cases, the available instrumental measurement series of mean temperature were long enough to permit both a calibration of ring-width with temperatures over one period, and the verification of this calibration over another quite distinct period. These series cover the period 1818-1960 for Grand St Bernard station, 1852-1980 for Marseille, 1811-1930 and 1951-1970 for Rome. The temperature grid spans the period 1851-1984. The series corresponding to NE Spain, established from ten meteorological stations (including Madrid, Cuenca, Una and Teruel) covers the period 1902-1984.

18.2.2 Precipitation reconstructions

Reconstructions of total precipitation over a given period of a year (Table 18.1) are far less abundant than temperature reconstructions. There are two major reasons for this. The first reason is the great variation in precipitation from one place to another, making it difficult to find a homogeneous response of ring-widths to this factor. The second reason is that the growth of the trees that are used in reconstructions is often more frequently related to temperature than to precipitation, since these trees, especially the larch, often grow at an altitude and/or on a sub-stratum where precipitation is not a limiting factor. The three sets of precipitation reconstructions presented here involve species from regions and/or altitudes where the influence of precipitation is at least as important as that of temperature.

The reconstruction of precipitation, over seven monthly periods, made for Marseille (Serre-Bachet 1988; Serre-Bachet *et al* 1988) (Table 18.2) is based on 15 ring-width series of five different species from ten sites (mostly at low altitudes) in five French districts where Mediterranean climatic conditions prevail (Serre 1976, 1977; Serre-Bachet 1982, 1986; Tessier 1986a; Gadbin *et al*. 1988). The series that are of unequal length were all adjusted to a common period of 197 years (A.D. 1784 to 1980) the missing data of the shortest series being extrapolated from the complete series that are best correlated with them (Guiot 1985). The precipitation reconstructions in NE Spain (Richter 1988) (Table 18.2) are based on the same tree-rings series as were used for temperature reconstructions.

Lastly, for Morocco (Table 18.2) the reconstructions of three of the five climatic stages distinguished in this country by Till and Guiot (1990) i.e. humid, subhumid and semi-arid, have been retained. These reconstructions are based on 46 cedar master chronologies referring to 46 different sites in the High and Middle Atlas and in the Rif (Till 1985, 1988).

Instrumental precipitation records for Marseille, available back to A.D. 1748 (Serre-Bachet 1988) were used for the period 1861-1980. The instrumental data for precipitation in Spain come from the same stations as those defined above for temperature. For Morocco, the climatic regions were defined on the basis of records from 41 meteorological stations; the averaged records of stations characteristic of each region cover the 1924-1977 period.

18.2.3 Reconstruction methodology

Among the works mentioned previously, some illustrate methodological procedures used in climate reconstruction (Guiot 1984; 1985, 1989); others are the outcome of research focused on a well delimited geographical area (Richter 1988; Serre-Bachet *et al*. 1988; Till and Guiot 1990). In all cases, estimates have ,as far as possible, been compared with data originating from the same meteorological station(s) as were used for calibration, or from other stations where long sequences were available (for example, the Grand St Bernard record, the Basle record [Schüepp 1961] and the Central England record [Manley 1974]). In order to assess the probability of reconstructed variations over certain periods or years, other proxy data were in some cases taken into account. These include periods of glacier advance or retreat, grape-harvest dates, dates of processions to ask for rain, or to give thanks for the cessation of rain etc. Comparison has also been made between the reconstructions and other proxy climatic reconstructions.

Details on the above reconstructions, the procedure used to calculate them as well as the statistical results they provide are summarized in Table 18.2. The ring-width series involved in these reconstructions are either indexed series (see Bradley and Jones, this volume, Chapter 1) or residual series obtained after autoregressive modeling (Guiot 1984, Richter 1988). Exceptionally, the raw data have been used. The regressors matrix may involve both the data series at time "t" and at time "t+k" (with "k" varying from -2 to +2). Autocorrelation of the tree-ring data and delayed influence of climate on growth are thus taken into account (Guiot 1984). Some reconstructions are performed on the spectral decomposition of the data by digital filtering (Guiot 1984, 1985, 1989).

The reconstructed climatic parameters for a given annual period have sometimes been chosen on the basis of ring-width/climate relationships previously established mostly by response functions using at least 30 years of data (Serre-Bachet and Tessier 1989; Richter 1988). They have also been chosen from those monthly and/or seasonal climatic parameters, that gave the most significant results from the selected calibration period. Calibration was achieved using techniques related to multiple linear regression and to spatial regression as described by Bradley and Jones (this volume, Chapter 1). More specifically, orthogonal multiple regression (Richter 1988; Serre-Bachet and Guiot 1987) spectral canonical regression (Guiot 1984, 1985) and "boot-strapped" orthogonal multiple regression were used (Guiot 1989; Serre-Bachet 1988; Serre-Bachet *et al.* 1988; Till and Guiot 1990). Depending on the calibration methods, the significance of the results was tested using the correlation coefficient (R or R^2) between the reconstructed and independent climate data for both calibration and verification periods as well as on the reduction of error (RE) for the verification period. In the bootstrapped orthogonal multiple regression, the correlation coefficient (R e/o) is the mean of "n" correlation coefficients obtained from "n" simulations made on "n" random samples of the data (Guiot 1989).

In view of the diversity of regressors and methods, it is difficult to inter-compare the 16 temperature reconstructions and the 11 precipitation reconstructions summarized in Table 18.2. Therefore, these reconstructions have all been analysed in order to select those which are most representative of common criteria.

18.3 Analysis of the preceding reconstructions

18.3.1 Selection of representative and reliable series

In order to detect the most representative and reliable series, the 16 temperature reconstructions are compared to the actual temperatures presented by Jones *et al.* (1985) for the five gridpoints (35°N 10°W, 50°N 0°W, 40°N 0°W, 40°N 10°E and 50°N 20°E) that are the nearest to the meteorological stations concerned. The seven reconstructed combinations of months are considered at each gridpoint: June-July, July to August (the combination June-September being assimilated to the latter, see Table 18.1) August, December, September to November, November to March, and lastly the whole year.

In the absence of a grid such as that developed for temperature by Jones *et al.*, the 13 precipitation reconstructions are compared to the actual series of Marseille for the nine reconstructed combinations of months: the biological year from October to September, May to August, October to March, June-July, April to September, June to September, October-November, April, October to May.

All of the series, both the actual and the reconstructed ones, are smoothed using a filter which cuts off the periods that are shorter than four years; of course this filter has no effect on the series previously filtered with larger cut off period. Comparison between filtered reconstructed and actual series is made using canonical analysis (Bradley and Jones, this volume, Chapter 1). This analysis provides independent linear combinations, called canonical components, for which correlations between actual and reconstructed series are maximal (for more details, see Clark 1975). Correlation of both data from the grid of Jones *et al.* and reconstructed data with these canonical components enables one to choose the most reliable reconstructions with respect to the data from the grid.

Temperature   The common analysis period of the different series is 1851-1961. The three first canonical correlations are successively 0.96, 0.92 and 0.90; only the three corresponding components are retained. Figure 18.2 makes clear the correlations of each of them with the 16 reconstructed series on the one hand (bars) and the 16 actual series at the nearest gridpoints of Jones *et al.*, on the other hand (lines). For each component, the reconstructed series are selected on the basis of the most significant correlations within each set of months, i.e. 0W40, 2E50, SuRo, WiC1 for the first component, 0W50 for the second component, 1E40 and SuB2 for the third one (see also Figure 18.1 and Table 18.2).

Precipitation   When the same analysis is applied to precipitation series compared with the Marseille data over the period 1822-1979, it provides lower canonical correlation values in comparison with temperature, because the Spanish or Moroccan proxy-series upon which the reconstructions are based are poorly representative of the Marseille actual series. Therefore only the two first canonical components, with 0.65 and 0.45 correlations, are retained. The first component (Figure 18.2) leads to the choice of MHu. Other parameters reconstructed at Marseille, some of them redundant, can also be selected; only two complementary parameters are retained : OcMr and ApSp. The second component leads us to choose MaC2 (see also Figure 18.1 and Table 18.2).

18.4   Discussion of the reconstructions

Figures 18.3 to 18.7 show the 11 selected reconstructions over the period 1500-1980. Table 18.1 shows that many of them start long before A.D. 1500. As regards climate since A.D. 1500, these different reconstructions show marked coolings over the period ca 1550-1850, the so-called "Little Ice Age", followed by a warming which is still persisting to-day (Guiot 1989). It seems that not only were summers cooler during the Little Ice Age (Serre-Bachet & Guiot 1987) but also cool winters (Richter 1988) affected both western Mediterranean regions and more northern regions. In the central Mediterranean area, temperatures (especially summer temperatures) during the Little Ice Age were not markedly lower on average, but rather showed only slight variations compared to previous or subsequent periods (Guiot 1985). In Morocco, where only precipitation could be reconstructed over a long time-interval (Till 1988, Till & Guiot 1990) the Little Ice Age coincided with severe drought periods which sometimes were synchronous with those recorded in the Sahara.

Annual temperature (Figures 18.3 and 18.4, and Figure 18.1)   The series 50°N 0°W (0W50 in Figure 18.4 and Tables) relating to northwestern France describes the best documented climate in the literature (Lamb 1977, LeRoy Ladurie 1971, Manley 1974, Flohn and Fantechi 1984 etc.). A "Little Ice Age" is recorded which is comparable to that characterized by a temperature fall exceeding 1°C reported by LeRoy Ladurie (1971) and Lamb (1977) at the beginning of the latter half of the 16th century. Most of the 17th century was relatively warm. A cool period began at the end of the 17th century and lasted some 150 years, with two minima at about 1700 and 1815, the latter corresponding to the Tambora eruption (Lamb 1977). As a whole, the first half of the 20th century was warm. The series 40°N 10°E (1E40) (Figure 18.3) relating to Italy shows the same temperature evolution as the preceding series,

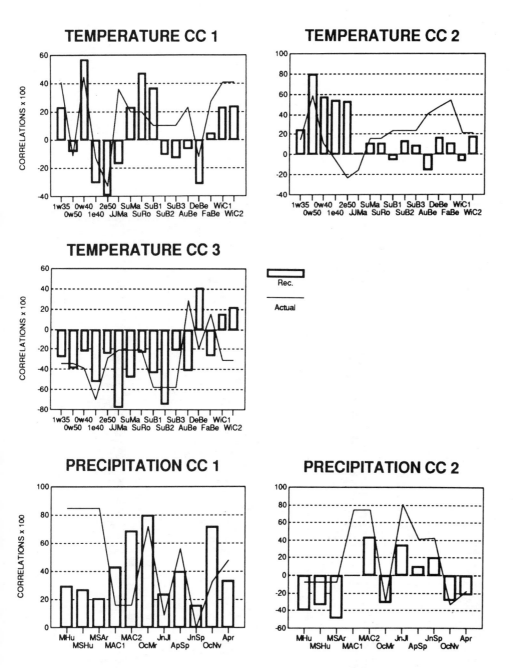

Figure 18.2 Results of the canonical analysis used to select the most representative and reliable reconstructions. Three canonical components are retained for temperature and two for precipitation. The *bars* represent the correlation coefficients between the canonical component and the reconstructed series. The *line* links the correlation coefficients of the same canonical component with the Jones *et al.* climatic actual data at the gridpoint closest to the reconstructed series. The meaning of the codes used for the variables is given in Table 18.2. For the temperature, the first five series concern the whole year, the following seven ones concern summer or a part of it, the 13th and 14th concern December and autumn and the last two winters. For the precipitation, the first three series refer to annual precipitation in Morocco, the following two concern summer temperature in Spain and the others the various annual periods reconstructed for Marseille. See also Figure 18.1 and Table 18.2.

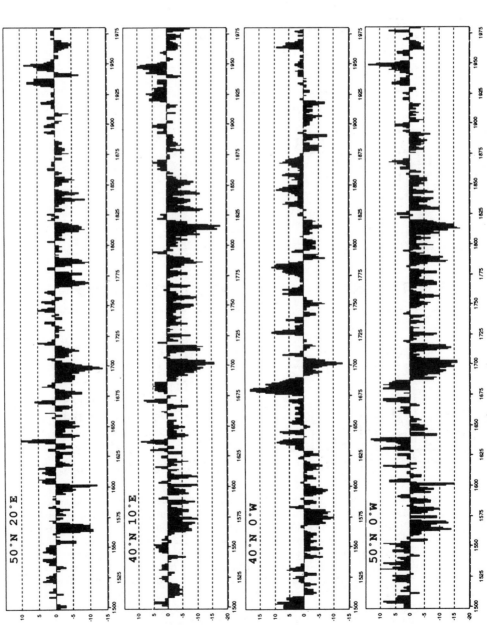

Figures 18.3 and 18.4 Annual temperature (°C × 10) at the four gridpoints of the Jones *et al.* network (1985) expressed as departures from the mean of the 1951-1970 period.

with a slightly colder 17th century. The series 50°N 20°E (2E50) (Figure 18.3) which is representative of the continental climate of Germany and Poland shows steady low temperatures and the episodes described above are far less marked, except the 20th century warming.

For Western Mediterranean 40°N 0°W (0W40) (Figure 18.4) a slightly different climatic evolution is obtained, compared with the three other regions. From A.D. 1500, very few long cold episodes are recorded in the western Mediterranean climate. There is no indication of a Little Ice Age; the 17th, 18th and 19th centuries were rather warm and the Tambora eruption of 1815 had practically no effect. The 20th century warming-up is not obvious in comparison with the previous warm episodes recorded in the same region.

The correlation analysis of the four series confirms the division between Western and Central Mediterranean (Italy): the correlation between series 40°N 0°W and series 40°N 10°E is only 0.13. In contrast, there exists an obvious correlation (0.68) between central Mediterranean 40°N 10°E and northwestern France 50°N 0°W and a lower one (0.48) between the regions North (50°N 0°W) and South (40°N 0°W) of the Pyrenees.

Summer temperature (Figure 18.5 and Figure 18.1)   Summer temperature curves at Rome (SuRo) and Grand St Bernard (SuB2) are based on a number of common dendrochonological series, which in part explains their parallelism. This parallelism also results from the climate itself, since the instrumental annual temperature series of the same geographical area (40°N 10°E and 50°N 0°W, Jones *et al.* 1985) are correlated. Except for several warm episodes during the 18th century, which do not appear in annual reconstructions, these summer curves show minima at the same periods as in the annual reconstructions for the central Mediterranean and northwestern France. The Little Ice Age period is mainly characterized by great variability.

There is no summer temperature reconstruction available for the region west of the Mediterranean Basin, thus no comparison can be made for this period between this region and the central Mediterranean region (Italy). However, the correlations established between the annual series can reasonably be also assumed for the summer series in view of the correlation between the reconstructed summer temperature at Rome and Grand St Bernard (0.63) both series being themselves closely correlated with their corresponding reconstructed annual series (0.63 between Rome and 40°N 10°E and 0.47 between Grand St Bernard and 50°N 0°W).

Winter temperature (Figure 18.5 and Figure 18.1)   The winter series for northeastern Spain (WiC1) is generally negatively correlated with all of the reconstructed annual series and particularly with the 40°N 0°W series, this suggesting that, in the Mediterranean climate, cold periods may include mild winters and, conversely, cold winters may exist during hot periods. There also exists a negative correlation (-0.36) between the reconstructed winter temperature of northeastern Spain and the reconstructed summer temperature of Grand St Bernard. However, these relations are not in agreement with the conclusions of Pfister (1981) and Manley (1974) who show that the Little Ice Age was mainly characterized by severe winters.

A possible hypothesis to explain this discordance could be that the winter temperature reconstruction can be distorted by the limiting action of precipitation on tree-ring widths during this period (Serre-Bachet 1988). An argument in favour of this idea is that in Jones *et al.* (1985) there appears no such inverse relation between winter and annual temperature in

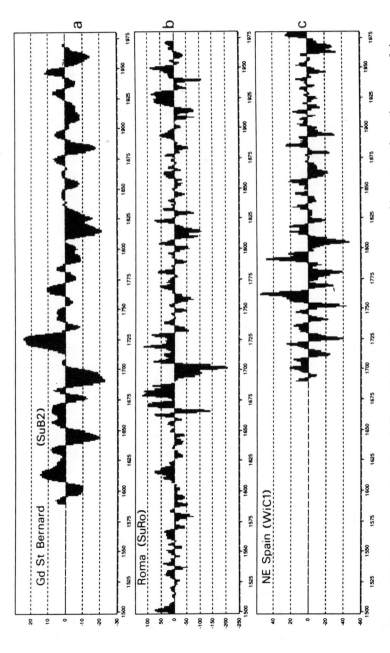

Figure 18.5 Reconstructed series of: Grand St Bernard summer temperature (departures from the mean of the 1851-1930 period in °C × 10); Rome summer temperature (departures from the mean of the 1811-1930 and 1951-1970 periods in °C × 10); Northeastern Spain winter temperature (departures from the reference period in °C × 10).

the same region; in fact, there is a positive correlation. Therefore it is clear that the winter temperature reconstruction in northeastern Spain may not be entirely reliable, as was also suggested Richter (1988, p.170) on the basis of comparisons between his estimates and various historical data.

Precipitation (Figures 18.6 and 18.7, and Figure 18.1)   Precipitation reconstructions are more difficult to interpret than temperature reconstructions. Annual precipitation (October to September) in the humid climatic region of Morocco (MHu) mainly in the Rif (Figure 18.7) shows a long dry period up to the beginning of the 20th century in comparison with the calibration period 1924-1977. The complete reconstruction made by Till and Guiot (1990) indicates that this drought period began long before A.D. 1500. Since A.D. 1500 some precipitation peaks coincide with low temperatures in the European area, particularly at the 40°N 0°W gridpoint.

The series from northeastern Spain (Figure 18.7) relating to summer (MaC2) can not be compared to the series from Morocco in which annual precipitation is exclusively centered on the period October to May. It is noteworthy, however, that in Spain summers are marked by a great variability in rainfall. This corroborates Lamb's remark (1982, p.224) for the Little Ice Age: "In southern Europe we find records which indicate an enhanced variability from year to year and decade to decade, particularly as regards rainfall, and the difficulties it caused. In Spain there were some runs of drought years and others characterized by flooding of the rivers". This variability also prevails at Marseille, although from the latter half of the 19th century, there has developed a marked trend towards a more stable and drier climate, more pronounced from October to March (OcMr) than from April to September (ApSp).

18.5   Conclusions

The eleven annual, summer and winter temperature and precipitation series selected as representative of South West Europe and North West Africa are the first elements of a synthesis concerning the climatic changes that have occurred in the Mediterranean region since A.D. 1500. They make it possible to compare these changes with those known for northwestern Europe.

The climate characteristics derived from the four annual temperature reconstructions make clear the difference between the Western Mediterranean region in comparison with other geographical areas. Any similarity, which was not perceived in the first analyses, appears here probably because of the joint utilization of ring-widths series and other proxy data as regressors. As a matter of fact, there exists a stronger teleconnection between the non-Mediterranean climate of France and the Mediterranean climate of Italy than between the latter and the Mediterranean climate of Spain or Morocco. This is reflected in the appearance, in Italy, of a Little Ice Age such as it is described in Europe, while no such event is reported in Spain or Morocco. In the latter regions, the 20th century warming up to 1975 does not appear to be exceptional in view of the numerous warm periods that have occurred previously. This situation in the Western Mediterranean may be related to the screening effect of the Pyrenees against the polar air masses, and/or to greater anticyclonicity, Spain

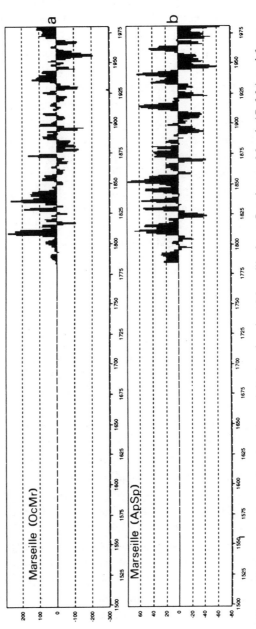

Figure 18.6 Reconstructed series of precipitation (mm) at Marseille from October to March (OcMr) and from April to September (ApSp) (departures from the mean of the 1861-1980 period).

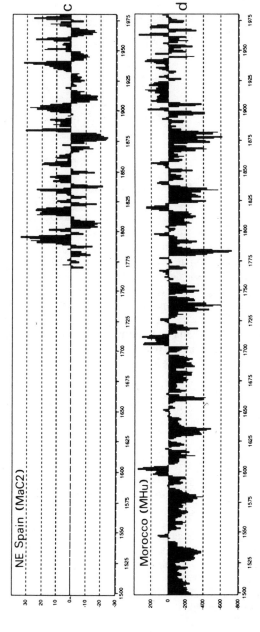

Figure 18.7 Reconstructed series of northeastern Spain summer (May to August) total precipitation (departures from the reference period mean in mm) and annual (October to September) precipitation (mm) for the humid climatic region of Morocco (departure from the mean of the 1924-1977 period).

and Morocco being thus spared the cooling which periodically affects northwestern European regions, the Alps and Italy.

Summer temperatures reveal the same teleconnections as annual temperatures. For the latter, in particular, it appears that, contrary to what was inferred on the basis of only one tree-population, the Little Ice Age was certainly characterized by a marked variability. This agrees with the variability observed in ring-widths only for the same period (Serre 1978, Serre and Guiot 1987, Tessier *et al.* 1986). The only winter temperature series available is not reliable enough for any inference to be drawn.

Annual precipitation in Morocco was generally lower from the 16th to 19th centuries than during the 20th century, though some humid episodes are recorded. More stable and/or drier conditions than before characterize the October to March period at Marseille, and the summer period in Spain, in the latter half of the 19th century.

References

Clark, D. 1975. *Understanding canonical correlation analysis*. Catmog 3, University of East Anglia, Norwich: Geo. Abstracts Ltd.

Flohn, H. and R. Fantechi (eds) 1984. *The climate of Europe: past, present and future*. Dordrecht: Kluwer.
Fritts, H. C. 1976. *Tree-rings and climate*. New York: Academic Press.

Gadbin, C., J. Guiot, F. Serre-Bachet and L. Tessier 1988. Croissance radiale de quelques résineux et feuillus en réponse aux précipitations mensuelles en milieu méditerranéen. In: *Time scales and water stress*, F. Di Castri, Ch. Floret, J. Rambal and J. Roy (eds) 401-414. Paris: I.U.B.S.
Guiot, J. 1984. Deux méthodes d'utilisation de l'épaisseur des cernes ligneux pour la reconstruction de paramètres climatiques anciens, l'exemple de leur application dans le domaine alpin. *Palaeogeography, Palaeoclimatology, Palaeoecology* 45, 347-368.
Guiot, J. 1985. The extrapolation of recent climatological series with spectral canonical regression. *Journal of Climatology* 5, 325-335.
Guiot, J. 1989. The climate of Central Canada and southwestern Europe reconstructed by combining various types of proxy-data: a detailed analysis of the 1810-1820 period. In: C. Harington (ed.) *1816: The Year without a summer?* Syllogeus Series, National Museums of Canada, Ottawa in press.
Guiot, J., A. L. Berger, A. V. Munaut and C. Till 1982. Some new mathematical procedures in dendroclimatology with examples from Switzerland and Morocco. *Tree-Ring Bulletin* 42, 33-48.
Guiot, J., L. Tessier, F. Serre-Bachet, F. Guibal, C. Gadbin and C. Till 1988. Annual temperature changes reconstructed in W. Europe and N.W. Africa back to A.D.1100. *Annales Geophysicae* Special Issue, 85.

Jones, P. D., S. C. B. Raper, B. D. Santer, B. S. G. Cherry, C. Goodess, R. S. Bradley, H. F. Diaz, P. M. Kelly and T. M. L. Wigley 1985. *A grid point surface air temperature data set for the Northern Hemisphere*, 1851-1984. DoE Technical Report TR022. Washington, DC: U.S. Dept. of Energy, Carbon Dioxide Research Division.

Lamb, H. H. 1977. *Climate: present, past and future*, Volume 2. London: Methuen.
Lamb, H. H. 1982. *Climate History and the the Modern World*. London: Methuen.
Le Roy Ladurie, E. 1971. *Times of feast, times of famine*. New York.

Manley, G. 1974. Central England temperatures: monthly means 1659 to 1973. *Quarterly Journal of the Royal Meteorological Society* 100, 389-405.

Pfister, C. 1981. An analysis of the Little Ice Age climate in Switzerland and its consequences for agriculture production. In: *Climate and History*, Wigley *et al* (eds) 214-248. Cambridge: Cambridge University Press.

Richter, K. 1988. *Dendrochronologische und Dendroklimatologische Untersuchungen an Kiefern* (*Pinus sp.*) *in Spanien*. Unpublished PhD. dissertation, University of Hamburg.

Serre, F. 1976. Les rapports de la croissance et du climat chez le pin d'Alep (*Pinus halepensis* Mill.) I et II. *Oecologica Plantarum* 11, 143-173 and 201-224.

Serre, F., 1977. A factor analysis of correspondences applied to ring widths. *Tree-Ring Bulletin* 37, 21-32.

Serre, F. 1978. The dendroclimatological value of the european larch (*Larix Decidua* Mill.) in the French Maritime Alps. *Tree-Ring Bulletin* 38, 25-34.

Serre-Bachet, F. 1982. Analyse dendroclimatologique comparée de quatre espèces de pins et du chêne pubescent dans la région de la Gardiole près Rians (Var, France). *Ecologia Mediterranea* 8, 167-183.

Serre-Bachet, F. 1985. Une chronologie pluriséculaire du sud de l'Italie. *Dendrochronologia* 3, 45-66.

Serre-Bachet, F. 1986. Une chronologie maitresse du sapin (*Abies alba* Mill.) du Mont Ventoux (France). *Dendrochronologia*, **4**, 87-96.

Serre-Bachet, F. 1988. La reconstruction climatique à partir de la dendroclimatologie. *Publications de l'Association Internationale de Climatologie* 1, 225-233.

Serre-Bachet, F. and J. Guiot 1987. Summer temperature from tree-rings in the mediterranean area during the last 800 years. In: *Abrupt Climatic Change*, W. H. Berger and L. D. Labeyrie (eds) 89-97. Dordrecht: Reidel.

Serre-Bachet, F. and L. Tessier 1989. Response function analysis for ecological study. In: *Methods of Dendrochronology: Applications in the Environmental Sciences*, E. Cook and L. Kairiukstis (eds) 247-258. Dordrecht: Kluwer.

Serre-Bachet, F., L. Tessier, J. Guiot and C. Gadbin 1988. Rainfall reconstructions from tree-rings in the French Mediterranean region. *Annales Geophysicae* Special Issue, 87.

Tessier, L. 1981. Contribution dendroclimatique à la connaissance écologique du peuplement forestier des environs des chalets de l'Orgère (Parc National de la Vanoise). *Travaux Scientifiques du Parc de la Vanoise* 11, 29-61.

Tessier, L. 1986a. Approche dendroclimatologique de l'écologie de *Pinus silvestris* L. et *Quercus pubescens* Willd. dans le sud-est de la France. *Acta Oecologica, Oecologica Plantarum* 7, 339-355.

Tessier, L. 1986b. Chronologie de mélèzes des Alpes et Petit Age Glaciaire. *Dendrochronologia* 4, 97-113.

Tessier, L., M. Coûteaux and J. Guiot 1986. An attempt at an absolute dating of a sediment from the last glacial recurrence through correlations between pollen-analytical and tree-ring data. *Pollen et Spores* 28, n°1, 61-76.

Till, C. 1985. Recherches dendrochronologiques sur le cèdre de l'Atlas (*Cedrus Atlantica* (Endl.) Carrière) au Maroc. Unpublished Ph.D. dissertation, Louvain-la-Neuve: Université Catholique de Louvain.

Till, C. 1988. Reconstitution of precipitation in Morocco since A.D.1100 according to tree-ring series. *Annales Geophysicae* Special Issue, 84.

Till, C. and J. Guiot 1990. Reconstruction of precipitation in Morocco since A.D..1100 based on *cedrus Atlantica* tree-ring widths. *Quaternary Research*, in press.

19 Recent dendroclimatic evidence of northern and central European summer temperatures

K. R. Briffa and F. H. Schweingruber

19.1 Introduction

This paper is concerned with recent tree-ring based reconstructions of summer temperatures in the northern and central regions of western Europe. The paper is divided into two primary sections. The first describes the reconstruction of a single mean Fennoscandinavian temperature series representing the late summer months of July and August back beyond A.D. 1600. The second section describes the reconstruction of spatial patterns of temperature for a longer summer season (April to September) across much of western Europe back to A.D. 1750.

19.2 Reconstruction method

We use the orthogonal spatial regression technique (OSR) based on linear least squares regression of tree-ring principal components on climate data. This is similar to the canonical correlation approach developed by Fritts and co-workers (see references in Fritts 1976 and Lofgren and Hunt 1982). A general discussion of this approach is given in Bradley and Jones (Chapter 1, this volume). The specific technique used here is fully described in Briffa *et al.* (1986). For a fuller description of the mathematical details and a comparison of the applications of the OSR and canonical regression techniques, see Cook *et al.* (1988).

In the first application described in this paper, a single climate series is reconstructed from a network of ring-width and densitometric chronologies. In the second application, year-by-year patterns of summer temperature are reconstructed using a network of densitometric chronologies. In both cases, the climate data are modelled as a simple function of the appropriate tree growth parameter(s) as follows

$$C_t = f(G_{tk})$$

where C_t is the climate in year t (one series in the first application and a two-dimensional array in the second) and G_{tk} is the chronology value at the various sites (k = 1, 2 . . . etc.) in year t.

Having established the prediction equations by fitting over a pre-selected calibration period, some feeling for the likely quality of any past estimates of climate produced by applying these equations to past tree-ring data can be gleaned by comparing the climate estimates with actual climate data. This comparison involves calculating a range of verification statistics over an independent period (i.e. one not used to calibrate the regression equations).

19.3   Mean July August Fenno-Scandinavian temperatures

Studies on the relationships between tree-growth and climate have a long history in Fenno-scandia. Early reviews were produced by Schulman 1944; Eklund 1954; Hoeg 1956 and Mikola 1956. The dominant influence of summer temperature as a control on the annual ring widths of northern conifers is well documented (for example, see Schove 1954; Eklund 1957; Sirén 1961; Mikola 1962; Jonsson 1969 and Aniol and Eckstein 1984). The following section describes recent work involving the use of ring-width and densitometric data to reconstruct summer temperatures on a regional scale.

Briffa *et al.* (1988a) assembled a group of 21 ring-width and two maximum latewood density chronologies, all *Pinus sylvestris*, spread mainly across northern Norway, Sweden, Finland and the western U.S.S.R. After experimenting with various lagged tree-ring predic-tors and different seasonal predictands, they used this network to reconstruct mean July-August temperatures for northern Fennoscandinavia back to A.D. 1700. The predictand temperature series was an average of gridded data at the intersections between the two latitude lines 65°N and 70°N and three longitude lines 10°, 20° and 30°E. These six gridpoint series had previously been constructed from selected series of homogenized monthly anom-aly data (1951-1970 reference period) from 19 stations spanning a latitudinal range between 62.6° and 70.4°N and a longitudinal range between 9.6° and 34.8°E (Jones *et al.* 1985).

The A.D. 1700 starting point for the earlier tree-ring reconstruction was dictated by the length of a number of the chronologies. Here we present a reconstruction of the same northern Fennoscandinavian temperature data based on a smaller network of sites but one in which the chronologies allow us to reconstruct back as far as A.D. 1580 (see Figure 19.1 and Table 19.1). The new network contains two total ring width (TRW) and two maximum latewood density (MXD) chronologies which were not included in the earlier reconstructions (i.e. numbers 4, 6, 8 and 9 in Table 19.1) and five individual site chronologies located in the region around Lake Inari which have now been averaged to form a single (and extended) 'Finnish Lapland' chronology.

19.4   Standardization of the raw data and chronology quality

The chronologies used in earlier reconstructions were standardized using 60-year Gaussian filters. This somewhat restrictive approach was adopted because many of the data were available only as averaged chronologies and so it was not possible to measure the degree of correspondence, especially of medium- to long-timescale fluctuations, in the raw data from different trees. It was therefore felt that a circumspect approach was warranted. Obviously, as a consequence of this, temperature fluctuations on timescales of about 60 years and above could not be reconstructed. We have attempted to rectify this shortcoming here.

The raw data were available for all but two of the chronologies listed in Table 19.1 (numbers 2 and 3, for which the early raw data were not available). The remaining chrono-

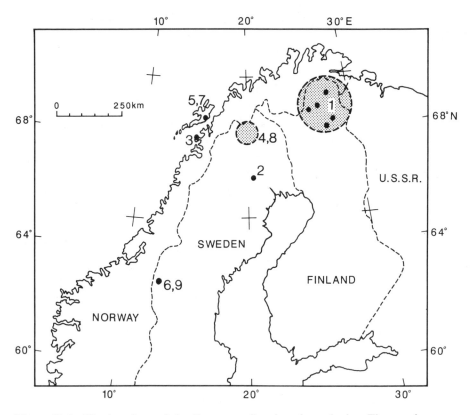

Figure 19.1 The locations of the Fennoscandinavian chronologies. The numbers are identified in Table 19.1. The gridpoints for which data were averaged to produce the Fennoscandinavian temperature series are shown as six crosses.

logies have been standardized using smoothing splines (Cook and Peters 1981). The flexibility of the splines was such as to remove 50 per cent of the variance of a cycle with a period equal to two thirds of the length of the particular series being processed. In other words, a 300-year series of raw ring-width measurements would be transformed into indices by calculating the quotients of the raw over the estimated data where the estimated data are the values from a 200-year spline fit through the raw data. Similarly, a 150-year series would use only a 100-year spline. This approach allows the data to dictate how much low-frequency variability is left in the standardized series, while at the same time ensuring that any age-related trends are effectively removed. This is a much less stringent approach to the removal of low-frequency variance than was used in the earlier work cited above.

The average length of standardization splines used in the production of each chronology is shown in Table 19.1. The values range from 118 years for Öst Fröstsjöåsen to 203 for Lofoten. Figure 19.2 illustrates the medium- to long-timescale variability present in two groups of individual site ring-width chronologies, one in the east (Figure 19.2a) and one in the west (Figure 19.2b) of Lapland following this '67 per cent' spline standardization. The two regionally-averaged chronologies formed from these data are also shown (Figure 19.2c). They are approximately 450km apart. The two lower curves (Figures 19.2d and 19.2e) show

Table 19.1 The chronologies used as predictors in the new northern Fennoscandinavian temperature reconstruction. Numbers on the left identify the chronologies shown in Figure 19.1.

Ring-width chronologies

Site	Lat.(N)	Long.(E)	Start	End	Mean spline[a]	r̄[b]	SSS>0.85[c]
1. Finnish Lapland mean of	~68°45'	~27°20'	1486	1983	202	0.38	1570
Sompio/Riukuselkå mean	68°15'	27°15'	1486	1983	191	0.36	1630
Morgammaras/Jurmarova mean	68°45'	26°30'	1536	1983	213	0.43	1629
Suojanperå	69°19'	28°08'	1532	1983	210	0.49	1624
2. Muddus	66°47'	20°08'	1532	1983	-	0.45	1720
3. Steigen	67°55'	15°05'	1396	1981	-	0.32	1454
4. Torneträsk	~68°15'	~20°15'	441	1980	194	0.35	551
5. Lofoten	68°29'	16°02'	1485	1978	203	0.33	1686
6. Öst Fröstsjöåsen	62°20'	12°48'	1580	1983	118	0.46	1617

Maximum latewood density chronologies

Site	Lat.(N)	Long.(E)	Start	End	Mean spline	r̄	SS>0.85
7. Lofoten	68°29'	16°02'	1485	1978	203	0.49	1670
8. Torneträsk	~68°15'	~20°15'	441	1980	194	0.40	551
9. Öst Frostsjöåsen	62°20'	12°48'	1580	1983	118	0.47	1617

Data sources: Chronologies 4, 8   Bartholin and Karlén (1983), Schweingruber et al. (1988).

                  5, 7   Schweingruber (pers. comm.).

                 Details of the other chronologies can be found in Briffa et al. (1988a).

[a]   The mean length of all individual splines used to standardize the raw data. For chronologies 2 and 3 a 60-year filter was used.

[b]   The mean correlation coefficient calculated using the maximum overlap between individual pairs of mean-tree series (see text). The values for chronologies 4 and 8 are based on one-core-per tree.

[c]   The earliest year when the SSS exceeds 0.85 (see text).

the correlations between these two regional chronologies at relatively high (periods <10 years) and low frequencies (periods >10 years) calculated over running 50-year periods (plotted at the end of the period). Except for the very early section before about A.D. 1580, the high frequency correspondence is generally very good with correlations for the most part at, or above, 0.7. The poor correlations in the 16th century data are a product of poor replication in the Lake Inari data (c.f. Table 19.1). The correspondence in the low-frequency curves is more variable but is still generally high for much of the 1600s and from the mid-18th to the mid-20th centuries. There are two periods of lower correlations (~0.4) corresponding to runs of comparison periods ending between about 1650 to 1700 and 1890 to 1910. The

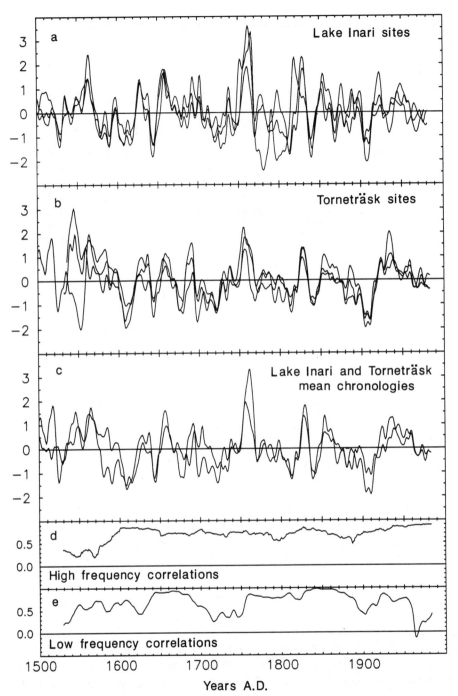

Figure 19.2 Low-frequency (i.e. variability >10 years) plots of ring-width chronologies at several sites around Lake Inari in northern Finland (a) and in the Torneträsk region of northern Sweden (b). The two regional-average curves are also plotted (c). Running 50-year correlation coefficients (plotted at the end of each 50-year period) are shown for the 10-year high-pass (d) and low-pass (e) comparisons of the Lake Inari and Torneträsk mean series. The chronology data are plotted as normalized departures from the mean of the whole data series. We thank Thomas Bartholin and Dieter Eckstein for allowing us to use their unpublished Torneträsk ring-width data which were used to produce Figure 19.2b.

similarity between the individual low-frequency site curves within each of the two regions during these same periods implies that these lower correlations probably reflect differences in inter-regional forcing and are not the result of retaining spurious (i.e. not regionally representative) low-frequency variability in the mean regional chronology. The sharp drop in the correlations plotted at about 1950-60 implies that the low frequency regional chronologies are out of phase between 1910 and 1960. In this instance, the individual site curves suggest that there may not be a common intra-regional signal within the Lake Inari data in this period. Taken as a whole, Figure 19.2 offers convincing evidence that the use of the 67 per cent spline is a valid means of standardizing these northern coniferous series where the replication is reasonable.

Averaging the correlation coefficients calculated between all possible standardized mean-tree series (where each comparison involves the maximum overlap between the two series) gives a value which represents the strength of common forcing (r in Table 19.1) within each chronology (Briffa and Jones 1989). This value, in conjunction with the numbers of constituent series throughout the length of the chronology, can be used to calculate the Subsample Signal Strength (SSS). This measures the statistical reliability of a part or parts of a chronology in relation to the uncertainty inherent in another, better replicated, section (Wigley *et al.* 1984; Briffa and Jones 1989). By calculating early SSS values relative to the recent sections of the chronologies (against which the chronology/climate links are calibrated) one can show the reliability or otherwise of the early data. Table 19.1 shows that only four of the nine chronologies used as predictors here have SSS values greater than 0.85 for the whole period from 1580. A value of 0.85 is equivalent to a maximum uncertainty of 15 per cent. This is an arbitrary but nonetheless useful figure for general comparison (Wigley *et al.* 1984). Two other chronologies exceed this value after 1617 but the other three do not reach this value until after 1670, 1686 and 1720. We shall return to this point later.

19.5 Results

The performances of alternative regression equations (calibrated over the two 50-year sub-periods 1876-1925 1926-1975) are summarized in Table 19.2. This shows the variance accounted for by the fitted equations and includes the results of a number of verification tests involving comparisons of the regression-based temperature estimates with independent observational data. These results indicate that both sub-period prediction equations are performing well. The results also compare favourably with those achieved in the earlier published work (Briffa *et al.* 1988a; also summarized in Table 19.2). Both verification period r^2 values (i.e. the square of the simple correlation coefficient between the actual and the estimated values) produced here are higher than in the previous work. In addition, variance is now apparently being faithfully reconstructed at both high and low frequencies. The r^2 values calculated after applying a 10-year high-pass filter to the actual and estimated data are comparable with the equivalent result achieved in the earlier work. Previously, however, the low-pass verification r^2 value was only 0.29, probably due to the removal of too much of the low frequency variance during chronology standardization. Now, following the 67% spline approach, higher values are achieved using both early and late-period low-pass verification comparisons: 0.69 in the late-calibration/early verification case. The other verification tests

Table 19.2 Independent verification period statistics for the early (1876-1925) and late (1926-1975) calibrated reconstructions. For comparison, the results for the previously published reconstruction calibrated over 1891-1964 and verified over 1852-1890 are also shown.

	Early Calibration	Late Calibration	Briffa et al. (1988a)
Calibration R^2	0.45	0.60	0.56
Verification r^2	0.55***	0.48***	0.45***
Verification r^2 (frequency variations < 10 years)	0.56***	0.43***	0.56***
(frequency variations > 10 years)[a]	0.52	0.69*	0.29
Reduction of Error[b]	0.57	0.63	0.45
Coefficient of Efficiency[c]	0.41	0.43	0.45
1st Difference Sign test[b]	42 correct*** 7 incorrect	35 correct* 14 incorrect	30 correct* 8 incorrect
Product Means Test[b]	4.4***	3.2**	3.7*

*p = 0.01   **p = 0.001   ***p = 0.0001

[a]   degrees of freedom corrected for autocorrelation when calculating significance. (Mitchell et al., 1966)
[b]   see Fritts (1976) and Gordon and Le Duc (1981).
[c]   see Briffa et al. (1988a).

also indicate that reconstruction performance is good on both year-to-year and longer timescales.

Having established this, and in order to maximize the range of frequencies against which we could fit the final prediction equation, we recalibrated the model over the whole 1876-1975 period. This equation accounted for 58 per cent of the temperature variance over this period. The regression weights for this and the two sub-period calibrations are plotted in Figure 19.3. The equations are generally stable in as much as only the weight on the Steigen ring-width chronology changes noticeably between sub-periods. The reconstructions based on all three equations therefore correlate highly at both high and low frequencies as is demonstrated in Table 19.3. This also demonstrates how the final reconstruction presented here (based on the 1876-1975 calibration) differs from the earlier published reconstruction most noticeably on longer timescales. The most significant predictors are the ring-width and the maximum latewood density chronologies at Öst Fröstsjöåsen and the density chronology from Torneträsk. The statistical quality of these most important predictor chronologies is reasonable from around the beginning of the 17th century (c.f. Table 19.1) and we consider the reconstructions to be generally reliable after this date. Table 19.4 shows the results of an additional verification exercise where early fragmentary station records in Finland and

Sweden (Bradley *et al.* 1985) are compared with our regional temperature reconstruction. Considering the disparity between the spatial representativeness of the regional and various local records these correlations are encouraging.

The actual and the estimated data for the period 1876-1975 are shown in Figure 19.4 and their variance spectra calculated over the same period, are shown in Figure 19.5a. Over 20 per cent of the variance of the actual data is associated with periods at and above 66 years (the trend accounts for over 11 per cent and 10 per cent is associated with a cycle at about 66 years).

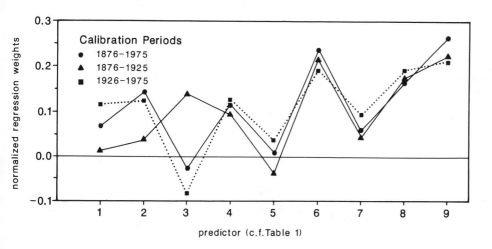

Figure 19.3 The standardized regression weights for each of the chronologies (see Table 19.1) calculated using different periods of dependent data.

Table 19.3 Comparison of the reconstructions based on equations variously calibrated over the periods 1876-1925 (early) 1926-1975 (late) and 1876-1975 (overall). The figures shown are correlation coefficients calculated between the raw series and between the 10-year high- and low-pass filtered reconstructions over the period 1580-1975. Comparisons with the earlier published reconstruction (Briffa *et al.* 1988a) over the period 1700-1964 are also shown.

		Early			Late			Overall		
		raw	high	low	raw	high	low	raw	high	low
Early	raw				0.92			0.97		
	high					0.96			0.98	
	low						0.88			0.97
Late	raw							0.98		
	high								0.99	
	low									0.97
Briffa et al. (1988a)	raw	0.60			0.65			0.63		
	high		0.70			0.74			0.73	
	low			0.45			0.51			0.47

Table 19.4 Comparison of final reconstruction with early fragmentary data for stations in Sweden and Finland.

Station	Lat.(°N)	Long.(°E)	Period (years)	r
Karesuando	68.5	22.5	1830-1838(9)	0.64*
Umeå	63.8	20.3	1797-1803(7)	0.48
			1797-1802(6)	0.86**
Övertorneå[1]	66.4	23.8	1802-1832(29)	0.43**
Hailuoto Carlö[2]	65.0	24.7	1817-1836(20)	0.50**
Oulu	64.9	25.4	1776-1784(9)	0.42
Vöyri	63.2	22.0	1800-1824(25)	0.66***

* p = 0.1   ** p = 0.05   *** p = 0.001

[1] Does not include 1814 and 1815 which are obviously in error in the station data

[2] Inhomogeneity in 1828-36 corrected by addition of + 3.1°C

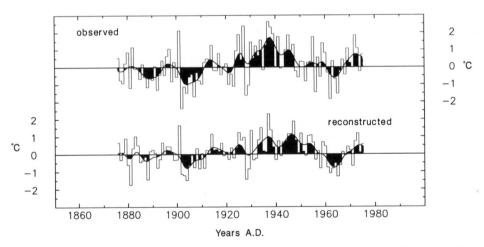

Figure 19.4 Actual and estimated July/August temperatures for northern Fennoscandia. The estimates are based on the regression equation calibrated over 1876-1975. The smoothed curve shows 10-year filtered values.

The other significant peak in the spectrum corresponds to a period of 2.36 years. This explains 4 per cent of the temperature variance. There are also insignificant peaks at 2.0 and 6.6 years. The spectrum of the estimated data again shows power at low frequencies (~66 years and above) and at frequencies corresponding to periods of 2.2-2.3 and 6 years. However, the only significant peaks in this spectrum represent a range of periods from 3.3 to 3.9 years. The coherency spectrum for the actual and estimated data is shown in Figure 19.5b. The coherency is generally good across the whole spectrum. Only in one area, between 2.7 and 2.9 years is there relatively low coherence. The coherence at periods of 2.3, 3.3 to 3.7, 9.4

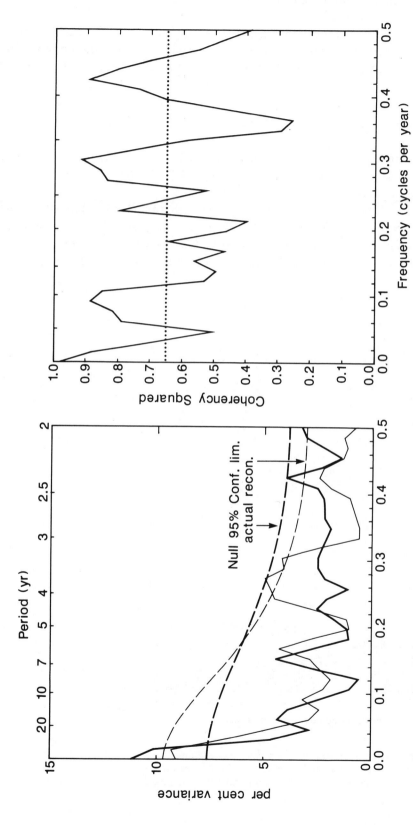

Figure 19.5   The variance spectra for the actual Fennoscandinavian temperature data (thick line) and the estimated data (thin line) calculated over the long calibration period, 1876–1975. The spectra were calculated using 33 lags of the autocovariance function. The raw spectral estimates were filtered with the Hamming window and each estimate has 7 degrees of freedom. The coherency spectrum for the two series is also shown. The dotted line shows the 95 per cent confidence limit. For *a priori* chosen frequencies with values above this level, the null hypothesis that the true coherence is zero must be rejected (e.g. see Jones 1985).

to 13 and above 66 years is particularly high. Interestingly, the second of these (3.3 to 3.7 years) is the band where the estimated data showed a significant concentration of variance (i.e. at periods between 3.3 and 3.9 years) not shown in the actual data.

19.6   Discussion of the reconstruction

The variance spectrum of the reconstructed data from A.D. 1580 to 1978 is shown in Figure 19.6. Significant peaks correspond to periods of 2.33, 2.89, 2.99, 3.02, 3.13, 3.59, 3.97, 4.16, 33.25, 38.0 and 88.7 years.

Sirén and Hari (1971) analysed tree-ring data from northern Finland and found that they contained major cycles at ~3.3, 3.6, 7.4, 22.8, 32 and 80-96 years. They also analysed a 780-year series of summer-deposited lake varves from Estonia which were said to data from circa 1200 B.P. The spectrum for these contained major peaks at periods of 2.2, 3,6, 4.3, 5.9,

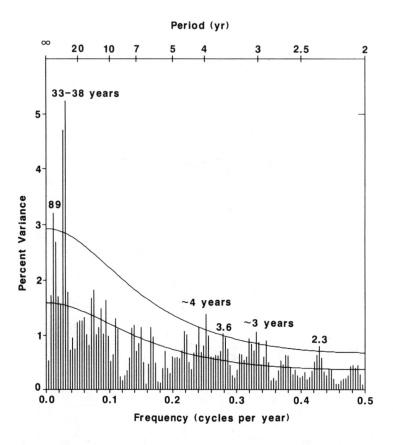

Figure 19.6   The variance spectrum of the reconstructed Fennoscandinavian temperature series calculated over 1580-1978. The estimates are calculated from 133 lags of the autocorrelation function and have 7 degrees of freedom.

7.0, 9.1, 10.7 and 32.0 years. This series also contained a periodicity at about 80-96 years though the authors considered that it was not statistically significant. Concentrations of variance corresponding to cycles with periods of about 3.6, ~33.0 and 80-96 years are therefore notable in both of these series and in our reconstructed temperature data. The longest of these cycles corresponds to the length of the much discussed Gleissberg cycle which has a period of about 88 years (for example see Attolini *et al.* 1988). A significant cycle of this length has also been identified in the power spectrum of ^{14}C production rate at the Earth's surface, itself extracted from the C^{14} tree-ring record (Stuiver and Kra 1986). Stuiver and Braziunas (1989) have apparently also identified a near-88 year cycle in records of alpine glacier fluctuations produced by Röthlisberger (1986) (see also, Stuiver and Braziunas, Chapter 30, this volume). By far the most significant concentration of variance in our temperature reconstructions, however, corresponds to a period band between 33 to 38 years. Renberg *et al.* (1984) report a strong cyclic variation with a period of between 30 and 40 years in the thicknesses of varves in Lake Judesjön, east central Sweden. Their series spans the years A.D. 230-1532 and is thought to reflect year-to-year changes in organic matter production resulting from variations in summer (June-August) insolation.

The reconstructed July-August temperatures back to A.D. 1580 are plotted in Figure 19.7. The smoothed curve joins the 10-year low-pass filtered values. The data are plotted in degrees Celsius as anomalies from the mean for 1951 to 1970. It is apparent that a number of periods since the beginning of the 17th century have been equal in degree and duration to the relative warmth of the 1930s and 1940s. They include the periods encompassing the 1650s and 1660s, the 1680s and 1690s, the 1750s and 1760s and much of the 1850s and 1860s. Other shorter warm periods include the 1620s, mid-1720s to mid-1730s, and the ten years from about 1825 to 1835. In relation to the relatively cool base period of 1951-1970, there are fewer cool periods in the reconstruction. The most significant by far is the period from 1590 to 1609. Excluding two years reconstructed as near the base period mean (in 1598 and 1599) the mean for the other 18 years of this period is -0.81°C. Other relatively cold periods reconstructed prior to the start of the instrumental data include the 1670s, mid-1700 to 1720 and the early part of the decade beginning in 1800. Between A.D. 1580 and 1875, the reconstructed data contain four positive and two negative values that are more than 2°C away from the

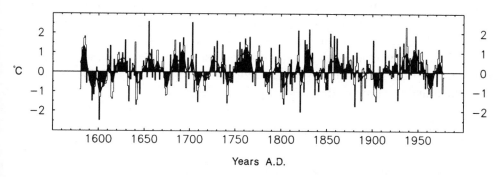

Figure 19.7 The reconstructed July/August temperatures for northern Fennoscandia plotted from 1580 to 1975. The values are °C anomalies from the mean for the period 1951-1970. The smooth curve represents 10-year low-pass filtered data.

base-period mean. The positive values are in 1655 (2.6°C) 1703 (2.6°C) 1819 (2.2°C) and 1831 (2.2°C. The largest negative values occurred in 1601 (-2.5°C) and 1821 (-2.1°C).

19.7   Central European temperatures from A.D. 1750

Summer temperatures were recently reconstructed over Europe from A.D. 1750 using a 37-chronology network comprised only of maximum latewood density chronologies (Briffa *et al.* 1988b). Data at these sites were chosen on the basis of length from a larger network assembled at the Swiss Federal Institute for Forest, Snow and Landscape Research, Birmensdorf (Bräker 1987). Summary response function analyses showed the presence of a widespread common climate response in these chronologies: enhanced density being associated with anomalous warmth, particularly during the spring months of April and May and the late summer months of August and September. A similar, but generally less ubiquitous response was apparent in the mid-summer months of June and July. These results indicated the potential for reconstructing summer half-year temperatures (April-September) over western Europe using this network of site data.

19.8   Climate and tree-ring data

19.8.1   Temperature Data

The predictand temperature grid is shown in Figure 19.8. There are 25 points covering a range of latitudes from 40° up to 70°N and a maximum range of longitudes from 0° to 30°E (there are no data at 70°N 0°E; 70°N 10°E and 65°N 0°E). These data are part of the hemispheric compilation of monthly mean data interpolated from station series and expressed as degree Celsius anomalies with respect to the period 1951-1970 (Jones *et al.* 1985 1986). The effective starting date for the gridded temperature data is 1875.

19.8.2   Tree-Ring Data

The predictors are a set of 37 chronologies each extending back at least to A.D. 1750 (Figure 19.8). They are a subset of the longest series from a larger European site network constructed from trees selected for maximum potential temperature sensitivity (Schweingruber 1987). The sites tend to be at high elevation or high latitude and are primarily located north of 65°N or south of 50°N. Between these latitudes there are only six sites and four of these lie west of the Greenwich meridian, beyond the western edge of the predicted grid. Details of the precise altitudes and locations of the sites are given in Table 19.5.

19.9   Results and verification analyses

19.9.1   An initial cross calibration-verification analysis

As with the previously described Fennoscandinavian work, a cross-calibration/verification scheme was used to gauge the strength of the tree-growth climate link and its stability in time.

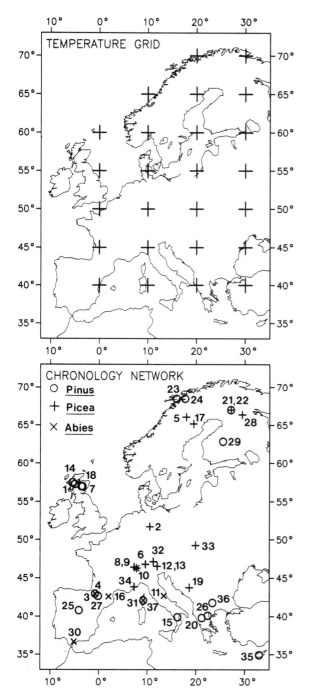

Figure 19.8 The locations of the 37 maximum latewood density chronologies. The chronologies are identified and further details are given in Table 19.5. The 25 temperature grid points which make up the dependent grid are shown as crosses.

This involved splitting the temperature data into two 50-year periods, 1876-1925 and 1926-1975, and then first calibrating the spatial transfer function using the earlier data and verifying the calibrated equations by comparing the estimated and actual data over the later period. The periods were then reversed and the process repeated. The results can be expressed in terms of the major principal component (PC) amplitude series of the predictand network or

379

Table 19.5 Details of the maximum latewood density chronologies shown in Figure 19.8. PISY = *Pinus sylvestris*; PIMU = *P. mugo*; PILE = *P. leucodermis*; PINI = *P. nigra*; PCAB = *Picea abies*; ABAL = *Abies alba*; ABPI = *Abies pinsapo*.

	Chronology Name	Lat(^{O}N)	Long	Alt(m)	Species
1	Glen Affric	57°17'	4°55'W	300	PISY
2	Andreasberg	51°43'	10°34'E	900	PCAB
3	Pic d'Anie	42°58'	0°44'W	1750	PIMU
4	Arette, Col St.Martin	42°59'	0°45'W	1500	ABAL
5	Arjeplog	66°04'	17°59'E	600	PCAB
6	Arosa	46°48'	9°41'E	1940	PCAB
7	Ballochbuie	56°59'	3°19'W	380	PISY
8	Bürchen Bielwald	46°17'	7°50'E	1740	PCAB
9	Bürchen Bielwald	46°17'	7°50'E	1740	ABAL
10	Lauenen Brüchli	46°25'	7°19'E	1500	PCAB
11	Ceppo	42°41'	13°26'E	1700	ABAL
12	Cortina d'Ampezzo	46°32'	12°04'E	1820	PCAB
13	Cortina d'Ampezzo	46°32'	12°04'E	1900	PCAB
14	Coulin	57°32'	5°21'W	250	PISY
15	Sierra da Crispo	39°54'	16°14'E	2000	PILE
16	Formigueres	42°36'	2^{O}04'E	1700	ABAL
17	Gallejour	65°10'	19°28'E	480	PCAB
18	Inverey	57°00'	3°35'W	500	PISY
19	Jahorina	43°45'	18°38'E	1700	PCAB
20	Katara Pass	39°48'	21°13'E	1750	PILE
21	Laagennus	67°00'	27°07'E	320	PISY
22	Laagennus	67°00'	27°07'E	270	PCAB
23	Lofoten	68°29'	16°02'E	200	PISY
24	Narvik	68°29'	17°44'E	50	PISY
25	Navacerrada	40°48'	4°02'W	2050	PISY
26	Olympos	40°05'	22°25'E	2250	PILE
27	Ordessa	42°40'	0°07'W	1870	PIMU
28	Ouluangan	66°22'	29°26'E	260	PCAB
29	Pyhän Häkin	62°51'	25°29'E	185	PISY
30	Torrecilla	36°40'	5°05'W	1650	ABPI
31	Col de Sorba	42°04'	9°12'E	1400	PINI
32	Stubaital	47°08'	11°17'E	1850	PCAB
33	Swistowko	49°15'	19°55'E	1500	PCAB
34	Le Tournairet	43°52'	7°20'E	2050	PCAB
35	Troodos	34°55'	32°55'E	1600	PINI
36	Vihren	41°46'	23°23'E	1920	PILE
37	Vizzavona	42°05'	9°12'E	1500	ABAL

in terms of the individual grid point data (Briffa *et al.* 1986). Table 19.6 summarizes the results of the reconstructions of the major climate PCs. These are defined here as those PCs which together explain about 80 per cent of the climate variance: five in the early period and three in the later period. Overall, somewhat superior results are achieved when the transfer function is calibrated over the later period. Nevertheless, the early and late verification results indicate that when averaged over the whole grid, some 30 per cent of the temperature variance is being reliably estimated. However, reconstruction performance varies across the grid. The link between temperature and maximum ring density is strongest in the north and west of the grid. This is illustrated in the pattern of explained variance in the calibration period (Figure 19.9a) and, more importantly (as this map gives the best picture of the likely reliability of the reconstructions from 1750 onwards) in the verification period (Figure 19.9b). In the verification map over 50 per cent explained variance is apparent over central Fennoscandia. Over southern Scandinavia and much of Germany, Switzerland, northern Italy and

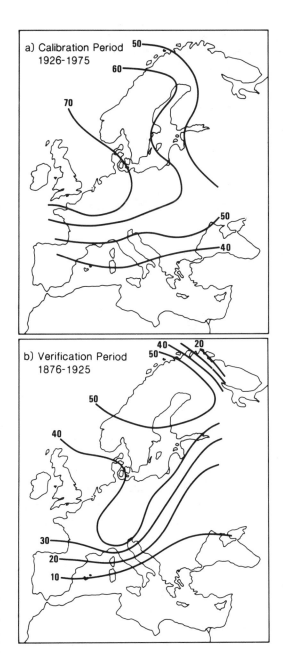

Figure 19.9 Variance explained when calibrating (a) and verifying (b) the transfer function for April-September temperatures.

eastern France the explained variance is over 40 per cent. With the exception of these regions the explained variance north of 45°N (including the British Isles and west and central France) is still above 30 per cent. Over the southern Mediterranean and the Balkans however, explained variance falls to below 10 per cent and little confidence can be placed in the temperature estimates for the extreme southern and southeastern grid points.

Table 19.6 Cross calibration/verification results in terms of explained variance over the whole grid and in terms of the individual climate PCs. See Table 19.2 for references to RE and CE.

Early calibration period: 13 candidate predictor PCs
Total Grid Variance Explained 0.47 in Calibration 0.30 in Verification

Climate PC	% Variance	No of Predictors	Calibrated Variance	Verified Variance	RE	CE
1	29.4	7	0.77	0.27	0.24	-0.13
2	20.4	9	0.59	0.48	0.43	0.38
3	15.7	4	0.52	0.54	0.53	0.53
4	8.4	4	0.17	0.11	0.09	-0.32
5	5.0	4	0.32	0.13	0.14	-0.01

Late calibration period: 14 candidate predictor PCs
Total Grid Variance Explained 0.55 in Calibration 0.33 in Verification

Climate PC	% Variance	No. of Predictors	Calibrated Variance	Verified Variance	RE	CE
1	32.6	10	0.72	0.51	0.29	-0.21
2	23.4	5	0.62	0.43	0.32	0.31
3	21.7	9	0.67	0.55	0.53	0.54

19.10 A recalibrated transfer function

Following the cross-calibration exercise the final transfer function was recalibrated using all 100 years of gridded temperature data from 1876-1975. This was done in order to maximise the potential variance of the calibration data and to include potential variability at longer wavelengths than can be represented in a 50-year period. The variance explained by the recalibrated transfer function averages 44 per cent across the whole grid but the spatial pattern conforms to that shown in Figures 19.9a and 19.9b with greater variance calibrated in the northwest and central regions and poorest calibration in the extreme southeast. Figure 19.10 illustrates the similarity between the gridpoint temperature estimates based on the recalibrated transfer function and those based on the two separate 50-year calibrations verified formally in Table 19.6 and Figure 19.9b. These maps are based on comparisons made over the period 1750-1875 (i.e. outside any of the fitting periods). Common-variance generally above 80 per cent is apparent over most of the reconstruction region, indicating that the spatial pattern of verification performance shown in Figure 19.9b provides a reasonable guide to the reliability of our ultimate reconstructions which are based on the recalibrated transfer function.

Additional support for the veracity of these reconstructions is provided in Table 19.7. This contains correlation coefficients between nine long instrumental temperature series (Bradley *et al.* 1985) and the nearest of our reconstructed grid point series. The correlations are all significant at the 0.01 level of probability. They are higher in the northern half of the grid which is again consistent with the Figures 19.9a and 19.9b. These correlations should be viewed alongside those calculated between the actual station data and the predictand gridpoint data (also shown in Table 19.7) calculated over 1876-1975. The correspondence is not

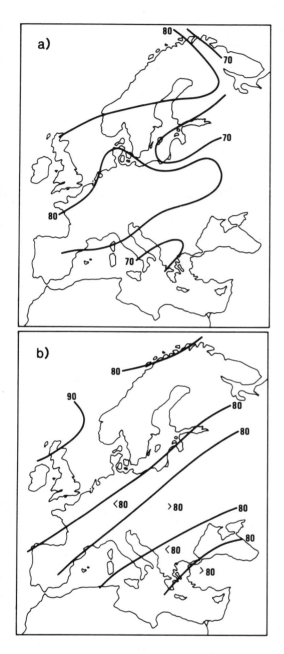

Figure 19.10 Variance in common between the reconstruction based on the 1876-1975 calibration and each of the sub-period calibrations: (a) 1876-1925 and (b) 1926-1975. The contours are the squared correlation coefficients calculated at each grid point over the period 1750-1875.

perfect, even between the station data and the actual gridpoint data, as the gridded series represent inverse-distance weighted averages of data from a varying number of stations (Jones *et al.* 1986) and some (non-common) variance from each station series is inevitably lost in the averaging. Viewed in this light the early reconstruction/station correlations demonstrate the general fidelity of the dendrochronologically-based temperature estimates.

Table 19.7 Comparisons of early station temperature records (April-Sept) and the nearest reconstructed grid point series. All correlations are significant at the 0.01 level of probability.

Station (Lat.Long.)	Grid Point[1]	Distance[2] (km)	Years	r_{recon}[3]	$r_{station}$[4]
Trondheim (63.4°N 10.5°E)	(65°N 10°E)	180	1761-1875(115)	0.57	0.74
Stockholm (59.4°N 18.1°E)	(60°N 20°E)	120	1756-1875(120)	0.71	0.78
Edinburgh (55.9°N 3.2°W)	(55°N 0°E)	240	1764-1875(112)	0.67	0.84
Paris (48.8°N 2.5°E)	(50°N 0°E)	220	1757-1875(119)	0.64	0.90
Berlin (52.5°N 13.4°E)	(55°N 10°E)	360	1750-1875(126)	0.46	0.76
Milan (45.5°N 9.2°E)	(45°N 10°E)	90	1763-1875(113)	0.52	0.76
Rome (41.7°N 12.5°E)	(40°N 10°E)	280	1811-1875(65)	0.58	0.58
Leningrad (60.0°N 30.3°E)	(60°N 30°E)	90	1750-1875(126)	0.38	0.68
Central England (~52.5°N ~1.0°W)	(55°N 0°E)	290	1750-1875(126)	0.72	0.96
Root mean square				0.59	0.79

[1] Nearest grid-point.

[2] Approximate distance between the station location and the grid point.

[3] Correlation coefficient between the grid-point reconstruction and the station data.

[4] Correlation coefficient between predicted gridpoint temperature data and the individual station temperature series calculated over 1876 to 1975.

19.11   Assessing the role of the chronology predictors

One measure of the overall importance of each chronology is given by the root mean square (MW) of their standardized regression weights over all individual gridpoint reconstruction equations i.e.

$$MW = \left[\sum_{i=1}^{I} d_{in}^2/I \right]^{1/2}$$

where the d_{in} are the regression weights for $I=25$ gridpoints and n is the nth chronology of the 37-site network. The values range from 0.093 to 0.024. The five chronologies with largest MW values are Swistowko, Coulin Formigueres, Pyhän Häkin and Sierra da Crispo (c.f. Figure 19.8 and Table 19.5). For each of these in turn, the regression coefficients in all grid-point equations are contoured in Figures 19.11a to 19.11e. These more important chronologies are

Figure 19.11 Maps showing the standardized regression weights of the five most influential chronologies (a to e) as defined in the text. The chronologies are shown as large dots. The weight on the particular chronology in each individual gridpoint regression equation is contoured to demonstrate the areas of the grid where that chronology is most influential in predicting summer temperature. The final map (f) illustrates the influence of the least significant chronology.

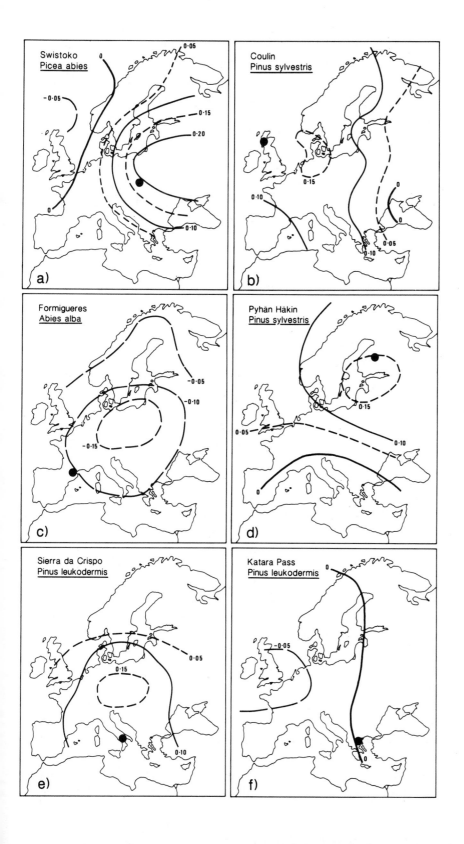

seen to be generally well separated and spaced evenly across most of the map. It is also apparent that their major spheres of influence encompass the whole central band between 45° and 65°N in which the density of chronologies in the original network is very low. All but one of the major chronologies have positive MW values. The exception is Formigueres in the French Pyrenees. Ring density at this site displays a negative relationship with warm temperatures over central Europe, probably reflecting the effect of increased water stress brought about during warmer summers. Seven of the chronologies with smallest MW values lie south of the Alps. It is likely that they are influenced by climate conditions prevailing over the Mediterranean which are not necessarily systematically associated with conditions over our more northerly grid. An example from this group, Katara Pass at Metsovon in Greece, is shown in Figure 19.11f. It has the lowest MW value of the chronology network. The pattern of regression weights for this chronology has little definition and its largest weights are for grid points over Great Britain. These are almost certainly spurious.

19.12   Regional average reconstructions

Figure 19.12 shows three series of reconstructed half-year temperatures. Each series represents a different region of our temperature grid. The 'United Kingdom' series is the average of the reconstructions at each of three grid points, all on the Greenwich meridian: at 50°, 55° and 60°N. The 'Scandinavia' series is made up of eight series: two at 70°N at 20° and 30°E, three at 65°N at 10°, 20° and 30°E and three at 60°N at 10°, 20° and 30°E. The 'Central Europe' series is the mean of eight grid point series: three at 55°N at longitudes of 10°, 20° and 30°E, three at the same longitudes at 50°N and two at 45°N at 10° and 20°E. These three regional series therefore represent 19 out of the 25 series which make up our whole grid. The poor reconstruction performance in the southern part of the grid does not warrant the production of a 'Mediterranean' series.

All of these regional data are plotted as anomalies in degrees Celsius with respect to the mean for the period 1951-70. The filtered values are the 10-year low pass data. There are four periods during the reconstruction when temperatures appear to be in phase throughout Europe and Scandinavia. Two are cold excursions, one from around 1810 to 1817 and the other from around 1832 to 1838. Between these periods, from about 1819 to 1828 relatively warm conditions were widespread. Apparently, general warmth also prevailed from about 1868 to 1875. There were no other periods between 1750 and 1875 when the same conditions prevailed over the whole of the grid. The early 1860s were cool in the U.K. and in Scandinavia but relatively warm in central Europe, whereas the 1840s and 1850s were warm in Scandinavia and central Europe but conditions were close to the 1951-70 norm in Britain during this time.

For quite a long period (some 80 years from 1760 to 1840) the smoothed temperature curves for the United Kingdom and Scandinavia appear to run in parallel with both series showing periods of relative warmth centred around 1760 and 1779 and during the decade of the 1820s. They also both show cooler periods occurring in 1767-1772 and to some extent during the early 1780s as well as during the 1810s and 1830s. During this whole period (i.e. from 1760-1840) with the one exception of the 1810s to which we have already referred, summers in central Europe remained firmly above the mean for 1951-70. Indeed, but for the

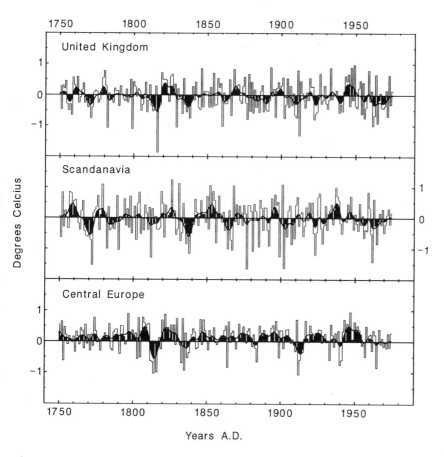

Figure 19.12 Regionally averaged reconstructions of April-September temperature. The regions are defined in the text. The filtered curves are 10-year low-pass values.

1810s and 1830s, the smoothed anomaly curve representing central European summers is consistently above zero for the whole period of reconstruction for 1750 to 1875.

One final point worthy of mention relates to the Scandinavian curve. The period between 1790 and 1810 is marked by strong interannual variability, to a degree which is not matched at any other period in the Scandinavian reconstruction or indeed at any time in either of the reconstructions for the U.K. or central Europe.

19.13   Annual maps

Besides the mean regional summer temperature curves shown in Figure 19.12 we have also produced a series of year-by-year April-September summer temperature maps. The full set of maps will be published elsewhere (Schweingruber *et al.*, in preparation). However we present some examples here. Figure 19.13 shows observed and reconstructed temperature anomalies contoured at 0.5°C intervals for the eight years from 1932 to 1939. These were

Beobachtete (obere Reihe) und rekonstruierte Temperaturen (untere Reihe) 1932–1939
Observed (upper maps) und Reconstructed Temperatures (lower maps) 1932–1939

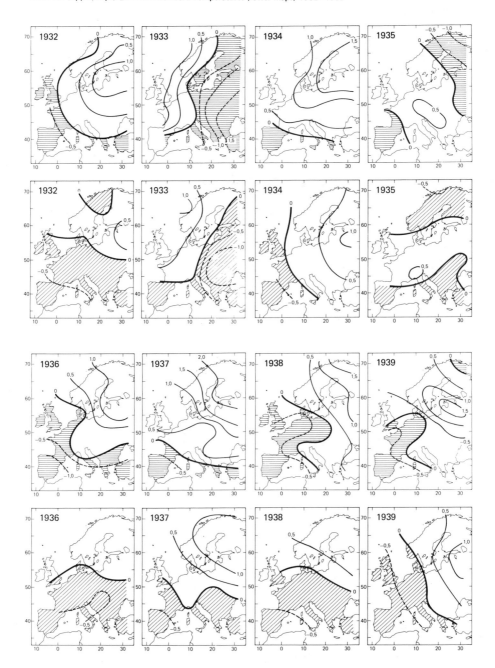

Figure 19.13   An example set of actual and estimated summer temperature maps.

chosen at random to give some indication of the level of correspondence between the observed and estimated patterns and to give some idea of the degree of definition and spatial coherence between the actual and the estimated data. Figure 19.14 is a similarly random

Rekonstruierte Temperaturen 1814–1829
Reconstructed Temperatures 1814–1829

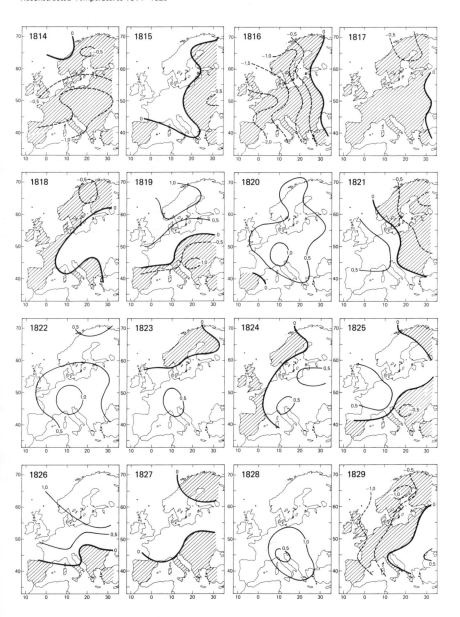

Figure 19.14 A selection of the summer temperature reconstructions from a period prior to the start of the gridded temperature data.

example showing 16 reconstructed maps (1814-1829) for a period prior to the start of the gridded temperature data in 1875. This is an interesting period as it contains much of the decade of the 1810s, a period when conditions were generally cold throughout Europe and the 1820s which as we have said earlier were generally warm. This warmth is particularly apparent in 1820, 1822 and 1828.

389

19.14 Conclusions

We have described the reconstruction of past summer temperatures in the north and central regions of western Europe and presented time series of late summer temperatures for northern Fennoscandinavia extending back to before A.D. 1600, and three regional series representative of April-September temperatures for the whole of Scandinavia, the United Kingdom and central Europe from A.D. 1750. In addition we have shown examples drawn from a set of year-by-year summer temperature maps covering much of western Europe between 40° and 70°N and 0° and 30°E.

We consider that these results are a vindication of the dendroclimatic sampling strategy proposed by Schweingruber and colleagues; namely the collection of material suited for the measurement of densitometric as well as ring-width data especially from cool moist sites so as to maximize the potential value of the resultant data for the reconstruction of past temperature variability. Already further collections have been made in western North America and across northern Canada. Summer temperature reconstructions have been made back to A.D. 1600 using the American data and these will be published shortly. The Canadian data, collected in collaboration with Gordon Jacoby, Lamont Doherty Geological Observatory, New York are currently being processed. Field work to collect samples from the northern parts of the U.S.S.R. is currently planned for the summers of 1990 and 1991.

References

Aanstad, S. 1938: Die Jahresringbreiten einiger seltener Kiefern in Steigen, Nordland. *Nytt Magasin for Naturvidenskapene 79*, 127-140.

Aniol, R. W. and Eckstein, D. 1984: Dendroclimatological studies at the northern timberline. In Mörner, N.-A. and Karlén, W. (eds.), *Climatic Changes on a Yearly to Millenial Basis*. Dordrecht: Reidel, 273-279.

Attolini, M. R., Galli, M. and Nanni, T. 1988: Long and short cycles in solar activity during the last millennia. In *Secular Solar and Geomagnetic Variations in the Last 10,000 years* (F. R. Stephenson and A. W. Wolfendale (Eds.) 49-68. Kluwer, Dordrecht.

Bartholin, T. S. and Karlén, W. 1983: Dendrokronologi: Lapland. *Dendrokronologiska Sallskapets Meddelanden 5*, 3-16.

Bradley, R. S., Kelly, P. M. Jones, P. D., Goodess, C. M. and Diaz, H. F. 1985: *A Climatic Data Bank for Northern Hemisphere Land Areas, 1851-1980*. U.S. DoE Technical Report TRO17, U.S. Department of Energy Carbon Dioxide Research Division, Washington, D.C.

Bräker, O. U. 1987: The European data base at the EAFV. In *Methods of Dendrochronology-I* (L. Kairiukstis, Z. Bednarz, and E. Feliksik, Eds.), pp. 133-136. IIASA/Polish Academy of Sciences, Warsaw.

Briffa, K. R. and Jones, P. D. 1989: Basic chronology statistics and assessment. In *Methods of Dendrochronology* (E. R. Cook and L. Kairiukstis Eds.) 137-152. Kluwer, Dordrecht.

Briffa, K. R., Jones, P. D., Wigley, T. M. L., Pilcher, J. R. and Baillie, M. G. L. 1986: Climate reconstruction from tree rings: Part 2, Spatial reconstruction of summer mean sea-level pressure patterns over Great Britain. *Journal of Climatology 6*, 1-15.

Briffa, K. R., Jones, P. D., Pilcher, J. R. and Hughes, M. K. 1988a: Reconstructing summer temperatures in northern Fennoscandinavia back to A.D. 1700 using tree-ring data from Scots Pine. *Arctic and Alpine Research 20*, 385-394.

Briffa, K. R., Jones, P. D., and Schweingruber, F. H. 1988b: Summer temperature patterns over Europe: A reconstruction from 1750 A.D. based on maximum latewood density indices of conifers. *Quaternary Research 30*, 36-52.

Cook, E. R. and Peters, K. 1981: The smoothing spline: a new approach to standardizing forest interior tree-ring width series for dendroclimatic studies. *Tree-Ring Bulletin 41*, 45-53.

Cook, E. R., Briffa, K. R. and Jones, P. D. 1988: *Spatial regression methods for dendroclimatology: a review and comparison of two techniques.* Research Report Submitted to the Scientific Affairs Division North Atlantic Treaty Organization. 59pp.

Eklund, B. 1954: Årsringsbreddens klimatiskt betingade variation hos tall och gran inom norra Sverige åren 1900-1944. *Meddelanden från Statens Skogsforskninstitut, 44(8)*. 150pp.

Eklund, B. 1957: Om granens arsringsvariationer inom mellersta Norrland och deras samband med klimatet. *Meddlelanden från statens Skogsforskningsinstitut 47*, 1-63.

Feynman, J. and Fougere, P. F. 1984: Eighty-eight year periodicity in solar-terrestrial phenomena confirmed. *Journal of Geophysical Research 89* (A5), 3023-3027.

Fritts, H. C. 1976: *Tree Rings and Climate*. Academic Press, London. 567pp.

Gordon, G. A. and LeDuc, S. K. 1981: Verification statistics for regression models. In *Preprints Seventh Conference on Probability and Statistics in Atmospheric Sciences* 129-133. American Meteorological Society, Monterey.

Hoeg, O. A. 1956: Growth-ring research in Norway. *Tree-Ring Bulletin 21* 2-15.

Jones, R. H. 1985: Time series analysis – Frequency domain. In *Probability, Statistics, and Decision Making in the Atmospheric Sciences*. (A. H. Murphy and R. W. Katz Eds.) 189-221. Westview Press, Boulder, Colorado.

Jones, P. D., Raper, S. C. B., Bradley, R. S., Diaz, H. F. Kelly, P. M. and Wigley, T. M. L. 1986: Northern Hemisphere surface air temperature variations, 1851-1984. *Journal of Climate and Applied Meteorology 25*, 161-179.

Jones, P. D., Raper, S. C. B., Santer, B. D., Cherry, B. S. G., Goodess, C., Bradley, R. S., Diaz, H. F., Kelly, P. M. and Wigley, T. M. L. 1985: *A Grid Point Surface Air Temperature Data Set for the Northern Hemisphere: 1851-1984.* U.S. DoE Technical Report TR022, U.S. Department of Energy Carbon Dioxide Research Division, Washington, DC.

Jonsson, B. 1969: Studier över den av väderleken orsakade variationene i årsringsbredderna hos tall och gran i Sverige. *Institution for skogsproducktion skogshögskolan Rapporter och Uppsatser 16*, Stockholm. 297pp.

Lofgren, G. R. and Hunt, J. H. 1982: Transfer functions. In *Climate from tree rings* (M. K. Hughes, P. M. Kelly, J. R. Pilcher and V. C. LaMarche Jr. Eds.) 50-58. Cambridge University Press, London.

Mikola, P. 1956: Tree-ring research in Finland. *Tree-Ring Bulletin 21*, 16-20.

Mikola, P. 1962: Temperature and tree growth near the northern timberline. In Kozlowski, T.T. (ed.), *Tree Growth*, 265-274. Ronald Press, New York.

Mitchell, J. M., Jr., Dzerdzeevskii, B., Flohn, H., Hofmeyr, W. L., Lamb, H. H., Rao, K. N. and Wallén, C. C. 1966: *Climatic Change*. W.M.O. Technical Note 79. W.M.O. Geneva. 79pp.

Renberg, I., Segerström, U. and Wallin, J-E. 1984: Climatic reflection in varved lake sediments In *Climatic Changes on a Yearly to Millennial Basis* (N.-A. Mörner and W. Karlén, Eds.) 249-256. Reidel, Dordrecht.

Röthlisberger, F. 1986: *10000 Jahre Gletschergeschichte der Erde*. Verlag Sauerländer. Aarau.

Schulman, E. 1944: Tree Ring work in Scandinavia. *Tree-Ring Bulletin 11*, 2-6.

Schweingruber, F. H. 1987: Site selection and sampling strategy. In: *Methods of Dendrochronology – 1* (L. Kairiukstis, Z. Bednarz and E. Feliksik, Eds.) pp.15-22. IIASA/Polish Academy of Sciences, Warsaw.

Schweingruber, F. H., Bartholin, T., Schär, E. and Briffa, K. R. 1988: Radiodensitometric-dendroclimatological conifer chronologies from Lapland (Scandinavia) and the Alps (Switzerland). *Boreas 17*, 559-566.

Schove, J. D. 1954: Summer temperatures and tree-rings in north-Scandinavia A.D. 1461-1950. *Geografiska Annaler 36*, 40-80.

Sirén, G. 1961: Skogsgränstallen som indikator för klimafluktuationerna i norra Fennoskandien under historisk tid. *Communicationes Instituti Forestalis Fenniae 54*, 66pp.

Sirén, G. and Hari, P. 1971: Coinciding periodicity in recent tree rings and glacial clay sediments. *Report of the Kevo Subarctic Research Station 8*, 155-157.

Stuiver, M. and Braziunas, T. F. 1988: The solar component of the atmospheric ^{14}C record. In *Secular Solar and Geomagnetic Variations in the last 10,000 Years* (F. R. Stephenson and A. W. Wolfendale, Eds.) 245-266. Kluwer, Dordrecht.

Stuiver, M. and Kra, R. Eds. 1986: *Radiocarbon (Calibration Issue) 28*, 2B.

Wigley, T. M. L., Briffa, K. R. and Jones, P. D. 1984: On the average value of correlated time series, with applications in dendroclimatology and hydrometeorology. *Journal of Climate and Applied Meteorology 23*, 201-213.

20 Dendroclimatic evidence from the northern Soviet Union

D. A. Graybill and S. G. Shiyatov

20.1 Introduction

We make a reliable reconstruction of average June-July temperature for A.D. 961-1969 using tree-ring width variation of *Larix sibirica* from the north Polar Urals (Graybill and Shiyatov 1989). It is of considerable interest because it is the first millennial-aged proxy record of seasonal temperatures in the Soviet sub-arctic. Some of the longer term trends (century scale) in the reconstruction are similar in direction and timing to those found in other summer temperature proxy records from the European Arctic, Greenland and the Americas that are commonly referred to as "the Medieval Warm Epoch" and the "Little Ice Age" (Lamb 1977; Williams and Wigley 1983; Grove 1988). We cautiously note that our record is only considered one of many that will be required to test rigorously hypotheses about the spatial and temporal generality of such temperature changes.

Additionally, our reconstruction may hold information about trends and long-term variation in temperature over large areas. The tree-ring chronology was derived from an area just south of the Kara Sea, bounded by latitudes 66°45', 66°55'N and longitudes 65°15', 66°05'E (Figure 20.1) (Shiyatov 1986) The region around the Kara Sea is thought to be particularly sensitive to long-term trends and variations in temperature over the arctic and possibly the Northern Hemisphere (Kelly *et al.* 1982).

Because of the unique nature of this record, and the strong possibility that others will want to use it for various scientific purposes, much of our discussion concerns analytical procedures used in its production. We then evaluate the quality of the reconstruction from several perspectives. An understanding of this background is crucial to subsequent comparisons or interpretations that might involve this record.

20.2 Data and data pre-treatment

20.2.1 Tree-ring widths

Tree-ring samples are from various localities in a research area on the eastern slope of the Polar Ural Mountains (Figure 20.1). This is a subset of the collections that includes the longest available series. They were selected to preserve as much low frequency variance as possible in a final chronology. Forty-eight series are from different living individuals of Siberian larch (*L. sibirica*). These were growing between 150 and 300m asl., the upper elevational limit for tree growth in 1970 when most data were collected. In addition, 23

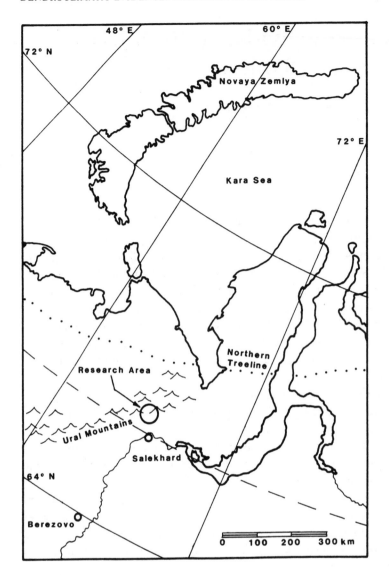

Figure 20.1 Region of research.

samples from one of the localities represent dead larch individuals found at elevations ranging up to 80m above the current treeline.

Ring width series were cross-dated with each other and then each ring was assigned a calendar year based on the known collection date of living tree samples. The total time span represented by the living series is A.D. 1541-1969. Samples from dead wood cover the time range of A.D. 960-1741 (Figure 20.2).

Inspection of time series plots of the measurements of each sample indicated that most had decaying biological growth trend that is best described as negative exponential. Removal of the biological growth trend and subsequent computation of dimensionless tree-ring indices is commonplace in dendrochronological studies (Fritts 1976; Graybill 1982; Briffa 1984; Cook

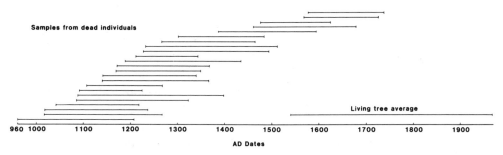

Figure 20.2   Tree-ring chronology components.

1985; Shiyatov and Mazepa 1987; Cook *et al.* 1990). A comparison of the Shiyatov 'Corridor Method' with some of the American procedures suggested that it was somewhat superior in preserving low frequency variance (Shiyatov, Fritts and Lofgren 1989). It was used with the 71 ring width series to remove age-related growth trend and to preserve other trends that appeared consistent across samples for any particular period. While the former is considered primarily biological, the latter are presumably climatic in origin. The data did not appear to have short-term (<10 years) disturbance signals that might reflect individual injuries. Because of rigorous growth conditions here, stand densities and competition are relatively low. The trees are not particularly susceptible to widespread disturbances such as fire or insect damage.

A final chronology was produced as a simple average of the 71 individual indexed series. This effectively reduced the unique error associated with each component series and maximized the common signal. Changes in the strength of the common signal in chronologies such as this are of obvious interest. As sample size per year decreases, usually near the early portion, generalizations about the reliability of a reconstruction may need to be tempered. A useful measure for consideration of this is the Subsample Signal Strength (SSS) (Wigley *et al.* 1984). This estimates the agreement between an average series made from a few samples with one made from an optimum or larger number of series. It was possible to make an estimate of SSS from 13 ring-width index series that spanned the interval of 1800-1960. When sample numbers are in the low range of two through four, a chronology derived from them explains respectively 83%, 89%, and 92% of the variance of one composed of the optimal number of series. Assuming that the living and dead trees had a relatively consistent growth response through time, these figures indicate the reliability of the early part of the record.

20.2.2   Climate data

Temperature data for 1881-1969 were selected from a gridded monthly data set for the Northern Hemisphere (Jones *et al.* 1985) at the grid point nearest the tree-ring data (65°N, 70°E). The values are expressed as departures from a 1951-1970 reference period. This grid point series represents data from two stations, Salekhard (WMO #233300) and Berezovo (WMO #236310). They are located respectively, about 75km and 350km from the center of the research area. For initial screening of climate-tree growth relationships we also used monthly precipitation totals (1883-1969) from the Salekhard records (Laboratory files, Institute of Plant and Animal Ecology, Sverdlovsk).

20.2.3   Climate-tree growth relations

Growing season temperature has been shown to be the dominant natural forcing function for tree growth at high latitudes, and in some cases for upper elevational treelines at mid-latitudes (Tranquilini 1979). This is reasonably well understood from physiological perspectives. In the European and North American sub-arctic it has been empirically demonstrated in many cases (Eklund 1954; Mikola 1962; Briffa *et al.* 1988; Giddings 1941; Garfinkel and Brubaker 1980; Jacoby *et al.* 1985; Jacoby and D'Arrigo 1989). Warm temperatures during the growing season, as well as through the year or so preceding growth, are found to have a positive causal effect on the width of annual growth increments. Conversely, colder periods place limitations on a variety of growth processes, resulting in diminished tree-ring widths. Evergreen genera such as spruce (*Picea sp.*) apparently integrate enough of the annual temperature variation that they can be successfully used for reconstruction of that signal (Jacoby and D'Arrigo 1989). Deciduous genera such as the larch used in this study integrate a seasonal signal. It reflects those temperatures most crucial to early initiation and maintenance of photosynthetic processes that lead to cell division. This may be for a shorter period than the actual growing season, that is, June and July *vs* June-August (Giddings 1954).

Soil moisture is also necessary for growth and can be a limiting or stressful factor. However, it has not commonly been found as limiting to growth as temperature by dendrochronologists working at high latitudes, and is not commonly reconstructed in these settings (Jacoby and Cook 1981). This disparity in response to temperature and precipitation is not so clear for tree growth at upper treeline in arid, mid-latitude settings (LaMarche 1974; Graybill 1987; Graumlich 1989; Hughes, Chapter 21, this volume).

A preliminary investigation of the relationship between climate and Polar Urals larch was undertaken. It first involved a review of simple correlations of the final index chronology with monthly temperature averages for the grid point data. Similar computations were performed using tree-ring indices and monthly precipitation totals from Salekhard. Somewhat as expected, given physiological considerations and results of similar studies cited above, there is a strong positive and significant response of tree growth to June and July temperatures (Table 20.1). There is a stronger correlation with the average for those two months ($r=0.72$, $p<0.001$). Significant correlations with temperature in the May and June of the year preceding growth probably reflect a biological dependence on stored food produced in prior years (Fritts 1976). No explanation is offered for the significance of the correlation with January temperatures. Limited but significant correlation of the index series with precipitation is present for various months as well as with a 12 month sum of July-June ($r=0.37$, $p=0.001$). The relationships with precipitation are not considered strong enough to warrant further consideration here. Continued discussion is focused on the average June-July temperature relationship to tree growth.

20.2.4   Time-series considerations

Time series and other statistical characteristics of the tree-ring and temperature data were evaluated to determine their suitability for use in simple linear regression. These evaluations focused on the period of common data overlap (1881-1969) as well as on the longer period of tree growth back to A.D. 960. Primary concerns were with the normality of the distributions

Prior Growth Year   Current Growth Year

Months	M	J	J	A	S	O	N	D	J	F	M	A	M	J	J	A
Data 65°N 70°E temperature departures with prewhitened indices	.29	<u>.37</u>	-.14	.00	.20	.17	.04	.09	.15	.02	.00	.19	.21	<u>.61</u>	<u>.67</u>	.02
65°N 70°E temperature departures with untreated indices	.32	<u>.55</u>	.10	.03	.22	.20	.07	.14	<u>.26</u>	-.08	-.09	.21	<u>.22</u>	<u>.60</u>	.56	.05
Salekhard precipitation with prewhitened indices	.01	-.08	-.19	.00	.13	.06	-.13	-.10	.05	-.01	.21	.10	.20	.07	<u>.30</u>	-.00
Salekhard precipitation with untreated indices	.17	.05	-.01	.09	<u>.32</u>	.19	.09	.16	<u>.25</u>	.17	<u>.34</u>	<u>.26</u>	<u>.28</u>	.17	<u>.32</u>	.11

Table 20.1 Pearson correlations of tree-ring series with monthly temperature and precipitation series (underlined value indicates $p < 0.05$).

and with the statistical independence of values in each series. Regression of series with non-normal distributions or with strong persistence can result in unreliable estimates of covariance and in problems with interpretation of the results (Wonnacott and Wonnacott 1981, Monserud 1986). Based on the results of one-sample Kolmogorov-Smirnov tests (Bradley 1968) all series are from normally-distributed populations.

Evaluation of the persistence in these series was accomplished with standard procedural software (Dixon 1985) by reviewing auto-and partial autocorrelations. If the persistence was significant then standard Box and Jenkins protocol (1976) was used to model this feature. The June-July grid point temperature series is best fit by an AR(0,2) model ($\phi_2 = 0.27$, %var. = 14.5). During the same period of 1881-1969 the best fit model for the larch indices is AR(1,2) ($\phi_1 = 0.41$, $\phi_2 = 0.34$, %var. = 47.5). The differences in the two models and the greater amount of variance in the latter model fit indicates that much of the persistence in the tree-ring indices is non-climatic in origin.

At this stage of decision making we do not find widespread agreement on all of the analytical processes that should follow in developing a reconstruction. Part of this stems from the fact that while autocorrelation in tree-ring series is normally recognized and pretreated before regression, or handled in other fashions during regression, autocorrelation in temporally correspondent climate series has not been so commonly discussed. Recent exceptions include the reconstruction of a regional drought index (Stahle and others 1988) and a study on methods of reconstruction (Meko 1981). The latter author suggests one approach where significant autocorrelation is removed from both climatic and tree-ring series. This is accomplished by pre-whitening each series after best fit ARMA models are found for their period of common overlap. Regression procedures are then used with the pre-whitened residual series and the climatic persistence is later added back into the reconstruction. One problem here is that the climate of the instrumented period may be somewhat unique or different from that of preceding centuries. If so, the persistence structure of either or both the climate and tree-ring series for that period may not be a good realization or estimate of longer term climatic processes. This does in fact appear to be the case with the larch tree-ring indices and is discussed further below. If this is true with respect to a climate series then spurious variation may be introduced into a reconstruction by reddening it with a model based on data from the past century. The obvious dilemma here is that increased understanding of centennial to millennial scale climatic processes can only be obtained from multiple and relatively consistent reconstructions by proxy records from the same or neighboring regions. These do not presently exist in the North Polar Urals. In the absence of such information we thus proceed to obtain what is possible with the current data and techniques.

In developing a time series model for the 1009 year larch chronology we considered the possibility that the model fitting results might be affected by changes in sample depth during the early years. Several trials were made. The beginning point for the modelling procedure was stepped along the points in the chronology where sample depth increased progressively from one to nine series. Dates for these points range from A.D. 960 to 1141. In all cases an ARMA(1,1) model was the best fit and the model coefficients were stable. The AR coefficients only ranged from 0.77-0.78 and the MA coefficients from 0.44-0.45. The amount of variance associated with the fits ranged from 21.4 to 22.2%. The second most parsimonious and best fit model in these trials was AR(1,2). Those coefficients were also stable with ϕ_1 ranging from 0.34-0.35 and ϕ_2 from 0.20-0.21. The amount of variance ranged from 21.2 to

22.0%. Following discussion above, it is notable that the coefficients for the AR processes are somewhat less in the longer term models than in those for 1881-1969. Also, the variance associated with the longer term fits is less than half that for the more recent period. The increasingly upward trend in growth from 1881 to 1923 (Figure 20.3) and the lack of such a predominant trend over the longer period (Figure 20.4) are at least partially responsible for these differences.

One concern that surfaces in these circumstances is whether the use of time series models for pre-whitening tree-ring data actually removes trends that are assumedly or demonstrably driven by a climatic factor (Jacoby and D'Arrigo 1989:45). We do not have an immediate reference but personal experience indicates that this procedure is sometimes referred to in jargon as 'detrending'. There may be a semantic problem here in addition to the statistical issues. We suggest that the term 'detrending' should be expunged from these discussions. However, to examine this issue in more detail, we developed reconstructions of June-July temperature using both untreated and pre-whitened tree-ring indices. The ARMA (1,1)

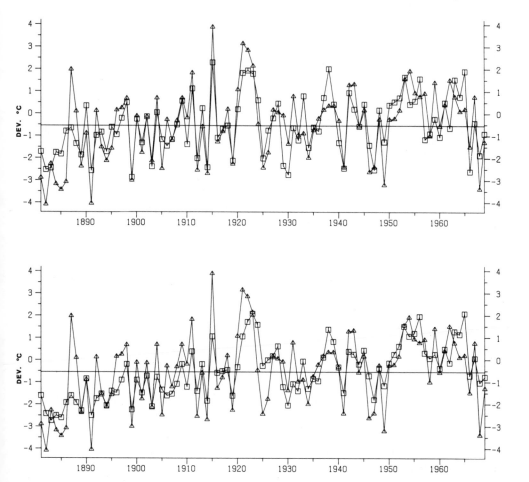

Figure 20.3 Actual and reconstructed average June-July temperature series for the Polar Urals region. (a. Reconstructed from pre-whitened tree-ring indices, b. reconstructed from untreated tree-ring indices.)

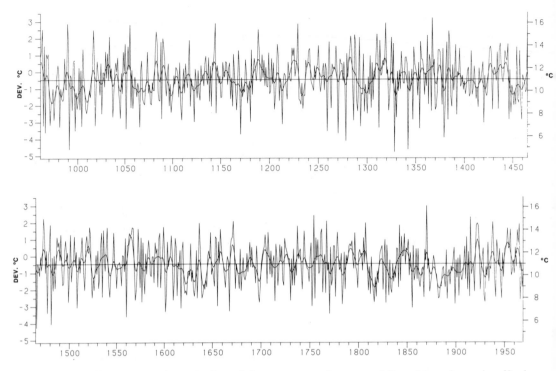

Figure 20.4 Reconstructed average June-July temperature from pre-whitened tree-ring series. (Scales on the left are for deviations from the 1951-1970 normal period at 65°N 70°E and scales on the right are for actual °C at Salekhard.)

model coefficients used for pre-whitening were 0.78 and 0.45 respectively. Pearson correlations were again computed between the pre-whitened indices and the same monthly climate data discussed above (Table 20.1). Note that the correlations with precipitation are decreased from those with the indices and that the correlation with January temperature is low and not significant. The strongest relationship of tree growth to monthly climate remains with the temperatures for the current June and July, and to a lesser degree with those of the prior May and June. The correlation with the current June-July temperature average is 0.78 (p <0.001).

20.3   Calibration and verification procedures

Several calibration and verification procedures were conducted to evaluate the stability of the temperature-tree growth relationships at different times (Table 20.2). In each calibration either the pre-whitened tree ring series, or the untreated index series, was used as the independent variable in linear regression with the June-July temperature average. The equations derived from these regressions were used to reconstruct temperature values during the corresponding verification periods. Each reconstruction was then compared with the actual temperature values. Evaluation of the calibration regressions utilized standard goodness-of-fit criteria (Draper and Smith 1981). Comparison and evaluation of the recon-

	Periods cal.	Periods ver.	r^2_a (1)	r^2 (2)	S_d (3)	S_m (4)	T (5)	W (6)	F_c (7)	RV_c (8)	F_v (9)	RV_v (10)	RE (11)	CE (12)
I	1926-1969	1881-1925	.40	.76	a	a	a	a	r	.42	r	.20	.57	.41
	1881-1925	1926-1969	.75	.42	a	a	a	a	a	.76	a	1.54	.18	.05
	1881-1969		.60						r	.60				
II	1926-1969	1881-1925	.44	.59	a	a	a	a	r	.45	r	.29	.53	.50
	1881-1925	1926-1969	.58	.45		a	a		r	.59	a	.89	.16	.04
	1881-1969		.51						r					
III	1912-1969	1883-1911	.58	.56	a	a	a	a	r	.59	r	.30	.61	.51
	1883-1911, 1941-1969	1912-1940	.54	.66	a	a	a	a	r	.55	a	.74	.65	.55
	1883-1940	1941-1969	.62	.51	a	a	a	a	r	.62	a	.76	.52	.48
	1881-1969		.59						r					
IV	1912-1969	1883-1911	.50	.59	a	a	a	r	r	.51	r	.26	.45	.30
	1883-1911, 1941-1969	1912-1940	.49	.49	a	a	a	a	r	.50	r	.40	.51	.48
	1883-1940	1941-1969	.50	.56	a	a	a	r	r	.50	r	.45	.44	.39
	1881-1969		.49						r					

1. Calibration period regression covariance adjusted for degrees of freedom, alpha = .01.

2. Covariance of independent and dependent series during the verification period, alpha = .01.

3. Sign test of first differences, see text, alpha = .01.

4. Sign test of direction from mean, see text, alpha = .01.

5. Students T-test, alpha = .05.

6. Wilcoxon matched pairs signed ranks test, alpha = .05.

7. F test for equality of variances, actual and reconstructed temperature, calibration period, alpha = .05.

8. Ratio of reconstructed to instrumented temperature variance, calibration period.

9. F test for equality of variances, actual and reconstructed temperature, verification period, alpha = .05.

10. Ratio of reconstructed to instrumented temperature variance, verificationb period.

11. Reduction of Error.

12. Coefficient of Efficiency.

a = null hypothesis accepted

r = null hypothesis rejected

Table 20.2 Results of calibration and verification procedures of tree-ring data with June-July temperature departures from 65°N 70°E (I, III used pre-whitened larch indices; II, IV used untreated larch indices).

structed and actual temperature values utilized standard parametric and non-parametric statistics. Other not-so-standard statistics (CE, RE) that are sometimes found in the hydrologic and dendroclimatic literature were also included in the analyses (Nash and Sutcliffe 1971; Fritts 1976; Gordon and LeDuc 1981).

In the first set of trials the data were divided into two equal and sequential halves that were alternatively used for calibration and verification. There is nothing particularly sacrosanct about this kind of data division although it is simple and is common practice. One kind of problem that can occur in this circumstance, and did with this data set, is that one or more of the basic distributional characteristics of the climate or tree-ring data may actually change from one time period to the next. This makes comparison of the actual and reconstructed series during the verification period somewhat difficult. A notable change of this nature is a decrease in the variance of the temperature data. It is about 50% less in the post-1925 period than in earlier years. The tree-ring indices exhibit this to a lesser extent with a decrease in variance near 30% over the same periods. The pre-whitened series show only limited change in that value. Other grid point temperature data in the region bounded by 50°N, 65°N and 50°E, 90°E were reviewed to see if they had similar temporal patterns of change in average June-July temperature variance. This pattern is restricted to the northeastern sector of that region, appearing in grid point series along the meridian at 70°E from 55°N to 65°N and at 65°N, 90°E. Grid point data along the meridian at 80°E are too limited in length for these considerations. This kind of change appears to correspond temporally with a shift from more meridional to more zonal flow in the Northern Hemisphere. The former period was more continental and cooler than the latter (Lamb and Johnson 1959, 1961; Dzerdzeevskii 1962). Sometime near 1960 there appears again to be a shift to meridional flow and higher variance in the same sector just described. This was apparent in our computations (1881-1984) and is also indicated for much of the Soviet Union by Borisenkov (Chapter 9, this volume).

Given this situation a second series of calibration-verification trials was undertaken to try to avoid such potentially dichotomous results. Three trials successively used the first, middle, and final two-thirds of the data for calibration and the remaining portion for verification. In reviewing the results of these exercises (Table 20.2) the highest amount of calibration variance in common between tree-growth and temperature is found in early periods that are the coldest and have the highest climate variance. The converse is also true, particularly in the analyses where the data are only split into two parts. However, all correlations (and by extension-covariances) are highly significant with associated probabilities being less than 0.001. Similar kinds of results were reported for conifers in the North American sub-Arctic which were being used to reconstruct annual temperatures from 1880-1973 (Jacoby and D'Arrigo 1989:46). Calibration period covariance was highest during the earlier and colder portion of the temperature record. They infer, as we do, that the warmer temperatures of the mid to late 20th century were not as limiting to growth as cooler temperatures of earlier years. As those earlier stresses decreased, other factors (noise at this time) have recently intervened, which somewhat obscure the growth response.

Other information presented in Table 20.2 provides further perspectives on the quality of the Polar Urals tree-ring data for reconstruction purposes. A basic consideration is how well the mean temperature of the verification period data is reproduced. To evaluate this, the non-parametric Wilcoxon matched-pairs signed-ranks test (Bradley 1968) was used in addition to the T-test because some of the tripartite schemes involved limited sample sizes. In most cases the mean values were reconstructed, although the Wilcoxon test resulted in two more rejections than the T-test. Models based on regression using the residual white noise series only failed to reconstruct one of the means. Models based on untreated indices could not do this in four cases, given results of the Wilcoxon test.

Results of F tests for the equality of variances in the actual and reconstructed climate series (F_c, F_v) are not commonly included in these exercises. To speculate, this may be due to the fact that not all of the variance in a climate series can normally be reconstructed; a large and significant F value is commonplace, and may not therefore be of interest. Even with an alpha level of 0.05, it is apparent that rejection of the hypothesis of equal variances is more common than acceptance. However, in a few of the calibrations and several of the verifications this is not the case. Another consideration is how the amount of variance in the reconstructed temperature values, relative to that of the actual values, varies from the calibration to the verification periods. Those ratios are labelled RV_c and RV_v in Table 20.2. One would commonly hope to find limited differences, assuming at least some stationarity in the mean and the variance of both series. Given discussion above, the largest differences in those ratios are expected to be found where the calibration is based on the least variable temperature data and the verification on the most variable temperature data. Differences in those ratios are somewhat diminished in other trials where the temperature variances of the two periods are not so dissimilar. These patterns are similar for both the pre-whitened and untreated larch indices.

Two of the other results in Table 20.2 (S_m, S_d) are from non-parametric sign tests designed to evaluate two hypotheses. (1) The locations of the annual values of the reconstructed and dependent series either above or below the mean of the dependent series are not significantly different (S_m). (2) the first differences of the actual and reconstructed values do not differ significantly in sign (S_d). In all cases these hypotheses could not be rejected at the 0.001 alpha level. Values for the reduction of error (RE) and coefficient of efficiency (CE) are also reported in Table 20.2. A discussion of the similarities and differences in these can be found in Briffa *et al.* (1988). The RE statistic was originally designed for evaluation of errors in predictions of the distributional characteristics of short-term changes (hours, days) in weather (Lorenz 1956). Its applicability to longer-term phenomena can be questioned, and there are no significance tests associated with either the RE or CE statistic. Their inclusion here is solely for comparative purposes because they are now presented frequently in literature on climatic reconstruction. Values of the two statistics can range from negative infinity to 1.0. Any positive values are considered a reflection of good calibration model performance. Given those criteria, all the calibration models developed here were successful to varying degrees. The lowest RE and CE values are from trials where the data were split into two portions and the calibrations were performed on the sector of temperature data with the highest variance. The highest values are found with some of the threefold data divisions where the differences in temperature variance are smallest.

Final calibrations of the pre-whitened tree-ring indices and the untreated indices with the June-July temperature record used the full span of available data to obtain the fullest sample of covariation in each series. Tree-ring covariation with the temperature series was highest for the pre-whitened indices (Table 20.2). These linear regressions provided equations that were used to compute reconstructions (R_1, R_2). The data sets and derived equations are:

(1) Pre-whitened tree-ring indices (PI) grid point average June-July temperature departures

$$R_1 = (PI * 3.6858) - 0.4633$$

(2) Untreated tree-ring indices (UI) grid point average June-July temperature departures

$$R_2 = (UI * 2.6143) - 2.9914$$

20.4 Comparison of reconstructions

Although the quality of these reconstructions can to some degree be judged by reference to Table 20.2 and the discussion above, it is pertinent to consider other aspects of them. The following discussion compares the two reconstructions, indicating various strengths and weaknesses of each.

One of the first exercises we undertake upon completion of a reconstruction is to consider the distributional characteristics of the actual and reconstructed data during their period of overlap, and of the longer period reconstruction. A lack of similarity in the moments of the instrumented and reconstructed series can signal various limitations on inferences that might be drawn from the reconstruction. This is aided by reference to Table 20.3 and to time series plots for some of the values (Figures 20.3-6). An eight year low pass filter was used to produce the smooothed line through the central portion of Figures 20.4-6. The temperature scale on the left of the long period reconstructions is departure values derived from the grid point reconstructions. On the right side, this axis is annotated with temperature values for the June-July season at Salekhard. This was included in anticipation of physiological and ecological research in the region that might require estimates of instrumented values rather than deviations. To reduce the visual dominance of high frequency variation in Figure 20.4, and to aid in recognition of other important trends and patterns, we computed 20 year non-overlapping averages of the reconstructed values (Figure 20.6a). They were positioned such

Table 20.3 Statistics for actual and reconstructed June-July temperature data, grid point 65°N 70°E (I. – from pre-whitened larch indices; II – from untreated larch indices).

I.	Actual	Reconstructed	
Dates	1881-1969	1881-1969	1061-1969
Mean	-.54	-.54	-.46
Std. dev.	1.66	1.29	1.35
Skewness	.01	.21	-.15
Kurtosis	2.76	2.31	2.80
II.			
Mean	-.54	-.54	-.39
Std. dev.	1.66	1.19	1.08
Skewness	.01	.30	.08
Kurtosis	2.76	2.38	2.57

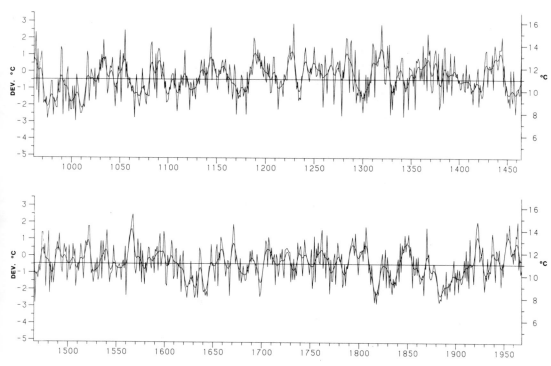

Figure 20.5 Reconstructed average June-July temperature from untreated tree-ring series. (Scales on the left are for deviations from the 1951-1970 normal period at 65°N 70°E and scales on the right are for actual °C at Salekhard.)

that the final period includes the last 20 years of data (1950-1969). The first eight years of the reconstruction were not used.

During the full calibration period the mean values of the actual and reconstructed series are the same, by definition of the regression process. The slightly warmer long-term mean values are somewhat a product of the higher values through the period of about A.D. 1120-1450 (Figures 20.4 and 20.6a). The variance of the actual temperature data is best reconstructed by the pre-whitened index series. The relatively high amount of variance in the long period reconstruction from the pre-whitened index series is partially due to higher variance from the late 1200s to the mid-1400s (Figure 20.4). This is more apparent in a plot of 20 year non-overlapping averages of the sums of absolute first differences (Figure 20.6b). These values were used instead of variance to avoid any scale dependence on either local or global means.

Skewness of the temperature values is near the expected value in a normal distribution and is best reconstructed by the pre-whitened indices. The slight negative skew of the long period reconstruction from pre-whitened indices is probably due in part to the presence of many extreme low values in the period of about A.D. 1250-1390. The kurtosis values for the actual and reconstructed series are not remarkably different from each other or from an expected value of 3.0 in a normal distribution.

We now consider the question of whether the quality of reconstruction is somewhat different for low and high climatic values. It is commonly found that lower values are best

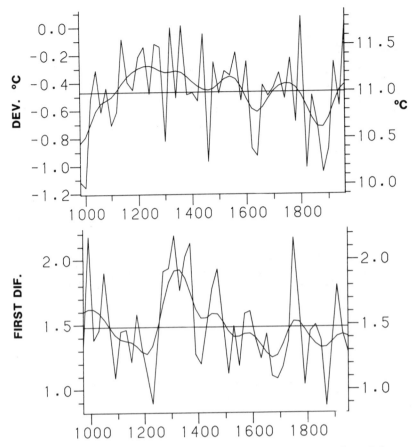

Figure 20.6 Twenty year averages of (a) reconstructed average June-July temperature and (b) of absolute first differences of average June-July temperature. (The scale on the left of (a) is for deviations from the 1951-1970 normal period at 65°N 70°E and the scale on the right of (a) is for actual °C at Salekhard.)

reconstructed because they represent conditions of highest biological stress. Most trees from a stand will respond in a more similar fashion under these conditions and in a less similar fashion when stress is not so strong (Fritts 1976; Graybill 1989). In review of this issue, two linear regressions of tree-ring data and June-July temperature series were conducted using (1) values above the means, and (2) values below the means of the actual temperature data sets (Table 20.4). In all cases the lower values were better explained than the higher. Covariances of the pre-whitened indices and untreated indices with values below the mean of the grid point data are about the same. In general, the untreated indices do the poorest job of reconstructing high temperature values as well as the full range of values (Table 20.2). This is also visually apparent in Figure 20.3. These findings suggest that generalizations about the highest temperatures that may be represented in the reconstructions must be limited in nature and made with recognition that the highest reconstructed values are likely underestimations.

Although no accurate statistical generalizations can be made about the probable amounts of actual temperature underestimations in times prior to instrumented records, we did

Table 20.4 Regressions of tree-ring series with temperature data above and below mean levels. Data sets are: I grid point June-July temperature, pre-whitened indices, II grid point June-July temperature, untreated indices.

Location Relative to Temperature Mean

	Below			Above		
	r^2_a	dif_b	n	r^2_a	dif_a	n
	(1)	(2)	(3)	(1)	(2)	(3)
I	.38	.65	40	.25	.67	49
II	.42	.75	40	.14	.78	49

1. Regression covariance adjusted for degrees of freedom.
2. Average value of underestimated temperature departures (I,II).
3. Number of cases in computation of 2.

compute those values for the instrumental period. When temperature values were both above the mean and also underestimated by the reconstructions, the differences were summed and averaged (dif_a, Table 20.4, these equate to about 1.4-1.2°C at Salekhard). Since some observers may also be interested in the average underestimate of values below the mean, this value was also computed (dif_b, Table 20.4, these equate to about 1°C at Salekhard). It represents the average of differences for only those years where the actual temperature values were less than the estimates from the tree-ring data.

Another consideration of the two reconstructions is focused on their ability to reproduce the increasing temperature trend from the late 19th into the early quarter of the 20th century. Trends over this period were computed by simple linear regression of incremental annual values with the June-July data of the climate series and with the reconstructions from the tree-ring data (Table 20.5). The trends represented by these slopes are indistinguishable statistically given the range of the standard errors of the regression coefficients. This indicates that other trends of at least this length (45 years) and of similar slope may be reasonably well represented throughout all sectors of the reconstructions.

One other aspect of the records concerns their ability to reproduce the autocorrelation pattern in the temperature series. It will be recalled that this exhibits higher autocorrelation at lag two than one (0.19, 0.24) and the second value is significant at the 0.05 level. During the instrumental period, both the pattern and amount of autocorrelation in the reconstruction from the pre-whitened indices are similar to that of the temperature data (0.15, 0.30). The untreated index series reconstruct their own pattern of changes and levels in those lag values and do not adequately model the persistence in the temperature series. The lag one and two values in this reconstruction are respectively 0.50 and 0.48, and both values are highly significant ($p < 0.001$).

A cross-spectral analysis (Jenkins and Watts 1968) of the temperature data and the reconstruction based on pre-whitened indices indicated strong coherence at most periods less than the 20 year frequency range. These 89 year records are somewhat limited for high resolution of spectral densities but there are some significant similarities at periods of about

2.0-2.1 and 2.8 years. We do not speculate on possible reasons for these periodicities. Spectral analysis of the reconstruction based on untreated indices is not appropriate given the high amounts of autocorrelation present (Chatfield 1980).

Because of the results of the comparisons and analyses presented here, we think that the most reliable generalizations will be based on use of the temperature reconstruction from the pre-whitened index series. Further discussion below is accordingly restricted. This decision, as well as the discussion above, should not however be construed as polemics. Seldom have the relative merits of these kinds of approaches to data manipulation been discussed or compared in detail with full presentation of results in the professional literature. Continuing investigation of these issues is fully warranted in dendroclimatic research. We also look forward to similar, detailed examination of the quality of climatic reconstructions from other proxy sources by our colleagues in disciplines such as palynology and ice core studies. Only when this is accomplished can we collectively attempt to make reliable generalizations about long-term regional, hemispheric and global patterns and processes of climate variability.

20.5   Variation in the reconstruction

Our time series plots (Figures 20.4 and 20.5) of the reconstruction extend back to A.D. 961 as a matter of completeness of record. However, comments here will be restricted to variation in that record after A.D. 1086 when five or more component tree-ring series are usually present. The overall similarity of longer term trends to those found in many other summer temperature proxy records outside of this region was earlier noted. Somewhat warmer than normal temperatures in medieval times are followed by a general decline into the late 1800s, with an obvious increase again into the 20th century. Now, most of the constituent series forming the chronology are from about 220-400 years in length, although a few are somewhat shorter. Given this, and the fact that standardization can remove trends that are the same or greater length than each series, the ability of this reconstruction to mirror some of these longer term trends may be questioned. There are, however, important field observations that bear on this issue. Series from the dead individuals were found at elevations up to 80m above the current treeline, although the majority were found in the first 20 to 40m of that incline. All the dead series that were living in the early to mid-1600s, a period of inferred temperature decrease, experienced dramatic growth decline at that time. Their demise occurred then or shortly thereafter. This is also the time period when many of the oldest living trees found near the current elevational treeline germinated, although some date to the mid-1500s. Therefore, it is possible that the longer term trends in temperature are at least reasonably estimated. The two different elevational groupings of trees provided continuous monitors of temperature variation at their respective locations and sequentially their records may have captured those trends.

Early medieval times from about A.D. 1100-1250 experienced warming above the long term mean and stable interannual variation (Figures 20.4 to 20.6). June-July average temperatures generally remained above that mean until the late 1500s, with two major exceptions. They are cool periods centered on A.D. 1300 and 1460 that represent two of the three most rapid temperature decreases between adjacent 20 year periods (1° and 1.4°C at Salekhard). On even shorter time scales there are some remarkable changes in the variability of reconstructed temperatures as measured by absolute first differences (Figures 20.4 and 20.6b). This is particularly evident through later medieval times from A.D. 1259 to 1379. It is not thought to be an artefact of sample depth because one of us (SGS) noted extensive variability in the ring widths of most samples from these times. Overall, this 121 year period is relatively warm and contains five of the twelve years that are greater than two standard deviations above the long term mean. However, it also contains nine of the seventeen years with values that are more than two standard deviations below the mean, as well as the only two values that are more than three standard deviations in either direction (below the mean, in 1328 and 1342).

The predominant trend of decreasing temperature from the late 1500s to 1880 is primarily due to three periods with one or more strong excursions below the long-term mean. These are from about 1610-1640, 1810-1820 and 1850-1880. The relatively warm recent period of 1950-1969 (11.74°C) is only exceeded by that of 1790-1809 (11.80°C). The decline from the 20 year period centered on 1800 to that centered on 1820 is the greatest in the reconstruction (1.62°C). In terms of possible arguments about volcanic forcing of tree growth in this time period, via restrictions of sunlight from aerosols, it is noted that tree growth began to decline after 1807, reaching a low in 1818. Increases in growth are found from 1819-1824. Whether the eruption of Tambora in Indonesia, April of 1815, may have had some or any cooling effect on summer temperatures at this high latitude, thereby limiting tree growth in that or subsequent years, is a difficult issue to resolve (Sear et al. 1987; Bradley 1988).

Reconstructed interannual temperature variation after A.D. 1500 is highest from 1750-1769, with a lesser peak again in the late 19th and early 20th century. Changes in circulation types that were noted above, in connection with variance changes in the instrumental record, may also have been involved at earlier times. The extreme nature of most of these, as well as the duration of high values in medieval times, are clearly outside the range of variation during the instrumental period. Improved understanding of the spatial pattern of such changes, and of the changes themselves, will require a larger spatial array of reconstructions than is now available.

The autospectrum of the larch reconstruction has three significant peaks that occur at 2.1, 3.8 and 8.6 cycles per year. These are mentioned only for the sake of documentation. We have no strong *a priori* convictions or hypotheses to test about the physical meaning of such periodicities and they are not discussed further.

20.6 Comparisons with other temperature reconstructions

Other estimates of past summer temperatures from northern Europe and Asia have been made with ice core data from the arctic islands of Svalbard (Spitsbergen) Franz-Josef Land, Novaya Zemlya and Svernaya Zemlaya (Tarussov, Chapter 26, this volume). It is thought by Tarussov that the most reliable record of temperatures is for a seasonal period of June-August. This is based on annual melting percentage (AMP) values in a core from the Austfonna region (Spitsbergen) some 1800km to the northwest of Salekhard. The data have a temporal resolution of about 4-5 years. Some trends that are similar to those in our larch reconstruction include cooling from about 1550 to 1750, and warming events near 1780 and 1820. The sharp cooling between the latter two events that is found in the tree-ring record, and in other ice core data, is not apparent in this series. Tarussov suggests there are physical reasons for this. In some additional agreement, mid-19th century cooling and late to post-19th century warming are apparent in both records. Given the differences in temporal resolution of the ice core and tree-ring records, these are interesting coincident results. Further ice core and tree ring records from this sector of the Arctic will be needed to validate these correspondences.

Other AMP records from high north latitudes that are thought to provide summer temperature records over the past millennium are from Dye 3 in Greenland (Herron *et al.* 1981) and from Devon Island in the Canadian Arctic (Fisher and Koerner 1981). Those localities are respectively about 4300km and 4000km from Salekhard. The 1979-1980 series from Dye 3 has some correspondence with the larch reconstruction in the earlier years. Both show increases to relatively high temperatures by the first half of the 1300s. This is followed by a sudden decrease to near long-term mean values in the ice core series, while the larch series has a more gradual decline in the same succeeding five centuries. The Devon Island record indicates a warming trend from 1300 to 1450, followed by a substantial warming event in the first half of the 1500s. This is somewhat different from the Polar Urals reconstruction. The two series share an overall cooling trend in the succeeding 200 years, as well as warming events in the late 1700s, the mid-1800s, and after about 1880.

Dendroclimatic reconstruction of late summer temperatures in northern Fennoscandia (July-August, A.D.1700-1964) are presented in another chapter of this volume by Briffa and Schweingruber. The center of that study region is about 1880km west of Salekhard. The simple correlation coefficient of the Polar Urals reconstruction with theirs is relatively low and nonsignificant ($r=-0.05$, $p=0.30$). Comparison of the respective time series plots does not indicate any remarkable similarities in the estimated temperature records for single years or for short term (10-20 years) variation. It can be questioned whether temperature of the two areas would demonstrate strong synchroneity of change on short time-scales, based on studies of instrumental records in the Arctic for the period of 1881-1980 (Kelly *et al.* 1982). Briffa and Schweingruber's principal component analyses of annual and seasonal temperature variation suggest some differences in the two regions, and they note that summer temperature patterns are less well organized spatially than patterns in other seasons. Additionally, the seasonal period of strongest climatic sensitivity varies by one month, and the two genera (*Larix vs Pinus*) may be responding somewhat differently to seasonal as well as to annual temperatures due to biological differences.

20.7 Conclusions

After extensive experimentation we conclude that the highest quality dendroclimatic reconstructions of June-July average temperatures in the Polar Urals from *Larix sibirica* are best obtained by using pre-whitened tree-ring indices rather than untreated tree-ring indices. This empirical finding is only one of such studies and may not be generalizable to others, but it remains an issue for continuing investigation.

Early summer temperature variation in the Polar Urals evidences long-term trends on the scale of centuries or more that are broadly similar to changes derived from many other proxy temperature records over the past millennium from diverse areas that include the European

High Arctic, Greenland and the Americas (see especially Williams and Wigley 1983). These include warming to above long-term means from near A.D. 1000 to 1200 or 1300. Subsequently there is overall cooling until the 19th century that usually includes some regional episodes of intervening warm periods. In the Polar Urals the post-A.D. 1500 period is generally one of cooling until the early 1880's, the coldest part of the record. The relatively sustained rise in temperature from the 1880's to the 1960's has the greatest magnitude in the record for such an extended period (1.6°C difference between the 20 year periods centered on 1870 and 1960). Other extremes in the post-1500 period include the highest 20 year mean value for the period centered on 1800 (11.8°C) the greatest decrease (1.6°C) from one 20 year period to the next 1790-1810, and the second warmest value in the entire reconstruction (1870: 16.1°C).

Comparison of the larch temperature reconstruction with AMP records in ice cores from Svalbard, Devon Island and southern Greenland indicates that the greatest correspondence since 1500 is with Svalbard which is the closest and the northernmost record. This suggests that a regional signal might be forthcoming if more ice core records could be developed from the remaining archipelagos to the north of Siberia. The Polar Urals reconstruction and the two other AMP records collectively suggest that the period from about 1000-1550 was one with various regional episodes of warming well above long term means. Subsequent centuries underwent overall cooling until the late 1800s and all three records indicate a brief but obvious warming just before 1800.

Comparison of the early and late summer dendroclimatic reconstructions from the Polar Urals and Fennoscandia did not reveal strong similarities. Probable reasons for this are biological differences in the climatic responses of larch and pine as well as in the short-term climates of the two regions. A larger network of tree-ring chronologies that includes series from the intervening sector of the northwestern Soviet Union and climatic reconstruction from longer series (Schweingruber et al. 1988) is desirable. In conjunction with other new proxy records such as ice cores, we would then be better able to evaluate the strength, coherency, and history of climate signals across these regions.

These kinds of endeavors are important beyond the level of academic exercise. They can inform on climatic extremes or states that we have not yet experienced but which may potentially occur. The Early Medieval warm period is a strong illustration of this and contains some parallels to recent climatic history which are pertinent in this regard. Early Medieval warming was accompanied by relatively low and stable interannual summer temperature variation, much as the Polar Urals experienced from about 1930 to 1960. Although later medieval times remained relatively warm, they were characterized by a long period of exceedingly high interannual variance that is unparalleled in the rest of the record. If widespread, changes of this kind may have had important economic and social implications. For example, reliance upon subsistence agriculture in marginal environments was and still is a precarious enterprise at best. Major changes in the interannual variability of a crucial factor such as growing season temperatures at high latitudes could result in substantial decreases in crop production, which would probably also be accompanied by exhaustion of any stored reserves. Barring a set of social and political mechanisms to cope with these circumstances, famine and population displacement were not unlikely. Is it possible that some of those events during the thirteenth and fourteenth century in European Russia that are described by Borisenkov (Chapter 9, this volume) were driven by such climate changes, and, is the recently increasing variance of Soviet Arctic climate a portent of things to come?

References

Bradley, J. V. 1968. *Distribution-free statistical tests*. Englewood Cliffs, New Jersey: Prentice-Hall.

Bradley, R. S. 1988. The explosive volcanic eruption signal in Northern Hemisphere continental temperature records. *Climate Change* 12, 221-243.

Briffa, K. R. 1984. *Tree-Climate Relationships and Dendroclimatological Reconstruction in the British Isles*. Ph.D. Thesis, University of East Anglia, Norwich, England.

Briffa, K. R., P. D. Jones, J. R. Pilcher and M. K. Hughes 1988. Reconstructing summer temperatures in northern Fennoscandinavia back to A. D. 1700 using tree-ring data from Scots Pine. *Arctic and Alpine Research* 20(4) 385-394.

Box, G. E. P. and G. M. Jenkins 1976. *Time series analysis: forecasting* and control. San Francisco: Holden-Day.

Chatfield, C. 1980. *The analysis of time series: an introduction*. New York: Chapman and Hall.

Cook, E. R. 1985. *A time-series approach to tree-ring standardization*. Ph.D Thesis, University of Arizona, Tucson.

Cook, E., K. Briffa, S. Shiyatov and V. Mazepa 1990. Tree-ring standardization and growth-trend estimation. In *Methods of Dendrochronology: Applications in the Environmental Sciences*, E. R. Cook and L. A. Kairiukstis (eds) 104-123. Boston: Kluwer.

Dixon, W. J. (ed.) 1985. *BMDP statistical software*. Berkeley: University of California Press.

Draper, N. and H. Smith 1981. *Applied regression analysis*. New York: Wiley.

Dzerdzeevskii, B. 1962. Fluctuations of climate and of general circulation of the atmosphere in extra-tropical latitudes of the Northern Hemisphere and some problems of dynamic climatology. *Tellus* 14, 328-336.

Eklund, B. 1954. Arsingsbreddens klimatiskt betingade variation hos tall och gran inom norra Sverige aren 1900-1944. *Meddelanden fran Statens Skogsforskninstitut* 44(8) 1-150.

Fisher, D. A. and R. M. Koerner 1981. Some aspects of climate change in the high Arctic during the Holocene as deduced from ice cores. In *Quaternary Paleoclimate*, W. C. Mahaney (ed) 249-271. Norwich: Geo Abstracts.

Fritts, H. C. 1976. *Tree rings and climate*. London: Academic Press.

Garfinkel, H. L. and L. B. Brubaker 1980. Modern climate-tree-growth relationships and climatic reconstruction in subarctic Alaska. Nature 268, 872-874.

Giddings, J. L. 1941. Dendrochronology in northern Alaska. *University of Arizona Bulletin* 12(4) 1-107.

Giddings, J. L. 1954. Tree-ring dating in the American Arctic. *Tree-Ring Bulletin* 20, 23-25.

Gordon, G. A. and S. K. Leduc 1981. Verification statistics for regression Models. In *Preprints seventh conference on probability and statistics in atmospheric sciences* 129-133. Monterey: American Meteorological Society.

Graumlich, L. J. 1989. Interactions between climatic variables controlling subalpine tree growth: implications for climatic history of the Sierra Nevada, California. In *Proceedings of the sixth annual pacific climate (paclim) workshop*, compiled by J. L. Betancourt and A. M. MacKay 102-105. Asilomar, California.

Graybill, D. A. 1982. Chronology development and analysis. In *Climate from tree rings*, M. K. Hughes, P. M. Kelly, J. R. Pilcher and V. C. LaMarche, Jr. (eds) 21-28. New York: Cambridge University Press.

Graybill, D. A. 1987. A network of high elevation conifers in the Western U.S. for detection of tree-ring growth response to increasing atmospheric carbon dioxide. In *Proceedings of the international symposium on ecological aspects of tree-ring analysis*, 463-474. Springfield, Va.: National Technical Information Service.

Graybill, D. A. 1989. The reconstruction of prehistoric Salt River stream-flow. Chapter 3 in *The 1982-1984 excavations at Las Colinas, environment and subsistence*, by D. A. Graybill, D. A. Gregory, F. L. Nials, S. K. Fish, R. E. Gasser, C. H. Miksicek and C. R. Szuter. Archaeological Series 162, Volume 5, Part 1, Cultural Resource Management Division, Arizona State Museum, University of Arizona, Tucson.

Graybill, D. A. and S. G. Shiyatov 1989. A 1009 year tree-ring reconstruction of mean June-July temperature deviations in the Polar Urals. In *Air pollution effects on vegetation including forest ecosystems*, R. D. Nobel, J. L. Martin and K. F. Jensen (eds) 37-42. Proceedings of the Second U.S.-U.S.S.R. Symposium. U.S. Department of Agriculture, Forest Service, Northeastern Forest Experiment Station.

Grove, J. M. 1988. *The Little Ice Age*. New York: Methuen.

Herron, M. H., S. L. Herron and C. C. Langway, Jr. 1981. Climatic signal of ice melt features in southern Greenland. *Nature* 293, 389-391.

Jacoby, G. C. and E. R. Cook 1981. Past temperature variations inferred from a 400 year tree-ring chronology from Yukon Territory, Canada. *Arctic and Alpine Research* 13(4) 409-418.

Jacoby, G. C., E. R. Cook and L. D. Ulan 1985. Reconstructed summer degree days in Central Alaska and Northwestern Canada since 1524. *Quaternary Research* 23 18-26.

Jacoby, G. C. and R. D'Arrigo 1989. Reconstructed Northern Hemisphere annual temperature since 1671 based on high-latitude tree-ring data from North America. *Climate Change* 14, 39-59.

Jenkins, G. M. and D. G. Watts 1968. *Spectral analysis and its applications*. San Francisco: Holden-Day.

Jones, P. D., S. C. B. Raper, B. D. Santer, B. S. G. Cherry, C. Goodess, R. S. Bradley, H. F. Diaz, P. M. Kelly and T. L. Wigley 1985. *A grid point surface air temperature data set for the Northern Hemisphere 1851-1984*. U.S. DOE Technical Report, TR022, U. S. Department of Energy Carbon Dioxide Division, Washington D.C.

Kelly, P. M., P. D. Jones, C. B. Sear, B. S. G. Cherry and R. K. Tavakol 1982. Variations in surface air temperatures: Part 2, Arctic regions 1881-1980. *Monthly Weather Review* 110, 71-83.

LaMarche, V. C. Jr. 1974. Paleoclimatic inferences from long tree-ring records. Science 183 1043-1048.

Lamb, H. H. 1977. *Climate: present, past and future*, Vol. 2. London: Methuen.

Lamb, H. H. and A. I. Johnson 1959. Climatic variation and observed changes in the general circulation: Parts I and II. *Geografiska Annaler* 41, 94-134.

Lamb, H. H. and A. I. Johnson 1961. Climatic variation and observed changes in the general circulation: Part III. *Geografiska Annaler* 43, 363-400.

Lorenz, E. N. 1956. Empirical orthogonal functions and statistical weather prediction. In *Statistical forecasting project science report*, no. 1, 604-1566. M.I.T.

Meko, D. M. 1981. *Applications of Box-Jenkins methods of time series analysis to the reconstruction of drought from tree-rings*. Ph.D. Thesis, University of Arizona, Tucson.

Mikola, P. 1962. Temperature and tree growth near the nothern timberline. In *Tree growth*, T. T. Kozlowski (ed.) 265-274. New York: Ronald Press.

Monserud, R. A. 1986. Time series analyses of tree-ring chronologies. *Forest Science* 32(2) 349-372.

Nash, J. E. and J. V. Sutcliffe 1971. Riverflow forecasting through conceptual models. 1, A discussion of principles. *Journal of Hydrology* 10, 282-290.

Schweingruber, F. H., T. Bartholin, E. Schar and K. R. Briffa 1988. Radiodensitometric-dendroclimatological conifer chronologies from Lapland (Scandinavia) and the Alps (Switzerland). *Boreas* 17 ,559-566.

Sear, C. B., P. M. Kelly, P. D. Jones and C. M. Goodess 1987. Global surface-temperature responses to major volcanic eruptions. *Nature* 330, 365-367.

Shiyatov, S. G. 1986. *Dendrochronology of the upper forest boundary in the Urals*. Moscow: Nauka (in Russian).

Shiyatov, S. G., H. C. Fritts and R. G. Logfren 1989. Comparative analysis of the standardization methods of tree-ring chronologies. In *Air pollution effects on vegetation including forest ecosystems*, R. D. Nobel, J. L. Martin and K. F. Jensen (eds) 13-25. Proceedings of the Second U.S.U.S.S.R. Symposium. U.S. Department of Agriculture, Forest Service, Northeastern Forest Experiment Station.

Shiyatov, S. G. and V. S. Mazepa 1987. Some New Approaches in the Construction of More Reliable Dendrochronological Series and in the Analysis of Cycle Components. In *Methods of Dendrochronology I*. Proceedings of the Task Force Meeting on Methodology of Dendrochronology: East/West Approaches. Laxenberg, Austria: IIASA.

Stahle, D. W., M. K. Cleaveland and J. G. Hehr 1988. North Carolina climate changes reconstructed from tree-rings: A.D. 372 to 1985. *Science* 240 1517-1519.

Tranquillini, W. 1979. *Physiological ecology of the Alpine timberline*. New York: Springer-Verlag.

Wigley, T. M. L., K. R. Briffa and P. D. Jones 1984. On the average value of correlated time series with applications in dendroclimatology and hydrometeorology. *Journal of Climate and Applied Meteorology* 23, 201-213.

Williams, L. D. and T. M. L. Wigley 1983. A comparison of evidence for Late Holocene summer temperature variations in the Northern Hemisphere. *Quaternary Research* 20, 286-307.

Wonnacott, T. H. and R. J. Wonnacott 1981. *Regression: a second course in statistics*. New York: Wiley.

21 Dendroclimatic evidence from the western Himalaya

M. K. Hughes

21.1 Introduction

Instrumental and proxy records of climate with annual resolution are concentrated in the developed countries of the temperate zone (Bradley *et al* 1985; Hughes 1987a). The forests of the Western Himalaya offer a possibility for extending this coverage (Bhattacharya *et al* 1988). In addition to tree species and sites analogous to those that have yielded valuable dendroclimatic records in semiarid regions there are extensive subalpine conifer forests. These have phytogeographical and ecological similarities to forests in the European Alps (Aymonin and Gupta 1965; Meusel and Schubert 1971) where the power of wood densitometry as a dendroclimatological technique has been demonstrated (Schweingruber *et al* 1978).

The use of ring width and wood density from subalpine conifers as proxy climate records, as pioneered by Parker and Henoch (1971) and Schweingruber *et al* (1978) was the chosen approach in this project. The results reported here are for the most promising tree species (*Abies pindrow* [Royle] Spach.) in the most intensively sampled part of the region, the Vale of Kashmir and its surrounding mountains – the Pir Panjal and the western ranges of the Greater Himalaya (Figure 21.1).

21.2 Data sources

21.2.1 Meteorological and tree-ring data

Monthly temperature and precipitation data for Srinagar, Kashmir (34°5'N, 74°50'E, 1587m a.s.l.) are available from A.D. 1893 from World Weather Records.

The details of tree-ring sites for which replicated cross-dated chronologies of *Abies pindrow* have been established are given in Table 21.1. The distribution of these sites in the mountains surrounding the Vale of Kashmir, where the Srinagar meteorological station is found, is shown in Figure 21.1. Several of the chronologies extend back well into the seventeenth century, but replication is not adequate in some cases until the early eighteenth century. Emphasis was placed on collecting from a number of locations rather than establishing a definitive sample at each. The length and replication of chronologies was limited by the prevalence of butt rot in *A. pindrow* and difficulty in obtaining good density measurements from many cores or parts of cores that would be acceptable for ring width. Most of the remaining trees of moderate age (200 years or more) are found on steep slopes (Table 21.1) and so suffer greatly from distorted xylem structure.

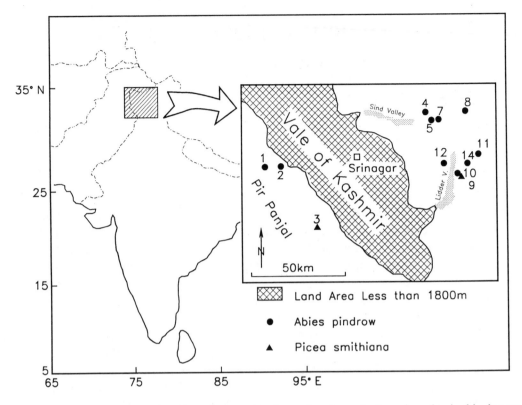

Figure 21.1 The location of replicated, cross-dated tree-ring chronologies referred to in this chapter.

Table 21.1 Site details of *Abies pindrow* chronologies.

Site	Abbrev-iation	Latitude ° ' N	Longitude ° ' E	Altitude metres	Aspect	Slope degrees	Time span
Gulmarg	GUL	35 05	74 18	2740	E	0-40	1620-1680
2. Khillanmarg	KHI	35 05	74 20	3125	ESE	55	1668-1980
4. West of Sonmarg	WSO	34 39	75 28	3200	NE	40	1682-1982
5. Sonmarg	SON	34 35	75 33	3125	N	60	1764-1980
7. Thijwas	THI	34 35	75 32	3400	NNE	60	1655-1982
8. Sarbal	SAR	34 30	75 45	3110	N	40	1644-1982
10. Pahlgam	PAH	34 02	75 42	2620	ENE	40	1656-1980
11. Chandanwani	CHA	34 15	75 55	3200	W	50	1777-1982

Site numbers refer to a series of collections, some of which have not yielded cross-dated chronologies.

All materials were cross-dated (Hughes and Davies 1987) and analyzed using the X-ray microdensitometric procedures described by Lenz *et al* (1976) with two principal modifications. These were the use of a moving carriage system to produce radiographs (Milsom and Hughes 1978) and the use of a linearising optical wedge in the densitometer (Hughes and Sardinha 1975). The measurements for each core were dated maximum latewood density

(MXD) minimum earlywood density (MND) earlywood width (EWW) latewood width (LWW) and total ring width (TRW). All series were standardized by fitting a polynomial curve and taking the quotient of the actual and fitted value for each year as the index. The fitted curves were inspected to insure that variation common to a majority of cores from a site was not removed in this process. The series of indices for maximum and minimum density, earlywood, latewood and total ring width for the cross-dated trees at each site were then combined to form the five chronologies for that site.

21.2.2   Tree-ring chronology statistics

In order to compare the nature of the five chronologies from each of the eight *Abies pindrow* sites, a range of chronology statistics were calculated for a common period of 60 years. Three statistics are presented in Table 21.2, namely the mean, the serial correlation, and the proportion of the variance of the individual series attributable to the common between-year pattern (%Y) (Fritts 1976). MND shows little common pattern (measured as %Y) for any combination of sites and so will not be discussed further. Of the remaining four variables LWW generally has the smallest %Y. It also tends to have a low serial correlation, a positive

Table 21.2   Chronology statistics 1921-1980 *Abies pindrow*.

SITE	1. GUL	2. KHI	4. WSO	5. SON	7. THI	8. SAR	10. PAH	11. CHA
Number of trees (cores)	15(14)	9(13)	17(17)	6(6)	25(25)	16(16)	9(9)	9(9)
Maximum density								
Mean (g/ml)	0.82	0.80	0.74	0.79	0.71	0.76	0.83	0.78
Serial r	0.25	0.32	0.27	0.16	0.45	0.45	0.49	0.35
%Y	31.0	26.3	17.9	23.8	14.3	16.0	28.9	37.2
Minimum density								
Mean (g/ml)	0.36	0.34	0.30	0.31	0.32	0.31	0.33	0.30
Serial r	0.36	0.41	0.19	0.31	0.41	0.34	0.43	0.34
%Y	3.5	3.3	7.1	1.7	8.3	4.8	8.4	8.2
Earlywood width								
Mean (mm.)	1.34	1.16	1.30	1.35	0.84	0.97	1.24	1.81
Serial r	0.64	0.59	0.59	0.63	0.43	0.63	0.59	0.51
%Y	24.3	11.2	26.4	28.0	14.6	27.3	19.6	30.7
Latewood width								
Mean (mm.)	0.22	0.24	0.18	0.17	0.21	0.20	0.20	0.26
Serial r	0.36	0.44	0.29	0.36	0.46	0.42	0.52	0.31
%Y	14.8	12.3	9.0	13.5	8.8	11.9	21.6	18.5
Total ring width								
Mean (mm.)	1.56	1.40	1.48	1.52	1.05	1.17	1.44	2.07
Serial r	0.66	0.67	0.60	0.64	0.46	0.65	0.62	0.50
%Y	24.3	14.3	34.2	31.5	17.5	27.4	25.1	30.0

%Y is the proportion of sample variance retained by the final chronology as determined from an analysis of the period 1921 to 1980.
A full explanation of the calculation and interpretation of chronology statistics is given by Fritts (1976).

attribute in terms of time series statistics. MXD has the lowest serial correlation in most cases. TRW usually has a much higher serial correlation. This persistence is manifest primarily in EWW. There appears to be no evidence of systematic variation in these statistics on a geographical or habitat basis.

The values of %Y and serial correlation presented here are in general similar to those presented by Schweingruber *et al* (1978) for sites in the Alps (mainly *Picea abies*) and by Hughes (1987b) for *Pinus sylvestris* in Scotland. Serial correlations are higher for MXD in several Kashmir *Abies* chronologies than for these other species and regions. Kienast (1985) reports a serial correlation of 0.31 for an *Abies alba* chronology in Switzerland. It may be that MXD in the genus *Abies* shows particularly great persistence as compared with this variable in other genera.

The eight chronologies for each variable (except MXD) were subject to an analysis of variance that showed strong regional coherence. Seven sites only were analyzed for MXD as there were measurement problems with MXD at one site (Pahlgam) in the last few decades. The other variables were unaffected by this. The percentage of variance attributable to the common regional pattern varied thus for the 90 years 1891-1980: MXD 61%; EWW 29%; LWW 47%; TRW 45%. Only MXD and TRW will be used in the work reported in this paper. TRW is used in preference to LWW because it is a more reliable measurement, being less dependent on the density level chosen to demarcate earlywood from latewood. EWW is not used because it has a weak regional signal.

21.3   Analysis methods

21.3.1   Calibration methods

Calibration tests were conducted in which TRW chronologies (TRW' series) for eight and MXD chronologies (MXD' series) for seven *Abies* sites were predictors and various combinations of monthly mean temperature and monthly total precipitation at Srinagar were predictands. The tree-ring data for years i-1 and i+1 were presented to the regression as well as those for year i to allow for autoregression in the tree-ring data and for lag effects. The calibrations were derived using an orthogonalized stepwise multiple regression procedure described by Hughes *et al* (1984) and based on that developed by Guiot (1980). The predictors were first screened using a stepwise multiple regression procedure in which values of F significant at p=0.1 and 0.15, respectively, were used to determine the entry and exit point of a predictor to and from the model. The principal components of the selected variables were then taken and the effective number of predictors further reduced by only retaining those components whose cumulative eigenvalue product exceeded unity. These components were entered into a final multiple regression producing the calibration equation in which Srinagar temperature or rainfall is expressed in terms of the tree-ring variables retained after the original screening.

21.3.2   Verification methods

Verification tests were applied for the 26 years (1943-1968) that were not used in the calibration. Four tests were applied in the case of each calibration.

1) The statistical significance of the correlation coefficient between the actual and estimated temperatures was calculated.

2) The reduction of error (RE) statistic was calculated. This is a measure of the residual error variance relative to that which would be obtained if the reconstructed values were compared with the mean for the calibration period. If RE is greater than zero there is some skill in the model compared with the trivial persistence estimate.

3) A sign test was applied which involves the comparison of the signs of first differences of the two series. This gives an indication of the extent to which the reconstruction reproduces the highest frequency component of the instrumental data.

4) A product means test was applied such that a significant positive product mean indicates a tendency for both actual and estimated departures from the average to be large when the sign is correctly estimated and to be small when the sign is incorrectly estimated.

A fuller account of these tests is given by Fritts (1976). In determining whether or not to further investigate a reconstruction, its performance in all four of these tests was considered.

21.4 Calibration and verification

21.4.1 Temperature

Calibration tests using the 15 tree-ring series (eight TRW' and seven MXD') to predict monthly mean temperatures at Srinagar yielded several statistically significant relationships. After examining these along with the correlation structure of the monthly temperatures for the 50 year period 1894-1943, the calibration procedure was repeated for groups of months on the grounds of either biological relevance or the structure of the seasons in Kashmir (Raza *et al*, 1978). Particularly promising results emerged for April-May, April-May-June-July and June-July mean temperatures (Table 21.3a).

Only those cases where a statistical significant relationship was established in the calibration were subject to verification tests for the period 1943 to 1968 (Table 21.3b). Good verification on all tests was found for reconstructions of temperature at Srinagar for April-May and August-September. The April-May reconstruction performed less well in the sign test indicating relative failure to capture high frequency variation. This is made clear when the time series are partitioned into high- and low-frequency bands by the use of a pair of reciprocal digital filters (Fritts, 1976) designed to pass variance with approximate wavelengths less or greater than eight years respectively (Figure 21.2). The August-September reconstruction performed well in all verification tests, but failed notably to capture one extreme event, the extraordinarily hot August of 1959. This reconstruction failed to capture low-frequency variation (Figure 21.3). These are the only temperature reconstructions derived from the eight TRW' and seven MXD' series that will be discussed further. In the case of April-May temperature, much the largest regression coefficient was for year i MXD' at Gulmarg (GUL). For August-September temperature the predictor with the largest regression coefficient was year i MXD' at the much higher elevation Chandanwani site (3200m at CHA as compared to 2740m at GUL).

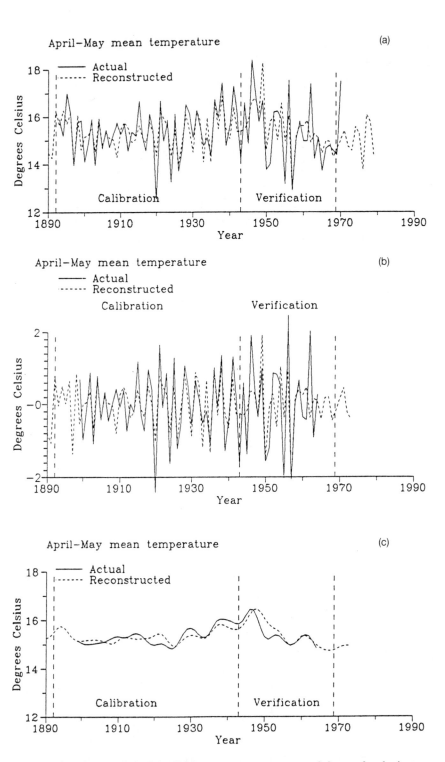

Figure 21.2 Time series of April-May mean temperature at Srinagar for the instrumental period for (top) unfiltered; (middle) high-pass filtered and (bottom) low-pass filtered data. Details of the filtering are given in the text. The reconstruction used the 15-series network described in sections 21.2 and 21.3.

Table 21.3 Performance of regression model with 15-series, (i-1, i, i+1) predictor set Monthly mean temperatures at Srinagar.

a) **Calibration period (1893-1942)**

Period	Multiple correlation coefficient	Predictors retained	F-value	Probability
Prior November- December	0.579	4	7.0	<0.0001
April-May	0.728	5	11.6	<0.0001
April-July	0.696	13	4.3	<0.005
June-July	0.642	4	9.8	<0.0001
August-September	0.496	3	7.0	<0.01

b) **Verification period (1943-1968)**

Period	Correlation coefficient	Reduction of error (RE)	Sign Test	Product means 't'
Prior November- December	0.415*	-0.068	15/25	1.021
April-May	0.663****	0.381+	15/25	2.385*
April-July	0.593***	0.102+	19/25*	1.397
June-July	0.335	-0.56	13/25	-0.447
August-September	0.478**	0.389+	19/25*	4.060****

+ - RE greater than 0

* - $p < 0.05$; ** - $p < 0.02$; *** - $p < 0.01$; **** - $p < 0.001$

21.4.2 Precipitation

Table 21.4a shows the results of calibration tests with total precipitation for several groups of months as predictands. Statistically significant relationships exist in all cases.

The reconstruction of total precipitation in the period from April through September performed well in all four verification tests (Table 21.4b). There appears to be no evidence of a difference in the ability to capture high and low extremes of total precipitation. Both high frequency (Figure 21.4b) and low frequency (Figure 21.4c) variations are reproduced in the verification period. The predictors with the largest regression coefficients are MXD'$_i$ and MXD'$_{i-1}$ at GUL. It is not entirely surprising that MXD emerges as a proxy for precipitation as well as for temperature, since there is a negative relationship between April-September precipitation and temperature in April-May ($r = -0.340$, $p<0.005$) April-May-June-July ($r = -0.483$, $p<0.005$) and August-September ($r = -0.179$, $p>0.05$).

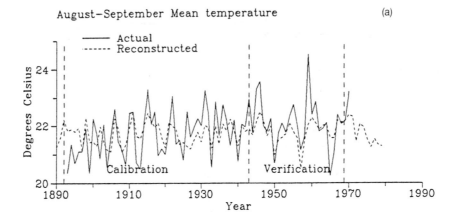

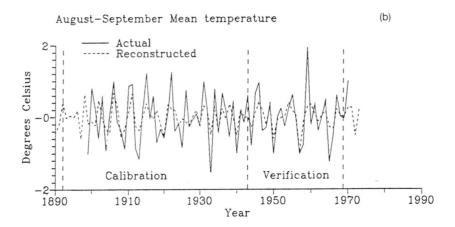

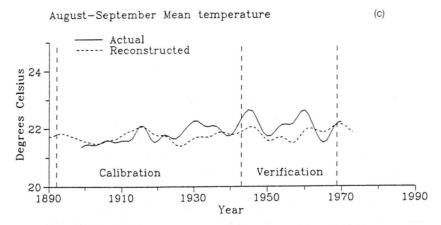

Figure 21.3 Time series of August-September mean temperature at Srinagar for the instrumental period for (top) unfiltered; (middle) high-pass filtered and (bottom) low-pass filtered data. Details of the filtering are given in the text. The reconstruction used the 15-series network described in sections 21.2 and 21.3.

Table 21.4   Performance of regression model with 15-series
(i-1, i, i+1) predictor set. Precipitation at Srinagar.

a) Calibration period (1894-1943)

Period	Multiple correlation coefficient	Predictors retained	F-value	Probability
Prior October to September (12 months)	0.607	3	12.56	<0.0001
Prior October to March (6 months)	0.556	4	6.25	<0.005
April-June	0.467	1	12.25	<0.0001
July-September	0.585	9	3.38	<0.005
April-September	0.685	5	12.37	<0.0001

b) Verification period (1943 1969)

Period	Correlation coefficient	Reduction of error (RE)	Sign Test	Product means 't'
Prior October to September (12 months)	0.467**	-0.513	13/25	0.881
Prior October to March (6 months)	0.499**	-1.029	13/25	0.039
April-June	0.587***	0.333+	12/25	1.401
July-September	0.538***	-0.274	11/25	0.748
April-September	0.731****	0.312+	18/25+	2.16*

* p < 0.05; ** p < 0.02; *** p < 0.01; **** p < 0.001

21.5   The reconstructions

21.5.1   Temperature

The April-May network-based reconstruction shows markedly greater variance in the period
before the mid-eighteenth century than thereafter. This is likely to be an artifact arising from
poorer replication in the earlier part of the chronologies most strongly weighted in the
calibration. This effect is greater in the case of the early part of the TRW' series than for
MXD' series. Hence only the later part (after A.D. 1780) of the April-May and August-
September temperature reconstructions based on the eight TRW' and seven MXD' series will
be discussed further (Figures 21.5a and 5b).

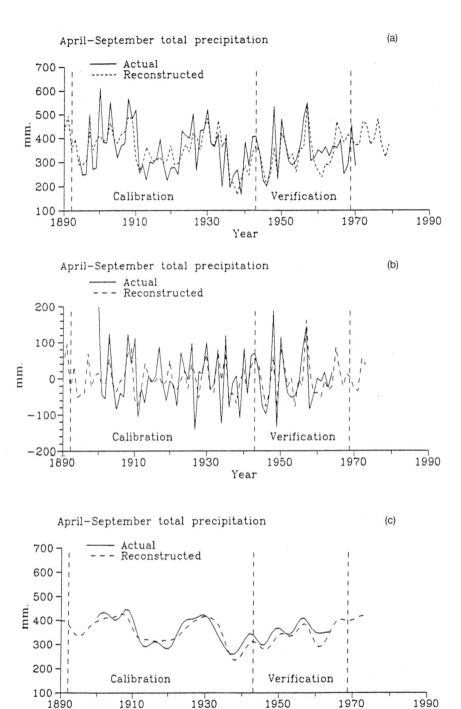

Figure 21.4 Time series of April-September total precipitation at Srinagar for the instrumental period for (top) unfiltered; (middle) high-pass filtered and (bottom) low-pass filtered data. Details of the filtering are given in the text. The reconstruction used the 15-series network described in sections 21.2 and 21.3.

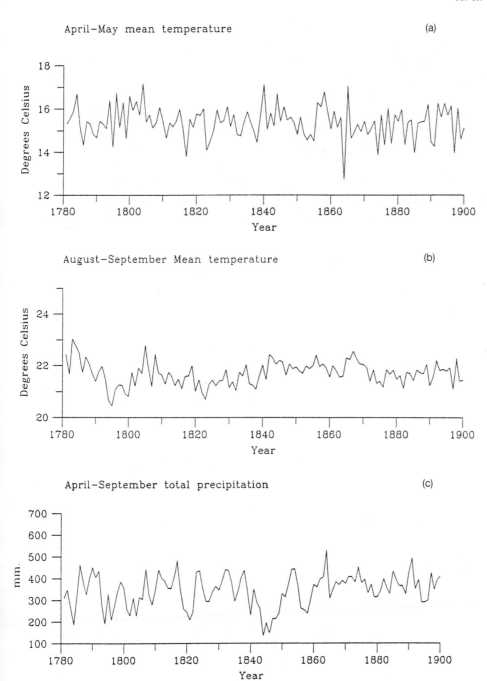

Figure 21.5 Reconstructions at Srinagar based on the 15-series network described in sections 21.2 and 21.3; (top) April-May mean temperature; (middle) August-September mean temperature; (bottom) April-September total precipitation.

The problem of inadequate replication in the early years of key chronologies may be circumvented by using a mean MXD' series for *A. pindrow* from all eight sites in Kashmir. The evidence for a strong regional signal in MXD has been given above and by Hughes and

425

Davies (1987). These results are consistent with those of Schweingruber (1985) on larger spatial scales in Europe. Calibration tests were conducted in which regional MXD'_i and MXD'_{i+1} (current and next year's maximum latewood density) were offered as predictors in a regression equation for April/May temperature. A strong relationship was demonstrated (Figure 21.6a) for the period 1893-1942 (multiple correlation coefficient 0.609, one predictor retained (MXD'_i) F value 26.55, $p<0.0001$). For the verification period (1943-1968) this model performed well in three of four tests (Correlation coefficient 0.536, $p<0.01$; reduction of error 0.399; product means 't' 2.330, $p<0.05$)). Performance in the sign test was moderately encouraging (17 agreements out of 25 possible, $p>0.05$). This reconstruction (Figure 21.6b) shows no great change in variance, at least back to A.D. 1690. Notably cold springs are reconstructed for 1694, 1755 and 1769. There is no apparent difference in the nature of this reconstruction between its early part and that after A.D. 1780.

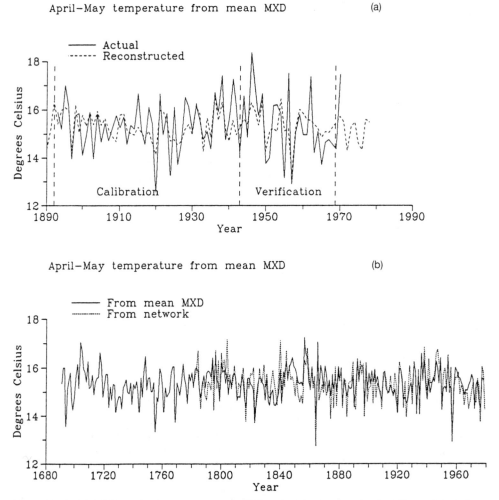

Figure 21.6 April-May mean temperature at Srinagar reconstructed using the regional mean MXD' series (see section 21.5.1); (above) for the instrumental period, with the actual record; (below) since A.D. 1690, with the record from the 15-series network also shown.

426

21.5.3   Precipitation

As for the temperature reconstructions also based on the eight-site tree-ring network, the summer precipitation reconstruction will only be discussed for the period after A.D. 1780 (Figure 21.5c). It is characterized throughout its length by multi-year periods above or below the long-term mean.

21.5.4   Reconstructed climate at Srinagar

The reconstructions presented here extend the record for spring temperature, late summer temperature, and growing season rainfall, back from the beginning of the instrumental record in 1894 to 1780 with a fairly high degree of confidence. Further information on spring temperatures between 1690 and 1779 has been derived from a mean regional MXD' series. There is no evidence of consistent trend on the multi-decade to century timescale in these reconstructions. This may indicate a genuine lack of trend or it may result from the method used to standardize the tree-ring data. Two major characteristics of climate at Srinagar in the last two centuries emerge from the reconstructions. First, the record is punctuated by years of markedly cold springs (1864, 1823, 1817) cold late summers (1823, 1800, 1795, 1794) and wet growth seasons (notably 1891, 1864 and 1817). The frequency of cold springs from 1700 to 1779 is similar to that in the following century, the springs of 1694, 1755 and 1769 being reconstructed as particularly cold. In addition to 1694, the springs of 1698, 1699 and 1700 constitute an unusual sequence of cold springs. No obvious or consistent relationship has yet been found between these extreme years and major climate effects such as the Southern Oscillation or the strength of the Indian Monsoon. Second, there are multi-year runs of markedly below-average April-September rainfall. Within the instrumental period these occurred, for example, in the late 1930s and the mid-1940s. Such dry periods were absent between 1860 and 1930, whereas several were reconstructed between 1780 and 1860.

No other calibrated reconstructions have been reported for this region at the time of writing. Kashmir had an extraordinarily complex history during the eighteenth and nineteenth centuries, with less direct British involvement than in most of the subcontinent. As a consequence, it is difficult to find documentary information on the history of the region's climate in European languages. Two sources yield information relevant to the verification of the reconstructions presented here (Koul, 1978; Bamzai, 1962) and indicate that a more systematic search by a climate historian could be of value. Both draw information from a variety of sources, making some reference to climate-related phenomena such as floods and famines. I have not subjected these reports to rigorous examination. Koul (1978) asserts that "Every great famine that occurred in Kashmir was caused, not by summer drought, but by a too mild winter or by heavy rains in harvest season which destroyed the crops". Even so, the reports of crop yields or famines seemed most likely to be of relevance to dendroclimatic reconstructions, since they represent the effect of conditions over several months rather than a single event of a few hours or days duration such as a flood. Each of these authors reports two crop-related events after 1780. Koul (1978) reports that crops did not ripen in 1813, whilst Bamzai (1962) reports famine in 1814. These may refer to the same event. Bamzai also reports that the rice crop remained unharvested in October of 1832, whilst Koul describes the harvest of 1864 as 'scanty'. There is nothing in the reconstructions to distinguish 1813, 1814 or

1832, although the latter was moderately cool and wet. 1864 was marked by the coldest April-May in the reconstruction and the second wettest summer.

21.6   Discussion

21.6.1   Uncertainties

Quantitative paleoclimatic reconstruction of the kind reported here is based on two key assumptions. First, that the relationship between the climate variable of interest and the proxy record is sufficiently well represented by the model used. Second, that this relationship is stable over the time period of the reconstruction. Uncertainty as to the reliability of the reconstruction increases to the extent that these assumptions are violated.

There is a considerable body of published evidence (Parker and Henoch 1971; Schweingruber *et al* 1978; Hughes *et al* 1984; Conkey 1986) that there is a strongly linear relationship between the maximum latewood density (MXD) of conifers that are not usually drought-stressed and seasonal temperatures in the growth season. Conkey (1986) has proposed a physiological model to explain the observation, repeated here, that there is often a strong link with spring as well as late summer temperatures. This pre-existing understanding, along with the calibration results, supports the contention that the climate-growth relationship is well represented by the calibrations reported above.

The second assumption concerns the temporal stability of the relationship between the climate variable and the proxy record. There are two components of this problem, namely the stability of the tree ring-climate relationship and the consistency of the tree-ring record itself. It is difficult to test the first part of this rigorously as it has been in Europe (e.g. Hughes *et al*, 1984; Hughes, 1987b) when only a few decades of instrumental data are available, as in this case. Even so, the verification results indicate that there is temporal stability in the climate-growth relationships. If the available historical climate information is brought together and subjected to critical analysis, it could, in addition to its innate worth, be invaluable in testing dendroclimatic reconstructions.

Even if the underlying relationship between tree-ring properties and the climate variable is strong and stable, the reconstruction will be rendered invalid if replication in the tree-ring series drops below some acceptable level (Wigley *et al*, 1984). This is most commonly relevant to the earliest years of a reconstruction. In addition, the risks associated with reconstructing conditions not found during the calibration period (i.e. extrapolating) should be borne in mind (Graumlich and Brubaker 1986). The truncation of all but one reconstruction at A.D. 1780 served to minimize these risks.

21.6.2   Potential

The aim of this work was to test the feasibility of reconstructing the climate of recent centuries in the western Himalaya using the techniques of dendroclimatology, particularly wood microdensitometry. The approach adopted was to seek out species, climate and ecological conditions analogous to those in which Parker and Henoch (1971) Schweingruber *et al* (1978) Conkey (1986) and Hughes *et al* (1984) had succeeded in other regions. Field work over two seasons yielded sets of chronologies (TRW', MXD', etc.) from 20 sites in three Indian states (Jammu and Kashmir, Himachal Pradesh and Uttar Pradesh) and representing three tree species. A number of the chronologies extend back to the seventeenth century. The eight *Abies pindrow* sites in Kashmir used in the reconstructions reported here constitute the only strong regional concentration of chronologies established so far. Even so, the other collections indicate an equivalent potential for dendroclimatic reconstruction in the other regions. There is also potential for improving and extending of the tree-ring dataset in Kashmir.

Bhattacharya *et al* (1988) conducted exploratory sampling with promising results, focussing particularly on species and situations where TRW' series might be drought sensitive. Ramesh *et al* (1985) working with material of *Abies pindrow* from the Gulmarg site, *inter alia*, demonstrated that stable isotopic ratios from the rings of this species have potential as a paleoclimatic record.

The results of the calibration and verification tests reported here confirm that reconstructions of at least equivalent quality to many of those from Europe and North America have been developed for Kashmir. It is inferred from this and the availability of similar materials in neighboring areas that it would be feasible to develop such reconstructions for the Western Himalaya from central Nepal to northern Pakistan. The successful reconstruction of precipitation reported here demonstrates that density chronologies from subalpine conifers may reflect precipitation as well as temperature. Notwithstanding the good quality of the reconstructions, their immediate applicability to problems of climatology is limited. This is because they represent conditions in a small area (diameter circa 150 km.) in an extremely mountainous region. There do not appear to be strong connections between the instrumental climate record of the Vale of Kashmir and that of the rest of Asia, or of larger scale circulation indices. Hence, it is unlikely that such teleconnections would be revealed in a Kashmir proxy record that is little more than double the length of the instrumental record. This does not detract from the value of this feasibility study. It has been demonstrated that dendroclimatological techniques can be applied successfully in this region. A Himalayan tree-ring network would make possible the reconstruction of synoptic scale interannual variation of several meteorological variables. These reconstructions could extend over three, perhaps four centuries.

21.7   Summary

a) A study of the feasibility of using tree-ring series to reconstruct the climate of recent centuries in the western Himalaya had a positive outcome. Not only were suitable materials found at locations in three Indian Himalayan states but preliminary results from the Vale of Kashmir are very encouraging.

b) Ring width series from eight and density series from seven subalpine *Abies pindrow* sites were used. The resulting reconstructions of spring temperature, late summer temperature and growth season rainfall at Srinagar extend the record back from the beginning of the instrumental record in 1894 to 1780. Further information on spring temperatures between 1690 and 1779 came from a mean regional maximum latewood density series.

c) The reconstructed record is punctuated by years of markedly cold springs (1864, 1823, 1817, 1769, 1755, 1694) cold late summers (1823, 1800, 1795, 1794) and wet growing seasons (notably 1891, 1864 and 1817). Multiyear periods of below average April-September rainfall occur repeatedly before 1860 and after 1930.

Acknowledgements

Preliminary work and the 1980 sampling trip were funded by the Royal Society of London. Subsequent work including the 1982 field work was supported by grant GR3/4705 from the U.K. Natural Environment Research Council. Thanks are due to the Chief Conservator of Forests of the State of Jammu and Kashmir, to the Physical Research Laboratory, Ahmedabad, India and to Liverpool Polytechnic where all the laboratory analyses were conducted. A. C. Davies, R. Ramesh, S. K. Bhattacharya, G. B. Pant, P. Mayes, V. Kaul and C. K. Varshney all rendered invaluable help.

References

Aymonin, G. G. and R. K. Gupta 1965. Étude sur les formations végétales et leur succession altitudinale dans les principaux massifs du 'système alpine' occidental. Essai de comparaison avec l'Himalaya. *Adamsonia*, 5, 49-94.

Bamzai, P. N. K. 1962. *A history of Kashmir*. Dehli: Metropolitan Book Co.

Bhattacharya, A., V. C. LaMarche, Jr. and F. W. Telewski 1988. Dendrochronological reconnaissance of the conifers of Northwest India. *Tree-Ring Bulletin*, 48, 21-30.

Bradley, R. S., P. M. Kelly, P. D. Jones, C. M. Goodess, and H. F. Diaz 1985. *A climatic data bank for Northern Hemisphere land areas, 1851-1980*. DOE Technical Report TR017. Washington: U.S. Department of Energy Carbon Dioxide Research Division.

Conkey, L. E. 1986. Red spruce tree-ring widths and densities in Eastern North America as indicators of past climate. *Quaternary Research*, 26, 232-243.

Fritts, H. C. 1976. *Tree Rings and Climate*. Academic Press, London.

Graumlich, L. J. and L. B. Brubaker 1986. Reconstruction of annual temperature (1590-1979) for Longmire, Washington, derived from tree rings. *Quaternary Research*, 25, 223-234.

Guiot, J. 1980. *Spectral multivariate regression in dendroclimatology*. Institute d'Astronomie et de Géophysique, Université Catholique de Louvain-la-Neuve, Contribution No.21.

Hughes, M. K. 1987a. Requirements for spatial and temporal coverage. In *Methods of dendrochronology: East-West approaches*. L. Kairiukstis, Z. Bednarz and E. Feliksik (eds.) 107-116. IIASA/Polish Academy of Sciences.

Hughes, M. K. 1987b. Dendroclimatology of Pinus sylvestris in the Scottish Highlands. In *Applications of tree-ring studies: current research in dendrochronology and related areas*, R.G.W. Ward (ed.) 91-106. Oxford: B.A.R. International series, 333.

Hughes, M. K. and Davies, A. C. 1987. Dendroclimatology in Kashmir using tree ring widths and densities in subalpine conifers. In *Methods of dendrochronology: East-West approaches* L. Kairiukstis, Z. Bednarz and E. Feliksik (eds.) 163-176. IIASA/Polish Academy of Sciences.

Hughes, J. F. and R. M. de A. Sardinha 1975. The application of optical densitometry in the study of wood structures and properties. *Journal of Microscopy*, 104, 91-103.

Hughes, M. K., F. H. Schweingruber, D. Cartwright, and P. M. Kelly 1984. July-August temperature at Edinburgh between 1721 and 1975 from tree-ring density and width data. *Nature*, 308, 341-344.

Kienast, F. 1985. *Dendroökologische Untersuchungen an Höhenprofilen aus verscheidenen Klimabereichen*. Doctoral dissertation of the University of Zurich. Zurich: Juris Druck Verlag.

Koul, A. 1978. *Geography of the Jammu and Kashmir State*. (revised by P. N. K. Bamzai). New Dehli: Light and Life Publishers.

Lenz, O., E.Schär and F. H. Schweingruber 1976. Methodische Probleme bei der radiographisch-densitometrischen Bestimmung der Dichte und der Jahrringbreiten von Holz. *Holzforschung*, 30, 114-123.

Meusel, H. and R. Schubert 1971. Beiträge zur Planzengeographie des Westhimalajas: 2. Teil: Die Waldgesellschaften. *Flora, Jena*, 160, 372-432.

Milsom, S. J. and M. K. Hughes 1978. X-ray densitometry as a dendrochronological technique. In *Dendrochronology in Europe*, J. Fletcher (ed.) 317-324. Oxford: B.A.R. International series, 51.

Parker, M. L. and W. E. Henoch 1971. The use of Engelmann spruce latewood density for dendrochronological purposes. *Canadian Journal of Forest Research*, 1, 90-98.

Ramesh, R., S. K. Bhattacharya and K. Gopalan 1985. Dendroclimatological implications of isotope coherence in trees from Kashmir Valley, India. *Nature*, 317, 802-804.

Raza, M., A. Aijazuddin and A. Mohammad 1978. *The Valley of Kashmir: Vol.I – the land*. New Dehli: Vikas.

Schweingruber, F. H. 1985. Dendroecological zones in the coniferous forests of Europe. *Dendrochronologia*, 3, 67-73.

Schweingruber, F. H., H. C. Fritts, O. U. Bräker, L. G. Drew and E.Schär 1978. The X-ray technique as applied to dendroclimatology. *Tree-Ring Bulletin*, 38, 61-91.

Wigley, T. M. L., K. R. Briffa and P. D. Jones 1984. On the average value of correlated time series with applications in dendroclimatology and hydrometeorology. *Journal of Climate and Applied Meteorology*, 23, 201-213.

22 Dendroclimatic studies in China

X. D. Wu

22.1 Introduction

In the 1930s and 1940s, a few Chinese scientists (Zheng 1935; Deng 1948) tried to use variations of tree-ring width to describe local climatic change in northern China. Tree-ring studies have been going on in certain regions ever since then, but the greatest advances were made in the 1970s because of increased emphasis on research into climatic change and the increased use of computers in China (Zheng, Wu and Lin 1982).

Tree-ring analysis was developed in many institutes and departments, in which researchers collected samples and performed analyses. Taking the work in the Institute of Geography, Academia Sinica as an example, in the 1970s we focused our efforts in three areas: northeast China, the Tibetan Plateau and the Hengduan Mountains. Our results indicate the local variations of temperature and precipitation in these areas. Scientists in meteorological bureaus and universities obtained many tree cross sections during the 1970s and the early 1980s. Most of the dendrochronological work was concentrated in the western part of China, and some of the results have been published (Li *et al.* 1977; Liu 1982; Zhuo *et al.* 1978; Wang *et al.* 1982).

Recently, there have been four main developments in dendrochronology in China. One is an increase in the number of samples taken at each site; another is collection in new areas; the third is the use of historical documents with tree ring analysis and the fourth is the adoption of new techniques of climate reconstruction using tree-ring data (Wu, Cheng and Sun 1987).

Before 1980, replicated samples were not taken, since it was difficult to obtain numerous cross sections because of labor requirements and transportation problems. In fact, it is obvious that this approach violates the basic principle of cross-dating which is fundamental to reliable paleoclimatic reconstruction. The importance of multiple samples has been emphasized in China since the Norwich meeting of 1980 (Hughes *et al.* 1982) and so the number of samples from each site has been greatly increased. For example, Yang *et al.* (1984) pointed out that a total of 140 cross sections were collected for 10 chronologies in the East Tianshan Mountains. A large number of increment cores have been collected by the Tree-Ring Laboratory of the Institute of Geography and other units in recent years. Generally, ten trees or more were sampled at each site. Sometimes, an entire cross-section can increase the chance of detection of absent and double rings and other local distortions, so one or two cross sections were taken along with cores from each site. Using multiple samples, more attention has been paid to cross-dating. When some chronologies from the Tibet Plateau were constructed, the "three-step dating" approach was used. As a result, the reliability and applicability of some dendrochronologies has been greatly increased.

There has been a significant increase in the number of tree-ring localities in China. Collection of samples in the 1980s has been carried out in the Tianshan Mountains, northwest China by local meteorological bureaus and the Institute of Geography, Academia Sinica.

More dendroclimatic work is being done on these collections now. Some scientists have been collecting samples in other areas, such as the middle reaches of the Yellow River, north China and the Qilianshan Mountains, northwest China.

There are other proxy climate records (historical documents) in many parts of China. It seems useful to judge missing or false rings in a new area where there is still not any chronology and to verify reconstructions of past climate using these records. Actually, many very dry or cold years recorded in local documents can be compared with the narrow rings in trees from the same area.

Using a combination of dendroclimatic data and climatic reconstructions from historical documents, large scale climatic patterns can be described. The combined reconstruction has often given a more reliable estimate of past climate than either type of record can do individually. In addition, some response function and transfer equation analyses have been adopted. In order to enhance the reliability of past-climate reconstruction, basic procedures of calibration and verification have also been emphasized in our recent projects.

22.2   Establishment of tree-ring chronologies

In order to produce reliable and useful tree-ring chronologies, it is necessary to describe the sampling in some sites of China and the process of building ring-width series based on important dendrochronological concepts and our own experiences.

22.2.1   Sampling sites

Based on fundamental principles of site selection (Fritts 1976) most tree-ring sites we have sampled can be classified into two types.

Dry sites   Most areas in Tibet are quite dry; the annual precipitation is less than 500 mm, except for a small region located in the southeastern Xizang. We have calculated aridity over the plateau according to Penman's formula (as the ratio of potential evaporation to precipitation) and find the aridities of most sites are higher than 2.0, and even larger than 20.0 in the desert areas (Lin and Wu 1981). It can be said that the Tibet plateau is one of the driest areas in China. The annual average total precipitation in most sites where we have collected samples is near 400 mm, and even as low as 300 mm a year at a few sites. Other arid areas where tree-ring samples have been collected are the East Tianshan Mountains, Northwest China and the Hengduan Mountains, Southwest China. Most sampling sites are located in the typical arid climate region, where the annual precipitation is not more than 400 mm. In these arid areas, the forests are limited in extent and are quite sparse, except for some woods located by lakes and rivers. If the trees are far from such water sources, most of them may be useful samples for reconstructing past precipitation (Wu and Lin 1977).

High altitude sites   The other kind of area we have sampled is near the upper treeline at high altitudes. The Tibet plateau is famous for its high altitude and cold climate. The upper limits of most trees in Tibet can extend to over 4000 m, and a few even reach beyond 4500 m. Trees are rarely found at such high elevations in other parts of the world. The air temperature due to the altitude is quite low; annual mean temperature is near 0°C, the mean of the warmest month (July) is often below 15°C and the mean of the coldest month (January) could be well below -10°C. That means the days with daily air temperature below 0°C would last more than half a year. The precipitation, meanwhile, is generally more than that at the base or the summit of the mountains because the height at which maximum precipitation occurs is often close to the altitude of upper tree line. Similar cases can be found at the upper tree line in the Hengduan and other mountains ranges, where tree growth may mainly respond to temperature variations. It is most likely that ring widths in trees at the tree line are limited by cold rather than drought, and this is supported by the positive relationship between variations in ring width and variations in air temperature.

A large number of samples have been collected in China since the 1970s. A total of 39 sites have been sampled and the distribution of sampling sites is shown in Figure 22.1. Some tree-ring chronologies which consist of multiple samples and have been used as a time series to reconstruct local past climate are listed in Table 22.1. All of them are located in the western part of China, especially in the Tibetan plateau, the Hengduan Mountains and the East Tianshan Mountain areas.

22.2.2 Dating procedures

Crossdating is one of the most important concepts in dendrochronology. Based on basic crossdating methods and our own practice, the "three-step dating" approach has been named

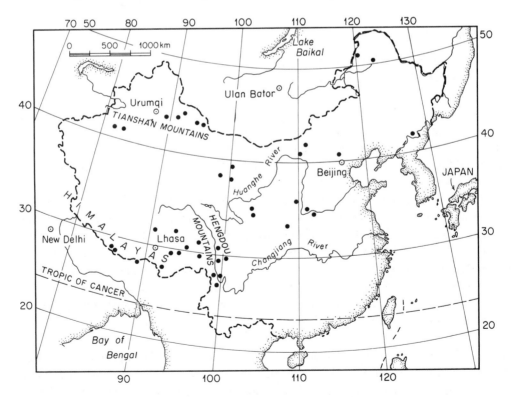

Figure 22.1   Tree-ring sample localities in China.

Table 22.1 Summary of some dendrochronologies in China.

I.D.	Lat. (°N)	Long. (°E)	Elevn. (m)	Species	Number of trees	Length (Yr.)
Li 2	27 10′	100 15′	3600	*Abies forrestii*	10	254
Li 1	27 15′	99 50′	3900	*Abies forrestii*	10	198
Zhong3	27 46′	99 41′	3300	*Picea likiangensis*	15	245
Zhong2	27 47′	99 41′	3900	*Abies forrestii*	18	378
Zhong1	27 48′	99 43′	3750	*Larix gmelini*	21	383
Zhong4	27 49′	99 43′	3850	*Sabina tibetica*	10	266
De 2	28 16′	89 58′	3500	*Picea likiangensis*	10	199
De 1	28 18′	89 57′	4200	*Sabina tibetica*	10	332
MAI1	29 12′	93 36′	3150	*Pinus densata*	15	244
Dao1	29 13′	100 10′	4200	*Sabina tibetica*	14	315
Ba 1	29 38′	96 48′	4000	*Sabina tibetica*	5	710
NYI1	29 42′	94 23′	3145	*Cypressus gigantea*	14	367
NYI2	29 44′	94 34′	4500	*Sabina saltuaria*	21	605
LHU1	30 14′	91 22′	4320	*Sabina wallichina*	14	392
Qam7	31 12′	96 52′	3800	*Sabina wallichiana*	10	125
Qam 301	31 12′	96 59′	3250	*Picea likiangensis*	8	227
BWW 1	40 48′	110 20′	1450	*Pinus tabulaeformis*	21	264
BWN1	40 50′	110 21′	1470	*Pinus tabulaeformis*	14	215
XM*1	42 21′	93 09′	3200	*Larix sibirica*	25	298
XM 3	42 55′	94 34′	2950	*Larix sibirica*	19	224
XM 4	43 01′	95 02′	2670	*Larix sibirica*	9	524
XM 5	43 01′	95 02′	2840	*Larix sibirica*	10	350
XM 2	43 21′	93 09′	2920	*Larix sibirica*	8	327
XM 9	43 29′	92 56′	2370	*Larix sibirica*	5	220
XM 10	43 33′	93 03′	2730	*Larix sibirica*	10	424
XM 6	43 43′	93 17′	2490	*Larix sibirica*	5	166
XM 7	43 44′	93 18′	2850	*Larix sibirica*	7	270
XM 8	43 47′	93 17′	2680	*Larix sibirica*	8	517

* All with XM are from Yang *et al.* (1984)

by Wu, Sun and Cheng (1988). This approach could be described as visual dating of cores, skeleton plotting (Stokes and Smiley 1968) and computer dating checks (Holmes 1983).

22.2.3 Measurement of tree-ring width

Although a modern incremental measuring system, has been installed in our Tree-Ring Laboratory and all ring width data are measured carefully, probable errors in judgment by workers might be difficult to avoid. In order to control the data quality in tree-ring measurement, each core must be independently measured at least twice. Each of the differences is squared, and all squares summed up are calculated. A value of 0.10 for the total sum of the squares can be considered as a precise criterion for acceptance or rejection. If the value is more than 0.10, a third measurement should be taken.

22.2.4 Standardization

Very low-frequency trends and changes in tree growth can appear in an appropriate curve fitted to the data for each core by least-squares techniques. For our samples, an exponential function or a straight line is often used. Sometimes, other functions (such as polynomial or cubic spline) might be used.

22.2.5 Calibration and verification

We used to check the quality of ring-width series in two ways. One was based on the calculated relationship between ring widths and climatic factors, either air temperature or precipitation, at the nearest meteorological station. The other was to find some local documents, in which severe cold or drought years have been recorded. These historical documents could be helpful in comparison with small rings from the site. For example, 1905 and 1937 were anomalously cold and drought years, respectively, in the historical writings around Lhasa. We find corresponding narrow rings in each chronology built from trees near the upper tree line and the arid area in Middle Tibet respectively (Wu and Lin 1987a).

Of course, the systematic calibration and verification must assess the strength of the growth-climate linkages. Response functions and transfer functions have become widely used to estimate the linkages in the reconstruction of past climate. Functions relating the tree-ring chronologies and climate will be described in the next part.

22.3   Reconstruction of past climate

22.3.1 Response function

In dendroclimatology, response functions have become widely used to estimate how ring-width growth responds to variations in monthly climatic conditions. This analysis is a multiple regression technique using the principal components of monthly climatic data to estimate indexed values of ring-width growth. It is obvious that it is necessary to establish the response function before reconstructing past climate. Fritts and Wu (1986) analyzed three different response-function programs. The results were described and compared to one another as well as those obtained from some other regression methods. Also, the rationale for the different response-function solutions was discussed. As an example, the response function for chronology LHU1 with Lhasa climatic data is shown in Figure 22.2. All of the eigenvectors (100%) were included as initial regressors in each program. The vertical lines designate the 0.95 confidence ranges; the significant elements are marked with an asterisk in the Figure.

It is clear that the tree ring response to air temperature is positive in most months. The values appear negative in a few months, but they never reach the 5% significance level. In fact, eight of the 12 months have a significant positive response. The multiple correlation coefficient (R) of the response regression for LHU1 is 0.86. The number of significant elements is 13. The variance explained by climatic factors is 59.2%. The prior growth and total variances are 14% and 73%, respectively. The statistics for this and another site are listed in Table 22.2.

436

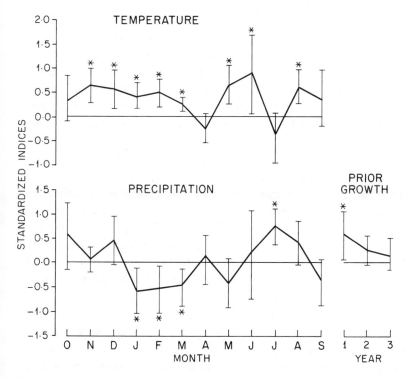

Figure 22.2 Response function of LHU1.

Table 22.2 Statistics for two response functions.

Chronology	NYI 1	LHU 1
Variance (%)		
Prior growth	16.59	14.15
Climate	60.06	59.24
Total	76.65	73.39
Multiple R	0.876	0.857
Total factors	19	27
Significant elements	10	13

Response function analysis demonstrates that these chronologies are positive in response to climate and coincident with the biological theory of tree growth at these collection sites. Consequently, it will be possible to use them for the reconstruction of local climatic change.

22.3.2 Transfer equation and test

In reconstructing climate from tree rings it is necessary to focus attention on understanding the statistical and physical relationships between climate and tree growth. Generally, a number of tree-ring chronologies are needed for reconstructing past climate. Taking LHU1 (from the timber line in Tibet) as an example, the variance in the chronologies explained by climate can reach 60% and the response functions are physically meaningful. Fortunately, we found the tree growth corresponds to the yearly mean temperature (the prior October through September) very well. With the adoption of the Empirical Orthogonal Function (EOF) which guarantees independence of all predictors, the yearly index and the prior growth (lag one to three years) in the regression could achieve better results than simple linear regression analysis (Wu, Sun and Zhan 1989). The observed and reconstructed temperature departures for LHU1 are shown in Figure 22.3. Similarly, The growing season (May through August) precipitation departures for NYI1 are shown in this figure. The reconstructed values of temperature and precipitation (broken line) derived from the transfer equation are close to the observed data (solid line). The difference between the observed data estimated values for each tree-ring site is given in Table 22.3. The average deviation of air temperature is near 0.2°C. The average deviation of precipitation is 24 mm. Reduction of error (RE) is commonly adopted in verification tests for dendroclimatic reconstructions. The statistic is a most rigorous test of the reliability of the estimated climate. The RE values reach 0.55 and 0.70 for LHU1 and NYI1, respectively. This also suggests that these reconstructions are fairly reliable. Of course, we often adopt more tree-ring chronologies to reconstruct past climate in an area, using multivariate regression or averaging several tree-ring series. In order to compare with documentary data which are formed in non-continuous grades (see section 3.3, below) the averaged series may be more convenient than using several series in multivariate regression in our tree-ring analyses.

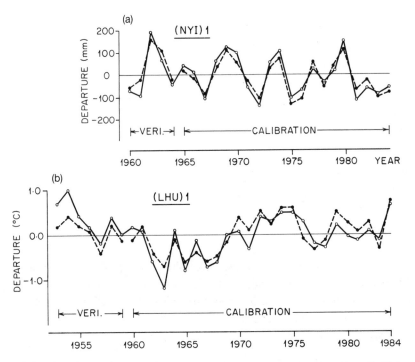

Figure 22.3 A comparison between reconstructed and observed values of (a) temperature and (b) precipitation for LHU 1 and NYI 1.

Table 22.3 Statistics for reconstruction at two sites.

I.D.	NYI 1	LHU 1
Predictand	Precipitation in growing season	Annual temperature
r	0.731	0.752
Average deviation	23.7 mm	0.22°C
RE	0.702	0.547
Calibration	1965-84	1960-84
Verification	1960-64	1953-59

We have also tried to find some local historical documents, in which severe cold or drought events have been recorded to test the reconstructions. For example, it was anomalously cold in Lhasa in 1905 according to the diary of a former top officer to Tibet, You Tai (Lin and Wu 1984). At the same time, the reconstructed temperature departure was -1.1°C, which indicates that 1905 was quite a cold year. Another example concerns drought. According to the Tibetan calendar (a special calendar of historical records in Tibet) an anomalous drought occurred in the summer of 1937 in mid-Tibet. The estimated precipitation values from NYI1 have large negative departures in 1937.

22.3.3 Combination of tree-ring and historical data

When tree-ring chronologiesare used to describe local climatic variations, other kinds of proxy data such as historical documents, are often used to supplement the paleoclimatic reconstructions. Historical documents, in which the historical weather or climate conditions were recorded, have been converted to five categories: 1=very dry or cold, 2=dry or cold, 3=normal, 4=wet or warm, and 5=very wet or warm. With similar treatment, the tree-ring series can also be converted into a new series with five corresponding grades. Four values: (-2.5s, -1.5s, +1.5s and +2.5s, where s = the standard deviation of a series) are adopted to divide each tree-ring series into five grades (Wu and Lin 1981a). Then, all of the graded documentary data can be merged with the tree-ring data for the same year.

Located in southwest China, the Hengduan Mountains form a transitional zone between the Tibetan plateau and the Yunnan-Guizhou plateau and Sichuan basin. The high mountains and deep gorges run in parallel from south to north in this area. Also, it has a rather rich flora and fauna, and some rare animals and plants (such as panda and dove trees) are still found here. Field investigations in this area, in the early 1980s, resulted in a large number of tree ring samples and other kinds of proxy data (Wu and Lin 1983; 1987b). Twelve tree-ring chronologies from the Hengduan Mountains have been constructed. Most of the tree-ring sites are located in the western part of the area. However, these are not well-spaced enough to reconstruct the past climate over the whole area. Fortunately, all 15 documentary data sites are located in the eastern part. By combining these two kinds of data it is possible to analyze climatic changes in the area during the last few hundred years. Recently, this combination approach has been widely adopted for climate reconstruction in western China. Wu and Lough (1987) have demonstrated how different types of proxy climate records, tree ring and documentary data, can be combined to give a more reliable estimate of past climate than either record can do individually. The basis for the reconstruction was that both Chinese dryness/wetness variations (as recorded in documentary records) and western North American tree-ring width variations are influenced by North Pacific summer sea-level pressure variations. These proxy climate records are used separately and together to estimate summer sea-level pressure variations over the North Pacific back to A.D. 1600. Although both sets of predictors calibrate most of the variance in the vicinity of the North Pacific subtropical anticyclone, the model which provides the best estimate is one of the combined models, demonstrating the potential of combining different proxy data sources to derive better estimates of past climate.

22.4   Climatic change

Based on each time series derived from tree-ring data, the local climatic change can be assessed. Considering climate classification in China, climatic changes of several regions are summarized below.

22.4.1   The Tibetan Plateau

Using tree ring and various other kinds of data, climatic change in Tibet during historical times has been analyzed by Wu and Lin (1978 1981b). Five periods: the "Climatic Optimum," "Neo-glacial Period", "Warmer Period", "Little Ice Age" and the "Last Warm Period", are characteristic of the last 7,000 years. The climate since A.D. 1500 encompasses the last two periods.

Taking 12 annual ring series from various parts of Tibet (showing yearly average temperatures) we converted these into a temperature anomaly series, in order to find the range of mean annual temperature in Tibet for the last 500 years (see Figure 22.4). From Figure 22.4a, it can be seen clearly that the climate of Tibet since A.D. 1500 has obviously fluctuated between warm and cold periods. There was a warmer period during the mid-16th century with 0.2°C or more positive temperature departures. The coldest period in the last 500 years occurred approximately in the mid-17th century, when the mean annual temperature was generally 0.5°C or more below the modern average value. At the same time, glacial advances occurred in many parts of Tibet. Both before and after this, it was warmer than in the 17th century. Since the 19th century, a warming tendency has been even more marked, with comparatively warm temperatures mostly being maintained despite several short periods of cooler temperatures. Generally speaking, relatively warmer periods came in the mid-16th century, at the beginning of the 18th century, in the first half of the 19th century and in the period following the 1970s. The relatively cooler periods were in the early 16th century, in the mid-17th century, in the mid-18th century, in the 1860s , and in the 1920s and 1960s. The

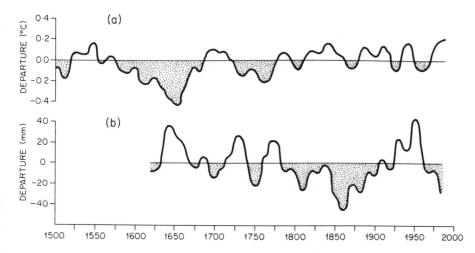

Figure 22.4 Fluctuation of (a) temperature and (b) precipitation in Tibet (0 means the average value during the last 30 years).

average temperature over the past 500 years is about 0.08°C lower than that calculated from meteorological records over the last 30 years.

Temperature trends in Tibet are essentially in agreement with the trend of the eastern region in China, which was derived by Zhu (1973) and other scientists. However, it should be noted that there remain comparatively major regional differences between the climates of Tibet and Eastern China.

With similar treatment, one can also obtain departures of annual rainfall in Tibet (Figure 22.4b). This series was derived from 7 tree ring series over the plateau. It can be seen that there were some marked wet and dry periods during the last 400 years. The major wet periods occurred in the middle to late 18th century and the first half of the 19th and 20th centuries, (especially in the 1920s and 1930s, with 40 mm or more positive departures). All the rainfall variations correspond closely to water-level fluctuations on the plateau since the 18th century.

22.4.2 *The Hengduan Mountains area*

Based on 12 tree-ring chronologies, which respond to the local variations of temperature or precipitation, two series from the Hengduan Mountains have been derived (Figure 22.5). Figure 22.5a, shows that a long cold period prevailed in the Hengduan Mountains from the early 17th century through the 1650s. This may have been the coldest period of the last 400 years. The temperature then increased rapidly and remained high until the 1740s, so that a long warm period occurred in the area. More obvious cold and warm periods came one after another, but the fluctuation amplitude is smaller and the successive duration of these periods is shorter than those before the 1740s. In the present century, two cold periods were obvious but there has been warming since the 1970s.

From Figure 22.5b, it can be clearly seen that several dry and wet periods appeared alternately during the last 400 years. The distinct dry periods are: 1610s-1620s, 1660s-1670s,

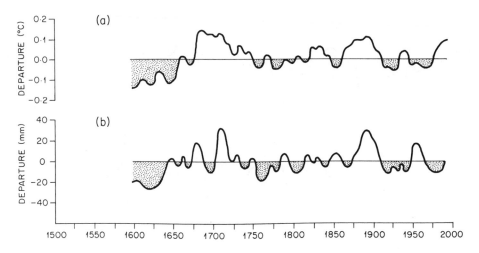

Figure 22.5 Fluctuation of (a) temperature and (b) precipitation in the Hengduan Mountains area (0 means the average value during the last 30 years).

1690s, 1750-1770s, 1790s-1820s, and 1910s-early 1940s. Other intervals were relatively wet. The length of each wet period, generally speaking, is about 10 years; the maximum seldom reaches 20 years.

Using tree ring chronologies and historical documentary data divided into five categories, five climate patterns have been derived from the two graded series. Each year could correspond with one of the five patterns. They are: cold and dry, cold and wet, normal, warm and dry, and warm and wet (Wu, Lin and Sun 1988). Looking at the area as a whole, the frequency of occurrence and percentage of each climate pattern in the Hengduan Mountains in fifty-year intervals, (except the most recent period of 32 years) is given in Table 22.4.

Obviously, the pattern of cold and dry conditions was quite common, accounting for 58% of all years during the first half of the 17th century. In the same period, the patterns related to a warm climate never appeared. In the second half of the century and in the successive 50 years, the climate became warmer and wetter. The warm and wet pattern occurred more than 20% of the time. From the 1850s through the 1940s, the cold and wet pattern which is important for glacier activity (Li *et al.*,1983) appeared frequently. In fact, this period does correspond to a time of active glacier advances during this hundred year period. It should be noticed that the warm and dry pattern has increased in frequency from 4% (1900-1949) to 15.5% (1950-1981) and that is an important trend in the climate of this area.

22.4.3   Northwest China

Zhang *et al.* (1981) analyzed one tree sampled from the Qilian Mt. (see Zhou *et al.* 1978). Three warm periods (1538-1621, 1741-96, 1871-1923) and four cold periods (1428-1537, 1622-1740, 1797-1870 1924-) were identified in the last 500 years. Kang (1983) collected eight tree-ring cross-sections in the Altai mountain region, Xinjiang, Northwest China, and analyzed the variation of cold and warm climate in the last 300 years, and of dry and wet conditions over the last 180 years. A total of five warm periods (1719-28, 1752-84, 1802-46,

Table 22.4 Frequency and percentage of each climate pattern in the Hengduan Mountains region (N = frequency; % = percentage).

Periods	Cold and Dry		Cold and Wet		Normal		Warm and Dry		Warm and Wet	
	N	%	N	%	N	%	N	%	N	%
1600-1649	29	58	8	16	13	26	0	0	0	0
1650-1669	4	8	3	6	25	50	6	12	12	24
1700-1749	4	8	6	12	23	46	7	14	10	20
1750-1799	2	4	8	16	36	72	1	2	3	6
1800-1849	5	10	7	14	36	72	1	2	1	2
1850-1899	7	14	10	20	22	44	1	2	10	20
1900-1949	9	18	12	24	23	46	2	4	4	8
1950-1981	7	22	4	12.5	14	44	5	15.5	2	6

1895-1937, 1954-69) and five cold periods (1729-51, 1785-1801, 1847-94, 1938-53, 1970+) were identified. There were also four wet periods (1820-38, 1866-88, 1913-27, 1943-53) and four dry periods (1839-65, 1889-1912, 1928-42, 1954+) in this mountain area. Glacier fluctuations could be related to the reconstructed climatic conditions (Wu, Lin and Sun, 1989).

Based on some tree-ring chronologies, Li (1985) described climatic fluctuations in the East Tianshan Mountains during the last 500 years. The cold periods were before the 1530s, the 1570s-1640s, 1670s-1720s and 1840s-1880s. The warm periods appear to have been the mid-17th century, the mid-18th century and after the 1880s. Similarly, there were two wet periods (1726-1812, 1890-1926) and three dry periods (1685-1725, 1813-1889 and 1927+) during the last 300 years.

Li *et al.* (1988) reconstructed the temperature sequence in the growing season for the last 200 years using one *Populus euphratica* chronology in the middle reaches of Talimu River, Xinjiang, Northwest China. According to the ten-year running mean plots of reconstructed temperature, three cold periods (1829-66, 1880-1930, 1949-69) and four warm periods 1807-28, 1867-79, 1931-48, 1970+) during the last 200 years could be identified.

In the future, there will be more dendroclimatic results from Northwest China because some excellent chronologies have been developed in recent years.

22.4.4 *Northeast China*

In the Jilin province, 13 trees from Mt. Changbai were analyzed to determine the relationship between tree rings and meteorology. Three warm periods (1821-41, 1862-95, 1919-52) and three cold periods (1842-61, 1896-1918, 1953-74) could be identified.

Gong *et al.* (1979) used two trees at 50°48'45"N, 121°21'40"E, (Northeast China) and analyzed temperature variations since the 17th century. From historical documents and other data, it is clear that the 17th century was the coldest (and the 1650s-60s were the "climax") of the so-called "Little Ice Age". During the 1650s, the growing season was more than one month shorter than today and the ice-thickness on the Heilongjiang River more than one meter greater.

Although some progress in dendroclimatic reconstruction has been made in recent years in China, we expect further advances in the near future with the development of new chronologies from North China and Middle China.

References

Deng Shugun, 1948. Tree-ring and climate in Kansu. Botanical *Bulletin of Academia Sinica*, 2 (3) pp. 211-214.

Fritts, H. C., 1976. *Tree Rings and Climate*. Academic Press, London.
Fritts, H. C. and Wu Xiangding, 1986. A comparison between response-function analysis and other regression techniques. *Tree-Ring Bulletin*, 46, pp. 31-46.

Gong Gaofa, Chen Enjiu and Wen Huanran, 1979. The climatic fluctuation in Heilongjiang Province, China. *Acta Geographica Sinica*, 34 (2) pp. 129-138. (in Chinese with an English abstract)

Holmes, R. L., 1983. Computer-assisted quality control in tree-ring dating and measurement. *Technical Note* No.28, U of A, pp. 1-19.
Hughes, M. K., Kelly, P. M., Pilcher, J. R. and Lamarche Jr, V. C., 1982. *Climate from Tree Rings*. Cambridge University Press.

Kang Xingcheng, 1983. Correlation between tree ring and climatic and glacial variations in the region of Mts. Altay. *Journal of Glaciology and Cryopedology*, 5 (4) pp.57-62. (in Chinese with an English abstract)

Li Jiangfeng, 1985. Climatic variation of Xinjiang in recent 3000 years. *Papers on Quaternary Research in Arid Area of Xinjiang*, pp. 1-7. (in Chinese)
Li Jiangfeng, Yuan Yujang and Wang Chengyi, 1988. Temperature sequence in the area of the Middle reaches of Talimu River and its changes in recent 2000 years. *Geographical Research*, 7(3) pp. 67-71. (in Chinese with an English abstract)
Li Jijun *et al.*, 1983. Investigation of glaciers on the Gongga Shan. *Study of Tibetan Plateau*, 140-153, Yunnan Press, Kunming. (in Chinese with an English abstract)
Li Zhaoyuan *et al.*, 1977. Tree-rings and climatic change in Shaanxi Province. *A Symposium on Climatic Change and Prediction*, pp. 72-76. Science Press, Beijing. (in Chinese)
Lin Zhenyao and Wu Xiangding, 1981. Climatic regionalization of the Qing-Xizang plateau. *Acta Geographica Sinica*, 36(1) pp.22-32. (in Chinese with an English abstract)
Lin Zhenyao and Wu Xiangding, 1984. The climate of Lhasa in the early 20th century. *Plateau Meteorology*, 3(4) pp.14-20. (in Chinese with an English abstract)
Liu Chuangzhi, 1982. Tree-ring in the Wolong region and climatic variation of west Sichuan Province in recent century. *Meteorological Monthly*, 5, pp. 18-20. (in Chinese)

Stokes, M. A. and Smiley, T. L., 1968. *An Introduction to Tree Ring Dating*. University of Chicago Press.

Wang Yuxi *et al.*, 1982. Relationship between the tree rings in the Qilian Mountain and climatic and glacial change in China. *Kexue Tongbao*, 27 (21) pp. 1316-1319. (in Chinese)

Wu Xiangding and Lin Zhenyao, 1977. Tree ring samples and the climatic change in Southern Tibet. *A Symposium on Climatic Change and Prediction*, pp. 68-71. Science Press, Beijing. (in Chinese)

Wu Xiangding and Lin Zhenyao, 1978. A preliminary analysis of climatic variation during the last hundred years and its outlook on Tibetan Plateau. *Kexue Tongbao*, 23 (12) pp. 746-750. (in Chinese)

Wu Xiangding and Lin Zhenyao, 1981a. Climatic change during the last 2000 years in Tibet. *Proceedings of Symposium on Climatic Change*, pp. 18-25. Science Press, Beijing. (in Chinese)

Wu Xiangding and Lin Zhenyao, 1981b. Some characteristics of the climatic change during the Historical Time of Qinghai-Xizang Plateau. *Acta Meteorologica Sinica*, 39 (1) pp. 90-97. (in Chinese with an English abstract)

Wu Xiangding and Lin Zhenyao, 1983. The climatic change and tree ring analysis in the Hsiao Zhongdian area of the Yunnan Province. *Study of Tibetan Plateau*, pp. 206-213. Yunnan Press, Kunming. (in Chinese with an English abstract)

Wu Xiangding, Cheng Zhigang and Sun Li, 1987. Status of dendrochronological work in China. *Dendrochronologia*, 5, pp.127-133.

Wu Xiangding and Lin Zhenyao, 1987a. Sampling in Tibet. Dendrochronological Methods. *Proceedings of Task Force Meeting on Methodology of Dendrochronology: East/West Approaches*. pp.23-33. eds. L. A. Kairiukstis *et al.* WOSI. Warsaw, Poland.

Wu Xiangding and Lin Zhenyao, 1987b. A preliminary study of the modern climatic change in *Hengduan Mountains*. Geographical Research, 6 (2) pp. 48-56. (in Chinese with an English abstract)

Wu Xiangding and Lough, J. M., 1987. Estimating North Pacific summer sea-level pressure back to 1600 using proxy climate records from China and North America. *Advances in Atmospheric Sciences*, 4 (1) pp. 74-84.

Wu Xiangding, Lin Zhenyao and Sun Li, 1988. A preliminary study on the climatic change of the Hengduan Mountains area since 1600 A.D. *Advances in Atmospheric Sciences*, 5(4) pp.437-443.

Wu Xiangding, Sun Li and Cheng Zhigang, 1988. Establishment of some tree-ring chronologies in Tibet. *Kexue Tongbao*, 33(15) pp. 1284-89.

Wu Xiangding, Lin Zhenyao and Sun Li, 1989. A preliminary analysis of climatic change in the arid area of Northwest China. *Chinese Journal of Arid Land Research*, 1(4) pp.341-348.

Wu Xiangding, Sun Li and Zhan Xuzhi, 1989. A preliminary study on reconstructing past climate using tree-ring data in the middle Tibetan plateau. *Acta Geographica Sinica*, 44(3) pp.334-342. (in Chinese with an English abstract)

Yang Guangxun *et al.*, 1984. Tree ring chronology in East Tianshan Mountains. *Meteorological Monthly*, 7, pp. 21-25. (in Chinese)

Zhang Xiangong, Zhao Zhen and Xu Ruizhen, 1981. The tree rings of the Qilian sabina and the climate trend in China. *Proceedings of Symposium on Climatic Change*, pp.26-35, Science Press, Beijing. (in Chinese)

Zheng Sizhong, Wu Xiangding and Lin Zhenyao, 1982. Asia. in *Climate from Tree Rings*, pp. 155-157. eds. M. K. Hughes *et al.* Cambridge University Press, Cambridge.

Zheng Zizheng, 1935. On the relation between tree rings and rainfall in Beijing. *Fangzhi Yueken*, 8(6) pp. 13-16. (in Chinese)

Zhou Zhenda *et al.*, 1978. The tree rings in the Qilian Mountains and climatic change in China during last 1000 years. *Journal of Lanzhou University*, 2, pp. 141-157. (in Chinese with an English abstract)

Zhu Kezhen, 1973. A preliminary study on the climatic fluctituation during the last 5000 years in China. *Scientia Sinica*, 16(2) 168-189. (in Chinese with an English abstract)

23 South American dendroclimatological records

J. A. Boninsegna

23.1 Introduction

Tree-ring work in South America has lagged greatly behind that in the Northern Hemisphere. The reasons for this lag include the scarcity of experienced people, the enormous diversity of forest-types, ranging from the Amazonian rain forest to the Subantarctic evergreen, and consequently, the need for screening a large number of tree and shrub species to assess their dendrochronological potential.

The relatively small size of the tree-ring data base in South America reflects the recent initiation of sampling and analysis programmes. The chronologies are, however, of high quality because modern sample size and replication requirements generally have been met, the dating is reliable, and modern computational techniques have been used in the processing and analysis of the data. The needs and prospects for further collections differ from region to region. At present, the major effort has been carried out in the development of chronologies in the temperate zone where 96 chronologies are available. Areas with virtually no coverage include the tropical parts of South America. A major obstacle has been the lack of clearly defined rings in most species and the presence of obvious intra-annual growth bands in others; these problems lead to major difficulties in dating. However, several promising species have been identified and some chronologies have been derived, especially in the northern part of Argentina. Future work could emphasize the development of more chronologies using these species and related taxa, the study of the climate response of the trees and the construction of suitable models to explain this relationship in the tropical regions.

Due to the limited development of the tree-ring data base, the approach to reconstruction of climate parameters beyond the local and regional scales departs somewhat from the schemes successfully applied in the Northern Hemisphere.

23.1.1 Biogeographical setting

South America has a land area of 17,800 km² lying mostly between 13°N and 55°S. The Cordillera de los Andes rises abruptly from the Pacific coast and runs the length of South America (8,000 km). The remainder of South America to the east consists of lowlands or low altitude plateaux. The Andes are the most important topographical obstruction to the hemispheric circulation in tropical and middle latitudes. The climatic context of South America can be found in Schwerdtfeger (1976) and Pittock *et al.* (1978).

According to F.A.O. (1985) the tropical forests of South America cover approximately of 788,500,000 ha, representing 44.3% of the total land surface of the continent, while the temperate forests (32,040,000 ha) account only for 1.8% of the total land surface. A comprehensive description of the forests of South America is presented by Hueck (1976).

23.1.2 Dendrochronological work

E. Schulman of the Laboratory of Tree Ring Research (University of Arizona) made the first successful collection of cores in Argentina and Chile (Schulman 1956). In early 1974, V. C. LaMarche, from the same Laboratory, collected samples in central Chile, produced a well-replicated site chronology and developed a first reconstruction of the annual rainfall at Santiago, Chile. (LaMarche 1975). From 1975-1978 LaMarche and others made extensive field trips in Argentina and Chile, collecting samples almost exclusively of the conifers *Araucaria araucana* and *Austrocedrus chilensis* species, and producing 21 chronologies in Argentina and 11 in Chile, between 36° and 44°S.

Boninsegna and Holmes (1985) and Villalba *et al.* (1990) developed several chronologies at the same latitudes using *Fitzroya cupressoides* and the southern beech Nothofagus pumilio. *Fitzroya cuppressoides*, found in the wetter areas of South central Chile and Argentina between 39° and 43°S, is the longest lived conifer in South America. It attains ages of at least 1800 years (Schulman 1956). However, the presence of groups of micro-rings make it very difficult to measure, and some rings which 'wedge-out' make the cross-dating occasionally problematical. The first successful cross-dating and chronology development for this species was reported by Boninsegna and Holmes (1985)

At 32°S, Roig and Boninsegna (1988a) were able to produce chronologies using the mountain shrubs *Adesmia horrida* and *Adesmia uspallatensis*. These chronologies have a strong signal related to climate. The wide, sparse geographic distribution of the genus *Adesmia* is of special interest to the expansion of the dendrochronological network in the Central and Northern Andes of Argentina and Chile as well as in southern Bolivia. Also Roig *et al.* (1988) produced a short chronology with the mountain shrub *Discaria trinervis* which has a strong temperature-related signal. At 24° to 26°S the tropical species *Cedrela lilloi*, *Cedrela angustifolia* and *Juglans australis* were used to develop tree ring chronologies by Villalba *et al* (1985, 1987). In the subantarctic region, at 54-55°S Boninsegna *et al.* (1989) constructed 21 chronologies using *Nothofagus pumilo* and *Nothofagus betuloides*.

A summary of the chronologies currently available and some chronology statistics are presented in Table 23.1 and Table 23.2. A map showing the chronology locations is presented in Figure 23.1. The longest chronologies are made with *Fitzroya cupressoides* ranging from 1120 to 1530 years, followed by the chronologies of *Austrocedrus chilensis* and *Araucaria araucana* with 800-1000 years, maximum. The *Nothofagus* species yielded chronologies of about 400 years while in the tropical region, one *Juglans australis* chronology attained 365 years.

Lamprecht (1984) investigated the application of X-ray densitometric methods to estimate the wood density in some angiosperm and native conifers of South America. Although the author failed to cross-date the material, the work represents the first description of density profiles in South American woods.

At present, no attempt has been made to produce floating chronologies with buried wood, so all the reconstruction are limited to the life span of living trees.

23.1.3 Methodological comments

In all of the dendroclimatic reconstructions discussed below, cross-dating accuracy was verified with the computer program COFECHA (Holmes, 1983) and chronology develop-

Table 23.1 Tree ring chronologies of South America. The latitudes are estimated from maps in the various references and should only serve as a guide to general locations.

Species	Number of sites	Latitudinal Range (L.S.)	Altitudinal Range (m. a.s.l.)	Maximum Length (yrs)	Source
Juglans australis	9	22°44'-27°42'	700-1850	1688-1985	Villalba, R. et al. (1988 a)
Cedrela angustifolia	2	24°-24°65'	1700-1900	1809-1981	Villalba, R. et al. (1985)
Cedrela lilloi	1	27°10'	1650	1729-1982	Villalba, R. et al. (1988 a)
Prosopis flexuosa	1	32°00'	585	1903-1986	Villalba & Boninsegna (1988)
Adesmia horrida	3	32°20'-32°43'	2560-3100	1747-1986	Roig, F. & J. Boninsegna (1988)
Adesmia uspallatensis	1	32°33'	2500	1859-1984	Roig, F. & J. Boninsegna (1988)
Discaria trinervis	1	32°35'	2300	1889-1985	Roig, F. et al. (1988)
Empetrum rubrum	1	52°10'	70	1905-1986	Roig, F. (1988)
Nothofagus pumilio	14	39°36'-54°55'	15-1700	1575-1986	Boninsegna et al. (1989)
Nothofagus betuloides	14	54°31'-54°55'	15-350	1504-1986	Boninsegna, J. et al. (1989)
Austrocedrus chilensis	14	32°40'-43°01'	650-1700	1012-1984	LaMarche, V. et al. (1979)
Araucaria araucana	20	37°41'-39°41'	890-1670	1140-1983	LaMarche, V. et al. (1979)
Fitzroya cupressoides	3	41°10'-42°30'	600-110	441-1987	Villalba, R. (1988)
Pilgerodendron uviferum	2	42°30'-43°00'	540-750	1489-1987	Roig, F. & J. Boninsegna (1989)

Table 23.2 Tree ring chronologies of South America. Summary of chronology statistics. M.S: Mean Sensitivity; M.AC.: Mean First Order Autocorrelation; M.ST.D.: Mean Standard Deviation; M.C.T.: Mean Correlation between Trees; M.V.F.EG.: Mean Variance explained by the First Eigenvector.

Species	Number of chronologies	Trees (range)	Radii (range)	M.S. (range)	M.AC. (range)	M.ST.D. (range)	M.C.T. (range)	M.V.F.EG. (range)
Juglans australis	9	15 (6-21)	22 (12-37)	.33 (.23-.41)	.39 (.12-.62)	.36 (.24-.48)	.37 (.15-.51)	48.6 (40.0-59.0)
Cedrela angustifolia	2	13 (13-13)	26 (26-26)	.32 (.30-.34)	.37 (.35-.42)	.39 (.33-.45)	.46 (.44-.49)	49.4 (47.6-51.2)
Cedrela lilloi	1	18	24	.29	.71	.54	.42	48.3
Prosopis flexuosa	1	36	71	.36	.57	.36	.44	46.2
Adesmia horrida	3	23 (17-32)	23 (17-32)	.27 (.25-.29)	.29 (.22-.35)	.27 (.24-.32)	.47 (.34-.70)	52.4 (41.7-70.6)
Adesmia uspallatensis	1	20	20	.23	.25	.28	.42	49.6
Discaria trinervis	1	36	36	.19	.67	.39	.44	49.8
Empetrum rubrum	1	14	18	.32	.19	.27	.42	40.5
Nothofagus pumilio	14	18 (9-38)	29 (12-54)	.21 (.14-.29)	.41 (.32-.56)	.25 (.19-.30)	.31 (.20-.48)	36.0 (21.3-48.6)
Nothofagus betuloides	14	20 (9-33)	32 (17-52)	.21 (.14-.35)	.55 (.39-.72)	.27 (.19-.37)	.27 (.14-.65)	34.0 (21.2-66.1)
Austrocedrus chilensis	14	14 (5-27)	30 (13-73)	.18 (.16-.23)	.61 (.51-.79)	.26 (.19-.33)	.36 (.26-.48)	44.4 (40.1-54.12)
Araucaria araucana	20	12 (5-18)	34 (7-93)	.14 (.10-.16)	.61 (.26-.75)	.19 (.13-.25)	.21 (.10-.45)	35.8 (31.4-44.7)
Fitzroya cupressoides	3	24 (21-28)	42 (33-49)	.16 (.14-.18)	.74 (.66-.84)	.29 (.24-.40)	.31 (.12-.56)	47.0 (24.7-68.4)
Pilgerodendron uviferum	2	20 (19-22)	41 (38-45)	.14 (13-.15)	.60 (.46-.71)	.22 (.16-.29)	.30 (.21-.41)	40.7 (33.0-48.6)

ment was carried out using the computer program ARSTAN (Cook and Holmes, 1985) as a routine procedure. The ARSTAN program allows the use of various curve-fitting techniques, two-stage trend removals, autoregressive modeling and arithmetic or biweight robust mean

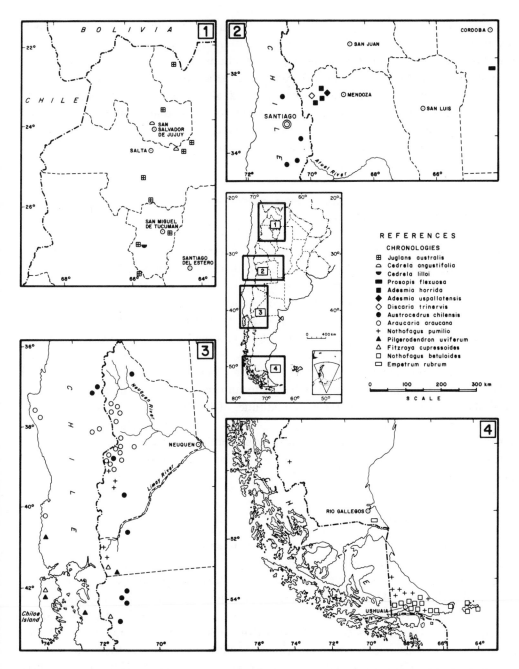

Figure 23.1 Map of the southern part of South America showing location of published tree ring chronologies, temperature, rainfall and river flow reconstruction records.

estimation to produce time series of ring-width indices representing common growth variations for the trees at a site (Cook, 1985).

A negative exponential function, followed by a cubic spline filter of 0.66 N (where N =

449

number of years in the series) were selected as the detrending procedure; arithmetic biweight robust mean estimation was used to calculate the chronology indices.

The chronologies used in the streamflow reconstructions (see Section 23.2.5, below) were those published by LaMarche *et al.* (1979). According to the authors, in the majority of cases, a negative exponential or straight-line curve was fitted, although the orthogonal polynomial option was occasionally used with considerable discretion to remove the trend of each measurement series. The individual radial ring-width index series were combined by simple arithmetic averaging to produce the site chronology. The reliability of each chronology was calculated according to Wigley *et al.* (1984).

23.2   Climate reconstructions

During the 1970's, dendroclimatology was largely concerned with the development of techniques to handle the data arising from sampling outside the semi-arid region of the southwestern Unites States. However, progress in the tropical regions is hampered at present by the lack of suitable techniques adapted and verified for the particular problems of those regions. Villalba *et al.* (1987) were able to establish the relationship between climate and tree growth in *Cedrela* and *Juglans* chronologies. Because of the relatively annual growing season in the subtropics, tree growth is better correlated with seasonal climate than with the climate of a particular month. This and other pioneering researches (Worbes, 1985 and 1989; Vetter and Botosso, 1989) indicate great potential in the quality and magnitude of climatic signal present in tropical trees (Jacoby, 1989). In the higher latitudes of South America, the relatively small land area means that the potential dendroclimatological coverage is less than in the Northern Hemisphere. It is likely that dendroclimatology in South America will be directed more towards the reconstruction of key climate indices rather than the reconstruction of dense grids of climate variables.

23.2.1   Southern oscillation

Lough and Fritts (1985) used 33 Southern Hemisphere chronologies and a set of 65 western United States chronologies to reconstruct the Southern Oscillation. The best estimates of the Southern Oscillation appear to be those from the western North American chronologies. Although the Southern Hemisphere information alone is capable of calibrating some Southern Oscillation variance, the resulting model is less reliable using independent data. According to the authors, the poorer performance of the Southern Hemisphere chronologies suggests that more work needs to be done with these chronologies before their full potential for large scale climate reconstruction can be realized. As indicated by Norton and Palmer (see Chapter 24, this volume) this may be a result of an inadequate selection of chronologies which have different climatic responses.

23.2.2   Subtropical Anticyclone Belt

Villalba (1989) was able to estimate the position of the Subtropical Anticyclone Belt through a set of chronologies in Argentina and Chile. The chronology set comprised 8 *Austrocedrus chilensis*, 7 *Araucaria araucana* and 2 *Fitzrova cupressoides chronologies*. The minimum number of samples necessary to provide reliable statistical estimates of the chronology indicated that 14 of the chronologies are reliable after the year 1550 and 17 after 1620.

The latitude of the subtropical high pressure belt is a parameter which has been widely used in paleoclimatic analysis (Flohn 1964) and in discussion of present and possible future climatic variations (Bryson 1974; Flohn 1969, 1984; Pittock 1971). The mean latitude of the subtropical high pressure belt (LSA) along the Chilean coast was determined by Pittock (1980) for the 1941-1962 period.

Eigenvector analysis of the 17 chronologies has revealed three dominant patterns of year-to-year tree-ring width variability. The second eigenvector, associated with *Austrocedrus chilensis* chronologies of central Chile, is correlated with the LSA index in winter ($r = 0.65$, significant at the 99% confidence level). The third eigenvector, grouping the *Austrocedrus chilensis* chronologies between 39° to 42°S east of the Andes, are mainly connected with the anticyclone position in summer ($r = 0.58$). Due to the short length of Pittock's series (1941-1962) a true reconstruction of the LSA index was not carried out. The amplitude values of the corresponding eigenvectors were used as estimates of the anticyclone position.

The latitudinal position of the Southeastern Pacific anticyclone's most extreme low- and high-index years for winter (1450 to 1972) and summer (1572-1974) are shown in Table 23.3. In the winter estimates, although the series suggests that events of similar magnitude have occurred in earlier times, extremely low index events are particularly concentrated in the last 100 years. Five of the lowest ten indices have occurred since A.D. 1863 (Figure 23.2b). On the other hand, the year-to-year variations indicate that the anticyclone reached its highest indices between 1560 and 1650, in coincidence with the main phase of the Little Ice Age registered in southern South America (Mercer 1976, 1982; Villalba 1990). The major historical summer low-indices were centered at approximately the following periods: late 1910s and early 1920s, early 1680s, early 1820s, mid-1620s and early 1630s, and early 1960s. (Figure 23.2c).

Table 23.3 Estimate Southeastern Pacific anticyclone belt: extreme low- high index years Winter: 1450 to 1972 period. Summer: 1572 to 1974 period.

Rank	Winter		Summer	
	Low	High	Low	High
1	1924	1593	1943	1579
2	1548	1657	1813	1652
3	1968	1600	1655	1797
4	1863	1942	1682	1945
5	1458	1589	1762	1940
6	1886	1950	1865	1926
7	1598	1561	1908	1878
8	1696	1654	1705	1941
9	1540	1843	1755	1878
10	1879	1652	1846	1808

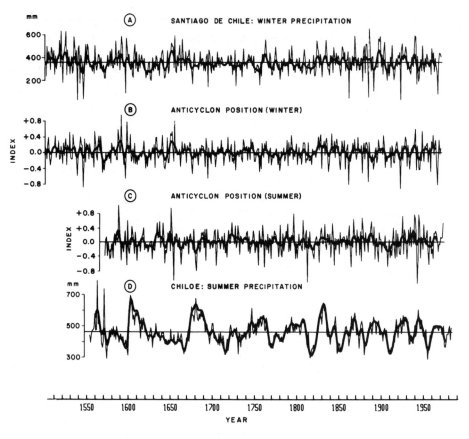

Figure 23.2 A: Estimated winter precipitation of Santiago de Chile; B and C: estimated winter and summer position of the Subtropical Anticyclone Belt respectively; D: estimated summer precipitation of Chiloe Island. Heavy line is the 7-year low-pass digitally filtered series.

Regimes of unusually low or high indices can become established and persist for two or more years. Periods of low-index condition as long as 9 years occur in the reconstructed records. The search for periodicities revealed that low- or high-indices have a marked rhythm of 3.4-3.7 years in winter anticyclone positions and rhythms of 5, 3.2 and 2 years in summer latitudinal shifts. An apparent increase in high-frequency variance in the latitudinal positions of the Pacific anticyclone has been noted since 1900-1910. It is probably an artifact of increased human disturbance in the forests, rather than a true climatic signal. However, since year-to-year variability is of critical importance to agriculture and water supply management, this result concerning variability is of great practical interest, and further study of the matter is needed.

23.2.3 Precipitation

Santiago de Chile LaMarche (1975) produced a first reconstruction of Santiago de Chile precipitation using an *Austrocedrus chilensis* chronology from El Asiento. Boninsegna (1988) reconstructed the winter precipitation of Santiago using an improved version of the same chronology which is highly reliable after the year 1250. The reconstruction was carried out using a simple correlation procedure (see Bradley and Jones, Chapter 1, this volume). Calibration and verification statistics are shown in Table 23.4.

Table 23.4 Summary of South American climate reconstruction models.

RECORD	CALIBRATION		VERIFICATION		RE
	Period (years)	% variance explained	Period (years)	% variance explained	
Southern High Pressure belt					
Winter	1941-1962	42.2			
Summer	1941-1962	33.6			
Precipitation					
Santiago de Chile	1870-1929	49.0	1930-1960	25.0	+.26
	1901-1960	42.2	1870-1900	43.0	+.43
	1870-1960	39.9			
Chiloe Island	1912-1947	36.6	1948-1968	21.0	+.15
	1933-1968	40.9	1912-1932	20.0	+.19
	1912-1968	32.0			
Temperature					
North Patagonia	1902-1938	46.0	1939-1974	40.9	+.36
	1939-1974	57.0	1902-1938	33.6	+.23
	1902-1974	52.0			
Rio Alerce	1908-1947	43.0	1948-1984	42.2	+.30
	1945-1984	46.0	1908-1944	37.2	+.22
	1908-1984	42.0			
Ushuaia	1901-1984	23.0			
Streamflows					
Rio Neuquen	1903-1960	53.3			
Rio Limay	1903-1960	53.3			
Rio Atuel	1922-1972	50.4			

The analysis of the reconstructed series (Figure 23.2a) shows that one long period of drought occurred during the years 1270-1450; another drought period occurred in the years 1600-1650. Several others of shorter duration are also apparent in the record; all are longer than any that have occurred since the beginning of instrumental measurements. Even if a major drought cannot be forecast, its future occurrence can be anticipated as highly probable.

No low frequency fluctuations seem important in the reconstructed series as indicated by the power spectrum estimates. According to J. Minetti (personal communication) very few long term variations were found in the central Chile precipitation records because Santiago usually lies close to the northern edge of storms travelling eastward from the Pacific Ocean and rainfall amounts are highly sensitive to shifts in the storm tracks.

Chile Island Roig and Boninsegna (1989b) made a preliminary reconstruction of the summer precipitation of Chile Island (44°S, 76°W) using a chronology of the conifer *Pilgerodendron uviferum* (Figure 23.2d). The reconstruction was carried out using lagged values of the chronology in t+1, t, t-1, t-2 and t-3 years. The amplitudes of the first and second eigenvectors were used as predictors, while the predicted was the combined summer precipitation data (December to March) of three meteorological stations. A multivariate regression procedure was used. Calibration and verification statistics are presented in Table 23.4.

Although the relation between annual growth and climate variations was studies using a rather short meteorological record, the reconstructed series allows inferences to be made about periods in which dry or wet conditions in 'summer' have occurred. Dry conditions seem predominant during 1700 to 1745, 1765 to 1785 and more frequently after 1800. Wet summers appear in 1675 to 1700, 1830 to 1840, alternating with droughts after this date. Perhaps the most interesting result is to show that *Pilgerodendron uviferum* is potentially a good source of paleoclimatic information. *Pilgerodendron uviferum*, known as the most austral conifer in the world, has a biogeographical distribution southward of 47°S and owing to its particular site-growth conditions. It is possible to find layers of fossil trunks in a state of good preservation beneath the surface of peat bogs, so that there is the possibility of extending the chronologies of living trees.

23.2.4 Temperature

Patagonia Villalba *et al.* (1989) studied the temperature sensitive *Araucaria araucana* chronologies and they were able to reconstruct the summer temperature (December to April) in Central Western Argentina back to A.D. 1491. The Araucaria araucana chronologies set is highly reliable after the year 1500. To derive a regression model relating tree-ring indices to temperature fluctuations, a regional temperature record (six weather stations) served as predicted and tree-ring chronologies from seven sites in Neuquen province served as predictors. Temperatures were reconstructed for the period December to May. This period was selected according to the explained variance for separate calibrations of tree-ring data on each monthly mean regional temperature. The pool of predictors was first screened using a stepwise multiple regression procedure in which values of F (significant at p=0.8 and 1.0, respectively) were used to determine the entry and exit point of a predictor to and from the model. The eigenvector scores of the selected variables were then entered in a best subset regression procedure. The primary criterion for choosing predictor variables was a minimization of the Mallow's CP statistics. Verification procedures followed Gordon and LeDuc (1981). Calibration and verification statistics are presented in Table 23.4.

The summer regional reconstruction (Figure 23.3a) pointed out the severity of the A.D. 1898-1903 cold period; the years 1898 and 1903 rank among the coldest 15 of the last 500 years. During the cold period around 1900, the Sub-Antarctic westerly belt extended northwards to about 48°S, its lowest known latitude for any period on record (Lamb, 1977). This displacement has been in parallel with shifts of the Antarctic ice limits. When conditions were averaged over periods of three or more years, the three coldest periods were from A.D. 1532

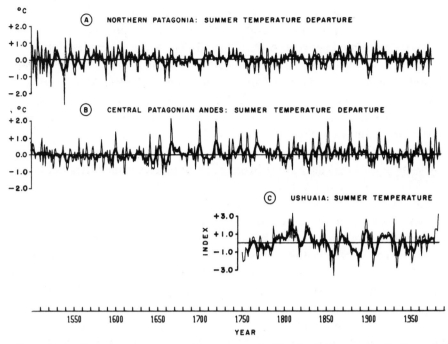

Figure 23.3 Estimated summer temperatures. A: Northern Patagonia; B: Central-West Patagonia; C: Ushuaia city (Tierra del Fuego). Heavy line shows the 7-year low-pass digitally filtered series.

to 1538, 1898-1903, and 1556 to 1562. On the other hand, the three warmest periods were between A.D. 1911 and 1920, 1859 to 1867, and 1563 to 1567. Although important cold intervals are reconstructed during the 16th and 17th centuries, the Little Ice Age cold episode is not as evident in the present reconstruction as in other tree-ring records located southward of 41°S (see following subsection). The influence of summer continental air masses of tropical origin at 37°S could be responsible for the observed differences.

Rio Alerce   Villalba (1990) built-up a long chronology using *Fitzroya cupressoides* at Rio Alerce (41°10'S, 71°46'W). This chronology is reliable after the year 1200. A detailed interpretation of the relation between climate and tree-growth indicates that the Rio Alerce chronology is strongly related to summer temperature over Patagonia, particularly from December to February. Best subset regression procedure (Drapper and Smith, 1981) was used to derive a model for predicting summer temperature deviations. A mean normalized summer temperature departure (December to February) involving three meteorological stations was used as predictand while the Rio Alerce chronology at time t and lagged one year backward (t+1) and one year forward (t-1) were the candidate predictor variables. Only the year t and t-1 of the chronology proved to be stable predictors of the temperature data. Calibration and verification statistics are shown in Table 23.4.

Four main climatic episodes can be distinguished in this proxy paleoclimatic record. The first, cold and moist, was from A.D. 900 to 1070, followed by the warm-dry period from A.D. 1080 to 1250. Afterwards, a long cold-moist period followed from A.D. 1280 to 1670, peaking around A.D. 1340 and 1650. These cold maximum episodes are contemporaneous with the two principal Little Ice Age events registered in the Northern Hemisphere (Lamb, 1977). Warmer conditions then resumed between A.D. 1720 and 1790. Since the cold period in the early 1800s, reconstructed temperatures have oscillated around the long-term mean, except for a warmer period from A.D. 1850 to 1890. Correlations between the Rio Alerce temperature reconstruction and the regional weather stations indicate a homogeneous summer weather pattern covering Patagonia east the Andes from 38° to 50°S. The reconstructed series from A.D. 1500 to 1980 is shown in Figure 23.3b.

Ushuaia Boninsegna *et al* (1989) have constructed 21 chronologies from Tierra del Fuego Island using *Nothofagus pumilio* and *Nothofagus betuloides*. As the southernmost extension of forested land in the world, the region has unique dendroclimatic potential. Response function studies have shown that increased growth is related to cooler temperatures in late winter and early spring. Correlation analysis between Ushuaia temperatures and the Tierra del Fuego chronologies revealed a positive correspondence between tree growth and growing season (November to February) temperatures at a number of sites. Four of the chronologies with the strongest correlation for the overlapping 1901-1984 period were selected and the first eigenvector scores of a principal components analysis of these four chronologies were used in a preliminary reconstruction of November to February (summer) temperatures.

The reliability of the four chronologies is reasonably acceptable after A.D. 1750. For the calibration period (1901-1984) the simple correlation coefficient between the first eigenvector scores and the seasonalized summer temperature is 0.48 and the resulting variance explained (adjusted for degrees of freedom) is 0.22. The reconstruction has yet to be verified and so verification statistics are not available. The reconstruction is a preliminary effort which will be improved with further modelling. The reconstructed summer temperatures show (Figure 23.3c) a period which was relatively warm in the late 1700s to early 1800s, and some cooler periods in the mid to late 1800s which are somewhat similar to a cooling during the last part of the Little Ice Age, observed in the Northern Hemisphere proxy data (Jacoby and D'Arrigo 1989) as well as low-latitude Southern Hemisphere ice core data (Thompson *et al*. 1986).

The reconstruction also shows a warming trend in recent decades, which appears to correspond to a similar positive trend in high latitude Southern Hemisphere temperature data (Hansen and Lebedeff 1987). Variance spectrum analysis for the reconstructed temperatures from 1760 to 1984 based on 60 lags reveals significant (95% level) concentration of variance at periods of 20-24, 3.1 and 2.1 years. The peak centered at around 24 years may be explainable based on the influence of wind. Another possible explanation is a correspondence with the Hale 22-year sunspot cycle. The peaks at about 3 and 2 years may correspond to the El Nino and Quasi-Biennal Oscillation phenomena, respectively.

23.2.5 Streamflows

Neuquen and Limay Rivers Holmes *et al*. (1982) made the reconstructions of the Rio Neuquen and Rio Limay streamflows using seven chronologies of *Araucaria araucana* and *Austrocedrus chilensis*. The Rio Neuquen and Rio Limay in the Patagonian Andes of Argentina drain an area extending from latitudes 36°S to about 44°S. These major rivers have good continuous gauging station records beginning in 1903. In the southern Andes, the prevailing air movement is from the west, thus the area drained by the two rivers is in the rainshadow of the Andes. Precipitation drops off steeply from over 2,000 mm per year at the base of the Andes to under 200 mm some 80 km to the east, out into the Patagonian steppe. Thus the tree-ring sites used in the reconstruction are close to the areas receiving most of the precipitation, but in areas sufficiently dry that moisture availability is frequently limiting growth processes in trees, so that dry years are reflected in narrower than normal rings.

The annual runoff of the rivers was reconstructed back to the year 1601 using seven tree ring chronologies and the multivariate technique of canonical analysis. The predicted variables were the gauged annual riverflow data from the Neuquen and Limay rivers and the predictor data included modified series from each of the seven sites. The correlation coefficient between the measured and estimated records was 0.73 in both cases. Verification statistics are not yet available. The reconstructed annual runoff series show important low flow periods between the years 1630 to 1750, and 1890 to 1925, while high flow periods appear during the years 1760 to 1830. (Figure 23.4b and 23.4c)

Atuel River Cobos and Boninsegna (1983) studied the glaciers of the upper Atuel river basin and reconstructed the streamflows of the river since 1534 using three *Austrocedrus* chronologies from Chile. Two of the three chronologies used are highly reliable after the year 1534, and the other after 1620. Stepwise multiple regression was carried out to calibrate the model (see Bradley and Jones, Chapter 1, this volume). During the calibration period (A.D. 1922-1972) the correlation between the measured and estimated streamflow accounted for the 50.4% of the total variance. This reconstruction also has not been verified.

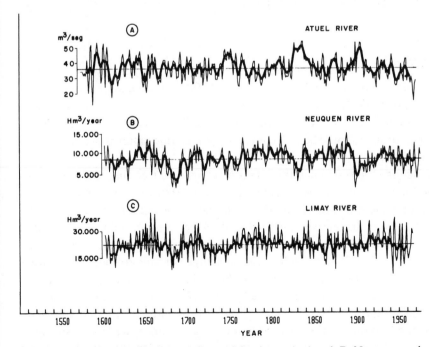

Figure 23.4 Reconstructed streamflows of the rivers: A: Atuel; B: Neuquen and C: Limay. Heavy line shows the 7-year low-pass digitally filtered series.

457

The reconstructed streamflow series for the Atuel River does not exhibit periods of greatly increased runoff and abundant precipitation, except for the period from 1820 to 1850. The trend of the series is slightly negative but not significant. In this series the proportion of years in which streamflow was greater than the general mean is one in two for the period from 1575 to 1850, one in three between 1850 and 1970 and one in four in the latter part of this period, from 1914 to 1970. Thus the proportion of years in which streamflow has been below the mean has increased (Figure 23.4a).

Recharge in the drainage basin likely depends not only on the quantity of snowfall, but also on its distribution in time, and on summer temperatures to provide enough energy for ablation. Data obtained and developed in Cobos and Boninsegna (1983) lead to the conclusion that in the last 400 years only one glacier has advanced and that this is probably a case of surging. Aside from this event, all evidence points to sustained recession of the glacier in the upper Atuel river basin.

23.3   Other related studies

23.3.1   Glacier movements

Rabassa *et al.* (1985) has studied the movements of Rio Manso and Castano Overo glaciers at Monte Tronador (42°S 70°W) and the moraines were dated using tree ring counts. Villalba *et al.* (1988b) reconstructed the oscillations of Frias glacier at Monte Trondador using local chronologies, ice scars in some particular trees and tree-ring counts to date the moraines. Both authors give the decade of 1670 as the maximum expansion of the ice fields after A.D. 1500.

23.3.2   Volcanic eruptions

Major explosive volcanic eruptions can have pronounced effects on climate that may persist for several years (Lamb 1972). It is generally agreed that a tropospheric cooling would be expected following sufficiently powerful volcanic eruptions (Taylor *et al.* 1980, Angell and Korshover 1985). Such cooling should be reflected in the annual ring patterns of temperature sensitive tree-ring records.

Villalba and Bonisegna (1989) signaled the relation between 50 tree ring chronologies of South America and the ten major volcanic eruptions from 1780 to 1970. The main findings from this study are that the magnitude, the duration and the geographical extent of the volcanically-induced tree ring decrease in southern South American chronologies are related to the eruption type (specifically, whether they are rich or not in sulphur gases) and to the hemispheric debris distribution. It is also indicated that a large volcanic episode is not always related to a uniform decrease of temperature in all the regions. It seems that tropical and extratropical warming in South America associated with El Nino events tended to mask the cooling due to volcanic eruptions. The El Nino events in 1814 and 1817 (Quinn and Neal, Chapter 32, this volume) around the time of the eruption of Tambora as well as the strong El Nino in 1884 following the Krakatoa eruption may have moderated the effects of these eruptions.

23.4 Conclusions

Despite several difficulties, dendrochronological work in South America has progressed during the last ten years. The number of chronologies is still small in relation to the total land surface, but some results are very encouraging. The chronologies show climatic signals more or less comparable in quality and quantity to the chronologies derived in Australasia and in the Northern Hemisphere.

The reconstruction of the movement of the Southeast Pacific Anticyclone Belt, the retraction and advances of glaciers, the reconstructions of temperature, precipitation and streamflow in the temperate zones seem to indicate the occurrence of the Little Ice Age at a more or less synchronous time with the Northern Hemisphere, culminating near A.D. 1670. The strength and extension of the phenomenon was probably very different, due to the relation between land and water masses, the atmospheric circulation and other biogeographical particulars of the continent. An improvement in the dendrochronological network will be necessary to study these relations. As more chronologies become available in Australasia (see Norton and Palmer, Chapter 24, this volume) and in South America, reconstruction of climate parameters at a regional scale, as well as teleconnection studies, provides a very promising new field that needs to be explored in the near future.

In the Northern Hemisphere, the human impact on natural forest has been far more noticeable that in South America, especially in the temperate region. Pollution problems such as acid rain are not widespread in South America. These facts make the South American forests especially valuable to assess other global problems such as the increase in CO_2 and the greenhouse effect.

The tropical forest seems to be a very promising new field for the development of "tropical dendroclimatology"; the extension of climatic records and a dendrochronological network in that area is highly desirable. Finally, South America has the southernmost extension of forested land in the world, so the region has unique dendroclimatic potential, with good prospects of producing centuries of climatic information useful for studies of climate on a hemispheric scale.

References

Angell, J. K. and J. Korshover. 1985 Surface temperature changes following the six major volcanic episodes between 1780-1980. *Journal of Climate and Applied Meteorology*, 24: 937-951.

Boninsegna, J. 1988. Santiago de Chile winter rainfall since 1200 as reconstructed by tree rings. *Quaternary of South America and Antartica Penninsula*. Balkema ed. The Netherlands (in press).
Boninsegna, J. and R. L. Holmes. 1985. Fitzroya cupressoides yields 1534-year long South American chronology. *Tree Ring Bulletin*, 45:37-42.
Boninsegna, J. A., J. Keegan, G. C. Jacoby, R. D. D'Arrigo and R. L. Holmes. 1989. Dendrochronological studies in Tierra del Fuego, Argentina. *Quaternary of South America and Antarctica Peninsula*. Balkema ed. The Netherlands (in press).
Bryson, R. A. 1974. A perspective on climatic change. *Science*, 184:735-760.

Cobos, D. R. and J. A. Boninsegna. 1983. Fluctuations of some glaciers in the upper Atuel river basin, Mendoza, Argentina. *Quaternary of South America and Antarctica Peninsula*. 1:61-82.

Cook, E. R. 1985. *A time series analysis approach to tree-ring standardization*. Ph.D. dissertation. University of Arizona. 171 pp.

Cook, E. R. and R. L. Holmes. 1985. *Program ARSTAN user's manual*. Lab of Tree-Ring Res. University of Arizona, Tucson, Arizona.

Draper, N. R. and H. Smith. 1981. *Applied Regression analysis*. 2nd Ed. Willey, N. York.

F.A.O. 1985. *International Year of the Forest*. F.A.O. Fact Sheet.

Flohn, H. 1964. Grundfragen der Palaoklimatologie im Lichte Einer Theoretischen Klimatologie. *Geol. Runds*. 54:504-515.

Flohn, H. 1969. *Climate and Weather*. Weidenfeld and Nicolson, London.

Flohn, H. 1984. Climatic belts in the case of a unipolar glaciation. In: *Climatic Changes on a Yearly to Millennial Basis*. N.A. Morner and W. Karlen (eds.). D. Reidel Publishing Company.

Gordon, G. A. and S. K. LeDuc. 1981. Verification statistics for Regression Models. *Proceedings of Conference on Probability and Statistics in Atmospheric Science*. Monterrey, California.

Hansen, J. and S. Lededeff. 1987. Global trends of measured surface air temperature. *Journal of Geophysical Research*. 92:13345-13372.

Holmes, R. L. 1983. Computer assisted quality control in tree-ring dating and measurement. *Tree Ring Bulletin*. 43:69-75

Holmes, R. L., C. W. Stockton and V. C. LaMarche. 1982. Extension of Riverflow records in Argentina. In: *Climate from tree rings*. M. K. Hughes, P. M. Kelley, J. R. Pilcher and V. C. Lamarche (eds). Cambridge Univ. Press. London.

Hueck, K. 1978. *Los Bosques de Sudamerica* (translated by R.Brun). Sociedad Alemana de Cooperacion Tecnica, Eschborn, West Germany.

Jacoby, G. C. 1989. Overview of tree-ring analysis in tropical regions. *IAWA Bulletin* 10(2):99-108.

Jacoby, G. C. and R. D. D'Arrigo. 1989. Reconstructed Northern Hemisphere annual temperature since 1671 based on high-latitude tree-ring data from North America. *Climatic change*. 14: 39-59.

LaMarche, V. C. 1975. Potential of tree rings for reconstruction of past climate variations in the Southern Hemisphere. *Proceedings of the WMO/IAMAP Symposium on Long Term Climatic Fluctuations*, 21-30, Norwich.

LaMarche, V. C., R. L. Holmes, P. W. Dunwiddie, and L. G. Drew. 1979. *Tree-Ring Chronologies of the Southern Hemisphere. Vol 1: Argentina and Volume 2: Chile*. Chronology Series V, University of Arizona, Tucson, Arizona.

Lamb, H. H. 1972. *Climate: Present, Past and Future*. Vol 1: Fundamentals and Climate Now. Methuen, London.

Lamb, H. H. 1977. *Climate: Present, Past and Future*. Vol 2: Climatic History and the Future. Methuen, London.

Lamprecht, A. M. 1984. Dendroklimatologishe Untersuchungen in Sudamerika. *Rapports No. 263 of the Swiss Federal Institute of Forestry Research*, Birmensdorf.

Lough, J. and H. C. Fritts. 1985. The Southern oscillation and tree rings: 1601-1961. *Journal of Climate and Applied Meteorology*. 24:952-965.

Mercer, J. H. 1976. Glacial History of Southernmost South America. *Quaternary Research*. 6, 125-166.

Mercer, J. H. 1982. Holocene Glacier Variations in Southern South America. In: *Holocene Glaciers*. W. Karlen (ed.) Striae, 48:35-40. Upsala.

Pittock, A. B. 1971. Rainfall and the general circulation. *Preprints Int. Conf. on Weather Modification*, Camberra, Amer. Meteor. Soc., 330-338.

Pittock, A. B. 1980. Patterns of climatic variation in Argentina and Chile. I: Precipitation, 1931-1960. *Monthly Weather Review* 108(9):1347-1361.

Pittock, A. B., L. A. Frakes, D. R. Jensen, J. A. Peterson and J. W. Zillman (eds.). 1978. *Climatic change and variability*. A southern perspective. Cambridge University press, London.

Rabassa, J., A. Brandani, J. Boninsegna and D. R. Cobos. 1985. Glacier fluctuations during and since the Little Ice Age and forest colonization: Monte Tronador and Volcan Lanin, Northern Patagonian Andes. *Proceedings of the International Symposium on Glacier Mass Balance, Fluctuations and Runoff*, Alma Ata, October 1985, U.S.S.R.

Roig, F. 1988. Growth conditions of *Empetrum rubrum* Vahl. ex Will. in the south of Argentine. *Dendrochronologia* 6:43-59.

Roig, F. A. and J. A. Boninsegna. 1989a. Ring-width Chronologies from Central west ranges of Argentina. *Tree Ring Bulletin* (in press).

Roig, F. and J. A. Boninsegna 1989b. Chiloe Island (Chile) summer precipitation since 1556 as reconstructed by a tree ring chronology of *Pilgerodendron uviferum*. *Unpublished Manuscript*.

Roig, F., R. Villalba and A. Ripalta 1988. Climatic factors affecting *Discaria trinervis* growth in Argentina Central Andes. *Dendrochronologia* 6:62-70.

Schulman, E. 1956. *Dendroclimatic change in Semiarid America*. University of Arizona Press. Tucson. Arizona.

Schwerdtfeger, W. 1976. *World Survey of Climatology*. Vol 12: South America. Elsevier. New York.

Taylor, B. L., T.Gal-Chen and S. E. Schneider. 1980. Volcanic eruptions and long-term temperature records: An empirical search for cause and effect. *Quarterly Journal of the Royal Meteorological Society*, 106:175-199.

Thompson, L. J., E. Mosley-Thompson, W. Dansgaard and P. M. Grootes. 1986. The Little Ice Age as recorded in tropical Quelcaya Ice cap. *Science* 234:361:364.

Vetter, R. E. and P. C. Botosso. 1989. Remarks on age and growth rate determination of Amazonian trees. *IAWA Bulletin*. 10(2):133-145.

Villalba, R. 1990. Climatic fluctuations in Northern Patagonia during the last 1000 years as inferred from tree ring records. *Quaternary Research* (in press).

Villalba, R. 1989. Latitude of the surface high-pressure belt over Western South America during the last 500 years as inferred from tree-ring analysis. *Quaternary of South America and Antarctica Peninsula*, Balkema (in press).

Villalba, R. and J. A. Boninsegna. 1988. Dendrochronological studies on *Prosopis flexuosa* DC. *IUFRO, All-Division 5 Conference*, San Pablo, Brazil.

Villalba, R. and J. A. Boninsegna. 1989. Changes in Southern South America Tree-Ring Chronologies following major Volcanic Eruptions between 1750 to 1970. In: *Climate in 1816*. C. Harrington (ed.). Syllogeus, National Museums of Canada. (in press).

Villalba, R., J. A. Boninsegna and R. L. Holmes 1985. *Cedrela angustifolia* and *Juglans australis*: two tropical species useful in dendrochronology. *Tree Ring Bulletin*, 45:25-35.

Villalba, R., J. A. Boninsegna and A. Ripalta. 1987. Climate, site conditions, and tree growth in subtropical northwestern Argentina. *Canadian Journal of Forest Research*, 17:1527-1539.

Villalba, R., J. A. Boninsegna and A. Ripalta. 1988a. Climate and tree growth on subtropical northwestern Argentina. *IUFRO, All-Division 5 Conference*, San Pablo, Brazil.

Villalba, R., J. C. Leiva, S. Rubulis, J.A. Suarez and L. Lensano. 1990. Climate, tree rings and glacier fluctuations in the Rio Frias valley, Rio Negro, Argentina. *Arctic and Alpine Research*. 22: (in press).

Villalba, R., J. A. Boninsegna and D. R. Cobos. 1989. A tree-ring reconstruction of summer temperature between A.D. 1500 and 1974 in western Argentina. *Third International Conference of the Southern Hemisphere Meteorology and Oceanography*, Buenos Aires, Argentina November 1989.

Wigley, T. M. L., K. R. Briffa and P. D. Jones. 1984. On the average value of correlated time series, with applications in dendroclimatology and hydrometeorology. *Journal of Climatology and Applied Meteorology*. 23:201-213.

Worbes, M. 1985. Structural and other adaptations to long-term flooding by trees in Central Amazonia. *Amazoniana* 9(3):459-484.

Worbes, M. 1989. Growth rings, increment and age of trees in inundation forest, savannas and mountain forest in the neotropics. *IAWA Bulletin*. 10(2):109-122.

24 Dendroclimatic evidence from Australasia

D. A. Norton and J. G. Palmer

24.1 Introduction

Few instrumental climate records in Australasia extend back before A.D. 1900. For example, the Dunedin temperature series (Salinger, 1977) starting in 1853, is the longest continuous New Zealand record. Consequently, for information on climate variations prior to the present century we must rely on indirect sources, from so called 'proxy' data. However, much of the proxy data for palaeoclimate studies has a rather coarse temporal resolution. Furthermore, the evidence is often contradictory and difficult to interpret in terms of climate (e.g. Burrows & Greenland 1979). Dendroclimatology has, however, been shown to be a promising tool for reconstructing annual to decadal fluctuations in palaeoclimates (Fritts 1976; Hughes *et al.* 1982) and in this chapter the use of dendroclimatic techniques to reconstruct Australasian climates since A.D. 1500 is reviewed.

The basic principles of dendroclimatology have been reviewed extensively elsewhere (e.g. Fritts 1976; Hughes *et al.* 1982; Schweingruber 1983) and are not discussed here. However, the specific techniques used to develop the different reconstructions presented are discussed.

24.2 Australasian dendrochronology

General reviews of the use of dendroclimatic techniques in Australasia are given by Ogden (1978a, 1981, 1982) Dunwiddie (1979) Dunwiddie and LaMarche (1980) Norton and Ogden (1987) and Norton (1990). Results from early attempts to use dendrochronological techniques were largely inconclusive, either because cross-dating was unsuccessful (e.g. Scott 1972, Wells 1972) because the trees were too short-lived (e.g. *Callitris*, Dunwiddie & LaMarche 1980) or because of uncertainties over the annual nature of growth ring production in some species (e.g. *Eucalyptus*, Ogden 1978a; *Diospyros*, Duke *et al.* 1981; Ash 1983). Since the mid-1970s, modern dendrochronological techniques, which have been successfully applied to mesic forest trees elsewhere in the world, have been used in Australia and New Zealand. The first properly cross-dated and replicated Australasian tree-ring chronologies were developed as part of a Southern Hemisphere sampling program initiated by the Laboratory of Tree-Ring Research, University of Arizona (LaMarche *et al.* 1979a, b). Thirty-eight chronologies were developed for 12 species, but no climate reconstructions were published. Subsequent research has seen the number of chronologies increase to 89, involving 14 tree species (Table 24.1). Both coniferous and angiosperm genera are represented.

In Australia success has been greatest with coniferous species. Although one chronology has been developed from the short-lived arid-zone conifer *Callitris* (LaMarche *et al.* 1979b)

Table 24.1 Synopsis of Australasian tree-ring chronologies at the time of publication. No., number of published chronologies. Lat., latitudinal range of chronologies. Alt., altitudinal range of chronologies in metres. Length, average chronology length. AC, average lag-one autocorrelation. MS, average mean sensitivity. R, average mean correlation between all radii in chronology.

Species	No.	lat.	alt.	length	AC	MS	R
a. Australia							
Arthrotaxis cupressoides[1]	3	41°45'-42°41'	1200	699	0.65	0.13	0.38
A. selaginoides[1]	1	41°38'	1000	778	0.62	0.14	0.27
Callitris robusta[1]	2	32°00'-32°07'	4-15	54	0.06	0.21	0.40
Nothofagus gunnii[1]	1	41°38'	1000	244	0.52	0.17	0.40
Phyllocladus aspleniifolius[1]	10	41°11'-43°28'	200-900	388	0.26	0.29	0.36
b. New Zealand							
Agathis australis[*,2,3]	12	35°11'-37°36'	75-468	316	0.31	0.26	0.31
Halocarpus biformis[2]	1	45°32'	305	410	0.75	0.10	0.15
Lagarostrobos colensoi[2]	2	39°21'-42°23'	244-1000	543	0.62	0.13	0.17
Libocedrus bidwillii[2]	11	39°15'-46°23'	244-1067	459	0.68	0.15	0.30
Nothofagus menziesii[4]	5	43°03'-45°18'	950-1275	347	0.42	0.31	0.35
N. solandri[4]	25	43°01'-45°18'	610-1400	222	0.50	0.31	0.42
Phyllocladus alpinus[2]	1	42°54'	915	260	0.57	0.13	0.19
P. glaucus[2,5]	5	37°30'-38°42'	520-1000	291	-0.36	0.44	0.46
P. trichomanoides[*,2,5]	10	38°18'-41°07'	15-640	289	0.13	0.26	0.33

[*] subfossil chronologies are not included
[1] LaMarche *et al.* 1979a
[2] LaMarche *et al.* 1979b
[3] Palmer 1982
[4] Norton 1983a,b,c
[5] Palmer 1989

the greatest potential appears to lie with the genera *Arthrotaxis*, *Phyllocladus*, and *Lagarostrobos* in Tasmania (Ogden 1978a,b; LaMarche *et al.* 1979b; Dunwiddie & LaMarche 1980; Francey *et al.* 1984). Preliminary work with *Arthrotaxis cupressoides* and *A. selaginoides* (Ogden 1978b) showed that cross-dating could be achieved and four chronologies have been published (LaMarche *et al.* 1979b). The longest chronology, for *A. cupressoides*, covers the period A.D. 1028-1974 and is the longest in Australasia. Ten chronologies have been developed for *Phyllocladus aspleniifolius* (LaMarche *et al.* 1979b) the longest extending back to A.D. 1310. Chronology development is also underway with the very long-lived *Lagarostrobos franklinii* (Francey *et al.* 1984) although no chronologies have yet been formally published. One relatively short chronology has been developed from the dwarf angiosperm tree *Nothofagus gunnii* (LaMarche *et al.* 1979b).

Initial sampling in New Zealand concentrated on seven coniferous species (*Libocedrus bidwillii*, *Phyllocladus alpinus*, *P. glaucus*, *P. trichomanoides*, *Lagarostrobos colensoi*, *Halo-*

carpus biformis, and *Agathis australis*) with 21 chronologies produced (LaMarche *et al.* 1979a; Dunwiddie 1979). Additional chronologies have been developed subsequently for *Agathis australis* (Palmer 1982; Ahmed 1984; Ahmed & Ogden 1985) *Phyllocladus glaucus* and *P. trichomanoides* (Palmer 1989) and *Libocedrus bidwillii* (Norton 1983c). Several of these conifer chronologies extend back prior to A.D. 1500, with the oldest from *Libocedrus bidwillii* extending to A.D. 1256. Good progress has also been made with the angiosperm genus *Nothofagus* in South Island, with thirty chronologies having been developed using two species, *N. menziesii* and *N. solandri* (Norton 1983a,b, 1985, 1987). The oldest, for *N. menziesii*, extends back to A.D. 1580.

Some success has also been achieved with developing subfossil chronologies in Australasia. In Australia, buried wood of *Lagarostrobos franklinii* and *Phyllocladus aspleniifolius*, dating from throughout the last 8000 years (McPhail *et al.* 1983) has been recovered and offers the potential for developing a long historical chronology comparable to those developed for oak and bristlecone pine in the Northern Hemisphere. In the North Island of New Zealand, subfossil chronologies of *Agathis australis* (Bridge & Ogden 1986) for the period 3500-3000 B.P. (before present) and *Phyllocladus trichomanoides* (Palmer 1989) for the period 2200-1800 B.P., have been developed. The presence of abundant buried wood at several swamp sites, particularly in the northern part of New Zealand, presents the opportunity for developing long sub-fossil chronologies.

24.3   Dendroclimatic reconstructions in Australasia

The six published Australasian palaeoclimate reconstructions (including stream-flow) based on tree-rings are now reviewed.

24.3.1   *Tasmanian temperature*

The only temperature reconstructions published for Australia (LaMarche & Pittock 1982) were based on a grid of 11 tree-ring chronologies from Tasmania (Figure 24.1; LaMarche *et al.* 1979b). The data set was not ideal, including three genera *Arthrotaxis*, *Phyllocladus* and *Nothofagus*) and three composite chronologies (obtained by averaging two or more individual chronologies together). The chronologies were developed using a mixture of standardization techniques (LaMarche *et al.* 1979b). The shortest chronology ended in 1776, setting the maximum length of the reconstructions. However, the early portion of the reconstructions must be interpreted with caution since no assessment was made of the strength and reliability of the climatic signal contained in the chronologies as the sample size of trees reduced back in time (c.f., the subsample signal strength statistic (SSS) of Wigley *et al.* 1984).

Response-function analyses indicated that temperature had a strong influence on tree growth, although this response differed between the two main genera. In *Arthrotaxis* there was a dominant positive response to higher than normal summer temperatures in the preceding and current years, while in *Phyllocladus*, the response to temperature was generally negative, especially during the previous season. *Phyllocladus* also showed a strong positive temperature response during the winter prior to growth. LaMarche and Pittock

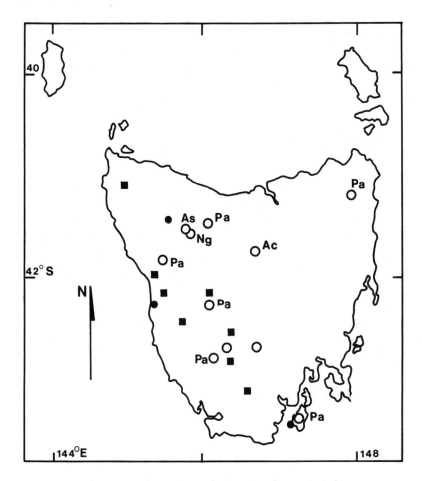

Figure 24.1 Location of tree-ring sampling sites (open circles) temperature stations (closed circles) and river-flow stations (closed squares) in Tasmania. PA, *Phyllocladus aspleniifolius*. AS, *Arthrotaxis selaginoides*. Ac, *A. cupressoides*. Ng, *Nothofagus gunnii*.

(1982) do, however, suggest that the climate responses observed may in fact reflect a relationship between solar radiation and growth. No information is given on the *Nothofagus* temperature response.

The eleven tree-ring chronologies together with data from 15 climate stations were then subjected to principal components analyses for a 29 year period with common climate data. The first three components of the temperature data (October-May) explained 87% of the total variance and reflected regional temperature anomalies. The first four chronology components explained 66% of the total variance, but did not reflect broad- scale growth anomalies. Rather the first two components were associated with differences between the *Phyllocladus* (component 1) and *Arthrotaxis* (component 2) chronologies. The third and fourth components did, however, reflect spatial growth anomalies, although their explained variance was low (17%).

These data were then used in a step-wise canonical regression (Fritts *et al*. 1979) to derive a transfer function for reconstructing past climates. The predictand data set comprised the three temperature principal components and the predictor variables were the amplitudes of the four most important chronology components, including current and two lagged values, giving 12 predictor variables in total. The canonical regressions yielded three canonical variates with canonical correlation coefficients of 0.87, 0.85 and 0.34. Only the first two were used in subsequent analyses. The canonical equations were solved to yield regression coefficients expressing the relationship between the 12 predictor variables and the three predictand variables (i.e. three equations were developed relating each temperature component to the 12 predictor variables).

From these equations, estimates of yearly values of October-May temperature were made for each of the 15 climate stations. These were obtained by first estimating the values of the amplitudes of the three temperature components using the canonical regression equations. The resulting yearly amplitudes were then multiplied by the elements of the corresponding temperature component for each station, and the respective products summed. The equations were verified for three stations that had adequate temperature records (Table 24.2, Figure 24.2a). For these stations, variance explained in calibrating the transfer functions ranged from 55% to 79% and in verification from 18% to 40%. However, none of the three verifications passed all non-parametric verification tests, while two passed them in calibration.

The reconstructions were interpreted as showing periods of above average temperatures between the early 1830s and the late 1850s, and the late 1890s to the mid-1910s. Cooler periods occurred during the mid-1810s to early 1830s, and the late 1850s to late 1870s.

Table 24.2 Summary of Australian climate reconstruction models.

record	Calibration period	Calibration % variance explained	Verification period	Verification % variance explained	RE
Temperature					
Waratah	1941-1969	81.0	1899-1940	39.7	0.12
Cape Sorell	1941-1969	77.4	1900-1940	18.5	-0.58
Cape Bruny	1941-1969	54.8	1924-1940	31.4	0.04
Riverflow					
Station 40	1958-1973	36.0	1922-1933 & 1951-1957	49.0	0.40
Station 46	1958-1973	59.3	-	-	-
Station 78	1958-1973	51.8	1924-1957	9.0	-0.38
Station 119	1958-1973	68.9	1949-1957	32.5	0.11
Station 154	1958-1973	53.3	-	-	-
Station 159	1958-1973	27.0	-	-	-
Station 183	1958-1973	59.3	-	-	-
Record 10087	1958-1973	60.8	-	-	-

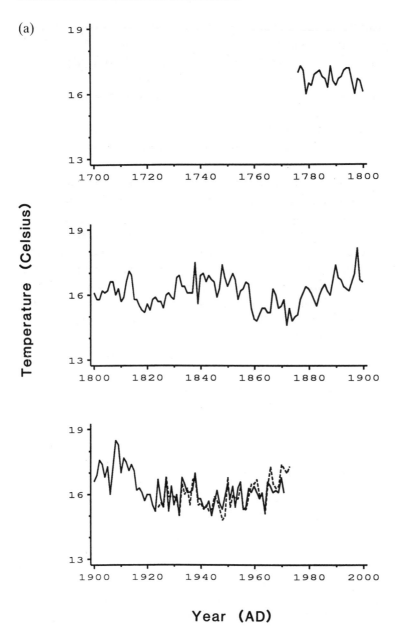

Figure 24.2(a)

24.3.2   Tasmanian stream-flow

Reconstructions of stream-flow for eight rivers in western Tasmania were developed (Campbell 1982) using the same grid of 11 tree-ring chronologies described in the last section (Figure 24.1). Response function analyses suggested that despite the high precipitation experienced at the chronology sites, the chronologies were sensitive to precipitation and were therefore potentially useful for reconstructing past stream-flow. Hydrological data were obtained from rivers with at least 16 years of data; the longest data set was 34 years. Transfer functions were developed using the canonical regression procedure described above and

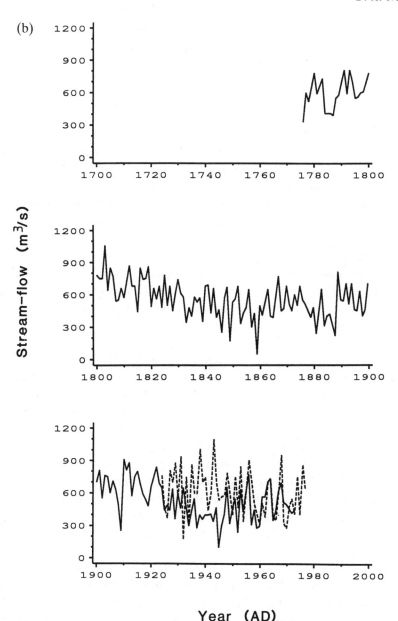

Figure 24.2 Reconstructed (solid line) and recorded (dashed line) temperature (a) and riverflow (b) records from Tasmania.

were calculated over the 16 year period (1958-1973) common to all stream-flow records. Verification was only possible for three of the stream-flow records (Table 24.2, Figure 24.2b).

Variance explained in calibrating the transfer functions ranged from 30% to 68% and in verification (three rivers only) from 9% to 49%. In verification, none passed all non-parametric tests. However, the results do suggest that some ability to reconstruct stream-flow is present in the tree-ring data set and examination of the one reconstruction presented

469

(Campbell 1982) indicates periods of higher than average flow from about 1790-1820 and 1890-1930, and below average flow since about 1930. No reconstructions of rainfall were presented.

24.3.3 New Zealand temperature (Nothofagus)

The first quantitative reconstruction of New Zealand palaeotemperature was based on ten *Nothofagus menziesii* and *N. solandri* chronologies from South Island (Figure 24.3; Norton *et al.* 1989). These chronologies (five of each species) were developed from trees growing at subalpine sites (Norton 1983a,b) where it was known that temperature had a strong influence on tree growth (Norton 1984a,b). Because many of the chronologies developed were short, the reconstruction did not extend prior to 1730.

Temperature records from seven New Zealand stations have been used to develop a regional New Zealand temperature series back to 1853 (Salinger 1980). The summer component (December-March) of this series was used in developing the temperature reconstruction. December to March are the months during which the majority of *Nothofagus* growth occurs (Norton 1984a) and are also the months in which growth is most strongly correlated with temperature (Norton 1984b).

The individual series of measured ring widths were standardized using the Gaussian filter method (Briffa 1984). In order to decide what degree of low-frequency variance should be removed from the raw ring-width data, the available instrumental temperature record was subject to a spectral analysis. This showed that about 17% of the variance of the temperature data was associated with variations longer then 30 years, 12% above 40 years, and about 10% above 60 years. Based on this, the tree-ring data were standardized using a filter designed to pass 25 per cent of the variance of a cycle with a period of 60 years. This effectively removed almost all variance associated with periods longer than 60 years while having little effect on shorter periods. Because little of the variance above 60 years is preserved in the standardized tree-ring chronologies, climate fluctuations on these time-scales were not reconstructed, but as the spectral analysis shows, this represented only a small part of the observed climate variance ($\approx 10\%$).

A principal components regression technique (Briffa *et al.* 1983, 1986; see also Bradley and Jones, Chapter 1, this volume) was used to relate the tree-ring chronologies with the temperature data. The temperature data set was divided into two equal periods for analysis, 1853-1915 and 1916-1979, with two regressions being developed for each period and verified on the other (independent) period. Results of the two regressions and their verifications are presented in Table 24.3 and Figure 24.4. The regression derived over the 1853-1915 period (model N1) explained 66% of the variance of the temperature data and in verification over the period 1916-1979, explained 34%. The periods were then reversed (model N2) giving a calibration value of 55% and a verification of 42%. The correlation coefficients for both verification periods were highly significant and one of the reduction of error (RE) values was positive. The regression weights showed that neither regression was dominated by one or two heavily weighted chronologies. The second transfer function model (N2) was chosen for reconstruction as it had a positive RE value and showed more consistent correlation coefficients during calibration and verification.

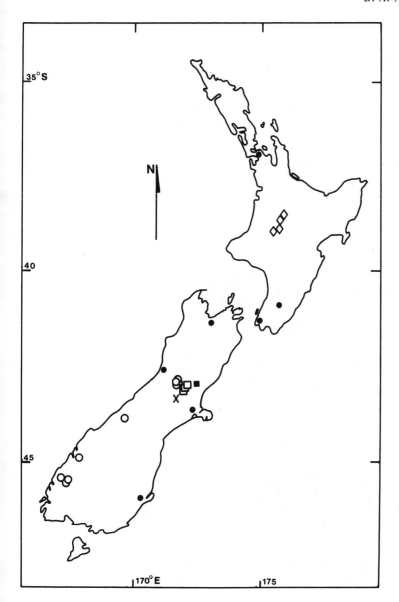

Figure 24.3 Location of tree-ring sampling sites, and temperature, rainfall and river-flow stations in New Zealand. Open circles, subalpine *Nothofagus* chronology sites. Triangles, *Phyllocladus* chronology sites. Open squares, rainfall/river flow *Nothofagus* chronology sites. Closed circles, temperature stations. Closed square, river flow gauging station. Cross, rainfall station.

The N2 regression equation was then applied to the tree-ring data for the period 1730-1852 and a reconstruction of summer temperature developed (Figure 24.4). Prior to about 1760 this reconstruction must be considered with some caution as the number of individual tree-ring time-series (radii) in some of the chronologies becomes small. Based on the SSS

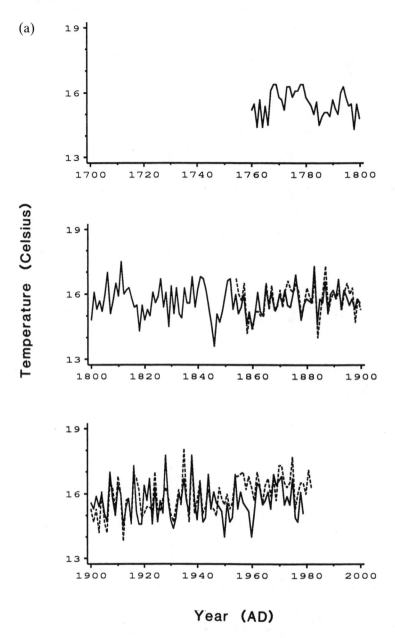

(a)

Temperature (Celsius)

Year (AD)

Figure 24.4(a)

criteria (Wigley *et al.* 1984) only six of the ten chronologies are reasonably reliable prior to 1760.

Variance spectra calculated over the common period 1853 to 1978 show a concentration of variance at periods above about 80 years, at about 20 years and for a range of values around 3 years for the instrumental data. The spectrum of the reconstruction shows similar relatively high variance at periods of 20 years, and around 3 years, and the expected lack of variance at periods of above about 60 years. Interestingly, it also shows concentrations of variance at periods of around 9.3 and 5 years which are not evident in the instrumental data. These results indicate that periodicities in the data up to about 30 years in length are reliably

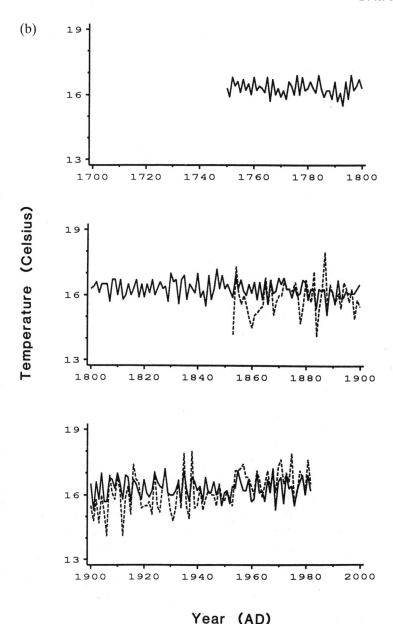

Figure 24.4 Reconstructed (solid line) and recorded (dashed line) temperature records from New Zealand [Nothofagus (a) and Phyllocladus (b)].

constructed. They also suggest that the reliability may be even greater at higher frequencies. Low common variance above cycles of 40 years is a consequence of the standardization technique we have used.

The reconstruction (Figure 24.4) was interpreted as showing that cooler summers were more common in the mid-1730s, 1760s, 1780s, and mid-1840s, while warmer summers

473

Table 24.3 Summary of New Zealand climate reconstruction models.

record	Calibration period explained	% variance	Verification period	% variance	RE
Temperature					
Nothofagus (N1)	1853-1915	65.6	1916-1979	33.6	-0.13
Nothofagus (N2)	1916-1979	54.8	1853-1915	42.3	0.06
Phyllocladus (P1)	1853-1917	46.9	1918-1982	24.6	0.14
Phyllocladus (P2)	1918-1982	32.0	1853-1917	32.1	0.28
Rainfall					
Coleridge	1913-1962	46.2	1963-1977	57.8	0.56
Riverflow					
Hurunui	1956-1977	51.8	-	-	-

appeared to have been more common in the early-1730s 1750s, 1770s, mid-1790s, 1805-1815, and the 1830s.

24.3.4   New Zealand temperature (Phyllocladus)

Unlike the more 'conventional' approach of selecting chronology sites where climate is likely to be an important factor limiting growth (i.e. environmentally limiting sites) the seven chronologies derived from *Phyllocladus trichomanoides* and two from *P. glaucus* were developed from closed canopy, relatively flat, mesic forest sites in the central North Island (Figure 24.3; Palmer 1989). As such, the sensitivity of the ring sequences and chronology variance may have been expected to be low. However, a high amount of common variance was contained within the chronologies, suggesting their suitability for climate modelling.

It has been hypothesized that trees growing on wet sites may be physiologically 'tuned' to wet conditions, and may therefore be more sensitive to, or limited by, dry conditions (Phipps 1982). However, response function analyses of the *Phyllocladus* chronologies showed only a clear association between ring width and temperature and not with precipitation. Climate reconstruction was therefore only attempted using temperature data. Regional temperature series for the central North Island were found to be very similar to the homogenized New Zealand temperature series (Salinger 1980) and the latter was used in the reconstructions. Transfer functions were developed using the principal components regression method described above.

Three transfer function models were tested using different months of temperature data, but only those using the 'summer' months of January, February and March produced significant correlations in calibration and positive RE values in verification (Palmer 1989). Results of the two regressions and their verifications are presented in Table 24.3 and Figure

24.4b. The regression derived over 1853-1917 (model P1) period explained 47% of the variance of the temperature data and in verification over the period 1918-1982, explained 25%. Reversing the periods (model P2) give a calibration value of 32% and a verification of 32%. The correlation coefficients for both verification periods were highly significant and both reduction of error (RE) values were positive. However, the drop in correlation from calibration to verification in the first regression (calibrated 1853-1917) may indicate some artificial predictability in the calibration regression. Because of this, the reciprocal model (P2) was chosen as the best to use for reconstructing summer temperatures since its correlation values remained almost identical over the two periods.

Summer temperature values were then reconstructed back to 1750 using the P2 regression equation (Figure 24.4b). Prior to 1750, the reliability of the chronology series becomes questionable as the number of individual tree-ring time-series in some chronologies becomes small (based on the SSS statistic of Wigley *et al*. 1984).

Cross-spectral analysis was used to assess how well the estimated summer temperatures portrayed the high and low frequency variation shown in the instrumental data over the common period (1853-1982). As expected (from standardizing with a 50-year Gaussian filter) the reconstructed series best modelled the high frequency variation. Differences were emphasised in a coherency diagram, where the most notable period with low common variance (after that above 50 years) was around 10 years and to a lesser extent near 4 years. The reason for this was not known but a factor might be the low (32%) amount of variance explained by the reconstructed series. However, one encouraging feature was that no periodicities occurred in the reconstructed series that were not already present in the observed summer temperature data.

Interpretation of the reconstructed temperature data was cautious because of the relatively low amount of variance explained in calibrating the transfer function models. However, the similarity between the *Phyllocladus* reconstruction and that developed by Norton *et al*. (1989) with *Nothofagus* was considered encouraging and is discussed in more detail below.

24.3.5 Canterbury precipitation and stream-flow

Four *Nothofagus solandri* tree-ring chronologies developed from trees growing at mid-altitude sites in Canterbury, on the east coast of South Island (Figure 24.3) were used to develop reconstructions of precipitation and stream-flow (Norton 1987). The chronologies were developed from trees growing in open stands on bluffs and talus slopes, and were found to be strongly correlated with growing season (November-January) precipitation (Norton 1987). However, because these sites were subject to frequent disturbance (rockfall) the chronologies were short, the longest extending back to A.D. 1787.

Chronology development used the same Gaussian filter standardization method described earlier. Although, the shortest chronology started in 1833, poor replication meant that the first reliable year that could be used for climate reconstruction (based on the SSS criteria of Wigley *et al*. 1984) was 1879. Precipitation data from Lake Coleridge, which lies in the same precipitation response area as the sampled sites, and stream-flow data for the Hurunui River, whose headwaters were adjacent to the study area, were used. The precipitation record spanned the period 1913-1977 and was divided into two for calibration (1913- 1962) and verification (1963-1977). The Hurunui stream-flow record was short (1956-1977) and it was not possible to retain an independent period for verification. Both reconstructions were for the summer (November-January) period.

Statistics describing the two regressions are presented in Table 24.3. The regression with precipitation explained 46% of the variance in calibration and 58% in verification, while the stream-flow regression explained 52% of the variation. As the stream-flow regression could not be directly verified, an independent test was used. A correlation coefficient of 0.85 was calculated between *observed* stream-flow and observed precipitation for the period 1956-1977 and a value of 0.62 between *estimated* stream-flow and observed precipitation for the period 1933-1955. The close agreement between these two values suggested that the stream-flow regression was valid.

Although the precipitation reconstruction is too short to identify any obvious trends with time, the stream-flow reconstruction shows a greater frequency and intensity of low flow events prior to the observed data (Figure 24.5).

24.4   Discussion

The transfer function models reviewed here have the potential to provide information on climate variation in Australasia over the last two hundred years. However, information for the full period back to A.D. 1500 is not presently available from dendroclimatic evidence (but see below). Of the six strictly climate models (3 Tasmanian temperature, *Nothofagus*-N2 temperature, *Phyllocladus*-P2 temperature, and Coleridge precipitation) variance explained in calibration ranges from 32% to 81% (mean of 57.7%) and in verification from 19% to 58% (mean of 37%). Four of the six models had positive RE values in verification. The riverflow reconstructions have a similar range of calibration variance explained (27-69%) although the general lack of verification suggests that these need to be interpreted cautiously. Longer observed records may alleviate this problem in the future.

The reconstructions presented are limited by the short length of some chronologies and by the small number of trees available during earlier centuries. In New Zealand, the reconstructions of temperature are not considered reliable prior to 1760 (*Nothofagus*) and 1750 (*Phyllocladus*) and for rainfall prior to 1879. The Tasmanian temperature reconstructions extend back to 1777, but no information is available on chronology reliability. At the present time, short chronology lengths and a lack of replication within chronologies is a serious limitation with the existing dendrochronological data base in Australasia.

The Tasmanian temperature reconstructions show considerably more low-frequency variation than the New Zealand temperature reconstructions. This is likely to be a result of differences in the standardization methods used. The New Zealand chronologies were standardized using either 50 or 60 year Gaussian filters, removing most lower-frequency variance from the series (Norton *et al.* 1989; Palmer 1989) while the Tasmanian chronologies were standardized using a mixture of curve fitting techniques including polynomial, negative exponential and straight lines (LaMarche *et al.* 1979a,b). With this latter approach there is no control over the amount of variance removed at different periodicities, with considerable low-frequency (and not necessarily climatic) variance remaining. Because of these uncertainties we do not interpret these reconstructions further here.

The two independently developed New Zealand reconstructions extend existing instru-

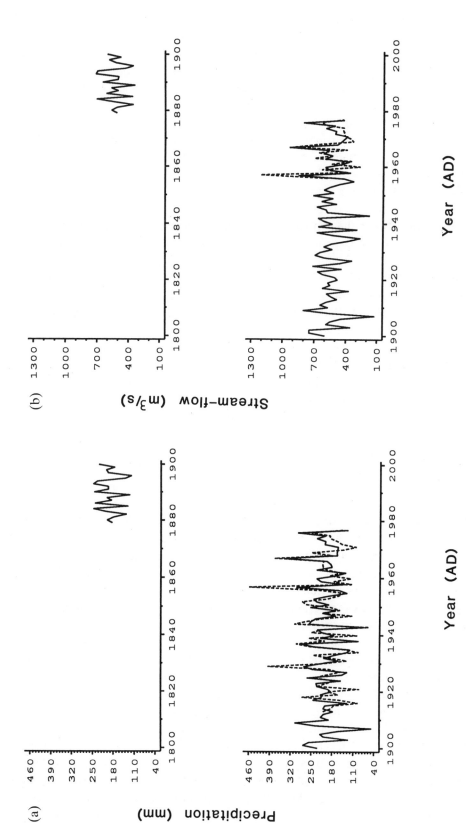

Figure 24.5 Reconstructed (solid line) and recorded (dashed line) rainfall (a) and river-flow (b) records from New Zealand.

mental records by 100 years. Although they differ in the temperature data used in their respective transfer function models (December-March in *Nothofagus* and January-March in *Phyllocladus*) correlation between the two reconstructions is significant (r=0.40, n=226, P<0.001) although only a low amount of variance is 'common' (16%). This correlation is encouraging, despite the low variance, and suggests that some confidence can be attached to the interpretation of these reconstructions. In part this low level of 'common' variance is not surprising given the diverse types of habitat the respective *Nothofagus* and *Phyllocladus* chronologies were derived from. The *Nothofagus* sites were all in South Island subalpine forests while the *Phyllocladus* sites were from lower altitude North Island forests.

Both New Zealand temperature reconstructions show fluctuating temperatures since 1750, although agreement between the two series is limited. However, both indicate that cool and warm periods usually do not last for more than about 10 years, but it would be unwise to over-interpret the precise magnitude or duration of these events.

The period since 1950 in New Zealand has been characterised by a mean temperature increase (Salinger 1982) occurring in two phases; an initial increase centred around 1955 and the second centred around 1971 (Salinger 1980). These are apparent in the observed summer temperature series shown in Figure 24.4. The first increase has been modelled by the *Phyllocladus* transfer function model (P2; Table 24.3) but the second increase has not been. The temperature model developed for *Nothofagus* (Table 24.3) shows the reverse situation. The inability to model both these periods of temperature increase have been accounted for by reference to synoptic climate conditions (Norton *et al.* 1989; Palmer 1989). The warming during the 1950s resulted from increased north to northwest airflow with maximum warming occurring in the eastern areas of New Zealand. Easterly airflow was more important during the second warming (centred on 1971) with increased cloud in eastern areas and a strong warming in the west. Both regional air-flow anomalies resulted in an increase in temperature but with different spatial emphasis. The warming associated with eastern areas seems to have been better modelled by the *Phyllocladus* chronologies from the central North Island, while the warming associated with western areas appears to have been better modelled by the *Nothofagus* chronologies from the western South Island mountains.

Comparisons of the temperature reconstructions presented here with other proxy palaeo-climate data allow some general patterns to be deduced. Lack of dating control or coarse temporal resolution makes comparison difficult in many cases (e.g. with palynological data) and only two sources of proxy temperature records appear comparable; data on the expansion of glaciers and records of the movements of icebergs into low latitudes. Both are said to respond to decadal or shorter time scale temperature fluctuations (Burrows & Greenland, 1979). Evidence for glacier expansion comes mainly from the central Southern Alps and is based on minimum moraine ages derived from lichen and rock weathering rind measurements and tree age estimates (Wardle 1973; Birkeland 1982; Burrows 1982; Gellatly 1985). The evidence suggests that glacier expansion occurred during the mid-18th century and mid-19th century. However, dating uncertainties (Burrows 1982) and differences in the response-time of individual glaciers to specific climatic events (Gellatly & Norton 1984) make comparison with the tree-ring reconstructions difficult. The mid-19th century period of glacial expansion does not directly follow any obvious period of cool summers although glacier advance need not necessarily reflect summer temperature. However, the pronounced cool period in the 1780s and early 1790s is evident in both the *Nothofagus* and *Phyllocladus* reconstructions (Figure 24.4) and may have initiated this advance. The cooler conditions initiating the mid-18th century advance may pre-date the tree-ring reconstructions. Records of iceberg sightings in low latitudes (Burrows 1976) suggest that unusually high numbers of icebergs were present in the early 1770s and early 1850s. Both these periods follow decades with apparently cooler summers in the *Nothofagus* reconstruction (Figure 24.4a) and may reflect similar synoptic situations. However, other decades with cool summers (e.g., 1780s) are not associated with iceberg sightings, although this may reflect a lack of ships in the southern oceans at this time (Burrows 1976).

What is clear from these comparisons is that the tree-ring reconstructions allow a far more detailed interpretation of palaeotemperatures than do other proxy evidence. They show a more complicated sequence of palaeotemperature variations than have previously been inferred and would suggest that interpretations based on evidence such as glacial expansion may underestimate the complexity of palaeoclimate variation. The period since 1750 has been characterised by fluctuating summer temperatures, with cool and warm periods usually lasting no more than about 10 years. Though it would be unwise to over-interpret the precise magnitude or duration of these events, the relative coolness of the 1780s to early 1790s and late 1840s and warmth of the 1770s to 1780 appear to be notable features of the reconstruction.

An attempt to reconstruct the Southern Oscillation using 33 Southern Hemisphere tree-ring chronologies, including some from Australasia, was unsuccessful (Lough & Fritts 1985) although they were able to reconstruct the Southern Oscillation using western North American chronologies. Although some variance was explained using the Southern Hemisphere chronologies, verification was unsuccessful. It has been suggested (Norton 1990) that this may be a result of an inadequate knowledge of the ecology and climatic response of some of the species used.

We believe that dendroclimatic techniques have considerable potential for palaeoclimatic reconstruction in Australasia, particularly because of the close location to major climatic phenomena such as the Southern Oscillation. A number of long-lived tree species are present including both conifers and angiosperms, although more information on the ecology and life-histories of some of these trees is required (Norton 1990). A systematic chronology development program is urgently needed to develop a network of long tree-ring chronologies from species with known climatic response. This program should draw on the very promising early work described in this chapter and has the potential to result in the development of high quality reconstructions of palaeoclimates over the last 500 years. This type of project clearly requires considerable collaboration between Australian and New Zealand scientists, and between plant ecologists and climatologists. Detailed reconstructions that will provide us with information on climatic variations in the recent past are unlikely to be developed until a coordinated project of this type is undertaken. Expansion to include South American chronologies (see Boninsegna, Chapter 23, this volume) could also be of considerable value for reconstructing larger scale climate patterns.

Acknowledgements

We would like to acknowledge the continued assistance of John Ogden with much of the work described here.

References

Ahmed, M. 1984. *Ecological and dendrochronological studies on Agathis australis Salisb. (kauri).* Unpublished Ph.D. thesis, University of Auckland.

Ahmed, M. & Ogden, J. 1985. Modern New Zealand tree-ring chronologies 3. *Agathis australis* (Salisb.) – kauri. *Tree-Ring Bulletin* 45, 11-24.

Ash, J. 1983. Note on paper 'Growth rings and rainfall correlations in a mangrove tree of the genus *Diospyros* (Ebenaceae)', by N.C. Duke *et al. Australian Journal of Botany* 31, 19-22.

Birkeland, P. W. 1982. Subdivision of Holocene glacial deposits, Ben Ohau Range, New Zealand, using relative dating methods. *Geological Society of America Bulletin* 93, 433-449.

Bridge, M. C. & Ogden, J. 1986. A sub-fossil kauri (*Agathis australis*) tree-ring chronology. *Journal of the Royal Society of New Zealand* 16, 17-23.

Briffa, K. R. 1984. *Tree-climate relationships and dendroclimatological reconstructions in the British Isles.* Unpublished Ph.D. thesis, University of East Anglia.

Briffa, K. R., Jones, P. D., Wigley, T. M. L., Pilcher, J. R. & Baillie, M. G. L. 1983. Climate reconstructions from tree-rings. Part 1. basic methodology and preliminary results for England. *Journal of Climatology* 3, 233-242.

Briffa, K. R., Jones, P. D., Wigley, T. M. L., Pilcher, J. R. & Baillie, M. G. L. 1986. Climate reconstructions from tree-rings. Part 2. spatial reconstruction of summer mean sea-level pressure patterns over Great Britain. *Journal of Climatology* 6, 1-15.

Burrows, C. J. 1976. Icebergs in the Southern Ocean. *New Zealand Geographer* 32, 127-138.

Burrows, C. J. 1982. On New Zealand climate within the last 1000 years. *New Zealand Journal of Archaeology* 4, 157-167.

Burrows, C. J. & Greenland, D. E. 1979. An analysis of the evidence for climatic change in New Zealand in the last thousand years; evidence from diverse natural phenomena and from instrumental records. *Journal of the Royal Society of New Zealand* 9, 321-373.

Campbell, D. A. 1982. Preliminary estimates of summer streamflow for Tasmania. In Hughes, M. K., Kelly, P. M., Pilcher, J. R. & LaMarche V. C., Jr. (eds) *Climate from Tree Rings.* Cambridge University Press, Cambridge. pp 170-177.

Duke, N. C., Birch, W. R. & Williams, W. T. 1981. Growth rings and rainfall correlations in a mangrove tree of the genus *Diospyros* (Ebenaceae). *Australian Journal of Botany* 29, 135-142.

Dunwiddie, P. W. 1979. Dendrochronological studies of indigenous New Zealand trees. *New Zealand Journal of Botany* 17, 251-266.

Dunwiddie, P. W. & LaMarche, V. C., Jr. 1980. Dendrochronological characteristics of some native Australian trees. *Australian Forestry* 43, 124-135.

Francey, R. J. & 18 co-authors. 1984. Isotopes and Tree Rings. *Technical Paper* 4. Commonwealth Scientific and Industrial Research Organisation, Australia. Division of Atmospheric Research

Fritts, H. C. 1976. *Tree-Rings and Climate.* Academic Press, New York.

Fritts, H. C., Lofgren, G. R. & Gordon, G. A. 1979. Variations in climate since 1602 as reconstructed from tree-rings. *Quaternary Research* 12, 18-46.

Gellatly, A. F. 1985. Glacial fluctuations in the central Southern Alps New Zealand: documentation and implications for environmental change during the last 1000 years. *Z. fur Gletscherkunde und Glazialgeol* 21, 259-264.

Gellatly, A. F. & Norton, D. A. 1984. Possible warming and glacial recession in the South Island, New Zealand. *New Zealand Journal of Science* 27, 381-388.

Hughes, M. K., Kelly, P. M., Pilcher, J. R. & LaMarche V. C., Jr. (eds). 1982. *Climate from Tree Rings.* Cambridge University Press, Cambridge.

LaMarche, V. C., Jr., Holmes, R. L., Dunwiddie, P. W. & Drew, L. G. 1979a. Tree-ring chronologies of the Southern Hemisphere. 3, New Zealand. *Chronology Series V*. Laboratory of Tree-Ring Research, University of Arizona, Tucson.

LaMarche, V. C., Jr., Holmes, R. L., Dunwiddie, P. W. & Drew, L. G. 1979b. Tree-ring chronologies of the Southern Hemisphere. 4, Australia. *Chronology Series V*. Laboratory of Tree-Ring Research, University of Arizona, Tucson.

LaMarche, V. C., Jr. & Pittock, A. B. 1982. Preliminary temperature reconstructions for Tasmania. In, Hughes, M. K., Kelly, P. M., Pilcher, J. R. and LaMarche V. C., Jr. (eds) *Climate from Tree Rings*. Cambridge University Press, Cambridge. pp 177-185.

Lough, J. M. & Fritts, H. C. 1985. The Southern Oscillation and tree rings: 1600-1961. *Journal of Climate and Applied Meteorology* 24, 952-966.

McPhail, S., Barbbetti, M., Francey, R., Bird, T. & Dolezal, J. 1983. 14C variations from Tasmanian trees – preliminary results. *Radiocarbon* 25, 797-802.

Norton, D. A. 1983a. Modern New Zealand tree-ring chronologies. I. *Nothofagus solandri. Tree-Ring Bulletin* 43, 1-17.

Norton, D. A. 1983b. Modern New Zealand tree-ring chronologies. II. *Nothofagus menziesii. Tree-Ring Bulletin* 43, 39-49.

Norton, D. A. 1983c. *A dendroclimatic analysis of three indigenous tree species, South Island, New Zealand*. Unpublished Ph.D. thesis, University of Canterbury.

Norton, D. A. 1984a. Phenological growth characteristics of *Nothofagus solandri* trees at three altitudes in the Craigieburn Range, New Zealand. *New Zealand Journal of Botany* 22, 413-424.

Norton, D. A. 1984b. Tree-growth-climate relationships in subalpine *Nothofagus* forests, South Island, New Zealand. *New Zealand Journal of Botany* 22, 471-481.

Norton, D. A. 1985. A dendrochronological study of *Nothofagus solandri* tree growth along an elevational gradient, South Island, New Zealand. *Eidgenossische Anstalt fur das Forstliche Versuchswesen*, Berichte 270, 159-171.

Norton, D. A. 1987. Reconstructions of past river flow and precipitation in Canterbury, New Zealand, from analysis of tree-rings. *Journal of Hydrology (New Zealand)* 26, 161-174.

Norton, D. A. 1990. Dendrochronology in the Southern Hemisphere dendrochronology. In, Cook, E. and Kairiukstis, L. (eds) *Methods of Dendrochronology: Applications in the Environmental Sciences*. Kluwer, Dordrecht, 17-21.

Norton, D. A., Briffa, K. R. & Salinger, M. J. 1989. Reconstruction of New Zealand summer temperature to 1730 AD using dendroclimatic techniques. *International Journal of Climatology* 9, 633-644.

Norton, D. A. & Ogden, J. 1987. Dendrochronology: A review with emphasis on New Zealand applications. *New Zealand Journal of Ecology* 10, 77-95.

Ogden, J. 1978a. On the dendrochronological potential of Australian trees. *Australian Journal of Ecology* 3, 339-356.

Ogden, J. 1978b. Investigations of the dendrochronology of the genus *Arthrotaxis* D. Don. (Taxodiaceae) in Tasmania. *Tree-Ring Bulletin* 38, 1-13.

Ogden, J. 1981. Dendrochronological studies and the determination of tree ages in the Australian tropics. *Journal of Biogeography* 8, 405-420.

Ogden, J. 1982. Australasia. In, Hughes, M. K., Kelly, P. M., Pilcher, J. R. and LaMarche V. C., Jr. (eds) *Climate from Tree Rings*. Cambridge University Press, Cambridge. pp 90-103.

Palmer, J. G. 1982. *A dendrochronological study of kauri (Agathis australis)*. Unpublished M.Sc. thesis, University of Auckland.

Palmer, J. G. 1989. *A dendroclimatic study of Phyllocladus trichomanoides D. Don (tanekaha)*. Unpublished Ph.D. thesis, University of Auckland.

481

Phipps, R. L. 1982. Comments on interpretation of climatic information from tree rings, eastern North America. *Tree-Ring Bulletin* 42, 11-22.

Salinger, M. J. 1977. Dunedin temperatures since 1853. *New Zealand Geographical Society, Proceedings of the Ninth Geography Conference 1977*, 106-109.
Salinger, M. J. 1980. The New Zealand temperature series. *Climate Monitor* 9, 112-118.
Salinger, M. J. 1982. On the suggestion of post-1950 warming over New Zealand. *New Zealand Journal of Science*, 25, 77-86.
Schweingruber, F. H. 1983. *Der Jahring. Standort, Methodik, Zeit und Klima in der Dendrochronologie*. Paul Haupt, Bern.
Scott, D. 1972. Correlation between tree-ring width and climate in two areas of New Zealand. *Journal of the Royal Society of New Zealand* 2, 545-560.

Wardle, P. 1973. Variations of the glaciers of Westland National Park and the Hooker Range, New Zealand. *New Zealand Journal of Botany* 11, 349-388.
Wells, J. A. 1972. Ecology of *Podocarpus hallii* in Cantral Otago, New Zealand. *New Zealand Journal of Botany* 10, 399-426.
Wigley, T. M. L., Briffa, K. R. & Jones, P. D. 1984. On the average value of correlated time series, with applications in dendroclimatology and hydrometeorology. *Journal of Climate and Applied Meteorology* 23, 201-213.

Section C: ICE CORE EVIDENCE

25 Ice core climate signals from Mount Logan, Yukon A.D. 1700-1897

G. Holdsworth, H. R. Krouse and M. Nosal

25.1 Introduction

The successful interpretation of high altitude mountain firn and ice cores as a source of paleoclimatic data depends very strongly on a knowledge of (1) the physical and meteorological characteristics of the site, (2) the history of the water vapor-condensate arriving at that location, and (3) the history of subsequent processes of evaporation, sublimation and snow scour occurring there. Some specific core site problems are discussed by Dansgaard (1964) Gow (1965) Giovinetto and Schwerdtfeger (1966) Dansgaard *et al.* (1973) Moser and Stichler (1974) Reeh *et al.* (1978) Whillans (1978) Fisher *et al.* (1983) and Grootes *et al.* (1989). Although these are mostly discussed in the context of ice sheets or caps, all are applicable to high mountain sites. Several processes may accordingly influence the quality and amount of information that is stored in an ice core.

Two main climate variables that may be extracted from ice core information are: (a) proxy temperature and (b) net accumulation of solid precipitation and drift at the site. A third parameter, the relative humidity of the air at the water vapor source, has been studied by Jouzel *et al.* (1982) but will not be discussed here because of insufficient hydrogen isotope data and because of the shortness of the record. Possible data distortion due to relative isotope changes in the snowpack (as discussed, for example, by Moser and Stichler 1974) are unknown, but believed to be relatively small.

Firstly, it is assumed that proxy temperature information is contained in the stable isotope ($\delta^{18}O$) time series. However, because of the influence of other factors on the $\delta^{18}O$ of precipitation, such as the shifting of storm track trajectories, (Hage *et al.* 1975) or due to large changes in vapor source meteorology (Covey and Haagenson 1984) it can be very difficult to extract the pure temperature signal. The $\delta^{18}O$ time series represent a form of proxy climate data but, in general, a temperature scale cannot be quantitatively attached to them. However, some ice core $\delta^{18}O$ data series seem to contain a coherent temperature signal which dominates over other signals (e.g. Thompson *et al.* 1986) whereas others, such as the one discussed here, do not.

Secondly, net accumulation data acquired from ice cores by direct measurement, do not, by definition, represent proxy data. This may appear to be an inferior method of obtaining precipitation data when compared to precipitation recording devices, but the efficiencies of these and their corresponding errors vary greatly both with the type of device and the local conditions (Sevruk 1986). In principle, the ice core data compare in quality to snow ruler data, except that the former represent *net* accumulation rather than total precipitation. It is important to realize that although the total snow precipitation at the core site is not retrievable from the core data, the relative changes in net accumulation can be shown to be of great climatic significance.

For regional hydrological studies, long term precipitation/accumulation time series are most relevant. In the case of Mt. Logan, it has been found that cross correlations with local station precipitation series from low altitudes, do not yield encouraging results, although some significant negative correlations are found. In contrast to this, we find some significant correlations with data from distant regions. These results belong to a class of phenomena known as teleconnections (Bjerknes 1969).

25.2   Data sources

25.2.1   Ice core analyses, data reduction and corrections

In 1980, an electromechanical drill (Holdsworth 1984) was used on Mt. Logan to retrieve a 103m firn and ice core sequence from a site on a saddle (60°35′N, 140°35′W; 5340m a.s.l) where the mean annual firn temperature at 10m depth is close to -29°C. Further pit sampling in 1986 and 1988 over the borehole site has enabled the earlier core results to be updated to 1987.

Several measurements have been performed on the core meltwater. Where these do not extend over the full length of the core they have been labelled *incomplete*:

1   stable isotopes of oxygen and hydrogen ($\delta^{18}O$; δD incomplete)
2   liquid electrolytic conductivity
3   pH (incomplete)
4   Total ß-activity (1943-1980) (Holdsworth *et al.* 1984)
5   $[NO_3^-],[SO_4^=],[Cl^-],[F^-]$ (incomplete except $[NO_3^-]$)
6   $[Na^+]$, $[K^+]$, $[Ca^+]$, $[Mg^{++}]$ (incomplete)

Chemical analyses, particularly for the major (acid) anions were preferentially performed across intervals in which elevated conductivity and depressed pH values occurred. These were initially assumed to be of volcanic origin. Specific ion analyses were then performed on samples of the melt water. When $[SO_4^=]$ was dominant, or in the few cases when $[Cl^-]$ values were of the same magnitude as $[SO_4^=]$ (as in the case of some island volcanoes) (Symonds etal. 1988) the compendium *Volcanoes of the World* (Simkin etal. 1981) was consulted. To help select the most likely volcanic event responsible for the chemical signature we used as a guide only, an approximate time scale based on a simplified two dimensional ice flow model. We also took into account the size of the eruption, the size, shape and phasing of the chemical signatures and the probable air mass trajectory. When the resulting sequence of volcanic events matched a monotonically decreasing slope of the depth-time curve (Figure 25.1) then this established confidence in the result. The resulting depth-time relationship is shown in Figure 25.1, which allows the computation of a steady state vertical strain rate profile.

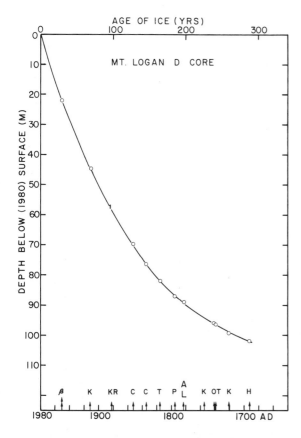

Figure 25.1 Depth versus time relation for the Mt. Logan core referenced to 1980. The double arrow (↕) marks the end of the total β-activity dating and the single arrows associated with a letter represent volcanic events identified by a chemical signature. K: Katmai 1912; KR: Krakatau 1884; C: Chikurachki 1853; C: Cosiquina 1835; T: Tambora 1816; P: Pogromni 1795; A,L: Asama, Laki 1783; K: Katla 1755; O: Oshima 1741; T: Tarumai 1739; K: Katla 1721; H: Hekla 1693.

Because this resembles theoretical profiles, we have available another (qualitative) check on the integrity of the overall time scale. Finally, the exercise is satisfactorily completed if the right number of annual increments can be fitted in between the volcanic events.

Annual increments have been determined by a number of methods: (a) Total ß-activity variations, (b) $\delta^{18}O$ oscillations and (c) nitrate concentration variations. In some sections, limited sodium ion concentration variations were also used. These latter methods replace the use of seasonal $\delta^{18}O$ variations, which cannot be unambiguously recognized below about 50m due to diffusion effects in the lower part of the core (Holdsworth *et al.* 1989). Two sections (A.D. 1693-1720 and A.D. 1729-1735) have not had annual increments assigned to them because of the lack of a reliable seasonal indicator. Thus, only mean net accumulation values for these sections, flanked by volcanic events, have been given. In order to convert observed annual increments, determined along the core, to water equivalent values that apply at the surface, it is necessary to perform some computations, already outlined in Holdsworth *et al.* (1989). The depth-density profile (Figure 25.2) shows the firn-ice transition to be at about 65m depth. This occurs at a density of 825 ± 5 kg cm^{-3} which practically defines the density at which all voids are completely sealed.

In principle, the conversion of observed annual increments in the core to equivalent surface values could be done by a numerical model employing the finite element method (FEM). Such a model would have to account for: (a) three dimensional geometry, (b) compressible firn to 65m depth and thereafter slowly compressible to 'incompressible' ice, (c)

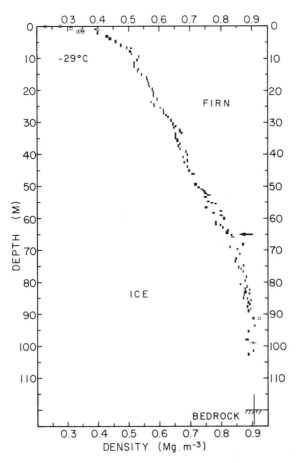

Figure 25.2 Depth-density curve for the 1980 core. Open symbols represent field measurements, filled symbols are laboratory measurements. The shape indicates the size of the sample and the error on the determination of density. The horizontal arrow at 65m depth marks the firn-ice transition. Bedrock is at 120±5m as determined by radar measurements at frequencies of 5 and 640 MHz.

an accumulation gradient across the col and (d) a time varying accumulation rate. Preliminary numerical experiments indicate that the model would also require more extensively known boundary geometry, ice property and ice flow information than we have at present. For these reasons our present approach to reducing the data has had to be over-simplified, semi-analytical and applied in discrete steps. The steps are as follows :

(i) Down to the firn-ice transition, the water equivalent (b_n) of an annual increment was determined from:

$$b_n(z) = a_n(z).\rho_n(z) \tag{1}$$

where n is an annual increment number, a_n is an individual annual increment, measured parallel to the core axis, ρ_n is the corresponding mean density of that increment and z is depth, measured along a vertical axis, originating at the top of the 1980 borehole.

(ii) Below the firn-ice transition (z >65m) a correction for layer creep thinning is applied. This is based on the assumption that ice flow processes are beginning to be significant once grain compaction processes have diminished and crystal fabric is starting to develop as shear stresses increase.

The vertical strain rate $\dot{\epsilon}_z(z)$ was computed numerically from the depth-time curve as d/dz(dz/dt). We have only the vertical strain rate distribution in one vertical column. However, between the firn-ice transition and the 300 year isochron, the vertical strain rate in upstream columns, to a horizontal distance of 50m should not be substantially different (Holdsworth *et al.* 1989, Figure 25.3, and FEM simulation). Therefore, we have modelled the creep thinning process as if ice flow was directed vertically down the borehole.

Denoting the firn-ice transition annual layer thickness by $a_n(O^*)$ and the annual layer thickness at z^* below the firn-ice transition by $a_n(z^*)$ it can be shown (Holdsworth *et al.* 1989) that

$$a_n(O^*)/a_n(z^*) = \{\exp[_0\!\int^t \dot{\epsilon}_z(z^*)dt]\}^{-1} \tag{2}$$

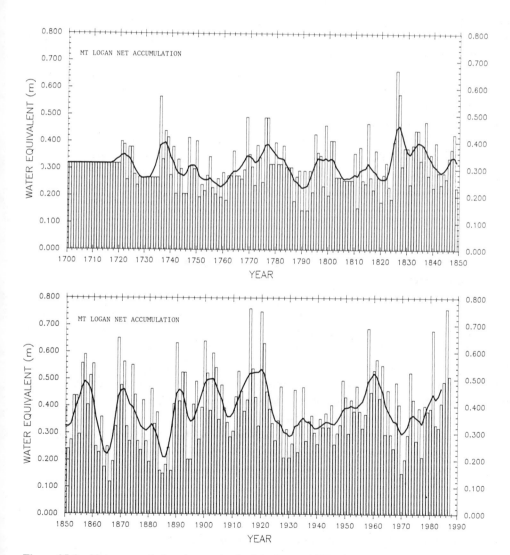

Figure 25.3 Net accumulation time series for Mt. Logan 1700-1987. Individual annual values represented as bars. Heavy line represents smoothing of data using a 7 year triangular filter.

where t is time and $65 < z^* < 102$ m. Thus, a correction, given by (2) with dimensionless values lying between one and five is applied to the $a_n(z^*)$ values. Then, the water equivalent value b_n for the layer below the firn-ice transition is calculated from

$$b_n = a_n(O^*).\rho(O^*) \tag{3}$$

where $\rho(O^*)$ is the density at the firn-ice transition (825 ± 5 kgm^{-3}).

(iii) The next correction to be applied is for the tilt of the isochrons, which is indicated by the progressive increase, with depth, of the inclination of thin (≤ 0.3 mm) glaze crusts which are assumed to have a common origin on the lee side of surface sastrugi, as presently observed. These crusts, evidently formed by exposure of the lee-side of sastrugi to wind and solar radiation, were observed in 1986 to have a mean slope of about 2°. The isochrons, in the vicinity of the borehole, should have a slope of $0° \pm 0.5°$ at the crest of the col to about $1.6° \pm 0.5°$ from where the lowest ice in the core is believed to be derived. The borehole tilt was measured by inclinometry in 1980 and found to be $<2°$ and to average $1.1°$, over a 96m depth range. The measured inclination of glaze crusts to a plane perpendicular to the core axis varied from $2° \pm 2°$ to $31° \pm 4°$ at 102m depth. Because there is no azimuth control on either the borehole tilt data or the glaze crust slopes, no correction for deviation of the borehole from the vertical can be applied. By subtracting 2° from all the measured glaze crust slopes we consider that the resulting $\alpha_n(z)$ series represents the slope of isochrons with depth to within ± 2 degrees. This obviates the need for making a rather trivial correction for the variation of isochron and glaze crust slopes along the surface away from the borehole site.

Thus the correction to previously computed $b_n(z)$ values is obtained from:

$$b_n'(z) = b_n(z) \cos \alpha_n(z) \tag{4}$$

where the error due to the uncertainty in α_n ranges from negligible at the surface to ± 0.013m at the base of the core.

(iv) A final correction must be applied to allow for the fact that firn and ice occurring in the core have different origins at the surface and hence originate from points of different accumulation rates. In this case, a knowledge of the flow-lines in a vertical flow plane through the borehole is required. This is best accomplished by simple 2-D finite element model simulation because of the strong relief in basal topography. The surface velocity data had shown that a displacement of nearly 90m exists between the topographic divide, where the borehole was situated, and the flow divide. The surface net accumulation gradient over this distance is known only for the period 1975-80.

If $a_n(x^*,0)$ is the value at distance x^* 'upstream' from the top of the borehole, (see Holdsworth *et al* 1989) then at the point $(0,z)$ where the flowline starting at $(x^*,0)$ intersects the borehole, the correction factor $C(z)$ to be applied to the "observed" $b_n(z)$ is:

$$C(z) = a_n(0,0) /a_n(x^*,0) \tag{5}$$

$C(z)$ was found to range between 1.00 and 1.176, and is applied as:

$$b_n''(z) = b_n'(z).C(z). \tag{6}$$

488

(v) A final step is the conversion from the depthseries $b_n''(z)$ to the time series $b_n''(t)$ using the experimentally derived relation $z=z(t)$ shown graphically in Figure 25.1. Figure 25.3 shows this series together with one in which the annual values have been smoothed by a seven year running triangular filter. The undated section prior to A.D. 1693 is estimated to be about 5 years long and is not used here. The $b_n''(t)$ series is truncated at A.D. 1700.

25.2.2   Data precision

The error for individual $b_n''(z)$ values is variable as a result of the different techniques used to delineate annual increments $a_n(z)$ and the depth dependent corrections that are subsequently applied to the data. A maximum estimated error on the mean $b_n''(0.35m)$ is $\pm 0.016m$ (varying between $\pm 0.005m$ at the surface to $\pm 0.025m$ at the base of the core. The standard deviation of the 250 year series is 0.12 meters.

25.2.3 $\delta^{18}O$ time series

The $\delta^{18}O$-depth series shown as Figure 25.1 in Holdsworth *et al*. (1989) has been converted to a time series (Figure 25.4) using the same annual depth increment identification derived previously. The number of $\delta^{18}O$ values per annual increment from A.D. 1860-1987 varied from 5 to 17 with a mean of 9.7. From A.D. 1736-1859 the corresponding values are 3 to 13 with a mean of 7.0. Although the details cannot be discussed here, we suspect that this series contains at least two major signals, one related to temperature changes and one related to water vapor (and thus air mass) trajectory changes. The maximum analytical error on individual $\delta^{18}O$ values is $\pm 0.3‰$. Because we do not yet have a knowledge of the seasonal distribution of snowfall at the site, the mean annual $\delta^{18}O$ value will probably be biased towards the months of July through December, which is the period of maximum precipitation at Yakutat, Alaska. Also, inter-annual changes in wind characteristics causing non-uniform removal or isotopic modification of precipitation and of deposited snow will build in other 'climatic signals', which must necessarily be regarded here as noise.

25.3   Calibration procedures

25.3.1   Net accumulation time series

Because the net accumulation time scale is established directly, there is no essential difference, in principle, between the method of determining accumulation amount in the core and that of using snow 'ruler' depth measurements made at some instrumental stations. Corrections that have to be applied to the core data are often not exact and may not completely restore the information to its proper condition.

25.3.2 $\delta^{18}O$ Time series

This series contains true proxy data, but so far, no satisfactory method has been devised to decompose the series into its component signals (*viz*. temperature, air mass trajectory and

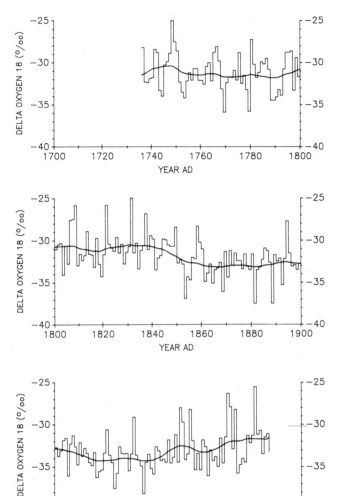

Figure 25.4 Mean annual $\delta^{18}O$ data, using the annual increments as employed in constructing Figure 25.3.

wind scour changes). Some attempts have been made to search for coherence between this time series and dendrochronologies in which temperature signals are believed to exist. So far, only sporadic in-phase correlations have been found between the mean annual $\delta^{18}O$ series and one northern Yukon tree-ring chronology (Jacoby and Cook 1981). Except for some sections, no significant correlation exists between Alaskan and Yukon instrumental temperature series and the $\delta^{18}O$ series, suggesting that other types of strong climate signals occur in the latter. Hence, no calibration has been carried out on the series.

25.4 Verification procedures

25.4.1 Cross-correlation of time series with instrumental data

Net accumulation time series We have carried out cross-correlations between the series shown in Figure 25.3 (1880-1987) and instrumental precipitation time series, both local and distant. We find that a number of significant long distance correlations exist. These 'tele-connections' evidently involve large scale air mass dynamics (Wallace 1985) especially over ocean regions. These results are too extensive to discuss further here and will be the subject of a separate publication. They are mentioned because they provide one important form of semi-quantitative verification of the snow accumulation time series. Figure 25.5 shows the smoothed time series of net accumulation for Mt. Logan from 1890 to 1987 and the smoothed time series of mean total precipitation for five stations in Japan for the same period. The Pearson product moment cross-correlation coefficient, r, is +0.6, significant at the 95% level. Although other significant correlations do exist, this is the best correlation established using the whole northern hemisphere precipitation data set (Bradley *et al*. 1987). The reasons for this 'teleconnection' may be related to dynamical aspects of north Pacific ocean-atmosphere interactions, discussed further in section 25.5.

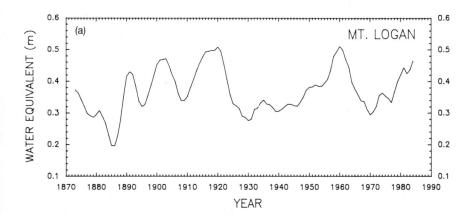

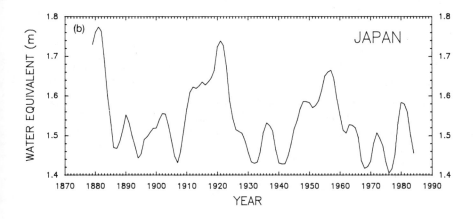

Figure 25.5 Time series of (a) Net accumulation for Mt. Logan (1870-1987). Annual data smoothed by a 7 yr triangular filter. (b) Total precipitation averaged for 5 stations in Japan (Nagasaki, Tokyo, Miyako, Akita, Nemuro). Annual data smoothed by a 7 yr triangular filter.

491

$\delta^{18}O$ time series In section 25.3.2, we discussed the reasons for being unable to assign any single climate parameter scale to this series because it evidently contains multiple climate signals. In section 25.5, a qualitative form of verification of the series is attempted using early U.S. meteorological data and non-instrumental accounts of weather during extreme events.

25.4.2   Time series analyses

Two tests were applied to both time series to determine if they were significantly different from white noise. These were the Bartlett's Kolmogorov-Smirnov Statistic and Fisher's Kappa test (Fuller 1976). Both series, according to these tests, are significantly different from white noise at the 99% level.

In addition, the coherence $(0 \leq C(\omega) \leq 1)$ between the two series was computed. Significant values of $C(\omega)$ occurred at several frequencies corresponding to periods of about 2.6 years (0.62) and 3.0 years (0.48) for the full series and of about 3.0 yr (0.56) for the first half and 2.6 yr (0.53) for the last half. Although both series individually show strong power between 8.5 and 12.5 years (thus containing the solar cycle) $C(\omega)$ only reaches 0.43 at 8.5 yr for the full series.

Net accumulation time series Power spectral analyses have been performed on the annual data series from 1736-1987 using the Fourier Transform, the Fast Fourier Transform and the Maximum Entropy Methods. Figure 25.6 shows power spectra derived by the FFT and MEM techniques. All three methods yielded power spectra with prominent spectral peaks at ~3.8 years, ~11 years and ~21 years. Some power is also seen at ~15 years for the full series and for a partitioned series running from 1736-1860. Whatever series partitioning was used, the ~3.8 year and ~11 year spectral peaks persisted and are therefore assumed to be stable. These correspond very closely with the mean El Niño recurrence interval (all severities) (Quinn *et al.* 1987) and with the mean sunspot cycle, respectively. Power at ~36 years was found in the spectrum of the first sub-series (1736-1860) but not elsewhere. The existence of physically significant spectral peaks in the time series implies a degree of 'robustness' in the data (see section 25.5).

$\delta^{18}O$ time series Power spectral analyses have been performed on this series. The spectrum shows significant power between 9.5 and 12.5 years, or essentially bracketing the periods of the sun-spot cycle. No significant power occurs at the 'preferred' El Niño-Southern Oscillation (ENSO) frequency but power is indicated at certain frequencies, some of which correspond approximately to the Quasi-Biennial Oscillation (QBO) and ENSO periods (2.1 ± 0.1 yr, 3.0 ± 0.1 yr, 3.5 ± 0.1 yr and 4.6 ± 0.1 yr). The FFT spectrum is shown in Figure 25.7.

25.5   Results and relationships to other studies

25.5.1   Overview of results

The net accumulation time series has the potential of being a useful source of climatic and hydrologic information. Evidently, any changes in wind regime at this site have not been

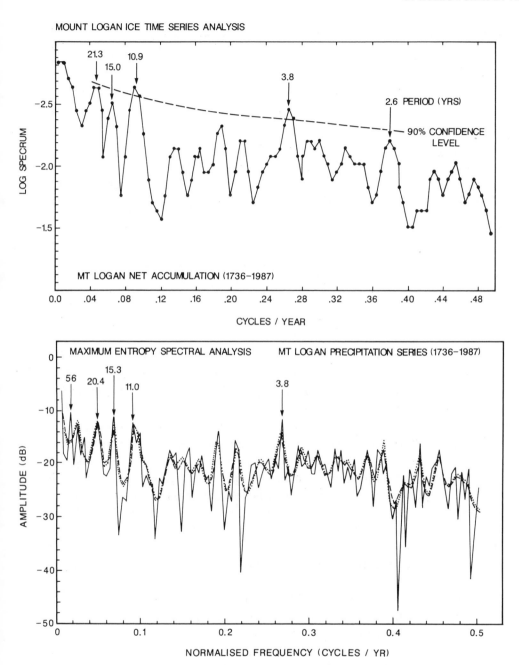

Figure 25.6   Power spectra for net accumulation time series (a) Fast Fourier Transform and (b) Maximum entropy method (Courtesy of S. Nichols). Periods corresponding to spectral peaks are indicated. Note that both ordinates represent logarithmic scales.

large enough to cause irregular snow scour with time and thus prevent significant correlation between this series and several precipitation time series derived by instrumental methods. Even if the Mt. Logan precipitation regime is biased strongly towards the fall-winter months

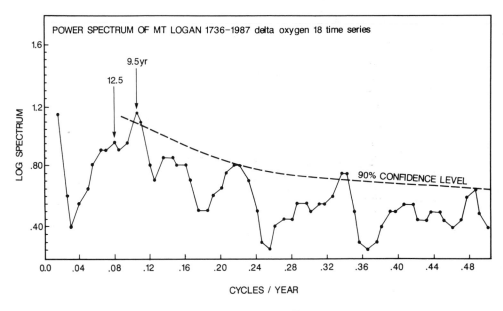

Figure 25.7 Fourier Transform Power spectrum for $\delta^{18}O$ time series.

(which is likely) this situation will not necessarily cause any difficulty in interpretation, since many station precipitation records when decomposed into seasonal components still cross-correlate very highly with the annual series. In addition to the Japanese example discussed in section 25.4.1, other regions that are teleconnected with the Mt. Logan series (with r ≥0.3 but over more restrictive time intervals) include parts of the North American prairies and the steppes of the Soviet Union (Budyko 1977). The latter region is a case where winter (November to March) precipitation correlates not only very highly with total precipitation there but also (restrictively) with the Mt. Logan net accumulation time series. An interesting observation is that prior to about 1915 and after about 1960, the composite time series from the steppes is out of phase with the Mt. Logan time series, whereas between those years (which includes the well-known agricultural drought period) it is clearly in phase. Rather than tending to negate the correlation, this could be a case of bistable phasing (Currie 1984, 1987).

Instrumental time series from coastal Alaska and the Yukon either do not correlate or correlate negatively with the Mt. Logan series. This result might be expected because of the influence of the extreme topography of the coastal mountain ranges on advecting surface Pacific air masses. There is evidence from $\delta^{18}O$ and δD profiles determined between altitudes of 1800m and 5920m on Mt. Logan that above about 5300m (51 kPa level) nearly geostrophic air flow conditions might prevail, and that a different air mass from the one below ~3300m exists with a distinctly different stable isotope fractionation history. Between these two air layers is a mixed layer with nearly homogeneous isotopic properties. The association between teleconnecting regions is probably associated with this observation but it is evidently very complex and remains to be fully elucidated (Wallace 1985).

Evidently the extra-tropical western Pacific (Hamilton 1988) is a key region in any attempt to explain the teleconnection between the Japanese precipitation series and the Mt. Logan net accumulation time series. Because the latter series contains a signal which is believed to be associated with ENSO events, it is logical to consider oceanic as well as atmospheric linkages between the southwestern and the northeastern Pacific ocean. The Kuroshio current is clearly a major component in the system. This current may be traced back to the region of the western tropical Pacific which Hamilton (1988) denotes as region I. When sea surface temperatures (SSTs) are high in region I, the extra-tropical response to an ENSO event is enhanced. The effects of the ENSO events must be propogated both by the Kuroshio current (Kawabe 1985) its extension, the North Pacific drift and the Alaska current, and by cyclones that originate in the region of these currents (Sander and Gyakum 1980). Rasmusson and Wallace (1983) have shown that there is a very high positive correlation between equatorial Pacific SSTs and rainfall.

A positive correlation between the above variables also holds for the Gulf of Alaska region (Royer 1985). In principle, this can be extended to other regions of the Pacific. Figure 25.8 shows isopleths of the Pearson product moment cross correlation coefficient, r, at zero lag between the Mt. Logan net annual accumulation time series and North Pacific mean annual SSTs for a 37 yr period. Values of r from +0.2 to +0.4 are given. In the region of the Bering Sea, r approaches +0.5. These latter values are statistically significant at the 95% level. The geometry of the isopleths exhibits an ordered pattern due to spatial correlation in the SST field. The isopleths shown in Figure 25.8 may be interpreted to imply that SSTs, particularly in peripheral regions of the North Pacific, have recently been in phase with the Mt. Logan precipitation. During times of higher sea ice incidence in the Bering Sea (and Sea of Okhotsk) which could occur either during times of lower SST or lower air temperature or lower wind speeds (which may not necessarily be synchronous) less moisture would be available for precipitation. Hence, during these times, generally lower snowfalls would be expected on

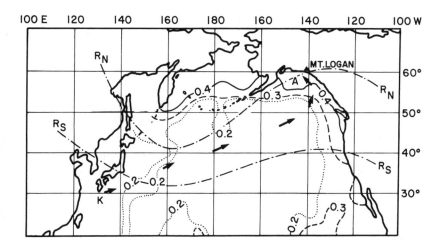

Figure 25.8 Cross-correlation coefficients between the Mt. Logan net annual accumulation time series and annual North Pacific sea surface temperatures: 1947-1984. (Data courtesy of D. Cayan, Scripps Institution of Oceanography). Isopleths of r values (Pearson product moment correlation coefficients) are shown for positive values at zero lag. K: Kuroshio current, A: Alaska current. The preferred tracks of Rossby Waves defined by R_N and R_S are also shown.

Mt. Logan. Whereas rather long and detailed sea ice records are available for Iceland, the same is not true for the North Pacific (Lamb 1977) therefore we are not able to extend the immediate discussion in more detail.

The preferred zone of upper level Rossby wave tracks is shown in Figure 25.8. Variations in the geometry of these waves, which are related to the position of the mid- to high-latitude quasi-stationary cyclones and anticyclones, influence the trajectories of storms and thus are a possible source for the nearly simultaneous modulation of Mt. Logan and Japanese precipitation.

Mt. Logan accumulation has apparently increased about 16% over the full time span of the core. Most of this has occurred after about A.D. 1850. This increase could be due to one or several factors and the result should be initially treated with caution. Possible causes for the trend are as follows:

(a) The creep thinning correction (Equation 2) is based on strain rate data derived from the core, whereas additional but unavailable data from immediately up the flowline should be incorporated in a more rigorous computation.

(b) The accumulation gradient across the col was measured only in recent years, and a condition of steady state is implicit in the data reduction. If this gradient has changed over the last 250 yr (due to shifts in prevailing winds) then the correction will contain an error which we have no basis for assessing. Nevertheless, another climatic signal will have been built into the data.

(c) If the amount of surface snow removal, which is related to the wind speed and frequency at the site, has changed significantly over the last 250 yr, this could produce an apparent trend in net accumulation. Variations of windiness appear to be pseudo-cyclic and may be related to shifts in the semi-permanent low pressure centers (Lamb 1972, 1977).

(d) The trend in accumulation may be a direct climatic one. Long-term trends in other snow accumulation as well as precipitation series do exist, as shown in Figure 25.9, although the data are not displayed to imply that the apparent correlation existing between any of these series has a physical basis. Nevertheless, comparisons have been made (Lamb 1977) between trends in spatially remote climate time series over time spans as long as the ones shown in Figure 25.9. Trends in shorter precipitation series derived from instrumental data and for specific latitude bands, have been studied by Diaz *et al*. (1989). That work demonstrated that decadal scale trends in precipitation do exist but that more study is required in order to identify the reasons for such trends. A component in the trend of the Greenland and South Pole series (3 and 4) shown in Figure 25.9, may be due to surface site characteristics and thus, individually, these curves should also be viewed with some caution. In spite of this, the fact that several long series derived from widely different sites show quite similar trends, suggests that the data could be reliable, if it is subsequently shown that the climatic regions which they occupy, are teleconnected.

Another observation that can be made about the net accumulation series from Mt. Logan is that before about A.D. 1820, the variance of the annual net accumulation series (as well as the filtered series) is significantly less than after that time. Caution must also be applied to any climatic interpretation of this observation. Because of layer thinning and diffusion effects,

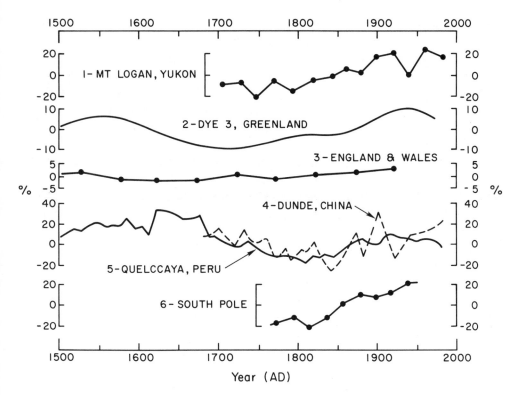

Figure 25.9 Time series of: (1) Mt. Logan net accumulation (as % deviation from the 1693-1987 average). Twenty year averages shown. (2) Total annual precipitation of England and Wales (as % deviation from the 1916-1950 average). Fifty year averages (from Lamb 1972). (3) Net annual accumulation for Dye 3, Greenland (as % of the 1250-1970 average). Data is low-pass filtered with a 120 yr cut-off period (Reeh *et al.* 1983). (4) Decadal averages of net annual accumulation on the crest of the Dunde Ice Cap (China) expressed as a % deviation from a 380 yr mean value (from Chapter 27). (5) Decadal averages of net annual accumulation on the summit of Quelccaya ice cap (Peru) expressed as a % deviation from a 600 yr mean (from Thompson *et al.* 1986). (6) Twenty year averages of 10 yr running means of net annual accumulation at South Pole (as % deviations from 1760-1957 average) (after Giovinetto and Schwerdtfeger 1966). Note scale changes for each series.

which are accelerated towards the bedrock, identification of annual layers becomes progressively less precise. It appears that using the present techniques, the practical limit of annual resolution has been reached at about 100m depth (~A.D. 1720). In the lower part of the core, only (assumed) seasonal oscillations of the nitrate ion concentration were used as an annual indicator. A certain amount of subjective smoothing of layers may have resulted from this procedure of fitting nitrate peaks between volcanic signatures. On the other hand, an increase in variability of precipitation may have started around the middle of the last century.

25.5.2 *Time series analysis*

Spectral analyses may provide information on the frequency content of a time series and on the phase of a waveform contained in the series (Currie 1984). If more than one physically significant frequency peak is found in the power spectrum, this provides a degree of reliability in the time scale of the series. The net accumulation time series contains significant information both in the time domain (1890 to 1987) (Figure 25.5) and the frequency domain (full series) (Figure 25.6). In particular, spectral power exists at frequencies corresponding to periods/cycles of ~3.8 yr, ~11 yr and ~21 yr in the complete series. When the series is partitioned, the ~3.8 yr and ~11 yr peaks still persist. An ~18.6 yr periodicity persistently found by Currie (1984, 1987) can be found in the accumulation series (R. Currie, personal communication) but it is very weak. Based on current knowledge, several of the peaks in the power spectra may be linked to physical mechanisms. According to Quinn et al. (1987) the average time between the onset of near moderate or stronger El Niño events between 1803 and 1987 is about 3.8 yr. For all events between 1700 and 1987 the average time between them is almost 4 yr. This is probably an upper limit since reliability in the older historical data can be questioned. El Niño should be referred to as a pseudo-periodic event (Quinn et al. 1987; Enfield 1989). We therefore assume that the persistent ~3.8 yr peak in the accumulation power spectra is due to modulation of precipitation by ENSO events. However, other cycles are involved in the El Niño time series (Hanson et al. 1989) and it is possible that the 15 yr spectral peak and the peaks between 5 and 7 yr (seen in Figure 25.6) are associated with ENSO events. It was shown (Section 5.1 and Figure 25.8) that cross-correlations between Mt. Logan accumulation series and North Pacific SSTs occurred for the western, northern and eastern regions, and therefore to find an ENSO signal is a reasonable result. A visual comparison between the El Niño chronology and the annual accumulation series has been made. There seems to be a tendency during times of meridional air flow for an increase in accumulation at the Mt. Logan site immediately following an ENSO event, (cf. Ropelewski and Halpert 1986) but the consistency is not uniform (see also Cannon et al. 1985).

An 11yr cycle has been found in many temperature and precipitation time series (Currie 1984, 1987). Some spectral power coinciding with this period is seen both in the accumulation and the $\delta^{18}O$ series although it is not a highly stable peak, as seen by changing the characteristics of the spectral window. The length of the nominal 11 yr solar spot cycle also varies. By generating the pseudo-11 year waveform component of the series in the time domain it may be shown that it holds a close bi-stable phase relationship with the solar spot cycle (Currie 1987 and personal communication). Recently, an hypothesis for indirect solar modulation of climate variables has been outlined by Tinsley (1988) based on the work of Labitzke and van Loon (1988). Although the 11 yr cycle was observed in north Atlantic winter storm track trajectories (influenced by the QBO) we suppose that the same solar modulation might also be present in the north Pacific. Figure 25.10 (from Lamb 1972) shows the amplitude of the variations of mean annual pressure with an approximate 11 yr cycle length over a 40 yr time span. The overall N. Hemisphere maximum amplitude occurs in a region centered over the Alaska-Yukon border, through which the locus of maximum frequency of aurorae also passes. Mt. Logan is within 1100 km of this region and therefore it is not unexpected that a signal corresponding to the solar cycle occurs in the ice core time series, if it is assumed that pressure variations lead to variations in precipitation and temperature through the redirection of air masses and storm track trajectories. This interpretation is consistent with the work of Bucha (1984).

Spectral peaks occurring between 15 and 22 yr (Figure 25.6) represent particular problems in interpretation which will largely be avoided here. Peaks within this interval may be seen in

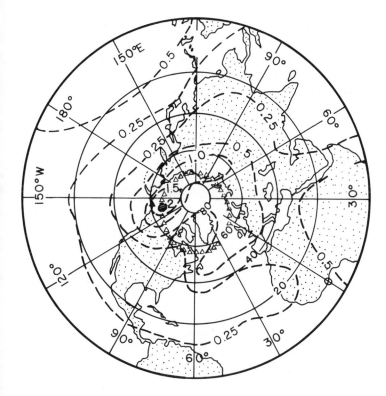

Figure 25.10 Amplitude of variations of annual mean N. Hemisphere pressure with approximately 11 yr periodicity. Data from 1899-1939. Iso-amplitudes (dashed lines) in millibars, reach a maximum (2mb) over Alaska (dashed region). The zone of maximum frequency of aurorae is shown by the ellipse of open triangles (after Lamb 1972). Mt. Logan is shown by a solid triangle (60°30′N 140°30′W).

other precipitation time series (e.g. Shao-Wu and Zong-Ci 1981; Clegg and Wigley 1984; Currie 1984 1987). The signal of the lunar nodal (M_N) tidal period occurs within this interval (Currie and O'Brien 1988). A strong 20 yr periodicity was found in the Greenland $\delta^{18}O$ time series by Hibler and Johnsen (1979) who proposed an astronomical theory for its existence. There is still controversy concerning the ~22 yr cycle (Currie and O'Brien 1988) but in order for a 'solar-weather' connection to exist at this frequency it may not be necessary for direct solar luminosity to vary. Tinsley (1988) has discussed a possible mechanism involving cosmic ray modulation (see Webber and Lockwood 1988) by solar activity and multistage amplification of the effects in the mesosphere, stratosphere, and troposphere through gas reaction changes and subsequent radiational variations.

25.5.3 *Comparison with other studies*

Tree ring series   We have also analyzed one of the ring-width chronologies (TT-HH) of Jacoby and Cook (1981; see also D'Arrigo and Jacoby, this volume) derived from northern Yukon tree line cores. This series evidently contains temperature information. For an A.D. 1700 to 1975 series there is weak power at 3.6 years (which is close to the 'ENSO period' of 3.8

499

years for the snow accumulation series) and at 10.9 years. There is also considerable power at the very low frequency end of the spectrum but this will not be discussed here.

The frequency coherence ($C(\omega)$) between the $\delta^{18}O$ series and the TT-HH tree ring width series was computed for (a) A.D. 1736-1975 and (b) A.D. 1860-1975. For series (a) coherence at zero lag occurred at 2.3 yr, 3.2 yr, 3.6 yr and 60yr. Values of $C(\omega)$ were respectively 0.44, 0.65, 0.54 and 0.74. For series (b) coherence occurred at 2.2 yr, 3.6 yr and 9.6 yr corresponding to $C(\omega)$ values of 0.44, 0.45, and 0.65.

Oceanographic data   In section 25.4.1, a connection between the ice core data and the North Pacific ocean-atmosphere system was discussed. The results of spectral analyses of N. Pacific oceanographic time series presented by Kawabe (1985) show some similarities to the present results. Although we have not accessed the original data to compute coherence between the time series, the published spectra indicate some striking similarities with the Mt. Logan spectra. For SST data, there is power at ~2 yr and ~3.6 yr. For the large meander of the Kuroshio current there is power at ~3.4 yr, 8.5 yr, and ~21 yr. All these cycles are close to ones seen in the Mt. Logan spectra. In addition, the 3.6 yr period is close to the period of strong coherency between the δ and tree ring chronologies.

Historical and early instrumental data   Records kept at Fort Snelling, Minnesota from 1819 to 1858 and at St. Paul from 1859-1870 may be useful, since there appears to be some parallels between the climate of this region and the Yukon. Cross correlation between the Mt. Logan accumulation series and Northern Hemisphere instrumental data (1901-1985) show some significant correlation coefficients for the North American mid-west (R. S. Bradley, personal communication). According to Ludlum (1968) from 1845 to 1857, winters were extremely cold with record coldest months for the instrumental period, 1819-1870. Although temperatures were recorded throughout the period, precipitation was only recorded in the last two decades of the period in the mid-west.

The tree ring-width chronology of Jacoby and Cook (1981) provides an interesting basis for an attempt to determine the probable date of a notably cold period last century in the Yukon. In particular, the late 1840s through the 1850s, was evidently a very cold period in the Yukon according to the tree ring-width time series. This is probably the time of the year of the "Two Winters", well known to Yukon Indians (Cruikshank 1981). Although we are not able to specify a temperature scale for our $\delta^{18}O$ curve, it is interesting to note that a low δ value is recorded after ~A.D. 1850 (see Holdsworth *et al.* 1989). Similar negative δ values are also recorded for 1715-1725 a period which corresponds rather closely with the other lowest ring width section for the TT-HH chronology given by Jacoby and Cook (1981).

Another source of proxy climate information comes from western North American Indian winter counts (e.g. Praus 1962; Dempsey 1964) several of which start before A.D. 1800. Although quite subjective, the many accounts which relate to weather conditions may be cross-verified and often seem to represent regional conditions. Winters of the 1850s are recorded as being particularly severe. Significantly, there is no written evidence for a climatic anomaly corresponding to the "year without a summer" (1816) (see Stommel and Stommel 1983). Neither the ring-width index series of Jacoby and Cook (1981) nor the $\delta^{18}O$ series from Mt. Logan show a significant anomaly corresponding to this climatic event. Such evidence helps to set limits on the spatial extent of that well documented climatic perturbation.

25.6 Summary

Two climate time series for the Mt. Logan ice core have been presented and analyzed for information content. Although not demonstrated here, the $\delta^{18}O$ series is known to contain more climatic information than just temperature. Power spectra determined for the complete series (1736-1987) and for its two halves indicate spectral power between 9 and 12.5 years. This includes the 10-11 yr peak usually ascribed to the sunspot cycle. This signal could have arisen through either the temperature influence on δ, through the 11 yr modulation of storm track trajectories or both.

A tree ring-width chronology from the northern Yukon was analysed by the same methods and yields a similar spectral peak at 10 ± 0.5yr. Another important spectral peak corresponding approximately to the ENSO quasi-periodic phenomenon is seen in this series. The frequency coherence between these two series indicates similarities in information content. From computing power spectra and coherence between the full series and partitioned series it is evident that both the δ and tree ring width series contain similar component waveforms and that the A.D. 1860-1987 δ-series is probably more reliable than the earlier half of the series. Two prominent minima (A.D. 1715-1725 and A.D. 1850 ± 5) occur in the tree ring series, which is considered to contain a strong 'proxy' temperature component. We consider that the δ-series may be showing a climatic perturbation after 1850. This time is thought to correspond to the Yukon Indian account of the year of the "Two winters", which seems to have been analogous to, but much later than the "year without a summer" (1816) an event that was apparently not experienced in the Yukon. For the series as a whole, however, we have not been able to isolate a consistent proxy temperature component, because of an incomplete understanding of the processes leading to a δ value in surface snow.

The net accumulation time series, when subjected to spectral analysis, can also be shown to contain some physically significant spectral peaks. For the full series (A.D.1736-1987) these give periods of 2.6 yr, 3.0 yr, 3.8 yr, 11 yr, 15 ± 1 yr and 20 ± 1.5 yr. For the series A.D.1736-1860 peaks exist with periods of about 2.2 yr, 3.8 yr and 36 ± 5 yr, whereas for the A.D. 1861-1987 series, peaks exist with periods of 2.6 yr, 3.8 yr, 11.5 yr and 21 yr. Several of these peaks are significant at the 90% level, but the occurrence of several peaks that also occur in other time series and that correspond to physical phenomena is of great importance here. Thus the 2.2-2.6 yr periodicity may be associated with the QBO, the $3.8\pm$ yr (and possibly also the 15 yr) periodicity with ENSO events and the 11 ± 2 yr periodicity with the quasi-periodic sunspot cycle.

Coherence between the spectra of $\delta^{18}O$ and net snow accumulation was determined for the full series and the sub-series. Both the full series and the first half had strong coherence at 2.9 yr (0.48) and 3 yr (0.56) respectively. The full series and the last half had strong coherences at 2.6 yr (0.62 and 0.53 respectively). Thus, this result might lead to the identification of processes which possibly modulated both the $\delta^{18}O$ and the net accumulation.

The section of the Mt. Logan accumulation series from A.D.1890-1987, which strongly correlates with some instrumental precipitation series, not only permits exploration of an interesting climatological phenomenon (teleconnections) but it also indirectly serves to help verify the more recent section of the ice core data. Finally, the possibility exists that there is a trend (or a step function) in the net accumulation time series, in the sense that mean values after about A.D. 1825 are higher (as also is the variance) than before that time.

Acknowledgements

Dr. G. C. Jacoby (Lamont-Doherty Geological Observatory) kindly supplied the TT-HH tree ring width time series used in this study. We thank Dr. D. Cayan (Scripps Institution) for supplying the North Pacific SST data set and Professor S. Nichols (University of Calgary) for producing the maximum entropy power spectrum shown in Figure 25.6(b). Professor R. S. Bradley kindly performed cross correlation analyses between the accumulation time series and the northern hemisphere precipitation data set for an extension of our study of teleconnections and Professor M. Giovinetto (University of Calgary) read and commented helpfully on the first draft of this chapter.

References

Bjerknes, J. 1969. Atmospheric teleconnections from the equatorial Pacific. *Monthly Weather Review* 97 162-172.

Bradley, R. S., H. F. Diaz, J. K. Eischeid, P. D. Jones, P. M. Kelly and C. M. Goodess 1987. Precipitation fluctuations over northern hemisphere land areas since the mid-19th Century. *Science* 237 171-75.

Bucha, V. 1984. Mechanism for linking solar activity to weather-scale effects, climatic changes and glaciations in the northern hemisphere. In *Climatic Changes on a Yearly to Millennial Basis*, N. A. Mörner and W. Kalen (eds.) 415-448. Amsterdam: D. Reidel Publ. Co.

Budyko, M. I. 1977. *Climatic changes*. AGU, Washington, D.C.

Cannon, G. A., R. K. Reed, P. E. Pullen 1985. Comparison of El Niño events off the Pacific Northwest. in *El Niño North*, W. S. Wooster and D. L. Fluharty (eds) 75-84. University of Washington.

Clegg, S. L. and T. M. L. Wigley 1984. Periodicities in precipitation in northeast China 1470-1979. *Geophysical Research Letters* 11(12) 1219-1222.

Covey, C. and P. L. Haagenson 1984. A model of oxygen isotope composition of precipitation: Implications for paleoclimate data. *Journal of Geophysical Research* 89 (D3) 4647-4655.

Cruikshank, J. 1981. Legend and landscape: convergence of oral and scientific traditions in the Yukon Territory. *Arctic Anthropology* XVIII-2, 67-93.

Currie, R. G. 1984. Periodic (18.6-year) and cyclic (11-year) induced drought and flood in western North America. *Journal of Geophysical Research* 89(D5) 7215-7230.

Currie, R. G. 1987. Examples and implications of 18.6- and 11-yr terms in world weather records. In *Climate history, periodicity and predictability*, M. R. Rampino, J. E. Sanders, W. S. Newman and L. K. Königsson (eds.) 378-403. New York: Van Nostrand Reinhold Company.

Currie, R. G. and D. P. O'Brien 1988. Periodic 18.6 and cyclic 10 to 11 year signals in northeastern United States precipitation data. *Journal of Climatology* 8, 255-281.

Dansgaard, W. 1964. Stable isotopes in precipitation. *Tellus* XVI, 4, 436-468.

Dansgaard, W., S. J. Johnsen, H. B. Clausen and N. Gundestrup 1973. Stable isotope glaciology. *Meddelelser om Gronland* 197(2). Copenhagen: C. A. Reitzels Forlag.

Dempsey, H. A. 1964. A Blackfoot Winter Count. *Occasional Paper* No 1. Calgary: Glenbow Foundation.

Diaz, H. F., R. S. Bradley and J. K. Eischeid 1989. Precipitation fluctuations over global land areas since the late 1800s. *Journal of Geophysical Research* 94(D1) 1195-1210.

Enfield D. 1989. El Niño, past and present. *Reviews of Geophysics* 27 (1) 159-187.

Fisher, D. A., R. M. Koerner, W. S. B. Paterson, W. Dansgaard, N. Gundestrup and N. Reeh 1983.

Effect of wind scouring on climatic records from ice core oxygen isotope profiles. *Nature* 301, 205-209.

Fuller, W. A. 1976. *Introduction to statistical time series*. New York: Wiley.

Giovinetto, M. B. and W. Schwerdtfeger 1966. Analysis of a 200-year snow accumulation series from the south pole. *Archives für Meteorology, Geophysik und Bioklimatology* A 15, 227-250.

Gow, A. J. 1965. On the accumulation and seasonal stratification of snow at the South Pole. *Journal of Glaciology* 5(40) 467-477.

Grootes, P. M., M. Stuiver, L. G. Thompson and E. Mosley-Thompson 1989. Oxygen isotope changes in tropical ice, Quelccaya, Peru. *Journal of Geophysical Research* 94(D1) 1187-1194.

Hage, K. D., J. Gray and J. C. Linton 1975. Isotopes in precipitation in north western North America. *Monthly Weather Review* 103, 958-966.

Hamilton, K. 1988. A detailed examination of the extratropical response to tropical El Niño-Southern Oscillation events. *Journal of Climate* 8 (1) 67-86.

Hanson, K., G. W. Brier and G. A. Maul 1989. Evidence of significant nonrandom behavior in the recurrence of strong El Niño between 1525 and 1988. *Geophysical Research Letters* 16 (10) 1181-1184.

Hibler, W. D. III. and S. J. Johnsen 1979. The 20-yr cycle in Greenland ice core records. *Nature* 280, 481-483.

Holdsworth, G. 1984. The Canadian Rufli-Rand electro-mechanical core drill and reaming devices. In *Special Report* No. 84-34, 21-32, U.S. Army Cold Regions Research and Engineering Laboratory, Hanover, New Hampshire.

Holdsworth, G., M. Pourchet, F. A. Prantl and D. P. Meyerhof 1984. Radioactivity levels in a firn core from the Yukon Territory, Canada. *Atmospheric Environment* 18(2) 461-66.

Holdsworth, G., H. R. Krouse, M. Nosal, M. J. Spencer and P. A. Mayewski 1989. Analysis of a 290 yr net accumulation time series from Mt. Logan, Yukon. In Proceedings IAHS Third International Assembly. Baltimore, Maryland, May 1989. *IAHS Publ.* no. 183, 71-79.

Jacoby, G. C, and E. R. Cook 1981. Past temperature variations inferred from a 400 year tree ring chronology from Yukon Territory, Canada. *Arctic and Alpine Research* 13(4) 409-000.

Jouzel, J., L. Merlivat and C. Lorius 1982 Deuterium excess in an east Antarctic ice core suggests higher relative humidity at the oceanic surface during the last glacial maximum. *Nature* 299, 688-691.

Kawabe, M. 1985 El Niño effects in the Kuroshio and western north Pacific. in *El Niño North*, W. S. Wooster and D. L. Fluharty (eds.) 31-43. University of Washington.

Labitzke, K. and H. van Loon 1988. Associations between the 11 year solar cycle, the QBO and the atmosphere. Part I: the troposphere and the stratosphere in the northern hemisphere in winter. *Journal of Atmospheric and Terrestrial Physics.* 50(3) 197-206.

Lamb, H. H. 1972. *Climate, present, past and future*. Vol. 1. London: Methuen.

Lamb, H. H. 1977. *Climate, present, past and future*. Vol. 2. London: Methuen.

Ludlum, D. M. 1968. *Early American Winters II 1821-1870*. Boston: American Meteorological Society.

Moser, H. and W. Stichler 1974 Deuterium and oxygen-18 contents as an index of the properties of snow covers. In Proceedings IAHS Assembly, Grindelwald Snow Mechanics Symposium, April 1974. *IAHS Publ.* no. 114 122-135.

Praus, A. 1962. The Sioux 1798-1922: A Dakota Winter Count. *Cranbrook Institute of Science Bulletin* No. 44.

Quinn, W. H., V. T. Neal and Santiago E. Antunez de Mayola. 1987. El Niño occurrences over the past four and a half centuries. *Journal of Geophysical Research* 92(C13) 14449-14461.

Rasmusson, E. M. and J. M. Wallace 1983. Meteorological aspects of the El Niño/Southern Oscillation. *Science* 222 1195-1202.

Reeh, N., H. B. Clausen, W. Dansgaard, N. Gundestrup, C. U. Hammer and S. J. Johnsen 1978. Secular trends of accumulation rates at three Greenland stations. *Journal of Glaciology* 20(82) 27-30.

Ropelewski, C. F. and M. S. Halpert 1986. North American precipitation and temperature patterns associated with the El Niño/Southern Oscillation (ENSO). *Monthly Weather Review* 114(12) 2352-62.

Royer, T. C. 1985. Coastal temperature and salinity anomalies in the northern Gulf of Alaska 1970-84. in *El Niño North*, W. S. Wooster and D. L. Fluharty (eds.) 107-115. University of Washington.

Sander, F., J. R. Gyakum 1980. Synoptic-dynamic climatology of the "bomb". *Monthly Weather Review* 108 1589-1606.

Sevruk, B. (Ed.) 1986. Correction of precipitation measurements. *ETH/IAHS/WMO Workshop*. Zurich 1-3 April 1985. Heft 23, Geographisches Institute ETH.

Shao-Wu, Wang and Zong-Ci, Zhao 1981. Droughts and floods in China 1470-1979. in *Climate and History*. T. M. L. Wigley, M. J. Ingram and G. Farmer (eds.) 271-287. Cambridge: Cambridge University Press.

Simkin, T., L. Siebert, L. McClelland, D. Bridge, C. Newhall and J. H. Latter 1981. *Volcanoes of the World*. Stroudsburg, PA: Hutchinson Ross Publishing Co.

Stommel, H. and E. Stommel 1983 *Volcano weather: The story of the year without a summer 1816*. Newport, R.I.: Seven Seas Press.

Symonds, R. B., W. I. Rose and M. H. Reed 1988 Contribution of Cl- and F- bearing gases to the atmosphere by volcanoes. *Nature* 334 (6181) 415-18.

Thompson, L. G., E. Mosley-Thompson, W. Dansgaard and P. M. Grootes 1986. The Little Ice Age as recorded in the stratigraphy of the tropical Quelccaya ice cap. *Science* 234, 361-364.

Tinsley, B. 1988. The solar cycle and the QBO influences on the latitude of storm tracks in the north Atlantic. *Geophysical Research Letters* 15(5) 409-12.

Wallace, J. M. 1985 Atmospheric response to equatorial sea surface temperature anomalies. in *El Niño North*. W. S. Wooster and D. L. Fluharty (eds.) 9-21. University of Washington.

Webber, W. R. and J. A. Lockwood 1988. Characteristics of the 22-year modulation of cosmic rays as seen by neutron monitors. *Journal of Geophysical Research* 93(A8) 8735-40.

Whillans, I. M. 1978 Surface mass balance variability near 'Byrd' Station, Antarctica and its importance to ice core stratigraphy. *Journal of Glaciology* 20(83) 301-310.

26 The Arctic from Svalbard to Severnaya Zemlya: climatic reconstructions from ice cores

A. Tarussov

26.1 Introduction

Four archipelagos are located in this area: Svalbard (including North-East Land) Franz-Jozef Land, Novaya Zemlya and Severnaya Zemlya. The northern island of Novaya Zemlya is covered by a large ice sheet with outlet glaciers; the glaciation of the other archipelagos consists of ice caps and valley glaciers, with areas varying from ten to thousands of square kilometers. The maximum altitude of the ice divides ranges from 500-1200m a.s.l. and ice thickness varies from 100 to 700m. These ice caps together with the ice sheet of Novaya Zemlya provide an excellent opportunity to obtain paleoclimatic data with high time resolution. We need, however, to take into account the difference between these glaciers and cold polar ice sheets: ice caps of the Eurasian Arctic are very "warm" not only in comparison with the Antarctic and Greenland ice sheets, but also compared to the ice caps of Canadian Arctic. Some of them are temperate by Ahlmann's classification (i.e. temperature is at the melting point throughout the glacier) but most of them are subpolar: they have negative internal temperatures though intensive melting occurs every summer on their surfaces. Some data necessary to estimate the surface conditions are given in Table 26.1.

The amount of summer melting is quite variable; in some cases, it accounts for more than 50% of accumulation and influences ice structure and temperature. The latter is almost 10°C higher than mean annual air temperature due to heat transfer by meltwater. Naturally, the vertical distribution of the soluble components is transformed in comparison with the original profile, and so it is not possible to use methods of annual stratification, based on chemical or $\delta^{18}O$ curves, to develop a meaningful chronology.

Table 26.1 Characteristics of ice caps discussed in text.

	Annual mean air temp. (°C)	Annual accumulation*	Summer melting*	10m ice temperature (°C)
Svalbard	-11	50-70	15-25	-2
Franz-Joseph Land	-13	25-40	15-20	-4
Novaya Zemlya	-9	80-100	60-80	0
Severnaya Zemlya	-15	15-30	10-15	-10

*cm water equivalent

26.2   Data available

A large amount of data on the glaciation of this area was obtained during the IGY in 1957-59, but only one 120m borehole was drilled at Churlionis ice cap (Franz-Jozef Land). However, this ice core was not processed properly to get any paleoclimatic information. Almost nothing was done in this area for a long time after IGY; only in the middle of the 1970s was drilling work begun on Svalbard by the Institute of Geography, and on Severnaya Zemlya by the Arctic and Antarctic Research Institute. From 1974 to 1987, seven boreholes were drilled on Svalbard, the deepest of which reached the bedrock of Austfonna, the largest ice cap of Svalbard, at a depth of approximately 566m. A few boreholes were drilled in Severnaya Zemlya, almost all of them on Vavilov ice cap and another one (720m deep) on the Academy of Sciences ice cap. The first borehole on the Churlionis ice cap and a few most interesting boreholes as sources of data for climatic reconstruction are shown in Figure 26.1.

Recently some climatic reconstructions have been produced using different indicators, such as $\delta^{18}O$, chloride content and ice stratigraphy (Gordienko *et al.* 1980; Vaikmyae and Punning 1982). The ice cores were not all processed in the same way; for example, we have isotopic curves for Vavilov ice cap and Lomonosovfonna, but at present only a stratigraphic record for Austfonna and the Academy of Sciences ice cap. The most complex data are available for Vestfonna (North-East Land) such as isotopic and chemical data, stratigraphy and crystalline structure, but the profile is too short and glaciological conditions are unsuitable for reliable conclusions. Nevertheless, all sources of data were compared and we know the advantages of each method and each glacier involved (this problem will be discussed later in detail). It is worth mentioning that almost all boreholes were drilled by a thermal drill and core quality is perfect; hence the cores were convenient for stratigraphic study.

Further discussion will concern three time series with the best time resolution. The first of them is the reconstruction of the accumulation rate for the last 300 years (from A.D. 1656)

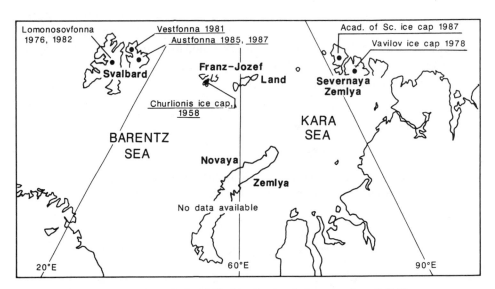

Figure 26.1   Locations of boreholes drilled in the Arctic ice caps since I.G.Y.

inferred from the Lomonosovfonna ice core (Westspitsbergen, 1982). This is the only accumulation record available from the area. Temperature reconstructions are not so rare as for accumulation, and we have selected two which we consider are the most reliable. These are unpublished data obtained from a stratigraphic study of Austfonna in 1987 and $\delta^{18}O$ data for Vavilov ice cap (Severnaya Zemlya) from Vaikmyae and Punning (1982).

26.3 Methods of reconstruction

It must be stressed that all glaciers in our area are subpolar with variable summer melting on their surfaces. The summer melting usually exceeds 10-15% of annual accumulation, and we agree with Paterson and Waddington (1984) that such an amount of meltwater smooths the original seasonal stratification of $\delta^{18}O$ and makes it impossible to determine annual layers by chemical or isotopic profiles. Nevertheless, the average isotopic ratio and chemical composition of the ice is not changed by melting providing there is no significant runoff. For such an assumption to be made, negative 10m temperatures and firn thickness of more than 5-6 m are required. This problem was investigated in detail by Arkhipov (1986) on the ice caps of Svalbard and Severnaya Zemlya. Therefore, we can use $\delta^{18}O$ values as climatic indicators if the necessary conditions are satisfied; we only have to estimate mean values for the percolation layer (usually 2-6m, or 3-10 annual layers) because the composition of the individual samples is stochastic due to mass exchange inside the percolation layer.

Summer melting smooths the isotopic stratification, but at the same time it creates the "melting features", usually in the form of layers of so-called infiltration ice. This type of glacier ice is relatively clear, with small amounts of air bubbles and high density (0.88-0.90). It is clearly visible throughout the ice core and is easily distinguishable from bubble-rich ice formed by firn densification. These layers are good indicators of surface conditions, as discussed by Koerner (1977) and Herron *et al.* (1981) and so we used them to reconstruct the summer temperature. These researchers correctly appreciated the importance of melting features as climatic indicators for the Devon Island ice cap and for Greenland in the percolation zone; we will try to demonstrate that the stratigraphic method is even more worthwhile and reliable for the cold saturation zone, where melting is much greater than in Greenland and where $\delta^{18}O$ and chemical methods are of limited use.

The main melting parameter measured in ice cores is the so-called annual melting percent (AMP). The procedure for measuring AMP is completely described by Koerner (1977); we used the same method. However, the distribution of melting features is quite random in the percolation zone where AMP is no more than 15-20%, and that part of the meltwater diffusing in wet snow is of the same order as the amount of meltwater forming the ice layers and lenses. The situation is rather different in saturation conditions (we mean the situation when meltwater penetrates much deeper than the first annual layer but all of it refreezes); the ice layers are then usually thick (5-50cm) and stable spatially and only a small part of the meltwater accumulates as ice glands and columns or diffuses in the upper snow layers. This conclusion was proved by comparing two cores from 200m boreholes drilled on Austfonna 2 kilometers apart in the same conditions (altitude and accumulation). These cores were almost identical, the AMP (averaged over 5 meters of core length) did not differ by more than 5%. Additionally, we drilled 20 more shallow holes to investigate the spatial variability and it was shown to be low. Of course, the depth resolution of our reconstruction cannot be better than the depth of water penetration which is generally 2-5 meters.

These results enabled us to calculate the mean summer temperature by means of the ice core stratigraphy. Summer temperature is the main factor controlling the mass balance of the glacier (Koerner 1977) and the melt features have undoubted advantages as climatic indicators, such as:

1) the ice core stratigraphy reflects surface conditions and is not affected by any complicated processes like evaporation of sea water. Herron *et al.* (1981) noted earlier that melt features are independent of the summer-winter precipitation ratio (unlike $\delta^{18}O$) and AMP is more sensitive to temperature changes; once formed, ice layers remain unchanged for thousands of years.

2) no large equipment and only a little time is required, because the registration of infiltration ice layer thickness may be carried out visually or by any optical sensor immediately after core recovery.

Annual melting is connected with the mean air temperature of the three summer months by the following relationship:

$$A = (T + 9.5)^3 \quad \text{(Krenke-Khodakov equation)}$$

where A is summer melting in mm of water and T is the mean June-August air temperature (°C). This ratio was developed by the analysis of a large set of data from different regions (Krenke 1982) and is accurate enough for our purposes.

Summer melting (and, naturally, ice content) is very sensitive to any changes of summer temperature due to this cubic relationship. Of course, climatic inferences are correct only if no drastic changes of glaciological conditions took place in the past (for example, changes of glaciological zones from percolation to saturation conditions or from negative to zero ice temperature). All such changes are clearly seen in the ice structure and we can observe them while studying the core.

The possibility of quantitative reconstruction of temperature conditions using stratigraphic data is very important because of the difficulty of using $\delta^{18}O$ to estimate past temperatures (relating to short-term and weak climatic changes during past centuries) due to uncertainties in the formation process of the isotopic composition of precipitation. We also stress that the Krenke-Khodakov equation is reliable enough for our purposes, and the problem is to calculate melting by the amount of infiltration ice (AMP) measured in the ice core. Some attempts to calibrate the curves by correlation with the nearest meteorological stations were not successful (Zagorodnov 1985); in any case, this way has many limitations. We propose a simple model to evaluate summer melting. The model connects infiltration ice percent with surface melting and is based on the following assumptions:

1) all meltwater is divided into two parts: a small part of it diffuses into the snow, and the rest goes to the formation of infiltration ice layers;

2) approximately 0.5 cm of water per cm of snow is necessary to increase initial snow density (0.40) to ice density (0.90) and make one cm of infiltration ice; hence, one cm of meltwater produces two centimeters of infiltration ice.

3) Only the upper annual layer increases its density by meltwater; this contribution is negligible for the deeper layers.

Figure 26.2 shows that fast densification occurs during the first year after deposition of snow due to refreezing of water; later the densification continues slowly only by recrystalization (0.02g/cc per year). So, underlying firn layers do not take up water for densification, but only for the formation of infiltration ice layers. The explanation of this fact is shown in Figure 26.3. Meltwater does not diffuse into the firn as it does into the snow, but penetrates in deeper layers by thin "channels" because firn permeability is much less than in snow. This process was observed in a pit wall at Lomonosovfonna in June 1982 after adding some color dye to the snow.

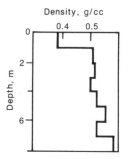

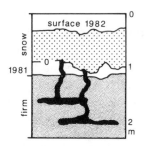

Figure 26.2 Densification curve, Lomonosovfonna, West Spitsbergen.

Figure 26.3 Meltwater patterns in the upper two meters. The 1981 summer surface is a boundary between snow (wet in its upper part) and firn, already influenced by melting and refreezing. Wet snow is shown by the dotted area, and water channels and lenses by thick black lines.

Water capacity of wet snow is close to 10% of its weight (according to Krenke et al. 1988) hence we calculate that the part of the meltwater which diffuses into snow as 0.1(C-A-R); (C= accumulation, A= melting in water equivalent, R= sum of the ice layers thicknesses in the annual layer. R= AMP*C).

4) Part of the meltwater refreezes in deeper annual layers of firn, but in the next few years meltwater from the subsequent annual layers will compensate for this loss of water, so the average accumulation over a few years is not changed by melting, if assumptions made earlier (no runoff, negative ice temperature) are satisfied.

Using 1 to 4, we obtained a few very simple equations. These equations give the "melting-infiltration ice content" relationship as:

$$A = 0.41R + 0.09C.$$

We used this simple model to calculate summer melting and then (using the Krenke-Khodakov equation) mean summer temperature. Accumulation is assumed to be independent of time; we consider this assumption to be acceptable because ice stratigraphy is much more sensitive to temperature than to accumulation changes. For example, a decrease of temperature from -4.0 to -4.2°C is equal to 30% accumulation growth.

This model was checked by calculation of the R-T relationship for a few Svalbard shallow cores, where both parameters are known from direct measurements. The error is less than 0.3°C for ten-year periods; the boundary conditions for the cold saturation zone, too, were exactly predicted by the model, in good agreement with Shumskiy (1955). These conditions are as follows: infiltration ice layers can form while melting is more than 9% of accumulation; if it exceeds 45-50% of accumulation, meltwater fills all firn pores. We can conclude that the method described above is suitable for the investigation of short-term changes in local conditions, whereas $\delta^{18}O$ is a better method for investigating global changes.

The dating of ice cores from subpolar glaciers is a serious problem due to smoothing of seasonal layers in $\delta^{18}O$ and other profiles; any climatic reconstruction is useless without dating. The best methods of dating are absolute dating by ^{14}C using AMS and direct counting of seasonal cycles in microparticle concentration, but our researchers did not have the opportunity to use them. Therefore a completely new method of measuring the thinning of annual layers according to depth, useful for glaciers with summer melting on the surface, was developed (Zagorodnov 1985). It is based on the fact that the melting-refreezing process produces natural markers for the measurement of ice deformation with depth, namely, the infiltration ice layers. Later we improved this method by choosing not the ice layers, but the intervals between them (or the layers of white ice) as thinning markers, because they are less dependent on surface changes and are normally distributed.

We approximated the relationship between ice layer thicknesses and depth by a second-order polynomial equation: $[L_h = ah^2 + bh + c]$ and then used the mean ice layer thickness near the surface L_0 and recent accumulation C_0 to calculate the relationship between annual layer thickness t and depth $[t_h]$ from the ratio:

$$t_h/C_0 = L_h/L_0$$

hence,

$$t_h = L_h * C_0/L_0.$$

This method was used successfully for dating of the Austfonna (1987) ice core (see below). Naturally, it assumes that accumulation is independent of time, but it does not lead to big errors because any short-scale changes would compensate for themselves. At the same time it has one important advantage in that it does not assume a steady state glacier, because the real cumulative deformation of layers is measured and all past changes of glacier dynamics are automatically taken into account.

26.4 Accumulation changes since the 17th century.

The infiltration ice layers described above as temperature indicators may sometimes serve as a tool for determining the annual cycles in the ice core. The basis of this approach is described in Benson (1960) and Hattersley-Smith (1963). They mentioned that infiltration ice content is usually increased in the lower part of the annual layer, and the groups of infiltration ice layers in a shallow pit or in an ice core (not the individual ice layers) mark the annual boundaries. Of course, the distribution of infiltration ice layers is rather random, therefore this method is not accurate for each annual layer, but for the mean thickness of three or five layers (depending on percolation depth) it is accurate enough.

The annual stratigraphy shown by the non-uniform distribution of infiltration ice in the vertical profile may be registered by crystal size variations, too, because the infiltration ice has coarse-grained structure in comparison with ice originating from the firn by recrystallization. All these methods are efficient on arctic glaciers where seasonal stratification in $\delta^{18}O$ curves is smoothed by melting. Only the peaks of microparticle concentration may be better seasonal indicators, but we have no such data for Svalbard and Severnaya Zemlya. The only source of accumulation data in our area is the Lomonosovfonna ice core obtained in 1982. The borehole was drilled at the ice divide of this ice cap in the central part of West Spitsbergen at 1040m a.s.l.; the surface is relatively flat and the spatial accumulation pattern is not disturbed by mountains. The ice thickness is 135m; all drilling procedures and previous results are described by Zagorodnov *et al.* (1984).

We determined the thicknesses of more than 300 annual layers in this ice core by ice stratigraphy. All details are in Tarussov (1988); we give only the main principles here. The correction for glacier flow could not be made accurate enough, and we prefered to eliminate the trend from the data. This procedure prevented us from studying long-term variability (over a period of more than 100 years) but it did not affect studies of shorter periods.

Figure 26.4 shows the accumulation values for Lomonosovfonna and for Milcent station in Southern Greenland (from Reeh *et al.* 1978) after corrections for density changes and thinning of layers with depth. The mean accumulation is almost equal at both sites, approximately 56cm in ice equivalent, and a comparison of these curves is very interesting. The Svalbard curve shows that some periods or epochs of lower accumulation took place during cooler periods in the eighteenth and nineteenth centuries, and accumulation was evidently higher in warmer periods (for example, since the 1930s and during a sudden warm event in the 1820s, which is well-known from many sources (see Groveman and Landsberg 1979). The standard deviation of the Svalbard accumulation is 9.5cm (the coefficient of variation, $Cv = 17\%$); such variability is usual for the majority of coastal areas. The mean accumulation of relatively warm periods is 10% higher than during cold periods. Thus, accumulation changes

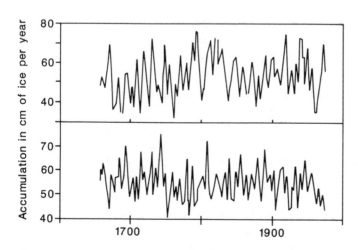

Figure 26.4 Accumulation changes measured in ice cores from Lomonosovfonna (upper plot) and Milcent, Greenland (lower plot).

in the Arctic tend to compensate for warming effects on the mass balance of glaciers, but, of course, a 10-15% increase of accumulation rate is insufficient to completely compensate for the effect of a 1-1.5°C rise of summer temperature. The temporal variability of accumulation in Greenland is almost two times less than at Svalbard; Cv at Milcent station is only 10%. Paterson and Waddington (1984) also mentioned this characteristic of Greenland accumulation series: Cv is usually 9-12% at ice sheet stations while it is 17-22% or more for almost any point outside the Greenland ice sheet. This difference could be explained by the smoothing influence of a large ice sheet on the short-term climatic changes; it suggests that any inferences about climate based on Greenland (and, probably, the Antarctic) data for surrounding areas may lead to mistakes.

No correlation exists between the two curves shown in Figure 26.4; the linear correlation coefficient, r= -0.13. Comparison of the Milcent curve with other accumulation and temperature data from Scandinavia and Iceland shows no connection; even other Greenland stations have different temporal patterns of accumulation (see Hammer *et al.* 1978) defined, probably, by the particular circulation system existing over the Greenland ice sheet.

The strong increase of the accumulation rate in Svalbard during 1780-1820 is very interesting because it is similar to one or two drastic peaks in the temperature curves of this period. These temperature events will be discussed later in the section concerning temperature reconstructions. We tried to find some periodicities in the accumulation records using Fourier analysis, but no significant periods were found in either series. The method of processing the Svalbard curve excludes periods shorter than 4-6 years, and we can only conclude that no periodicities with frequencies of 0.1-0.01 per year were apparent.

26.5 Temperature changes

26.5.1 Svalbard

We also mentioned that some ice cores from Svalbard have been studied since 1974. All results are in agreement concerning general features, but the ice core from Austfonna (1987) provided the most reliable and detailed information so we preferred to use it as a basis for the temperature reconstruction. This ice core has the best temporal resolution: no worse than 4-5 years (equal to the water percolation depth). The dating of the core was made by the stratigraphical method described above. Figure 26.5 demonstrates the thinning of ice layers from the surface downwards used as the basis for age calculations. The age of the ice near the bottom is tentatively 3-4 ka B.P., hence the last 500 years coincide with the upper 250m.

Isotopic samples are presently being processed, and we can use only stratigraphic data as a climatic indicator. As shown above, infiltration ice content is even more sensitive to temperature changes than $\delta^{18}O$, and it correlates well with other proxy data. We measured annual melting percent (AMP) i.e. the sum of infiltration ice layers thicknesses in some unit of core length (usually one meter) and then calculated the surface summer temperature using the melting-refreezing process model described above. Figure 26.6 demonstrates the whole profile of AMP versus age. The constant cooling from 3 ka B.P. to the ninth century A.D. was caused, probably, not by climatic processes, but by the relatively fast growth of the ice cap and the rising of its surface. Austfonna became steady in size only after the advance of its edge

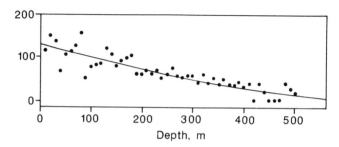

Figure 26.5 Thinning of white ice layers with depth in Austfonna ice core (each point represents a 10m mean).

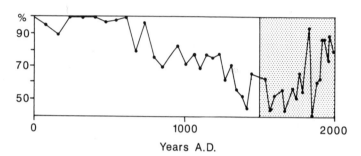

Figure 26.6 The AMP profile for Austfonna ice core from the surface to 400m depth versus age, calculated by the stratigraphic method. The last 500 years are shown by the shaded area.

to the tidewater position (otherwise it would have continued to grow during the cold time of the "Little Ice Age").

The part of the temperature profile representing the last 500 years is shown in Figure 26.7. It looks like the majority of European climatic series for this time: the Little Ice Age is seen from 1550 to 1920, and the first half of the 16th century is 0.5°C warmer. The 20th century is warmer than any other time in the last 500 years; only the period from 1800 to 1830 had such conditions, a strong sharp warming, of great interest for us. It can be found in many records (see Groveman and Landsberg 1979); in some it consists of two warming events near 1780 and 1820, but in others only one of them is pronounced. We cannot even estimate the temperature of this period because infiltration ice content is close to 100% and some unknown runoff undoubtedly occurred; we can only conclude that temperature was no lower than -2.5°C, as in the 1930s. A similar peak of accumulation has already been mentioned above. However, we have to take into account possible overestimation of the real duration of this event; this is possible because the large amount of meltwater produced by such warming could saturate the 10-20 metres of firn deposited before; in the same way any cold breaks in it would be smoothed out. Hence, the real duration of the warm event near A.D. 1800 could be less than it seems from the stratigraphic data and it might be divided into several short peaks.

The results of a statistical study of those data are almost the same as in the case of accumulation data; no periodicities were found by spectral analysis, but some epochs with

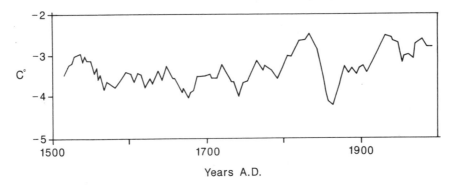

Figure 26.7 Mean summer temperatures at Austfonna, calculated by the melting model (smoothed by an 11-year filter).

cool or warm conditions can be seen in the original plot. The duration of such epochs does not exceed 150 years. A slight warming trend is observed from the end of 18th century. This trend cannot be explained by the lowering of the ice surface, because all glaciological evidence suggests a stagnation of Austfonna during the last centuries, due to the stabilizing effect of tidewater outlet glaciers and positive mass balance. Mean summer temperature was -4.0°C in the second half of the 16th century, -3.7°C at the end of the 19th century, and -2.9°C for the last 50 years. Temperatures have been much more variable in the last 180 years than in earlier times. This growth of variability seems to begin during the pronounced warm event near A.D. 1800. We have no explanation for this feature.

26.5.2   Severnaya Zemlya

These islands are the easternmost place in the Eurasian Arctic where large ice caps exist and where ice core investigations have been carried out. The 556m ice core from Vavilov ice cap (79°N, 96°E) was obtained in 1978 by the Arctic and Antarctic Research Institute. The results of $\delta^{18}O$ analysis of this core are described in Vaikmyae and Punning (1982). Unfortunately, there are some uncertainties in the dating of the ice core; the ice age is evidently overestimated, even in the upper part of the profile. The interpretation of this core is problematical due to the strange absence of correlation even between the chloride and $\delta^{18}O$ curves of the same core. The authors suppose, that such disagreement is connected with strong surface melting in the past.

Nevertheless, we can make some conclusions: slight warming evidently occurred during the last 250 years; the temperature rose fast in the last 120-140 years. Some periods of lower and higher variability are seen in the curve, and the last two centuries have anomalously high variability of Cl- and $\delta^{18}O$. One feature of this profile is worth mentioning. The overall amplitude of $\delta^{18}O$ variations in the Holocene time is 5-8‰, while the same value in Greenland and in Antarctica does not exceed 2-3‰. The same variability was described by Paterson and Waddington (1984) for ice caps of the Canadian Arctic, and it is usual for all subpolar ice caps. We suppose that it is partly due to surface melting, but two other effects involved are climatic: smoothing of short-term changes over the ice sheets (but not over relatively small ice caps because their feedback on climate is negligible) and feeding of the ice caps by precipitation from lower condensation levels with greater temperature variability.

Preliminary study of the stratigraphy of the ice cores from the Academy of Sciences ice cap (Savatyugin and Zagorodnov 1987) shows that the 20th century is the warmest time of the last millennium, but the recent warming began here earlier than in Svalbard and Europe, nearly 130-140 years ago, as in the North American Arctic (Hansen and Lebedeff 1987). All the warm periods in the past such as in the 16th century and in the Viking time are clearly seen and synchronous with the same events in Svalbard, but very weakly pronounced. We are sure that such a result is not produced by lowering of the ice surface during the last century, because the warming record in 1890-1930 was too fast to be produced by any possible retreat of the ice cap. Of course, our knowledge of the climatic history of this region is very limited and incomplete; we hope to improve it after the publication of the results of the Academy of Sciences core study.

26.6   Conclusions

The main climatic results of ice cores studies in the Eurasian part of Arctic are the following:

1) the main climatic events in the last 500 years were synchronous for all the Arctic (to within 10 years) but their amplitudes were different; individual warm or cool periods could be more or less pronounced in different areas. We suppose that reports about time lags in some regions were due to incorrect dating of proxy data.

2) we cannot estimate ranges of temperature changes accurately enough, but they undoubtedly were similar in the whole area. At Lomonosovfonna (Svalbard) the coldest summers were near 1630 and 1850 (-4.2°C) the warmest near 1820 and in the 1930s (-2.5°C) when the mean for the whole period is -3.3°C and the standard deviation is 0.72°C for the 7-year mean.

3) the warming in the 20th century did not start suddenly; some rising trend is seen in the curves from the end of the 18th century. It is probable that significant changes in atmospheric processes occurred between 1780 and 1820, resulting in increased variability and a strong warming event at about this time.

4) no regular periodicities are found in time series of temperature and accumulation. However, some anomalous periods, which are pronounced in all climatic parameters, undoubtedly occurred. Maybe, these periods are connected with changes in zonal and meridional circulation patterns; the observed increase in accumulation rates during warm epochs supports this inference.

5) Some differences exist between the time series of accumulation (and, probably, temperature) in Greenland and other territories. The most reliable cause of such differences is the strong influence of a large ice sheet on local climate.

6) Climatic reconstructions based on ice stratigraphy in subpolar glaciers are more reliable than isotopic methods for the study of short-term changes of surface conditions.

References

Arkhipov S. M., 1986. *Chemical and isotopical composition of glacier ice as indicators of ice formation conditions.* Sc.D. thesis, Institute of Geography, Academy of Sciences, Moscow. (In Russian)

Benson C. S., 1960. Stratigraphic studies of the snow and firn of the Greenland ice sheet. *USA SIPRE Research Report* 70, 31 pp.

Gordienko F. G., Kotlyakov V. M., Punning J-M. K. and Vaikmyae R. A., 1980. Study of the 200m ice core from Lomonosovfonna, Spitsbergen, and paleoclimatic implications. *Izvestiya Vsesoyuznogo Geograficheskogo Obshestva*, vol.112, no. 5, 394-401 (in Russian).

Groveman B. S. and Landsberg H. E., 1979. Reconstruction of the Northern Hemisphere temperature: 1579-1880. *Publication No. 79181/182*, Department of Meteorology, University of Maryland.

Hammer, C. U., Clausen, H. B., Dansgaard, W., Gundestrup, N., Johnsen, S. J. and Reeh, N., 1978. Dating of Greenland ice cores by flow models, isotopes, volcanic debris and continental dust. *J. Glaciology*, 20, 3-26.

Hansen J. and Lebedeff S., 1987. Global trends of measured surface air temperature. *Journal of Geophysical Research*, 92, 13,345-13,372.

Hattersley-Smith G., 1963. Climatic inferences from firn studies in Northern Ellesmere Island. *Geografiska Annaler*, Vol. 45, No. 2/3, 139-151.

Herron M. M., Herron S. L., and Langway C. C. Jr., 1981. Climatic signal of ice melt features in southern Greenland. *Nature*, 293 (5831) 389-391.

Koerner R. M., 1977. Devon Island Ice Cap: Core stratigraphy and paleoclimate. *Science*, 196 (4285) 15-18.

Krenke A. N., 1982. *Mass exchange in glacier systems of the USSR*. Gidrometeoizdat, Leningrad, 288 pp. (In Russian)

Krenke A. N. et al., 1988. *Marukh Glacier*. Gidrometeoizdat, Leningrad, 254 pp. (In Russian)

Paterson W. S. B. and Wadddington E. D., 1984. Past precipitation derived from ice core measurements: methods and data analysis. *Reviews of Geophysics and Space Physics*, vol. 22, 2, 123-130.

Reeh N., Clausen H. B., Dansgaard W., Gundestrup N., Hammer C. U. and Johnsen S. J., 1978. Secular trends of accumulation rates at three Greenland stations. *Journal of Glaciology*, vol. 20, No. 82, 27-30.

Savatyugin L. M., and Zagorodnov V. S., 1987. Glaciological studies at Academy of Sciences Ice Cap. *Materialy Glaziologicheskikh Issledovaniy*, 61, 228. (In Russian)

Shumskiy P.A., 1955. *Fundamentals of ice structure study*. Moscow, 492 pp. (In Russian)

Tarussov A. V., 1988. Accumulation changes on Arctic glaciers during 1656-1980 A.D. *Materialy Meteorologicheskikh Issledovaniy*, 14, 85-89 (In Russian)

Vaikmyae R. A. and Punning J.-M. K., 1982. Isotope and geochemical studies of the Vavilov Ice Dome, Severnaya Zemlya. *Materialy Glyaziologicheskikh Issledovaniy*, 44, 145-149 (In Russian)

Zagorodnov V. S., 1985. Ice formation and inner structure of glaciers. In: *Glaciology of Spitsbergen* (ed. by V.M.Kotlyakov) Moscow, 119-144 (In Russian)

Zagorodnov V. S., Samoylov O. Y., Raikovskiy Y. V., Tarussov A. V., Kuznetzov M. P. and Sazonov A. V., 1984. Inner structure of the Lomonosovfonna, Westspitsbergen. *Materialy Glaziologicheskikh Issledovaniy*, 50, 119-126. (In Russian)

27 Ice core evidence from Peru and China

L. G. Thompson

27.1 Introduction

Ice sheets and ice caps are recognized as being the best of only a few sources of atmospheric history from which past climatic and environmental conditions may be extracted. Much of the climatic activity of significance to humanity may not be strongly expressed (or may not affect) the polar ice caps. Ice core records can be recovered from polar ice sheets, as well as from a select few high altitude, low- and mid-latitude ice caps. In addition, the high accumulation on these ice caps and ice sheets makes it possible to recover high temporal resolution records of particulates, chemical constituents and oxygen isotope ratios ($\delta^{18}O$) for the last 1000 years. The aerosol records (soluble and insoluble) records serve as indicators of drought, wind strength, volcanic activity, and net accumulation, while $\delta^{18}O$ can, in some cases, serve as a proxy for temperature. Dust plays a deterministic role in the atmospheric radiation balance. Thus, the variety of chemical and physical data extracted from ice caps and ice sheets provide a multi-faceted record of both the climatic and environmental history of the earth and may allow assessment of the relative importance of potential forcing functions such as volcanic activity, greenhouse gas concentrations, atmospheric dust, and solar variability.

One of the important needs in addressing global change issues which the ice core records can provide are long time series, that is, a frame of reference against which present and future changes can be compared. The longer-term perspective available from polar cores is well documented. Recent evidence (Thompson *et al.* 1989a) indicates that longer-term glacial-interglacial records from non-polar regions can be obtained from the Qinghai-Tibetan Plateau. Such records provide a more global perspective of climate which is needed to understand fully the earth's climate system, and specifically to determine how well polar ice cores reflect climate variability in the subtropics. This chapter presents ice core evidence of climatic and environmental variability since A.D. 1500 with emphasis on those records from the tropical Quelccaya ice cap, Peru, and the subtropical Dunde ice cap, China (Figure 27.1).

The Tibetan Plateau and Bolivian-Peruvian altiplano have similar mean elevations of approximately 4000 meters. The regional weather patterns are driven by the sensible heat flux over these plateaus and the subsequent latent heat release during precipitation. Gutman and Schwerdtfeger (1965) studied the role of latent and sensible heat for the development of the Bolivian High and concluded that similar conditions exist there as in the highlands of Tibet (Flohn 1965). In both cases the upper troposphere anticyclone is a quasi-stationary system, persisting throughout the warm season. A typical weather pattern over the plateau during the wet season is a clear morning during which the plateau is heated, followed by an afternoon thunderstorm. Over 80% of the annual snowfall on the Quelccaya ice cap, Peru and the Dunde ice cap, China falls during the wet season (November-April) and (April-August) respectively. Since the Tibetan High is an integral part of the Asiatic monsoon circulation during the northern summer, it is instructive to compare not only the Bolivian and

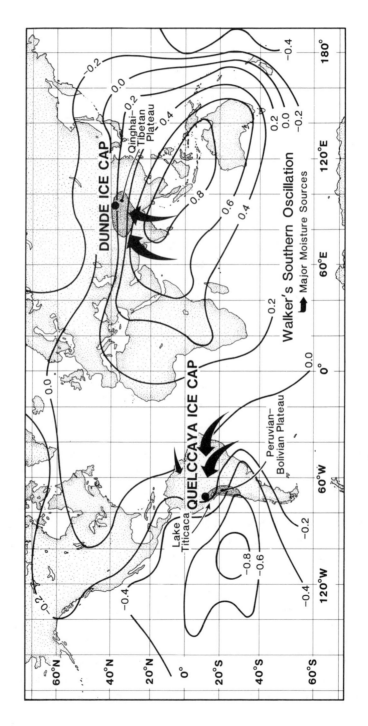

Figure 27.1 Location map showing the position of the Quelccaya ice cap, south of the equator on the Bolivian-Peruvian altiplano and the Dunde ice cap, north of the equator on the Tibetan Plateau. Arrows indicate the major moisture sources for both locations. The Walker circulation pressure correlations illustrate how teleconnections could link these two widely separated ice caps.

Tibetan Highs, but also other major features of both the upper and lower level circulation over South America and the Asiatic summer monsoon circulation.

Virji's (1979) comparison between the features of the Asiatic summer circulation and that over South America during the southern summer, suggests that the systems are similar in a number of important respects, and that the character of the summertime circulations over tropical and subtropical South America is monsoonal. Virji notes that the circulation system over South America is relatively more compact, confined largely to the continental zone. In contrast, the Asiatic monsoon system embraces a much larger area including the Indian Ocean and the adjacent continental region of Africa. Despite obvious differences in areal extent and intensity of the various features, the two circulation systems resemble each other as follows:

(a) The quasi-stationary Bolivian and Tibetan Highs have maximum amplitudes near the 300mb level (Dean 1971; Krishnamurti and Bhalme 1976). Both are localized regions of intense Hadley-type meridional, as well as seasonal, east-west circulations.

(b) The well-marked upper level easterly flow around 10°S latitude over South America occasionally produces high speeds, analogous to the tropical easterly jet of the Asiatic summer monsoon system around 10°N latitude.

(c) The continental heat low over Gran Chaco and the Pampean Sierras between 15° and 30°S and 60° and 68°W just east of the Andes is a relatively weaker feature compared with the monsoon trough over northern India.

27.2 Quelccaya ice cap

Research programs were conducted on the tropical Quelccaya ice cap in the Peruvian Andes (13°56′S; 70°50′W) between 1974 and 1984. This relatively large ice cap is characterized as follows: summit elevation, 5670m; total area 55 km^2; mean annual temperature, -3°C; maximum summit ice thickness, 164m; flat bedrock topography; and net annual accumulation, 1.15m of water per year. The annual cycle in precipitation is characterized by 80-90% of the snow falling from November to April (Thompson et al. 1984a). This produces the distinct seasonality in precipitation which is subsequently preserved in the ice stratigraphy.

Because the site is so remote and too high for use of a conventional drill system, a newly designed, portable, lightweight, solar-powered drill was used in 1983 to recover two ice cores (163.6m to bedrock and 154.8m) without contaminating the pristine environment or the core samples. This was the first major ice core drilling project using solar power.

The annual layers for the past 1500 years were counted using a combination of visible seasonal dust layers and the seasonal variations of $\delta^{18}O$, microparticles and liquid conductivity, thus allowing a very precise time scale to be established. Total particle (diameter >0.63μm) and large particle (diameter >1.59μm) concentrations, liquid conductivity, $\delta^{18}O$, net accumulation and pollen have been measured (Thompson et al. 1986; 1988a).

Figure 27.2 illustrates profiles of total particle concentrations, liquid conductivity and $\delta^{18}O$ measurements for the Quelccaya ice cap, plotted with depth. Each of the parameters show significant variations with depth. In order to document changes in conductivity which may occur due to storage time in which water is bottled before analysis in the laboratory, 1000

Quelccaya Summit Core

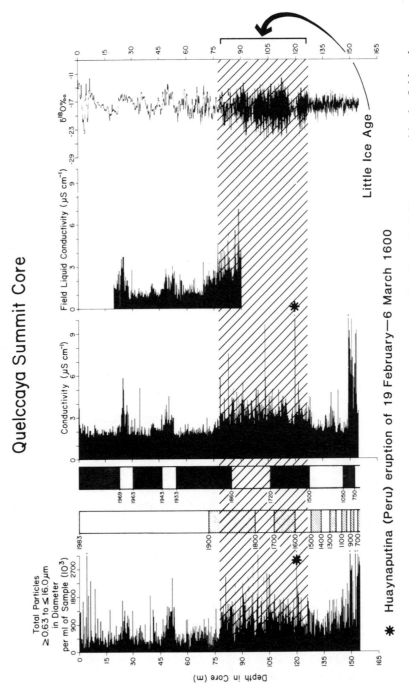

* Huaynaputina (Peru) eruption of 19 February—6 March 1600

Figure 27.2 Overview of Quelccaya core 1 results; total microparticles, liquid conductivities measured in the field and oxygen isotopes plotted with depth in core. The second column is the stratigraphic time scale illustrating the compression of time with depth. The third column indicates the wet and dry periods as reported in Thompson *et al.*, (1985). These data provide a continuous record of the environmental changes since ∼A.D. 470. The hatched area indicates the Little Ice Age (LIA).

conductivity measurements were made in the field immediately after the ice core was drilled, and then remeasured several months later when the samples arrived at the laboratory. The results illustrated in Figure 27.2 demonstrate that although background levels increase due to dissolution of some particles and absorption of atmospheric CO_2 with time, the major features are reproducible. The second column in Figure 27.2 illustrates how the rate of change in age increases with depth due to compression and thinning of the annual layers as a result of ice flow. The third column illustrates wet periods (dark shading) and dry periods (light shading) as calculated from the measured annual layer thickness. These data illustrate that there is no constant relationship between dry periods and increased concentration of either soluble or insoluble dust.

27.3 Dunde ice cap

Research programs have been conducted on the subtropical Dunde ice cap (38°06′N; 96°25′E) from 1984 to present. This ice cap is characterized as follows: summit elevation, 5325m; total area, 60 km²; mean annual temperatures, -7°C (as determined from measuring 10m temperatures and is considered a maximum); maximum summit ice thickness, 140m; flat bedrock topography; and net snow accumulation, 0.2m of water per year. The record of dust is very promising as the Asian continent is a major source area for continental dust (e.g., Gobi Desert) which contributes substantially to the North Pacific atmospheric dust burden today and in the past.

As with Quelccaya, over 80% of the snow falls in the wet season (monsoon season, May-August) producing the marked visible dust stratigraphy. In 1987 three ice cores, 136.6, 138.4 and 139.8m in length, were drilled to bedrock at the summit of the ice cap. Borehole temperatures indicate that even though the ice cap is located 38°N, it is a "polar" glacier with a bottom temperature of -4.7°C, well below the pressure melting point (Thompson *et al.* 1988b; Thompson *et al.* in press).

27.4 Annually resolvable ice core records

Fortunately, the high altitude regions in the low and mid latitudes, as well as the polar regions, contain ice caps and ice sheets which have continuously recorded climatic and environmental changes over the past several thousand years, often with annual resolution (Dansgaard 1954; Johnsen *et al.* 1970, 1972; Thompson 1977; Thompson *et al.* 1986; Mulvaney and Peel 1988; Hammer 1989). The records can be used to address problems which are of concern to a wide sector of the scientific community, governments, industries, and the general public. These include: (a) global scale climatic events such as the Little Ice Age (LIA) and El Niño/Southern Oscillation (ENSO) and monsoonal variability; (b) abrupt climatic change and evidence for changes in the amplitude of the annual cycle over the last 1000 years; (c) impact of past low and high frequency climatic changes on human activities, and (d) documentation of climate changes in the 20th century relative to long time-perspectives provided by ice core records.

The construction of a time scale for an ice core is generally based upon the integrated high resolution (8 to 12 samples per year) records of $\delta^{18}O$, microparticle concentrations, conductivity, and ionic concentrations which often exhibit a distinct seasonal variation in flux (Thompson *et al.* 1986; Hammer 1989). When possible, as is the case for many high elevation non-polar ice caps, visible dust layers allow rapid and accurate dating (Thompson *et al.* 1985). The measurement of total Beta radioactivity makes it possible to identify the major global time-stratigraphic horizons associated with known atmospheric thermo-nuclear tests (Picciotto and Wilgain 1963; Lambert *et al.* 1977, Thompson *et al.*, in press). Two or more ice cores should be drilled at the same site and similarly analyzed to provide an independent verification of the time scale and to ensure an uninterrupted physical and chemical stratigraphic record. For most of the ice core records presented here, at least two, and in some cases three, cores are available from each site. The utility of ice core analyses hinges upon accurate dating of cores which requires the use of multiple stratigraphic features exhibiting a seasonal variation and/or visible annual dust layers which are preserved within the ice. Figure 27.3

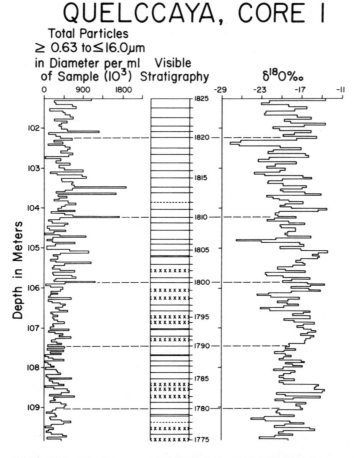

Figure 27.3 The 50-year period from A.D. 1775-1825 illustrates stratigraphic parameters used to date the Quelccaya cores. Annual signals are recorded in microparticle concentrations (diameters $\geq 0.63 \leq 16.0\mu$m per ml of sample) oxygen isotope ratios and visible stratigraphy. For stratigraphy, a single solid line represents a normal dry-season dust layer; a single dashed line and double dashed line represent very light and light dust layers, respectively. Series of X's symbolize diffuse dry season layers (Thompson *et al.*, 1986).

illustrates a 50-year period A.D. 1775-1825, from the Quelccaya Ice Cap. Distinct annual layers are preserved in particles, visible stratigraphy and $\delta^{18}O$ and permit very accurate dating of the ice cores (Thompson *et al.* 1986). On the Quelccaya ice cap the least negative oxygen isotopes, highest concentration of insoluble and soluble dust and the visible dust layers occur during the dry season – May through October. Figure 27.4 illustrates the detailed stratigraphy preserved in particle concentrations and $\delta^{18}O$ in snow pits from the summit of the Dunde ice cap, China. The average measured accumulation rate from 31 stakes in the summit strain network for the 1986-1987 accumulation year was 80cm of snow or 0.34m of water equivalent. On the Dunde Ice Cap high dust concentrations occur during the late fall to early spring, prior to the onset of the monsoon snowfall, while the least negative $\delta^{18}O$ values occur in August and September after the monsoon season. The low average $\delta^{18}O$ values and large seasonal variations in $\delta^{18}O$ in snow accumulating on the Quelccaya ice cap can be quantitatively explained as a result of atmospheric processes between the oceanic source of water vapor and Quelccaya that determine the isotopic composition of snow deposited on the ice cap, and local conditions determining how the depositional isotope signal is modified during firnification (Grootes *et al.* 1989). However, this explanation for the observed $\delta^{18}O$ variations is probably not unique as it does not consider the very large vertical height of convective thunderstorms in the Amazon Basin which are often two to three times higher than the Andes. Thus, the observed $\delta^{18}O$ in precipitation is the result of condensation at many different temperatures throughout the vertical column. The comparison of the measured accumulation with the dust and oxygen isotope records confirms the annual cycle in these parameters.

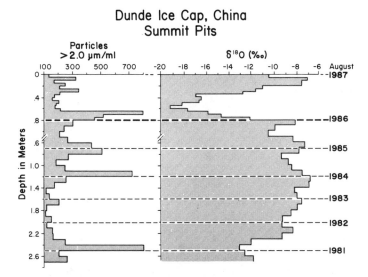

Figure 27.4 The stratigraphic parameters used to date the Dunde cores are illustrated for two snow pits, sampled in 1986 and 1987. The 1986-87 accumulation was determined by measuring accumulation stakes. Annual signals are recorded in microparticle concentrations (diameters ≥ 2.0 and $\leq 60 \times \mu m$ per ml of sample) and oxygen isotope ratios. The $\parallel$ symbol indicates where the 1987 and 1986 snow pit records are joined.

Table 27.1 presents the correlations of decadal averages for Quelccaya ice cap core 1 and summit core for accumulation, $\delta^{18}O$ and dust for the period A.D. 1570-1980. The $\delta^{18}O$ records are most strongly correlated and the dust records are least well correlated. Table 27.1 illustrates that the reproducibility of any given chemical constituent of the record is partially a function of the parameter being measured. Stable isotopes are more highly correlated in part because they undergo smoothing due to vapor and molecular diffusion during the firnification process. Thus, some of the "noise" or detail in the record is smoothed out and lost in the process. A comparison of the annual oxygen isotope data between core 1 and summit core for the period A.D. 1500-1983 yields an r value of 0.87, significant at the 99.9% level only slightly less than the decadal correlation of 0.95. Particles, on the other hand, represent physical entities within the ice and are not subject to diffusion. Moreover, the position of the drill site upwind or down wind of the summit could greatly influence the concentration of dust. Differences may also result from the fact that the measured annual layers at the core 1 site average 19cm over the past 1000 years versus 18cm at the summit core site. The difference in average layer thickness is a result of the different flow regimes and vertical strain rates at the two sites (Thompson *et al.* 1985). Thicker individual annual layers contain a greater volume of water equivalent for the same year which produces a lower particulate count per unit volume per sample.

Table 27.1 Statistical correlations of decadal averages for Core 1 and Summit Core ice core parameters for the period A.D. 1570-1980.

	r	r^2
Accumulation	.89*	79%
Oxygen isotope ratio	.95*	90%
Microparticle concentration	.56*	31%

*significant at 99.9% level

In areas where it can be demonstrated that little or no mass loss occurs due to melting or removal by wind, ice cores provide the very best record of the past variations in precipitation available. The annual accumulation rate can be determined by measuring layer thickness changes based on annually varying ice core parameters such as dust, $\delta^{18}O$ and ionic concentrations. Records covering 1500 years have been produced from the Greenland ice sheet (Reeh *et al.* 1978) and from the tropical Quelccaya ice cores (Thompson *et al.* 1985). The annual records of annual layer thickness from Quelccaya are compared in Figures 27.5-27.7

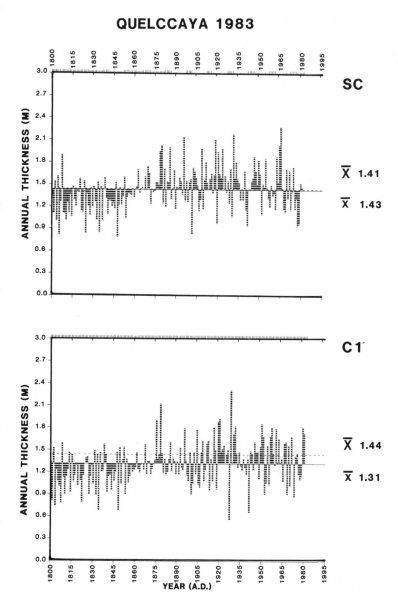

QUELCCAYA 1983

Figure 27.5 Annual layer thicknesses were determined in the summit core (SC) and core 1 (C-1) for the period A.D. 1800 to 1980. Large $\bar{X}$ is the average for entire core, small $\bar{x}$ is the average for the 180 year period illustrated.

which illustrate substantial annual variability, due in part to the effects of drifting and to drill site position on the ice cap.

On the Quelccaya ice cap the annual net accumulation record of A.D. 1915-1984 compares well with annual changes in Lake Titicaca water levels and with annual precipitation at El Alto (La Paz, Bolivia) suggesting that the 1500-year net accumulation record from Quelccaya may serve as a proxy for water level changes in Lake Titicaca (Thompson *et al.* 1988a). The

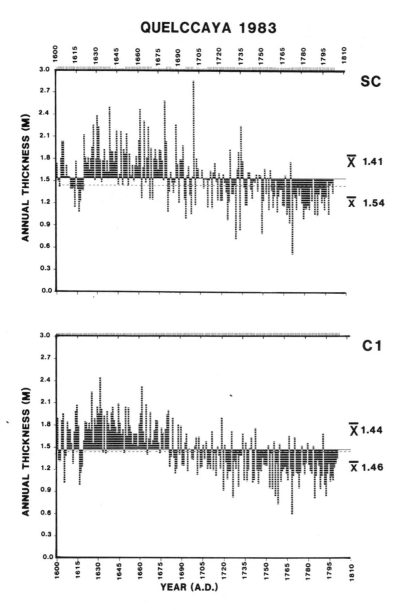

Figure 27.6 Identical to Figure 27.5, but representing the period from A.D. 1600 to 1800, except the small x̄ is the average for the 200-year period illustrated.

ice-core record suggests that climatic variability, reflected in lake level changes, has strongly influenced fluctuations in agricultural activity in southern Peru (Thompson *et al*. 1988a). As historic and prehistoric highland civilizations in Peru, Ecuador and Bolivia were largely agrarian based, and since the high plateau areas (being at the upper limits of agriculture) are climatically sensitive (Cardich 1985) it is likely that climate played an important role in the survival and economic well being of these cultures (Thompson and Mosley-Thompson 1989). The accuracy of the precipitation reconstructions, particularly in the deeper core sections,

526

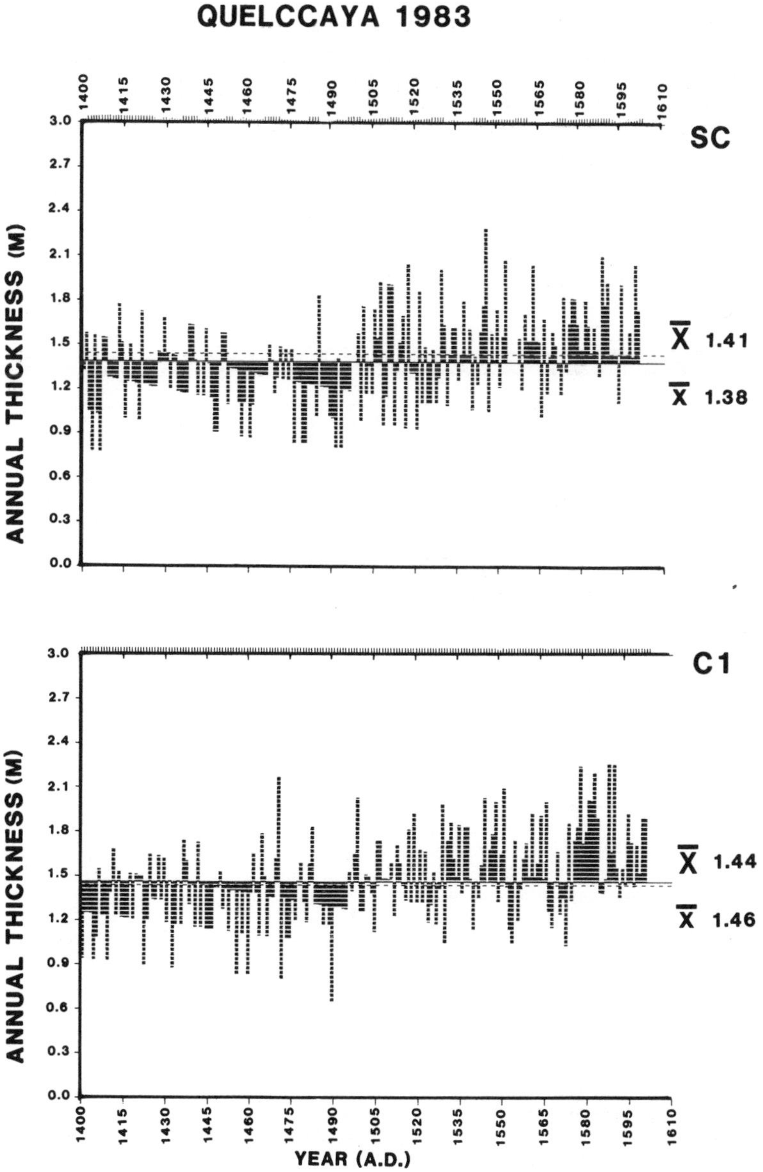

QUELCCAYA 1983

Figure 27.7 Identical to Figure 27.5, but representing the period from A.D. 1400 to 1600, except the small x̄ is the average for the 200-year period illustrated.

depend on the application of appropriate constraints to the ice flow model (Paterson and Waddington 1984). Comparisons between accumulation (Figures 27.5-7) $\delta^{18}O$ (Figures 27.8-10) and particulates (Figures 27.11-13) illustrate that caution must be exercised in the interpretation of climatic and environmental history from a single ice core record. Figures 27.5, 27.6 and 27.7 illustrate that in both Quelccaya ice cores (1) A.D. 1400 to 1500 was a period of reduced annual layer thickness, (2) starting ~A.D. 1500 and lasting until

527

~A.D. 1720 was a period of enhanced annual layer thickness, (3) A.D. 1720 marks the beginning of a long period of reduced annual layer thickness which lasted until about A.D. 1860 and (4) the period of transition between this period of reduced annual layer thickness which characterized the 18th and 19th century and the slightly above average annual layer thickness of the twentieth century is marked by a transitional period A.D. 1860 to 1875 characterized by very little interannual variability in layer thickness. Figures 27.8, 27.9 and

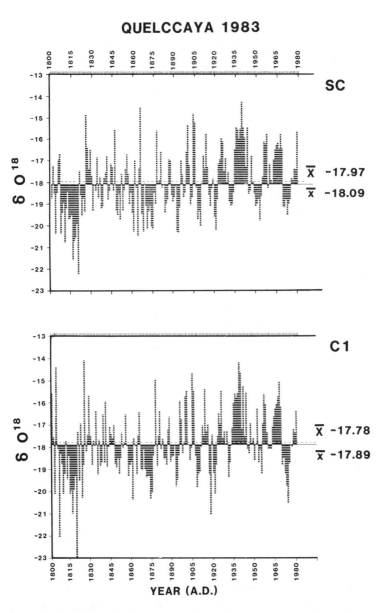

Figure 27.8 Annual $\delta^{18}O$ values determined in the summit core (SC) and core 1 (C-1) for the period from A.D. 1800 to 1980. Large $\bar{X}$ is the average for the entire core, small $\bar{x}$ is the average for the 180 year period illustrated.

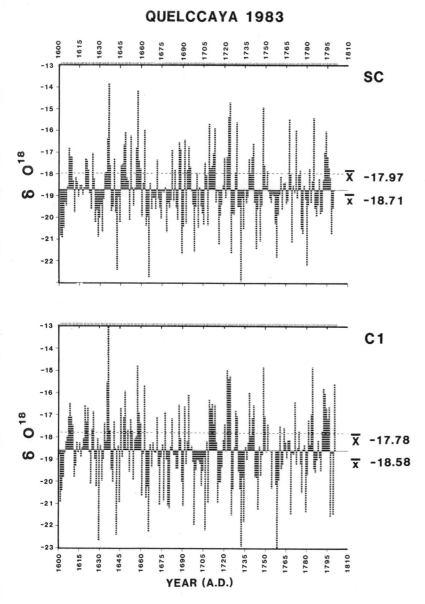

Figure 27.9 Identical to Figure 27.8, but representing the period from A.D. 1600 to 1800, except the small x̄ is the average for the 200-year period illustrated.

27.10 show that for both Quelccaya ice cores: (1) A.D. 1400 to 1530 was a period of average interannual isotopic variability; (2) around A.D. 1530 the annual averages of $\delta^{18}O$ become more negative and remain generally more negative until around A.D. 1880 and (3) the annual averages of $\delta^{18}O$ of the twentieth century are generally less negative. These annual oxygen isotopic values indicate a greater interannual variability during the LIA which is particularly pronounced in Figure 27.9 representing the period A.D. 1600 to 1800. Figures 27.11, 27.12

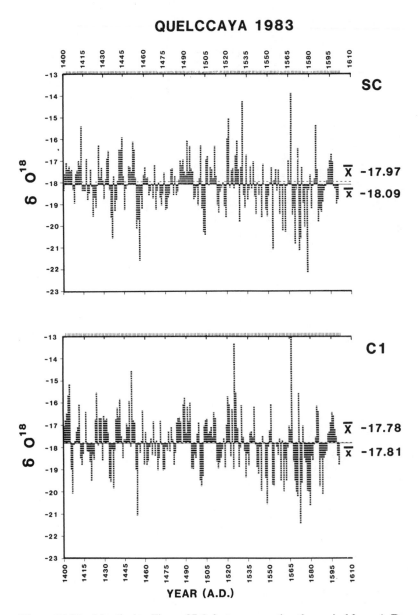

Figure 27.10 Identical to Figure 27.8, but representing the period from A.D. 1400 to 1600, except the small x̄ is the average for the 200-year period illustrated.

and 27.13 illustrate the interannual variability in total particles (insoluble dust). These data indicate a greater variability both on an interannual comparison and between cores throughout the A.D. 1400 to 1984 period than seen in either the accumulation or oxygen isotope records. Total particle concentrations increase slightly during the LIA period as can be seen in Table 27.2. These figures demonstrate that it is essential that spatial, as well as temporal, variability of the physical and chemical stratigraphic records be determined.

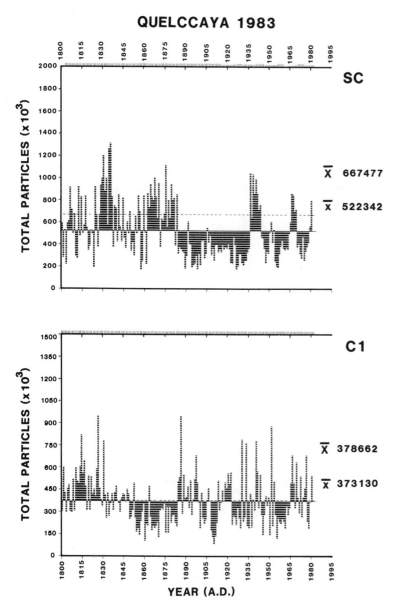

Figure 27.11 Mean annual particulate concentration (diameters from ≥0.63 and ≤16.0μm per ml) in the summit core (SC) and core 1 (C-1) for period A.D. 1800 to 1980. Large X̄ is the average for the entire core, small x̄ is the average for the 180 year period illustrated.

Figure 27.14 shows the annual records of total particles (diameter >0.63μm) conductivity, δ^{18}O and net accumulation for the last 500 years as reconstructed from the Quelccaya summit ice core. The A.D. 1600 eruption of Huaynaputina, the most explosive event recorded in the central Andes of Peru, is very distinct in the particulate and conductivity records (Thompson *et al.* 1986). However, no marked increase in insoluble or soluble dust occurs in the 1815 to

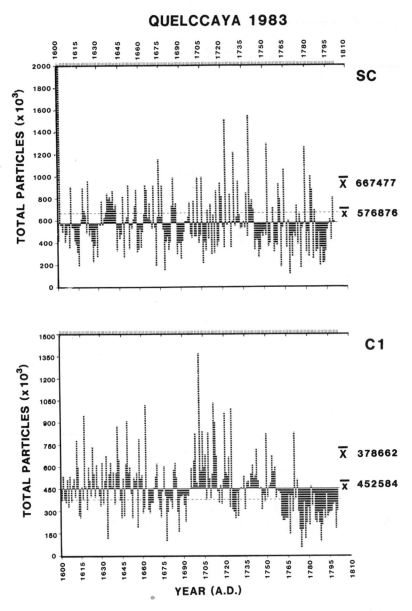

Figure 27.12 Identical to Figure 27.11, but representing the period from A.D. 1600 to 1800, except the small x̄ is the average for the 200-year period illustrated.

1820 period associated with the 1815 eruption of Tambora. This is not unexpected in view of predominance of the local Altiplano as the major dust source which may mask more distant dust events. The $\delta^{18}O$ record, however, shows a prominent cooling which reaches a minimum in A.D. 1819-1820. The most negative $\delta^{18}O$ values in the entire 1500-year record occur in the wet season snowfall of this year as illustrated in Figure 27.3.

During the LIA in southern Peru, A.D. 1490-1880, microparticles and conductivities were

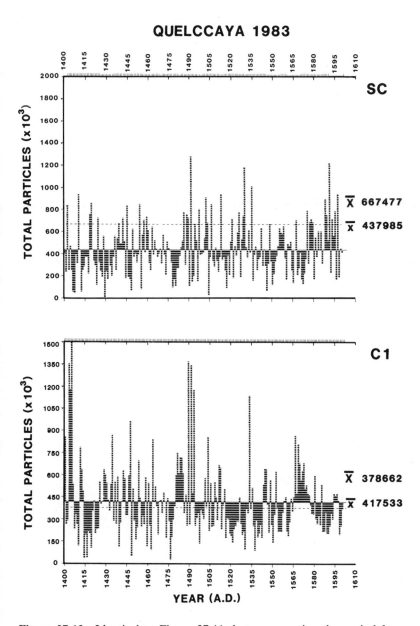

Figure 27.13 Identical to Figure 27.11, but representing the period from A.D. 1400 to 1600, except the small x̄ is the average for the 200-year period illustrated.

20 to 30% above the respective averages for the entire core. Increases in particulates may result from increased atmospheric impurities and/or decreased accumulation. The accumulation history (Figures 27.5, 27.6, 27.7 and 27.14) from Quelccaya is well documented which suggests that the increase in particulates must be due to increased atmospheric loading, not decreased accumulation. Figure 27.14 illustrates that the period A.D. 1500 to 1720 was the wettest interval in the 600 years. Similarly, A.D. 1720 to 1860 was very dry and yet both

Table 27.2 Quelccaya summit and Core 1 data summary.

Time (A.D.)	$\delta^{18}O$ ($^o/_{oo}$)		Total Particles ≥ 0.63 to ≤ 16.0 μm in diameter per ml of sample		Conductivity (μS cm-1)	
	$\bar{x}$	(Δ)	$\bar{x}$	(Δ)	$\bar{x}$	(Δ)
			Summit Core			
1880-1983	-17.69		430,000		1.95	
		0.98		158,000		0.85
1530-1880	-18.67		588,000		2.80	
		0.85		105,000		0.63
1250-1530	-17.82		483,000		2.17	
% change: 1880-1983 1530-1880	5.3%		26.8%		30.4%	
% change: 1250-1530 1530-1880	4.6%		17.9%		22.5%	
			Core 1			
	$\bar{x}$	(Δ)	$\bar{x}$	(Δ)		
1880-1983	-17.85		373,000			
		0.66		47,000		
1530-1880	-18.51		420,000			
		0.65		31,000		
1250-1530	-17.86		389,000			
% change: 1880-1983 1530-1880	3.6%		11.2%			
% change: 1250-1530 1530-1880	3.5%		7.4%			

particulates and conductivities remain unchanged from previous wet period values. Thus, the variations in microparticles and conductivity concentrations cannot be explained simply by changes in the rate of snow accumulation. Preliminary SEM and light microscope analyses of insoluble particles show no significant changes in the types of particles deposited during the LIA. Therefore, it is most likely that increased particulate deposition resulted from increased wind velocities across the high altiplano of southern Peru. Additional support for this interpretation is found in that the number of large particles (diameter > 1.59 μm) also increase during the LIA period (Thompson *et al.* 1986). Between A.D. 1530 and 1880 the

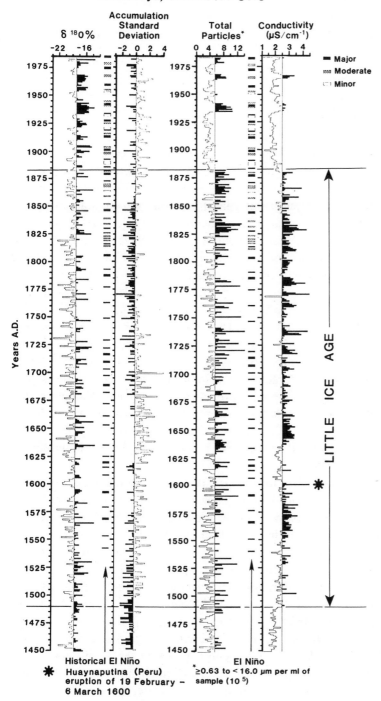

Figure 27.14 Annual variations in microparticle concentrations (total particles) conductivity, oxygen isotope ratios and accumulation (in standard deviation units, based on the last 500 years). 1(sigma) is equivalent to 34cm of accumulation. The "Little Ice Age" stands out clearly and is characterized by increased soluble and insoluble dust and decreased (more negative) $\delta^{18}O$. It appears to have been a major climate event in tropical South America. The large dust event centered on A.D. 1600 was produced by the February 19-March 6 A.D. 1600 eruption of Huaynaputina, Peru. Also shown is the historical El Niño record from Quinn *et al.*, 1987 (cf. Quinn and Neale, this volume).

average oxygen isotope ratios are 0.9‰ lower than in preceding or subsequent periods (see Table 27.2).

27.5   Abrupt climatic changes

The total ecosystem, and particularly the future well-being of man, is clearly as affected by the "rate of climatic change" as by the magnitude of climate change which actually occurs. In general, abrupt changes in climate denote a rupture of the established range of experience, and would be an unexpected and surprisingly fast transition from one state to another (Berger and Labeyrie 1987). Recent research (Dansgaard *et al*. 1989; Jouzel *et al*. 1987; Oeschger *et al*. 1984; Thompson and Mosley-Thompson 1987, 1989b) demonstrates the variety of parameters which can be measured for these periods and how ice core records can be used to define very precisely the rate of climate change.

These abrupt changes seem to occur on a spectrum of time scales. For example, in Greenland ice cores the transition from the last glacial to the present interglacial (~10,000 years ago) occurred in less than a hundred years (Thompson 1977) or perhaps in less than thirty years (Herron and Langway 1985) while the termination of the Younger Dryas climate event may have occurred in less than twenty years (Dansgaard *et al*. 1989). In the Dunde ice cores (China) the glacial-interglacial transition in dust concentration is completed within a 30cm section of ice representing approximately forty years (Thompson *et al*. 1989a; Thompson *et al*. 1990). These terminations may reflect steep gradients across the site in question, such as frontal zones in the climatic system, however, when they occur at the same time at many points on the globe, then they must reflect abrupt changes in the climate system which are not just site specific.

The annual record of climate variations for the last 500 years from the upper portion of the 1500-year Quelccaya, Peru, ice core record (shown in Figure 27.14) shows abrupt changes in tropical South America (Thompson and Mosley-Thompson 1987; 1989) with the transition from the LIA to the warmth of the current century occurring over a two to three year period centered on A.D. 1880. It was marked by a sharp change from the high amplitude seasonal oxygen isotope variations which characterized the Little Ice Age (LIA) period (Figure 27.15). The seasonal $\delta^{18}O$ range which averages 2‰ for the period A.D. 1880 to 1980 is twice as large, averaging 4‰, for the LIA period (A.D. 1520 to 1880) with a maximum of 13‰ in A.D. 1819-1820. Both cores illustrate a period during the LIA from A.D. 1590 to 1630 when the large seasonal $\delta^{18}O$ oscillations abruptly decrease. The large annual range in $\delta^{18}O$ during the LIA may reflect increased seasonality during this period, a feature noted in the historical records of Europe (Gribbin and Lamb 1978; Lamb 1984a, b). Alternatively, the annual signal may have been better preserved under the climatic conditions which existed during much of the LIA period. Regardless of the cause, the change in ice core records clearly illustrates that rapid alternation of the climate or environmental conditions occur around A.D. 1520, 1590, 1630 and at the termination of the LIA in A.D. 1880. Thus, during the LIA not only was there an apparent increase in the interannual variability as seen in Figures 27.8, 27.9 and 27.10, but also an increase in intrannual variability as seen in Figure 27.15.

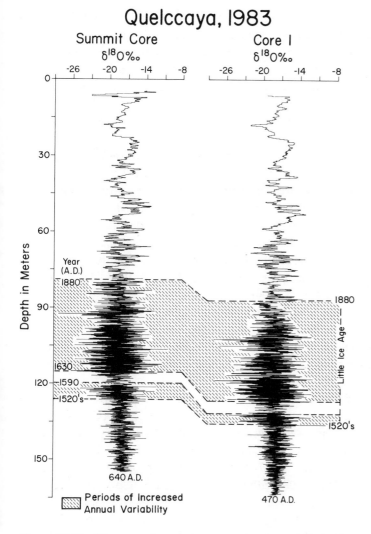

Figure 27.15 The annual range in oxygen isotope values in the Quelccaya summit core and core 1. The annual range in $\delta^{18}O$ increases by 100 percent during the Little Ice Age, reaching the largest values between A.D. 1650 to 1710. Low ranges occur in both pre- and post-Little Ice Age periods as well as in the unusual period from A.D. 1590 to 1630. Periods of high oscillations are shaded.

27.6   Comparison of Quelccaya and Dunde Ice core records

The Qinghai-Tibetan Plateau is a major heat source for the lower atmosphere and thus, greatly affects the hemispheric circulation. The present-day northern summer-monsoon circulation is driven by the relative warmth of the African-Asian landmass in comparison with the cooler surrounding ocean (Fein and Stephens 1987).

The interannual fluctuations in the summer monsoon rains over southeastern Asia have a

537

profound socio-economic impact on China and India. Weak monsoons are associated with drought, crop failure, and, in extreme cases, famine. Strong monsoons are associated with devastating floods and the accompanying loss of life, property and crops. The importance of the Qinghai-Tibetan Plateau as a heat source has been noted by many authors (e.g., Flohn 1957, 1965, 1968; Koteswaram 1958; Murakami 1958, 1981; Virji 1979; Yeh 1981; Luo and Yanai 1983; Reiter 1983; Lau and Li 1984; Barnett *et al.* 1988; Hansen *et al.* 1988).

Eastern Asia is directly influenced by the thermal and dynamical forcing enhanced by this huge elevated landmass. Sensible heat flux over the semi-arid western region of the Plateau and latent heat release above the Himalayas, contribute to a strong tropospheric heat source which maintains the large-scale Asian monsoon circulation. Lau and Li (1984) point out that not only does the Plateau determine the large-scale circulation, but also strongly influences synoptic scale monsoon events over China. Unfortunately, there are no long historical records of climatic variability from the far western sections of the Plateau. However, the detailed historical records of drought, flood and dust storms from central and eastern China provide independent ties for assessing the spatial significance of events on the Plateau.

The present-day glaciers in China are widely scattered, mainly across the Qinghai-Tibetan Plateau, and cover a large area of about 57,000km^2 (Shi and Wang 1979). The snow line elevations range from 3000m in Altay Shan in northwestern China, to 6200m in southern and western areas of the Plateau. The annual precipitation ranges between 0.8 to 2.5m in parts of the outer mountains, and declines to only 0.2 to 0.3m on the inner Plateau. The greater absorption of solar radiation over the vast expanse of the Plateau than on the surrounding lowlands also raises the snowline elevations.

From about A.D. 1400 to 1900, temperatures in China were 1 to 2°C lower than present and resulted in prominent advances of glaciers. According to Zhu Kezhen (Chu Ko-Chen) (1973) the lowest temperatures of the last 500 years in central and eastern China occurred around A.D. 1700, clearly within the LIA. The Dunde oxygen isotope records reveal a cold period in western China from A.D. 1670 to 1740. The comparison of the Dunde ice cap decadal average $\delta^{18}O$ from A.D. 1580 to 1980 with Quelccaya (Figure 27.16) shows a general correspondence to Landsberg's first approximation for decadal Northern Hemisphere temperature departures for this period (Landsberg 1985; Groveman and Landsberg 1979). The correlation coefficients between decadal oxygen isotope records from Dunde and Quelccaya records and Landsberg's northern hemisphere temperature departures are presented in Table 27.3. All ice core records in Figure 27.16 show isotopic values below the A.D. 1880-1980 mean for most of the LIA, while all of the records illustrate the warming of the twentieth century.

The study of climate change on the time scale of centuries to millennia provides a valuable time perspective for interpreting climatic patterns of the present century and assessing the potential future effects of climate change due to a greenhouse warming. For example, there has been considerable debate regarding when and if a "greenhouse" signal will rise above the "climate noise" or variance within the unperturbed climatic system on time scales of decades to centuries. However, to make more confident projections about future climate change, it is necessary first to describe and understand the sources of that variance on the same time scale (Zhang and Crowley 1989).

In the monsoon region of China the precipitation has a marked annual cycle with 80% of the rain or snow (higher elevation sites) generally occurring in the summer half-season. The

Table 27.3 Statistical correlations (r) among the decadal averages of $\delta^{18}O$ for Dunde and Quelccaya ice caps and Landsberg's decadal temperature departures. Deviations are calculated with respect to the 1881 to 1980 mean. r is the linear correlation coefficient, r^2 is the coefficient of determination, and p is the level of statistical significance.

	Quelccaya Summit Core $\delta^{18}O$	Northern Hemisphere Temperature Departures
Dunde Core D-1 $\delta^{18}O$	r = 0.433 r^2 = 18.7% p = 99.5%	r = 0.482 r^2 = 23.2% p = 99.8%
Northern Hemisphere Temperature Departures	r = 0.632 r^2 = 39.9% p = 99.9%	

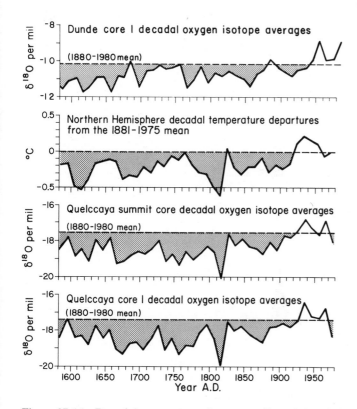

Figure 27.16 Decadal temperature departures (from 1881-1975 mean) in the Northern Hemisphere from A.D. 1580 to 1975 compared with decadal average $\delta^{18}O$ values for both Quelccaya ice cores and the Dunde D-1 core. The dashed line is the A.D. 1881-1980 mean for the $\delta^{18}O$ records.

rainy season is closely associated with the position of the polar front, which is generally located at the northern periphery of the subtropical high. The retreat of the polar front accompanies the advance of the summer monsoon. The intensification of the Hadley circulation is a result of intense summer heating on the Eurasian landmass (e.g., Lau and Li 1984).

Lau *et al.* (1988) summarize the development of the summer monsoon in three stages: (1) the late spring position of the rainbelt is in south China, and its early summer (mainly June and early July) position is in the Yangtze Valley (the so-called *Mei-Yu* or "Plum Rains"); (2) the northernmost positions reach China and inner Mongolia in July-August; (3) in autumn the rainbelt retreats to the south and persistent rain is prominent only in southwest China, although typhoon rains peak at this time and can significantly influence rainfall totals. In winter the whole country is dry except in the Yangtze Valley, where protracted small rains are frequent and under the influence of a (subtropical) southwesterly jet stream.

Figure 27.17 illustrates the initial comparison of decade-averaged accumulation in meters per year for the period A.D. 1610 to 1980 for Quelccaya and Dunde ice core records. These records show a general similarity in net balance for the last 400 years for these two widely separated areas. The correlation between the decadal averaged net accumulation for Dunde and Quelccaya increase from r = 0.29 with no lag to r = 0.57 (significance = 99%) when Dunde record is lagged 50 years. It has been recognized since the turn of the century that teleconnections link these two widely separated sites. The El Niño-Southern Oscillation

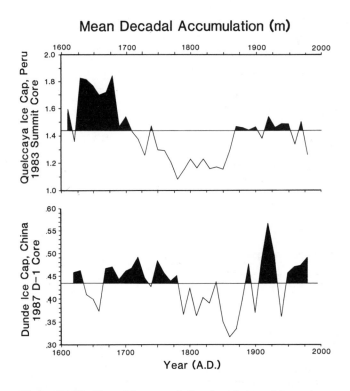

Figure 27.17 Decadal accumulation (net balance) in meters for the Quelccaya, Peru summit core compared with the Dunde, China core D-1 for the period A.D. 1610 to 1980.

phenomenon perturbs the ocean-climate system of the Pacific basin episodically and is related to anomalous weather patterns covering much of the globe (Rasmussen and Wallace 1983; Nicholls 1987; Enfield 1989). The longer-term similarity of patterns in accumulation illustrated in Figure 27.17 suggest that teleconnections may exist between these widely separated areas not only for high frequency events such as El Niño-Southern Oscillation phenomenon, but also for low frequency events lasting centuries. Further evidence for large scale teleconnections are also suggested by the comparison of the Quelccaya and Dunde ice cap accumulation records with those from Mt.Logan, Canada (Holdsworth *et al.*, this volume).

27.7 Climate, man and the environment in South America

The historical record of man's activities in pre-Spanish South America is scant and incomplete, and the process of piecing it together is hampered by the lack of written language. Archaeologists, through excavations of sites of pre-Hispanic civilizations in Ecuador and Peru, have been able to better define the chronology and activities of civilizations in these regions. Additional pieces of the historical puzzle can be provided by non-archeological sources. For example, the Quelccaya ice cores have provided records of climatic events which may have played a major role in prehistoric man's activities.

The archeological sites in Peru may be assigned to either coastal or highland cultures. The accumulation (meters of ice) record for the southern Peruvian highlands presented in Figure 27. 18 is a composite of the decadal averages from both core 1 and summit core. Four very marked dry periods are found: A.D. 540 to 610, A.D. 650 to 730, A.D. 1040 to 1490 and A.D. 1720 to 1860, and two distinct wet periods occur from A.D. 760 to 1040 and A.D. 1500 to 1720. It is important to note that under the present climate regime, coastal Ecuador and coastal Peru suffer from heavy rains in association with El Niño-Southern Oscillation events (ENSO) while the highlands of southern Peru, where the Quelccaya ice cap is located, often experience drought conditions (Thompson *et al.* 1984b; Lam and Del Carmen 1986). Paulsen (1976) reported a similar relationship in the longer term archeological records establishing the rise and fall of coastal cultures of Peru and Ecuador. For comparison, the cultural record (dating based largely on highly refined ceramic sequences and some ^{14}C measurements) from Peru and Ecuador are also presented in Figure 27. 18. Paulsen reported that highland and coastal cultures seemed to flourish out of phase with each other, i.e., highland cultures flourish when coastal cultures decline and *vice versa*. Since both the prehistoric coastal and highland civilizations were largely agrarian and both the coastal areas (due to dependence on a limited water supply) as well as the high plateau areas (being the upper limits of agriculture) are very climatically sensitive (Cardich 1985) it is certain that climate played an important role in the survival of these cultures. However, the exact nature of that role of climatic variability as a dominant independent variable in prehistoric Andean cultural changes is debated (Stahl 1984; Paulsen 1984).

The comparison of the records of accumulation from the Quelccaya ice cores and the archeological history demonstrate that highland cultures flourished during wet periods on the Bolivian-Peruvian plateau and conversely, coastal cultures flourished when the highlands were dryer. Assuming that the same seesaw relationship which currently exists during ENSO

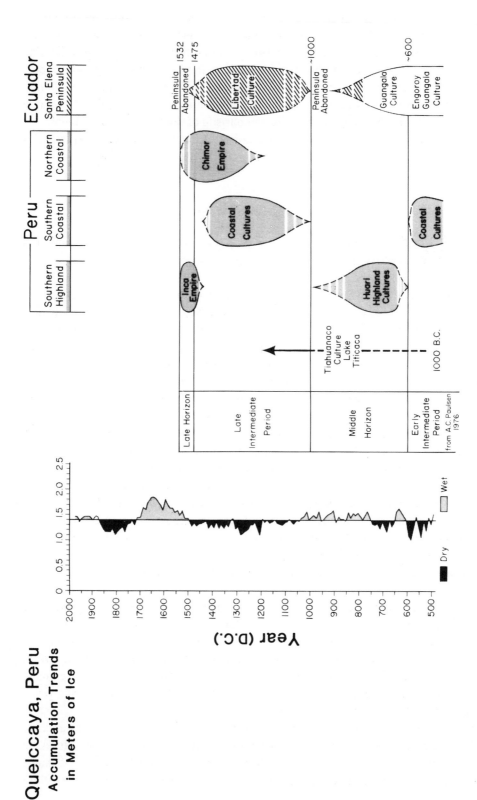

Figure 27.18 Quelccaya decadal accumulation trends in meters of ice are presented as a composite of core 1 and summit core records. Wet and dry periods are noted. On the right the periods of the rise and fall of coastal and highland cultures of Ecuador and Peru are illustrated (taken mainly from Paulsen, 1976).

events (Thompson *et al.* 1984b) was valid over the longer periods of El Niño-like ocean and atmospheric circulation patterns, then wetter coastal conditions would be expected during periods of highland drought. The fact that coastal cultures flourished during the dryer highland periods implies that this seesaw relationship existed over extended time periods.

Correlations have been established between the shifts of cultural activity between highland and coastal civilizations in the region and the accumulation record obtained from the ice cores. This record may also provide a key to reconstructing the succession of habitation and abandonment of sites of so called "lost" civilizations uncovered by archaeologists, such as the Gran Pajaten. However, a correlation between past environmental events and cultural events does not necessarily imply a direct connection. When there is a connection, it involves that culture's adaptation to new conditions and the mode of the adaptation depends on the culture.

27.8 The global perspective

The assimilation of multifaceted ice core data sets from polar and non-polar regions makes possible the determination of both temporal and geographical variability of any or all the individual ice core parameters, i.e., dust, chemistry, gas content, isotopes, etc. The $\delta^{18}O$ records are discussed on a global scale using examples in Figure 27.19. The $\delta^{18}O$ records reflect in varying degrees (1) air temperature at which condensation takes place, (2) atmospheric processes between the oceanic source of water vapor and the site of deposition, (3) local conditions under which the isotope signal is modified during firnification, (4) the surface elevation of the depositional site and (5) the latitude of the site (see Dansgaard *et al.* 1973 and Bradley 1985, for review). Although the correlation of atmospheric temperatures with oxygen isotope ratios and its spatial representativeness is still under discussion, the method continues to be used widely as a proxy for climate and in particular for temperature (Dansgaard *et al.* 1973; Jouzel *et al.* 1983; Thompson *et al.* 1986 and Peel *et al.* 1988). Figure 27.19 provides a preliminary global perspective of the decadal and centennial variations for the periods ranging between 550 to 1000 years B.P. from five sites across the globe. The sites are, from north to south, Camp Century, Greenland (Johnsen *et al.* 1970); Dunde ice cap, China (this paper); Quelccaya ice cap, Peru (Thompson *et al.* 1986) and South Pole and Siple Station, Antarctica (Mosley-Thompson *et al.*, in press). In Figure 27.19, isotopically heavier, and hence warmer, periods are illustrated by shaded areas and isotopically lighter, and hence colder, periods are unshaded. The records show a large diversity in detail not unlike that which would be found if only five widely dispersed meteorological stations were used to reconstruct global temperatures. However, several features stand out, such as the similarity of variability of isotopes in the two northern hemisphere sites with a rather pronounced 180-year oscillation in the China decadal oxygen isotopic record. In the southern hemisphere there is a strong similarity in the oxygen isotopic records from the tropical Quelccaya ice cap, Peru and South Pole, Antarctica. The LIA stands out as a period of more negative isotopic values from ~A.D. 1530 to 1900, while before and after, both locations were generally isotopically less negative. Of these five records, the Siple Station record is unique because during the LIA this site exhibits less negative oxygen isotopes and hence presumably warmer conditions (see Mosley-Thompson, this volume). All sites show isotopically less negative

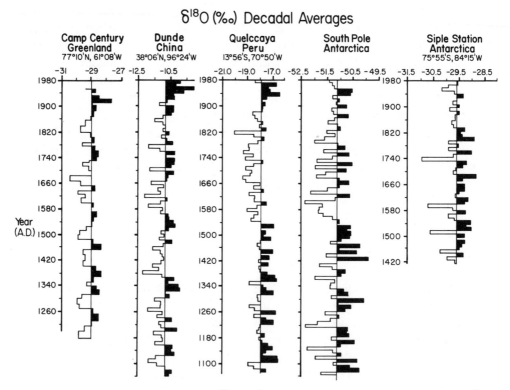

Figure 27.19 Decadal averages of the $\delta^{18}O$ records in a north-south global transect from Camp Century, Greenland to the South Pole, Antarctica. The shaded areas represent isotopically less negative, or warmer, periods and the unshaded areas represent isotopically more negative, or colder, relative to the respective means of the individual records.

conditions during the twentieth century, except Siple Station which is, in general, isotopically more negative for the last 100 years. However, for the last thirty years the isotopic evidence (Peel *et al.* 1988; Mosley-Thompson *et al.*, in press) indicates a general trend to less negative values and thus warmer conditions which is consistent with the warmer overall, atmospherically-measured temperatures in the Antarctic Peninsula from 1960-1980.

A most striking feature of the $\delta^{18}O$ records in Figure 27.19 is the extreme warmth in central China during the last 60 years, with the warmest decades being the 1940s, 1950s, and 1980s. Using the $\delta^{18}O$ record as a proxy for temperature, the last 60 years constitute one of the warmest periods in the entire record, equalling levels of the Holocene maximum 6000 to 7000 yr B.P. (Thompson *et al.* 1989a,; Thompson *et al.*, in press). Model results of Hansen *et al.* (1988) suggest that the central part of the Asian continent may be one of the first places to exhibit an unambiguous signal of the anticipated "greenhouse warming." Certainly the Dunde ice core results suggest that the recent warming on the Tibetan Plateau has been substantial. Recent radiosonde data from Southern India (Flohn and Kapala 1989) show that, in fact, the average tropospheric temperature has increased nearly 1°C since 1965. It must be cautioned that more robust temperature-isotope transfer functions must be developed for the Tibetan Plateau and indeed for all regions of the earth where isotopic ice core records are being used as proxy indicators of temperature.

In conclusion, the ice core climatic and environmental records from low, middle and high latitudes can greatly increase our knowledge and understanding of the course of climatic events of the past. This is essential to predict future climatic oscillations, which may or may not be dominated by increasing greenhouse gas concentrations. The forcing factors, internal and external, which have operated in the past to prevent climatic stability will continue to operate and influence the course of events (Grove 1988). The cores from the Dunde ice cap, China cores provide the first ice core record of the Holocene/late Pleistocene climate from the subtropics. The stable isotope record indicates that the last 60 years on the Tibetan Plateau constitutes one of the warmest periods in the entire record. Events such as the LIA are global in scale and thus result from large scale climatic forcing which influences the entire earth system. The manifestation of the LIA in any given record is quite variable and more distinct in the higher elevation sites, such as Dunde, Quelccaya and South Pole, than in the lower elevation sites of Camp Century and Siple Station. This may imply importance of climatic change as a function of elevation as well as of the more subtle changes in climate like the LIA. The latter are more clearly recorded farther from the mitigating influence of the oceans. The high temporal resolution available from ice cores illustrates that the transition from climatic "norms" may be abrupt on the scale of the LIA and glacial/interglacial events. The tropical and subtropical ice core records provide the potential to establish long records of El Niño-Southern Oscillation events and monsoonal variability. Information of variations in their magnitude and frequency through time potentially may provide information on the causes of these global-scale events. Moreover, a global array of cores can be used to establish changes in precipitation over the last 1000 to 2000 years.

References

Barnett, T. P., L. Dümenil, V. Schlese, E. Roeckner, 1988. The effect of Eurasian snow cover on global climate. *Science*, 239, 504-507.
Berger, W. H. and L. D. Labeyrie, 1987. *Abrupt Climate Change, Evidence and Implications*. NATO Advanced Science Institutes Series C: Mathematical and Physical Sciences, Vol. 216, D. Reidel Publishing Co., Dordrecht, 425 pp.
Bradley, R. S., 1985. *Quaternary Paleoclimatology*, Allen and Unwin, Boston, 472 pp.

Cardich, A., 1985. The fluctuating upper limits of cultivation in the Central Andes and their impact on Peruvian prehistory. *Advances in World Archaeology*, Academic Press, Inc., New York, 4, 293-333.

Dansgaard, W., 1954. Oxygen-18 abundance in fresh water. *Nature*, 174, 234.
Dansgaard, W., S. J. Johnsen, H. B. Clausen and N. Gundestrup, 1973. *Stable Isotope Glaciology*. Meddelelser om Gronland, 197, 1-53.
Dansgaard, W., J. W. C. White and S. J. Johnsen, 1989. The abrupt termination of the Younger Dryas climate event. *Nature*, 339, 532-534.
Dean, G. A., 1971. The three dimensional wind structure over South America and associated rainfall over Brazil. *Report LAFE-164*, Dept. Meteorology, Florida State University.

Enfield, D. B., 1989. El Niño, past and present. *Reviews of Geophysics*, 27, 159-187.

Fein, J. S. and P. L. Stephens (Eds.) 1987. *Monsoons*, Wiley, New York, 632 pp.
Flohn, H., 1957. Large-scale aspects of the summer monsoon in South and East Asia. *Journal of the Meteorological Society of Japan*, 75th Anniversary Volume, 180-186.

Flohn, H., 1965. Thermal effects of the Tibetan Plateau during the Asian monsoon season. *Australian Meteorological Magazine*, 49, 55-58.

Flohn, H., 1968. Contributions to a meteorology of the Tibetan highlands. *Report No. 130*, Department of Atmospheric Sciences, Colorado State University, Ft. Collins, 120 pp.

Flohn, H. and H. Kapala, 1989. Changes of tropical sea-air interaction processes over a 30-year period. *Nature*, 338, 244-246.

Gribbin, J. and H. H. Lamb, 1978. Climatic change in historical times. In: *Climatic Change*. J. Gribben, (Ed.) Cambridge University Press, 68-82.

Grootes, P. M., M. Stuiver, L. G. Thompson and E. Mosley-Thompson, 1989. Oxygen isotope changes in tropical ice, Quelccaya, Peru. *Journal of Geophysical Research*, 94(D1) 1187-1194.

Grove, J. M., 1988. *The Little Ice Age*, Methuen, London, 498 pp.

Groveman, B. S. and H. E. Landsberg, 1979. Simulated northern hemisphere temperature departures 1579-1880. *Geophysical Research Letters*, 6, 767-769.

Gutman, G. J. and W. Schwerdtfeger, 1965. The role of latent and sensible heat for the development of a high pressure system over the subtropical Andes, in the summer. *Meteorol. Rundsch.*, 18(3) 69-75.

Hammer, C. U., 1989. Dating by physical and chemical seasonal variation and reference horizons. In: *The Environmental Record in Glaciers and Ice Sheets*, H. Oeschger and C. C. Langway, Jr., (eds)., Wiley and Sons, Chichester, 99-121.

Hansen, J., I. Fung, A. Lacie, D. Rind, S. Lebedeff, R. Ruedy, G. Russell and P. Stone, 1988. Global trends of measured surface air temperature. *J. Geophys. Res.*, 93(D8) 9341-9364.

Herron, M. M. and C. C. Langway, Jr., 1985. Chloride, nitrate, and sulfate in the Dye 3 and Camp Century, Greenland ice cores. In: *Greenland Ice Core: Geophysics, Geochemistry, and the Environment*, C. C. Langway, Jr., H.

Johnsen, S. J., W. Dansgaard, H. B. Clausen, C. C. Langway, 1970. Climatic oscillations 1200-2000 A.D., *Nature*, 322, 430-434.

Johnsen, S. J., W. Dansgaard, H. B. Clausen and C. C. Langway, Jr., 1972. Oxygen isotope profiles through the Antarctic and Greenland ice sheets. *Nature*, 235, 429-434.

Jouzel, J., L. Merlivat, J. R. Petit and C. Lorius, 1983. Climatic information over the last century deduced from a detailed isotopic record in the South Pole snow. *J. Geophys. Res.*, 88(C4) 2693-2703.

Jouzel, J., C. Lorius and L. Merlivat, 1987. Abrupt climate changes: the Antarctic ice core record during the Late Pleistocene. In: *Abrupt Climate Change, Evidence and Implications*, W. H. Berger and L. D. Labeyrie (eds.) NATO Advanced Science Institutes Series C: Mathematical and Physical Sciences, Vol. 216, D. Reidel Publishing Co., Dordrecht, 235-245.

Koteswaram, P., 1958. The easterly jet stream in the tropics. *Tellus*, 10, 43-87.

Krishnamurti, T. N. and H. N. Bhalme, 1976. Oscillations of a monsoon system. Part I. Observational aspects. *Journal of the Atmospheric Sciences*, 33, 1937-1954.

Lam, J. A., and C. Del Carmen, 1986. The evolution of rainfall in northern Peru during the period January 1982 to December 1985, in the coastal, mountain and jungle regions. *Chapman Conference on El Niño*, Guayaquil, Ecuador, April 27-31, 1986.

Lamb, H. H., 1984a. Climate and history in northern Europe and Elsewhere. In: *Climatic Changes on a Yearly to Millennial Basis*. N. A. Morner and W. Karlen (Eds.) D. Reidel Publishing Company, 225-240.

Lamb, H. H., 1984b. Some studies of the Little Ice Age of recent centuries and its great storms. In: *Climatic Changes on a Yearly to Millennial Basis*, N. A. Morner and W. Karlen (Eds.) D. Reidel Publishing Company, 309-329.

Lambert, G., B. Ardouin, J. Sanak, C. Lorius and M. Pourchet, 1977. Accumulation of snow and radioactive debris in Antarctica: a possible refined radiochronology beyond reference levels, *IAHS* 48, 146-158.

Landsberg, H. E., 1985. Historical weather data and early meteorological observations. In: *Paleoclimate Analysis and Modeling*, A. D. Hecht (Ed.) John Wiley and Sons, 27-70.

Lau, K. M. and M. T. Li, 1984. The monsoon of east Asia and its global associations – a survey. *Bulletin of the American Meteorological Society*, 65(2) 114-125.

Lau, K. M., G. J. Yang and S. H. Shen, 1988. Seasonal and intraseasonal climatology of summer monsoon rainfall in East Asia. *Monthly Weather Review*, 116, 18-37.

Luo, H. and M. Yanai, 1983. The large scale circulation and heat sources over the Tibetan Plateau and the surrounding area during the early summer of 1979. Part I: precipitation and kinematics. *Monthly Weather Review*, 111, 922-944.

Mosley-Thompson, E., L. G. Thompson, P. Grootes and N. Gundestrup, 1989. Little Ice Age (Neoglacial) paleoenvironmental conditions at Siple Station, Antarctica, *Ann. Glaciol.*, in press.

Mulvaney, R. and D. A. Peel, 1988. Anions and cations in ice cores from Dolleman Island and the Palmer Land Plateau, Antarctic Peninsula. *Ann. Glaciol.*, 10, 121-125.

Murakami, T., 1958. The sudden change of upper westerlies near the Tibetan Plateau at the beginning of the summer season. *Journal of the Meteorological Society of Japan*, 36, 239-247 [in Japanese].

Murakami, T., 1981. Orographic influence of the Tibetan Plateau on the Asiatic winter monsoon circulation. Part I: large-scale aspects. *Journal of the Meteorological Society of Japan*, 59, 66-84 [in Japanese].

Nicholls, N., 1987. Prospects for drought prediction in Australian and Indonesia. In: *Planning for Drought: Toward a Reduction of Societal Vulnerability*, D. A. Wilhite and W. E. Easterling (Eds.) Westview, Boulder, Colorado, 61-72.

Oeschger, H., J. Beer, U. Siegenthaler, B. Stauffer, W. Dansgaard, C. C. Langway, 1984. Late glacial climate history from ice cores. In: *Climate Processes and Climate Sensitivity*, J. E. Hansen, T. Takahashi (eds.) American Geophysical Union, Geophysical Monograph no. 39, Maurice Ewing Series, 5, 299-306.

Oeschger, H. and W. Dansgaard (eds.) American Geophysical Union Monograph No. 33, 77-84.

Paterson, W. S. B. and E. D. Waddington, 1984. Past precipitation rates derived from ice core measurements, methods and data analysis. *Reviews of Geophysics and Space Physics*, 22(2) 123-130.

Paulsen, A. C., 1976. Environment and empire: climatic factors in pre-historic Andean culture change. *World Archaeology*, 8(2) 121-132.

Paulsen, A. C., 1984. Discussion and criticism, reply. Current Anthropology, 25(3) 352-355.

Peel, D. A., R. Mulvaney and B. M. Davison, 1988. Stable isotope/air-temperature relationships in ice cores from Dolleman Island and the Palmer Land Plateau, Antarctic Peninsula. *Ann. Glaciol.*, 10, 130-136.

Picciotto, E. E. and S. E. Wilgain, 1963. Fission products in Antarctic snow: a reference level for measuring accumulation. *J. Geophys. Res.*, 68, 5965-5972.

Quinn, W., V. Neal, S. Antunez de Mayolo, 1987. El Niño occurrences over the past four and half centuries. *J. Geophys. Res.*, 92(C13) 14449-14461.

Rasmussen, E. M. and J. M. Wallace, 1983. Meteorological aspects of the El Niño/Southern Oscillation. *Science*, 222, 1195-1202.

Reeh, N., H. B. Clausen, W. Dansgaard, N. Gundestrup, C. U. Hammer, and S. J. Johnsen, 1978. Secular trends of accumulation rates at three Greenland stations. *J. Glaciol.*, 20, 27-30.

Reiter, E. R., 1983. Teleconnections with tropical precipitation surges. *Journal of the Atmospheric Sciences*, 40(7) 1631-1647.

Shi, Y-F. and J. Wang, 1979. The fluctuations of climate, glaciers and sea level since Late Pleistocene in China. *Sea Level, Ice and Climatic Change*, IAHS-AISH Publication No. 131, 281-293.

Stahl, P., 1984. Discussion and criticism, on climate and occupation of the Santa Elena Peninsula: implications of documents for Andean pre-history. *Current Anthropology*, 25(3) 351-352.

547

Thompson, L. G., 1977. Microparticles, ice sheets and climate. *Institute of Polar Studies Report* 64, The Ohio State University, 148 pp.

Thompson, L. G., E. Mosley-Thompson, P. M. Grootes, M. Pourchet and S. Hastenrath, 1984a. Tropical glaciers: potential for ice core paleoclimatic reconstructions. *J. Geophys. Res.*, 89(3) 4638-4646.

Thompson, L. G., E. Mosley-Thompson and Benjamin Morales Arnao, 1984b. Major El Niño/ Southern Oscillation events recorded in stratigraphy of the tropical Quelccaya Ice Cap. *Science*, 226(4670) 50-52.

Thompson, L. G., E. Mosley-Thompson, J. F. Bolzan and B. R. Koci, 1985. A 1500-year record of tropical precipitation in ice cores from the Quelccaya Ice Cap, Peru. Science, 229, 971-973.

Thompson, L. G., E. Mosley-Thompson, W. Dansgaard and P. M. Grootes, 1986. The Little Ice Age as recorded in the stratigraphy of the tropical Quelccaya Ice Cap. *Science*, 234, 361-364.

Thompson, L. G. and E. Mosley-Thompson, 1987. Evidence of abrupt climate change during the last 1,500 years recorded in ice cores from the tropical Quelccaya Ice Cap, Peru. In: *Abrupt Climate Change. Evidence and Implications*, W. H. Berger and L. D. Labeyrie (eds.) NATO Advanced Science Institutes Series C: Mathematical and Physical Sciences Vol. 216, D. Reidel Publishing Co., Dordrecht, 99-110.

Thompson, L. G., M. Davis, E. Mosley-Thompson and K. Liu, 1988a. Pre-Incan agricultural activity recorded in dust layers in two tropical ice cores. *Nature*, 336, 763-765.

Thompson, L. G., E. Mosley-Thompson, X. Wu, and Z. Xie, 1988b. Wisconsin/Wurm glacial stage ice in the subtropical Dunde ice cap, China. *GeoJournal*, 17(4) 517-523.

Thompson, L. G., E. Mosley-Thompson, M. E. Davis, J. F. Bolzan, J. Dai, T. Yao, N. Gundestrup, X. Wu, L. Klein and Z. Xie, 1989a. Pleistocene climate record from Qinghai-Tibetan Plateau ice cores. *Science*, 246, 474-477.

Thompson, L. G. and Mosley-Thompson, 1989b. One-half millennia of tropical climate variability as recorded in the stratigraphy of the Quelccaya Ice Cap, Peru. In: *Climate Change in the Eastern Pacific and Western Americas*, D. Peterson (ed.) Monograph 55, American Geophysical Union, 15-31.

Thompson, L. G., E. Mosley-Thompson, M. E. Davis, J. Bolzan, J. Dai, N. Gundestrup, T. Yao, X. Wu, L. Klein and Z. Zichu, 1990. Glacial stage ice core records from the subtropical Dunde Ice Cap, China. *Ann. Glaciol.*, 14, in press.

Virji, H., 1979. *Summer circulation over South America from satellite data*. Ph.D. dissertation, University of Wisconsin, 146 pp.

Yeh, T. C., 1981. Some characteristics of the summer circulation over the Qinghai-Xizang (Tibet) Plateau and its neighborhood. *Bulletin of the American Meteorological Society*, 62, 14-19.

Zhang, J. and T. J. Crowley, 1989. Historical climate records in China and reconstruction of past climates. *Journal of Climate*, 2(8) 835-849.

Zhu, K., 1973. A preliminary study on the climate fluctuations during the last 5000 years in China. *Scientia Sinica*, 16(2) [in Chinese].

28 Ice core evidence from the Antarctic Peninsula region

D. A. Peel

28.1 Modern perspective

28.1.1 Present-day climate distribution

The Antarctic Peninsula is of special climatological interest because it transects the southern circumpolar trough, forming a partial bridge between the Antarctic ice sheet and South America. Moreover it lies close to one of the most important sea-ice producing areas of Antarctica – the Weddell Sea. In this sense it offers a unique observational platform to evaluate the role played by these features in climate change of the Southern Hemisphere. Climate records from the region also will help to forge a connection between records from the interior of the Antarctic ice sheet and those from lower latitudes.

Instrumental weather observations are available at several stations since 1947 (Figure 28.1) a relatively dense network in comparison with other parts of Antarctica. One record at Orcadas (60°44′S, 44°44′W) extends back to 1903. Analysis of these records has highlighted the role of the Antarctic Peninsula as a major climate divide (e.g. Schwerdtfeger 1984). The presence of an unbroken, steep mountain chain of 1400m-2000m altitude, south of 64°S, is a barrier to air-flow in the lower atmosphere. It limits advection of mild, maritime air from the South Pacific across the Peninsula from west to east. At the same time it blocks the prevailing eastward flow of cold, strongly stable air that flows around the edge of the Antarctic ice sheet. These factors give rise to frequent, strong, south-westerly ('barrier') winds along the east coast of the Peninsula, and a more continental climate regime. Schwerdtfeger estimates that at a given latitude annual mean temperatures are 4-6°C colder on the east coast sector. This finding has been confirmed by more recent studies based on extensive surveys of temperatures measured in the snowpack (Martin and Peel 1978).

28.1.2 Spatial variability of modern climate

Temperature time-series from Peninsula stations show a large interannual variability (with a standard deviation ranging from ±1.0°C at Orcadas to ±1.7°C at Faraday) in comparison with that obtained in the interior of the Antarctic ice sheet (±0.5-0.8°C). The exceptionally large variability at Faraday probably arises from topographic factors and local variations in the distribution of sea ice and open water (Pepper 1954). Analysis of the long series of ice observations from Laurie island, Scotia Bay (Schwerdtfeger 1975) has shown that some 64% of the variance in the annual temperature series from Orcadas may be linked to variations in

Figure 28.1 Distribution of the principal ice-core drilling sites and weather stations in the Antarctic Peninsula region.

the annual duration of ice cover. But Raper *et al.* (1984) find no firm evidence that such variations are directly linked to large-scale changes in ice extent.

Detailed analyses of the year-to-year temperature records by Schwerdtfeger (1975) and confirmed by the more detailed statistical analysis of Raper *et al.* (1984) shows that these records are well correlated north of about 65°S, but the correlation with Orcadas deteriorates southwards. There is no significant correlation between Orcadas and stations along the south coast of the Weddell sea, and a significant negative correlation with coastal stations to the east of longitude 10°W. The Orcadas temperature series shows no significant correlation with the mean temperature across Antarctica for the period 1957-82. Such findings have led Raper *et al.* to warn that "the possibility of extending Antarctic mean temperatures back before 1957 on the basis of temperatures at single stations is remote". Nevertheless in the period from about 1960 to the present there has been a significant (5% level) warming throughout the Antarctic Peninsula, in line with the trend in the annual mean Antarctic temperature anomaly series.

28.2   The glaciological regime

The Antarctic Peninsula has several features that offer advantages for ice core drilling to reconstruct climate change through the past several thousand years. Most significant is the relatively high annual accumulation rate, averaging around 0.8 m water equivalent (Peel and Clausen 1982) which ensures that in many places seasonal cycles in proxy climate parameters can be recognized for at least several hundred years and potentially up to several thousand years at the optimal locations. Highly resolved records can be achieved, which can be dated accurately by analysing several independent, stratigraphic parameters (e.g. $\delta^{18}O$, sulphate, and chloride) that show strong seasonal variations (Figure 28.2). In addition, the availability of relatively long, reliable instrumental weather observations from stations in different parts of the region provides a useful basis for exploring the relationship between ice core parameters and climatic controls. Such studies are needed to establish reliable transfer functions.

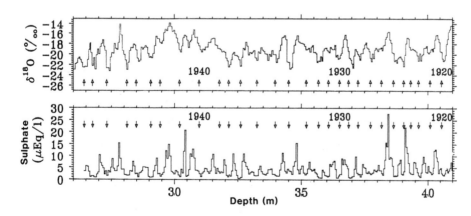

Figure 28.2   Variations in sulphate and $\delta^{18}O$ along a section of the Dolleman Island ice core. Sulphate shows a clear annual cycle, with a maximum concentration during the late austral summer.

28.2.1 Location of ice core drilling sites

A variety of sites is available for drilling. Table 28.1 summarises the main characteristics of the ice core sites that will be considered here. Ice dome or divide positions are preferred for deeper drilling because they have a simple ice-flow pattern. Several ice rises occur along the east coast, although they are fairly small features (ice thickness 200-500m); one of these, Dolleman Island has been drilled to 133m. The ice cap on James Ross Island, off the north-east tip of the Peninsula, has been drilled to 154.3m. The longest records will be found in the thickest ice (1000-1500m deep) found at sites along the spine of the Peninsula. One of these sites ('Gomez') has been drilled to 87m depth. No deeper drilling has yet been undertaken on the west coast although several dome sites are available, mainly on islands off the north-west coast of the Peninsula. Owing to the higher snow accumulation rate (1.5-2.5m/a) and thinner ice, records from these sites are likely to be limited to several hundred years. Generally, disturbance of the ice core records by meltwater percolation does not present a significant problem at the sites considered here, although some melting during summer was reported from James Ross Island (Aristarain *et al.* 1982). On the east coast ice rises, melt layers are very limited and rarely exceed 1cm thickness. Even mm-thick ice layers are a rare occurrence on the Peninsula plateau.

28.2.2 Proxy climate parameters

The best developed ice core proxy-climate parameter is the stable isotope (^{18}O or D) content of the ice. The mean isotopic composition of snow and ice ($\delta^{18}O$ or δD, expressed in units of

Table 28.1 Drilling site characteristics.

	Gomez	Dolleman Is	James Ross Is (Dalinger Dome)
Position	74°01'S, 70°38'W	70°35'S, 60°56'W	64°13'S, 57°41'W
Site type	Flank	Dome	Dome
Elevation	1130 m	398 m	1640 m
Ice thickness	~1500 m	~460 m	~300 m
Accumulation rate	0.88 m water/a	0.42 m water/a	~0.4 m water/a
10m temperature	-17.32°C	-16.75°C	-14.3°C
Drilling date	Jan 1981	Jan 1986 Jan 1976	Mar 1981
Depth drilled	30.5 m & 83 m	133 m,32m 11 m & 11 m	154.3 + many shallow cores
Data span	1942-80	1795-1986	1850-1980
Data sources	Peel & Clausen (1982), Peel et al (1988), Mulvaney & Peel (1988), Peel (this work)		Aristarain (1980), Aristarain & Delmas (1981), Aristarain et al (1986), Aristarain et al (1987)

per mil relative deviation from SMOW) is primarily dependent on the temperature of formation of the precipitation, the proportion of the heavy isotope decreasing with decreasing temperature (Dansgaard 1964). However it is influenced also by other factors, including distance from the source of moisture, and the rate processes of evaporation and condensation (Jouzel and Merlivat 1984). Before considering the climate implications of the long-term ice core isotope records, the evidence supporting use of a simple isotope:temperature transfer function to interpret these data must be considered. Observation of the seasonal cycles in δ also provides one method for dating the ice core sections, and in turn allows the annual snow accumulation rate to be reconstructed.

28.3   Calibration of stable isotope records

28.3.1   Dating

Primary dating of all cores considered here has been carried out by identifying the well-established reference horizons in total ß-activity (principally ^{90}Sr and ^{137}Cs) corresponding to two major series of nuclear weapons' tests (Pourchet et al. 1983, Peel and Clausen 1982). The principal peaks in Antarctic snow strata appear in summer 1954-55 and summer 1964-65. The dating is carried back in time by stratigraphic analysis of species that exhibit a regular seasonal variation. Stable isotope composition, using either δ^{18}O or δD, generally shows a clear seasonal cycle, with no significant mixing of snows of different seasonal origin evident at any of the sites considered here. The profiles show no serious deterioration with depth due to diffusion, even at the greatest depths drilled so far. However, typically 10-15% of seasonal cycles from Peninsula sites show double peaks (Peel et al. 1988, Aristarain et al. 1986) and without additional evidence, dates based on isotope stratigraphy alone may be subject to a similar absolute error.

The following is our assessment of the accuracy of dating of the cores listed in Table 28.1:

James Ross Island (A.D. 1850-1890)   Aristarain et al. (1986) consider that an error of 'several years' cannot be eliminated. However they note the general correspondence of the smoothed time series of δD with the temperature record from Orcadas back to 1904, where the principal cold and warm events match within ±2 years throughout.

Palmer Land plateau (A.D. 1942-1980)   The stratigraphic dating was supported by a procedure in which the δ^{18}O profile (converted to a time-series) was matched to find its optimum statistical fit with the temperature time-series from Faraday (Peel et al. 1988). The procedure assumes *a priori* that temporal variations in δ and air temperature are correlated, furthermore that a significant part of these variations can be recognized throughout the region. In support, Limbert (1974) has indeed reported a good correlation between the principal temperature anomalies recorded at weather stations lying between 61 and 68°S. The statistical match achieved supports an estimated ±1 year dating accuracy to 1947.

Dolleman Island (A.D. 1795-1986)   Procedures as in 3.1.2 were adopted to give a dating accuracy of ±1 year down to 1947 (Peel *et al.* 1988). Sulphate, which shows a strong annual cycle at this site (Mulvaney and Peel 1988) was analysed along the core at the same frequency as $\delta^{18}O$. In most cases, where there was a possible ambiguity in the assignation of an annual horizon in one of the species, this could be resolved by inspection of the second parameter. A maximum absolute error of ± 5 years down to 1795 is estimated (this work).

28.3.2   Isotope: temperature-spatial relationship

The problem of relating stable isotope variations in polar ice to absolute temperature change has been considered by many authors including Robin (1983) and specifically related to the Antarctic Peninsula region by Peel and Clausen (1982) Aristarain *et al.* (1986) and Peel *et al.* (1988). In the absence of reliable series of air temperature data close to ice core drilling sites, the δ:T relation is frequently determined from studies of the spatial variations of the mean annual parameters in the neighbouring region. In this case the annual mean temperature is taken as the temperature measured at 10m depth in the snowpack, and δ is averaged over several years. These studies, which are based on the analysis of more than 30 shallow cores spread across the Antarctic Peninsula region (compiled in Figure 28.3) show that spatially there is a well defined δ:T relation, with a gradient (0.9±0.06‰/°C) compatible with a simple Rayleigh model of the isotope fractionation process (Dansgaard 1964). In this model the 'annually-averaged air mass' respon-

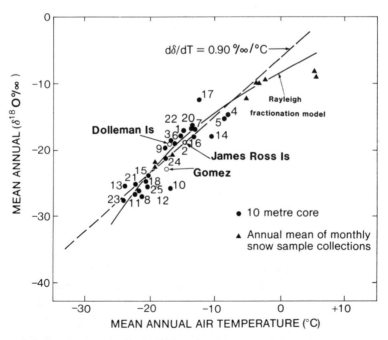

Figure 28.3   Mean annual stable isotope ratio vs mean annual air temperature (10m temperature in snowpack or instrumental average) for sites in the Antarctic Peninsula (64°-74°S, 59°-75°W).

sible for precipitation throughout the region is assumed to originate in the same general oceanic area. As the air mass is advected towards cooler regions, isotopic differentiation between different deposition sites arises as a result of progressive condensation, a process primarily controlled by temperature.

However, this approach cannot detect the effect of changes in the initial isotopic composition of the air mass (e.g. due to changes in the effective source area, temperature or humidity) possible influence of inter-annual variations in the extent of sea ice as indicated by Kato(1978) or systematic biasing arising from an uneven seasonal pattern of snow accumulation. Over a fixed time period such factors may exert a similar influence on the isotopic composition over a wide area, and hence may not contribute significantly to the spatial δ:T gradient. On the other hand, these factors may play a significant role in determining the effective δ:T gradient controlling isotope fluctuations at a single site over time. Direct comparisons between isotope and temperature records are needed to address this problem.

28.3.3 Isotope: temperature-temporal relationship

Whilst several relatively long instrumental temperature records are available from the Peninsula region (Jones and Limbert 1987) most are separated by several hundred km from sites that are suitable for ice core drilling. There are major contrasts in climate regime across the region, in addition to potentially very local influences on both the temperature (e.g. sea ice variations) and ice-core records (e.g. variations in the snow deposition pattern). Consequently, it is important to establish the general level of correlation between T and δ at the available sites and to assess their likely spatial representativeness. In this way it is possible to identify optimum sites, between which δ:T gradients should be estimated.

Local noise in the δ profiles   Time series of annually-averaged parameters derived from two cores drilled close together will differ considerably simply owing to local spatial variations in the snow accumulation pattern (e.g. Robin 1983, Fisher 1985) leading to 'deposition noise'. An additional error (definition noise) arises because the horizons (in the case of Gomez and Dolleman Island, the δ minima) dividing successive snow accumulation years do not fall on a constant calendar date. The deposition noise factor has been estimated from replicated shallow cores drilled at Gomez and Dolleman (Figure 28.4). Definition noise has been estimated assuming that the annual δ cycle follows a similar course to that observed at Faraday and Halley, where δ data are available for snowfall collected on a monthly basis over 12 years (International Atomic Energy Agency 1964-78). In this series, the standard deviation of the timing of the isotopic minimum is ± 1.5 months. Calculated contributions of both noise factors at Gomez and Dolleman are summarised in Table 28.2, which includes comparative estimates for other sites in Antarctica. Mainly because of the high snow accumulation rate, noise contributions to the true climate signal are much smaller than elsewhere in Antarctica. Data averaged over a 5-year span should reveal true climate-related shifts in δ of order 0.2‰ ($\sim$0.3°C).

For this reason, to correlate the δ and temperature time series, 5-year binomially-smoothed averages have been used throughout. This procedure also reduces the influence of 1-2 year errors in absolute dating of the ice core records, which cannot be excluded.

Table 28.2 Comparative local noise contributions to annual $\delta^{18}O$ at Antarctic drilling sites.

SITE	Accum.rate (m water/a)	Variance in $\delta^{18}O$ signal			AVG* (years)
		Deposition noise	Definition noise	Total	
GOMEZ	0.89	0.068	0.057	0.35	4
DOLLEMAN	0.45	0.212	0.285	2.28	6
LAW DOME	0.6	0.67	?	1.23	40
S POLE	0.092	3.2	?	3.5	100
DOME C	0.037	8.4	?	9	210

AVG* : Length of average needed to reduce
local noise level to ± 0.2⁰/₀₀

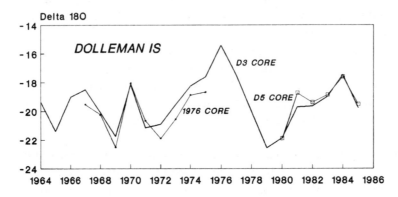

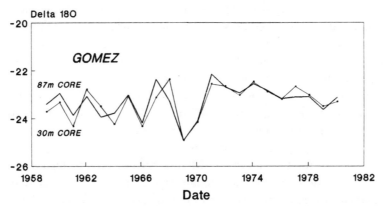

Figure 28.4 Comparison of mean annual stable isotope values from replicated cores at Gomez and Dolleman Island. The two cores from Gomez and the D3 and D5 core at Dolleman Island were drilled a few m apart. The '1975' core from Dolleman was drilled several hundred m from the D3/D5 cores.

Correlation of δ and temperature records since 1943 Unsmoothed time series of mean annual $\delta^{18}O$ from the three ice cores are shown in Figure 28.5. To ease comparison, the data for James Ross Island which were measured as δD have been converted to equivalent $\delta^{18}O$ values using the relationship $\delta D = (7.6\delta^{18}O - 2.9)$ determined by Aristarain *et al.* (1986) for this site. Table 28.3 summarises the observed correlations between the three isotope records, and between these records and temperature records selected on the basis of length and representation of the principal climate zones. Faraday, chosen to represent the west coast regime, is also the closest instrumental record to the Gomez and Dolleman Island sites. Esperanza, representing the east coast regime (correlation between Esperanza and Marambio, r=0.93) lies about 100km from the James Ross drilling site. As this record is taken from a point very close to the ice edge, and is likely to be very sensitive to local sea ice conditions, the record from Halley has been included to represent larger-scale conditions in the Weddell Sea sector. Finally an averaged time series for the Antarctic Peninsula region (MAP) is included, which was computed by Limbert (1984) as an average of records from Faraday, Frei-Bellingshausen, Hope Bay-Esperanza, and Adelaide-Rothera. Orcadas is of

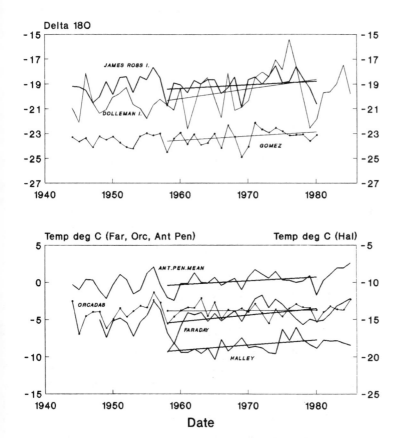

Figure 28.5 Time-trends, through the period of instrumental weather observations, of mean annual stable isotope content at Dolleman Island, James Ross Island and Gomez, in comparison with mean annual air temperature at Faraday, Orcadas, Halley, and a regional average for the Antarctic Peninsula. Data are un-smoothed.

Table 28.3 Correlation matrix for stable isotope (δ) and mean annual temperature (T) series (1943-1985) using 5 year binomially-weighted averages.

	FAR(T)	MAP(T)	ORC(T)	ESP(T)	HAL(T)	GOM(δ)	JRI(δ)	DI(δ)
FAR(T)	-	***0.90	*0.38	[-0.04]	[0.13]	***0.69	***0.60	***0.66
MAP(T)	-	-	***0.53	[0.21]	[0.20]	***0.58	***0.67	***0.56
ORC(T)	-	-	-	***0.71	[0.15]	[-0.01]	**0.43	[0.02]
ESP(T)	-	-	-	-	*0.33	[-0.20]	[0.08]	[-0.28]
HAL(T)	-	-	-	-	-	[0.24]	[0.28]	*0.42
GOM(δ)	-	-	-	-	-	-	***0.58	***0.60
JRI(δ)	-	-	-	-	-	-	-	***0.64

```
Significance level : *** - <0.001        Stations:    Faraday (FAR)
                      ** - <0.01                       Antarctic Peninsula mean (MAP)
                       * - <0.05                       Orcadas (ORC)
                                                       Esperanza (ESP)
                                                       Halley (HAL)
                                                       'Gomez' (GOM)
                                                       Dolleman Is (DI)
                                                       James Ross Is (JRI)
```

interest specifically because it is the longest temperature series available from the region (starting 1903) hence offering the potential to extend the ice core calibrations further back in time.

It is clear that correlations between the eastern and western sides have weakened markedly since about 1970 (cf. Schwerdtfeger 1976) and there is now no significant correlation between Faraday and Esperanza or Halley. Moreover the correlation between the Orcadas and Faraday records is now only just significant (95% level). The correlation between Orcadas and Esperanza remains very highly significant, and whilst there is still no correlation between Orcadas and Halley there is a weak connection between Esperanza and Halley.

Two of the ice core δ records are from sites that lie clearly within the east coast regime. Nevertheless the most significant correlations for all three cores are with the Faraday temperature series. In the case of the Gomez and Dolleman Island series the level of correlation deteriorates slightly when they are related to the MAP temperature record, which incorporates data from around the northern tip of the region: these sites show no correlation with the Orcadas record and slightly negative (although not significant at the 5% level) correlation with Esperanza. There is, however, a significant (5% level) correlation between Dolleman Island and Halley. It appears that in terms of inter-annual variability, Gomez is almost totally responding to influences from the western side. At Dolleman Island these influences overall are also dominant, but this site is also sensitive to conditions in the Weddell sector. Both sites are essentially de-coupled from short-term variations around the northern tip of the Peninsula. Whilst James Ross Island also shows very strong correlation with the

Faraday record, an improved correlation with the MAP series, and good (1% level) correlation with Orcadas demonstrate some sensitivity to conditions around the northern tip of the Peninsula. Surprisingly, we find no significant correlation with Esperanza, which despite its proximity evidently experiences important local, short-term fluctuations. There is, however, some connection, albeit not quite significant at the 5 % level, with Halley. Some influence from the Weddell sector is therefore apparent, this influence becoming increasingly important further south along the east coast.

The ice core data, although widely dispersed geographically, are better correlated than the temperature records, suggesting that they may be less sensitive to local topographic and sea ice conditions, and hence more representative of regional-scale climate change, than are the individual temperature records.

δ and Temperature trends and anomalies   Limbert (1974) has identified several characteristic periods of pronounced temperature anomalies (broadly defined by departures of more than ±2°C at Orcadas) which can be identified in all of the instrumental records from 1947 to 1973. He suggests that these features may be used as reference epochs which should be observed in records throughout the region. The principal features, sustained over two or more years, are listed in Table 28.4, which indicates the strength of corresponding anomalies in the ice core records. It is clear that in most cases, corresponding δ anomalies are observed. However, the magnitudes of the δ anomalies corresponding to the same temperature event generally vary markedly between the cores. Usually, the Gomez record shows around 60% smaller standard deviation in both the size of these features and in its overall inter-annual variation. This may be a real reflection of reduced inter-annual temperature variation in comparison with lower

Table 28.4 Principal mean annual temperature anomalies for the Antarctic peninsula (Limbert 1974) compared with the strength of corresponding isotopic signals in ice cores from Gomez, Dolleman, and James Ross.

Temperature anomaly	Sign(strength) in temp record	Isotopic signal strength Gomez	Dolleman	J Ross
1970-72	+increase (*)	**	**	**
1959-62	+increase (*)	*	**	*
1956-59	-decrease (**)	**	**	**
1954-57	+warm (**)	**	*	**
1947-49	-decrease (*)	-	**	*
1943-45	-decrease (*)	**	**	**
1929-36	+increase (**)	-	**	**
1927-30	-cold (**)	-	**	**

** - Strong anomaly
* - Moderate anomaly
- - Weak/absent anomaly

latitude Peninsula stations, as seen at stations of similar latitude elsewhere in Antarctica (Raper *et al.* 1984).

Additionally there are features in the ice core records from both Dolleman Island and James Ross Island of similar magnitude but which do not match a temperature anomaly. Thus it seems that, although these features may help to confirm the chronology of cores collected from the region, short-period (i.e. over a few years) trends alone cannot be used to infer an absolute temperature shift unambiguously. This is especially true for the two east coast sites, where for example the largest isotope anomaly in both records since 1948 – an apparently extreme warm event between 1973-77 – corresponds to a period of minor decrease in temperature at Orcadas and Esperanza whereas to the west it spans the final part of a period of a pronounced positive temperature anomaly, including the final cooling stage (Schwerdtfeger 1976). The significance of this event will be discussed later in relation to the interpretation of longer term isotope variations. A similar event can be seen in the longer time series, Figure 28.6a, during the period 1933-38.

Fortunately there is greater consistency in longer-term trends observed in both the δ and temperature series. Linear trends through the period since 1958-80 have been fitted to the profiles shown in Figure 28.5; the results are summarised in Table 28.5. This period corresponds to that for which a statistically significant (5% level) trend (~0.03°C/a) has been demonstrated for the mean Antarctic temperature anomaly series (Raper *et al.* 1984).

With the exception of Orcadas, where both short and longer-term variations appear to be strongly moderated by the surrounding ocean, the increasing temperature trend through this period has been greater in the Peninsula than over continental Antarctica. However, the gradients throughout the Antarctic Peninsula and across the Weddell sea sector are similar and appear relatively insensitive to short-term local influences which perturb the inter-annual climate signal. Correspondingly, there is reasonably close agreement in the gradient of $\delta^{18}O$ increase detected in the three ice cores.

Isotope: temperature transfer function   The most conservative temperature signal available for comparison with the δ records is the long-term temperature gradient. Data in Table 28.5 yield an average δ:T gradient of 0.66‰/°C. Ratios computed from the correlations between mean annual parameters are generally rather smaller than this. In the case of Gomez, as mentioned earlier, this may arise in part simply because the interannual variability of temperature is lower at this site than at Faraday. Non-correlated local noise in either the isotope or temperature data may also be an important factor. For Dolleman Island and James Ross Island, the ratios achieved between the best-correlated mean annual δ and temperature series are generally close to the 'long-term' estimate. Aristarain *et al.* (1986) have suggested that a similar $\delta^{18}O$ ratio (~0.59‰/°C should be used to interpret climatic shifts at James Ross Island, on the basis of inter-annual correlations with combined temperature records from Faraday and Esperanza.

The temporally-derived δ:T gradient (0.66‰/°C) is approximately 30% smaller than the spatially derived value. This problem has been discussed by Aristarain *et al.* (1986) and Peel *et al.* (1988). Aristarain *et al.* argue that this could still be consistent with the simple Rayleigh model of isotope fractionation, provided that similar temperature variations occur in both the moisture source region and at the precipitation site. Peel *et al.* (1988) felt that this condition

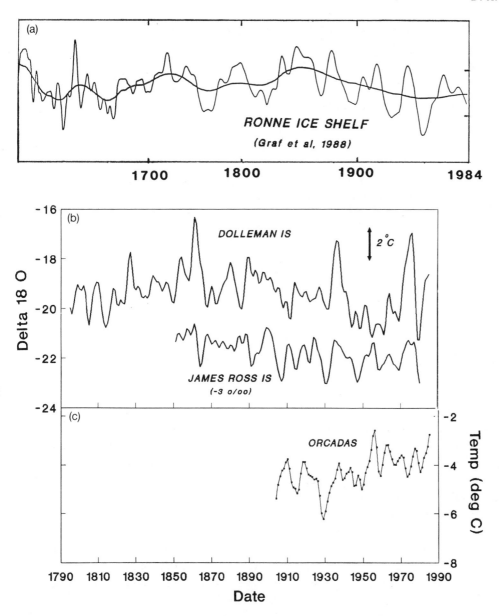

Figure 28.6 (a) 5-Year binomially-weighted averages of annual mean $\delta^{18}O$ content at Dolleman Island and James Ross Island in comparison with the temperature record at Orcadas for the period 1904-86. The arrow bar within the isotope record represents the equivalent temperature shift, assuming that present-day isotope:temperature gradients applied in the past. Since 1904 the general trend in isotope content has opposed the overall temperature trend reflected in the record from Orcadas. (b) Smoothed $\delta^{18}O$ profile from ice core T340 drilled on the Ronne Ice Shelf in 1984 (Graf *et al* 1988). The profile has been first-order corrected for the effects of ice flow. Note the general maximum $\delta^{18}O$ value in the mid-19th century as seen also in the Dolleman Island profile.

was unlikely to be met at sites farther south in the Antarctic Peninsula where both the interannual variability and longer term trends are much greater than over the Southern

Table 28.5   Isotope and temperature trend lines for the period 1958-1980.

Station	Standard Dev (°C)	Trend (°C/a)	r^2	
Faraday	1.49	.087 ±.044	0.16	(sig 5% level)
Orcadas	0.90	0.00±.029	0.00	
MAP	1.06	.050±.032	0.10	
Halley	1.05	.071±.030	0.21	(sig 5% level)

Average = 0.070 °C/a

Station	Standard Dev (‰)	Trend (‰/a)	r^2
Gomez	0.69	.034±.021	0.11
J Ross	0.87	.029±.027	0.053
Dolleman	1.9	.076±.058	0.078

Average = 0.046 ‰/a

Ocean. Using data obtained from monthly snowfall collections at Faraday between 1964-75, they showed that the reduced gradient was most probably a direct result of biasing in the isotope profiles, which only 'record' temperature during periods of snowfall. Snowfall tends to occur when temperatures are higher than the seasonal average – the discrepancy increasing as temperature decreases. Very cold periods, frequently associated with temperature inversions during the winter months, are not 'recorded' in ice cores. In the isotope records, winter temperatures are effectively over-estimated, and both the annual range of temperature and longer-term trends are under-estimated by ~36%.

Peel *et al.* (1988) also examined the possibility that further biasing might arise owing to a non-uniform seasonal snow deposition pattern. Analysis of the daily meteorological register at Faraday, using days per month on which precipitation was reported as an index of snow accumulation rate, showed a strongly bimodal distribution with maxima during October/November (principal) and March/April, corresponding to the equinoctial periods. The analysis showed that the mean annual isotopic composition was relatively insensitive to the seasonal accumulation pattern.

28.4   Ice-core evidence since 1795

28.4.1   Stable isotopes

The δ records for Dolleman Island and James Ross Island are presented in Figure 28.6a together with the Orcadas temperature record since 1904. The general level of correlation between the James Ross Island and Orcadas records (r=0.38, 0.1% significance) and between the James Ross Island and Dolleman Island records (r=0.50, 0.1% significance) is sustained over the longer period, which suggests that there is no important residual dating error in the ice-core profiles. An inferred temperature scale has been added to the graphs, assuming the overall δ:T gradient estimated from behaviour in the post-1958 period, applied previously. There are strong reasons to doubt this assumption for the east coast sites. It is clear that there is a series of major positive isotope anomalies in both records, especially marked at Dolleman Island. Two of the largest features, mentioned earlier (1973-77 and 1933-38) are not associated with any significant deviation in the Orcadas record. The amplitudes of these anomalies are comparable with the major long-term trend visible in both profiles. The major feature is a long-term decrease in δ up to 1960, from a broad maximum in the mid-19th century. Aristarain *et al.* (1986) have suggested that this might indicate a general climatic cooling in the Antarctic Peninsula region.

There is evidence that the feature is not a local perturbation in the climate round the east coast of the Peninsula. Figure 28.6b shows a smoothed δ profile from an ice core drilled on Ronne Ice Shelf in 1984 (Graf *et al.* 1988) which has been corrected for the effects of ice flow. The same broad picture of major decadal-scale fluctuations, and broad maximum δ values in the mid-19th century is apparent. This suggests that it may be a general signal around the Weddell sea region. If the signal does reflect a temperature shift, this would apparently conflict with other evidence for climate change over this period both within Antarctica and in the sub-Antarctic, which at least qualitatively conforms with the large-scale air temperature trend in the Southern Hemisphere. Compiled surface air temperature data for the Southern Hemisphere show a more or less sustained warming of 0.49°C through the period 1881-1984 (Jones *et al.* 1986).

28.4.2 *Snow accumulation rate*

5-year binomially smoothed average snow accumulation rates for Dolleman Island, James Ross Island and Gomez are presented in Figure 28.7. In all cases there has been a significant increase in snow accumulation rate since 1950 (Dolleman Island, 0.8%/a; James Ross Island 1.8%/a; Gomez, 0.6%/a) in parallel with the regional temperature increase (0.07°/a) during the same period. Accumulation rates at Dolleman Island for the period 1955-1980 are on average 17% greater than the average for 1805-1940, and 25% greater at James Ross Island. The average rate of accumulation increase since 1950 of 1.16%/a corresponds to a relative change with temperature of 15.2%/°C. This may be compared with a relative change in water vapour pressure at -19°C (the mean air temperature at the drilling sites) of 10.0%/°C. The general link between snow accumulation rate and air temperature has been considered by several authors (e.g. Robin 1977, Young *et al.* 1982) on the basis of spatially-derived data across the Antarctic ice sheet. Although no causal relationship has been proposed, generally the accumulation rate seems to be controlled by the amount of water vapour that can be carried in the atmosphere. A detailed study near Dome C of snow accumulation rate changes between the radioactive horizons of 1965 and 1955 (Petit *et al.* 1982) indicates that the snow accumulation rate has increased by about 30% since 1965 in comparison with the 1955-65 period. A similar increase is apparent at Law Dome (Morgan 1985). Hence it seems that

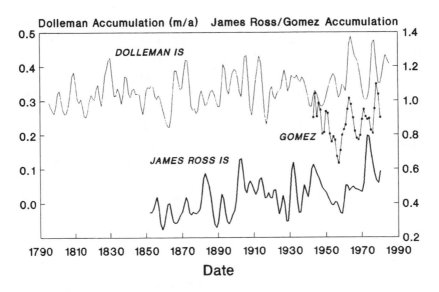

Figure 28.7 5-year, binomially-weighted average annual snow accumulation data from Gomez, Dolleman Island and James Ross Island. At all sites, marked increases have occurred since 1950, relative to the long-term average, in parallel with the increasing temperature trend in the Antarctic Peninsula region. The signal appears to be most marked at James Ross Island, where values appear to have increased generally since around 1850.

important increases in accumulation rate have occurred on a large scale in Antarctica during the past 30 years.

28.5   Relationship with other studies

28.5.1   Evidence from Antarctica

Isotope data from three ice cores drilled in continental Antarctica show evidence of negative δ anomalies associated with the 'Little Ice Age' event, albeit possibly a less pronounced climate feature than in northern latitudes (see Mosley-Thompson, this volume). A well-dated ice core record (473m deep) from Law Dome (66.7°S, 112.8°E; 1395m asl) shows a minimum δ between A.D. 1790 and 1850, indicating temperatures lower than at any time during the past 2000 years. The δ value has subsequently shown a general increase up to 1977, with secondary fluctuations (Morgan 1985; Wishart 1985). Evidence from the 905m deep core at Dome C (74°39′S, 124°10′E; 3240m asl) which cannot be dated so precisely, also suggests that there was a cool period between A.D. 1200 and 1800 (Benoist *et al.* 1982). Finally, a discontinuous record from Mizuho station (70°42′S, 44°20′E) also shows a strong minimum δ around 1800, a secondary minimum around 1860, followed by a general increase up to 1960 (Watanabe *et al.* 1978).

Comparisons between instrumental records from Antarctica and from lower latitudes, although limited by paucity of data, suggest some de-coupling of short-term trends (Jones *et al*. 1986). Nevertheless evidence over longer time scales, mainly from glacier fluctuations (Clapperton and Sugden 1988) indicates broadly synchronous behaviour during the past 3000 years. In the South American Andes, Patagonia, South Georgia and the sub-Antarctic islands of South Shetlands and South Orkneys, glacier advances during the Little Ice Age culminated during the 18th and 19th centuries with moraines marking extremes of this event dated to the mid-late 19th century in the sub-Antarctic. With this evidence it becomes increasingly difficult to account for a corresponding period of warmth in adjacent areas of the Antarctic Peninsula. Clues which enable another explanation for the positive isotope anomaly during the 19th century can be gained from closer examination of the positive anomaly during the period 1973-77.

28.5.2   *The 1973-77 isotope anomaly*

During the period 1973-77 there was an exceptionally strong positive δ anomaly at Dolleman Island. A similar, but smaller feature, is evident in the δ record from James Ross Island. Records from stations on the western side of the Peninsula show a significant positive temperature anomaly for the period 1969-75, but temperatures returned to the 1947-86 average by 1976, at the maximum of the δ anomalies. Records from Esperanza, the station closest to James Ross Island, show a significant negative temperature anomaly for the period (Jones *et al*. 1987). Clearly,the isotope records are not yielding a faithful representation of mean annual temperature change in the region during this period.

Schwerdtfeger (1976) noted that there had been significant changes in the surface temperature field of the Peninsula during the five-year period up to 1976, and gave evidence that there had been an increase in advection of warm air towards the western side of the Peninsula, associated with greatly reduced ice cover in the Marguerite Bay area (68.3°S). He suggested that the contrasting, lower temperatures recorded on the eastern side could be related to abundant ice production in this sector. This raises the question as to whether such significant changes in atmospheric circulation pattern can lead to changes in the effective source areas for moisture deposited on the ice sheet, and in turn to changes in the isotopic composition of the ice not necessarily related to the temperature at the deposition site.

According to the pure Rayleigh model of the isotope fractionation process, which ignores kinetic effects, $\delta^{18}O$ and δD vary in parallel, with a $\delta D/\delta^{18}O$ gradient close to a factor 8 (Dansgaard 1964). However, kinetic effects during phase changes will cause deviations from this relationship. During the condensation stage, kinetic effects due to supersaturation of vapour with respect to ice, have been shown to be not significant at the temperatures occurring in the Antarctic Peninsula (Jouzel and Merlivat 1984). However deviations arise if there are variations in the rate of evaporation, which is related to temperature, in the source areas of the moisture. Modelling studies in Greenland (White and Dansgaard 1988, Dansgaard *et al*. 1989) have shown that the 'deuterium excess' ($d = \delta D - 8\delta^{18}O$) is mainly determined by the sea surface temperature in the source region, d increasing by about 0.4‰/°C.

Unfortunately δD and $\delta^{18}O$ have not been measured together on a continuous basis along a core from the Antarctic Peninsula. As a first approximation we assume that the δ data from both Dolleman Island and James Ross Island mainly respond to the same regional climatic signal. A semi-quantitative index of the deuterium excess (called here the 'relative deuterium excess', d') is then calculated using the $\delta^{18}O$ data from Dolleman Island and δD data from James Ross Island. The derived profile (Figure 28.8a) although suffering from excessive noise due to local perturbations, shows a pronounced negative anomaly in d' in the mid-1970s compared with the average for the period 1948-86, implying that there was significant input of moisture from sources at much lower temperature than the average. In turn this situation would contribute to elevated δ values during this period, due to the reduced temperature gradient (and therefore less fractionation of oxygen isotopes) between the source and deposition site. Without measurements of absolute values for d' it is not possible to estimate whether this could account completely for the isotopic maxima, but the general magnitude of the shift in d' suggests that this may be possible.

During the period 1974-76, the central part of the isotope anomaly, there was exceptional potential for local moisture advection from the Weddell Sea – the Weddell Polynya. In this period a large open-water area ($1-3 \times 10^5$ km^2) in the middle of the southern ocean sea ice off Queen Maud Land, was sustained throughout the winter months. It was centred at about 0°E, 67°S in 1974 and shifted westwards to about 20°W, 68°S in 1976 (Carsey 1980). Both heat and water vapour would have entered the atmosphere at enormous rates from the exposed ocean surface which is relatively warm, albeit some 8-10°C colder than the probable dominant source areas that lie to the north of the Antarctic Convergence.

Thus it seems possible that the positive δ anomaly observed during this period along the east coast of the Peninsula is consistent with lower temperatures in the same general area, and the δ signal can be dominated by the effects of influx of local moisture for an extended period. Clearly confirmation is now needed from a single site under the influence of the Weddell Sea zone.

28.5.3 Implications for long-term climate change in the Antarctic Peninsula

Figure 28.8b shows the profile of relative deuterium excess computed in the same way for the period since 1850. It can be seen that the 1973-77 isotope anomaly was but one of a series. The strong positive isotope anomaly in 1933-38, referred to previously, is also associated with a sharply lower relative deuterium excess. Moreover this feature and the earlier periods of sustained low d' values were also associated with severe ice conditions reported off the northern tip of the Peninsula, in turn probably linked to high ice discharge rates from the Weddell Sea (the major periods reported by Schwerdtfeger (1970) are marked on Figure 28.8b). It is tempting to speculate that during these periods there was also either major polynya development or some other mechanism operating that enhanced local water vapour production in the Weddell Sea sector. Moreover the average d' value was significantly lower (4.7‰) during the period 1850-1940 than in 1940-85, implying sustained lower temperatures in the moisture source region during the former period.

Whilst the explanation for the climatic significance of the positive δ anomaly during the mid-late nineteenth century must remain ambiguous until data for both δD and $\delta^{18}O$ are obtained on the same core, present evidence suggests that this finding may be primarily a

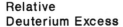

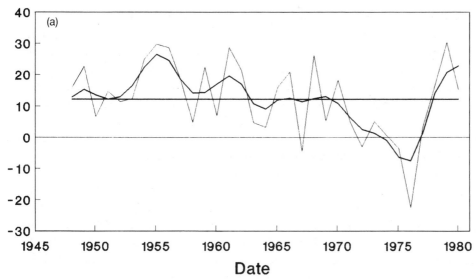

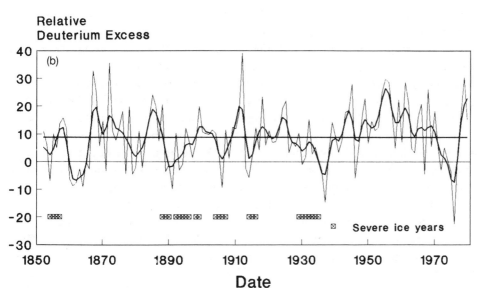

Figure 28.8 (a) Mean annual values (un-smoothed and 5-year binomially weighted) for the 'relative deuterium excess' (d') computed using δD data from James Ross Island and δ¹⁸O data from Dolleman Island. Strong negative anomalies in d' occurred during 1973-77, when a large polynya persisted in the Weddell Sea. (b) Trends in 'relative deuterium excess' (d') since 1850, computed as in Figure 8a, show that d' was generally lower than average during the 19th century. The major sustained periods with negative anomalies in d' generally appear to match periods where severe sea-ice conditions were reported in the Atlantic sector (Schwerdtfeger 1970).

result of enhanced local moisture input from the Weddell Sea. The possibility that temperatures may have been cooler during the mid-19th century than at present cannot be excluded.

28.6 Summary

The Antarctic Peninsula spans a major climate divide, separating a maritime regime to the west from a colder, more continental regime to the east, where conditions are dominated by influences from the Weddell Sea. Recent climatic trends, from instrumental records at several weather stations since 1947, show a very large inter-annual variability in comparison with other parts of Antarctica. Nevertheless, most major temperature anomalies in this period, and an overall increasing temperature trend since the mid-1950s, can be identified in all the records, with the single exception of Orcadas, where the record appears strongly moderated by the surrounding ocean. Such findings imply that there is a significant regional climatic signal that should be detected in ice cores drilled across the region.

Major features and trends in the temperature records are reflected in stable isotope profiles from the three ice cores drilled so far that span the post-1947 period: Site 'Gomez' (period spanned, 1942-80) on the southern Palmer Land plateau lies mainly within the west-coast regime, whereas Dolleman Island (1795-1986) and James Ross Island (1850-1980) lie along the east coast. A combination of relatively large snow accumulation rates in the Peninsula combined with strong seasonal cycles in the isotope content (δ) has allowed these isotope records to be dated precisely. Hence it has been possible to make a detailed study of the mean annual δ:temperature gradient on the basis of temporal variations. The snow accumulation rate has increased by about 20% since 1955 at both Dolleman Island and James Ross Island compared with the long-term average (1805-1940) in parallel with the increasing temperature trend. This now seems to be a very widespread climate signal in Antarctica.

The derived mean $\delta^{18}O$:T gradient of 0.66‰/°C, is about 30% less than that obtained by analysis of spatial variations in the mean annual values. This discrepancy can be accounted for by a systematic biasing of the ice core records, which 'record' the temperature only when snow is falling. When this factor is taken into account, the general characteristics of the δ:temperature relation appear to match fairly well with the predictions of a pure Rayleigh model of the isotope fractionation process based on the hypothesis that, on average, snow accumulation throughout the region is derived from the same general oceanic source areas for the moisture. Nevertheless there are sustained periods (e.g. 1973-77) in both of the east-coast records when clearly the isotope profiles are not giving, even qualitatively, a realistic representation of temperature change in the region. However, site Gomez, which seems to be mainly influenced by the west coast regime, appears to have behaved normally during this period.

Making the assumption the James Ross Island and Dolleman Island are responding on an annual basis to a similar climate signal, a relative deuterium excess parameter (d'= δD - 8$\delta^{18}O$) has been calculated by combining the δD data measured at James Ross Island with the $\delta^{18}O$ data measured at Dolleman Island. d' has been shown to be mainly sensitive to the sea surface temperature in the moisture source region. A very pronounced negative anomaly in d' at James Ross Island/Dolleman Island during 1973-77 implies low source region temperatures during this period. This could account also for the observed anomalously high δ values during this period, due to the reduced temperature gradient between the source and deposition areas. During 1973-77 a large polynya developed in the Weddell sea, during the winter months, This is proposed as the most likely cause of the δ anomaly.

The most significant feature of the longer-term records from Dolleman Island and James

Ross Island is a general decrease in δ from a broad maximum in the mid-19th century up to 1960. Associated values for d' show that this parameter was generally lower during 1850-1940 compared with 1940-85. A series of events showing anomalously low d' values generally match reported periods of severe ice conditions (high ice discharge rates from the Weddell Sea) in the Drake Passage area. Whilst an unambiguous statement of the climatic significance of the isotopic warm feature cannot yet be made, the evidence suggests that there may have been sustained polynya development during this period. Air temperatures may have been cooler than now, in line with trends inferred from other evidence, both elsewhere in Antarctica and in the sub-Antarctic.

Isotopic data from ice cores collected in any area influenced by air masses passing over the Weddell Sea may be similarly affected. Parallel data on δD and δ^{18}O are needed to separate quantitatively the effects of changing source temperature on the isotope records so that the underlying temperature trends can be isolated. For the Antarctic Peninsula region it is now important to obtain records from high altitude sites on the plateau region (e.g. Gomez) that should be much less perturbed by input of moisture from local sources.

References

Aristarain, A. J. 1980. *Etude Glaciologique de la Calotte Polaire de l'Ile James Ross (Peninsula Antarctiqe)*. Publ. 322, CNRS – Lab. de Glaciol. et Geophys. de l'Environ., Grenoble, France.

Aristarain, A. J. and Delmas, R. 1981. First Glaciological Studies on the James Ross Island Ice Cap, Antarctic Peninsula. *Journal of Glaciology* 27, (97) 371-379.

Aristarain, A. J., Delmas, R. J. and Briat, M. 1982. Snow Chemistry on James Ross Island (Antarctic Peninsula). *Journal of Geophysical Research* 87, (C13) 11004-11012.

Aristarain, A. J., Jouze, J., Pourchet, M. 1986. Past Antarctic Peninsula Climate (1850-1980) deduced from an Ice Core Isotope Record. *Climatic Change* 8, 69-89.

Aristarain, A. J., Pinglot, J. F., and Pourchet, M. 1987. Accumulation and temperature measurements on the James Ross Island ice cap, Antarctic Peninsula, Antarctica. *Journal of Glaciology* 33, (115) 357-62.

Benoist, J. P., Jouzel, J., Lorius, C., Merlivat, L., and Pourchet, M. 1982. Isotope climatic record over the last 2.5 ka from Dome C, Antarctica, ice cores. *Annals of Glaciology* 3, 17-22.

Carsey, F. D. 1980. Microwave observation of the Weddell Polynya. *Monthly Weather Review* 108, 2032-44.

Clapperton, C. M., and Sugden, D. E. 1988. Holocene glacier fluctuations in South America and Antarctica. *Quaternary Science Reviews* 7, 185-98.

Dansgaard, W. 1964. Stable isotopes in precipitation. *Tellus* 16, 436-68.

Dansgaard, W., White, J. W. C., and Johnsen, S. J. 1989. The abrupt termination of the Younger Dryas climate event. *Nature* 339, (6225) 532-4.

Fisher, D. A. 1985. Stratigraphic noise in time series derived from ice cores. *Annals of Glaciology* 7, 76-83.

Graf, W., Moser, H., Oerter, H., Reinwarth, O., and Stichler, W. 1988. Accumulation and ice-core studies on Filchner-Ronne ice shelf, Antarctica. *Annals of Glaciology* 11, 23-31.

International Atomic Energy Agency 1964-78. *Environmental isotope data Nos 2-6. World survey of isotope concentration in precipitation (1964-75)*. Vienna: International Atomic Energy Agency (Technical Reports Series 117, 129, 147, 165, 192).

Jones, P. D., Raper, S. C. B., and Wigley, T. M. L. 1986. Southern hemisphere surface air temperature variations: 1851-1984. *Journal of Climate and Applied Meteorology* 25, 1213-1230.

Jones, P. D. and Limbert, D. W. S. 1987. *A data bank of Antarctic surface temperature and pressure data*. Publ. TR038. US Dept. Energy, Washington, D.C: Office of Energy Research.

Jouzel, J. and Merlivat, L. 1984. Deuterium and oxygen 18 in precipitation: modelling of the isotopic effects during snow formation. *Journal of Geophysical Research* 88, (C4) 2693-703.

Kato, K. 1978. Factors controlling oxygen isotopic composition of fallen snow in Antarctica. *Nature* 272, (5648) 46-8.

Limbert, D. W. S. 1974. Variations in the mean annual temperature for the Antarctic Peninsula. *Polar Record* 17, (108) 303-6.

Limbert, D. W. S. 1984. West Antarctic temperatures, regional differences, and the nominal length of summer and winter seasons. In: *Environment of West Antarctica: potential CO_2-induced changes*. Report of a workshop held in Madison, Wisconsin, 5-7 July 1983. Washington, DC: National Academy Press, 116-139.

Martin, P. J. and Peel, D. A. 1978. The spatial distribution of 10 m temperatures in the Antarctic Peninsula. *Journal of Glaciology* 20, (83) 311-17.

Morgan, V. I. 1985. An oxygen isotope – climate record from the Law Dome, Antarctica. *Climate Change* 7, 415-26.

Mulvaney, R. and Peel, D. A. 1988. Anions and cations in ice cores from Dolleman Island and the Palmer Land plateau, Antarctic Peninsula. *Annals of Glaciology* 10, 121-125.

Peel, D. A. and Clausen, H. B. 1982. Oxygen-isotope and total beta-radioactivity measurements on 10 m ice cores from the Antarctic Peninsula. *Journal of Glaciology* 28, (98) 43-45.

Peel, D. A., Mulvaney, R. and Davison, B. M. 1988. Stable-isotope/air temperature relationships in ice cores from Dolleman Island and the Palmer Land plateau, Antarctic Peninsula. *Annals of Glaciology* 10, 130-136.

Pepper, J. 1954. *The Meteorology of the Falkland Islands and Dependencies* 1944-50. London: Falkland Islands Dependencies Survey.

Petit, J. R., Jouzel, J., Pourchet, M., and Merlivat, L. 1982. A detailed study of snow accumulation and stable isotope content in Dome C (Antarctica). *Journal of Geophysical Research* 87, (C6) 4301-8.

Pourchet, M., Pinglot, F. and Lorius, C. 1983. Some meteorological applications of radioactive fallout measurements in Antarctic snows. *Journal of Geophysical Research* 88, (C10) 6013-20.

Raper, S. C. B., Wigley, T. M. L., Mayes, R. R., Jones, P. D. and Salinger, M. J. 1984. Variations in Surface Air Temperatures. Part 3: The Antarctic, 1957-82. *Monthly Weather Review* 112, 1341-53.

Robin, G de Q. 1977. Ice cores and climatic change. *Philosophical Transactions of the Royal Society of London Ser B* 280, (972) 143-68.

Robin, G de Q. 1983. The climatic record from ice cores. In: Robin G de Q (ed) *The climatic record in polar ice sheets*. Cambridge: Cambridge University Press 180-95.

Schwerdtfeger, W. 1970. The climate of the Antarctic. In Orvig, S. (ed) *Climates of the Polar Regions*, Amsterdam: Elsevier, 253-355.

Schwerdtfeger, W. 1975. Relationship of temperature variations and ice conditions in the Atlantic sector of the Antarctic. *Antarctic Journal of the US* 10, 237-8.

Schwerdtfeger, W. 1976. Changes of temperature field and ice conditions in the area of the Antarctic Peninsula. *Monthly Weather Review* 104, 1441-3.

570

Schwerdtfeger, W. 1984. *Weather and Climate of the Antarctic*. Developments in Atmospheric Science, 15. Amsterdam: Elsevier.

Watanabe, O., Kato, K., Satow, K., and Okuhira, F. 1978. Stratigraphic analyses of firn and ice at Mizuho station. *Memoirs of National Institute of Polar Research. Special Issue* 10, 25-47.

White, J., and Dansgaard, W. 1988. the origin of arctic precipitation as deduced from its deuterium excess (Abstract). *Annals of Glaciology* 10, 219-20.

Wishart, E. R. 1985. Evidence of southern hemisphere warming from oxygen isotope records of Antarctic ice. *ANARE Research Notes* 28, 36-44.

Young, N. W., Pourchet, M., Kotlyakov, V. M., Korolev, P. A., and Dyugerov, M. B. 1982. Accumulation distribution in the IAGP area, Antarctica: 90°E-150°E. *Annals of Glaciology* 3, 333-338.

29 Paleoenvironmental conditions in Antarctica since A.D. 1500: ice core evidence

E. Mosley-Thompson

29.1 Antarctic ice core records

The previous chapter described the paleoclimatic records available from ice cores in the Antarctic Peninsula region. This paper synthesizes the records available for the East and West Antarctica Ice Sheets (Figure 29.1). The histories summarized in this chapter come

Figure 29.1 The map illustrates core sites and meteorological stations discussed in the text.

from different areas which are characterized by a broad spectrum of net balances, mean annual temperatures, surface climatologies and ice flow regimes. Section 29.1.2 focuses upon the different ice cores and discusses the strengths and limitations of each record for the synthesis of environmental conditions since A.D. 1500. The following section provides an overview of the dating of ice core records which is the first, critical step in paleoclimatic reconstruction.

29.1.1 Dating the ice cores

As discussed in Chapter 1, ice cores may be dated using numerous techniques (Hammer 1978) ranging from seasonally varying parameters such as dust and $\delta^{18}O$, identification of discrete well-dated events, and modeling techniques. To examine the last 500 years as recorded in Antarctic ice cores, annual to decadal time resolution is essential and the precision of the time scale is a major consideration. Data drawn from other authors are presented as faithfully as possible with respect to time, and any annual or decadal averages presented were calculated from the time series as originally published.

Johnsen (1977) demonstrated that the seasonal $\delta^{18}O$ signal will be smoothed gradually by diffusion during the firnification process and may not be preserved at depth when accumulation rates are below ~250 mm H_2O equivalent. In these regions other seasonally varying constituents such as insoluble particulates (MPC) sulfate (SO_4^{2-}) nitrate (NO_3^-) and direct current (D.C.) conductivity may be used. Ultimately, the net annual accumulation and temperature of a site limit the time resolution.

This may be illustrated using records from two very different places: Siple Station (75°55'S; 84°15'W; 1054m asl) and Amundsen-Scott South Pole Station (90°S; 2835m asl). Microparticle (insoluble) concentrations were analyzed in 5218 samples representing the entire length of a 101 meter ice core drilled in 1974 at South Pole. A 911 year time scale was produced using a single parameter, insoluble dust concentrations (Figure 2 in Mosley-Thompson and Thompson 1982). The time scale error at 101 meters was estimated as ~90 years or ± 10%. The low net annual accumulation at South Pole (~80mm H_2O equivalent) precludes the use of seasonal $\delta^{18}O$ and δD variations for dating at depths exceeding about 20 meters (Figure 29.2 in Jouzel et al. 1983). In contrast, the high net annual accumulation at Siple (560 mm H2O equivalent) and moderately low temperatures (MAT:-24°C) preserve the seasonal $\delta^{18}O$ signal which exhibits strong spatial reproducibility (Figure 29.2). Sulfate concentrations also exhibit a well preserved and spatially reproducible annual signal (Dai et al. 1990). Thus, using a combination of $\delta^{18}O$ and SO_4^{2-}, coupled with SO_4^{2-} identification of several volcanic events which serve as time-stratigraphic markers (e.g. Figure 29.3) the estimated accuracy of the time scale for the 302 meter Siple core is ± 10 years at A.D. 1417 or ± 2%. This 5-fold increase in precision illustrates the importance of net annual accumulation for obtaining highly resolved time scales.

The precision of the time scale also depends upon the sample size selected for individual analyses. The cores used in this synthesis were drilled at different locations (Figure 29.1) and were analyzed by different groups using a variety of techniques and sampling schemes. Therefore, the reader is encouraged to review the original records if more specific information is desired.

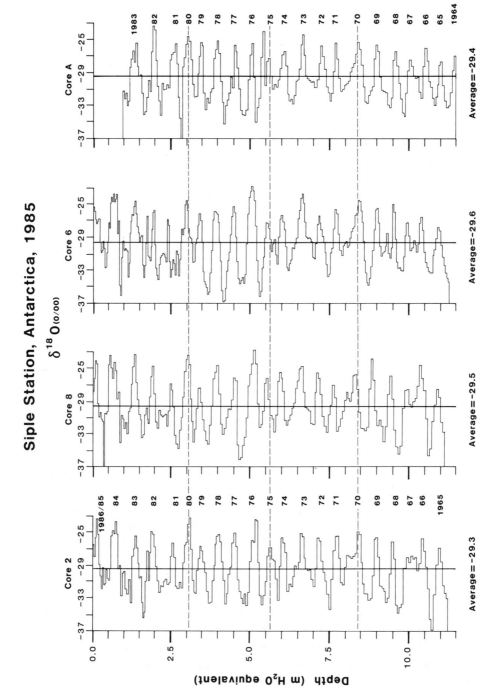

Figure 29.2 Four oxygen isotope records representing A.D. 1965-1985 at Siple Station illustrate the high degree of reproducibility of the annual signal used to date the deeper drill hole (see Mosley-Thompson *et al.*, 1990 for map).

SIPLE STATION, ANTARCTICA 1985

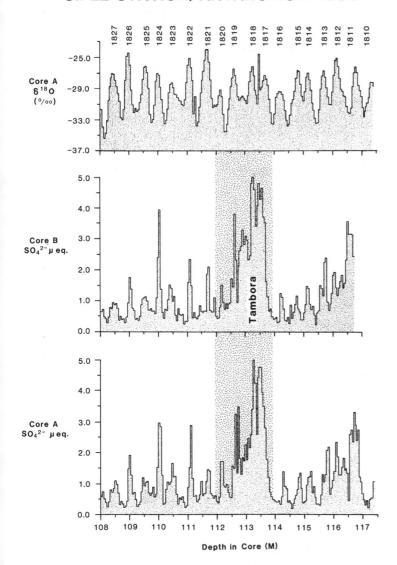

Figure 29.3 The sulfate concentrations (μeq. l⁻¹) in two parallel cores (A and B) from Siple Station exhibit an excellently preserved seasonal signal which is combined with the $\delta^{18}O$ record to produce the time scale. The SO_4^{2-} records reveal the eruption of Tambora (A.D. 1815) which serves as a time-stratigraphic marker, further confirming the time scale.

29.1.2 Description of the ice core records

Siple Station (75°55'S; 84°15'W; 1054m asl)   A 550 year record of the concentrations of dust, $\delta^{18}O$, and SO_4^{2-} was obtained from a 302 meter core drilled in 1985/86 at Siple Station (Figure 29.1). Siple lies between the Antarctic Peninsula region which is characterized by a complex near-surface wind regime (Schwerdtfeger and Amaturo 1979) and the high inland polar plateau. To the east of the Peninsula the continental character of the meteorological regime leads to a very cold Antarctic coastal belt while maritime conditions prevail to the west. It is likely that the Siple region is not dominated consistently by a single meteorological regime, but is a sensitive region of transition. This will be explored in more detail in Section 29.3.

The 302 meter Siple core was cut into 5757 samples each for microparticle concentrations and $\delta^{18}O$ and into 3492 samples for SO_4^{2-} analyses. The small sample size (and thus large number of samples) was necessary to isolate seasonal signals for establishing the best possible time scale. Both $\delta^{18}O$ and SO_4^{2-} records exhibit excellent seasonality (Figure 29.3) throughout the entire 302 meters and were used to produce the time scale previously discussed (Section 29.1.1). To create the annual records the $\delta^{18}O$ and particulate concentration values were averaged over individual annual layer thicknesses. The weight of each sample in the annual average was a function of its contribution to the thickness of the annual unit.

South Pole Station (90°S; 2835m asl)   A 101 meter core drilled at South Pole in 1974 was cut into 5218 samples for the analysis of microparticle concentrations (Mosley-Thompson and Thompson 1982) which were used to establish a 911 year record. The core was also cut into 1024 samples for $\delta^{18}O$ analyses at the University of Copenhagen. The $\delta^{18}O$ samples were cut to approximate a single year as defined by the current accumulation rate coupled with a steady state calculation of layer thinning with depth. Therefore, the $\delta^{18}O$ record does not contribute to the refinement of the time scale. The $\delta^{18}O$ data have been converted into a time series using the time-depth relationship derived from the particulate record. The $\delta^{18}O$ data from 1974 to 1982 were obtained by averaging over the annual layers in a pit 4 kilometers from the station (Mosley-Thompson *et al.* 1985). The net annual accumulation at South Pole is ~80 mm H_2O equivalent.

A second isotopic record is available from South Pole. Jouzel *et al.* (1983) produced a very detailed (~900 samples) continuous Deuterium (δD) record for the last 100 years with an estimated accuracy of ±5 years. The annual averages for A.D. 1887-1977 and the smoothed curves used in this paper are reproduced from Jouzel *et al.* (1983).

Dome C (124°10′E; 74°39′S; 3240m asl)   The low net annual accumulation (~37 mm H_2O eq.) at Dome C (Figure 29.1) precludes establishing an annually resolved record. The most detailed $\delta^{18}O$ records of the last 1000 years for Dome C (Figure 29.1) are from the upper 100 meters of two cores drilled in 1978 and 1979 (Benoist *et al.* 1982). Due to the large variability in net accumulation, a high level of smoothing was required to reduce the noise. Smoothing with filter band widths of 512 and 170 years precluded extraction of a detailed record. Benoist *et al.* (1982) conclude that the smoothed curves suggest generally cooler conditions from A.D. ~1200-1800.

T340: Filchner-Ronne Ice Shelf (~78°60′S; 55°W)   A 100 meter core was drilled in 1984 at site T340 on the Filchner-Ronne Ice Shelf (Figure 29.1) by the German Antarctic Research Program (Graf *et al.* 1988). Net annual accumulation at T340 is ~155 mm H_2O equivalent. The core was dated using the seasonal variations in $\delta^{18}O$ preserved in much of the core. The quality of the $\delta^{18}O$ record, and thus the time scale, was compromised by partial melting in the upper part of the core. Essentially, 479 annual layers were identified by $\delta^{18}O$ and of these 80 were expressed as small maxima or shoulders on larger peaks. In addition, 5 meters of core were unavailable. Extrapolating from surrounding sections lead to the addition of 41 years, representing this 5m section. Thus, a total of 520 years were estimated for the core which gives an age of A.D. 1460 for the bottom (Graf *et al.* 1988). No estimate of accuracy was given for the dating of T340. Due to the movement of the ice shelf, the ice in the core did not accumulate at a single location. Graf *et al.* (1988) attempted to correct the $\delta^{18}O$ record for the increasing continental effect down the length of the core since the ice at depth originated in a less maritime location.

Law Dome (66°44'S; 112°50'E; 1390m asl)   The Australian National Antarctic Research Expedition recovered a 473 meter ice core (BHD) in 1977 from the summit of the Law Dome. The net annual accumulation at the site of core BHD is ~800mm H_2O and the annual layers thin to approximately 110mm H_2O eq. at 450 meters. Pit studies and total Beta radioactivity profiles confirm the annual character of the well-preserved $\delta^{18}O$ signal. The upper 28 meters (1950-1977) were cut into roughly 10 samples per year to verify the seasonality of the $\delta^{18}O$ record. Below 28 meters, $\delta^{18}O$ was measured in selected sections and the results were extrapolated over intervening core sections. Recognizing that this introduces some uncertainty in the dating, Morgan (1985) suggests a dating accuracy of ± 10%.

Mizuho (70°41.9'S; 44°19.9'E; 2230m asl)   A 150 meter core was drilled at Mizuho Station by Japanese Antarctic Research Expeditions between 1970 and 1976 (Watanabe *et al.* 1978). Mizuho is situated in the Antarctic coastal zone (Figure 29.1) in a region dominated by katabatic winds. The mean annual accumulation is ~450 mm H_2O equivalent, but removal of material by wind produces hiatuses in the annual record. These hiatuses make the reconstruction of a continuous $\delta^{18}O$ record from the 150 meter core impossible. No obvious seasonal cycles in $\delta^{18}O$ were found. Principally, the core was dated by matching prominent isotope features to similar features in the upper part of the Camp Century, Greenland core which were assumed to be correlative. Thus, it is impossible to assess the quality of the time scale, but the error is likely to be higher than for other cores considered here.

29.2   The contemporary setting

Despite its zonal symmetry, the Southern Hemisphere atmospheric circulation is characterized by interannual variability larger than that in the Northern Hemisphere (Trenberth 1984). The distribution of storm tracks and preferred regions of blocking are tied to the planetary waves, and thus to the position of the mean jet stream. Rogers (1983) found that interannual temperature variability at a site reflects similar variability in the longitudinal positions of the upper level waves and associated surface cyclones. Positions of these features are controlled partially by the distribution of Antarctic sea ice and hence, by sea surface temperatures (Carleton 1984; Trenberth 1984). Thus, the high interannual temperature variability results from large-scale changes from year to year in the position of the jet stream and preferred storm tracks which control the penetration of warm air to the Antarctic interior.

Figures 29.4 and 29.5 illustrate the longest and most complete mean annual Antarctic surface temperature records expressed as departures from their respective time series means.

ANTARCTIC PENINSULA SURFACE TEMPERATURES

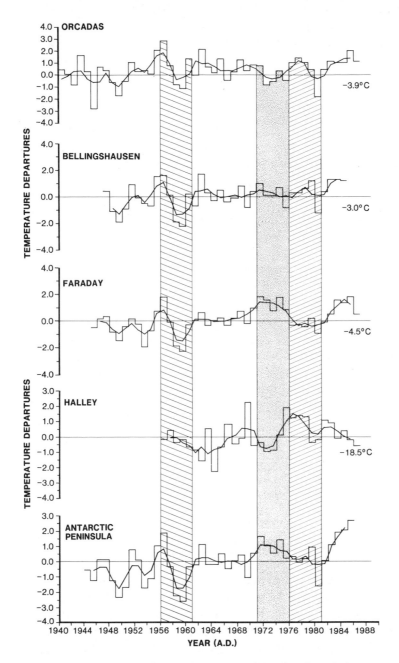

Figure 29.4 Annual surface temperatures (histogram) and three-year unweighted running means (solid line) are illustrated for Halley Bay and the following Antarctic Peninsula stations: Islas Orcadas, Bellingshausen, and Faraday. Composite temperatures for the Antarctic Peninsula are included (Limbert, 1984; Peel *et al.*, 1988).

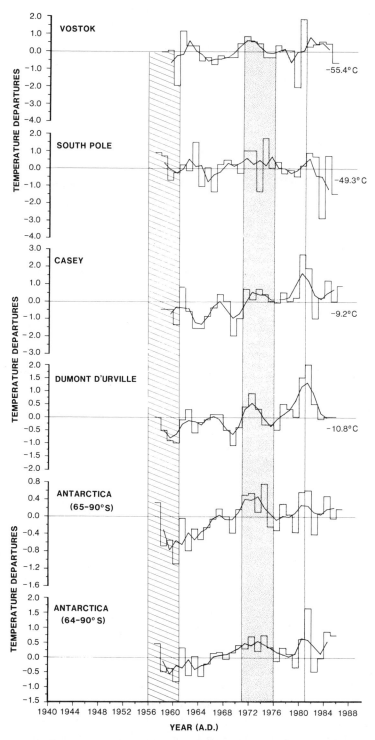

Figure 29.5 Annual surface temperatures (histogram) and the three-year unweighted running means (solid line) are illustrated for the following Antarctic stations: Vostok, Amundsen-Scott South Pole, Casey, and Dumont D'Urville. Two continental composites are included: Antarctic (65°-90°S) temperature trends (Raper *et al.*, 1984); Antarctic (64°-90°S) temperature trends (Hansen and Lebedeff, 1987).

To highlight trends in each record the three-year unweighted running mean is shown as a solid line and three periods of persistent multi-year temperature trends are shaded for later discussion. In Figure 29.4 stations from the Antarctic Peninsula region (Figure 29.1) are shown with the Antarctic Peninsula composite compiled by Limbert (1984) and updated by Peel *et al.* (1988). Islas Orcadas and Bellingshausen represent the northern end of the Peninsula while Faraday is characteristic of the western coastal region. Halley Bay, on the extreme eastern edge of the Weddell Sea, characterizes the colder, more continental regime to the east of the Peninsula. The temperature regime at Halley Bay is generally out of phase with that on the western side of the Ronne-Filchner Ice Shelf.

Figure 29.5 shows the most complete temperature records from East Antarctica. On the high polar plateau are Vostok and South Pole while Casey and Dumont D'Urville are along the Wilkes Land Coast. Included is a synthesis for (65°-90°S) derived from areal averaging (Raper *et al.* 1984; DOE 1987) and a spatially averaged trend analysis for 64° to 90°S (Hansen and Lebedeff 1987).

The main trends in the continental-scale composites (Figure 29.5) are reflected in the Peninsula composite (Figure 29.4) with the exception of the prominent warming trend since 1980 in the Peninsula region (excluding Halley Bay). This similarity arises partially because the continental-scale composites contain some areal bias toward the Peninsula region where the longer and more complete records exist (see Figure 29.2 in both Raper *et al.* 1984 and Hansen and Lebedeff 1987). Further, they are biased toward coastal conditions as only three inland stations (Amundsen-Scott, Vostok and Byrd) have long records and observations ended at Byrd in 1970. The greater similarity to trends at Casey and Dumont D'Urville illustrates this bias. If South Pole and Vostok are characteristic of the high polar plateau (above 2500 meters) then much of the areal extent of East Antarctica may not be represented well in the composite records. All three composites suggest a broad warming trend from the late 1950s to the early 1970s followed by moderate cooling. Since 1980 there has been no strong trend except in the Peninsula where a marked warming is evident at all stations (Figure 29.4) except Halley Bay.

The only long temperature record, Islas Orcadas (1903-1985) has been shown to be unrepresentative of both the Antarctic mean temperature series and the Southern Hemisphere (0°-60°S) composite for 1957-1983 (Jones *et al.* 1986; Raper *et al.* 1984). Much of the dissimilarity arises from a significant change in the temperature field in 1970. From 1957 to 1970 the temperature relationship between Orcadas and Faraday was consistent with colder temperatures at Faraday. However, this relationship reversed (Figure 29.4) in 1970. An analysis of the sea level pressure field led Schwerdtfeger (1976) to conclude that stronger and more frequent winds from the northwest increased the advection of warmer maritime air from the southeastern Pacific along the west coast of the Peninsula. Rogers' (1983) analysis suggested that a shift in the preferred location of surface cyclones and upper-level waves occurred. Comparison of the records in Figures 29.4 and 29.5 reveals that this warm event of the early 1970s affected a much larger area than the west coast of the Peninsula.

Antarctic temperature records reveal two other periods of persistent (multi-year) and geographically extensive temperature trends: 1975-1980 and 1955-1960. The spatial patterns of these trends are not simple and appear to change with time; that is, the temperature relationship between specific station pairs is not temporally consistent. For example, the 1970-1975 warm period, discussed above, which was so prominent at Faraday (Figure 29.4;

stipple) is present in all the composites, as well as Dumont D'Urville, Casey and Vostok (Figure 29.5). It is nearly absent at South Pole and Bellingshausen. Alternately, Halley Bay and Islas Orcadas exhibit cooler temperatures. It is interesting that the cooling trend in the late 1970s was prominent only where warming was pronounced earlier in the decade. This consistent spatial pattern may indicate that the consecutive warming and cooling throughout the decade was part of a large-scale circulation pattern which exhibited long-term persistence.

The spatial characteristics of the temperature pattern for the cooler period from 1955-1960 (Figure 29.4, hatched pattern) are different from those of the 1975-1980 cool period. Along the entire north-south axis of the Peninsula cooler temperatures were prevalent, but did not extend across the ice shelf to Halley Bay. The cooling was modest at Amundsen-Scott and Vostok, but more pronounced at Dumont D'Urville.

This discussion highlights the large-scale spatial differences in surface air temperatures which should be reflected in paleorecords reconstructed from $\delta^{18}O$ variations in ice cores. Such spatial differences may result from minor shifts in preferred locations for large-scale circulation features (Rogers 1983). Winter mean surface temperature trends in Antarctica have been linked to slow (multi-year) variations in atmospheric long waves (van Loon and Williams 1977) suggesting that mid-latitude large-scale circulation plays a significant role in the spatial variability of temperature over the continent. A more detailed discussion of the characteristics of the Antarctic meteorological regime is given by Schwerdtfeger (1984) and the precipitation regime is reviewed by Bromwich (1988).

29.3 Surface temperature and $\delta^{18}O$: A.D. 1945-1985

Annual $\delta^{18}O$ averages, like surface temperatures, exhibit interannual variability in response to large-scale circulation changes which control the frequency, duration, intensity, and seasonality of precipitation from cyclonic storms. In addition, ice core records contain glaciological noise superimposed upon the input signal by both surface and post-depositional processes. The climatological utility of an ice core record as an environmental proxy depends upon whether or not it reflects larger or regional-scale climatic trends. This assessment for Antarctic ice core records is hindered by the poor availability of long meteorological observations (DOE 1987) as previously illustrated (Figures 29.4 and 29.5). The only long and complete interior record is from South Pole. Vostok is of equal length but has three missing annual averages (Figure 29.5).

Comparison of $\delta^{18}O$ and surface temperatures provides a crude estimate of the larger-scale representativity of the ice core record although there are weaknesses in this approach (Peel, this volume). For example, the implicit assumption that accumulation occurs evenly throughout the year may introduce a bias. Precipitation falls throughout the year at Siple in association with persistent cyclonic activity, but the distribution throughout the year is unknown. On the other hand, winter (March to October) is the principal accumulation season at South Pole. The limitations of $\delta^{18}O$ as a proxy of condensation temperature are recognized as other factors such as distance from the source (e.g., sea ice extent, Bromwich and Weaver 1983) storm track trajectory and isotopic composition of the source also contribute to $\delta^{18}O$ at the deposition site. The relationships among these controlling factors are complex and their relative importance varies with geographic location.

Multi-year field programs are required to quantify the δ^{18}O-air temperature relationship. For the ice cores discussed here, such studies have been conducted only at South Pole. Aldaz and Deutsch (1967) collected precipitation for δ^{18}O analyses between November 1964 and October 1965 and combined these results with upper air (50 kPa) temperature observations. They formulated an empirical mean annual temperature-δ^{18}O relationship (δ^{18}O = 1.4(T°C)+4.0) which encompassed the majority of their observations. Using a similar approach, but with δD derived from pit studies for 1957-1978, Jouzel *et al.* (1983) found that mean annual deuterium values were best correlated (r= 0.57) with temperatures just above the inversion. This is consistent with the observation that precipitation forms just above the surface inversion (Miller and Schwerdtfeger 1972). For the purposes of this paper the δ^{18}O record is assumed to provide a proxy history for the condensation temperature of the precipitation at each core site. Thus, the classical interpretation of δ^{18}O for polar ice sheets is adopted, that is more (less) negative ratios imply cooler (warmer) condensation temperatures.

Figure 29.6 illustrates the δ^{18}O records for the only Antarctic ice cores (exclusive of the Peninsula, see previous chapter) analyzed in sufficient detail to provide meaningful annual averages for the period of overlap with meteorological observations. These are the South Pole δD record (Jouzel *et al.* 1983) the Law Dome δ^{18}O record (Morgan 1985) and the Siple δ^{18}O record (Mosley-Thompson *et al.* 1990, in press). The time scale accuracy of these cores has been discussed (Section 29.1.2). The 1974 South Pole δ^{18}O record is not included as it was cut into samples only approximating individual years (in the section on the South Pole Station) and thus, the 'annual' values lack the precision necessary for comparison with measured annual average temperatures; however, the record is quite appropriate for examination of longer trends. Figure 29.6 also illustrates two of the previously discussed temperature composites, the Peninsula and 65-90°S, respectively.

The large interannual variability makes it impossible to compare individual years among the three ice core records (Figure 29.6); however, the use of a three-sample unweighted running mean (solid line) highlights multi-year trends. Precise comparison of these records may be complicated by minor dating errors which could shift features several years in either direction. South Pole and Law Dome show remarkable similarity for sites of such disparate physical climatology (high plateau versus coastal). Nevertheless, both records reflect the same broad trends: an 'isotopic warm event' from 1955-1958, a subsequent cooling trend with little interannual variance throughout the early 1960s, and a warming trend since 1965. On the other hand, 2-3 year warm and cool events at Siple appear out of phase with those at South Pole and Law Dome, particularly prior to 1965. Only two longer trends characterize both South Pole and Siple: (1) the warming of the later 1960s which was also characteristic of the Peninsula (Figure 29.6) and a modest cooling from the mid-1940s to the mid-1950s. The latter was not characteristic of the Peninsula which experienced modest warming.

The large scale features in the Siple δ^{18}O record compare well with isotopic histories from James Ross Island, Dolleman Island, and Gomez Nunatak (Mosley-Thompson *et al.* 1990; Peel, this volume). This suggests that conditions at Siple may reflect those prevailing in the Peninsula area more frequently than over the high plateau.

There is observational evidence that temperature trends in the Peninsula area are generally 'out of phase' with those over the East Antarctica Plateau which is consistent with the South Pole and Siple records. Using factor analysis Rogers (1983) examined the spatial variability of

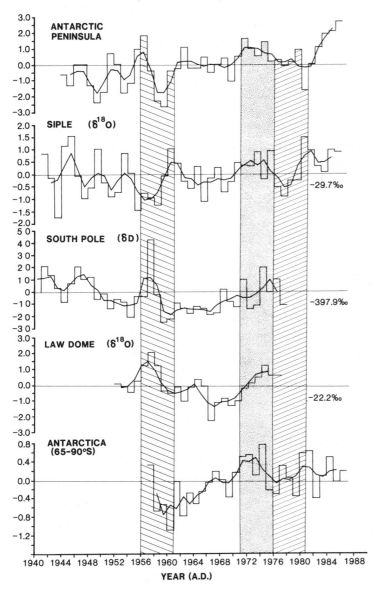

Figure 29.6 Stable isotope records for Siple (δ^{18}O) South Pole (δD) and Law Dome (δ^{18}O) are compared with the Antarctic Peninsula and continental (65-90°S) composites for A.D. 1940-1985.

seasonal mean temperature departures and reported an opposition in temperature anomalies between continental stations and those on or near the Antarctic Peninsula in all seasons but spring. For 1958 to 1980 the strength of the zonal westerlies (estimated from height differences across six pairs of mid-latitude and Antarctic stations) was strongly tied to the Peninsula-

583

continent temperature opposition pattern in winter and summer. Rogers found that in years when zonal westerlies are strongest, temperatures are anomalously cold at South Pole and anomalously warm in the Peninsula area. Previously, Swanson and Trenberth (1981) reported an opposition in the long-term temperature trends (1957-1979) between the northeast sector (roughly 0°-90°E; including South Pole) and the rest of the continent including much of West Antarctica and the Peninsula.

A principal components analysis (Raper *et al.* 1984) of the spatial characteristics of Antarctic annual and winter temperatures (1957-1983) support the Swanson and Trenberth results: that is, negative loadings (a cooling trend) in the sector between 40°E and 30°W and positive loadings (warming trend) for the rest of the continent including the Peninsula. These data illustrate the regional differences that exist over Antarctica and explain why no single meteorological record provides a consistent picture of Antarctic temperature trends.

29.4 The records since A.D. 1500

The most recent widespread Neoglacial episode (approximately A.D. 1500-1880) evident in both reconstructed Northern Hemisphere temperatures (Groveman and Landsberg 1979) and proxy records (Lamb 1977; Grove 1988) is commonly referred to as the Little Ice Age (LIA). Figure 29.7 illustrates the five Antarctic ice core $\delta^{18}O$ histories with sufficient time resolution and precision to examine environmental conditions over continental Antarctica during the last 480 years. A record from the Quelccaya ice cap (Thompson *et al.* 1986) located at 14°S at 5670 meters on the Altiplano of the southern Peruvian Andes, is included (Figure 29.7) as it closely resembles Northern Hemisphere temperatures reconstructed by Groveman and Landsberg (1979).

For the records in Figure 29.7, the Mizuho time scale is the least precise while that for Siple is the most precise. Partial melting makes assessing the Filchner-Ronne T340 time scale difficult; however, if approximately 20 years were missing from the upper part of T340, the major warm and cool events would correspond fairly well with those at Siple. Such errors are possible as the upper part of the core was affected by melting and contained most of the missing core sections for which extrapolations were used. In addition, the T340 $\delta^{18}O$ record (Figure 29.7) was adjusted for increasing continentality (^{18}O depletion) with depth in the core due to northward ice shelf movement and was finally smoothed with an unspecified filtering function (Graf *et al.* 1988).

Only selected sections of the Mizuho and Law Dome cores were analyzed. This discontinuous sampling results in a smoothed appearance. By contrast, the South Pole, Siple, and Quelccaya records were continuously analyzed, are annually resolved, and thus, exhibit a higher degree of variability. To facilitate comparison, a 48-point Gaussian filter was used to smooth the annual data (Figure 29.7). The horizontal line is the time series average for each core and values below the mean, inferred as cooler than average temperatures, are shaded.

The records from East Antarctica suggest cooler conditions during much of the LIA while the Siple record indicates warmer conditions for much of that period. Core T340 also suggests warmer conditions from A.D. 1650 to 1830 with a brief cool event at ~A.D. 1760. Clearly, in the last 300 years the T340 record most closely resembles that from Siple, particularly the downward trend in the last century.

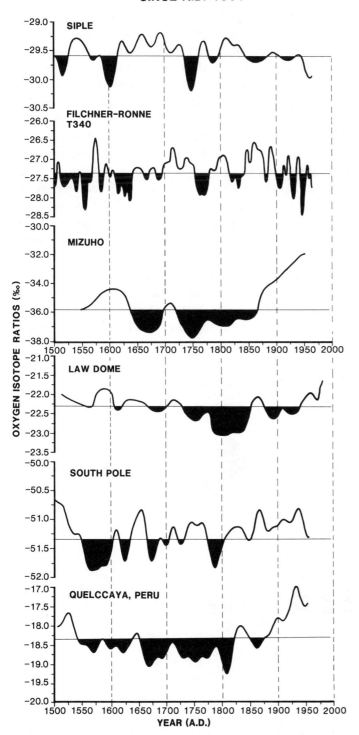

Figure 29.7 The Antarctic isotope records for A.D. 1500 to the present: Siple, T340, Mizuho, Law Dome, and South Pole. A comparable record for Quelccaya Ice Cap, Peru is included. Each time series mean is illustrated by the horizontal line. Isotopic values below the mean suggest cooler than normal temperatures and are shaded.

Figure 29.7 reveals several interesting spatial differences. First, Mizuho and Law Dome show the strongest similarity with coldest conditions between A.D. 1750 and 1850. Although conditions were cooler than average at South Pole from A.D. 1550 to 1800, this period is punctuated by warmer and cooler events with the coldest period in the mid to late 1500s. Using the empirical $\delta^{18}O$-temperature relationship of Aldaz and Deutsch (1967) the 'isotopically inferred' temperature depression in the late 1500s may have been ~0.5°C. A smoothed δD history from Dome C (not shown) also suggests cooler conditions from A.D. 1200 to 1800; however, because significant noise necessitated high level smoothing, further time resolution is impossible (Benoist et al. 1982). These records indicate that a warming trend has prevailed in East Antarctica since A.D. 1850 while cooling has clearly dominated at Siple and T340. The T340 record supports the suggestion that the longer-term trends in the Siple $\delta^{18}O$ history may reflect similar conditions for much of the Peninsula region. The opposition between the $\delta^{18}O$ records at Siple and those in East Antarctica is consistent with the currently observed opposition in surface temperatures (Section 3.). Since A.D. 1975 these trends appear to have reversed with cooling dominating over the Plateau and warming over the Peninsula (Figure 29.6).

The dust concentrations in cores from Siple and South Pole suggest further differences. Figure 29.8 illustrates the 10-year unweighted averages of particulate concentrations (diameters ≥ 0.63 μm) per milliliter sample for both cores. Concentrations above the time series mean for each core are shaded. From A.D. 1630 to 1880 dust concentrations at Siple are

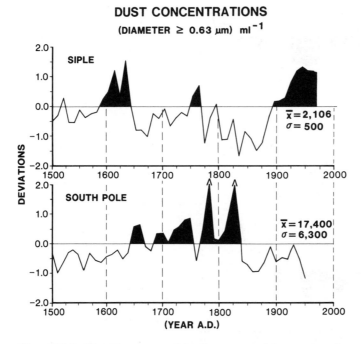

Figure 29.8 The 10-year unweighted averages of dust content for Siple and South Pole ice core are compared. Microparticle (diameter ≥ 0.63 μm) concentrations per ml are shown as standardized deviations from their respective time series means.

below average. A brief dust event around A.D. 1750 is associated with a negative (cooler) excursion in $\delta^{18}O$ (Figure 29.7). From A.D. 1880 to the present dust concentrations at Siple have increased while a cooling trend has prevailed (Figure 29.7). In contrast, at South Pole dust deposition was higher from A.D. 1650-1850. Note that the coldest temperatures at South Pole preceded the increase in dust by 100 years.

Every ice core extending back to the last glacial stage has exhibited a positive relationship between increased dust deposition and cooler temperatures as inferred from $\delta^{18}O$ (Thompson and Mosley-Thompson 1981; DeAngelis *et al.* 1987). A similar relationship has been demonstrated for a prominent Neoglaciation event on the Quelccaya Ice Cap in Peru (Thompson *et al.* 1986). The dust concentration-$\delta^{18}O$ relationship at Siple and South Pole supports this; e.g., reduced dust deposition during warmer conditions (less negative average $\delta^{18}O$) and increased dust deposition during cooler conditions. Thus, the insoluble particulate concentrations further support an inverse relationship between environmental conditions at Siple and South Pole during the last 500 years. Most of the broad temperature trends in the Peninsula region are reflected in the Siple $\delta^{18}O$ record (note the A.D. 1945-1955 exception) suggesting that the Siple region, possibly including much of the Antarctic Peninsula area, was characterized by warmer conditions from A.D. 1620-1830 than in the current century.

As previously discussed, the $\delta^{18}O$ record is interpreted strictly in terms of temperature and other factors may account for, or contribute to, the ^{18}O enrichment. For example, increased storm frequency during winter might enrich the annual $\delta^{18}O$ average (making it less negative) as storms tend to be associated with warmer than average temperatures. More frequent and/or intense cyclonic activity could increase warm air advection to the continent and possibly suppress sea ice extension. Parkinson (1990) examined sea ice limits using ship reports from early exploratory voyages to the Antarctic. The records, which are admittedly scanty and temporally discontinuous, showed no definitive evidence of sea ice extension. An increase in Cl^- and SO_4^{2-} concentrations could reflect an enhanced oceanic contribution as might be expected with reduced sea ice extent. However, from A.D. 1600 to 1830, the interval of greater warmth (least negative $\delta^{18}O$) the Cl^- and SO_4^{2-} concentrations in the Siple core are not elevated above the 480-year average. Figure 29.9 illustrates the 10-year averages of net accumulation for the 580-year record. The annual layers were converted to water equivalent using the measured depth-density relationship and were corrected for thinning with depth using a simple steady state model discussed by Bolzan (1984). A.D. 1600-1830 was not characterized by increased net balance although shorter intervals (several decades) of increase and decrease are evident.

29.5   Conclusions

The opposition in the Siple-South Pole $\delta^{18}O$ and dust records during much of the last five centuries may reflect an increase in the persistence of atmospheric and oceanic conditions which are responsible for the temperature opposition observed in the instrumental records. Rogers (1983) demonstrated a statistically significant relationship between intensification of the zonal westerlies, cooling at South Pole, and warming in the Peninsula region. It is unknown whether the intensification of the westerlies leads to cooling over the high polar plateau or vice versa.

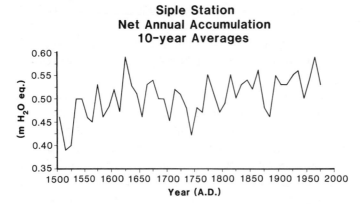

Figure 29.9 The 10-year unweighted averages of annual layer thicknesses for Siple are based upon the $\delta^{18}O$ time scale and converted to water equivalent. The thicknesses, which have been adjusted for thinning with depth, do not reveal major changes in net accumulation over the 480 year record.

The dissimilar dust concentrations probably reflect two different transport pathways from lower latitudes where the South American Altiplano and the South African Desert are the principal sources of dust. Observations at South Pole (Hogan *et al.* 1984) indicate the terrestrial (Al and Si) component of the aerosol mass is associated with an upper tropospheric or lower stratospheric source layer. Intensified westerlies at lower latitudes could entrain more material higher into the atmosphere and thus, increase the source of dust for the South Pole.

No similar aerosol studies have been conducted at Siple (accumulation rate $0.56m^{-1}$ H_2O eq.) where frequent and severe storms dominate throughout the year. However, it is more likely that particulates deposited at Siple have a lower tropospheric pathway associated with the passing cyclonic systems. Precipitation is an excellent mechanism for removal of entrained dust and thus lower tropospheric air reaching this region should be very clean (Hogan 1975). The microparticle analyses support this (Mosley-Thompson *et al.* 1990). If increased cyclonic activity along the periphery of the continent accompanied the postulated stronger westerlies, then the lower atmosphere should be cleansed further, leading to the low concentrations characterizing A.D. 1630 to 1880 (Figure 29.8). If this postulated increase in cyclonic activity resulted in more frequent storms at Siple, and hence increased net balance, the decrease in particulates could actually be accompanied by a net accumulation increase. However, as Figure 29.9 illustrates, the 480-year net accumulation record does not show an increase between A.D. 1650 and 1890, the interval of lowest dust deposition.

The similarity between the $\delta^{18}O$ records from South Pole and Quelccaya is intriguing. The excellent correspondence between the Quelccaya $\delta^{18}O$ record and Northern Hemisphere reconstructed temperatures has been demonstrated (Thompson *et al.* 1986). The similarity between the South Pole and Quelccaya $\delta^{18}O$ records, as well as the elevated dust concentrations, suggests the possibility of large-scale upper atmospheric teleconnections between the South American Andes and the high East Antarctica Plateau which warrants further investigation beyond the scope of this paper.

The 480-year records of $\delta^{18}O$ and dust concentrations from Siple suggest warmer and less dusty atmospheric conditions from A.D. 1600 to 1830 which encompasses much of the Northern Hemisphere Neoglacial period, the Little Ice Age. Dust and $\delta^{18}O$ data from South Pole, supported by the $\delta^{18}O$ results from Law Dome and Mizuho, indicate that opposite conditions (e.g., cooler and more dusty) were prevalent over the East Antarctica Plateau.

Meteorological data from 1945 to 1985 show that the Peninsula-East Antarctica Plateau temperature opposition prevailing during much of the last five centuries is consistent with the present spatial distribution of surface temperatures. There is some observational evidence suggesting that under present conditions stronger zonal westerlies are associated with cooler conditions on the polar plateau and warmer conditions in the Peninsula region. The physical processes controlling these spatial relationships must identified and better understood; however, the observational data base necessary for this assessment is currently lacking. These regional differences demonstrate that a suite of spatially distributed, higher resolution ice core records will be necessary to characterize more fully paleoenvironmental conditions since A.D. 1500 in Antarctica.

Acknowledgements

We thank W. Dansgaard and C. C. Langway, Jr. for providing the $\delta^{18}O$ results from the 1974 South Pole core. Larry Klein and Mary Davis conducted the particulate analyses and Jihong Dai conducted the sulfate measurements for the Siple core. Lonnie G. Thompson conducted the particulate analyses on the 1974 South Pole core. Pieter Grootes and Niels Gundestrup made the $\delta^{18}O$ measurements for the Siple core. Tom Johnstone and Kevin Herminghuysen organized the meteorological data and digitized the published ice core records. John Nagy, Susan Smith, Traci Temple, and Beth Daye produced the illustrations. This work was supported by NSF grant DPP-841032A04 to The Ohio State University, including a Research Experience for Undergraduates (REU) supplement. This is Contribution Number 714 of the Byrd Polar Research Center.

References

Aldaz, L. and S. Deutsch. 1967. On a relationship between air temperature and oxygen isotope ratio of snow and firn in the South Pole region. *Earth and Planet. Sci. Lett.*, 3, 2667-2674.

Benoist, J. P., J. Jouzel, C. Lorius, L. Merlivat and M. Pourchet. 1982. Isotope climatic record over the last 2.5 KA from Dome C, Antarctica. *Ann. Glaciol.*, 3, 17-22.

Bolzan, J. 1984. *Ice Dynamics at Dome C, East Antarctica*. Institute of Polar Studies Report No. 85, The Ohio State University, Columbus, Ohio.

Bromwich, D. H. 1988. Snowfall in high southern latitudes. *Rev. Geophys.*, 26(1) 149-168.

Bromwich, D. H. and C. J. Weaver. 1983. Latitudinal displacement from main moisture source controls delta-O18 of snow in coastal Antarctica. *Nature*, 301(5896) 145-147.

Carleton, A. M. 1984. Associated changes in west Antarctic cyclonic activity and sea ice. In: *Environment of West Antarctica: Potential CO₂-Induced Changes*. National Research Council, National Academy Press, 96-106.

Dai, J., E. Mosley-Thompson, L. G. Thompson and J. K. Arbogast. 1990. Chloride, sulfate and nitrate in snow at Siple Station, Antarctica, 1965-1985. *J. Glaciol.*, submitted.

DeAngelis, M., N. I. Barkov and V. N. Petrov. 1987. Aerosol concentrations over the last climatic cycle (160kyr) from an Antarctic ice core. *Nature*, 325, 318-321.

Department of Energy. 1987. *A Data Bank of Antarctic Surface Temperature and Pressure Data.* DOS Technical Report 038.

Graf, W., H. Moser, H. Oerter, O. Reinwarth and W.Stichler. 1988. Accumulation and ice-core studies on the Filchner-Ronne Ice Shelf, Antarctica. *Ann. Glaciol.*, 11, 23-31.

Grove, J. M. 1988. *The Little Ice Age.* London, Methuen, 498pp.

Groveman, B. S. and H. E. Landsberg, 1979. Simulated Northern Hemisphere temperature departures: 1579-1880. *Geophys. Res. Let.*, 6(10) 767-769.

Hammer, C. U., H. B. Clausen, W. Dansgaard, N. Gundestrup, S. J. Johnsen, and N. Reeh. 1978. Dating of Greenland ice cores by flow models, isotopes, volcanic debris and continental dust. *J. Glaciol.*, 20(82) 3-26.

Hansen, J. and S. Lebedeff. 1987. Global trends of measured surface air temperature. *J. Geophys. Res.*, 92(D11) 13,345-13,372.

Hogan, A. W. 1975. Antarctic Aerosols. *J. Appl. Meteor.*, 14(4) 550-559.

Hogan, A., K. Kebschull, R. Townsend, B. Murphey, J. Samson and S. Barnard. 1984. Particle concentrations at the South Pole, on meteorological time scales; Is the difference important? *Geophys. Res. Let.*, 1(9) 850-853.

Johnsen, S. J. 1977. Stable isotope homogenization of polar firn and ice. Isotopes and Impurities in Snow and Ice. *IAHS-AISH Publication* 118, 210-219.

Jones, P. D., S. C. B. Raper and T. M. L. Wigley. 1986. Southern Hemisphere surface air temperature variations: 1851-1984. *J. Clim. Appl. Meteorol.*, 25, 1213-1230.

Jouzel, J., L. Merlivat, J. R. Petit and C. Lorius. 1983. Climatic information over the last century deduced from a detailed isotopic record in the South Pole snow. *J. Geophys. Res.*, 88(C4) 2693-2703.

Lamb, H. H. 1977. *Climate: Present, Past and Future.* Volume 2: Climatic History and the Future. London, Methuen, 835pp.

Limbert, D. W. S. 1984. West Antarctic temperatures, regional difference and the nominal length of summer and winter seasons. In: *Environment of West Antarctica: Potential CO_2-Induced Changes.* National Research Council, National Academy Press, 116-139.

Miller, S. and W. Schwerdtfeger, 1972. Ice crystal formation and growth in the warm layer above the Antarctic temperature inversion. *Antarct. J. U.S.*, 7(7) 170-171.

Morgan, V. I. 1985. An oxygen isotope-climatic record from Law Dome, Antarctica. *Climatic Change*, 7(3) 415-426.

Mosley-Thompson, E. and L. G. Thompson. 1982. Nine centuries of microparticle deposition at the South Pole. *Quat. Res.*, 17, 1-13.

Mosley-Thompson, E., P. D. Kruss, L. G. Thompson, M. Pourchet and P. Grootes. 1985. Snow stratigraphic record at South Pole: potential for paleoclimatic reconstruction. *Ann. Glaciol.*, 7, 26-33.

Mosley-Thompson, E., L. G. Thompson, P. M. Grootes and N. Gundestrup. 1990. Little Ice Age (Neoglacial) paleoenvironmental conditions at Siple Station, Antarctica. *Ann. Glaciol.*, 14, in press.

Parkinson, C. L. 1990. Search for the Little Ice Age in southern ocean sea ice records. *Ann. Glaciol.*, 14, in press.

Peel, D. A., R. Mulvaney and B. M. Davison. 1988. Stable isotope/air-temperature relationships in ice cores from Dolleman Island and the Palmer Land Plateau, Antarctic Peninsula. *Ann. Glaciol.*, 10, 130-136.

Raper, S. C. B., T. M. L. Wigley, P. R. Mayes, P. D. Jones and M. J. Salinger. 1984. Variations in surface air temperatures. Part 3: The Antarctic, 1957-82. *Mon. Wea. Rev.*, 112, 1341-1353.

Rogers, J. C. 1983. Spatial variability of Antarctic temperature anomalies and their association with the southern hemispheric circulation. *Ann. Assoc. Am. Geog.*, 73(4) 502-518.

Schwerdtfeger, W. 1976. Changes of temperature field and ice conditions in the area of the Antarctic Peninsula. *Mon. Wea. Rev.*, 104(9) 1441-1443.

Schwerdtfeger, W. 1984. *Weather and Climate of the Antarctic*. Elsevier, Amsterdam, 261 pp.

Schwerdtfeger, W. and L. R. Amaturo. 1979. *Wind and weather around the Antarctic Peninsula.* Department of Meteorology, University of Wisconsin, Madison.

Swanson, G. S. and K. E. Trenberth. 1981. Trends in the Southern Hemisphere tropospheric circulation. *Mon. Wea. Rev.*, 109(9) 1879-1889.

Thompson, L. G. and E. Mosley-Thompson. 1981. Microparticle concentration variations linked with climatic change – evidence from polar ice cores. *Science*, 212(4496) 812-815.

Thompson, L. G. and E. Mosley-Thompson. 1982. Spatial distribution of microparticles within Antarctic snowfall. *Ann. Glaciol.*, 3, 300-306.

Thompson, L. G., E. Mosley-Thompson, W. Dansgaard and P. M. Grootes. 1986. The Little Ice Age as recorded in the stratigraphy of the tropical Quelccaya ice cap. *Science*, 234, 361-364.

Trenberth, K. E., 1984. The atmospheric circulation affecting the West Antarctic region in summer. In: *Environment of West Antarctica: Potential CO_2-Induced Changes.* National Research Council, National Academy Press, 73-87.

van Loon H. and J. Williams. 1977. The connection between trends of mean temperature and circulation at the surface: Part IV. Comparison of the surface changes in the Northern Hemisphere with the upper air and with the Antarctic in winter. *Mon. Wea. Rev.*, 105, 636-647.

Watanabe, O., K. Kato, K. Satow and F. Okuhira. 1978. Stratigraphic analyses of firn and ice at Mizuho Station. *Memoirs of the National Institute of Polar Research*. Special Issue 10, 25-47.

Section D: FORCING FACTORS

30 Evidence of solar activity variations

M. Stuiver and T. F. Braziunas

30.1 Introduction

For the investigation of a Sun-weather relationship, climate as well as solar change history has to be known in detail. A notable feature of solar change is the spottiness of the Sun which can be expressed at any time as the number and area of visible sunspots. Another expression of solar change is found in a variable distribution of charged particles in the interplanetary regions occupied by the solar wind plasma. The magnetic fields associated with these particles deflect a portion of the galactic cosmic ray flux directed towards the Earth. Because changes in the magnetic properties of the plasma originate at the Sun's surface, the intensity of cosmic rays measured on our planet depends on the surface conditions of the Sun. The relationship is such that cosmic ray intensity near our planet decreases when sunspot numbers increase. By monitoring the variable production rate of cosmogenic isotopes such as ^{14}C in our atmosphere, a time history of solar change can be obtained.

30.2 The tree-ring ^{14}C record

Trees, through $^{14}CO_2$ assimilation, lay down a ^{14}C record from which a proxy record of solar change can be derived. The derivation of the solar influence is complicated, however, because the ^{14}C record reflects additional processes influencing atmospheric $^{14}CO_2$ levels. These processes are (1) changes in geomagnetic dipole intensity (again changing ^{14}C production rate) and (2) climate-induced redistribution of ^{14}C between the various carbon reservoirs (atmosphere, world ocean and biosphere).

Determinations of atmospheric ^{14}C levels are made by the precise counting of the radioactivity of large samples, the sample activity being compared with the absolute National Bureau of Standards (NBS) oxalic acid standard activity. For tree-ring materials of known age a correction for ^{14}C decay is applied. The ^{14}C results are also normalized on a fixed $\delta^{13}C$ value of -25 per mil relative to the PDB standard. The results are expressed as $\Delta^{14}C$, which represents the ^{13}C-normalized ^{14}C activity at the time of formation, expressed as the deviation in parts per thousand from the oxalic acid standard activity (Stuiver and Polach 1977).

Tree-ring $\Delta^{14}C$ values reported in the *Radiocarbon* calibration issue (Stuiver and Kra 1986) were used to compile the 9,600 yr bi-decadal $\Delta^{14}C$ record given in Figure 30.1. The data of the

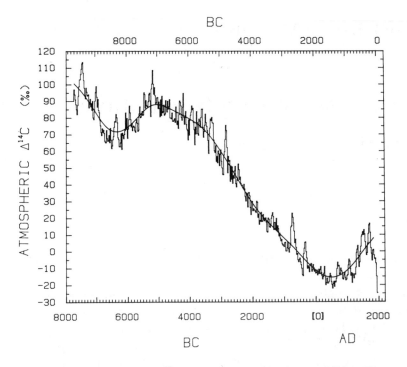

Figure 30.1 Atmospheric D^{14}C values in per mil for the past 9600 yr. The time scale (in yr A.D. or B.C.) was derived from dendrochronologically dated wood. The smoothed long-term trend represents a spline approximating a 400 yr moving average.

Seattle (Stuiver and Pearson 1986; Stuiver and Becker 1986) and Belfast (Pearson and Stuiver 1986; Pearson *et al* 1986) laboratories extend back to 5,200 yr B.C., whereas a mixture of results from Seattle (Stuiver *et al* 1986) La Jolla (Linick *et al* 1985) Tucson (Linick *et al* 1986) and Heidelberg (Kromer *et al* 1986) was used for the interval 5,200-7,750 B.C. The Seattle-Belfast portions of the data set have a typical standard deviation of 1.5-2.0 per mil for bi-decadal wood. For pre-5,200 B.C. data, the average standard deviation derived from the scatter of results within 20-yr blocks is 5 per mil.

The long-term Δ^{14}C trend in Figure 30.1 is usually attributed to either the influence of the geomagnetic dipole moment on ^{14}C production rate, or changing carbon reservoir parameters (induced by long-term climatic change) or a combination of both. The century scale oscillations in the Δ^{14}C record can be attributed to solar modulation of ^{14}C production. Arguments for a helio-magnetic origin of these oscillations take into account the Δ^{14}C maximum found during the A.D. 1654-1715 Maunder Mimimum in sunspot numbers (Stuiver and Quay 1980; Eddy 1976) the compatibility of the ^{14}C history with auroral evidence and naked-eye sunspot observations (Stuiver and Quay 1980; Stuiver and Grootes 1980; Krivsky and Pejml 1985) and the similarity (in timing as well as magnitude) of the 10Be cosmogenic record in ice cores and the atmospheric Δ^{14}C record (Beer *et al* 1988). Figure 30.2, obtained from the Figure 30.1 record by subtracting a spline approximating a 400 yr moving average, represents the solar component.

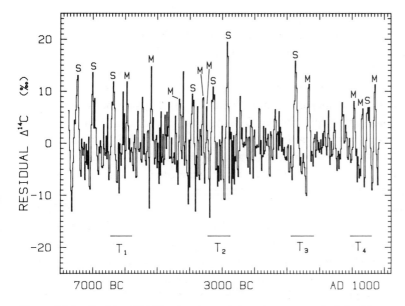

Figure 30.2 Residual D14C values remaining after removal of the Figure 30.1 spline. S and M variations indicate perturbations of the Spörer and Maunder variety. Triplet episodes, discussed in Stuiver and Braziunas (1989) are denoted by T_1-T_4.

The Earth carbon reservoir system transfers the ^{14}C production rate (Q) into an atmospheric ^{14}C signal. The Q record is more directly related to the Sun's surface changes than the Δ^{14}C record, and we therefore decode the atmospheric Δ^{14}C signal through carbon reservoir modeling. The carbon reservoir model (Stuiver and Quay 1980; Oeschger *et al* 1975) describes the terrestrial carbon exchange between the atmosphere, ocean and biosphere. To calculate the ^{14}C production rates from the record of atmospheric ^{14}C levels, the fluxes of ^{14}C between the carbon reservoirs (atmosphere, biosphere, oceanic mixed layer and deep sea) must be determined. Carbon and ^{14}C exchange is described by a finite- difference approximation of the differential equations governing the time rate of change of carbon and ^{14}C activities in the reservoirs. The transfer of carbon and ^{14}C between reservoirs is calculated over the time step of the iteration (0.04 year). The ^{14}C production rate is calculated from a mass balance of atmospheric ^{14}C. Over each time step the ^{14}C production rate is calculated to balance the change in atmospheric ^{14}C level interpolated from the Δ^{14}C record, plus the gain or loss of ^{14}C between the atmosphere and the other reservoirs. The Q values are averaged over a decade (or bi-decade) to represent the same time interval as the Δ^{14}C measurements. The calculations were made for a box diffusion model without direct contact between atmosphere and ocean. The basic properties of Q change, derived from the simple box-diffusion model, may undergo second order modifications when 'outcrop' and other models are used.

The Figure 30.2 Δ^{14}C variations were used to calculate ^{14}C production rate changes ΔQ around the long-term Qt values of the spline curve. Detailed information on calculated ΔQ/Qt values was presented in Stuiver and Braziunas (1988). There are several Spörer and Maunder type oscillations, identified by the M and S nomenclature, in the Figure 30.2 record.

Nine oscillations of the Maunder, and eight of the Spörer variety (Stuiver and Braziunas 1988) yield the prototype Maunder and Spörer production rate changes of Figure 30.3.

The changes in ^{14}C production rate Q (Stuiver and Braziunas 1988) observed at the Earth's surface since A.D. 1930 can be related to sunspot number S of the 11 yr cycle (Stuiver and Quay 1980). The gradient dQ/dS derived in this manner is too small to explain the full enhancement in Q during the Maunder minimum (Figure 30.3) when sunspot numbers are approaching zero. About one half of the Maunder Q change can be explained by the 11 yr Q-S relationship. The remaining one half appears to relate to solar wind modulation still occurring when sunspot numbers are near zero (Stuiver and Quay 1980). Using the above relationships, it is possible to convert the Δ^{14}C record into a paleo-record of sunspot numbers (Stuiver and Quay 1980).

For Sun-climate relationships the question of solar 'constant' change is of crucial importance. Changes of up to 0.06% have been measured for the recent part of the Hale cycle (Schatten 1988; Willson and Hudson 1988). The increase in solar constant E occurred when sunspot numbers changed from 20 to 140. On a longer timescale, the absence of sunspots during the A.D. 1645-1715 Maunder Minimum can be compared to an average sunspot number of 51 for the recent A.D. 1880-1965 interval. By analogy, the solar constant change between these periods would only be $51/120 \times 0.06 = 0.026\%$.

The above reasoning relates E change to changes in S. Another approach would be to

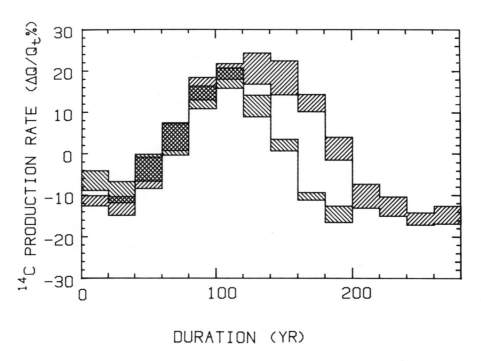

Figure 30.3 ^{14}C production rate variations of the proto-type Maunder and Spörer oscillations. The shaded error denotes the standard deviation in the mean bi-decadal values of nine M and eight S oscillations (Stuiver and Braziunas 1989). The ^{14}C production rate units are percent $\Delta Q/Q_t$ where ΔQ is the difference between the time-specific ^{14}C production rate Q and the concurrent value Q_t for the long-term trend in ^{14}C production as represented by the Figure 30.1 spline.

relate E change to changes in production rate Q. Here we expect a Maunder Minimum E change of $2 \times 0.026 = 0.052\%$ because solar modulation for this interval is twice that derived from the 11 yr sunspot cycle (see above). The 0.052% calculation assumes that the relationship between Q and E is also valid when solar activity diminishes after the sunspot signal "runs out" at zero (Stuiver and Quay 1980).

These calculations suggest that the magnitudes of solar constant change during the Maunder Minimum and the last solar cycle are comparable, but of different durations. A 0.05% change in E, however, corresponds to an equilibrium global temperature forcing of about 0.04°C ($\Delta E/E \sim 4 \Delta K/K$ with K the temperature in °K). Given our limited knowledge of climate sensitivity actual global temperature change could be larger (up to ~0.12°C) but this is still far short of the drop in temperatures during the "Little Ice Age". If caused by solar constant change, a mechanism of different origin is needed during the Little Ice Age. Such mechanisms have been postulated (Eddy et al 1982).

External factors, such as solar and volcanic forcing, leave a climatic imprint that is difficult to distinguish from that resulting from internal variability. The variance attributed to the solar forcing is small relative to the internally generated variance (for a large solar contribution the relationship between Sun and climate would have unequivocally been proven a long time ago). Correlation coefficients between climate and Q records usually are small, and when auto-correlation of both climate and Q are taken into account proof of a climate-Q relationship becomes difficult (Stuiver 1980).

As a minimum requirement – but not proof – for solar forcing of climate one demands similarities between the periodicities derived from the climate and Q records. Evidence for such similarity was given by Sonett and Suess (1984) who compared spectral properties of atmospheric $\Delta^{14}C$ and ring-widths of bristlecone pine grown in the White Mountains, Nevada. Another study suggesting a climate-Sun relationship is that of Wigley (1988) who noticed statistically significant correlations between century type variations of the atmospheric $\Delta^{14}C$ record and Röthlisberger's (1986) record of glacial advances. Eddy (1976) also notes similarities between climate and $\Delta^{14}C$. Pittock (1978, 1983) on the other hand, in critical reviews of sun-weather literature finds little support for a sun-weather relationship.

30.3   Spectral properties of the ^{14}C production rate record

The Q record, derived through carbon reservoir modeling from the Figure 30.1 $\Delta^{14}C$ record, has distinct spectral properties. Unfortunately, these spectral properties depend to some extent on the technical parameters used in deriving spectra. Figure 30.4 gives the spectral distribution obtained from the Burg MEM (maximum entropy method) for AR (autoregressive) order 20 and 120. The MEM time series model equation is a linear autoregression one in which each value is a weighted sum of M past data points together with random noise, where M is the autoregressive (AR) order. For the lower AR order a distinct set of harmonics is evident (periods 416, 215, 143, 85 etc., modified from Stuiver and Braziunas (1989) using an improved atmospheric $\Delta^{14}C$ record). For higher AR orders several additional peaks appear. The influence of AR order on the spectrum is given in Table 30.1 and depicted in Figure 30.5 for periodicities between 77 and 1000 yr. Figure 30.6 gives the more detailed plot of peak splitting ("split-plot") of the periodicities important for the current millennium. Derived

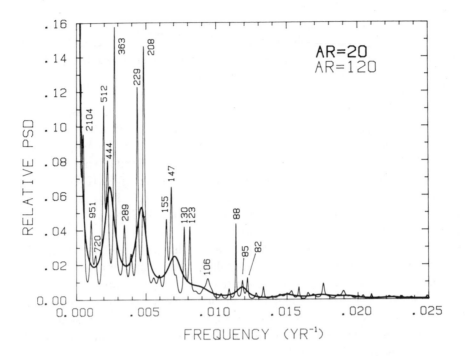

Figure 30.4 Power spectral density (PSD) normalized on the largest peak, obtained from MEM using the Burg algorithm. Autoregressive (AR) orders are 20 and 120. The periods are given in yr for each individual peak. The peaks present at AR of 20 are 416, 215, 143 and 85 yr.

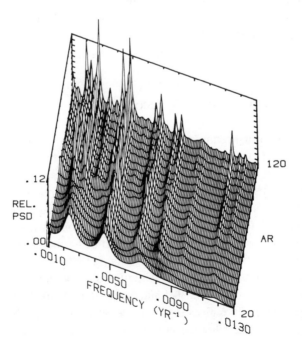

Figure 30.5 Power spectral density (PSD) normalized on the largest peak, versus AR order and frequency (yr⁻¹). The spectrum was derived from MEM analysis of the ^{14}C production rate Q calculated from carbon reservoir modeling. Frequency range is equivalent with periods ranging from 77 to 1000 yr.

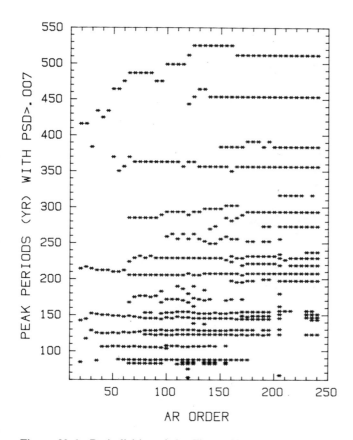

Figure 30.6 Periodicities of the Figure 30.5 peaks above a PSD threshold of 0.007 plotted versus AR order ("split-plot"). The lack of periodicities in the lower right corner is an artifact of the normalization on the highest peak. The split-plot of the record DQ/Qt obtained by removing the long-term trend does not exhibit a lower right corner devoid of periodicities.

periodicities clearly depend on AR order used, and differences of opinion are to be expected when discussing the relevance of a certain set of periodicities. Although the harmonic set of periodicities at AR = 20 fully represents, in our opinion, the basic features of the Q record it is also possible to claim that the larger number of periodicities found at higher AR order represents the Q record in finer detail.

In the power spectrum approach (Blackman and Tukey 1959, Mitchell *et al* 1966) the autocovariance function of the time series is calculated for all lags up to a maximum lag that represents a fraction of the total data length. A "lag window" of 10-30% of the data length is often recommended (eg. Mitchell *et al* 1966; Marple 1987) although successful results on artificial data have been shown for lags up to 80% as well (Kane 1977). Using half the number of data points as maximum lag, we obtained the spectrum in the upper part of Figure 30.7. And when using the often recommended maximum lag of 20% (Kane 1977) of number of data points (Figure 30.7, lower part) the 410 yr period that was observed for AR order 20 is represented also (although at much lower significance level than the three following harmonics).

Table 30.1 Periods (in years) identified in the ^{14}C production record.

AR=	20	60	120	180	240
			12395	9995	9995
			2104	2221	2161
			951	975	975
			720	740	
	416	475	512	512	512
			444	454	454
				391	384
		356	363	356	356
			289	294	294
			256	253	273
				232	238
			229		229
				222	219
	215	212	208	208	208
				198	198
			180		
			168		
			155	155	151
	143	148	147	149	147
				145	144
		126	130	130	123
			123	123	
		106	106		
	85	88	88		
			85		
			82		
			75		
			63		
			57		

The periodicities derived from the MEM and power spectrum method are compared in Figure 30.8. The bottom axis displays the AR order selected for the Burg analysis of the maximum lag chosen for the power spectrum analysis of the 480-point Q record. These choices correspond to the data fractions denoted on the upper axis. Circles and squares are the periodicities identified in the power spectrum record, with the squares being periodicities exceeding the 2 sigma significance level according to Mitchell's (1966) criteria. Asterisks represent periodicities derived from MEM analysis. Although differences exist (e.g. the splitting of the 416 yr periodicity) there is substantial agreement between the periods derived from both techniques. Q split-plot distributions of periodicities will depend on the time interval investigated. The split-plot distribution can be used as a fingerprinting method for matching the spectral properties of climate records covering the same interval. The somewhat subjective choice of AR order (or maximum lag) can be avoided when using the split-plot.

Users of spectral techniques disagree on MEM frequency resolution. Kay and Marple (1981) noted that the ability to distinguish the spectral response of two or more signals is

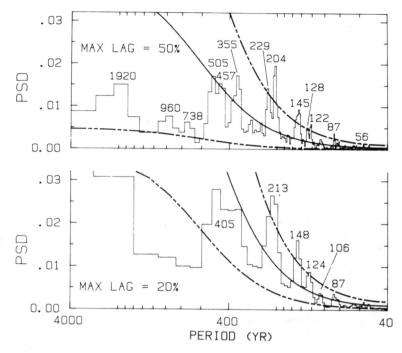

Figure 30.7 Power spectrum of the 9600 yr long ¹⁴C production record Q. Two sigma significance levels (dashed lines) above and below red noise (solid line) were calculated according to Mitchell *et al* (1966).

roughly the reciprocal of the time interval T over which sampled data are available. Thus with f the frequency, the resolution $\Delta f = 1/T$. Ulrych and Bishop (1975) estimated the Burg MEM resolution Δf to be twice as good ($\Delta f = 1/2T$). As can be seen from Figure 30.5, resolution depends on AR order. Marple (1987) gives Δf as a function of AR order and signal to noise ratio. The formula yields for our Q record (and signal to noise ratio of 1) a value of $1/1.25T$ for AR = 120, and $1/2.5T$ for AR = 240. Feynman and Fougere (1984) derived an 88.4 ± 0.7 yr period from aurorae reported per decade in Europe and the Orient from A.D. 450 to A.D. 1450. The quoted uncertainty corresponds to a Δf value of about $1/5T$. A less optimistic evaluation of MEM resolution is given by Sonett and Suess (1984) who estimated Δf to be about $3/T$.

Although the above resolution estimates differ by more than an order of magnitude we consider a Δf value of $1/T$ to be a conservative estimate of the MEM spectral resolution of the Q record. Figure 30.9 plots the "bandwidth" of the various periodicities t in which peak separation is not possible according to the $\Delta f = 1/T$ relationship. For a 1000 yr periodicity and a ¹⁴C record spanning 9600 yr one obtains $1/t = 1/1000 \pm 1/9600$ or t = 906 to 1116 yr. The percentage error in resolution increases drastically with length of the period (Figure 30.9, inset) and the longer periodicities (about 2100, 950 and 720 yr, Figure 30.4) have large uncertainties.

The length of the Q record has to exceed substantially the post-A.D. 1500 theme of this book when requesting a reasonable precision for the spectral features. The results are, nevertheless, also valid for the most recent portion of climate and solar history. Whereas the

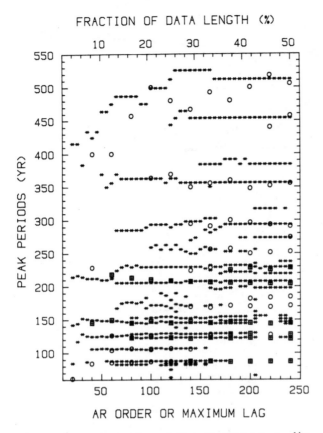

FRACTION OF DATA LENGTH (%)

Figure 30.8 A comparison of the periodicities in the ^{14}C production rate record derived from MEM (asterisks) and power spectrum analysis (squares when significant beyond the 2 sigma level, open circles when significance levels are below 2 sigma).

periodicities in the Q record below 500 yr can be attributed with fair confidence to solar influence, the longer periodicities may well be of different origin. Possibilities are production rate changes related to global changes in geomagnetic intensity, or a climate-induced redistribution of ^{14}C in the Earth carbon reservoirs. As our discussion is concerned mainly with solar forcing, we have not included climate-Q comparisons. These will be made elsewhere.

Acknowledgements

Important contributions to the spectral analysis were made by Travis L. Saling and Paula J. Reimer. The National Science Foundation supports the ^{14}C measurements of the Quaternary Isotope Laboratory at the University of Washington.

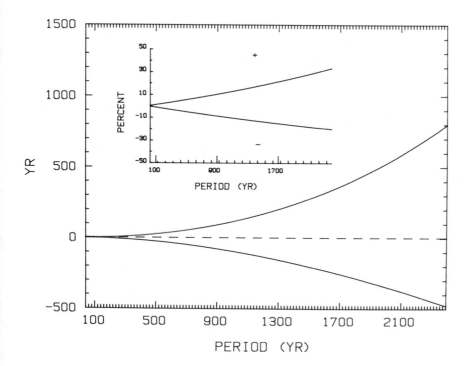

Figure 30.9 Estimated MEM spectral resolution of the 9600 yr ^{14}C production record for PSD peaks at particular periods. The inset gives the uncertainty in spectral resolution as a percentage of the period lengths.

References

Beer, J., U. Siegenthaler, G. Bonani, R. C. Finkel, H. Oeschger, M. Suter and W. Wölfli 1988. Information on past solar activity and geomagnetism from ^{10}Be in the Camp Century ice core. *Nature* 331, 675-679.

Blackman, R. B. and J. W. Tukey 1959. *The Measurement of Power Spectra*. New York, Dover.

Eddy, J. A. 1976. The Maunder Minimum. *Science* 192, 1189-1202.

Eddy, J. A., R. L. Gilliland and D. V. Hoyt 1982. Changes in the solar constant and climatic effects. *Nature* 300, 689-693.

Feynman, J. and P. F. Fougere 1984. Eighty-eight year periodicity in solar-terrestrial phenomena confirmed. *J. Geophys. Res.* 89, 3023-3027.

Kane, R. P. 1977. Power spectrum analysis of solar and geophysical parameters. *J. Geomag. Geoelectr.* 29, 471-495.

Kay, S. M. and S. L. Marple Jr 1981. Spectrum analysis – a modern perspective. *Proceedings of the IEEE 69*, 1380-1419.

Krivsky, L. and Pejml, K. 1985. Solar activity,aurorae and climate in central Europe in the last 1000 years. *Travaux Géophysiques XXXIII de L'academie Tchécoslovaque des sciences*, 77-151.

Kromer B., M. Rhein, M. Bruns, H. Schoch-Fischer, K. O. Münnich, M. Stuiver and B. Becker 1986. Radiocarbon calibration data for the 6th to the 8th Millennia B.C.. *Radiocarbon* 28, 954-960.

Linick, T. W., A. Long, P. E. Damon and C. W. Ferguson 1986. High-precision radiocarbon dating of bristlecone pine from 6554 to 5350 B.C.. *Radiocarbon* 28, 943-953.

Linick, T. W., H. E. Suess and B. Becker 1985. La Jolla measurements of radiocarbon in South German oak tree-ring chronologies. *Radiocarbon* 27, 20-32.

Marple Jr., S. L. 1987. *Digital Spectral Analysis with applications*. Englewood Cliffs, New Jersey: Prentice Hall, 492 pp.

Mitchell Jr., J. M., B. Dzerdzeevskii, H. Flohn, W. L. Hofmeyr, H. H. Lamb, K. N. Rao and C. C. Wallen 1966. *Climatic change*. World Meteorol. Organisation Tech. Note no 79, 79 pp.

Oeschger, H., U. Siegenthaler, U. Schotterer and A. Gugelmann 1975. A box-diffusion model to study the carbon dioxide exchange in nature. *Tellus* 27, 168-192.

Pearson, G. W., J. R. Pilcher, M. G. L. Baillie, D. M. Corbett and F. Qua 1986. High-precision ^{14}C measurements of Irish oaks to show the natural ^{14}C variations from A.D. 1840-5210 B.C.. *Radiocarbon* 28, 911-934.

Pearson, G. W. and M. Stuiver 1986. High-precision calibration of the radiocarbon time scale, 500-2500 B.C.. *Radiocarbon* 28, 839-862.

Pittock, A. B. 1978. A critical look at long-term sun-weather relationships. *Rev. of Geophys. and Space Phys*. 16, 400-420.

Pittock, A. B. 1983. *Quart. J. R. Met. Soc*. 109, 23-55.

Röthlisberger, F. 1986. *1000 Jahre Gletschergeschichte der Erde*. Aarau: Verlag Sauerländer.

Schatten, K. H. 1988. A model for solar constant secular changes. *Geophys. Res. Lett*. 15, 121-124.

Sonett, C. P. and H. E. Suess 1984. Correlation of bristlecone pine ring widths with atmospheric ^{14}C variations: a climate-Sun relation. Nature 307, 141-143.

Stuiver, M. 1980. Solar variability and climatic change during the current millennium. *Nature* 286, 868-871.

Stuiver, M. and B. Becker 1986. High-precision decadal calibration of the radiocarbon time scale, A.D. 1950-2500 B.C.. *Radiocarbon* 28, 863-910.

Stuiver, M. and T. F. Braziunas 1988. The solar component of the atmospheric ^{14}C record. In: *Secular Solar and Geomagnetic Variations in the last 10,000 years*, F. R. Stephenson and A. W. Wolfendale (edts) 245-266. Dordrecht: Kluwer.

Stuiver, M. and T. F. Braziunas 1989. Atmospheric ^{14}C and century-scale solar oscillations. *Nature* 338, 405-408.

Stuiver, M. and P. M. Grootes 1980. Trees and the ancient record of heliomagnetic cosmic ray flux modulation. In *The Ancient Sun*, R. O. Pepin. J. A. Eddy and R. B. Merrill (edts) 165-173. New York: Pergamon.

Stuiver, M. and R. Kra (eds.) 1986. Calibration issue. *Radiocarbon* 28, 805-1030.

Stuiver, M., B. Kromer, B. Becker and C. W. Ferguson 1986. Radiocarbon age calibration back to 13,300 years BP and the ^{14}C age matching of the German oak and U.S. bristlecone pine chronologies. *Radiocarbon* 28, 980-1021.

Stuiver, M. and G. W. Pearson 1986. High-precision calibration of the radiocarbon time scale, A.D. 1950-500 B.C.. *Radiocarbon* 28 805-838.

Stuiver, M. and H. S. Polach 1977. Discussion – reporting of ^{14}C data. *Radiocarbon* 19, 355-363.

Stuiver, M. and P. D. Quay 1980. Changes in atmospheric carbon-14 attributed to a variable sun. *Science* 207, 11-19.

Ulrych, T. J. and T. N. Bishop 1975. Maximum entropy spectral analysis and autoregressive decomposition. *Rev. of Geophysics and Space Physics* 13, 183-200.

Wigley, T. M. L. 1988. The climate of the past 10,000 years and the role of the sun. In *Secular Solar and Geomagnetic Variations in the last 10,000 years*, F. R. Stephenson and A. W. Wolfendale (edts) 209-224.

Willson, R. C. and H. S. Hudson 1988. Solar luminosity variations in solar cycle 21. *Nature* 332, 810-812.

31 Records of explosive volcanic eruptions over the last 500 years

R. S. Bradley and P. D. Jones

31.1 Introduction

Although explosive volcanic eruptions have long been suspected of having important effects on climate, the actual chronology of explosive events and their magnitude is not well known. Eruptions often occur in remote locations where, even today, they may go unrecorded (e.g. Sedlacek *et al.*, 1981; Mroz *et al.*, 1983). Only since the development of lidar observing stations in the 1970s have routine measurements of atmospheric aerosol loads been made, enabling specific aerosol clouds to be linked to preceding volcanic events (Reiter and Jager, 1986). Furthermore, the chemical composition of volcanic emissions is rarely well-documented and yet this is thought to be an important factor in determining what the consequent climatic effects might be (Rampino and Self, 1982; 1984; Devine *et al.*, 1984).

In this chapter we do not examine the climatic effects of explosive eruptions, but focus on the various chronologies of eruptions that have been constructed. The literature on climatic effects is large and there is little doubt that some eruptions in the past have affected climate over very large regions (for further discussions, see *inter alia*, Budyko, 1969; Lamb, 1970; Spirina, 1971; Yamamoto *et al.*, 1975; Oliver, 1976; Taylor *et al.*, 1980; Self *et al.*, 1981; Kelly and Sear, 1984; Angell and Korshover, 1985; Sear *et al.*, 1987; Kondratyev, 1988; Schonwiese, 1988; Bradley, 1988; Mass and Portman, 1989). The magnitude of any climatic effect depends on the volume of material ejected and the ejection height, the prevailing (and ensuing) stratospheric circulation pattern, and the chemical composition of the gas and tephra, particularly the amount of sulphur dioxide emitted. It appears that (at least for the major eruptions of the last century) effects on large-scale temperature averages were undetectable (that is, indistinguishable from noise) after 2-3 years (Kelly and Sear, 1984; Sear *et al.*, 1987; Bradley, 1988). However, others claim that eruptions may have had a significant impact on longer-term (lower frequency) changes of temperature (e.g. Budyko, 1969; Bryson and Goodman, 1980); indeed, there is some evidence that glacier advances over the last few centuries are closely linked to the cumulative atmospheric aerosol loading from volcanic eruptions (Bray, 1974; Porter, 1981, 1986).

Several attempts have been made to reconstruct the history of explosive volcanic eruptions over the last few centuries and four principal chronologies have been published. These are: a Dust Veil Index (DVI) (Lamb, 1970); a Volcanic Explosivity Index (VEI) (Simkin *et al.*, 1981; Newhall and Self, 1982); records of electrolytic conductivity or excess sulfate in ice cores (Hammer, 1977; Hammer *et al.*, 1980; Legrand and Delmas, 1987) and estimates of atmospheric optical depth (Pollack *et al.*, 1976; Bryson and Goodman, 1980). In addition, studies of frost damage in trees, historical and archeological records, and lunar eclipse data supplement and provide additional insight into these chronologies (LaMarche and

Hirschboeck, 1984; Stothers and Rampino, 1983; Keen, 1983). Several attempts have been made to improve on these approaches (e.g. Hirschboeck, 1980; Robock, 1981; Schonwiese, 1988) but the basic chronology of important eruptions is generally the same in each list. Here we discuss the derivation of these indices and their limitations; it will be apparent that none are ideal for climatic purposes.

31.2   Historical and geological records: the Dust Veil Index

Almost all studies of the climatic effects of volcanic eruptions have relied on Lamb's Dust Veil Index (DVI) chronology (Lamb, 1970). This was the first comprehensive effort to assess the probable climatic impact of volcanic eruptions and to construct a chronology of explosive eruptions and their magnitudes back to A.D. 1500. Unfortunately, the DVI assigned to individual eruptions is quite subjective and very dependent on observations and effects in mid-latitudes. Lamb based the DVI on historical and geological estimates of eruption magnitude, or on estimates of the volume of material dispersed into the atmosphere, or on volcanic effects on direct radiation receipts and surface temperature. These estimates were derived using one of the following formulae:

$$DVI = 0.97R.E.t, \quad or = 52.5T.E.t, \quad or = 4.4q.E.t$$

where R is the greatest percentage depletion of direct radiation, as registered by monthly averages in mid-latitudes of the hemisphere concerned,

T is the estimated lowering of annual temperature (in °C) over the mid-latitude zone of the hemisphere affected "for the year most affected",

q is the estimated total volume (in km^3) of solid matter dispersed as dust in the atmosphere,

and t is the total time in months between the eruption and last observation of the dust veil, or its effect on monthly radiation, or temperature values in mid latitudes.

The value of E ranges from 0.3 to 1.0, depending on the maximum extent attained by a dust veil. It is determined by the latitude of the eruption; for low latitude eruptions (20°N-20°S) E=1, for eruptions in the sub-tropics (20-35°N) E=0.7, for lower mid-latitude eruptions (35-42°N) E=0.5 and for higher latitude eruptions (>42°N) E=0.3.

The different multipliers used in each formula were derived empirically from a consideration of R, T and q for the 1883 eruption of Krakatau, so that the final DVI derived by each method would equal 1000 for this event. Lamb's 'final' estimate of the DVI for an individual eruption was based on an average of as many of these estimates as could be assembled. Global values of the DVI are given, as well as for each hemisphere separately, partitioning the dust between the hemispheres according to latitude of the eruption (see caption to Figure 31.1).

Values of R, T and t all depend on observations in mid-latitudes, though the effects of eruptions at different latitudes may be confined to the zone nearest to the eruption (Bradley, 1988). Furthermore, the derivation of the DVI using T (and an estimate of t based on temperature lowering) may lead to circular reasoning in any climatic analysis of the DVI. Lamb was cogniscent of this problem and made it clear which DVI values were derived wholly or partly in this way.

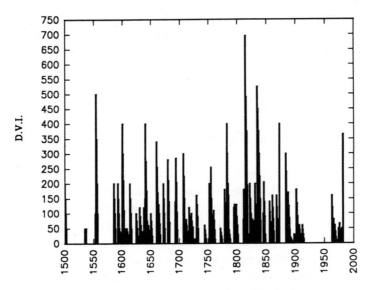

Figure 31.1 Cumulative DVI for the northern hemisphere, assuming the dust from an individual eruption is apportioned over four years with 40% of each DVI assigned to year 1, 30% to year 2, 20% to year 3 and 10% to year 4. Thus, the 1883 eruption of Krakatau (DVI=1000) results in values of 400 in 1883 declining to 100 in 1886. It is further assumed that all dust from eruptions poleward of 20°N remained in the northern hemisphere. For eruptions equatorward of 15°, the dust was assigned equally between the two hemispheres and for eruptions between 15 and 20°N and S, it was assumed that two thirds of the material remained in the hemisphere of the eruption and one third was dispersed to the other hemisphere (DVI values from Lamb, 1970, 1977, 1983).

When Lamb was unable to estimate a DVI based directly on any of the three formulae he assigned a DVI equal to the average for eruptions of similar size, using geological estimates of explosive eruption magnitude (largely based on Sapper, 1927). These were then adjusted (by E) for the latitude of the eruption. The averages were based on explosive eruptions which occurred after 1750 for which 'independent' DVI estimates (using the formulae above) could be made. Almost all of the DVI values before 1750 (and the majority of values from 1750 to 1900) were derived in this manner. When everything else failed, Lamb used historical records of unusual atmospheric phenomena ('volcanic sunsets'), plus his experience and knowledge of individual eruptions to produce a 'best estimate' of the eruption DVI.

In summary, Lamb's D.V.I. chronology is based on a number of different criteria, depending on the information available for a particular event. It is only partly objective and is biased towards the effects of eruptions on mid-latitudes. Nevertheless, Lamb's chronology is extremely useful if the methods used in its construction are clearly understood (cf. Kelly and Sear, 1982).

Figure 31.1 shows a time series of Lamb's cumulative DVI for the northern hemisphere, assuming the dust from an individual eruption is apportioned over four years, to simulate the gradual fall-out of dust from the atmosphere. The record is thus an estimate of the overall yearly volcanic aerosol loading of the atmosphere. It appears that dust loading was above

20th century levels for most of the preceding four centuries; highest levels prevailed in the early 19th century (1810s to 1830s) and in the 1780s, 1660s, 1630s and 1640s and in the 1600s. It is, of course, increasingly likely that as one goes back in time the cumulative values are only minimum estimates since many eruptions may not have been recognised.

Several attempts have been made to refine or improve upon Lamb's DVI. For example, Mitchell (1970) made a few minor changes to Lamb's estimates for the period 1850-1970 and Robock (1981) re-assessed the chronology from 1600 onwards. Robock eliminated as far as possible DVI values based partly or solely on temperature estimates and applied the time-distance decay model of Cadle *et al.*, 1976 (based on observed dust veil dispersal after the eruptions of Agung and Fuego) to produce a revised assessment of volcanic dust loading in time and space (Figure 31.2). The principal difference between Robock's estimates and

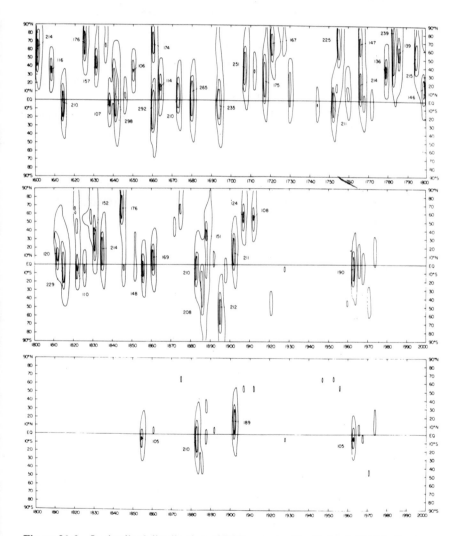

Figure 31.2 Latitudinal distribution of DVI as revised by Robock (1981). Contours are at index values of 20, 60 and 100 with the maximum value at the center also plotted (Robock, 1981).

609

Lamb's involve significantly lower DVI values for Krakatau (1883) Coseguina (1835) and Tambora (1815) and higher values for several events in the 17th and 18th centuries (Figure 31.3). However in all cases, the relative absence of major eruptions in the period after ~1920 is quite apparent.

31.3   Geological records: the Volcanic Explosivity Index

An alternative method of classifying explosive volcanic eruptions is that proposed by Newhall and Self (1982). They rank eruptions using only volcanological criteria to assess the magnitude, intensity, dispersive power and destructiveness of an event. The index does not rely on any assessment of temperature depression, atmospheric effects or reduction in radiation receipts, and is not weighted by climatic observations in mid-latitudes (as with the DVI or optical depth record) or high latitudes (as with the ice core record). It is thus a climatically independent estimate of explosivity, based primarily on geological criteria.

Each eruption is ranked from 1 to 8 (8 being the largest) and a chronology of volcanic events spanning the last 8,000 years has been constructed (Simkin *et al.*, 1981). The classification clearly identifies those events which are thought to have injected material into the stratosphere; such eruptions are assigned a Volcanic Explosivity Index (VEI) of 4 or more. These eruptions (described as Plinian or ultra-Plinian eruptions) produced at least 10^8 m^3 of ejecta and had column heights of 10-25km or more. Over the past 500 years there have been more than 110 eruptions with a VEI of 4 or more. These are listed in Table 31.1, divided into approximately equal area latitude bands in each hemisphere. Most major eruptions have occurred at high latitudes (>50°N) or within 10° of the Equator. There are relatively few records of large explosive eruptions in the southern hemisphere beyond the equatorial zone.

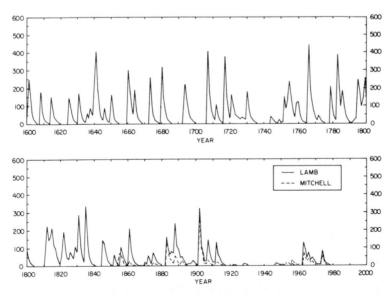

Figure 31.3 Annual average northern hemisphere volcanic dust veil indices as re-assessed by Robock (1981).

Table 31.1 Major Explosive Eruptions with VEI of 4 or more: 1500-1981 (after Simkin *et al.*, 1981).

Location	Elev. (m)	Lat.	Long.	Starting Date	VEI
A) Major High Latitude Eruptions (> 45°N)					
Alaid, Kurile Is.	2339	50.8	155.5E	04 1981	4
Gareloi, Aleutians	1573	51.8	178.8W	08.1980	4
Mt.St.Helens, U.S.A.	2549	46.2	122.2W	05 1980	5
Bezymianny,Kamchatka	2800	56.1	160.7E	02 1979	4
Augustine, Alaska	1227	59.4	153.4W	01 1976	4
Plosky Tolbachik	3085	55.9	160.5E	07 1975	4
Sheveluch, Kamchatka	3395	56.8	161.6E	11 1964	4
Bezymianny,Kamchatka	2800	56.1	160.7E	03 1956	5
Spurr, Alaska	3374	61.3	152.3W	07 1953	4
Hekla, Iceland	1491	61.0	17.7W	03 1947	4
Sarychev, Kurile Is.	1497	48.1	153.2E	11 1946	4
Kliuchevskoi,Kamchatka	4850	56.2	160.8E	01 1945	4
Kliuchevskoi,Kamchatka	4850	56.2	160.8E	03 1931	4
Raikoke, Kurile Is.	551	48.3	153.3E	02 1924	4
Katla, Iceland	1363	63.6	19.0W	10 1918	4
Katmai, Alaska	841	58.3	155.2W	06 1912	6
Ksudach, Kamchatka	1079	51.8	157.5E	03 1907	5
Thordarhyna, Iceland	1659	64.3	17.6W	05 1903	4
Augustine, Alaska	127	59.4	153.4W	10 1883	4
Askja, Iceland	1510	65.0	16.8W	03 1875	5
Grimsvotn, Iceland	1719	64.4	17.3W	01 1873	4
Sinarka, Kurile Is.	934	48.9	154.2E	? 1872	4
Sheveluch, Kamchatka	3395	56.8	161.6E	02 1854	5
Chikurachki, Kurile Is.	1817	50.3	155.5E	12 1853	4
Hekla, Iceland	1491	64.0	19.7W	09 1845	4
Isanotski, Aleutian Is.	2446	54.8	163.7W	03 1825	4
Beerenberg, Jan Mayen	2277	71.1	8.2W	? 1818	4
St. Helens, U.S.A.	2549	46.2	122.2W	? 1800(D)	4
Pogromni, Aleutians Is.	2002	54.6	164.7W	? 1795	4
Alaid, Kurile Is.	2339	50.8	155.5W	* 1793	4+
Laki, Iceland	500	64.1	18.3W	* 1784	4
Raikoke, Kurile Is.	551	48.3	153.3E	? 1778	4
Hekla, Iceland	1491	64.0	19.7W	04 1766	4
Katla, Iceland	1363	63.6	19.0W	10 1755	5
Oraefajokull, Iceland	2119	64.0	16.7W	08 1727	4
Katla, Iceland	1363	63.6	19.0W	05 1721	4
Chirpoi, Kurile Is.	624	46.5	150.9E	12 1712	4?
Hekla, Iceland	1491	64.0	19.7W	02 1693	4
Chikurachki, Kurile Is.	1817	50.3	155.5E	? 1690(T)	4
Hekla, Iceland	1491	64.0	19.7W	07 1510	4
St. Helens, U.S.A.	2549	46.2	122.2W	? 1500(D)	5
B) Major Mid Latitude Eruptions (20-45°N)					
Tiatia, Kurile Is.	1822	44.4	146.3E	07 1973	4
Komaga-take, Japan	1140	42.1	140.7E	06 1929	4
Sakura-jima, Japan	1118	31.6	130.7E	01 1914	4
Tarumai, Japan	1024	42.7	141.4E	03 1909	4
Suwanose-jima, Japan	799	29.5	129.7E	10 1889	4
Bandai, Japan	1819	37.6	140.1E	07 1888	4
Nasu, Japan	1917	37.1	140.0E	07 1881	4
Suwanose-jima, Japan	799	29.5	129.7E	? 1877	4

Table 31.1 – *continued*

Location	Elev. (m)	Lat.	Long.	Starting Date	VEI
Komaga-take, Japan	1140	42.1	140.7E	09 1856	4
Usu, Japan	725	42.5	140.8E	04 1853	4
Usu, Japan	725	42.5	140.8E	03 1822	4
Asama, Japan	2550	36.4	138.5E	05 1783	4
Komage-Take, Japan	1140	42.1	140.7E	? 1765	4
Oshima-O-Shima, Japan	714	41.5	139.4E	08 1741	4
Tarumai, Japan	1024	42.7	141.4E	08 1739	5
Fuji, Japan	3776	35.4	138.7E	12 1707	4
Iwate, Japan	2041	39.9	141.0E	02 1686	4
Tarumai, Japan	1024	42.7	141.4E	08 1667	5
Usu, Japan	725	42.5	140.8E	08 1663	5
Agua de Pau, Azore Is.	948	37.8	25.5W	06 1563	4

C) Major Low Latitude Eruptions, (0° to 20°N)

Location	Elev. (m)	Lat.	Long.	Starting Date	VEI
Mt.Pagan,Mariana Is.	570	18.3	145.8E	05 1981	4
Fuego, Guatemala	3763	14.5	90.9W	10 1974	4
Awu, Indonesia	1320	3.7	125.5E	08 1966	4
Taal, Philippines	400	14.0	121.0E	09 1965	4
Fuego, Guatemala	3763	14.5	90.9W	01 1932	4
Agrigan, Mariana Is.	965	18.8	145.7E	04 1917	4
Colima, Mexico	4100	19.4	103.7W	01 1913	4?
Taal, Philippines	400	14.0	121.0E	01 1911	4
Santa Maria,Guatemala	3772	14.8	91.6W	10 1902	6
Soufriere,West Indies	1178	13.3	61.2W	05 1902	4
Pelee, West Indies	1397	14.8	61.2W	05 1902	4
Dona Juana,Colombia	4250	1.5	76.9W	11 1899	4
Purace, Columbia	4600	2.4	76.4W	10 1869	4
Purace, Colombia	4600	2.4	76.4W	12 1849	4
Coseguina, Nicaragua	859	13.0	87.6W	01 1835	5
Colima, Mexico	4100	19.4	103.7W	02 1818	4
Mayon, Philippines	2462	13.3	123.7E	02 1814	4
Soufriere, St. Vincent	1178	13.3	61.2W	04 1812	4
San Martin, Mexico	1550	18.6	95.2W	03 1793	4
Jorullo, Mexico	1330	19.0	101.7W	* 1764	4
Tongkoko, Indonesia	1149	1.5	125.2E	? 1680	4
Gamkanora, Indonesia	1635	1.4	127.5E	05 1673	4
San Salvador, El Salvador	1850	13.7	89.3W	? 1671	4
San Salvador, El Salvador	1850	13.7	89.3W	? 1575	4?
Arenal, Costa Rica	1552	10.5	84.7W	? 1525	4

D) Major Southern Hemisphere Eruptions: Low Latitudes (0-20°S)

Location	Elev. (m)	Lat.	Long.	Starting Date	VEI
Ulawun, New Britain	2300	5.0	151.3E	10 1980	4
Negra, Galapagos Is.	1490	0.8	91.2W	11 1979	4
Fernandina,Galapagos	1495	0.4	91.6W	06 1968	4
Lengai, E. Africa	2886	2.8	35.9E	08 1966	4
Kelut, Indonesia	1731	7.9	112.3E	04 1966	4
Agung, Indonesia	3142	8.3	115.5E	03 1963	4
Bagana, Solomon Is.	1702	6.1	155.2E	02 1952	4
Ambryn, New Hebrides	1334	16.3	168.1E	* 1951	4
Lamington, New Guinea	1780	8.9	148.2E	01 1951	4

Table 31.1 – *continued*

Location	Elev. (m)	Lat.	Long.	Starting Date	VEI
Rabaul, New Britain	229	4.3	152.2E	05 1937	4
Manam, New Guinea	1725	4.1	145.1E	08 1919	4
Tungurahua, Ecuador	5016	1.5	78.5W	04 1918	4
Krakatau, Indonesia	813	6.1	105.4E	08 1883	6
Galunggung, Indonesia	2168	7.4	108.1E	10 1822	5
Tambora, Indonesia	2851	8.3	118.0E	04 1815	7
Papandayan, Indonesia	2665	7.4	107.7E	08 1772	4
Cotopaxi, Ecuador	5897	0.8	78.4W	04 1768	4
Cotopaxi, Ecuador	5897	0.8	78.4W	11 1744	4
Long Is., New Guinea	1304	5.4	147.1E	? 1700(C)	6
Quilotoa, Ecuador	3914	0.9	78.9W	11 1600	4?
Kelut, Indonesia	1731	7.9	112.3E	? 1586	4
Cotopaxi, Ecuador	5897	0.8	78.4W	06 1534	4

E) Major Southern Hemisphere Eruptions: Mid and High Latitudes (> 20°S)

Location	Elev. (m)	Lat.	Long.	Starting Date	VEI
Nilahue, Chile	400	40.4	72.1W	07 1955	4
Azul, Chile	3810	35.7	70.8W	04 1932	6
Puyehue, Chile	2240	40.6	72.1W	12 1921	4
Tarawera, New Zealand	1111	38.2	176.5E	06 1886	5
Peteroa, Chile	4090	35.3	70.6W	12 1762	4

* Continuous eruptions over one or more years.

A letter in parenthesis or ? after date indicates date uncertain; eruption dated dendrochronologically (D) or tephrochronologically (T) or by radiocarbon (C).

However, there is no doubt that this catalog omits or underestimates the size of many climatically significant eruptions which must have occurred in remote areas such as the Aleutians, and Kamchatka, or in the Andes, New Zealand and Antarctica. Those volcanoes which have been studied most carefully (e.g. Hekla in Iceland and Taupo in New Zealand) account for a large fraction of the major eruptions which are known to have occurred (Simkin *et al.*, 1981) and, no doubt, as further geological studies are carried out the list of explosive eruptions will expand, particularly for the period before about 1850.

Of particular relevance to the climatic effects of explosive eruptions is the total volume of sulfur-rich volatiles (such as SO_2 and H_2S) which is emitted (Rampino and Self, 1984). These gases result in sulphuric acid aerosols being produced in the stratosphere and such aerosols are known to be important in reducing solar radiation receipts at the surface (Toon and Pollack, 1984). Various estimates have been made of the H_2SO_4 'yield' from major eruptions; Table 31.2 lists recent results based on petrologic studies of glass inclusions in tephra from a number of the largest eruptions of the last 500 years, together with their corresponding VEI (Devine *et al.*, 1984; Rampino and Self, 1984; Palais and Sigurdsson, 1989). This provides a quite different perpective on the climatic significance of these eruptions. In terms of H_2SO_4 yield, the most important eruption of the last 500 years was the fissure eruption of Laki in Iceland (1783). This is the well-known event which produced 'dry fogs' over Europe, as described by Franklin (1784) and discussed at length by Lamb (1970). The Laki eruption

Table 31.2 Petrologic estimates of volatile emissions based on glass inclusion analysis (data from Rampino and Self, 1984; Symonds *et al.*, 1988 and Palais and Sigurdsson, 1989).

Eruption	Date	Lat.	VEI	H_2SO_4 (metric tons)
St. Helens, U.S.A.	1530	46°N	5	2.30×10^5
Laki, Iceland	1783	64°N	4	9.03×10^7
St. Helens, U.S.A.	1800	46°N	4	3.50×10^3
Tambora, Indonesia	1815	8°S	7	5.24×10^7
Coseguina, Nicaragua	1835	13°N	5	0
Krakatau, Indonesia	1883	6°S	6	2.94×10^6
Tarawera, New Zealand	1886	38°S	5	5.00×10^6
Santa Maria, Guatemala	1902	15°N	6	1.80×10^5
Soufriere, St. Vincent	1902	13°N	4	2.40×10^5
Katmai, Alaska	1912	58°N	6	7.90×10^6
Bezymianny, Kamchatka	1956	56°N	5	6.00×10^6
Agung, Indonesia	1963	8°S	4	2.84×10^6
St. Helens, U.S.A.	1980	46°N	5	7.90×10^4
El Chichon, Mexico	1982	17°N	4	7.00×10^4

(VEI=4) was comparable in H_2SO_4 production to Tambora which has a VEI rating of 7. Tambora is considered to have been the most violent explosive eruption of the Holocene; about $50 km^3$ of magma erupted within 24 hours, some of which probably reached the upper stratosphere (Self and Rampino, 1984; Stothers, 1984). By contrast, Laki was a very extensive ($>560 km^2$) non-explosive fissure eruption which continued for about 8 months. It is possible that intense convection cells associated with an eruption of this size could have led to stratospheric injection of gases and aerosols leading to similar climatic consequences as a more explosive eruption (Wolff *et al.*, 1984; Stothers *et al.*, 1986). In contrast to Tambora and Laki, Coseguina tephra reveals little evidence of sulfur-rich volatile emissions, though very large quantities of HCl were produced (1×10^7 metric tons) (Palais and Sigurdsson, 1989). The implications of such cloride-rich eruptions for climate are as yet unclear, but it seems likely that they have a major influence on stratospheric ozone concentrations (Johnson, 1980; Symonds *et al.*, 1988)

Table 31.2 also shows that the eruptions of Tarawera (1886) Katmai (1912) Bezymianny (1956) and Agung (1963) all produced more H_2SO_4 than Krakatau (1883) which explains why these events are so clearly noticeable in continental temperature records (Bradley, 1988). Indeed, there is a power relationship between petrologic estimates of sulfur yield and estimated temperature decrease over the northern hemisphere following major explosive eruptions over the last few centuries (Palais and Sigurdsson, 1989) (Figure 31.4).

31.4 Glaciological records

Large variations in the electrolytic conductivity of Greenland ice cores were first shown to be related to the deposition of acidic snow following large explosive eruptions, by Hammer (1977) and Hammer *et al.* (1980). Volcanic eruptions may produce large quantities of sulfur

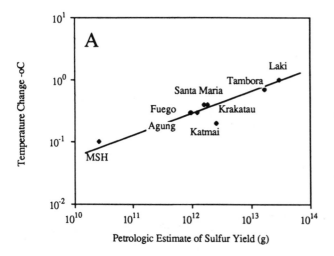

Figure 31.4 Relationship between petrologic estimates of sulfur yield (in grams) and estimates of northern hemisphere continental temperature decreases (°C) following the explosive eruptions indicated (Palais and Sigurdsson, 1989).

and chlorine gases which are converted to acids in the atmosphere. The resulting acidic snowfall (and dry deposition of acidic particles directly on the ice sheets) produces high levels of conductivity or 'spikes' above natural background levels. In most cases these acidity spikes result from excess sulphuric acid events (Figure 31.5). Hammer's original studies showed a remarkable similarity between electrolytic conductivity in the Greenland Crete ice core and Lamb's DVI, indicating that the elevated acidity (above background levels) could be used as an index of volcanic explosivity. This concept has now been extensively investigated in other cores from Greenland, Antarctica and elsewhere, with careful analyses to determine the precise chemistry of the acidity spikes (e.g. Holdsworth and Peake, 1984; Mayewski *et al.*, 1986; Legrand and Delmas, 1987; Lyons *et al.*, 1990). This enables more precise 'fingerprinting' of individual eruptions, some of which produce large amounts of HCl, for example, rather than H_2SO_4 (see discussion above).

Recently, Legrand and Delmas (1987) proposed that a Glaciological Volcanic Index (GVI) be compiled from ice core measurements in different parts of the world. Several problems face such a development. Firstly, the ice core records are primarily from high latitudes and are thus strongly biased towards high latitude eruptions, particularly (in the case of Greenland ice cores) those of Icelandic volcanoes (Hammer, 1984). Secondly, the eruption signal may differ significantly in magnitude from one ice core to another and important events may not appear at all in some ice core records (Delmas *et al.*, 1985). This reflects the fact that the spatial distribution of acidic snowfall varies after volcanic eruptions and hence the record of deposition will vary from one ice coring site to another (Clausen and Hammer, 1988). Thirdly, in some low latitude ice cores, deposition of alkaline aerosols may neutralise the volcanic acids and hence eliminate the eruption signal. These problems do not seem to be insuperable; more ice cores from lower latitudes (cf. Thompson, this volume) will help in constructing a complete catalog of explosive eruptions and in resolving latitudinal effects on

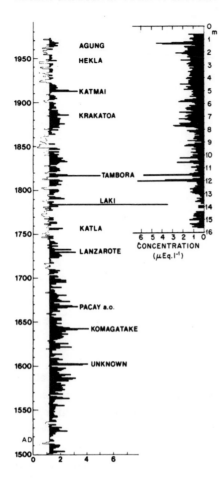

Figure 31.5 Mean acidity of annual layers from A.D. 1500 to 1972 in the ice from Crete, Greenland (left) and the excess sulfate record from Dome C, Antarctica for the last ~200 years (right). Acidity values in the Greenland core in excess of 1.2 equiv H^+ per kg ice are considered to be due to the fallout of volcanic acids (Greenland record from Hammer *et al.*, 1980; Dome C record from Legrand and Delmas, 1987).

the dispersal of volcanic material. The collection of many short cores from a wide area of the larger ice sheets, and along polar/alpine transects (e.g. from the South Pole to Peru) will enable a better assessment of the spatial pattern of acid deposition to be made (e.g. Clausen and Hammer, 1988; Mulvaney and Peel, 1987).

More detailed chemical studies thus hold the promise of a significant improvement in our understanding of volcanic explosivity through time. However, at present there is no comprehensive GVI catalog for eruptions, and extreme values in the published ice core conductivity profiles do not differ substantially from extremes in the DVI and VEI for the last 500 years.

31.5 Tree ring records

Major explosive eruptions may bring about changes in the atmospheric circulation that are sufficiently anomalous to seriously disrupt normal growing conditions for plants. In the case of trees, such anomalies may result in a sequence of extremely thin growth layers, or (in the most extreme cases) actual tissue damage to the growing cells.

In a study of frost ring damage to trees in the western United States, LaMarche and

Hirschboeck (1984) identified 17 times in the last 500 years when frost damage occurred in two or more regions of the western U.S. These are listed in Table 31.3. Many of these correspond to (or immediately follow) years in which major explosive eruptions occurred, according to Lamb (1970). Less correspondence is observed with the VEI chronology of Simkin *et al.* (1981). Table 31.2 also lists years with extremely narrow rings, or ring width sequences, in temperature-sensitive Pinus balfouriana from the Sierra Nevada of California (Scuderi, 1990). These years are assumed to be related to anomalous climatic conditions following major eruptions. Some correspondence with the frost-damaged tree ring record is apparent (e.g. 1601, 1640) but additional years of severe growth reduction are also identified.

When widespread conditions of frost-damage or narrow growth increments correspond to other independent records of explosive eruptions, the tree rings provide additional proxy evidence for the eruption and indicate the importance or magnitude of the explosive events

Table 31.3   Years with notable frost ring damage to trees in the western United States (1500-1970)* (LaMarche and Hirschboeck, 1984) and years with extremely narrow rings or ring width sequences in Sierra Nevada *Pinus balfouriana* (Scuderi, 1990).

Frost rings	Narrow rings
1965	
1941	
	1913
1912	
1902	
	1884
1866	
1837	
1831	
1828	
1817	
	1815
1805	
	1784
1761	
1732	
	1730
	1725
	1666
1660	
1640	1640
1601	1601

*Notable frost rings are those occurring in 50% or more of sampled trees in any one location and at two or more sites. Sites range from eastern California to New Mexico and Colorado.

through their impact on the general circulation. This information can be useful in refining the chronology of explosive eruptions and perhaps point to significant gaps in other volcanic chronologies. However, it is clear that not all growth damage or extremely narrow growth increments can be expected to result from volcanic eruptions; conversely, not all eruptions will produce reduced growth or tissue damage. Indeed, in some areas, circulation anomalies may result in *enhanced* growth (*cf.* Lough and Fritts, 1987). Hence, the records from trees must be considered as of only limited use in constructing a comprehensive chronology of explosive volcanic eruptions. However, assembling records of frost-damaged trees from different regions, together with studies of extremely narrow tree ring sequences (e.g. Baillie and Munro, 1988) would make it possible to determine which volcanic events were of greatest ecological significance. Attention can then be focused on the chemistry and dynamics of these events, and of the subsequent changes in atmospheric circulation which they brought about.

31.6   Instrumental records

Ideally, the optimum index of volcanically-induced turbidity would be a time series of direct radiation receipts at a set of well-distributed, high altitude sites. Unfortunately, such a data set does not exist. There have been actinometric (solar radiation) measurements in various locations around the world since the 1880s, but changes in instrumentation and the lack of fixed, long-term observations limit the usefulness of the records. Nevertheless, the available data provide an interesting comparison with the DVI, VEI and glaciological records discussed above. Figure 31.6 shows a composite record of solar radiation receipts at 20 observing stations in the northern hemisphere (between 32° and 62°N) (Asaturov *et al.*, 1986). This is similar to the optical depth record derived by Bryson and Goodman (1980) from 42 actinometric and pyrheliometric records between 20° and 65°N (Figure 31.6). Optical depth is a measure of the size and number of particles in a column of air (Toon and Pollack, 1980). Both compilations show sharp decreases in solar radiation receipts following major explosive eruptions such as Krakatau (1883) Santa Maria (1902) and Katmai (1912). Interestingly, both records also indicate a decrease in solar radiation (increase in optical depth) since ~1940, accentuated by the eruptions of Agung (1963) and El Chichon (1982). Soviet analysts (e.g. Budyko, 1969; Pivovarova, 1977) have argued for many years that the post-1940s cooling of the northern hemisphere parallels this decline in direct radiation. However, the reasons for the gradual increase in optical depth since 1940 are not clear; no comparable increase in volcanic activity is indicated by the DVI or VEI catalogs, though there is a small increase in Greenland ice core acidity over this period. Changes in tropospheric turbidity due to anthropogenic activities may be a factor in the optical depth increase.

Further research is needed to determine if these composite records accurately indicate volcanic aerosol loading over the last century. However, there is no propect of extending the record further back in time.

31.7   Conclusions

The record of large explosive eruptions since A.D. 1500 is probably quite incomplete, making it difficult to assess their overall impact on climate There is no single index of volcanic

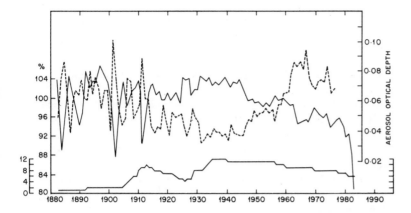

Figure 31.6 *Upper solid line*: variations in annual solar radiation receipts 1883-1983 at 20 actinometric radiation stations between 32° and 62°N as a percentage of the long-term mean for all records. *Lower solid line* shows the number of stations operating in each year; before 1905 only 2 stations were available (Pavlovsk/Leningrad and Montpelier). Similarly, after 1970, the average is based on 7 Soviet stations and Madison, Wisconsin (after Asaturov *et al.*, 1986).

Dashed line: mean annual residual aerosol optical depth, 1883-1975, based on a network of 42 stations between 20° and 65°N. This is the optical depth obtained after adjusting the data for the effects of clean air, water vapor, ozone etc., which amounts to ~0.212 in the northern hemisphere (Bryson and Goodman, 1980).

explosivity which is ideal for climatic studies since every index has its own limitations. In all catalogs, it is likely that some eruptions have been missed and the magnitudes of others have been mis-classified. However, there is little disagreement between the various catalogs about which were the biggest eruptions of the last 500 years, though geochemical studies reveal important differences in the potential climatic significance of explosive eruptions; not all large eruptions had the same potential for climatic effects. Only by combining geological studies with glaciological and other lines of evidence can a more complete assessment of volcanic explosivity through time be obtained.

References

Angell, J. K. and Korshover, J., 1985. Surface temperature changes following the six major volcanic episodes between 1780 & 1980. *J. Climate and Applied Meteorology*, 24, 937-951.

Asaturov, M. L., Budyko, M. I., Vinnikov, K. Ya., Groisman, P. Ya., Kabanov, A. S., Karol, I. L., Kolomeev, M. P., Pivovarova, Z. I., Rosanov, E. V. and Khmeletsov, S. S., 1986. *Volcanoes, stratospheric aerosols and Earth's climate*. Gidrometeoizdat, Leningrad. 256pp (in Russian).

Baillie, M. G. L. and Munro, M. A. R., 1988. Irish tree rings, Santorini and volcanic dust veils. *Nature*, 332, 344-346.

Bradley, R. S., 1988. The explosive volcanic eruption signal in northern hemisphere continental temperature records. *Climatic Change*, 12, 221-243.

Bray, J. R., 1974. Glacial advance relative to volcanic activity since A.D. 1500. *Nature*, 248, 42-43.

Bryson, R. A. and Goodman, B. M., 1980. Volcanic activity and climatic changes. *Science*, 207, 1041-1044.

Budyko, M. I., 1969. The effect of solar radiation variations on the climate. *Tellus*, 21, 611-619.

Cadle, R. D., Kiang, C. S. and Louis, J. F., 1976. The global scale dispersion of the eruption clouds from major volcanic eruptions. *J. Geophysical Research*, 81, 3125-3132.

Clausen, H. B. and Hammer, C. U., 1988. The Laki and Tambora eruptions as revealed in Greenland ice cores from 11 locations. *Annals of Glaciology*, 10, 16-22.

Delmas, R. J., Legrand, M., Aristarain, A. J. and Zanolini, F., 1985. Volcanic deposits in Antarctic snow and ice. *J. Geophysical Research*, 90, D7, 12901-12920.

Devine, J. D., Sigurdsson, H., Davis, A. N. and Self, S., 1984. Estimates of sulfur and chlorine yield to the atmosphere from volcanic eruptions and potential climatic effects. *J. Geophysical Research*, 89, B7, 6309-6325.

Franklin, B., 1784. Meteorological imaginations and conjectures. *Manchester Literary and Philosophical Society Memoirs and Proceedings*, 2, 375.

Hammer, C. U., 1977. Past volcanism revealed by Greenland ice sheet impurities. *Nature*, 270, 482-486.

Hammer, C. U., 1984. Traces of Icelandic eruptions in the Greenland ice sheet. *Jokull*, 34, 51-65.

Hammer, C. U., Clausen, H. B. and Dansgaard, W., 1980. Greenland ice sheet evidence of post-glacial volcanism and its climatic impact. *Nature*, 288, 230-255.

Hirschboeck, K. K., 1980. A new world-wide chronology of volcanic eruptions. *Palaeogeography, Palaeoclimatology, Palaeoecology*, 29, 223-241.

Holdsworth, G. 1985. Acid content of snow from a mid-tropospheric sampling site on Mount Logan, Yukon Territory, Canada. *Annals of Glaciology*, 7, 153-160.

Holdsworth, G. and Peake, E., 1985. Acid content of snow from a mid-troposphere sampling site on Mount Logan, Yukon Territory, Canada. *Annals of Glaciology*, 7, 153-160.

Johnson, D. A., 1980. Volcanic contribution to the stratosphere: more significant to ozone than previously estimated? *Science*, 209, 491-493.

Keen, R. A., 1983. Volcanic aerosols and lunar eclipses. *Science*, 222, 1011-1013.

Kelly, P. M. and Sear, C. B., 1982. The formulation of Lamb's Dust Veil Index. In: *Atmospheric Effects and Potential Climatic Impact of the 1980 Eruptions of Mount St. Helens*. A. Deepack (ed.). Conference Publication 2240, Scientific and Technical Information Branch, NASA, Washington D. C., 293-298.

Kelly, P. M. and Sear, C. B., 1984. Climatic impact of explosive olcanic eruptions. *Nature*, 311, 740-743.

Kondratyev, K. Ya., 1988. *Volcanoes and Climate*. WCP-54, W.M.O./T.D. 166. World Meteorological Organisation, Geneva, 103pp.

LaMarche, V. C. and Hirschboeck, K. K., 1984. Frost rings in trees as records of major volcanic eruptions. *Nature*, 307, 121-126.

Lamb, H. H., 1970. Volcanic dust in the atmosphere; with a chronology and assessment of its meteorological significance. *Philosophical Transactions of the Royal Society of London*, A266, 425-533.

Lamb, H. H., 1977. Supplementary volcanic dust veil index assessments. *Climate Monitor*, 6, 57-67.

Lamb, H. H., 1983. Update of the chronology of assessments of the volcanic dust veil index. *Climate Monitor*, 12, 79-90.

Legrand, M. and Delmas, R. J. 1987. A 220-year continuous record of volcanic H_2SO_4 in the Antarctic Ice Sheet. *Nature*, 327, 671-676.

Lough, J. M. and Fritts, H. C., 1987. An assessment of the possible effects of volcanic eruptions on North American climate using tree-ring data, 1602-1900 A.D. *Climatic Change*, 10, 219-239.

Lyons, W. B., Mayewski, P. A., Spencer, M. J., Twickler, M. S. and Graedel, T. E., 1990. A northern hemisphere volcanic chemistry (1869-1984) and climatic implications using a South Greenland ice core. *Annals of Glaciology*, 14, 176-182.

Mass, C. F. and Portman, D. A., 1989. Major volcanic eruptions and climate: a critical assessment. *J. Climate*, 2, 566-593.

Mayewski, P. A., Lyons, W. B., Spencer, M. J., Twickler, M., Dansgaard, W., Koci, B., Davidson, C. I., and Honrath, R. E., 1986. Sulfate and nitrate concentrations from a South Greenland ice core. *Science*, 232, 975-977.

Mitchell, J. M., Jr., 1970. A preliminary evaluation of atmospheric pollution as a cause of global temperature fluctuations of the past century, in: *Global Effects of Environmental Pollution* (ed. S. F. Singer) D. Reidel, Dordrecht, pp. 139-155.

Mulvaney, R. and Peel, D. A., 1987. Anions and cations in ice cores from Dolleman Island and the Palmer Land plateau, Antarctic Peninsula. *Annals of Glaciology*, 10, 121-125.

Mroz, E. J., Mason, A. S. and Sedlacek, W. A., 1983. Stratospheric sulfate from El Chichon and the mystery volcano. *Geophysical Research Letters*, 10, 873-876.

Newhall, C. G. and Self, S., 1982. The Volcanic Explosivity Index (VEI): an estimate of explosive magnitude for historical volcanism. *J. Geophysical Research*, 87, C2, 1231-1238.

Oliver, R. C., 1976. On the response of hemispheric mean temperature to stratospheric dust: an empirical approach. *J. of Applied Meteorology*, 15, 933-950.

Palais, J. M. and Sigurdsson, H., 1989. Petrologic evidence of volatile emissions from major historic and pre-historic volcanic eruptions. In: *Understanding Climate Change*. A. Berger, R. E. Dickinson and J. W. Kidson (eds.) Geophysical Monograph 52, American Geophysical Union, Washington D.C., 31-53.

Parker, D. E., 1985. Climatic impact of explosive volcanic eruptions. *Meteorological Magazine*, 114, 149-161

Pivovarova, Z. I., 1977. Utilization of surface radiation data for studying atmospheric transparency. *Meteorologia i Gidrologia*, No. 9, 24-31.

Pollack, J. B., Toon, O. B., Sagan, C., Summers, A., Baldwin, B. and Van Camp, W., 1976. Volcanic explosions and climatic change: a theoretical assessment. *J. Geophysical Research*, 81, 1071-1083.

Porter, S. C., 1981. Recent glacier variations of volcanic eruptions. *Nature*, 291, 139-142.

Porter, S. C., 1986. Pattern and forcing of northern hemisphere glacier variations during the last millennium. *Quaternary Research*, 26, 27-48.

Rampino, M. and Self, S., 1982. Historic eruptions of Tambora (1815), Krakatau (1883) and Agung (1963), their stratospheric aerosols and climatic impact. *Quaternary Research*, 18, 127-143.

Rampino, M. and Self, S., 1984. Sulfur-rich volcanic eruptions and stratospheric aerosols. *Nature*, 310, 677-679.

Reiter, R. and Jager, H., 1986. Results of 8-year continuous measurements of aerosol profiles in the stratosphere with discussion of the importance of stratospheric aerosols to an estimate of effects on the global climate. *Meteorology and Atmospheric Physics*, 35, 19-48.

Robock, A. 1981. A latitudinally dependent volcanic dust veil index, and its effect on climate simulations. *J of Volcanological and Geothermal Research*, 11, 67-80.

Sapper, K., 1927. *Vulkankunde*. Engelhorn, Stuttgart.

Schonwiese, C-D., 1988. Volcanic activity parameters and volcanism-climate relationships within the recent centuries. *Atmosfera*, 1, 141-156

Scuderi, L. A., 1990. Tree ring evidence for climatically effective volcanic eruptions. *Quaternary Research*, 34, 67-86.

Sear, C. B., Kelly, P. M., Jones, P. D., and Goodess, C. M., 1987. Global surface temperature responses to major volcanic eruptions. *Nature*, 330, 365-367.

Sedlacek, W. A., Mroz, E. J. and Heiken, G., 1981. Stratospheric sulfate from the Gareloi eruption, 1980: contribution to the "ambient" aerosol by a poorly documented volcanic eruption. *Geophysical Research Letters*, 8, 761-764.

Sedlacek, W. A., Mroz, E. J., Lazrus, A. L. and Gandrud, B. W., 1983. A decade of stratospheric sulfate measurements compared with observations of volcanic eruptions. *J. Geophysical Research*, 88, 3741.

Self, S., Rampino, M. and Barbera, J. J., 1981. The possible effects of large 19th and 20th century volcanic eruptions on zonal and hemispheric surface temperatures. *J. Volcanology & Geothermal Research*, 11, 41-60.

Self, S., Rampino, M. R., Newton, M. S. and Wolff, J. A., 1984. Volcanological study of the great Tambora eruption of 1815. *Geology*, 12, 659-663.

Simkin, T., Siebert, L., McClelland, L., Bridge, D., Newhall, C. and Latter, J. H., 1981. *Volcanoes of the World*. Hutchinson Ross, Stroudsberg, Penn. 233 pp.

Spirina, L. P., 1971. On the effect of volcanic dust on the temperature regime of the Northern Hemisphere. *Meteorologia i Gidrologia*, No. 10, 38-43.

Symonds, R. B., Rose, W. I. and Reed, M. H., 1988. Contribution of Cl⁻ and F⁻-bearing gases to the atmosphere by volcanoes. *Nature*, 334, 415-418.

Stothers, R. B., 1984. The great Tambora eruption in 1815 and its aftermath. *Science*, 224, 1191-1198.

Stothers, R. B. and Rampino, M. R., 1983. Volcanic eruptions in the Mediterranean before A.D. 630 from written and archaeological sources. *J. Geophysical Research*, 88, 6357-6371.

Stothers, R. B., Wolff, J. A., Self, S., and Rampino, M. R., 1986. Basaltic fissure eruptions, plume heights and atmospheric aerosols. *Geophysical Research Letters*, 13, 725-728.

Taylor, B. L., Gal-chen, T. and Schneider, S. H., 1980. Volcanic eruptions and long-term temperature records: an empirical search for cause and effect. *Quarterly J. Royal Meteorological Society*, 106, 175-199.

Toon, O. B. and Pollack, J. B., 1980. Atmospheric aerosols and climate. *American Scientist*, 68, 268-278.

Wolff, J. A., Self, S. A. and Rampino, M. R., 1986. Basaltic fissure eruptions, fire fountains and atmospheric aerosols. *Eos*, 65, 1148-1149.

Yamamoto, R., Iwashima, T. and Hoshiai, M., 1975. Change of the surface air temperature averaged over the northern hemisphere and large volcanic eruptions during the years 1951-1972. *J. Meteorological Society of Japan*, 53, 482-485.

32 The historical record of El Niño events

W. H. Quinn and V. T. Neal

32.1 Introduction

The El Niño is defined and discussed to some extent in Quinn (1987) where it is considered to be the regional manifestation of a recurring large-scale Southern Oscillation (SO)-related ocean-atmosphere fluctuation which is primarily noted over the Indo-Pacific area. There has been some confusion in terminology as a result of using the term ENSO (El Niño/Southern Oscillation) to represent the large-scale relative. Some authors refer to this large-scale short-term climate change as an equatorial Pacific warm event (e.g., Rasmussen *et al.*, 1983) and many others refer to the whole near-global-scale climatic change associated with these developments as El Niño. For the purposes of this chapter, it is most suitable to accept the El Niño as a regional manifestation of the parent large-scale SO-related climatic fluctuation which for the sake of simplicity we will refer to here as the ENSO. The region directly affected by the El Niño is southwestern Ecuador, northwestern Peru and their coastal waters. The following background information will provide some insight into the nature and development of this phenomenon.

Troup (1965) defined the SO as an exchange of air between the eastern and western hemispheres, principally in tropical and subtropical latitudes. He associated this exchange of air with a mean toroidal circulation driven by temperature differences between the eastern tropical Pacific and the western tropical Pacific-Indonesian area. He found that there was usually a flow in the upper troposphere from the Indonesian equatorial low pressure area across the Pacific to its eastern region which is compensated by a return flow in the lower troposphere from the eastern Pacific to the equatorial low. There is descent over the eastern Pacific due to subsidence over the cooler waters, ascent in the equatorial low region due to convergence and convection, and a toroidal circulation in the zonal-vertical plane is maintained. Variations in this circulation and the resulting locations and intensities of the associated atmospheric centers of action (principally the Indonesian equatorial low, the South Pacific subtropical high and the North Pacific subtropical high) determine the nature and strength of the Pacific trade winds and their westward extension in the equatorial easterlies. During the prolonged period of strong trades and equatorial easterlies, the equatorial currents are intensified coinciding with an east to west rise in sea level and an accumulation of warm water in the western tropical Pacific. When there is a general relaxation of the easterly wind stress, the westward-flowing equatorial currents will be weakened and the accumulated warm water above the thermocline in the western tropical Pacific will tend to return to the east: (1) in the form of an internal equatorial Kelvin wave, (2) through intensified north and south equatorial countercurrents and the equatorial undercurrent (Wyrtki *et al.* 1976). The result will be an accumulation of warm water above a depressed thermocline in the region off

Ecuador and Peru which essentially represents the El Niño. Hydrographic data off the coasts of Ecuador and Peru confirm the thermal structure depression and poleward spreading during El Niño (Enfield 1981) and, although coastal upwelling may continue, it is from the accumulated warm water above the thermocline. From a meteorological standpoint, as the ENSO sets in and the warm waters spread eastward, the equatorial low pressure center moves eastward and we note an eastward shift of the heavy equatorial Pacific precipitation from Indonesia and the westernmost equatorial Pacific to the central and at times into the eastern equatorial Pacific. Also, along with the warm offshore El Niño waters, southwestern Ecuador and the northwestern Peruvian desert region often experience anomalously heavy precipitation when the intertropical convergence zone (ITCZ) along with the southward diverted surficial north equatorial countercurrent water move south of the equator over the El Niño-affected region during the southeast trade relaxation.

In most cases the relaxational shift is from strong easterlies to weak or insignificant easterlies, but there are cases, such as the 1982-83 ENSO, where the shift is from weak-moderate easterlies to fairly strong equatorial westerlies. In either case we get a large circulation shift that provides similar results. It is this shift that is considered as the forcing factor for the El Niño.

When studied over a long time-history, the prominent circulation shifts are noted to vary considerably in period and amplitude. Berlage (1966) states that the SO period varies roughly between one and five years, but it is usually between two and four years. We often use SO indices (differences in sea level atmospheric pressure between sites along the South Pacific subtropical ridge and sites in the equatorial low pressure region) to represent the SO. The rising and high indices represent the build-up or anti-El Niño phase, the falling and low indices represent the relaxational or El Niño phase. For further information on El Niño and its related features, reference is made to Quinn (1987). For discussions on related climatic activity, reference is made to Tyson (1986) Rasmussen and Wallace (1983) and many others.

Although unusually strong ocean-atmosphere circulation shifts over the Indo-Pacific tropics and subtropics may be of sufficient magnitude to cause a cascade of regional climatic changes on a near global scale through teleconnective processes (e.g., the very strong 1877-78, 1891, 1925-26, and 1982-83 ENSO events) no two of these events exhibit identical characteristics and the regions they affect, and the degree to which they affect them, may vary greatly from one another. Even the regional El Niño manifestation may vary considerably from its parent ENSO event in time of onset, duration and intensity. For example, consider the three most recent events. With regard to the 1976-77 ENSO, the El Niño only occurred in 1976 yet, on the large-scale, the equatorial Pacific activity occurred in both years and was more significant in 1977. In the case of the 1982-83 ENSO, the large scale effects were first noted over the central and western parts of the ENSO-affected Indo-Pacific area between April and July 1982, yet the El Niño wasn't fully recognized off northwestern Peru until October 1982. Considering the 1986-87 ENSO, signs of its development were apparent over various parts of the tropical Pacific September-November 1986 (based on sea surface temperature analyses). It showed broad coverage December 1986-December 1987 and started breaking down in January 1988. This ENSO was only weak-moderate in intensity since the equatorial Pacific sea surface temperature anomalies were generally in the 0.5-2.0°C. range. The regional El Niño did not set in off northwestern Peru until February 1987; it lasted through May 1987 and was considered to be moderate in intensity. In many other cases the regional El Niño set in several months prior to the activity over the central and western equatorial Pacific. With regard to intensity, the strong 1917 El Niño did not show significant activity on the larger scale. The late 1904-05 ENSO event would warrant a strong intensity based on its wide spread activity yet the related El Niño was only a weak-moderate intensity. These differences are brought out to the reader, since our evaluations pertain to the regional El Niño and, although there is general agreement between the ENSO and its regional manifestation (El Niño) particularly when events are strong or very strong, the reader should be aware of possible deviations in time of onset, duration and intensity. Since data and information related to the large-scale ENSO event is limited to about 148 years, it was essential that we focus our attention on the regional El Niño if we wished to obtain the longest possible record on the SO-related ocean-atmosphere activity. Vasco Nunez de Balboa discovered the Pacific Ocean in 1511 (Raimondi 1876) and our study of the regional El Niño extends from shortly thereafter to the present.

32.2   Obtaining the record

Ecuador and Peru were parts of a land endowed with large quantities of gold, silver, gems and other valuable commodities avidly sought by at first Spain and later the other seafaring nations. As a result many detailed reports prepared from sailing vessel logs, diaries of conquistadors and their entourages, and early explorers became available from the early 1500s on. Later on there were many reports from missionaries, pirates, privateers, historians, geographers, engineers, geologists, hydrologists, newspapers, and scientists from the various other disciplines which provided a rather continuous source of information over both those areas directly affected by the El Niño and those areas affected by other unusual conditions related to the El Niño (e.g.,the droughts that often affect southeastern Peru and adjacent parts of Bolivia during an El Niño).

Our approach for identifying, evaluating and determining the strength of El Niño events was derived from a detailed study of what happened prior to, during and after the El Niños that occurred over the past 140-150 years when more data were available. Also, for further insight we had the SCOR Working Group 55 (1983) criteria and findings available for a quantitative evaluation of the events that occurred over recent years. The SCOR Group defined El Niño as: the appearance of anomalously warm water along the coast of Ecuador and Peru as far south as Lima (12°S) during which a normalized sea surface temperature (SST) anomaly exceeding one standard deviation occurs for at least four consecutive months at three or more of five coastal stations (Talara, Puerto Chicama, Chimbote, Isla Don Martin and Callao). The definition identified El Niños for 1957-58, 1965, 1972-73 and 1976, based on 1956-81 data. However, this basic definition only gives the minimal criteria for determining when an El Niño occurred and nothing concerning the strength of an event. We know that stronger events showed larger thermal anomalies and some extended further south than Lima. It has generally been considered that the 1957-58 and 1972-73 El Niños were strong events and that the 1965 and 1976 El Niños were of moderate intensity. We also believe that most investigators would concur in a very strong classification for the 1982-83 El Niño since its peak-month SST anomalies at the coastal sites were at least three times the minimal criteria. Obviously, to obtain a long historical record, we would have to rely on a subjective approach to identify and evaluate El Niños from the contents of descriptive historical, geographical, geological, hydrological, meteorological and oceanographic literature of the distant past. To accomplish this, we selected what we considered to be typical events (as role models) over the relatively recent past (when large quantities of data and descriptive information were available) to represent the strength categories. Although the temperature of the invading coastal waters is the desirable variable, we must now rely on associated meteorological, hydrological, oceanographic and other environmental alterations associated with the invading waters for this purpose. For the very strong (VS) events we used the available data and descriptive information for the 1877-78, 1891, 1925-26 and 1982-83 El Niños to get a comprehensive picture of activity to be expected with such events. Likewise, we used the data and descriptive information available for the 1932, 1940-41, 1957-58 and 1972-73 El Niños to obtain a fairly comprehensive picture of the range in activity that might be expected with a strong (S) El Niño. Considering events that would most likely just meet minimal SCOR criteria, we selected the 1918-19, 1923, 1930-31, 1976 and 1987 El Niños to represent the moderate (M) intensity.

The VS El Niños display most or all of the following characteristics over the coastal regions and/or adjacent offshore waters of southwestern Ecuador and northwestern to west-central Peru:

(1) very high sea and air temperatures with SST anomalies reaching 6-12°C above normal in peak months;

(2) presence of aguaje (red tide);

(3) thunderstorms, torrential rainfall, floods and erosion of the normally arid coastal lowlands;

(4) significant rises in sea level along the coast;

(5) invasion of northern and central Peruvian coastal waters by tropical nekton;

(6) destruction of housing areas, large buildings and sometimes whole cities by river inundations and flood waters;

(7) interruption of transportation as a result of destruction of bridges, roadways and railroad facilities by hydrological forces;

(8) departure of guano birds from coastal islands;

(9) mass mortality of various marine organisms, including guano birds, often with subsequent decomposition and a great stench from the release of hydrogen sulfide;

(10) destruction of agricultural crops and livestock;

(11) conditions causing the spread of tropical diseases;

(12) drastic reduction in coastal anchoveta fishery catches and fishmeal production.

The strong (S) El Niños will exhibit features similar to many of those listed for the VS events but to a significantly smaller degree. For example, the highest monthly SST values will be in the 3-5°C above normal range. Items (2) (4) (5) (8) (11) and (12) are likely to pertain to the S El Niño. Also, some heavy rainfall, flooding and agricultural setbacks may be noted.

The moderate (M) El Niños may exhibit several of the listed characteristics to a minor degree and to a lesser extent (in time and space). The highest monthly SST values will be in the 2-3.5°C above normal range. Items (2) (4) (8) and (12) are likely to pertain to the M El Niño.

While attempting to fit all of our El Niño events into the VS, S or M categories, we found many cases that fell significantly below or significantly exceeded the general characteristics of these categories. The differences were sufficient to cause us to add the S+ and M+ categories.

The S+ category represents those El Niños showing activity significantly stronger than the run-of-the-mill S events, particularly with regard to hydrological conditions, degree of destruction and/or effects on sea life. However, we felt that additional evidence would be required to raise such events to the VS category.

The need for the M+ category became particularly apparent as we attempted to extend our record on moderate events back in time beyond the 1800s. We found that the increased hydrological activity that accompanied the M+ events (e.g. the 1939, 1943, 1953 and 1965 El Niños) made them much more visible in the older literature than the minimal moderate event. They differed from the S events since their activity was more localized and generally extended over a shorter period time.

Although we eliminated the weak and very weak events from consideration, we included some events in Table 32.1 which are judged to fall somewhat below SCOR Working Group 55 criteria (e.g., 1862, 1904-05, 1951 and 1969) in an M-category since they were referenced in the literature and showed some significant activity.

The confidence ratings on the El Niño events of Table 32.1 range from 2 to 5 as they did in Quinn *et al*. (1987) with the lowest rating of 2 based primarily on circumstantial evidence which we will discuss later in the chapter.

Table 32.1 includes our additional investigative determinations since the publication of Quinn *et al*. (1987). The list of strong and very strong events has been refined to a minor degree, the moderate events have been extended back from the early 1800s to the early 1500s, and additional references are cited. The compound type of activity over the distant past where we are most likely covering two events of different intensities are represented by M/S (e.g., 1539-41, 1558-61, 1589-91, 1707-09, 1844-46) or M-/M+ (1887-89) since we lack the detailed information to further refine them. In more recent years we note similar developments which can be broken down in more detail due to the increased availability of reports and data. In some cases there may be an incomplete or staggered relaxation after a large anti-El Niño buildup; in some other cases there may be a secondary buildup to a higher level after an earlier premature relaxation; in still other cases there may just be two separate events with a one-year buildup between them. In the 1939-43 situation after the large buildup in 1938, there was an incomplete relaxation in early 1939, a completed relaxation late 1940-41, and a separate buildup in 1942 prior to the 1943 relaxation. In late 1973-early 74 there was a strong buildup, an incomplete relaxation in late 1974-early 75 resulting in the early 1975 aborted El Niño, and a strong secondary buildup in mid-late 1975 with a relaxation during 1976 resulting in the 1976 El Niño (the 1976-77 ENSO). Many similar cases over the past 80 years could be shown which might resemble those early periods involving two events with only about 15-20 months between them.

Typically the moderate events set in sometime between January and March and are essentially over by June or July; however, they at times show a small secondary thermal peak sometime during the following November-March period. The strong and very strong events usually set in between January and March, show a peak in sea temperature between February and June and a secondary peak between the following November and March. Of course there

Table 32.1 El Niño events of moderate, strong and very strong intensities, their confidence ratings and information sources.

El Niño Events	Event Strength	Confidence Rating	Information Sources
1525-26	S	3	Xeres (1534).
1531-32	S	4	Xeres (1534) Prescott (1892) Murphy (1926).
1535	M+	2	Mackenna (1877) Taulis (1934).
1539-41	M/S	3	Montesinos (1642) Cobo (1653) Raimondi (1876).
1544	M+	4	Albenino (1549) Montesinos (1642) Mackenna (1877) Gormaz (1901) Taulis (1934).
1546-47	S	4	Albenino (1549) Benzoni (1565) Raimondi (1876).
1552	S	4	Humboldt (1804) Moreno (1804) Palma (1894).
1558-61	M/S	3	Montesinos (1642) Martinez y Vela (1702) Garcia Rosell (1903)
1565	M+	2	Montesinos (1642).
1567-68	S+	5	Oliva (1631) Cobo (1639) Montesinos (1642) Gormaz (1901) Labarthe (1914) Portocarrero (1926).
1574	S	4	Garcia Rosell (1903) Taulis (1934)
1578	VS	5	Cabello Balboa (1586) Acosta (1590) Cobo (1639, 1653) Garcia Rosell (1903) Labarthe (1914) Brunning (1922-23) Portocarrero (1926) Huertas (1984).
1582	M	3	Montesinos (1642) Taulis (1934).
1585	M+	2	Montesinos (1642).
1589-91	M/S	3	Montesinos (1642) Martinez y Vela (1702) Barriga (1951).
1596	M+	3	Montesinos (1642).
1600	S	3	Gormaz (1901) Barriga (1951).
1604	M+	3	Montesinos (1642) Andrade (1948) Brooks (1971).
1607-08	S	5	Cobo (1639) Martinez y Vela (1702) Alcedo y Herrera (1740) Palma (1894) Labarthe (1914) Portocarrero (1926) Taulis (1934) Brooks (1971).
1614	S	5	Cobo (1653) Haenke (1799) Labarthe (1914) Portocarrero (1926) Andrade (1948) Brooks (1971).
1618-19	S	4	Vasquez de Espinosa (1629) Cobo (1653) Mackenna (1877) Taulis (1934).
1624	S+	5	Montesinos (1642) Cobo (1653) Labarthe (1914) Portocarrero (1926).
1634-35	S	3	Montesinos (1642) Puente (1885) Palma (1894).
1640-41	S	2	Martinez y Vela (1702) Mackenna (1877) Taulis (1934)
1647	M+	3	Mackenna (1877) Gormaz (1901) Taulis (1934).
1652	S+	4	Cobo (1653) Labarthe (1914) Portocarrero (1926).
1655	M	3	Alcedo y Herrera (1740) Taulis (1934).
1660	S	3	Labarthe (1914) Portocarrero (1926).
1671	S	3	Martinez y Vela (1702) Labarthe (1914). Portocarrero (1926).

Table 32.1 – *continued*

El Niño Events	Event Strength	Confidence Rating	Information Sources
1681	S	3	Rocha (1681).
1684	M+	2	Martinez y Vela (1702) Taulis (1934)
1687-88	S+	4	Unanue (1806) Melo (1913) Remy (1931) Taulis (1934).
1692-93	S	3	Martinez y Vela (1702) Andrade (1948) Brooks (1971).
1696-97	M+	3	Mackenna (1877) Palma (1894) Taulis, (1934).
1701	S+	5	De Sosa (1763) Bueno (1763) Haenke (1799) Humboldt (1804) Unanue (1806) Paz Soldan (1862) Palma (1894) Labarthe (1914) Portocarrero (1926) Nials *et al.* (1979).
1707-09	M/S	3	Cooke (1712) Alcedo y Herrera (1740).
1715-16	S	3	Gentil (1728) Labarthe (1914) Portocarrero (1926).
1718	M+	3	Bueno (1763) Barriga (1951).
1720	VS	5	Shelvocke (1726) De Sosa (1763) Bueno (1763) Alcedo (1786-89) Haenke (1799) Humboldt (1804) Moreno (1804) Unanue (1806) Paz Soldan (1862) Raimondi (1876) Palma (1894) Adams (1905) Labarthe (1914) Bachmann (1921) Portocarrero (1926) Nials *et al.* (1979) Huertas (1984).
1723	M+	3	Mackenna (1877) Taulis (1934) Andrade (1948) Brooks (1971).
1728	VS	5	Anson (1748) De Sosa (1763) Bueno (1763) Alcedo (1786-89) Humboldt (1804) Unanue (1806) Paz Soldan (1862) Spruce (1864) Palma (1894) Eguiguren (1894) Garcia Rosell (1903) Labarthe (1914) Portocarrero (1926) Nials *et al.* (1979).
1736	S	2	Andrade (1948) Brooks (1971).
1740	M	3	Juan and Ulloa (1748) Gormaz (1901).
1744	M+	3	Mackenna (1877) Taulis (1934) Andrade (1948) Brooks (1971).
1747	S+	5	Llano Zapata (1748) De Sosa (1763) Humboldt (1804) Moreno (1804) Palma (1894) Labarthe (1914) Portocarrero (1926) Taulis (1934) Nials *et al.* (1979).
1750	M+	4	Puente (1885) Labarthe (1914) Portocarrero (1926) Taulis (1934).
1755-56	M	3	Garcia Rodriguez (1779) Garcia Rosell (1903) Andrade (1948).
1761	S	5	Bueno (1763) Garcia Rodriguez (1779) Alcedo (1786-89) Haenke (1799) Ruschenberger (1834) Labarthe (1914) Portocarrero (1926) Andrade (1948).
1764	M	2	Mackenna (1877) Taulis (1934).
1768	M	2	Garcia Rodriguez (1779) Mackenna (1877) Taulis (1934).
1775	S	3	Puente (1885) Labarthe (1914) Portocarrero (1926).

continued

Table 32.1 – *continued*

El Niño Events	Event Strength	Confidence Rating	Information Sources
1778-79	M+	4	Garcia Rodriguez (1779) Mackenna (1877) Puente (1885) Labarthe (1914) Portocarrrero (1926) Andrade (1948) Brooks (1971).
1783	S	3	Mackenna (1877) Taulis (1934) Andrade (1948) Brooks (1971).
1786	M+	3	Labarthe (1914) Portocarrero (1926) Estrada (1977).
1791	VS	5	Unanue (1806) Ruschenberger (1834) Paz Soldan (1862) Spruce (1864) Hutchinson (1873) Eguiguren (1894) Garcia Rosell (1903) Adams (1905) Labarthe (1914) Leguia y Martinez (1914) Bachmann (1921) Portocarrero (1926) Andrade (1948) Brooks (1971).
1803-04	S+	5	Humboldt (1804) Moreno (1804) Unanue (1806) Stevenson (1829) Paz Soldan (1862) Spruce (1864) Eguiguren (1894) Palma (1894) Labarthe (1914) Portocarrero (1926) Petersen (1935) Lastres (1937) Andrade (1948) Brooks (1971).
1806-07	M	3	Unanue (1815) Stevenson (1829) Remy (1931).
1812	M+	3	Palma (1894) Gonzales (1913).
1814	S	4	Spruce (1864) Eguiguren (1894).
1817	M+	5	Eguiguren (1894) Labarthe, (1914) Portocarrero (1926) Taulis (1934) Andrade (1948).
1819	M+	5	Eguiguren (1894) Gormaz (1901) Taulis (1934).
1821	M	5	Eguiguren (1894) Fuchs (1925) Remy (1931) Taulis (1934).
1824	M+	5	Spruce (1864) Basadre (1884) Eguiguren (1894) Andrade (1948) Brooks (1971).
1828	VS	5	Ruschenberger (1834) Paz Soldan (1862) Spruce (1864) Hutchinson (1873) Eguiguren (1894) Middendorf (1894) Adams (1905) Sievers (1914) Labarthe (1914) Bachmann (1921) Portocarrero (1926) Taulis (1934) Brooks (1971).
1832	M+	5	Spruce (1864) Mackenna (1877) Basadre (1884) Eguiguren (1894) Bachmann (1921) Taulis (1934) Andrade (1948) Brooks (1971).
1837	M+	5	Mackenna (1877) Eguiguren (1894) Labarthe (1914) Portocarrero (1926) Taulis (1934).
1844-46	M/S+	4	Spruce (1864) Mackenna (1877) Basadre (1884) Eguiguren (1894) Adams (1905) Labarthe (1914) Portocarrero (1926) Taulis (1934) Andrade (1948) Brooks (1971).
1850	M	4	Mackenna (1877) Eguiguren (1894) Gormaz (1901) Fuchs (1925) Taulis (1934).
1852	M	4	Spruce (1864) Eguiguren (1894).
1854	M	4	Spruce (1864) Eguiguren (1894).

Table 32.1 – *continued*

El Niño Events	Event Strength	Confidence Rating	Information Sources
1857-58	M	5	Mackenna (1877) Eguiguren (1894) Gormaz (1901) Labarthe (1914) Gaudron (1925) Portocarrero (1926) Zegarra (1926) Taulis (1934).
1860	M	4	El Comercio (01/07/1860;02/04/1860) Gormaz (1901) Labarthe (1914) Portocarrero (1926) Taulis (1934).
1862	M-	4	Spruce (1864) Eguiguren (1894).
1864	S	5	Spruce (1864) Mackenna (1877) Eguiguren (1894) Taulis (1934).
1866	M+	5	Eguiguren (1894) Adams (1905) Labarthe (1914) Bachmann (1921) Portocarrero (1926).
1867-68	M+	4	El Comercio (01/10/1872) Mackenna (1877) Eguiguren (1894) Raimondi (1897) Gormaz (1901) Bachmann (1921) Taulis (1934).
1871	S+	5	Hutchinson (1873) Middendorf (1894) Eguiguren (1894) Adams (1905) Tizon y Bueno (1907) Leguia y Martinez (1914) Sievers (1914) Labarthe (1914) Bachmann (1921) Anonymous (1925) Gaudron (1925) Portocarrero (1926).
1874	M	4	LaPatria (02/09/1874) Gormaz (1901) Adams (1905) Bachmann (1921).
1877-78	VS	5	Mackenna (1877) Basadre (1884) Eguiguren (1894) Palma (1894) Adams (1905) Melo (1913) Sievers (1914) Labarthe (1914) Leguia y Martinez (1914) Bachmann (1921) Anonymous (1925) Portocarrero (1926) Murphy (1926) Taulis (1934) Andrade (1948) Brooks (1971) Kiladis and Diaz (1986).
1880	M	4	Puls (1885) Eguiguren (1894) Taulis (1934).
1884	S+	5	Eguiguren (1894) Gormaz (1901) Sievers (1914) Labarthe (1914) Bachmann (1921) Murphy (1925) Anonymous (1925) Portocarrero (1926).
1887-89	M-/M+	4	Eguiguren (1894) Gormaz (1901) Bravo (1903) Labarthe (1914) Portocarrero (1926) Taulis (1934) Andrade (1948) Brooks (1971).
1891	VS	5	Carranza (1891) Eguiguren (1894) Gormaz (1901) Adams (1905) Fuchs (1907) Labarthe (1914) Leguia y Martinez (1914) Sievers (1914) Bachmann (1921) Anonymous (1925) Zegarra (1926) Murphy (1926) Portocarrero (1926) Taulis (1934) Andrade (1948) Brooks (1971) Nials *et al.* (1979).
1897	M+	4	El Comercio (02/03/1897, 02/22/1897) Gormaz (1901) Bravo (1903) Bachmann (1921) Jones (1933) Andrade (1948) Brooks (1971).
1899-1900	S	5	El Comercio (2/10/1899) Gormaz (1901) Labarthe (1914) Bachmann (1921) Murphy (1923) Portocarrero (1926) Jones (1933) Taulis (1934) Schott (1938) Andrade (1948) Hutchinson (1950) Brooks (1971).

continued

Table 32.1 – *continued*

El Niño Events	Event Strength	Confidence Rating	Information Sources
1902	M+	4	El Comercio (02/17/1902) Bachmann (1921) Jones (1933) Taulis (1934) Schott (1938) Andrade (1948) Brooks (1971).
1904-05	M-	4	Bachmann (1921) Jones (1933) Taulis (1934) Schott (1938) Andrade (1948) Brooks (1971).
1907	M	3	Paz Soldan (1908) Remy (1931) Jones (1933) Andrade (1948) Brooks (1971).
1910	M+	3	El Comercio (02/10/1910) Labarthe (1914) Bachmann (1921) Portocarrero (1926) Jones (1933).
1911-12	S	5	Forbes (1913, 1914) Bowman (1916, 1924) Barclay (1917) Lavalle y Garcia (1917) Garcia (1921) Ballen (1925) Jones (1933) Schott (1938) Vogt (1940) Hutchinson (1950) Schweigger (1961).
1914-15	M+	5	Labarthe (1914) Garcia (1921) Bowman (1924) Zegarra (1926) Portocarrero (1926) Jones (1933) Taulis (1934) Petersen (1935) Schott (1938) Andrade (1948) Schweigger (1961) Brooks (1971) Markham (1972).
1917	S	5	Lavalle y Garcia (1917) Garcia (1921) Murphy (1923) Ballen (1925) Portocarrero (1926) Zegarra (1926) Jones (1933) Petersen (1935) Maisch (1936) Hutchinson (1950) Schweigger (1961).1918-19 M 5 Fuchs (1918, 1937) Garcia (1921) Murphy (1923) Portocarrero (1926) Jones (1933) Taulis (1934) Maisch (1936) Schott (1938) Vogt (1940) Andrade (1948) Hutchinson (1950) Brooks (1971) Markham (1972).
1923	M	5	Lavalle y Garcia (1924) Ballen (1925) Zegarra (1926) Jones (1933) Gunther (1936) Hutchinson (1950) Schweigger (1961).
1925-26	VS	5	Fuchs (1925, 1937) Murphy (1926) Zegarra (1926) Berry (1927) Sheppard (1930) Jones (1933) Petersen (1935) Maisch (1936) Vogt (1940) Mears (1944) Hutchinson (1950) Rudolph (1953) Nials *et al.* (1979) Mugica (1983).
1930-31	M	5	Jones (1933) Petersen (1935) Andrade (1948) Hutchinson (1950) Schweigger (1961) Brooks (1971) Miller and Laurs, (1975) Woodman (1984).
1932	S	5	Sheppard (1933) Jones (1933) Petersen (1935) Schott (1938) Vogt (1940) Mears (1944) Andrade (1948) Hutchinson (1950) Rudolph (1953) Brooks (1971) Mugica (1983) Woodman (1984).
1939	M+	5	Schweigger (1940) Vogt (1940) Mears (1944) Hutchinson (1950) Sears (1954) Mugica (1983) Woodman (1984).

continued

Table 32.1 – *continued*

El Niño Events	Event Strength	Confidence Rating	Information Sources
1940-41	S	5	Lobell (1942) Mears (1944) Hutchinson (1950) Sears (1954) Wooster (1960) Schweigger (1961) Brooks (1971) Mugica (1983) Woodman (1984) Quinn and Zopf (1984).
1943	M+	5	Schweigger (1961) Miller and Laurs (1975) Caviedes (1975) Mugica (1983) Woodman (1984).
1951	M-	5	Garcia Mendez (1953) Schweigger (1961) Brooks (1971) Wooster and Guillen (1974) Miller and Laurs (1975).
1953	M+	5	Rudolph (1953) Avilla (1953) Sears (1954) Wooster and Jennings (1955) Merriman (1955) Schweigger (1961) Mugica (1983) Woodman (1984).
1957-58	S	5	Wooster (1960) Schweigger (1961) Bjerknes (1961) Brooks (1971) Idyll (1973) Miller and Laurs (1975) Caviedes (1975) Mugica (1983). Hastenrath *et al.* (1984).
1965	M+	5	Guillen (1967, 1971) Wooster and Guillen (1974) Miller and Laurs (1975) Caviedes (1975) Mugica (1983) Woodman (1984).
1969	M-	5	Wooster and Guillen (1974) Miller and Laurs (1975) Quinn *et al.* (1978) Rasmusson and Hall (1983).
1972-73	S	5	Idyll (1973) (Wooster and Guillen (1974) Miller and Laurs (1975) Ramage (1975) Caviedes (1975) Nials *et al.* (1979) Mugica (1983) Woodman (1984).
1976	M	5	Quinn (1977, 1980) Ceres (1981) Smith (1983) Mugica (1983) Rasmusson and Hall (1983) Quinn and Neal (1983) Woodman (1984).
1982-83	VS	5	Mugica (1983) Rasmusson and Hall (1983) Rasmusson and Wallace (1983) Quiroz (1983) Smith (1983) Le Comte (1984) Quinn and Zopf (1984) Woodman (1984) Caviedes (1984) Canby (1984) Quinn and Neal (1984).
1987	M	5	Quinn *et al.* (1987) Le Comte (1988) reports of anomalous storm activity and heavy rainfall over subtropical and northern Chile (from V. T. Neal, during his 1987 assignment in Chile).

are occasionally deviations from this time-table, as exemplified by the strong late 1940-41 El Niño and the very strong late 1982-83 El Niño. Those El Niños that set in during the Southern Hemisphere summer are augmented by a strong seasonal input.

The foregoing details have been provided since they may be useful to investigators working with proxy data.

32.3   Additional evidence

The large number of references on early events was often essential for verification purposes. Sometimes the additional sources pinned down the departure time of a sailing vessel leaving Panama or its arrival time at Callao for verification of a reported rapid passage cited in another source as being based on the northerly wind and/or currents occurring during an El Niño. At another time, it may be the presence of a strong pre-event anti-El Niño that completes the evidence for the following El Niño onset. The whole idea is to come up with a sufficient amount of evidence both direct and circumstantial to reasonably assure the occurrence and strength of an event. Therefore, in addition to the 14 types of evidence, as listed in Quinn *et al.* (1987) for determining the occurrence of an El Niño, we have noted many other indications which have also been useful for strengthening our El Niño evaluations, some of which are as follows:

(1)   large scale starvation due to drought over southeastern Peru and adjacent parts of Bolivia;

(2)   widespread epidemics of disease and insect plagues due to excessive heat, pools of stagnant water, and at times huge quantities of putrid dead sea life in the northern and north central coastal regions of Peru; and also due to the extreme drought conditions such as those mentioned in (1) above for the highlands;

(3)   incidence of rabies in quadrupeds, particularly dogs, as the excessive heat caused them to immerse themselves in pools of stagnant water along the coast of Peru during an El Niño (In one case a rabid dog bit 14 people in one night, 12 of whom died.);

(4)   occurrence of anomalously heavy winter (Southern Hemisphere) precipitation in subtropical and northern Chile (Quinn and Neal 1983) (This occurs as the result of a breakdown in the southeast Pacific of the subtropical ridge during an ENSO, which allows the westerly storms to penetrate further to the north than usual along the coast of Chile.);

(5)   often very heavy anti-El Niño precipitation, floods and landslides occurring over eastern or southeastern Peru and adjacent parts of Bolivia about 6-15 months prior to the onset of El Niño [This was brought to our attention by Prof. A. Cornejo, Universidad Nacional Agraria, La Molina, Lima, Peru. He noticed that in several cases, when he had rainfall records for mountain stations, the heavy rainfall occurred during the peaks in our Easter-Darwin SO index and prior to the El Niños. We tested this on recent cases and noted that unusually heavy rainfall occurred January-April 1982 (Le Comte 1983) prior to the onset of the El Niño in October 1982; also, unusually heavy rainfall occurred January-April 1986 (Le Comte 1987) prior to the onset of El Niño in February 1987.];

(6)   years with an increased frequency of shipwrecks along the coast of central Chile due to adverse weather conditions (Caviedes 1985 referring to Gormaz 1901) [The rationale here is that the increase in ship losses off central Chile due to adverse weather conditions is most likely the result of the winter storms more frequently reaching lower latitudes along the coast of Chile, as in (4) above, due to the equatorward shift of westerly storm tracks as the southeast Pacific high breaks down during an ENSO.];

(7)   frequent agreement in times of occurrence between the northeast Brazil droughts (Secas) and the El Niño as noted by Caviedes (1973, 1985) [This agreement appears to

be quite substantial when we compare our recent El Niño record to that for the Seca. Although the frequency of occurrence for Secas is less than that for El Niños, in most cases when a Seca was reported there was also an El Niño. If we had reliable rainfall records for the same locations over a reasonably long period of time, it is estimated that Secas and El Niños might relate to one-another over 70% of the times that they occurred. The record of Brooks (1971) which agrees in general with that of Andrade (1948) reports that 4 Secas occurred in the 17th century, 11 in the 18th, 13 in the 19th and 8 in the 20th as recorded from 1603 to 1958. In our opinion the Secas of the distant past were based on more drastic conditions which caused massive evacuations by northeast Brazilian populations. This would account for the increased frequency of occurrence with time. Considering the global zonal-vertical Walker cells of Flohn (1971) this activity over Brazil would be associated with the Atlantic Cell in which the air of the southeast trades rises on the east side of the Andes and sinks off the coast of west Africa. The intermittent Secas are attributed to the fact that in drought years the ITCZ remains north of the equator and does not move southward over northeast Brazil (Caviedes 1973; Ratisbona 1976). In further noting similarities for occurrences of Secas and El Niños, Freise (1938) reports the rapid succession of dry periods (Secas) in the first quarter of the 19th century while we noted a similar rapid succession of El Niños over the same period. Also, the extreme Seca of 1877-79 agrees well with the occurrence of the very strong El Niño of 1877-78. In several cases the Seca lasted much longer than the El Niño. Both of these events are generally most prominent in the Southern Hemisphere summer and fall seasons, when rainfall is more likely to occur in significant amounts. The variations between listed Seca years over the past century by various authors are probably due to the use of different evaluation criteria, inadequate data sources or reliance on data from different parts of the drought region; Hastenrath *et al.* (1984) in their Figure 32.2 indicate that in some years there are large deviations in the amounts of rainfall over the northern northeast and southern northeast parts of the drought affected region. Many figures in Markham (1972) show the occasional far northward penetration of frontal weather conditions over the South American continent during Southern Hemisphere winter which may cause significant variations in the rainfall over the southern part of the northeast Brazil drought region. A good tree ring analysis over this drought region might provide us with a reasonably weighted record of drought activity over our whole period of interest.];

(8) reductions in tribute from Peruvian provinces for years adversely affected by El Niño-related activity, as reported for example in Huertas (1984) and Carnero y Pinto (1983).

(9) information relating to conditions associated with El Niño, as recorded in the historical logs of the large Peruvian haciendas.

32.4 The record

Using the contents of Table 32.1 and Figure 32.1, and considering the three to four year periods of compound activity to in each case represent two separate events of the indicated intensities, we come up with a total of 115 El Niño onsets between 1525 and 1987. Considering that there are 114 intervals between event onsets over a period of 462 years, we find

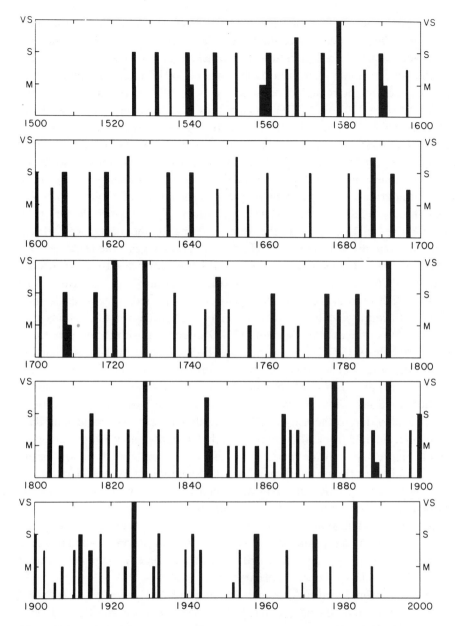

Figure 32.1 A graphic representation of the 1525-1987 record of El Niño events, as reported in Table 32.1 of the text, is shown here by century, with time of occurrence on the horizontal scale and strength on the vertical scale.

that there is an average of about four years between event onsets for the M-VS categories. If we just consider the S-VS events, we come up with 52 onsets between 1525 and 1982. This provides 51 intervals between onsets over a period of 457 years, and a long term average of about nine years between onsets of the stronger events. The record is too short for any statistical evaluation of VS events; their occurrences in 1578, 1720, 1728, 1791, 1828, 1877-78,

Table 32.2. El Niño onsets in the various intensities by century and over the period 1925-1987 when more data were available.

Century	M*-VS	M+-VS	S-VS	S+	VS
1525-1599	19	15	10	1	1
1600-1699	18	17	13	3	0
1700-1799	22	17	11	2	3
1800-1899	32	20	10	4	3
1900-1987	24	15	8	0	2
Totals	115	84	52	10	9
1925-1987	15	10	6	0	2

*Includes the M- cases noted in Table 32.2.

1891, 1925-26, and 1982-83 provide intervals of 142, 8, 63, 37, 49, 14, 34, and 57 years between onsets. However, they do reflect great irregularity in frequency of occurrence. Table 32.2 provides a record of events by century and also gives a breakdown over the period from 1925 to the present when we had a reasonably large amount of data and descriptive information available. Over this latter period we come up with the following intervals between M-VS onsets: 5, 2, 7, 1, 3, 8, 2, 4, 8, 4, 3, 4, 6, and 5 years. This gives an average of about 4.4 years between onsets. If we just consider the S and VS events, the intervals between onsets are: 7, 8, 17, 15, and 10 years; and, we get an average of 11.4 years between onsets for stronger events over the 1925-1982 period. When we compare the recent record with the past, we note that El Niño events haven't become more frequent in occurrence over recent years nor has there been a preponderance of stronger events over this period, as one might gather from our news sources. The same irregularity in time between event onsets prevails.

32.5 Long-term climatic changes

Extended periods of climatic change (near-decadal or longer) were noted throughout the record. The first period was 1539-1578, during which there was a large number of strong events culminating in the very strong 1578 El Niño. The next period of unusual activity was 1600-1624; here, Brunning (1922-23) refers to a report of very abundant rainfall in the province of Lambayeque during the second decade of the 17th century. The third such period was 1701-1728, during which two very strong El Niños (1720 and 1728) occurred in very close temporal proximity. The fourth unusual period was 1792-1802 during which there was an extended drought over northwestern Peru and no events. The fifth such period was 1812-1832 during which the M+ -VS events were unusually frequent. The sixth period, 1864-1891, was exceptional since the events occurred frequently and many of them were strong or very strong. The period 1897-1919 had frequent occurrences of significant El Niño activity. The unusual 1925-1932 period with its generally above normal sea surface temperatures and rainfall was discussed by Petersen (1935). The recent case which fostered the exceptionally strong 1982-83 El Niño (Quinn and Zopf 1984, Quinn et al. 1987) resulted in the Easter-Darwin SO index anomalies being generally about two millibars below average over the period April 1976-March 1988. A similar lowering took place in all SO indices over this period, and a general increase in the sea surface temperatures over the tropical Pacific and subtropical South Pacific appeared to be the cause (Quinn and Neal 1983, 1984).

Certainly the two most unusual periods of climatic change were 1701-1728 and 1864-1891. In both of these periods two of the highly unusual very strong El Niños occurred in close proximity timewise. During the 1864-1891 period the rainfall increased so much over northwestern Peru that the Sechura, a notoriously dry and barren desert region, became covered with trees and heavy vegetation, the likes of which were never seen before or afterward. Why this unusual development occurred when it did is questionable, but as indicated in Quinn *et al.* (1987) it may have been due to the general warming as we emerged from the "Little Ice Age" (LIA) in the latter part of the 19th century.

32.6   The Little Ice Age (LIA)

The time confines of the LIA vary considerably from author to author, but most determinations are based on Northern Hemisphere records and reports. Fairbridge (1987) provides a good coverage of this subject. We will assume that the LIA sets in during the first half of the 16th century and that it ends during the latter half of the 19th century. Fortunately this record of ocean-atmosphere activity spans the LIA period. The most significant changes that we noted were in the subtropical Chilean rainfall data. The first historic Chilean drought occurred 1636-1639 (Mackenna 1877, Taulis 1934). Precipitation was generally low during the 18th century, and drought conditions became particularly severe 1770-1803. The only ENSO-related climatic change that showed up with anomalously high rainfall in subtropical Chile between 1769 and 1816 was the 1783 event. As mentioned above, the increase in precipitation over northwestern Peru 1864-1891 may have been caused by emergence from the LIA.

32.7   Discussion

In several cases the historian obtains excellent descriptive material on an event by interviewing those who experienced it, but if the interview was many years after the event, dates of reference may be in error. For example, Cabello Balboa (1586) visited Trujillo in 1586 and got an excellent report on the north winds, rainfall deluge and flooding in an event for which the damage was still evident during his visit. It was obviously the VS 1578 El Niño which was being referred to, but he obtained a 1576 date from his informant. Our other sources helped us to correct the date of occurrence, but Cabello Balboa's descriptive information was still very valuable.

In an earlier section we mentioned the heavy anti-El Niño precipitation over the mountains of eastern Peru that is often noted prior to an El Niño. However, cases have also been noted in the past when extremely heavy precipitation occurred over southeastern Peru following an El Niño. In particular, we noted that very heavy precipitation and floods were experienced over the Andahuaylas region by President Gasca and his forces in early 1548, following the 1546-47 El Niño (Montesinos 1642). Also, there was very heavy rainfall and flooding over the Puno-Lake Titicaca region in 1748 (Paz Soldan 1862) after the 1747 El Niño. Of course, we have noted throughout this study that each event is accompanied by its own peculiar time-table of related activities. Various investigators have at times assumed the recurrent El

Niño events to be cyclic in occurrence (e.g., Berlage on page 25 of his 1957 publication, after referring to some ambiguous sources, lists the Peru heavy rains as occurring every seven years between 1728 and 1798; yet, of those listed, only the two he got from Eguiguren (1894) 1728 and 1791, actually occurred). The record of Table 32.1 and the discussion on it show very clearly the variability in frequency of occurrence of El Niño events.

The relationship between the El Niño and the anomalously heavy winter (Southern Hemisphere) precipitation over subtropical Chile, although they usually occur in the same year, is indirect in that they are both regional manifestations of the large-scale ENSO. They are both the result of a breakdown in the southeast Pacific subtropical high pressure cell (which of course triggers the relaxation process in the southeast trades and equatorial easterlies). However, in the case of the El Niño the invasion is of warm waters from the north and west and the atmospheric ITCZ from the north; whereas, the heavy precipitation over subtropical Chile is the result of a northward swing of the westerly storm tracks along the coast of Chile when the high breaks down.

At times there are other significant changes to the atmospheric circulation patterns which as yet have not been fully investigated; e.g., the occasionally substantial contribution from the southeast trades of the Atlantic to the meteorological pattern over northern Peru during some of the events (e.g., the 1982-83 El Niño). This input was noted about a century or more ago by Raimondi (Eguiguren 1894) and was shown quite clearly in the 1982-83 satellite photos and the work of Goldberg and Tisnado (1985). In the satellite photos dark, heavy clouds were noted to flow from east to west through the mountain passes of Peru. Goldberg and Tisnado noted from GOES satellite image cloud patterns, periods when lee wave structures indicative of easterly flow across the Andes prevailed during the 1982-83 event.

We feel that the relationship between the El Niño/ENSO and the Seca is substantial and warrants further study. The fact that both of these phenomena are primarily prominent in the Southern Hemisphere summer and fall is fortunate but to be expected. In most cases when there is a Seca, it can be associated with an El Niño, but the reverse relationship is not as frequent. However, the record on Secas is much less complete than that for the El Niño events and the basis for recognition is not consistent for Secas. In the associations either one may lead the other, but the El Niño appears to lead more frequently (e.g., the 1914-15 El Niño leads the 1915 Seca, the 1918-19 El Niño leads the 1919 Seca, the 1957-58 El Niño leads the 1958 Seca). Also, as explained earlier, the heavy rainfall over eastern Peru is often a forerunner of the El Niño. These relationships between the western part of the Atlantic Walker cell and the eastern part of the Pacific Walker cell require further investigation, but they also require more reliable and better quality meteorological data over the involved regions.

A question which is likely to arise with regard to this investigation is: "How many events (moderate to very strong in intensity) may have been overlooked?" We are quite sure that all of the VS events that occurred over our period of study have been recorded. Also it is quite likely that we have noted all of the S and S+ events. However, even though our search has been quite comprehensive, some of the moderate events with their less extensive effects could have been overlooked, particularly during periods when our information sources were limited. A review of the Table 32.1 record and the Figure 32.1 plot makes the period between the S+ 1624 El Niño and the S 1681 El Niño look suspect in this regard. Based on the overall record we rarely go beyond eight years without an event in the M-VS range, unless there were extenuating circumstances such as the persistent drought of 1792-1802. Ordinarily the time between events ranges from two to six years, with an overall average near four years. However, there were three periods between 1624 and 1681 when the time between events exceeded eight years: 1625-1633, 1661-1670 and 1672-1680. The average time between all events over the period 1624-1681 was about 7.1 years, which is far greater than our overall average near four years. Nevertheless, the average time between the S and S+ events over this same period was 9.5 years, which is very close to the overall average of nine years between S-VS events. It appears that the available information is insufficient in detail to detect all of the moderate events over this period; and, therefore, there may be as many as five moderate events missing from the record over this time span. If this happens to be the case, the overall average time between El Niños would be about 3.9 years.

With regard to the previous discussion, we have to some extent improved our capability for event detection and evaluation by finding additional direct, indirect and circumstantial sources of evidence and information to support our determinations. Of those sources, 23 are now listed between the contents of Quinn *et al.* (1987) and this chapter. However, this primarily holds for the stronger M+ -VS events. The M+ intensity events, which are at times difficult to separate from the S events, have now essentially become the lower limit for the stronger events; and, down to this level our event coverage appears to be quite thorough. Below this level, as mentioned earlier, literature sources available at this stage of our research are not sufficiently detailed over some periods to assure the detection of all moderate intensity events. A review of meteorological, hydrological and oceanographic records in relation to events below the M+ level over recent years will show why the weaker events would be more difficult to detect from contents of the earlier literature sources. We prepared Figure 32.2 for use by those investigators that only wish to consider the stronger events in relation to their proxy data, global/regional climatic change features, etc.

Considering El Niño onsets over the period 1525-1982, we find 84 M+ -VS onsets over 457 years and, therefore, an overall average of 5.5 years between onsets. Over the more recent 1925-82 period, we find 10 onsets in 57 years or an average of 6.3 years between onsets in the M+ -VS intensities. Of course, this M+ cutoff leaves off several of the recent events, such as 1951, 1969, 1976 and 1987. Also one finds an 18-year period, 1846-1863, without any stronger events (Figure 32.2) as a result of the M+ cutoff. However, this would be in line with Eguiguren's (1894) findings that there were none of the heavier class 3-4 rainfalls over Piura during this same 18-year period in his 100-year record (1791-1890) of rainfall estimates. It is interesting that this also relates to a long period without significant northeast Brazil droughts.

32.8 Summary

It was within the recurring large-scale ocean atmosphere circulation fluctuations over the Indo-Pacific area, which are represented by the Southern Oscillation (SO) that the forcing factor for the El Niño development was noted. It is the shift from a strong southeast trade and equatorial easterly buildup phase to a weak easterly (or westerly) breakdown (relaxation) phase that is the forcing factor which caused these developments to take place. By referring strictly to this regional manifestation (the El Niño) of the large-scale ENSO (El Niño/Southern Oscillation) which primarily affects southwestern Ecuador, northwestern

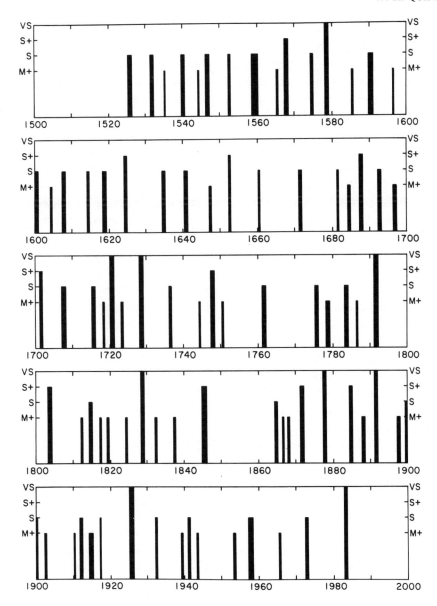

Figure 32.2 A graphic presentation of the 1525-1983 record of stronger El Niño events. The events shown here by century are limited to the M+-VS intensity levels of the entries in Table 32.1 of the text.

Peru and their adjacent ocean waters we can obtain the longest possible record of ocean-atmosphere activity (about 465 years) from existing historical records.

A list of moderate (M) strong (S) and very strong (VS) El Niño events that occurred between 1524 and 1987, along with their related references, is provided. In so far as possible, an attempt was made to bring to bear every available source of evidence for identifying the El Niño occurrences and evaluating their intensities. Unfortunately the amount of evidence depends on happenstance (whether or not a chronicler or his observational source was in the affected region at the time an event occurred). Indirect and circumstantial evidence such as droughts over southeast Peru and adjacent parts of Bolivia, northeast Brazil droughts, anomalously heavy subtropical and northern Chilean precipitation, and significant ship losses due to adverse weather conditions off central Chile (the latter two situations being due to an equatorward displacement in the westerly storm tracks along the coast of Chile when the southeast Pacific subtropical high pressure cell breaks down during an ENSO) added weight to our determinations. Anomalously heavy anti-El Niño precipitation over eastern or southeastern Peru and adjacent parts of Bolivia 6-15 months prior to a suspected El Niño occurrence likewise often adds positive weight to the El Niño determination. The observed absence of El Niño type activity over vulnerable areas was also of great value in constructing our record of events. It appears that a good tree ring analysis over northeast Brazil might be of tremendous climatological value in general and in particular to this investigation.

Long-term changes in climatic activity (near decadel or longer) were noted for 1539-78, 1600-24, 1701-28, 1792-1802, 1812-32, 1864-91, 1897-1919, 1925-32, and the recent case which at this time appears to be 1976-87 (April 1976-March 1988) based on SO index anomalies being significantly below average over this period. Considering the Little Ice Age (LIA) to set in during the first half of the 16th century and to end during the latter half of the 19th century, it appears that its greatest effects were a decrease in Chilean precipitation and an increase in the length and strength of Chilean droughts over the 17th, 18th and early part of the 19th centuries. The large increase in precipitation over northwestern Peru 1864-91 may have been caused by emergence from the LIA.

Considering the moderate, strong and very strong El Niños (a total of 115 events) we arrive at a long-term average of about four years between events. Considering just the strong and very strong El Niños (a total of 52 events) we get a long-term average of about nine years between events.

In addition to a figure that contains all events M-VS, we also provide a second figure with a lower level cutoff at the M+ intensity, since at this stage in our research there are some periods during which our literature sources may not have detected all of the weaker moderate events. Down to the M+ level our coverage appears to be quite thorough. Overall, the average time between M+ -VS onsets is about 5.5 years. Those investigators that are only interested in the stronger events may prefer to use this second figure.

Acknowledgments

We thank Dr. Santiago E. Antunez de Mayolo, Prof. Emeritus, Universidad Nacional Mayor de San Marcos, Lima, Peru for his copious notes on El Niño-related activity, Dr. Ramon Mujica, Dean of Faculty of Engineering, Universidad Piura, Piura, Peru for his copious data on the Department of Piura, Peru, the Interlibrary Loan Service of Oregon State University for excellent support, Florence Beyer for processing our information into readable form, and the National Science Foundation for our research support through NSF Grants ATM-85-15014 and ATM-8808185.

References

Acosta, Joseph de. 1590. *Historia natural y moral de las Indias*. Sevilla. (In: *Obras del Padre Jose de Acosta*, Biblioteca de Autores Espanoles, Madrid, 1954.)

Adams, J. I. 1905. Caudal, Procedencia y Distribucion de Aguas de la Provincia de Tumbes y los Departamentos de Piura y Lambayeque *Boletin de Ingenieros de Minas del Peru*, No. 27, Lima.

Albenino, Nicolao de. 1549. *Verdadera relacion delo sussedido en los Reyes e prouincias del Peru desde la yda a ellos de Virey Blasco Nunes Vela hasta el desbarato y muerte de Goncalo Picarro*. Sevilla.

Alcedo, Antonio de. 1786-89. *Diccionario geografico-historico de las Indias occidentales a America*. Madrid, 5 volumenes.

Alcedo y Herrera, Dionisio. 1740. *Aviso historico, politico, geografico con las noticias mas particulares del Peru, Tierra-Firme, Chile y Nuevo Reyno de Granada*. Madrid, Miguel de Peralta.

Andrade, Lopez de. 1948. *Introducao a sociologia das Secas*. Editora A. Noite, Rio de Janeiro.

Anonymous. 1925. Lluvias e inundaciones. *La Vida Agricola*, Vol. II. (No. 16) 238-245, Lima (Abril 1925).

Anson, G. 1748. *A voyage round the world in the years MDCCXL, I, II, III, IV*. London, Oxford University.

Avila, E. 1953. "El Niño" en 1953 y su relacion con las aves guaneras: problemas basicos referentes a la anchoveta. *Bol. Comp. Admin. del Guano*, Lima, 29(5):13-19. Bachmann, Carlos J. 1921. Departamento de Lambayeque, monografia historico-geografica. Lima, Imp. Torres Aguirre.

Ballen, F. 1925. Sobre la mortandad de las aves guaneras. *La Vida Agricola*, Vol. II: 301, Lima (April 1925).

Barclay, W. S. 1917. Sand dunes in the Peruvian desert. *Geographical Journal*, 49, 53-56.

Barriga, V. M. 1951. *Los terremotos en Arequipa 1582-1868*. Arequipa: La Colmena, S.A.

Basadre, Modesto. 1884. *Riquezas Peruanas*. Lima, Imp. de "La Tribuna."

Benzoni, Girolamo. 1565. *Historia del Mondo Nuovo*. Venice. Translated and edited by W. H. Smyth, New York, Burt Franklin, 1857.

Berlage, H.P. 1957. Fluctuations of the general atmospheric circulation of more than one year, their nature and prognostic value. *K. Ned. Meteorol. Inst. Meded. Verh.* 69, 152 p.

Berlage, H.P. 1966. The Southern Oscillation and world weather. *K. Ned. Meteorol. Inst. Meded. Verh.* 88, 152 p.

Berry, E. W. 1927. Meteorological observations at Negritos, Peru, December 1924 to May 1925. *Mon. Wea. Rev.*, 55 (2):75-78.

Bjerknes, J. 1966. Survey of El Niño 1957-58 in its relation to tropical Pacific meteorology. *Inter.-Amer. Trop. Tuna Comm. Bull.*, 12(2):1-62.

Bowman, I. 1916. *The Andes of southern Peru: Geographical reconnaissance along the seventy-third meridian*. New York, published for the Amer. Geog. Soc., Henry Holt and Co.

Bowman, I. 1924. Desert trails of Atacama, Ed. by G. M. Wrigley, *Amer. Geog. Soc., Spl. Pub.* No. 5, New York.

Bravo, Jose. J. 1903. Los huaycos. Inf. y Mem. Bol. Soc. de Ing. Tomo 5, No. 5: 13-21.

Brooks, R. H. 1971. Human response to recurrent drought in northern Brazil. *The Professional Geographer*, 23(1):40-44.

Brunning, Enrique. 1922-23. *Estudios monograficos del Departamento de Lambayeque*. Chiclayo, Libreria e Imprenta de Dionisio Mendoza.

Bueno, Cosme. 1763. *Geografia del Peru virreynal*. Ed. Daniel Valcarcel. Lima, D. M. Azangaro 858, 1951.

Cabello Balboa, M. 1586. *Miscelanea Antarctica-Una Historia del Peru Antiguo*. Universidad Nacional Mayor de San Marcos, Lima, 1951.

Canby, T. Y. (ed.). 1984. El Niños ill wind, *National Geographic* 165(2):144-183.

Carnero, N. y M. Pinto. 1983. *Diezmos en el Arbispado de Lima*. Seminario de Historia Rural Andina, Lima, Universidad Nacional Mayor de San Marcos.

Carranza, L. 1891. Contracorriente maritima observada en Payta y Pacasmayo. *Bol. Soc. Geogr., Lima*, 1:344-345.

Caviedes, C. N. 1973. Secas and El Niño: Two simultaneous climatical hazards in South America. *Proceedings of the Association of American Geographers* 5, 44-49.

Caviedes, C. N. 1975. El Niño 1972: Its climatic, ecological, human, and economic implications, *Geogr. Rev.*, 65:493-509.

Caviedes, C. N. 1984. El Niño 1982-83. *Geogr. Rev.*, 74:267-290.

Caviedes, C. N. 1985. South America and world climatic history. In: *Environmental history: Critical issues in comparative perspective*. Kendall E. Bailes (ed.) 135-152. Washington, D.C., University Press of America.

Ceres. 1981. New methods help in assessing fishery resources FAO *Rev. Agric. Dev.* 14(3):8-9.

Cieza de Leon, Pedro de. 1864. *The travels of Pedro de Cieza de Leon, A.D. 1532-1550, contained in the first part of his Chronicle of Peru*. Clements R. Markham, editor and translator. London, The Hakluyt Society, 340 pp. (Originally published in Spanish as *Cronica del Peru* in 1553.)

Cobo, P. Bernabe. 1639. Fundacion de Lima, escripta por El P. Bernabe Cobo de la Compania de Jesus. Ano de 1639. Mexico. (In: *Obras del Padre Bernabe Cobo*, Biblioteca de Autores Espanoles, Madrid, 1956).

Cobo, P. Bernabe. 1653. Historia del Nuevo Mundo. Lima. (In: *Obras del Padre Bernabe Cobo*, Biblioteca de Autores Espanoles, Madrid, 1956).

Cooke, Edward. 1712. *A voyage to the South Sea and around the world (1708-1711)*. London, 2 vols.

Eguiguren, D. V. 1894. Las Lluvias de Piura *Bol. Soc. Geogr. Lima* 4(7-9):241-258.

El Comercio A newspaper published in Lima.

Enfield, D. B. 1981. El Niño-Pacfic eastern boundary response to interannual forcing. In: *Resource management and environmental uncertainty: lessons from coastal upwelling fisheries*, M. H. Glantz and J. D. Thompson (eds.) 213-254. New York, Wiley Interscience, John Wiley and Sons.

Estrada Icaza, Julio. 1977. *Regionalismo y migracion*. Guayaquil. Pub. Archivo Historia.

Fairbridge, R. W. 1987. Little Ice Age. In: *The encyclopedia of climatology*. J. E. Oliver and R. W. Fairbridge (eds.) 547-550. New York, Van Nostrand Rheinhold Co.

Feijoo de Sosa, Miguel. 1763. *Relacion descriptiva de la ciudad y provincia de Trujillo del Peru*. Imprenta de Real y Supremo Consejo de las Indias, Madrid, 81 pp.

Flohn, H. 1971. Tropical circulation pattern, *Bonner Meteorologische Abhandlungen*, Heft 15, p. 24. Meteorologische Institut der Universitat Bonn.

Forbes, H. O. 1913. The Peruvian Guano Islands. *Ibis*, Series 10, 1:709-712.

Forbes, H. O. 1914. Notes on Molina's pelican (Pelicanus thagus). *Ibis*, Ser. 10, 2:403-420.

Freise, F. W. 1938. The drought region of northeastern Brazil. *Geographic Review*, 28, 363-378.

Fuchs, Federico G. 1907. Zonas lluviosas y secas del Peru. Inf. y Mems. *Bol. de la Soc. de Ingenieros de Lima*. Tomo VII:270-297.

Fuchs, F. G. 1918. Meteorologia del Peru. *Boletin de la Sociedad Geografica de Lima*, 34, 1-22.

Fuchs, Federico G. 1925. Las ultimas lluvias y la ciencia meteorologica. *La Vida Agricola*, Lima. Julio 1925, pp. 521-524.

Fuchs, F. G. 1937. Los cambios de clima en el mundo-Posibles causas del cambio de clima, en la costa del Peru. *Boletin de la Sociedad Geografica de Lima*, 54, 160-169.

Garcia, Pedro. 1921. Observaciones hidrologicas 1911-1920. *Boletin del Cuerpo de Ingenieros de Minas del Peru*, No. 102, 1-156.Garcia

Garcia Rodriguez, V. J. 1779. *Geografia de el Peru*, Tomo II, Lima.

Garcia Rosell, Ricardo. 1903. Monografia historica del departamento de Piura. *Bol. Soc. Geogr. de Lima*, Tomo 13: Trim. 2:193-242, Trim. 3:310-351, Trim. 4:419-462.

Gaudron, Julio. 1925. Las lluvias en la costa y la periodicidad de los fenomenos meteorologicos. *La Vida Agricola*, Lima. Mayo, 1925, pp. 361-368.

Gentil, Le. 1728. *Nouveau voyage au tour du monde*. Amsterdam, P. Mortier.

Goldberg, R. A. and G. Tisnado 1985. Andean lee waves on the western slopes during El Niño. In: *Ciencia, technologia y agression ambiental: El fenomeno El Niño*, p. 229-239. CONCYTEC, Lima, Peru.

Gonzales, Benito. 1913. *Ligero estudio sobre la meteorologia de los vientos en Lima*. Lima, San Marti.

Gormaz, P. V. 1901. *Naufrajios occurridos en las costas de Chile desde su descubrimiento hasta nuestros dias*. Imprenta Elzevier, Santiago.

Guillen, O. 1967. Anomalies in the waters off the Peruvian coast during March and April 1965. *Stud. Trop. Oceanogr.* Miami. 5:452-465.

Guillen, O. 1971. The "El Niño" phenomenon in 1965 and its relations with the productivity in coastal Peruvian waters. In: *Fertility of the Sea*, 1, p. 187-196.

Gunther, E. R. 1936. Variations in behavior of the Peru Coastal Current with a historical discussion. *Jour. Roy. Geogr. Soc.*, 88:37-65.

Haenke, Tadeo 1799. *Descripcion del Peru*. Reprint of a British Museum manuscript. The reprint is by Imprenta El Lucero, Lima, 1901.

Hastenrath, S., Ming Chin Wu and Pao Shin Chu. 1984. Towards the monitoring and prediction of northeast Brazil droughts. *Quarterly Journal of the Royal Meteorological Society*, 110, 411-425.

Huertas, L. 1984. *Tierras, diezmos y tributos en el obispado de Trujillo*. Lima, Universidad Nacional Mayor de San Marcos.

Humboldt, Alejandro von. 1804. La corriente de agua fria a lo largo de la costa occidental de Sudamerica. *Revista del Instituto de Geografia*, Universidad Nacional Mayor de San Marcos, Vol. 6:7-22. Lima, Peru, 1960.

Hutchinson, G. E. 1950. Survey of existing knowledge of biogeochemistry. 3. The Biogeochemistry of vertebrate excretion. *Bull. Am. Mus. Nat. Hist.* 96, 554 pp.

Hutchinson, T. J. 1873. *Two years in Peru, with exploration of its antiquities*. London, Samson Low, Marston, Low and Searle, 2 vols.

Idyll, C. P. 1973. The Anchovy crisis. *Scient. Amer.* 228(6):22-29.

Jones, H. H. 1933. Notas sobre la meteorologia y agricultura del departamento de Piura. *Bol. Soc. Geogr. de Lima*, Vol. 50, Trim. 1:16-18.

Juan, Jeorge and Antonio de Ulloa. 1748. *Relacion historica del viaje a la America Meridional*. Madrid. 4 vols. (translated by John Adams and printed for John Stockdale, London 1807).

Kiladis, G. N. and H. F. Diaz. 1986. An analysis of the 1877-78 ENSO episode and comparison with 1982-83. *Mon. Wea. Rev.*, 114:1035-1047.

Labarthe, Pedro A. 1914. Las avenidas extraordinarias en los rios de la costa. *Inf. y Mems. de la Soc. de Ingenieros del Peru*. Vol. 16, Nos. 11 y 12, p. 301-329.

La Patria - a newspaper published in Lima. Lastres, J. B. 1937. Hipolito Unanue y "El clima de Lima." *Boletin de la Sociedad Geografica de Lima*, 54, 75-87.

Lavalle y Garcia, J. A. de. 1917. Informe preliminar sobre la causa de la mortalidad anormal de las aves ocurrida en el mes de marzo del presente ano. *Mem. Comp. Adm. del Guano*, Lima. Vol. 8:61-88.

Lavalle y Garcia, J. A. de. 1924. Estudio de la emigracion y mortalidad de las aves guaneras ocurridas en los meses de mayo y junio del ano 1923. *Mem. Comp. Adm. del Guano*, Lima, Seccion Tecnica.

Le Comte, D. 1983. World weather 1982. *Weatherwise*, 36, 14-17.

Le Comte, D. 1984. Worldwide extreme floods and droughts. *Weatherwise*, 37, 9-18.

Le Comte, D. 1987. Highlights around the world-water, water almost everywhere... *Weatherwise*, 40, 9-10.

Le Comte, D. 1988. Global highlights. *Weatherwise*, 41, 10-13.

Leguia y Martinez, G. 1914. *Diccionario geografico, historico, estadistico, etc. del Departmento de Piura*. Lima, Tipografia El Lucero.

Llano Zapata, Joseph Eusebio. 1748. *Observacion diaria critico-historico meteorologica, contiene todo lo acaecido en Lima desde primero de Marzo de 1747 hasta 28 Octubre del mismo ano, etc.* Con licencia; impreso en Lima, ano de 1748.

Lobell, M. G. 1942. Some observations on the Peruvian coastal current. *Trans. Am. Geophys. Union*, 23:332-336.

Lorente, S. 1861. *Historia de la Conquista del Peru.* Lima.

Mackenna, B. V. 1877. *El clima de Chile.* Primera edicion, Santiago. Segunda edicion 1970, Buenos Aires, Compania Impresora Argentina S.A.

Maisch, Carlos. 1936. El clima de la costa Peruana-Sus causas y sus consecuencias. *Boletin de la Sociedad Geografica de Lima,* 53, 253-280.

Markham, C. G. 1972. *Aspectos climatologicos da Seca no Brasil-nordeste.* Recife, SUDENE Assessoria Tecnica.

Martinez y Vela, Bartolome. 1702. *Anales de la Villa Imperial de Potosi.* La Paz, Artistica, 1939.

Mears, E. G. 1944. The ocean current called "The Child." *Ann. Rep. Smithson. Inst.*(1943) p. 245-251.

Melo, Rosendo. 1878. *Apuntes para la irrigacion del Valle de Chira, Lima.* Imp. del Universo, 1888.

Melo, Rosendo. 1913. Hydrografia del Peru. *Bol. de la Soc. Geogr. de Lima*, Tomo 29, Trim. 1 and 2, p. 141-159.

Mendez, C. A. 1953. La corriente maritima del Peru y su replica en el clima de su litoral durante el ano de 1951. *Boletin Sociedad Geografica de Lima*, 70, 87-89.

Merriman, D. 1955. The El Niño brings rain to Peru. *Amer. Science* 43:63-76.

Middendorf, E. W. 1894. *Peru-Beobachtungen und studien uber das land und seine bewohner. Band II, Das kustenland von Peru.* Berlin, Robert Oppenheim (Gustav-Schmidt).

Miller, F. R. and R. M. Laurs. 1975. The El Niño of 1972-73 in the Eastern Tropical Pacific Ocean. *Inter-Amer. Trop. Tuna Comm. Bull.* 16(5):403-448.

Montesinos, Fernando de. 1642. *Anales del Peru.* Two vols. Published in 1906 in Madrid by Victor M. Maurtua.

Moreno, Gabriel. 1804. *Almanaque Peruano y Guia de Forasteros. Para el ano 1800.* Imp. Real del Telegrafo Peruano.

Mugica, R. 1983. *El fenomeno de El Niño Piura 1983.* Universidad Piura, Piura, Peru. 51 p.

Murphy, R. C. 1923. The oceanography of the peruvian littoral with reference to the abundance and distribution of marine life. *Geogr. Rev.*, 13:64-85.

Murphy, R. C. 1925. *Bird islands of Peru. The record of a sojourn on the west coast.* Putnam, New York. 362 p.

Murphy, R. C. 1926. Oceanic and climatic phenomena along the west coast of South America during 1925. *Geogr. Rev.*, 16:26-54.

Nials, Fred L., Eric E. Deeds, Michael E. Mosley, Shelia G. Pozorski, Thomas G. Pozorski, and Robert Feldman. 1979. El Niño: The catastrophic flooding of coastal Peru. *Field Museum of Natural History Bulletin.* Vol. 50, No. 7: p. 4-14, and No. 8: p. 4-10.

Oliva, Anello. 1631. *Historia del Peru y Varones insignes en Santidad de la Compania de Jesus.* Ed. Juan Francisco Pazos Varela y Luis Varela y Orbegozo. Lima, Imp. y Liv. de San Pedro, 1895.

Palma, Ricardo. 1894. *Tradiciones Peruanas.* Barcelona.

Paz Soldan, Eduardo. 1908. Inf. y Mems. *Bol. Soc. Ing.,* Vol. 10, No. 5, p. 187-189.

Paz Soldan, Mateo. 1862. *Geografia del Peru.* Paris. Libreria de Fermin Didot Hermanos.

Petersen, G. 1935. Estudios climatologicos del noroeste Peruana. *Bol. Soc. Geol. del Peru*, 7(2):1-141.

Portocarrero, Juan. 1926. Contribucion al estudio hidrologico del territorio Peruana. *Inf. Mem. Bol. Soc. Ing.* Vol. 28, No. 2, Lima, Febrero 1926. p. 68-93 y 1 grafico.

Prescott, W. H. 1892. *History of the conquest of Peru.* Philadelphia, J. B. Lippincott Co. Vol. 1 of 2 vols., 469 p.

Puente, Augustin de la. 1885. *Diccionario de la legislacion de aguas y agricultura del Peru.* Lima. Imp. Francisco Solis.

Puls, C. 1895. Oberflachtentemperaturen und stromungsverhaltnisse des aequatorial gurtels des Stillen Ozeans. *Arch. Deutsch Seewarte,* 18(1):37 p. + 3 charts.

Quinn, W. H. 1977. Diagnosis of the 1976-77 El Niño. *Proc. Second Annual Climate Diagnostics Workshop.* p. 21-1 to 21-14. [NTIS PB282151].

Quinn, W. H. 1980. Monitoring and predicting short-term climatic changes in the South Pacific Ocean. *Investigaciones Marinas* 8(12):77-114.

Quinn, W. H. 1987. El Niño. In: *The encyclopedia of climatology.* J.E. Oliver and R. W. Fairbridge (eds.) 411-414. New York, Van Nostrand Reinhold.

Quinn, W. H. and V. T. Neal. 1983. Long-term variations in the Southern Oscillation, El Niño, and Chilean subtropical rainfall. *Fish. Bull.,* U.S. 81:363-374.

Quinn, W. H. and V. T. Neal. 1984. Recent long-term climatic change over the eastern tropical Pacific and its ramifications. *Proc. Ninth Ann. Clim. Diag. Workshop.* U.S. Department of Commerce, Washington, D.C., p. 101-109 [NTIS-PB 85-183911].

Quinn, W. H. and D. O. Zopf. 1984. The unusual intensity of the 1982-83 ENSO event. *Tropical Ocean-Atmosphere Newsletter,* No. 26, D. Halpern, Ed, NOAA Pacific Marine Environmental Laboratories, Seattle, WA., p. 17-20.

Quinn, W. H., V. T. Neal and S. E. Antunez de Mayolo. 1987. El Niño occurrences over the past four and a half centuries. *Journal of Geophysical Research,* 92. No. C13: 14,449-14,461.

Quinn, W. H., D. O. Zopf, K. S. Short and R. T. Kuo Yang. 1978. Historical trends and statistics of the Southern Oscillation, El Niño, and Indonesian droughts. *Fish. Bull.,* U.S. 76:663-678.

Quiroz, R. S. 1983. The climate of the "El Niño" Winter of 1982-83. A season of extraordinary climatic anomalies. *Mon. Wea. Rev.,* 111:1685-1706.

Raimondi, A. 1876. *El Peru, Tomo II, Historia de la geografia del Peru.* Lima, Imprenta del Estado, Cale de la Rifa, Num. 58, por J. Enrique del Campo. 475 p.

Raimondi, A. 1897. Geografia fisica. *Bol. Soc. Geogr. de Lima,* Vol. 7, Nos. 7, 8 and 9, p. 268-278.

Ramage, C. S. 1975. Preliminary discussion of the meteorology of the 1972-73 El Niño. *Bull. Amer. Meteor. Soc.* 56:234-242.

Rasmusson, E. M. and J. M. Hall. 1983. El Niño: The great equatorial warming. *Weatherwise,* 36:166-175.

Rasmusson, E. M. and J. M. Wallace. 1983. Meteorological aspects of the El Niño/Southern Oscillation, *Science* 222:1195-1202.

Rasmusson, E. M., P. A. Arkin, L. H. Carpenter, J. Koopman, A. F. Krueger and R. W. Reynolds. 1982. Equatorial Pacific warm event. *Tropical Ocean-Atmosphere Newsletter (Special Issue)* No. 16, D. Halpern, Ed., NOAA Pacific Marine Environmental Laboratories, Seattle, WA, p. 1-3.

Ratisbona, L. R. 1976. The climate of Brazil. In: *World survey of climatology,* Vol. 12, 219-269, New York, Elsevier.

Remy, F. E. 1931. De la lluvia en Lima. *"El Comercio,"* Lima, 8/21/1931.

Rocha, Diego Andres. 1681. *Tratado unico y singular del origen de los Indios del Peru, Mexico, Santa Fe y Chile.* Lima, Manuel de los Olivos.

Rudolph, W. E. 1953. Weather cycles on the South American west coast. *Geogr. Rev.* 43(4):565-566.

Ruschenberger, W. S. W. 1834. *Three years in the Pacific.* Carey, Lea and Blanchard, Philadelphia, 441 pp.

Schott, G. 1938. Klimakunde der Sudsee Inseln. In: Vol. 4, part T of *Handbuch der Klimatologie.* Herausgegeben von W. Koppen, Graz and R. Geiger, Munchen. Berlin, Verlag von Gebruden Borntraeger 1930.

Schweigger, E.H. 1940. Studies of the Peru coastal current with reference to the extraordinary summer of 1939. *Proc. Sixth Pacific Sci. Cong.* 3:177-195.

Schweigger, E. H. 1961. Temperature anomalies in the eastern Pacific and their forecasting. *Soc. Geogr. Lima, Bol.* 78:3-50.

Scientific Committee on Oceanic Research (SCOR) Working Group 55. 1983. Prediction of "El Niño." *SCOR Proceedings* Vol. 19, Paris, September 1983, pp. 47-51.

Sears, M. 1954. Notes on the Peruvian coastal current. 1. An introduction to the ecology of Pisco Bay. *Deep Sea Res.*, 1:141-169.

Shelvocke, George. 1726. *A voyage round the world by the way of the Great South Sea*. London. Published 1971 by Da Capo Press, 227 West 17th St., New York.

Sheppard, G. 1930. Notes on the climate and physiography of southwestern Ecuador. *Geographic Review*, 20, 445-453.

Sheppard, G. 1933. The rainy season of 1932 in southwestern Ecuador. *Geogr. Rev.* 23:210-216.

Sievers, Wilhelm. 1914. *Reise in Peru and Ecuador ausgefurt 1909*. Munchen and Leipzig. 411 pp. e ilustraciones. Verlag von Duncker und Humboldt.

Smith, R. L. 1983. Peru coastal currents during El Niño 1976 and 1982. *Science* 221:1397-1398.

Spruce, R. 1864. *Notes on the valleys of Piura and Chira in northern Peru and on the cultivation of Cotton therein*. Eyre and Spottswoodie, London. 81 p.

Spruce, R. 1908. *Notes of a botanist on the Amazon and Andes,* 2 vols. Alfred R. Wallace, editor, Macmillan.

Stevenson, M., O. Guillen, and J. Santoro. 1970. *Marine atlas of the Pacific coastal water of South America*. Univ. of Calif. Press, Berkeley and Los Angeles, 23 p. plus 99 charts.

Stevenson, W. B. 1829. *A historical and descriptive narrative of twenty years residence in South America*. Vol. II, 434 p. London, Longman Rees, Orme, Brown and Green.

Taulis, E. 1934. De la distribution des pluies au Chile. In *Materiaux pour l'etude des Calamites*, Part 1, p. 3-20. Societe de Geographie de Geneve.

Tizon y Bueno, Ricardo. 1907. Descripcion sintetica de las condiciones hidrologicas de la quebrada del Rimac. *Inf. y Mem. Bol. Soc. Ing.* Vol 9(5):97-119.

Troup, A. J. 1965. The 'Southern Oscillation'. *Quarterly Journal of the Royal Meteorological Society*, 91, 490-506.

Tyson, P. D. 1986. *Climatic change and variability in southern Africa*. Capetown; New York: Oxford University Press.

Unanue, J. Hipolito. 1806. *El clima de Lima*. Lima. Second edition, Madrid, 1815.

Vasquez de Espinoza, Antonio. 1629. *Compendium and description of the West Indies*. Translated by Charles Upson Clark and published by the Smithsonian Institution, Washington, D.C., September 1, 1942 (*Publication* 3646).

Vogt, W. 1940. Una depresion ecologica en la costa Peruana. *Bol. Comp. Admin. del Guano*, 16(10):307-329.

Woodman, R. F. 1984. Recurrencia del fenomino El Niño con intensidad comparable a la del ano 1982-1983. In: *Proceedings of the Seminario Regional, Ciencia, Tecnologia y Agression Ambiental; El Fenomeno El Niño*. CONCYTEC, Lima-Peru, pp. 301-332.

Wooster, W. S. 1960. El Niño. *Calif. Coop. Oceanic Fish. Invest. Rep.* 7:43-45.

Wooster, W. S. and O. Guillen. 1974. Characteristics of El Niño in 1972. *J. Mar. Res.* 32 (33):387-404.

Wooster, W. S. and F. Jennings. 1955. Exploratory oceanographic observations in the eastern tropical Pacific January to March 1953. *Calif. Fish Game.* 41(1):79-90.

Wyrtki, K., E. Stroup, W. Patzert, R. Williams and W. Quinn. 1976. Predicting and observing El Niño. *Science*, 191, 343-346.

Xeres, Francisco de. 1534. *Verdadera relacion de la conquista del Peru. Seville.* (Reports on the discovery of Peru. Translated and edited by C. R. Markham; Burt Franklin, Publisher, New York, 1872)

Zegarra, J. M. 1926. Las lluvias y avenidas extraordinarias del verano de 1925 y su influencia sobre la agricultura del departamento de La Libertad. *Inf. y. Mems. Bol. de la Soc. de Ingenieros de Lima.* Vol. 28(1):1-46.

Section E: SUMMARY

33 Climatic variations over the last 500 years

P. D. Jones and R. S. Bradley

33.1 Introduction

The period since A.D. 1500 has seen dramatic changes in the fortunes of human life on this planet. At the turn of the 16th century the New World and tropical and southern Africa had just been discovered by European colonists, and only the Australian and Antarctic continents remained unknown to the outside world. At the beginning of the 16th century scientific understanding was extremely limited; the earth was believed to be only about 6000 years old and the Polish astronomer, Mikolaj Kopernik (Copernicus) was developing his theory that the earth and the other planets move around the sun. Almost 500 years on, at the dawn of a new millennium, satellites have reached every planet in our solar system except Pluto and the world's population is linked by instant communications. From a population of only about 400 million in A.D. 1500, the world now supports over 5 billion people.

Over this time, scientific understanding of the environment of our planet has increased to such an extent that we can now model the physics that underlies most of the processes of nature. In the field of climatology this increase in understanding has been dramatic over the last 25 years but there is still much that remains unknown. This is particularly true about the record of past climatic variations and their causes. The first attempts to estimate average hemispheric and global mean temperatures were made about a century ago (Köppen, 1873) and possible explanations were postulated for the changes in temperature which were seen since 1750. Despite these temperature series showing large interdecadal variability, most meteorologists during the first half of the 20th century believed that climate did not change by significant amounts. Climatology was generally considered to be the least fashionable aspect of meteorology (Lamb, 1977).

At present, the possibility that climate is changing and that this may lead to unprecedented warming, is at the top of the political agenda (Bromley, 1990). The increase in global mean temperatures by 0.3-0.6°C, since the middle of the 19th century (Folland *et al.*, 1990) is often cited as evidence that anthropogenic increases in greenhouse gases are affecting the climate. While the extent to which human activities have contributed to global warming over the last century is debateable, this uncertainty has highlighted the greater understanding of climate and climatic change that is required.

Greater understanding of climatic fluctuations is necessary, not only of the last century, but also of earlier times. The longer the record of global climate is, the more confidence we will have in determining how unusual recent events have been. In this volume we have provided a comprehensive assessment of climatic change covering the period since A.D. 1500. We are, therefore, in an excellent opposition to review both what has happened since that time and also to consider what some of the causes of the changes might have been. We begin, however, with a reconsideration of some of the proxy climatic reconstruction techniques and their limitations. These comments provide a caveat to those who would over-interpret the reconstructions, and serve to focus attention on where further methodological research is needed.

33.2   Reconstruction techniques

33.2.1   Seasonal limitations

In any proxy climatic reconstruction it is likely that the parameter being used (e.g. tree ring growth, ice core isotopic value) reflects climate during a particular season of the year. For example, historical records of variables such as grape flowering and harvest dates or snow cover duration are seasonally specific. Stating that the 17th century was cooler than the 20th based on greater numbers of days of snow lying may not be correct if the non-winter seasons were warmer. Similarly, ice core parameters such as $\delta^{18}O$ only represent conditions during periods of snowfall. Extremely cool periods during intense inversions, when there is no snowfall, will not be represented in annual time series. By contrast, ice core melt records reflect summer temperatures. Tree growth indices are largely indicative of conditions during the growing season, but they may also reflect preceeding conditions which influence soil moisture (or the physiological condition of the tree) prior to the growing season. Determining the correct seasonal signal in a particular proxy record is often one of the most difficult problems in paleoclimatic reconstruction. Much of the evidence presented in this book (particularly the concluding discussion which follows below) relates to conditions during the growing season, or the snowfall season over ice caps and glaciers. The temptation to extrapolate conditions in one season to the year as a whole must be resisted. Where multi-season reconstructions are available, as in Switzerland or the western Soviet Union, important differences are often apparent (cf. Figure 6.4 in Pfister, Chapter 6 and Figure 9.2 in Borisenkov, Chapter 9).

33.2.2   Temporal and spatial limitations

Limitations in the temporal resolution of some climate reconstructions are not always recognised. For example, in dendroclimatic reconstructions long chronologies have often been derived from a combination of living and standing dead trees, stumps and sub fossil material. If all the overlapping samples cover a relatively short period of time, it would be impossible to consider variations on longer time scales than the average length of the samples. Perhaps more significantly for studies the last 500 years, all tree-ring time series require some form of standardization to remove biologically-related age trends. The degree to which low-frequency aspects of chronologies are altered by this process is a matter for much debate (see Chapters 15, 19 and 24 by D'Arrigo and Jacoby, Briffa and Schweingruber, and Norton and Palmer). It is also true that the magnitude of yearly and decadal anomalies in any final chronology will be determined by the type of curve fitting procedure adopted (Cook and Kariukstis, 1990).

Historical records may also contain a temporal bias. Observers base descriptive comments about climate on their prior experience. Many early North American pioneers considered their first few winters harsh after their experience of much milder winters in Western Europe. With time, their opinions of the unusual or severe may change leading to a bias in what might otherwise appear to be a long and continuous record. Because of all these uncertainties, it should be apparent that no one record can be entirely relied upon to represent a faithful reconstruction unless there is corroborating evidence from another, independent source. The more supporting lines of evidence, the stronger will be the basis for confidence in the reconstructions. Each supporting line of evidence provides a further building block to which subsequent work can be added. In this way, a complex inferential pyramid of information is gradually constructed leading to a true and meaningful picture of climatic conditions in the past.

With ice cores, the climatic interpretation of certain parameters is problematical, particularly for the relatively small amplitude changes observed over the last 500 years. In polar regions, values of $\delta^{18}O$ and δD in snowfall decrease with decreasing condensation temperature in clouds, but the quantitative nature of these relationships are generally derived from the spatial (geographic) dependence between isotopic ratios in snowfall and mean annual temperature. Unfortunately temporal relationships may not be the same as the spatial relationship (see Section 28.3.2 in Peel, Chapter 28) and this may lead to incorrect paleoclimatic interpretations. Furthermore, different moisture sources can lead to apparent differences in reconstructed temperature, even when none occurred. Peel (Chapter 28) shows that it is essential to consider changes in moisture source when trying to understand the isotopic record from the Antarctic Peninsula region, and similar considerations are required in other regions of the world (Johnsen et al., 1989).

Much effort in recent years has gone into reconstructing large scale (hemispheric or globally averaged) changes of temperature and precipitation from instrumental records (e.g. Jones et al.,1986a, 1986b, 1991; Bottomley et al., 1990; Hansen and Lebedeff, 1988; Bradley et al., 1987a; Diaz et al., 1989; Vinnikov et al. 1990). Most of these records only extend back into the mid or late 19th century. What can we say about large-scale climatic changes further back in time? This is an extremely difficult problem since data from the oceans (70% of the globe) peters out in the mid-19th century and vast areas of the tropics (and much of the Southern Hemisphere) are similarly devoid of observations. At the present time it is simply not possible to derive a 'global' time series beyond the mid-19th century. Indeed such a record may be tenuous beyond the early part of this century (Bradley, 1991).

Can a hemispheric record be produced? There are currently far too few records to derive a meaningful long-term Southern Hemisphere record. Even today, instrumental records are only available for about 75% of the hemisphere and this coverage rapidly deteriorates back in time. In the Northern Hemisphere we are approaching the point where it may be possible to derive a long time series, combining tree ring, historical and ice core data. However, the problems of seasonal representation, discussed above, suggest that such an effort would have to focus on only one season, probably the summer. This may not be representative of the year as a whole. For example, in the Northern Hemisphere instrumental record there have been negligible changes in temperature since the mid-ninteenth century. Most of the long-term warming has occurred in the other three seasons (Jones and Bradley, Chapter 13).

Problems still prevail concerning large areas of the Tropics and much of interior Asia for which we have virtually no high resolution information. Nevertheless, with the development of new chronologies and techniques of climatic reconstruction we anticipate that a fairly good northern hemispheric summer temperature reconstruction, spanning several centuries, may be assembled within a few years. Previous attempts at such a synthesis (Groveman and Landsberg, 1981) are based on too few records (many of which were not properly calibrated in terms of climatic response) to provide a reliable time series.

An alternative approach would be to select those regions of the world which are known to be highly correlated with the hemispheric record of the last century (Bradley, 1991). If this can be demonstrated, regional records might be considered as proxies of larger scale variations. This requires the assumption that the relationship observed in the last century has prevailed over longer periods but there is evidence that this may not have been the case (Jones and Kelly, 1983). Nevertheless, such an assumption is no different from that made in all paleoclimatic reconstructions, that the relationships observed during the period of instrumental records (the calibration period) remained constant during earlier periods. The use of a regional time series to represent the hemisphere uses the same assumption, substituting a spatial frame of reference for a temporal one. Further consideration of such an approach is recommended.

33.3 Regional evidence

33.3.1 Europe

The European region is the part of the world with the most detailed climate history. Instrumental records discussed by Jones and Bradley (Chapter 13) extend back to the late 17th and early 18th century. Because meteorological instruments were developed in Europe, detailed information is available for Europe for about 100 more years than for any other continent. Prior to the instrumental time there is a wealth of written historical information back several hundred years further. This depth of climatic information from both instrumental and historical sources has meant that many of the 'generally' understood climatic variations of the last 1000 years were first recognized in Europe. Many methods of climatic reconstruction have been tested in Europe where the long records enable detailed verification of the results to be undertaken.

From the evidence presented for Europe in Chapters 5-9 and 18-20, the climate since A.D. 1500 has varied between extremely warm and extremely cool decades, but few of these decades appear synchronous over the whole of the European continent, from the Iberian Peninsula to the Urals. Evidence for a period of protracted cool temperatures during the so-called Little Ice Age during the 16th to 19th centuries does not appear that convincing. The best documented and most widespread cool periods occurred during the 17th and 19th centuries.

From warm temperatures during some decades of the early 16th century, conditions gradually began to cool during the second half of the century. Over Western Europe tree ring evidence indicates cool temperatures during the 1560s and 1570s (Serre-Bachet *et al*, Chapter 18). The severity of winters in northern Italy increased in frequency during the period 1570 to

1614, with milder conditions later in the first half of the 17th century (Camuffo and Enzi, Chapter 7). The decades of the 1590s and the 1600s were cool over northern Europe (Briffa and Schweingruber, Chapter 19) and in the northern Urals (Graybill and Shiyatov, Chapter 20). However, cool winters and springs were more common in Iceland during the 1630s and 1690s (Ogilvie, Chapter 5). In western parts of the USSR, winter and spring seasons were cooler in the 17th century than the first half of the 16th century though summers and autumns showed little difference (Borisenkov, Chapter 9).

Cool conditions returned to the European region during the late 18th and early 19th centuries. The frequency of severe winters in northern Italy increased again in the late 18th century (Camuffo and Enzi, Chapter 7). The number of cooler winters and springs and greater sea-ice extent off Iceland also increased at the same time (Ogilvie, Chapter 5). The 1800s and 1810s were cold throughout most of western Europe, as shown by evidence from Switzerland (Pfister, Chapter 6) southwestern Europe (Serre-Bachet et al., Chapter 18) and northern Europe (Briffa and Schweingruber, Chapter 19). According to the Swiss reconstructions, all seasons were cold during this period. However, further east, in the western Soviet Union, temperatures were warmer than the average during the summer and autumn seasons. Indeed, this was the period of the warmest summers in the entire 500 year period (Borisenkov, Chapter 9). In the northern Urals, however, relatively cool conditions seem to have prevailed (Graybill and Shiyatov, Chapter 20). A lack of sea-ice off Iceland is apparent during the first half of the 18th century (Ogilvie, Chapter 5) though summer temperatures in Svalbard appear to have been low at this time (Tarussov, Chapter 26).

Following the 1810s many European records indicate warmth during the 1820s followed by a return to cool conditions during the 1830s. This cold-warm-cold oscillation is evident in many of the proxy records discussed here as well as in most European instrumental records that extend to the period (Jones and Bradley, Chapter 13). Conditions were cooler again during the second half of the nineteenth century, particularly over the western USSR where the coldest summers of the last 500 years were experienced (Borisenkov, Chapter 9; Graybill and Shiyatov, Chapter 20; Jones and Bradley, Chapter 13).

In between these two relatively cooler centuries, the 18th century shows evidence of warmer conditions. For example, the instrumental record for Central England (Manley, 1974) indicates that temperatures were generally warmer than during the 19th century particularly during the 1730s. Warmth at this time is also indicated in Switzerland (Pfister, Chapter 6) and during the 1750s and 1760s in northern Europe (Briffa and Schweingruber, Chapter 19) and in the northern Urals (Graybill and Shiyatov, Chapter 20). Summers were 1°C warmer than the long-term average in the western U.S.S.R during the late 18th century.

33.3.2 Asia

Long-term climatic reconstructions for Asia have largely been confined to the eastern Asian region encompassing China and Japan. Reconstructions elsewhere on the continent are extremely scarce. Although the area is renowned for long historical records, there are actually very few for western China and the Asian interior and, as yet, very little is known about historical climatic records from south and southeast Asia. Dendroclimatic studies may hold the key to a better unnderstanding of past climatic variations in the Asian interior.

Some of the first dendroclimatic reconstructions from the Himalayas and Tibet are presented in this volume (Hughes, Chapter 21; Wu, Chapter 22).

The relatively few reconstructions available for the continent show no evidence of any prolonged periods of anomalous temperatures. Indeed, from the reconstruction of *summer* temperatures in Beijing (Wang *et al*., Chapter 11) extending back to 1725, the coolest period occurred during the 1960s. This contrasts strongly with the isotopic record from the Dunde ice cap further west (see below). The long temperature reconstructions from maximum latewood density in Kashmir for the spring and late summer periods (Hughes, Chapter 21) and those for summer in western China (Wu, Chapter 22) also show no trends over the last three centuries. The coldest period of the last 400 years in western China was ~1600-1670 and in northeast China it was the 1650s and 1660s. In Kashmir, dendroclimatic records only extend to 1690; the coolest decades of the last 300 years were during the 1720s and 1730s and 1790-1840. Reconstructions indicate warm decades in Kashmir during the mid-1700s and in western China during the mid-16th century, the early 18th century and again in the early 19th century.

Oxygen isotope measurements from the Dunde ice cap in western China (Thompson, Chapter 27) seem to confirm dendroclimatic evidence of cooler conditions during the late 16th and early 17th centuries (particularly from ~1580-1650) and to a lesser extent during the 19th century. The 18th century was warmer. The years since 1920 are the warmest of the entire series.

Reconstructions of precipitation totals from the clear and rain day records in China and Japan (Wang and Zhang, Chapter 10; Murata, Chapter 12) show evidence of decadal scale variations during the 18th century but no long-term trends. At the Chinese sites, summer precipitation was low during the 1740s and again during the 1770s and 1780s with wetter conditions during the 1750s and 1760s. The Japanese reconstructions show quite complex features over different regions of southern Kyushu. Drier conditions are evident during the late eighteenth century (1780s, 1790s) with a tendency towards wetter conditions during the 1740s and 1750s. The reconstructions in the two countries therefore appear to be in phase only at certain times.

33.3.3 North America

Historical evidence from the North American region is confined to the time since the European settlement. Climate reconstruction on the year-to-year time scale further back in time is only possible using tree-ring and ice core evidence. Diaries kept by the early settlers in the northeastern United States reveal warmer conditions during the 1740s and cooler annual temperatures during the 1750s, 1760s and 1810s (Baron, Chapter 4). No prolonged periods of cool temperatures are evident between 1640 and 1820. Further north in Canada the historical archives kept by the Hudsons Bay Company have been used to estimate temperatures and sea ice severity around Hudsons Bay during the 18th and 19th centuries (Ball, Chapter 3; Catchpole, Chapter 2). The sea ice records reveal the worst sea ice severity years during the 1810s and 1840s.

Over the western third of the United States, Fritts and Shao (Chapter 14) have reconstructed temperatures from tree-ring information back to 1602 for five regions. All regions show little evidence of protracted cool periods with the coldest period occurring during the late 19th and early 20th centuries. Prior to this time the 1750s to 1770s were cool, as were the 1830s and 1840s. Warm conditions occurred during the 1930s and 1950s and during the 1850s and 1860s. The period from about 1650 to 1740 was also generally warm, particularly over northern parts of the western United States. Average conditions from 1602-1900 were warmer and drier over most of the western United States compared to the period since 1900.

D'Arrigo and Jacoby (Chapter 15) reconstructed annual temperatures for northern North America from trees growing along the Canadian tree line from the Yukon to Quebec. Cooler temperatures occurred during 1725 to 1800 and again during the second half of the 19th century. The present century is clearly the warmest century in their reconstruction.

Tree-ring information from the central and eastern United States (Meko, Chapter 16; Cook *et al.*, Chapter 17) has been used to assess drought frequency since the early 18th century. Both reconstructions show little evidence of prolonged periods of drought or of *widespread* droughts, except possibly during the 1750s and 1760s and from 1814 to 1822. Fritts and Shao (Chapter 14) examined their reconstructions for the period 1814-1822 and found below normal precipitation during winter, spring and autumn months and above normal temperatures during summer.

In many regions drought appears to have occurred more regularly during the eighteenth century and again during the period after 1920. However, droughts which are significant locally may not have been widespread. Meko (Chapter 16) notes that although the 1930s and 1950s experienced severe droughts, there were longer, more persistent droughts which occurred earlier though these were less geographically extensive. The major feature of a number of regions is a lower frequency of droughts during the second half of the 19th century. Fritts and Shao (Chapter 14) postulate more frequent storms between 1600 and 1900 than since 1901.

In northern Canada, ice core melt records suggest that summer temperatures were generally low from ~1570-1860, with coldest conditions from 1550-1620 and ~1680-1700 (Fisher and Koerner, 1983; Alt, 1985). At Camp Century, Greenland, coldest conditions appear to have been in the 17th century (Dansgaard *et al.*, 1975). Holdsworth *et al.*, (Chapter 25) using an ice core from Mt. Logan in the Yukon show $\delta^{18}O$ values were below the long-term average from ~1850-1950 and above average for the period 1740-1850. However, these changes may not be linked to temperature in the same way as in polar regions.

33.3.4 Southern hemisphere

The greater area of ocean in this hemisphere considerably limits the number and extent of climatic reconstructions available. Written historical evidence is limited to the period since European settlement. Records from the time of the Spanish conquistadors have been used by Quinn and Neal (Chapter 32) to reconstruct a record of El Niño occurrence. Elsewhere in this hemisphere evidence is generally only available from tree-ring and ice core sources.

Tree-ring evidence from the hemisphere has been confined to southern South America, New Zealand and Tasmania. There is potential in parts of continental Australia and southern Africa but as yet this has not been fulfilled. Climatic reconstructions in New Zealand are, at present, limited by the age of species so far used. Dendroclimatic reconstructions have yet to be made from the longer lived species in the region. The main feature of the temperature reconstructions back to 1750 (Norton and Palmer, Chapter 24) is a cool period during the

1840s and 1860s (1860s and 1870s in Tasmania) that agrees with the earliest instrumental records available (Salinger, 1981). Since the coming of Europeans to New Zealand in 1840 there has been a widespread retreat of most mountain glaciers (Gellatly and Norton, 1984; Gellatly, 1985).

In southern South America longer chronologies have been produced from both coniferous and deciduous species. Temperature reconstructions have been made for both Patagonia and Tierra del Fuego, back to 1500 and 1750, respectively (Boninsegna, Chapter 23). All records show a great deal of decadal-scale variability but the cooler summers are apparent from 1500-1650 in central Patagonia and the latter half of the 17th and 18th centuries in northern Patagonia. In Tierra del Fuego, lowest summer temperatures occurred in the late 1800s. Hence there is no dendroclimatic evidence for a protracted "Little Ice Age" period in Patagonia.

Ice core evidence in this hemisphere is confined to the Quelccaya ice cap in Peru and to Antarctica. The Quelccaya record (Thompson, Chapter 27) indicates lighter isotopic values from about A.D. 1540 to the 1810s. The 1810s appear to have been the coldest decade of the entire record (recording the greatest heavy isotope depletion). By contrast, the 1820s were the warmest decade of the 19th century and temperatures have increased considerably since then.

In Antarctica, different ice cores reveal markedly different isotopic records, indicating that either a simple temperature interpretation is not possible, or persistent regional variations in anomalies and trends occurred in the past (Mosley-Thompson, Chapter 29). For example, the Mizuho record clearly shows a period of low $\delta^{18}O$ values from ~1630-1870, whereas at the South Pole the lowest values are in the 16th century with a generally 'cold' (low $\delta^{18}O$) period from ~1540-1740 (interrupted by a warmer period from ~1630-1670). At Law Dome the coldest interval is from ~1730 to the early 1900s, whereas towards the Antarctic Peninsula (Siple) and on the Filchner-Ronne ice shelf there is no clear picture of prolonged low $\delta^{18}O$ values.

Over the Antarctic Peninsula the interpretation of the isotope record is considerably more complex than elsewhere in Antarctica (Peel, Chapter 28). Peel shows that it is essential to derive both $\delta^{18}O$ and δD ratios in order to assist with the interpretation of the isotope/temperature record because isotopically light features may be related to polynya formation in the Weddell Sea. The resulting moisture source changes affect the isotopic ratios. In this region, where records currently only extend back to 1800, the first tentative reconstructions show that cool conditions prevailed in the 1860s, 1890s and 1930s.

33.4   Evidence for changes in possible forcing factors

Three possible forcing factors were considered in this volume: solar irradiance changes (Stuiver and Braziunas, Chapter 30) volcanic events (Bradley and Jones, Chapter 31) and changes in the frequency of El Niño/Southern Oscillation (ENSO) events (Quinn and Neal, Chapter 32). Obviously, in understanding the last hundred or so years of climatic variations we also need to consider anthropogenic changes of greenhouse gas concentrations in the atmosphere.

The evidence for solar activity changes was assessed using the variations of ^{14}C in tree rings.

Stuiver and Braziunas (Chapter 30) show that ^{14}C variation can be used as a proxy for solar irradiance change. Their analysis reveals significant periodicities of about 416, 215, 143 and 85 years. Since 1500 the most important variation of solar irradiance was probably related to the Maunder Minimum period between 1645 and 1715. During this interval there were almost no sunspots. This compares with an average sunspot number of 51 over the period 1880 to 1965. Stuiver and Braziunas, estimate that the maximum change of global mean temperatures due to the reduction in sunspot numbers is 0.12°C. This is fairly small given that global mean temperatures have increased by 0.45°C since 1880 (Jones and Bradley, Chapter 13). Evidence for any major climatic change during the 1645 to 1715 period is equivocal. While cool conditions are generally apparent during the 1690s-1710s, warmth is apparent in many regions during other decades of the Maunder Minimum, particularly the 1650s. At present it is not possible to relate historical climatic variations to changes in solar variability with any confidence.

Evidence for changes in the frequency of explosive volcanic events is discussed by Bradley and Jones (Chapter 31). It is apparent that the record of explosive eruptions is quite incomplete, particularly the chemical characteristics of volcanic events. Several different catalogs are in agreement on which were the largest eruptions of the last 150-200 years, but prior to that there is considerable uncertainty about the magnitude and timing of eruption events. It is telling that almost all of the mid and high latitude eruptions in the Northern Hemisphere that we know about before 1900 are in Japan or Iceland where studies are most complete. No doubt many Alaskan, Kamchatkan and Aleutian eruptions remain undocumented. The Southern Hemisphere volcanic chronology is similarly flawed. In view of the many studies which show unequivocal relationships between surface air temperature and certain major explosive eruptions, the importance of a more complete chronology of eruption events can not be over-emphasised; it is an essential pre-requisite to a better understanding of the variability of climate over the last few centuries.

It has only recently been shown that changes in the frequency of ENSO events can have global ramifications (see for example, Bradley et al., 1987b; Ropelewski and Halpert, 1987; Diaz and Kiladis, 1989; Jones, 1989). Particularly in the Tropics, large-scale temperature and precipitation anomalies are often related to the phase and magnitude of the ENSO event. Quinn and Neal (Chapter 32) have produced an historical record of El Niño events back to the beginning of the 16th century. Since 1525 there have been 84 medium to very severe events with an average recurrence interval of 5.5 years. The period since 1925 has had slightly fewer events than the long-term average with events occurring every 6.3 years. Strong El Niños recur about every 9 years. Certain periods stand out because of a lack, or an excess, of events. For example, between 1846 and 1863 there were no severe events. Periods with a higher frequency of events like 1976-87 were 1539-78, 1600-24, 1701-28, 1812-32, 1864-91, 1897-1919 and 1925-1932. These intervals should be focused upon in future research to determine the larger scale significance of periods with above or below average ENSO frequency. It is also important to determine the relative significance of ENSOs when coincident with major volcanic eruptions, as in 1982 (El Chichon) (cf. Handler, 1984).

33.5   The "Little Ice Age"

The term "Little Ice Age" has been used frequently in this volume without any discussion of what the term means. The term originated with Matthes (1939) who stated,

We are living in an epoch of renewed but moderate glaciation – a "little ice age" that already has lasted about 4,000 years.

Thus, in its original useage the term was informal (not capitalized) and referred to what is now called the period of neoglaciation (Moss, 1951; Porter and Denton, 1967). However, Matthes (1940) also noted that,

. . . glacier oscillations of the last few centuries have been among the greatest that have occurred during the 4,000 year period . . . the greatest since the end of the Pleistocene ice age.

It is this latest and most dramatic episode of neoglaciation to which the term "Little Ice Age" is now generally applied, though there is considerable uncertainty about when this period began (and ended) and what its climatic characteristics were. For example, Porter (1986) indicates that the Little Ice Age began near the end of the Middle Ages at around A.D. 1250 and continued until about 1920, whereas Lamb (1977) confines the Little Ice Age to 1550-1850, with its main phase from 1550-1700. Grove (1988) in her comprehensive treatise on the Little Ice Age, seems to concur with Lamb but does not explicitly define the term. Clearly, if the term is to be useful, it must be universally understood. This is especially important in determining what caused the Little Ice Age; you can not explain something if you do not know what it is! Since the focus of this book is the period which many consider to be within the Little Ice Age, it is appropriate that we evaluate the various records presented to try and shed some light on this confusion.

Early evidence for the occurrence of the Little Ice Age came from Europe (Lamb, 1977). Glaciers tended to be more advanced than at present particularly during the 18th and 19th centuries. For example, a series of pictures from the Grindelwald glacier (Zumbühl, 1980; Messerli *et al.*, 1978) illustrates quite dramatically the advance and retreat of the glacier over the last few hundred years. A variety of glacier evidence from other alpine regions of the world also indicates that many glaciers were extensive in the 19th century and have retreated dramatically over the last century (Grove, 1988; Wood, 1988). However, few regions have the detailed historical documentary records which are available for the European Alps to determine glacier positions over time, particularly before A.D. 1850. In most regions, glacier fluctuations have been dated by ^{14}C and/or lichenometry (often calibrated by ^{14}C). [14]C dates in the range of 100 to 500 B.P. often provide non-unique calendar year ages (Stuiver and Pearson, 1986) which create considerable uncertainty in reconstructing glacier positions over this interval (Porter, 1981). Glacier advances are also episodic events which result from cumulative increases in mass balance and the interaction of these changes with each glacier's unique dynamic system. Hence, glacier position changes are not easily ascribed to specific changes in climate. Mass balance changes can be brought about by a variety of climatic perturbations (such as changes in snowfall and/or temperature and/or radiation, etc) (Oerlemans, 1988; 1989).

A more useful approach towards understanding the nature of the Little Ice Age and its cause(s) is to examine continuous climatic and paleoclimatic records from around the world to determine what the principal climatic characteristics were during the last 500 years, and thereby to determine what was so different about climate in the recent past. A survey of data presented in this volume and elsewhere reveals three important facts:

1 The last 500 years have not experienced a monotonously cold Little Ice Age; certain intervals have been colder than others.
2 The coldest periods in one region are often not coincident with those in other regions. There is geographical variability in climatic anomalies.
3 Different seasons may show different anomaly patterns over time. Thus, for example, the historical reconstructions of seasonal temperature anomalies in Switzerland (Pfister, Chapter 6) and in the western U.S.S.R. (Borisenkov, Chapter 9) show distinctly different seasonal time series of past climatic anomalies within each region.

A survey of the longest time series from each area reveals quite distinct temporal patterns of temperature anomalies. In Europe, the 19th century experienced the most widespread negative anomalies, generally from around 1820 to ~1915. In many cases the 17th century was also cold. However, most records indicate relatively warm conditions in the 16th and 18th centuries.

In eastern Asia, the 17th century was the coldest period; the late 18th and early 19th century was also cold, but there is little evidence for persistent low temperatures throughout the 19th century, as in Europe.

North American records show the 19th century was the coldest period. In northern regions, dendroclimatic evidence suggests that the 17th century was also cold, but in many parts of the western U.S. conditions at that time were warmer than in the 20th century.

Southern Hemisphere records are consistent in showing the main period of negative anomalies occurred earlier than in the Northern Hemisphere, with widespread cool conditions in the 16th and 17th centuries. In some records these anomalies continued into the mid-19th century.

These conclusions are obtained by making broad generalisations about diverse records which often represent different seasons and which may not all span the entire 500 year period. Furthermore, proxy evidence tends to highlight variations at higher frequencies. Variations at lower frequencies are much more likely to be obscured by the proxy source itself or by the methods used to produce the reconstruction. Nevertheless, as a first step towards a better understanding of the Little Ice Age we feel that generalisations such as those made here are justifiable. They clearly show that the last 500 years was a period of complex climatic anomalies, the understanding of which is not well-served by the continued use of the term "Little Ice Age" (cf. Landsberg, 1985). The period experienced both warm and cold episodes and these varied in importance geographically. There is no evidence for a world-wide, synchronous and prolonged cold interval to which we can ascribe the term "Little Ice Age". Only a few short cool episodes (lasting sometimes for up to 30 years) appear to have been synchronous on the hemispheric and global scale. These are the decades of the 1590s-1610s, the 1690s-1710s, the 1800s-1810s and the 1880s-1900s. Synchronous warm periods are less evident although the 1650s, 1730s, 1820s and the 1930s and 1940s appear to be the most important. As more research begins to fill in the gaps in our knowledge, a better understanding of the prevailing circulation patterns at these various times should emerge. In the meantime, we suggest that the term "Little Ice Age" be used cautiously.

33.6   Recommendations

In order to improve our understanding of this vital period of climatic history we see three important areas requiring further study.

33.6.1   Improvements in data coverage

It is apparent that any conclusions we can draw about climatic fluctuations in the recent past are constrained by the limited geographical coverage of existing high resolution data sets. Large gaps exist for the Tropics, interior Asia, the Middle East and the southern continents. Each of these areas is likely to have different types of resources with which to expand our knowledge of past climatic variations. For example, tree ring reconstructions can almost certainly provide a greatly expanded perspective on conditions across the vast interior of the Asian continent, from the Ukraine to Kamchatka. Considerably more information can be obtained from tree ring studies in the Southern Hemisphere by both expanding the geographic coverage and by examining the records densitometrically. This relatively new approach should also be applied more widely in the Northern Hemisphere. Significant increases in the amount of verifiable climatic variance in densitometric reconstructions clearly demonstrates that the additional effort in chronology production is worthwhile (Briffa *et al.*, 1990).

Southern and southeast Asia can provide a wealth of historical information, from religious and dynastic archives as well as from colonial records (Dutch, Portuguese, French, Spanish and British). Colonial records pertaining to South America may also be available (in Spanish and Portuguese colonial archives). Spanish missionary records from central America and the southwestern U.S. are another possible source of historical climatic data. Historical records from Australasia and southern Africa cover a shorter interval but are worthy of scrutiny (e.g. Vogel, 1989; Nicholls, 1989). North African and Middle Eastern sources should also provide valuable records. The historical data banks that are being developed in Switzerland (Pfister, Chapter 6) and Japan (Murata, Chapter 12) are important. As new information becomes available it can be routinely added, allowing for easy and objective re-evaluations. Material which is later found to be dubious or of non-contemporary origin can be removed. Extension of the techniques outlined in Chapter 6 should be made to other regions of the world where long written histories are known to exist.

Additional approaches to high resolution paleoclimatic reconstruction include studies of varved sediments, and studies of the chemistry and density of growth bands in corals. At present, the value of coral growth increments as a paleoclimatic proxy is somewhat equivocal. Claims of strong relationships between climate and growth bands (e.g. Isdale, 1984) may be premature as recent research has shown the complexity of the problem (Lough and Barnes, 1990a). Nevertheless, there is potential here for valuable (indeed unique) data to be obtained concerning low latitude climate conditions (Lough and Barnes, 1990b; Cole and Fairbanks, 1990).

Studies of varved lake sediments have a long history, yet they have received scant attention in terms of high resolution paleoclimatic reconstruction (O'Sullivan, 1983). Recent studies in Europe (e.g. Zolitschka, 1989) demonstrate their extraordinary potential. Varved lake sediments can be found in many different parts of the world and may provide data from regions where no other proxies exist (e.g. Halfman and Johnson, 1988). Varved sediments may also be found in certain ocean basins where upwelling occurs (e.g. the Santa Barbara Basin, off southern California; Soutar and Crill, 1977; Baumgartner, 1987; the Cariaco Basin off Venezuela; Overpeck *et al.*, 1989). Sedimentary records from these regions may be related to the prevailing circulation regime. Further research on varved sediments, wherever they are found, is strongly recommended.

Eventually, if data coverage can be improved enough, it should be possible to construct large-scale maps of climatic conditions for selected intervals of time (cf Figure 14.2 in Fritts and Shao, Chapter 14). These may still be seasonally specific, but even so they could provide insight into the prevailing larger scale general circulation. For example, it would be of considerable interest to construct a series of maps of climatic anomalies spanning the early part of the nineteenth century when conditions appear to have been quite anomalous in many parts of the world (cf. Figure 19.13 in Briffa and Schweingruber, Chapter 19). It has often been argued that these conditions relate to one or more major explosive eruptions during this time. However, there is evidence that cold conditions may have prevailed *before* the largest eruption of the millennium (Tambora, 1815) and that the Tambora eruption merely accentuated the unusual conditions (Harington, 1991). The occurrence of severe El Niños around this period also complicates the issue. A year-by-year reconstruction of the interval would help to clarify the relationship between volcanic eruptions, ENSO events and climate during this time interval and also shed light on larger scale teleconnections within the climate system.

33.6.2 *Improvements in the record of climate forcing factors*

Although the record of ENSO events over the last 500 years is now fairly well-resolved (Quinn and Neal, Chapter 32) other factors likely to perturb the climate system are far less well known. Of most significance in this regard is the record of explosive eruptions, particularly the nature and quantity of volatile emmissions produced. This is of critical importance to a better understanding of climate variability over the last few hundred years. Even with an incomplete historical record, it is apparent that the 20th century has experienced far fewer eruptions than earlier centuries. Our knowledge of the number, magnitude, geographic distribution and chemical characteristics of explosive eruptions is increasingly incomplete as we go back in time. Further geological, glaciological, historical and dendroclimatic research is needed to improve the record of climatically significant explosive eruptions.

33.6.3 *Improvements in calibration and interpretation*

The value of many proxy records could be improved if there was a better understanding of the climatic signal which the record contains. Often the interpretation is constrained by a lack of information for the recent period with which to calibrate the proxy record. For example, the ice core records from Mount Logan, Yukon and Quelccaya, Peru are remote from locations where contemporary climatic records have been kept. If on-site measurements were available, a better understanding could be obtained of what the isotopic and other records in the ice cores represent. Similarly, varved sediment records are often difficult to interpret in climatic terms because long-term limnological and hydrological data are not available. Coral studies are also constrained by the inadequacy of oceanographic data which may be relevant to paleoclimatic interpretations of coral growth and geochemical records. An expansion of environmental measurements, focused on the calibration of specific paleoclimatic proxies, would go a long way towards improving confidence in the past record of climate from environmental archives.

33.7   Conclusions

It is apparent from this volume that we already know a great deal about the climate of the last few hundred years. Many thoughtful and creative approaches have been used to extract information about past climates from obscure and obdurate sources. Nevertheless, there is much that we do not know and much that remains to be done before a clear picture of global climatic variations and their causes will be available. We hope that this book has generated new ideas for resolving the many questions which remain, and that it will spur new and innovative approaches to resolving the climatic record of the last 500 years.

References

Alt, B. T., 1985. 1550-1620: a period of summer accumulation in the Queen Elizabeth Islands. In C. R. Harington (ed.) Climatic Change in Canada 5. *Syllogeus*, No. 55. National Museums of Canada, Ottawa, 461-479.

Baumgartner, T. R., 1987. *High resolution paleoclimatology from the varved sediments of the Gulf of California*. Ph.D. thesis, Oregon State University, Corvallis. University Microfilms, Ann Arbor, Michigan. 287pp.
Bottomley, M., Folland, C. K., Hsuing, J., Newell, R. E. and Parker, D. E., 1990. *Global Ocean Surface Temperature Atlas (GOSTA)*. Her Majesty's Stationary Office, London, 24pp and 313 plates.
Bradley, R. S., Diaz, H. F., Eischeid, J. K., Jones, P. D. and Kelly, P. M. and Goodess, C. M., 1987a. Precipitation fluctuations over global land areas since the mid-nineteenth century. *Science*, 237, 171-175.
Bradley, R. S., Diaz, H. F., Kiladis, G. N. and Eischeid, J. K., 1987b. ENSO signal in continental temperature and precipitation records. *Nature*, 327, 497-501.
Bradley, R. S. 1991. In: M. E. Schlesinger (ed.) *Greenhouse Gas-Induced Climatic Change*. Elsevier, New York (in press).
Briffa, K. R., Bartolin, T. S., Ekstein, D., Jones, P. D., Karlen, W., Schweingruber, S. and Zetterberg, P., 1990. A 1,400-year tree-ring record of summer temperatures in Fennoscandia. *Nature*, 346, 434-439.
Bromley, D. A., 1990. The making of a greenhouse policy. *Issues in Science and Technology*, 7, 55-61.

Cole, J. and Fairbanks, R.G., 1990. The Southern Oscillation recorded in the $\delta^{18}O$ of corals from Tarawa atoll. *Paleoceanography*, 5, 669-683.
Cook, E. R. and Kariukstis, L. (eds.) 1990. *Methods of Tree-Ring Analysis: Applications in the Environmental Sciences*. Kluwer, Dordrecht.

Dansgaard, W., Johnsen, S. J., Reeh, N., Gundestrup, N., Clausen, H. B. and Hammer, C. U., 1975. Climatic changes, Norsemen and modern man. *Nature*, 255, 24-28.

Diaz H. F. and Kiladis, G. N., 1989. Global climatic anomalies associated with extremes in the Southern Oscillation. *J. Climate*, 2, 1069-1090.

Diaz H. F., Bradley, R. S. and Eischeid, J. K., 1989. Precipitation fluctuations over global land areas since the late 1800s. *J. Geophysical Research*, 94D, 1195-1210. *et al.* 1989.

Fisher, D. A. and Koerner, R. M., 1983. Ice core study: a climatic link between the past, present and future. In C. R. Harington (ed.) Climatic Change in Canada 3, *Syllogeus* 49, National Museums of Canada, Ottawa, 50-69.

Folland, C. K., Karl, T. R. and Vinnikov, K. Ya., 1990. Observed Climate Variations and Change. In: J. T. Houghton, G. J. Jenkins and J. J. Ephraums, (eds.) *Climate Change. The IPCC Scientific Assessment.* Cambridge University Press, Cambridge, p.194-238.

Gellatly A. F. and Norton, D. A., 1984. Possible warming and glacier recession in the South Island, New Zealand. *New Zealand Journal of Science*, 27, 381-388.

Gellatly, A. F., 1985. Glacier fluctuations in the central southern Alps, New Zealand: documentation and implications for environmental change during the last 1000 years. *Zeitschrift fur Gletscherkunde und Glazialgeologie*, 21, 259-264.

Grove, J. M., 1988. *The Little Ice Age*. Methuen, London. 498pp.

Groveman B. S. and Landsberg, H. E., 1979. Simulated Northern Hemisphere temperature departures, 1579-1880. *Geophysical Research Letters*, 6, 767-769.

Halfman, J. D. and Johnson, T. C., 1988. High resolution record of cyclic climatic change during the past 4000 years from Lake Turkana, Kenya. *Geology*, 16, 496-500.

Handler, P. 1984. Possible association of stratospheric aerosols and El Niño type events. *Geophysical Research Letters*, 11, 1121-1124.

Hansen and Lebedeff, 1987. Global trends of measured surface air temperature. *J. Geophysical Research*, 92, 13345-13372.

Harington, C. R., 1991. *The Year without a Summer? World climate in 1816*. National Museums of Canada, Ottawa.

Isdale, P. 1984. Fluorescent bands in massive corals record centuries of coastal rainfall. *Nature*, 310, 578-579.

Johnsen, S. J., Dansgaard, W. and White, J. W. C., 1989. The origin of Arctic precipitation under present and glacial conditions. *Tellus*, 41B,

Jones, P. D., 1989. The influence of ENSO on global temperatures. *Climate Monitor*, 17, 80-89.

Jones, P. D. and Kelly, P. M. 1983: The spatial and temporal characteristics of Northern Hemisphere surface air temperature variations. *Journal of Climatology*, 3, 243-252.

Jones, P. D., Raper, S. C. B., Bradley, R. S., Diaz, H. F., Kelly, P. M. and Wigley, T. M. L. 1986a: Northern Hemisphere surface air temperature variations: 1851-1984. *J. Climate and Applied Meteorology*, 25 161-179.

Jones, P. D., Raper, S. C. B. and Wigley, T. M. L. 1986b: Southern Hemisphere surface air temperature variations: 1851-1984. *J. Climate and Applied Meteorology*, 25 1213-1230.

Jones, P. D., Wigley, T. M. L. and Farmer, G. 1991: Marine and land temperature data sets: A comparison and a look at recent trends. In: M. E. Schlesinger (ed.) *Greenhouse-gas-induced climatic change*, Elsevier, New York (in press).

Köppen, W. 1873: Uber Mehrjahriger Perioden der Witterung, Insbesondere Uber die 11 Jahrige Periode der Temperatur. *Zeitschrift der Osterreichischen*

Landsberg, H. E., 1985. Historic weather data and early meteorological observations. In A.D. Hecht (ed.) *Paleoclimate Analysis and Modeling*. J. Wiley, New York, 27-70.

Lamb, H. H., 1977. *Climate, Present, Past and Future*. Methuen, London, 835pp.

Lough, J. M. and Barnes, D. J., 1990a. Intra-annual timing of density band formation of Porites coral from the central Great Barrier Reef. *J. Experimental Marine Biology and Ecology*, 135,35-57.

Lough, J. M. and Barnes, D. J., 1990b. Possible relationships between environmental variables and skeleton density in a coral colony from the central Great Barrier Reef. *J. Experimental Marine Biology and Ecology* (in press).

Matthes, F. 1939. Report of Committee on Glaciers. *Transactions American Geophysical Union*, 20, 518-523.

Matthes, F., 1940. Committee on Glaciers, 1939-40. *Transactions American Geophysical Union*, 21, 396-406.

Messerli, B., Messerli, P., Pfister, C. and Zumbuhl, H. J., 1978. Fluctuations of climate and glaciers in the Bernese Oberland, Switzerland, and their geoecological significance. *Arctic and Alpine Research*, 10, 247-260.

Moss, J. H., 1951. Early man in the Eden valley. *University of Pennsylvania Museum Monograph*, 9-92. Philadelphia.

Nicholls, N., 1988. More on early ENSOs: evidence from Australian documentary sources. *Bulletin American Meteorological Society*, 69, 4-6.

Oerlemans, J., 1988. Simulation of historical glacier variations with a simple climate-glacier model. *J. Glaciology*, 34, 333-341.

Oerlemans, J., 1989. On the response of valley glaciers to climatic change. In: J. Oerlemans (ed.), *Glacier Fluctuations and Climatic Change*. Kluwer Academic, Dordrecht, 353-371.

O'Sullivan, P. E., 1983. Annually laminated sediments and the study of Quaternary changes: a review. *Quaternary Science Reviews*, 1, 245-313.

Overpeck, J. T., Peterson, L. C., Kipp, N, Imbrie, J. and Rind, D. 1989. Climate change in the circum-North Atlantic region during the last deglaciation. *Nature*, 338, 553-557.

Porter, S. C., 1981. Glaciological evidence of Holocene climatic change. In: T. M. L. Wigley, M. J. Ingram and G. Farmer (eds.) *Climate and History*. Cambridge University Press, Cambridge, 82-110.

Porter, S. C., 1986. Pattern and forcing of Northern Hemisphere glacier variations during the last millennium. *Quaternary Research*, 26, 27-48.

Porter, S. C. and Denton, G. H., 1967. Chronology of Neoglaciation in the North American Cordillera. *American J. Science*, 265, 177-210.

Ropelewski C. F. and Halpert, M. S., 1987. Global and regional scale precipitation patterns associated with the El Niño/Southern Oscillation. *Monthly Weather Review*, 115, 1606-1626.

Salinger, M. J., 1979. New Zealand climate: the temperature record, historical data and some agricultural implications. *Climatic Change*, 2, 109-126.

Soutar, A. and Crill, P. A., 1977. Sedimentation and climatic patterns in the Santa Barbara Basin during the 19th and 20th centuries. *Geological Survey of America Bulletin*, 88, 1161-1172.

Shen, G. T. and Sanford, C. L., 1989. Trace element indicators of climate variability in reef-building corals. In: P. W. Glynn (ed.) *Global Ecological Consequences of the 1982-83 El Niño-Southern Oscillation*. Elsevier, New York. 1975?

Stuiver, M. and Pearson, G., 1986. High precision calibration of radiocarbon time-scale, A.D. 1950-500 B.C. *Radiocarbon*, 28, 805-838.

Vinnikov, K. Ya, Groisman, P. Ya. and Lugina, K. M., 1990. Empirical data on contemporary global climate changes (temperature and precipitation). *J. Climate*, 3, 662-677.

Vogel, C. H., 1989. A documentary-derived climatic chronology for South Africa, 1820-1900. *Climatic Change*, 14, 301-307.

Wood, F., 1988. Global alpine glacier trends, 1960s to 1980s. *Arctic and Alpine Research*, 20, 404-413.

Zolitschka, B., 1989. Jahreszeitlich geschichtete Seesedimente aus dem Holzmaar und dem Meerfelder Maar. *Zeitschrift der deutschen geologischen Gesellschaft* (Hanover), 140, 25-33.

Zumbühl, H. J., 1980. *Die Schwankungen der Grindelwaldgletscher in den historischen Bild- und Schriftquellen des 12 bis 19 Jahrunderts*. Denkschriften der Schweizerschen Naturforschenden Gesellschaft in Zurich, 92.

Index